An Introduction to the Geometrical Analysis of Vector Fields

with Applications to Maximum Principles and Lie Groups

An Introduction to the Geometrical Analysis of Vector Fields

with Applications to Maximum Principles and Lie Groups

Stefano Biagi
Università Politecnica delle Marche, Italy

Andrea Bonfiglioli
Alma Mater Studiorum - Università di Bologna, Italy

World Scientific

NEW JERSEY · LONDON · SINGAPORE · BEIJING · SHANGHAI · HONG KONG · TAIPEI · CHENNAI · TOKYO

Published by

World Scientific Publishing Co. Pte. Ltd.

5 Toh Tuck Link, Singapore 596224

USA office: 27 Warren Street, Suite 401-402, Hackensack, NJ 07601

UK office: 57 Shelton Street, Covent Garden, London WC2H 9HE

Library of Congress Cataloging-in-Publication Data
Names: Biagi, Stefano, author. | Bonfiglioli, Andrea, author.
Title: An introduction to the geometrical analysis of vector fields : with applications to maximum
    principles and Lie groups / by Stefano Biagi (Università Politecnica delle Marche, Italy),
    Andrea Bonfiglioli (University of Bologna, Italy).
Description: New Jersey : World Scientific, 2018. | Includes bibliographical references and index.
Identifiers: LCCN 2018043034 | ISBN 9789813276611 (hardcover : alk. paper)
Subjects: LCSH: Vector fields. | Maximum principles (Mathematics) | Lie groups.
Classification: LCC QA613.619 .B53 2018 | DDC 516/.182--dc23
LC record available at https://lccn.loc.gov/2018043034

British Library Cataloguing-in-Publication Data
A catalogue record for this book is available from the British Library.

First published 2019 (Hardcover)
Reprinted 2020 (in paperback edition)
ISBN 978-981-122-124-8 (pbk)

For any available supplementary material, please visit
https://www.worldscientific.com/worldscibooks/10.1142/11165#t=suppl

We dedicate this book to Ermanno Lanconelli

on the occasion of his becoming Emeritus Professor
of Alma Mater Studiorum, University of Bologna

With our gratitude for his invaluable teaching

Preface

VECTOR fields, and many of their applications, are the main subjects of this book. Usually, students of scientific disciplines first meet vector fields during their undergraduate courses, either in connection with Ordinary Differential Equations (ODEs, in the sequel), or within physics, studying conservative forces and potentials. Later in their studies, many students encounter the geometric meaning of vector fields in the context of differential geometry and manifold theory. Vector fields reappear in more advanced studies, such as in Lie group theory or throughout the literature of Partial Differential Equations (PDEs, in the sequel). It is manifest, owing to their interdisciplinary nature, that vector fields play a remarkable role not only in mathematics, but also in physics and in applied mathematics.

The aim of this book is to provide the reader with a gentle path through the multifaceted theory of vector fields, starting from the definitions and the basic properties of vector fields and flows, and ending with some of their countless applications. The building blocks of rich background material (Chaps. 1-to-7) comprise the following topics, just to name a few of them:

- commutators and Lie derivatives; the semi-group property and the equation of variation; global vector fields; C^0, C^k, C^ω-dependence;
- relatedness, invariance and commutability; Hadamard-type formulas;
- composition of flows: Taylor approximations and exact formulas;
- the algebraic Campbell-Baker-Hausdorff-Dynkin (CBHD, in the sequel) Theorem; the CBHD series and its convergence; the CBHD operation and its local associativity; Poincaré-type ODEs;
- iterated commutators and the Hörmander bracket-generating condition; connectivity; sub-unit curves and the control distance.

Once the background material is established, the applications mainly deal, according to our choice, with the following three settings (Chaps. 8-to-18):

  (I) *ODE theory*;
 (II) *maximum principles* (weak, strong and propagation principles);
(III) *Lie groups* (with an emphasis on the construction of Lie groups).

Before describing the separate contents, we make some overall comments on these applications, thus providing a general motivation for the book.

(I). First, we would like to focus on the leading role played by ODEs in this monograph. This role is twofold:

- we shall obtain applications of our theory *to* ODEs; and
- we shall use ODEs *as a tool* for the derivation of many results of our theory.

Thus ODEs intervene, in a manner of speaking, in a "circular way": as an instrument and, at the same time, as an object of applications. On the one hand, it is not surprising that ODEs provide a major tool in the theories of vector fields or of Lie groups: just think that the flow of a vector field (hence, the exponential map for a Lie group) is the solution to an ODE. On the other hand, some applications of the analysis of vector fields to ODEs, especially concerning the CBHD Theorem, are less known in the literature, and are therefore amongst our main achievements.

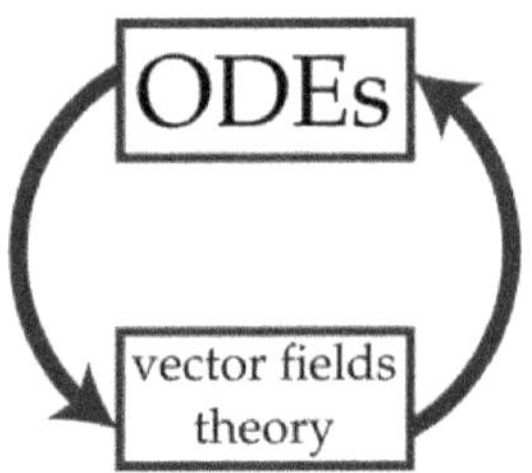

An example may help clarify this double role of ODEs: since the advent of modern Lie group theory, the theorem expressing the group-multiplication of two group-exponentials as another group-exponential, say

$$\mathrm{Exp}(X)\mathrm{Exp}(Y) = \mathrm{Exp}(Z(X,Y)), \tag{P.1}$$

is known, quintessentially, as *the* CBHD Theorem.[1] Most of its proofs (see e.g., the one in [Varadarajan (1984)]) *rely on an ODE* valid in the associated Lie algebra (see (P.8)), which we may trace back to none other than Poincaré (in the context of Lie groups of transformations). It does not escape the notice of experts that this ODE is valid not only in the Lie algebra of a Lie group, but also in other more general settings; however, buried as it tends to be in Lie group theory, the beautiful ODE argument may easily escape the attention of a student, or of a non-expert.

Most importantly, the ODE tool used in the proof of (P.1) can be formulated as *an independent ODE theorem* in a completely autonomous way, and so can be

---

[1]The fact that this Lie group result has a predecessor in non-commutative algebra which should be considered as the CBHD Theorem *par excellence* has already been discussed in [Bonfiglioli and Fulci (2012)]. In the literature, Dynkin's name is not often used as a label for this theorem; our choice is instead consistent with the historical presentation in [Bonfiglioli and Fulci (2012)] (see also [Achilles and Bonfiglioli (2012)] for more details).

viewed independently of any Lie group context. It can be thought of (and taught) as a result of ODE theory; it can even be exploited in the construction of Lie groups (as we do in this book), a procedure that is possible only after the ODE result has been liberated from the Lie group framework.

As a novelty with respect to the existing literature, this is our point of view with what we shall name 'the CBHD Theorem *for ODEs*'. In the choice of the ODE prerequisites as well, we shall take the liberty to make some non-standard choices. For example, even within the ODE theory itself there are results which are presented more and more rarely in ODE textbooks, and we hope that this monograph may be a good occasion (due to their subsequent applications) to bring them back to the attention of students. For instance, this is the case with the C^ω-dependence results (see App. B), or with the integral version of the equation of variation in the non-autonomous case (Chap. 12).

(II). We should now disclose our intent in introducing four chapters devoted to *Maximum Principles* (as applications of the vector-fields-and-flows machinery) in their many declinations: Weak and Strong Maximum Principles; Maximum Propagation along principal or drift vector fields.

We clearly have in mind the possible applications of Maximum Principles *to PDEs*, for example in obtaining an exhaustive Potential Theory for classes of partial differential operators. The theory presented here is general enough to comprise wide classes of operators, more general (for example) than the sub-Laplacians on Carnot groups considered in [Bonfiglioli *et al.* (2007)]. As a consequence, Chaps. 8–11 of the book may be useful also to young researchers in PDEs.

(III). As regards the Lie group theory, our applications are based on the multi-faceted use that one can make of the previously established machinery relating the CBHD Theorem and vector fields/flows:

(1) in the proof of (P.1), which reduces to a few lines once the ODE version of the CBHD Theorem is available;
(2) in the study of the CBHD series and of the CBHD operation on finite-dimensional Lie algebras;
(3) in implementing *Lie's Third Theorem* in some special but meaningful cases: in its local formulation, and also in the global one when dealing with nilpotent algebras; the tools mentioned in (2) prove to be particularly powerful and natural to use in solving this problem;
(4) in the *construction of Lie groups* starting from the exponentiation of Lie algebras of vector fields, without using the global version of Lie's Third Theorem;
(5) in a simple construction of Carnot groups (equipped with their homogeneous structure), starting from the stratified Lie algebras.

The Contents of the Book

With a view towards unveiling the unitary 'concept' of this monograph, let us describe more closely the topics it contains. In this introductory description, we shall keep the mathematical rigour at a minimum level, conveying the ideas rather than declaring the list of the assumptions behind any single result — there will be plenty of time for mathematical rigour throughout the main part of the book.

P.1. Basic facts on flows

Given a sufficiently regular vector field (v.f., for short) X on a domain Ω in $\mathbb{R}^N$, and fixing $x \in \Omega$, we denote by any of the following symbols

$$\gamma_{X,x}(t), \quad \gamma(t, X, x), \quad \Psi_t^X(x), \quad \exp(tX)(x)$$

the maximal solution (as a function of t), defined on its maximal domain $\mathcal{D}(X, x)$, of the Cauchy problem

$$\dot{\gamma}(t) = X(\gamma(t)), \qquad \gamma(0) = x.$$

As a function of x (or of (t, x), often in the literature), we shall refer to this function as *the flow of* X. We shall investigate $\Psi_t^X(x)$ as a function:

(A) of t (time),
(B) of x (the starting point), and
(C) of X (the vector field).

 (A). *As a function of t*, the flow admits the Taylor series expansion

$$\sum_{k=0}^{\infty} \frac{X^k(x)}{k!} t^k, \tag{P.2}$$

motivating the "exponential-type" notation $\exp(tX)(x)$, and giving a meaningful tool for the differentiation of a function f along a v.f. X:

$$\frac{\mathrm{d}^k}{\mathrm{d}t^k}\left\{ f(\gamma_{X,x}(t)) \right\} = (X^k f)(\gamma_{X,x}(t)).$$

The degree-two expansion (wrt t) of the flow furnishes a first remarkable interpretation of the commutator $[X, Y]$ of two v.f.s X and Y:

$$\lim_{t \to 0} \frac{\Psi_t^{-Y} \circ \Psi_t^{-X} \circ \Psi_t^{Y} \circ \Psi_t^{X}(x) - x}{t^2} = [X, Y](x). \tag{P.3}$$

 (B). *As a function of x*, particularly relevant is the Jacobian matrix

$$\mathcal{J}_{\Psi_t^X}(x) = \frac{\partial}{\partial x}\left\{ x \mapsto \Psi_t^X(x) \right\}.$$

For example, the Lebesgue measure of a set A evolves in the following way

$$\mathrm{meas}(\Psi_t^X(A)) = \int_A \exp\left(\int_0^t \mathrm{div}(X)(\Psi_\tau^X(x))\,\mathrm{d}\tau \right) \mathrm{d}x.$$

This result (Liouville's theorem for the flow of a v.f.), besides giving a meaningful interpretation of $\mathrm{div}(X)$ (the divergence of X), is a consequence of the *equation of variation* associated with the ODE defining the flow:

$$\frac{\mathrm{d}}{\mathrm{d}t} \mathfrak{J}_{\Psi_t^X}(x) = \mathfrak{J}_X(\Psi_t^X(x)) \, \mathfrak{J}_{\Psi_t^X}(x).$$

Again as a function of x, the composition of two flow maps (for the same v.f. X) gives rise to one of the most important features of the flow:

$$\textit{the semi-group property } \ \Psi_t^X \circ \Psi_s^X = \Psi_{t+s}^X.$$

Actually, the latter is so entrenched with v.f.s, that (roughly put) it characterizes the maps $F(t, \cdot) : \Omega \to \Omega$ which are flows of some v.f. Another story is the composition of two flows related to *different vector fields* X and Y (see Sec. P.2), the same as a different story is the product of two matrix-exponentials $e^A e^B$ when $A \neq B$.

(C). The analysis of $\Psi_t^X(x)$ *as a function of* X requires a lot of work, which is another of the main goals of this book. This work starts with the *non-autonomous equation of variation* ruling the following general (non-autonomous) Cauchy problem, also depending on the parameter ξ,

$$(\mathrm{CP}): \qquad \begin{cases} \dot{\gamma}(t) = f(t, \gamma(t), \xi) \\ \gamma(t_0) = x. \end{cases}$$

If $t \mapsto \Phi_{t,t_0}^\xi(x)$ denotes the solution of (CP), one can obtain the following *integral and non-autonomous* equation of variation of (CP):

$$\frac{\partial}{\partial \xi} \Phi_{t,t_0}^\xi(x) = \int_{t_0}^t \left(\frac{\partial}{\partial x} \Phi_{t,s}^\xi\right)(\Phi_{s,t_0}^\xi(x)) \cdot \left(\frac{\partial f}{\partial \xi}\right)(s, \Phi_{s,t_0}^\xi(x), \xi) \, \mathrm{d}s. \qquad (\mathrm{P.4})$$

Since this result is generally not[2] presented in ODE textbooks, we shall provide all the details. When the parametric f on the rhs of the ODE comes (in an autonomous way) from a vector field of the form

$$f(t, x, \xi) = \xi_1 X_1(x) + \cdots + \xi_m X_m(x),$$

where $X_1, \ldots, X_m$ are v.f.s forming the basis of a finite-dimensional Lie algebra $\mathfrak{g}$, one can obtain from (P.4) a formula for the differential of the map

$$\mathfrak{g} \ni X \mapsto \Psi_t^X(x).$$

This is the case, e.g., when $\mathfrak{g}$ is the Lie algebra $\mathrm{Lie}(\mathbb{G})$ of some Lie group $\mathbb{G}$.

But nothing forces us to be anchored to a pre-existing Lie group: for instance, $\mathfrak{g}$ may be *any finite-dimensional Lie algebra of smooth v.f.s*, and we may use the formula for the differential of $X \mapsto \Psi_t^X(x)$ to ask ourselves whether $\mathfrak{g}$ is the Lie algebra of a Lie group, without knowing it in advance.

We shall return to this question in Sec. P.7.

---

[2]Occasionally, the autonomous case when f does not depend on t is used in Lie group textbooks; see e.g., [Duistermaat and Kolk (2000)].

P.2. Composition of flows: the CBHD Theorem

As (P.2) shows, it is reasonable to expect that the composition $\Psi_t^Y \circ \Psi_s^X$ of the flows of two *different* v.f.s may be ruled by a formula analogous to the one behind the product of two formal exponentials series

$$e^{tY} e^{sX} = \sum_{j \geq 0} \frac{(tY)^j}{j!} \sum_{i \geq 0} \frac{(sX)^i}{i!}. \tag{P.5}$$

As a matter of fact, when passing from the formal exponential series $e^{tY} e^{sX}$ to the composition of flows of vector fields $\Psi_t^Y \circ \Psi_s^X$, one unexpectedly discovers, by taking the partial derivatives

$$\left. \frac{\partial^{i+j}}{\partial s^i \partial t^j} \right|_{(s,t)=(0,0)} \left\{ f\left(\Psi_t^Y \circ \Psi_s^X(x) \right) \right\} = (X^i Y^j f)(x), \tag{P.6}$$

that the order of tY and sX in (P.5) has to be reversed; yet the brilliant idea of the shift to the formal-power-series setting turns to be very fruitful. This idea traces back to the years between the nineteenth and twentieth centuries, with the works of Campbell, Baker and Hausdorff (see [Achilles and Bonfiglioli (2012); Bonfiglioli and Fulci (2012)]). Adding the name of Dynkin to those of the aforementioned mathematicians, we shall study in detail *the CBHD Theorem*, which states that

$$e^a e^b = e^{Z(a,b)}, \text{ where}$$

$$Z(a,b) = a + b + \frac{1}{2}\, [a,b] + \frac{1}{12}\, ([a,[a,b]] + [b,[b,a]]) + \cdots$$

is a series of Lie-polynomials in the non-commuting indeterminates a and b. The first explicit expression of the summands Z_h of $Z(a,b) = \sum_{h=1}^{\infty} Z_h(a,b)$ is due to [Dynkin (1947)] (whence our naming 'Dynkin polynomials').

In establishing the CBHD Theorem we make major use of *algebraic tools*; in this spirit, we follow the same pattern several times in the book:

- we establish an appropriate abstract algebraic setting;
- we derive the corresponding CBHD Theorem in this framework;
- we bring out an infinite family of identities by means of this theorem, identities that can profitably be used in many other different contexts.

Going back to the framework of the flows of v.f.s, and taking into account the needed reversing of X and Y visible in (P.6), we obtain the *Taylor approximation* of $\Psi_t^Y \circ \Psi_s^X$ up to arbitrary order n, in the following form

$$\exp(tY)(\exp(sX)(x)) = \exp\left(\sum_{h=1}^{n} Z_h(sX, tY) \right)(x) + \mathcal{O}\big(|s| + |t|\big)^{n+1}. \tag{P.7}$$

This type of formulas are particularly useful in the theory of linear PDEs; see e.g., [Bonfiglioli and Lanconelli (2012a); Christ *et al.* (1999); Citti and Manfredini (2006); Folland (1975); Magnani (2006); Morbidelli (2000); Nagel *et al.* (1985); Rothschild and Stein (1976); Varopoulos *et al.* (1992)].

Some natural questions arise: is it legitimate to let n go to ∞ in (P.7)? Is the series expected to converge? We shall answer these *non-trivial questions* in Sec. P.5, another goal of the book.

P.3. Our fundamental ally: ODEs ruling flows and the CBHD Theorem

A notable feature of the CBHD Theorem is that it rests on the validity (in an algebraic setting) of a suitable ODE, which traces back to [Poincaré (1900)], and for this reason we name it *'the Poincaré ODE'*: setting $Z(t) := Z(ta, tb)$, this is the ODE

$$\frac{\mathrm{d}}{\mathrm{d}t} Z(t) = \mathbf{b}\big(\operatorname{ad} Z(t)\big)(a) + \mathbf{b}\big(-\operatorname{ad} Z(t)\big)(b), \tag{P.8}$$

where $\mathbf{b}$ is the formal power series defined by

$$\mathbf{b}(z) = \frac{z}{e^z - 1} = \sum_{n=0}^{\infty} \frac{B_n}{n!} z^n,$$

which is the generating function of the Bernoulli numbers B_n. In its turn, (P.8) follows from the formal PDEs, each having its own interest and applicability,

$$\frac{\partial}{\partial s} C(s,t) = \mathbf{b}\big(\operatorname{ad} C(s,t)\big)(a) \quad \text{and}$$

$$\frac{\partial}{\partial t} C(s,t) = \mathbf{b}\big(-\operatorname{ad} C(s,t)\big)(b), \tag{P.9}$$

where $C(s,t) := Z(sa, tb)$. Just to give an idea of the depth of (P.8) and (P.9), we shall derive from them several applications to Lie group theory (Chaps. 14, 15, 17) and to ODE theory (Chap. 13).

Another indispensable tool in the analysis of v.f.s is the study of how a v.f. Y changes under the action of the flow of another v.f. X: this is given by $\mathrm{d}\Psi_{-t}^{X} Y$, the pushforward of the v.f. Y under the diffeomorphism Ψ_{-t}^{X}, the latter being the flow of X "running backward in time". We study this fundamental topic under the name of *Hadamard's Theorem for flows*. The reason for this (non-standard) naming is the analogy with the so-called Hadamard formula for formal power series in the indeterminates a, b:

$$e^a\, b\, e^{-a} = e^{\operatorname{ad} a} b := b + [a, b] + \tfrac{1}{2}\, [a, [a, b]] + \tfrac{1}{3!}\, [a, [a, [a, b]]] + \cdots.$$

This analogy will eventually become manifest, if we consider the formula

$$\mathrm{d}\Psi_{-t}^{X} Y = e^{\operatorname{ad} tX} Y := Y + t[X, Y] + \tfrac{t^2}{2}\, [X, [X, Y]] + \tfrac{t^3}{3!}\, [X, [X, [X, Y]]] + \cdots\,;$$

here, the higher order commutators of the v.f.s X and Y must satisfy some growth assumption for the series to converge: for example, if X and Y belong to a finite-dimensional Lie algebra of v.f.s, then this growth assumption is fulfilled.

As important as Poincaré's ODE (P.8), another ODE-like formula plays the leading role here, which we name *the Hadamard ODE for flows*:

$$\frac{\mathrm{d}}{\mathrm{d}t} \big(\mathrm{d}\Psi_{-t}^{X} Y\big) = \mathrm{d}\Psi_{-t}^{X} [X, Y], \tag{P.10}$$

an identity of time-dependent vector fields. When $t = 0$, (P.10) gives a remarkable interpretation of the commutator $[X, Y]$ (complementary to (P.3)), via the notion of the so-called *Lie derivative*.

Due to the invariance of a v.f. X under its flow, (P.10) becomes

$$\frac{\mathrm{d}}{\mathrm{d}t}\left(\mathrm{d}\Psi^{X}_{-t}Y\right) = [X, \mathrm{d}\Psi^{X}_{-t}Y].$$

In a finite-dimensional setting, this constant-coefficient, linear and homogeneous ODE can be integrated, to give the mentioned $\mathrm{d}\Psi^{X}_{-t}Y = e^{\mathrm{ad}\,tX}Y$.

P.4. Long commutators, connectivity and the control-distance

Let us consider a family $X = \{X_1, \ldots, X_k\}$ of (not necessarily distinct) smooth v.f.s. Like in (P.3), giving an approximation of the flow of $t^2\,[X_1, X_2]$ by means of $\Psi^{-X_2}_{t} \circ \Psi^{-X_1}_{t} \circ \Psi^{X_2}_{t} \circ \Psi^{X_1}_{t}$, we can *iterate this procedure* in order to approximate the flow of the long commutator

$$t^k\,[\cdots[[X_1, X_2], X_3], \cdots X_k], \tag{P.11}$$

by means of suitable compositions of the flows of $\pm X_1, \ldots, \pm X_k$ (usually referred to as the 'horizontal' directions wrt the family X).

The proof of the approximation of the flow of (P.11) is very laborious, but the "spirit" behind it is the same that we described in Sec. P.2:

- we first lift to an abstract level (formal power series in k indeterminates);
- we produce universal identities in this setting;
- we specialize these identities by putting v.f.s in place of the indeterminates.

This is the key ingredient for the remarkable *Connectivity Theorem* of Carathéodory, Chow, Hermann and Rashevskiĭ. This result states that, given a *Hörmander system* of smooth v.f.s X on a connected open set Ω, any pair of points $x, y \in \Omega$ can be connected by a continuous curve $\gamma_{x,y}$ in Ω, which piecewise is an integral curve of $\pm X_1, \ldots, \pm X_k$; the curve $\gamma_{x,y}$ is the typical example of a *X-subunit path*.

By definition, Hörmander systems of v.f.s satisfy the so-called *bracket-generating condition*, playing a remarkable role in Control Theory. Hörmander v.f.s are also fundamental in the study of hypoellipticity for linear PDEs, due to the celebrated result by [Hörmander (1967)] (see the survey [Bramanti (2014)]).

The notion of X-subunit path sets the basis for the definition of the *control distance* d_X associated with the family X: broadly speaking, the d_X-distance between x and y is obtained by minimizing the life-time of the X-subunit curves connecting x and y. The idea of modeling a more intrinsic geometry attached to X by means of the X-subunit paths proved to be one of the most fruitful ideas in the theory of sub-elliptic PDEs, as well as in Control Theory and in Geometric Measure Theory.

P.5. Applications: the CBHD Theorem for Lie algebras, and for ODEs

As anticipated at the end of Sec. P.2, a momentous problem for our investigation is the study of the convergence of the CBHD series in the realm of a *finite-dimensional Lie algebra* $\mathfrak{g}$. Due to its finite-dimensionality, $\mathfrak{g}$ supports a natural differentiable structure, where the Poincaré ODE-PDEs (P.8)-(P.9) are meaningful (besides being consequences of their abstract counterparts).

Hence, if x and y are close to 0 in $\mathfrak{g}$, we shall prove that
$$t \mapsto Z(t) := \sum_{n=1}^{\infty} Z_n(x, y)\, t^n$$
solves in $\mathfrak{g}$ (for $|t| \leq 1$) the Poincaré ODE (P.8). This fact has remarkable applications in Lie group theory: indeed, the *CBHD local operation*
$$(x, y) \mapsto x \diamond y := \sum_{n=1}^{\infty} Z_n(x, y) \tag{P.12}$$
has a "germ of associativity" on a small neighborhood Ω of 0 in $\mathfrak{g}$:
$$x \diamond (y \diamond z) = (x \diamond y) \diamond z, \quad \text{for every } x, y, z \in \Omega. \tag{P.13}$$
This result is a crucial point in solving Lie's Third Theorem in its local form (see Sec. P.7). Last but not least, the Poincaré PDEs (P.9) can be used to determine an enlarged domain of convergence for (P.12), even in infinite-dimensional contexts (see [Biagi and Bonfiglioli (2014); Blanes and Casas (2004); Mérigot (1974)]).

Deeply related to the convergence problem is the answer to our previous question, concerning the legitimacy of the limit "$\lim_{n \to \infty}$" in (P.7). A positive answer produces a non-trivial analog, *in the realm of ODEs*, of the CBHD Theorem. Roughly put, since the flow of a v.f. X is of "exponential-type", see (P.2), it is natural to presume that the composition of two flows $\exp(tY) \circ \exp(tX)$ may be ruled by a closed CBHD-type formula.

At the *finite* level of Taylor expansions, we have already stated that this is indeed the case, see (P.7), modulo suitable remainders. But much more is true if we deal with a finite-dimensional Lie-algebra V of smooth vector fields on a domain Ω. We shall derive *the CBHD Theorem for ODEs*:
$$\exp(Y)\big(\exp(X)(x)\big) = \exp(X \diamond Y)(x), \tag{P.14}$$
when $X, Y \in V$ are close to the zero vector field (and $x \in \Omega$ is fixed). In accordance with the spirit of the book, we prove (P.14) via an ODE argument: we show that
$$F(t) = \exp(tY)\big(\exp(X)(x)\big) \quad \text{and} \quad G(t) = \exp\big(X \diamond (tY)\big)(x)$$
solve (up to $t = 1$) the same Cauchy problem! In order to obtain $G'(t)$, we need many of our previous ingredients: Hadamard's Formula (P.10), Poincaré's ODE (P.8) in V, and the integral equation of variation (P.4) in the autonomous case.

As anticipated at the end of Sec. P.1, one can apply the CBHD Theorem for ODEs (P.14) when V is the Lie algebra of a real Lie group, but it is also possible to use this result in the absence of any background Lie group structure: this will be our approach for *the very construction of Lie groups* starting from Lie algebras V of smooth v.f.s satisfying minimal assumptions.

P.6. Applications to Maximum Principles

In Chap. 8 of the book we introduce an important class of linear PDOs, the *semielliptic operators* L of second order: we have chosen this name in place of the more usual (and longer) 'Picone elliptic-parabolic PDO' to mean that the second-order matrix of L is everywhere positive *semidefinite*. In this section, we denote by c the zero-order term of L.

We establish fundamental tools used for the study of these operators, of an undisputed independent interest in the PDE literature:

(a) *the Weak Maximum Principle* (WMP);
(b) *the Strong Maximum Principle* (SMP);
(c) *the Maximum Propagation Principle* (MPP) *for principal v.f.s;*
(d) *the MPP for the drift v.f.*

(a). Broadly put, we say that L satisfies the WMP on the set Ω if every $u \in C^2(\Omega)$ satisfying $Lu \geq 0$ must be non-positive on Ω whenever this is true on $\partial\Omega$ (in a suitable weak sense). Applications of the WMP are also provided, such as:

- comparison principles and *a priori* estimates;
- the uniqueness of the classical solution of the Dirichlet problem;
- the Green and Poisson operators related to L;
- the Maximum-Modulus Principle and the Maximum Principle.

The latter is important and it establishes that, when $c \equiv 0$, any L-subharmonic function (i.e., $Lu \geq 0$) takes its maximum on $\partial\Omega$, if L satisfies the WMP on Ω.

The semielliptic assumption on L is the natural hypothesis for the validity of the WMP: indeed, we show that (when $c \equiv 0$) the validity of the WMP on every bounded open set implies that L must be semielliptic. However, the semielliptic hypothesis is not sufficient as it stands, and some assumptions on the sign of c are needed. For example, $c < 0$ is a sufficient condition, as is $c \leq 0$ together with the existence of a so-called L-barrier for L.

Whereas our study of the WMP makes no use of any earlier result of the book, it serves as a bridge for the investigation of the SMP and of the MPP; the latter are indeed deeply connected with earlier topics and techniques of the book, such as the Connectivity Theorem and the use of integral curves.

(b). We say that L satisfies the SMP on the connected open set Ω if, whenever $u \in C^2(\Omega)$ satisfies

$$Lu \geq 0 \text{ and } u \leq 0 \text{ on } \Omega, \quad \text{and } F(u) = \{x : u(x) = 0\} \text{ is non-void,}$$

then $u \equiv 0$ on Ω. Actually, we shall prove the SMP for

$$L = \sum_{j=1}^{m} X_j^2 + X_0, \quad \text{where } X_1, \dots, X_m \text{ are Hörmander v.f.s.} \tag{P.15}$$

In studying the SMP, we are interested in how large $F(u)$ is (ideally, we aim at $F(u) = \Omega$); in particular, it is interesting to investigate those v.f.s whose integral curves "propagate" $F(u)$: the richer the class of these v.f.s, the larger $F(u)$; whence the validity of the SMP.

(c). It is clear from the above discussion that, in studying the SMP, the following classical approach (due to [Bony (1969)]) is the right path:

- given a set F, we are interested in those v.f.s X whose integral curves remain in F once they intersect F: in this case, we say that the set F is X-*invariant*;
- a theorem, due to Nagumo and Bony, tells us that F is X-invariant if and only if X is *tangent to* F in a weak sense (generalizing that of differential geometry);
- the so-called Hopf Lemma for L shows that any so-called *principal vector field X* for L is automatically tangent to $F(u)$.

Thus, if X is a principal v.f. for F, then $F(u)$ is X-invariant, so that

$$F(u) \text{ propagates along the integral curves of any principal v.f. } X \text{ for } L.$$

This is precisely the statement of the MPP for L.

Finally, the SMP for L in (P.15) follows due to the Connectivity Theorem, since the piecewise integral curves of $\pm X_1, \ldots, \pm X_m$ connect any pair of points of Ω, and the latter v.f.s are all principal for L.

(d). In dealing with PDOs of the form (P.15), we said nothing about the MPP along the v.f. X_0. This was not an oversight: the problem is that X_0 may not be a principal v.f. for L, as the Heat operator $H = \sum_{j=1}^{N} \partial_j^2 - \partial_t$ shows; as a matter of fact, H satisfies the WMP (on any bounded set) but it violates the SMP!

Nonetheless, we shall prove a redeeming fact: despite the possible lack of the X_0-invariance, still we have the *positive X_0-invariance* of $F(u)$. This means that, if $\gamma : [0, T] \to \Omega$ is an integral curve of X_0 with $\gamma(0) \in F(u)$, then we have $\gamma(t) \in F(u)$ for every $t \in [0, T]$. The proof of this fact is extremely delicate: we follow the approach by [Amano (1979)], where PDOs of the following form are considered (here $X_0, X_1, \ldots, X_N$ are regular enough v.f.s)

$$Lu = \sum_{i=1}^{N} \frac{\partial}{\partial x_i}(X_i u) + X_0 u; \tag{P.16}$$

also in this case we say that X_0 is the *drift* of L. Later, after we have studied PDOs of Amano form (P.16), we shall return to Hörmander PDOs (P.15).

P.7. Applications to Lie groups

In the realm of Lie groups, we show how the preceding flow theory produces *simple proofs* of the following results:

(i) the CBHD Theorem for Lie groups;

  (ii) the local Third Theorem of Lie;

 (iii) the global Third Theorem of Lie in the nilpotent case;

 (iv) the construction of Carnot groups;

  (v) the exponentiation of finite-dimensional Lie algebras of vector fields into Lie groups (under minimal assumptions).

Let us briefly analyze each topic.

(i). The CBHD Theorem for Lie groups $\mathbb{G}$ states that

$$\mathrm{Exp}(X)\mathrm{Exp}(Y) = \mathrm{Exp}(X \diamond Y), \tag{P.17}$$

for every $X, Y \in \mathrm{Lie}(\mathbb{G})$ sufficiently close to the null v.f.; here $\diamond$ is the CBHD operation in (P.12) and Exp is the Exponential Map for $\mathbb{G}$. The cornerstone for the proof of (P.17) is furnished by the following identity, valid for every $X, Y \in \mathrm{Lie}(\mathbb{G})$ and for time t close to 0,

$$\mathrm{Exp}(tX)\mathrm{Exp}(tY) = \Psi_t^Y\left(\Psi_t^X(e)\right),$$

where e is the identity element of $\mathbb{G}$. Thus the lhs of (P.17) is simply a composition of flows: this allows us to reduce the CBHD Theorem on a Lie group to the CBHD Theorem for ODEs, see (P.14).

(ii). The local (real) Third Theorem of Lie states that, for any finite-dimensional real Lie algebra $\mathfrak{g}$, there exists a local Lie group (on a neighborhood U of the origin of $\mathfrak{g}$) such that the smooth vector fields on U which are invariant under the (local) left-translations form a Lie algebra isomorphic to $\mathfrak{g}$. As a meaningful application of the results in Sec. P.5, we prove that this local Lie group is given by the CBHD operation $\diamond$ on U, see (P.12). The germ of associativity (P.13) enjoyed by this local operation is the main ingredient.

(iii). When $\mathfrak{g}$ is nilpotent, the result in (ii) can easily be globalized: the operation $\diamond$ is well defined throughout $\mathfrak{g}$, and $(\mathfrak{g}, \diamond)$ is a Lie group with Lie algebra isomorphic to $\mathfrak{g}$: this is the *global* version of Lie's Third Theorem for real and nilpotent Lie algebras. The global form of the Third Theorem of Lie holds for any finite-dimensional Lie algebra, not necessarily nilpotent. In this book we restrict to considering the latter case for two reasons: firstly, the general case requires a deep knowledge of differential geometry, which is beyond our scope; secondly the local case and the global nilpotent case are sufficiently interesting for our intent, since their proofs can be carried out in a constructive way as they exploit the CBHD series.

(iv). As a by-product of (iii), we shall derive a version of Lie's Third Theorem for finite-dimensional stratified Lie algebras $\mathfrak{s}$: a Lie algebra $\mathfrak{s}$ is stratified when it admits a decomposition of the form

$$\mathfrak{s} = V \oplus [V, V] \oplus [V, [V, V]] \oplus [V, [V, [V, V]]] \oplus \cdots,$$

where V is a subspace of $\mathfrak{s}$. If $\mathfrak{s}$ is finite-dimensional ($N = \dim \mathfrak{s}$, say), then $\mathfrak{s}$ is necessarily nilpotent. We can therefore equip $\mathfrak{s}$ with the Lie group structure $(\mathfrak{s}, \diamond)$

in (iii); it turns out that this Lie group can be further endowed with a homogeneous structure δ_λ, turning it into a so-called *homogeneous Carnot group* (HCG, for short) $\mathbb{G} = (\mathbb{R}^N, \diamond, \delta_\lambda)$, such that $\mathrm{Lie}(\mathbb{G})$ is isomorphic to $\mathfrak{s}$.

HCGs (and their sub-Laplacian operators) are studied in [Bonfiglioli *et al.* (2007)], by Lanconelli, Uguzzoni and one of us. The way HCGs are presented in the present book is more intrinsic than what is done in [Bonfiglioli *et al.* (2007)], in that we here shift the focus to the stratified Lie algebra as a *datum*, and the associated HCG is obtained by the constructive global Third Theorem of Lie. In this way, the many well-behaved properties of HCGs (mostly, the existence of dilations) are simple by-products of the Lie-algebra properties.

(v). Let us consider the following question:

(Q): *given a Lie subalgebra V of the smooth v.f.s on $\mathbb{R}^N$, is it possible to find a Lie group $\mathbb{G}$ whose manifold is $\mathbb{R}^N$ (with its usual differentiable structure) such that $\mathrm{Lie}(\mathbb{G}) = V$?*

Notice that we are requiring[3] the equality '$\mathrm{Lie}(\mathbb{G}) = V$', and not an equality 'up to an isomorphism' (as it happens with the global Third Theorem of Lie). We know that the following conditions are necessary for (Q) to have a positive answer:

- every $X \in V$ must be global;
- V must satisfy Hörmander's bracket-generating condition;
- the dimension of V must be equal to N.

We will show that these are also independent and sufficient conditions on V for a positive answer to (Q). Coherently within the spirit of the book, the following ingredients will be used: the CBHD Theorem for ODEs, in order to equip $V \equiv \mathbb{R}^N$ with a local-Lie-group structure; the use of a prolongation argument for ODEs in order to globalize this local Lie group.

How to read the book

About the Appendices. We assume that the reader is already acquainted with the basic notions of manifold theory and Lie groups; however, since we all know that notations and definitions may vary from book to book and from author to author, we equip our monograph with a short Appendix (App. C) containing the needed Lie group theory, where the reader can find all the definitions and the prerequisites clearly stated. The same is done with manifold theory in Chap. 4, whenever it is compulsory to fix the symbols and the nomenclature.

App. A, with some basic results of algebra and linear algebra, is functional to the reading of the rest of the book; once again it collects definitions and results which will avoid notational ambiguities and will spare the reader an endless search in the literature for those prerequisites which may not be standard for everybody.

---

[3]Here $\mathrm{Lie}(\mathbb{G})$ is meant as a Lie algebra of derivations, i.e., of vector fields in the sense of differential operators of order 1.

As regards App. B, devoted to background material on ODEs, this has another purpose: we hope that the reader has solid knowledge of basic ODE theory; but, as we already discussed, it is now increasingly rare to find an exhaustive treatment of parametric dependence in ODE textbooks. Thus, we provide an analysis of this subject (so important in our book), starting from the continuous dependence and ending with the C^ω case (the latter being usually omitted in textbooks); we skip most of the proofs, yet provide a complete presentation of all the relevant material.

How to use this book. Several parts of this monograph have been taught by the second-named author during his classes for the Master Degree and for the PhD in Mathematics at Bologna University. It was then our intention to write a book that could be used by students in their mathematic investigation as well as by teachers in giving their lectures; in this regard, the book contains:

- basic topics for an introductory course on ODEs (Chap. 1, App. B); and advanced topics in ODEs (Chaps. 12, 13);
- applications, suitable to a course on Lie groups (Chaps. 14–17 and 5);
- applications, suitable to PDE-oriented courses (Chaps. 8–11, 16, 17);
- introductory topics of Control Theory (Chaps. 6, 7);
- introductory topics of Differential Geometry (Chaps. 1–4).

Further material. We hope that students may benefit from the 182 exercises, and from the 58 figures. The interested reader can find some bibliographical references, grouped by chapter, at the end of the book, in the Further Readings section. The following figure describes the interdependence of the different chapters, and is therefore a guide through the reading of the book.

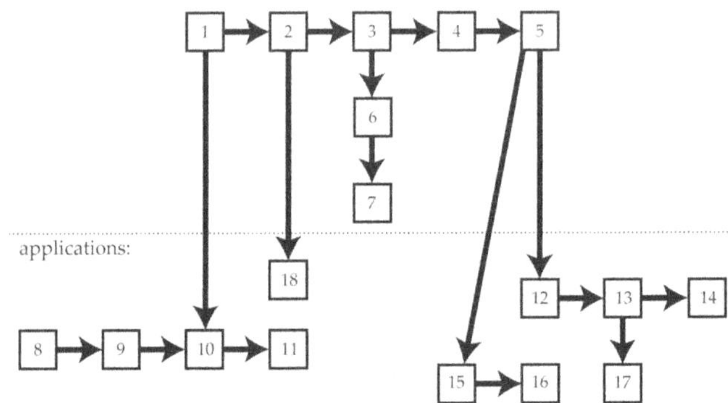

Bologna,
July, 2018

Stefano Biagi
Andrea Bonfiglioli

Contents

Chapter 1

Flows of Vector Fields in Space

THE aim of this chapter is to introduce one of the most important tools of this book: *the flow $\Psi_t^X(x)$ of a vector field X*. We shall investigate the flow as a function of t (time) and of x (the starting point). Later, in Chap. 4 we shall study it as a function of X (the vector field) as well.

In this introductory chapter, besides fixing notations that will return in every subsequent chapter, we prove the main facts about flows, only by using Ordinary Differential Equation (ODE, for brevity) theory and elementary Calculus. The "exponential-type" expansion of any flow allows to obtain meaningful expansions wrt time t, and an efficient method of *differentiating along integral curves*, besides offering a link to the contents of Chap. 2, where the so-called Exponential Theorem (for the composition of exponentials) of Campbell-Baker-Hausdorff-Dynkin will be investigated.

A useful tool is provided by the *equation of variation* naturally associated with the ODE system defining an integral curve. For example, this equation straightforwardly shows how the volume of a compact set evolves during the flow of a vector field (the so-called *Liouville's Theorem* on flows). Later in the book (Chap. 12), we shall also obtain from the equation of variation a formula for the differential of the flow *as a function of the vector field*.

The degree-two expansion (wrt time t) of the flow furnishes a first remarkable interpretation of the commutator $[X, Y]$ of two vector fields X and Y, via the study of the path $\Psi_t^{-Y} \circ \Psi_t^{-X} \circ \Psi_t^{Y} \circ \Psi_t^{X}$. Again on flows, in Chap. 3 we shall deal with the problem of longer compositions of flows (resulting in information on longer commutators).

As this is a short introductory chapter, our analysis is carried out in open subsets of ordinary space $\mathbb{R}^N$. Later on, we shall consider vector fields on arbitrary manifolds M and Lie groups $\mathbb{G}$. Clearly, this chapter furnishes results which can be applied in any local chart of M or $\mathbb{G}$.

The only prerequisite for this chapter is the knowledge of some elementary Analysis and basic ODE Theory. If need be, the dedicated App. B will provide the reader with background results of ODE Theory.

1

1.1   Notations for vector fields in space

In the sequel, Ω will always denote a non-empty open subset of $\mathbb{R}^N$, without the need to repeat it. If $k \in \{0, 1, 2, \ldots, \infty, \omega\}$, we denote by $C^k(\Omega)$ the vector space of the real-valued functions of class C^k on Ω; when $u \in C^\infty(\Omega)$, we say that u is smooth; as usual $C^\omega(\Omega)$ denotes the class of the real-analytic functions on Ω.

The points of $\mathbb{R}^N$ will be denoted by $x = (x_1, \ldots, x_N)$ (with $x_1, \ldots, x_N \in \mathbb{R}$). For $i \in \{1, \ldots, N\}$, any of the symbols

$$\partial_i, \quad \partial_{x_i}, \quad \partial/\partial x_i, \quad \frac{\partial}{\partial x_i}$$

will stand for the usual partial derivative operator wrt the x_i variable. Analogous notations will apply for higher order derivatives. In the sequel, if $f \in C^1(\Omega)$, we shall always denote its gradient as a $1 \times N$ *row-vector*:

$$\nabla f(x) = (\partial_1 f(x) \ \cdots \ \partial_N f(x)), \qquad x \in \Omega.$$

When f is vector-valued, say $f \in C^1(\Omega, \mathbb{R}^m)$, and $f_1, \ldots, f_m$ are its (real-valued) component functions, the row-notation for the gradient is of benefit for writing the Jacobian matrix $\mathcal{J}_f(x)$ of f at $x \in \Omega$ as follows

$$\mathcal{J}_f(x) = \begin{pmatrix} \nabla f_1(x) \\ \vdots \\ \nabla f_m(x) \end{pmatrix}. \tag{1.1}$$

When matrix computation arises, this implicitly suggests writing vector-valued functions as column-vectors, that will be our choice in a minute.

Throughout, we call **vector field** (v.f., for short) on Ω any linear partial differential operator (PDO, for brevity) X of the form:

$$X = \sum_{i=1}^{N} a_i \frac{\partial}{\partial x_i},$$

where $a_1, \ldots, a_N$ are real-valued functions on Ω. Thus, if $f \in C^1(\Omega)$, by Xf we mean the function

$$Xf : \Omega \longrightarrow \mathbb{R}, \qquad x \mapsto Xf(x) = \sum_{i=1}^{N} a_i(x) \frac{\partial f}{\partial x_i}(x).$$

We say that the v.f. X is of class C^k (with $k = 0, 1, 2, \ldots, \infty, \omega$) if the functions a_i are of class C^k in Ω. We say that X is a **smooth vector field** (s.v.f., for short) if the functions a_i are of class C^∞. Given $x \in \Omega$, the vector of $\mathbb{R}^N$

$$X(x) := (a_1(x), \ldots, a_N(x))$$

is called the **vector of the coefficients** (or the coefficient vector) of X at x. Frequently, in the literature, X is identified with the map $x \mapsto X(x)$. Instead, at this introductory level, we shall make the effort to distinguish between X (a PDO),

$X(x)$ (a vector), and $x \mapsto X(x)$ (a function from Ω to $\mathbb{R}^N$). In later chapters we shall feel free to pass from one notion to the other, under the same notation X, when confusion does not arise.

The only notational liberty that we take from the beginning is of identifying the N-dimensional vector $X(x)$ with the $N \times 1$ column-matrix

$$X(x) \equiv \begin{pmatrix} a_1(x) \\ \vdots \\ a_N(x) \end{pmatrix}. \tag{1.2}$$

As $\nabla f(x)$ is a row-matrix, this has the advantage of the compact notation

$$X f(x) = \nabla f(x) \cdot X(x), \qquad \forall f \in C^1(\Omega), \ x \in \Omega. \tag{1.3}$$

When X is smooth, we are allowed (and we systematically do it) to think of X *as a linear map of $C^\infty(\Omega)$ into itself*:

$$X : C^\infty(\Omega) \longrightarrow C^\infty(\Omega) \qquad f \mapsto Xf.$$

Hence a s.v.f. X can be thought of as being *an endomorphism of $C^\infty(\Omega)$*. We denote by $\mathrm{End}(C^\infty(\Omega))$ the vector space of all the endomorphisms of $C^\infty(\Omega)$.

Notation 1.1. The vector space of all the smooth vector fields on Ω will be denoted by $\mathfrak{X}(\Omega)$. Note that $\mathfrak{X}(\Omega)$ is a vector subspace of $\mathrm{End}(C^\infty(\Omega))$.

Remark 1.2. Throughout the book, $\mathfrak{X}(\Omega)$ *is implicitly equipped with the vector space structure of* $\mathrm{End}(C^\infty(\Omega))$. Some attention must be paid; indeed, with this vector space structure, the s.v.f.s $X_1, \ldots, X_m$ on Ω are linearly dependent iff there exist $c_1, \ldots, c_m \in \mathbb{R}$ such that $\sum_{i=1}^m c_i X_i$ is the null v.f., that is its component function are *identically vanishing*. For example, the v.f.s on $\mathbb{R}^2$

$$X_1 = \partial_{x_1} \quad \text{and} \quad X_2 = x_1 \, \partial_{x_2}$$

are linearly independent v.f.s in $\mathfrak{X}(\mathbb{R}^2)$; the reader must pay attention to the fact that this is not in contradiction to the fact that $X_1(x) = (1, 0)$ and $X_2(x) = (0, x_1)$ are linearly dependent when $x_1 = 0$, *as vectors of $\mathbb{R}^2$.* ♯

In the sequel, it will soon be convenient to let a v.f. X operate not only on real-valued functions, but on vector-valued functions as well.

Therefore, we shall allow any v.f. *operate component-wise on vector-valued functions*: namely, if $f : \Omega \to \mathbb{R}^m$ is C^1, by Xf we mean the function

$$Xf : \Omega \longrightarrow \mathbb{R}^m, \qquad x \mapsto (Xf_1(x), \ldots, Xf_m(x)).$$

Beware the following convention:

Convention. *When $f \in C^1(\Omega, \mathbb{R}^m)$ and X is a v.f. on Ω, if $Xf(x)$ occurs in matrix computations (for example after an $m \times m$ matrix), we opt for the convention that $Xf(x)$ be written as the column vector*

$$Xf(x) \equiv \begin{pmatrix} Xf_1(x) \\ \vdots \\ Xf_m(x) \end{pmatrix}. \tag{1.4}$$

This notation turns out to be consistent with (1.1) and (1.3), since we have

$$Xf(x) = \mathcal{J}_f(x) \cdot X(x), \qquad \forall\, f \in C^1(\Omega, \mathbb{R}^m),\ x \in \Omega. \tag{1.5}$$

This is an identity between the following matrices:

$$\begin{pmatrix} Xf_1(x) \\ \vdots \\ Xf_m(x) \end{pmatrix} = \begin{pmatrix} \nabla f_1(x) \\ \vdots \\ \nabla f_m(x) \end{pmatrix} \cdot \begin{pmatrix} a_1(x) \\ \vdots \\ a_N(x) \end{pmatrix}.$$

This is consistent with (1.3) since, for real-valued functions, the Jacobian matrix is simply the gradient row-vector. Occasionally and for the sake of clarity (a good example is given in identity (1.7)), we may want to denote by

$$I : \mathbb{R}^N \longrightarrow \mathbb{R}^N \qquad I(x) = \begin{pmatrix} x_1 \\ \vdots \\ x_N \end{pmatrix}$$

the identity function of $\mathbb{R}^N$, identified with an $N \times 1$ column vector, so that

$$XI(x) = \begin{pmatrix} a_1(x) \\ \vdots \\ a_N(x) \end{pmatrix} \tag{1.6}$$

is nothing but $X(x)$ under its column notation (1.2). This notation is particularly useful when we want to drop the notation of the point x and we need to consider $X(\cdot)$ as a vector-valued function. For example, we shall use the notation $\mathcal{J}_{XI}$ instead of the ambiguous[1] $\mathcal{J}_X$, to denote the Jacobian matrix of the function in (1.6); we shall also write $\mathrm{div}(XI)$ to denote the divergence function $\sum_{j=1}^N \partial_j a_j$.

For every multi-index $\alpha = (\alpha_1, \ldots, \alpha_N)$ (i.e., α is a vector with nonnegative integer components) we use this notation for higher order partial derivatives:

$$D_x^\alpha = \frac{\partial^{\alpha_1}}{\partial x_1^{\alpha_1}} \cdots \frac{\partial^{\alpha_N}}{\partial x_N^{\alpha_N}}.$$

If $\alpha = 0$, D_x^0 acts trivially:

$$D_x^0 f = f \quad \text{for every } f \in C^0(\Omega).$$

Moreover, we write $|\alpha| = \alpha_1 + \cdots + \alpha_N$.

We say that P is a **smooth linear PDO of order** n on Ω if

$$P = \sum_{|\alpha| \le n} a_\alpha(x)\, D_x^\alpha,$$

where, for every α with $|\alpha| \le n$, a_α is a smooth real-valued function on Ω. As for smooth v.f.s, we think of P as an endomorphism of $C^\infty(\Omega)$. The collection of

---

[1]Remember that X mainly denotes a PDO, not a vector-valued function.

the smooth linear PDOs (of any nonnegative order) on Ω will be denoted by $\mathcal{U}(\Omega)$. This is trivially a vector subspace of $\mathrm{End}(C^\infty(\Omega))$.

We remark that $\mathcal{U}(\Omega)$ *is closed under the operation of composition* (of endomorphisms). In other words, the composition of two smooth linear PDOs is again a smooth linear PDO. This proves that (following Def. A.1 in App. A) $(\mathcal{U}(\Omega), \circ)$ *is a unital associative algebra, subalgebra of* $(\mathrm{End}(C^\infty(\Omega)), \circ)$, *with unit element* D_x^0. Moreover, $\mathcal{U}(\Omega)$ is both a unital associative algebra (with the composition of operators) and a Lie algebra (with the associated bracket; see Rem. A.2). Note that $\mathcal{X}(\Omega)$ is *not* an associative subalgebra of $(\mathcal{U}(\Omega), \circ)$ since the composition of vector fields is not – in general – a vector field, but a PDO of order 2. What is instead true is that

$$\mathcal{X}(\Omega) \text{ is a Lie subalgebra of } \mathcal{U}(\Omega),$$

that is, the commutator of two smooth vector fields is a smooth vector field as the following computation shows: if $X = \sum_j a_j \partial_j$ and $Y = \sum_j b_j \partial_j$,

$$[X,Y] = \left(\sum_i a_i \partial_i \right) \circ \left(\sum_j b_j \partial_j \right) - \left(\sum_j b_j \partial_j \right) \circ \left(\sum_i a_i \partial_i \right)$$

$$= \sum_{i,j} \left(a_i(\partial_i b_j)\partial_j + \underline{a_i b_j \partial_{i,j}} - b_j(\partial_j a_i)\partial_i - \underline{b_j a_i \partial_{j,i}} \right)$$

(note that the underlined *second order* summands cancel out)

$$= \sum_j (X b_j)\, \partial_j - \sum_i (Y a_i)\, \partial_i = \sum_k (X b_k - Y a_k)\, \partial_k,$$

and the far rhs is clearly a smooth vector field. The mentioned cancelation follows from Schwarz's Theorem for mixed partial derivatives, since we understand that vector fields operate on C^∞ functions. The above computation still holds true whenever X, Y are C^1 vector fields operating on C^2 functions.

Remark 1.3. Using our notation (1.6), the above expression for $[X, Y]$ in terms of the component functions of X and Y can be rewritten as follows

$$[X,Y]I = X(YI) - Y(XI) \overset{(1.5)}{=} \partial_{YI} \cdot XI - \partial_{XI} \cdot YI; \tag{1.7}$$

Here v.f.s operate on vector-valued functions, written as column-vectors. ♯

We end the section with a warning: if X is a v.f. on $\mathbb{R}^N$ and if $f \in C^1(\mathbb{R}^N, \mathbb{R}^N)$, one must not make confusion between

$$X f(x) \quad \text{and} \quad X(f(x)).$$

Both are defined, but the former is the value at the point x of the $\mathbb{R}^N$-valued function $Xf = (Xf_1, \ldots, Xf_N)$, whereas the latter is the vector of $\mathbb{R}^N$ obtained by computing the coefficient vector of X at $f(x)$. Should confusion arise, the latter will be denoted by $XI(f(x))$, which is undisputable, due to (1.6).

1.2 The flow of a vector field

We next give the main definitions of this chapter: integral curves and flows. Again, Ω is understood to be a non-empty open subset of $\mathbb{R}^N$. We warn the reader that we shall use the word 'curve' in $\mathbb{R}^N$ with the meaning of a *function* $\gamma : I \to \mathbb{R}^N$ (with $I \subseteq \mathbb{R}$ an interval), whereas the set of the points $\gamma(I)$ will be referred to as the *image set* of the curve.

Definition 1.4 (Integral curve). Let X be a C^1 vector field on Ω. Any solution of the system of ODEs

$$\dot{\gamma}(t) = X(\gamma(t))$$

is called an **integral curve** of X. If $x \in \Omega$ and if $\gamma(t)$ is an integral curve of X such that $\gamma(0) = x$, we say that γ **starts at** x. In this case, the (unique) maximal solution of the Cauchy problem

$$\text{(CP)} \qquad \begin{cases} \dot{\gamma}(t) = X(\gamma(t)) \\ \gamma(0) = x \end{cases}$$

will be indifferently denoted by any of the following notations:

$$\gamma(t, X, x), \quad \gamma_X(t, x), \quad \gamma_{X,x}(t).$$

The domain of the maximal solution of (CP) will be denoted by $\mathcal{D}(X, x)$. The symbols X or x may be omitted when they are understood or irrelevant.

Finally, the following notation will be used

$$\mathcal{D}(X) := \Big\{ (t, x) \in \mathbb{R} \times \mathbb{R}^N : x \in \Omega, \ t \in \mathcal{D}(X, x) \Big\}. \tag{1.8}$$

See Fig. 1.1. If $X = \sum_i a_i \partial_i$, the ODE-system $\dot{\gamma}(t) = X(\gamma(t))$ in coordinates is

$$\begin{cases} \dot{\gamma}_1(t) = a_1(\gamma_1(t), \ldots, \gamma_N(t)) \\ \quad \vdots \\ \dot{\gamma}_N(t) = a_1(\gamma_1(t), \ldots, \gamma_N(t)). \end{cases}$$

We shall also write $\dot{\gamma} = X(\gamma)$; we see that this is an *autonomous* system of ODEs.

Obviously, the above definition makes sense under lower regularity assumptions on the coefficients of X, the hypothesis that X is C^1 being widely sufficient. For instance, it suffices that the components of X be locally Lipschitz continuous on Ω. In the sequel, we consider C^1 (C^2, occasionally) v.f.s for the sake of simplicity; the reader acquainted with basic ODE theory will recognize that much of the results to follow hold true under lower regularity assumptions.

Remark 1.5. We know that, when X is C^k, general results of ODE theory ensure that $\gamma_X(t, x)$ is of class C^k on the pair $(t, x) \in \mathcal{D}(X)$, and that the set $\mathcal{D}(X)$ in (1.8) is an *open set* (see App. B). We also know that the set $\mathcal{D}(X, x) \subseteq \mathbb{R}$ is a nonempty open interval containing 0.

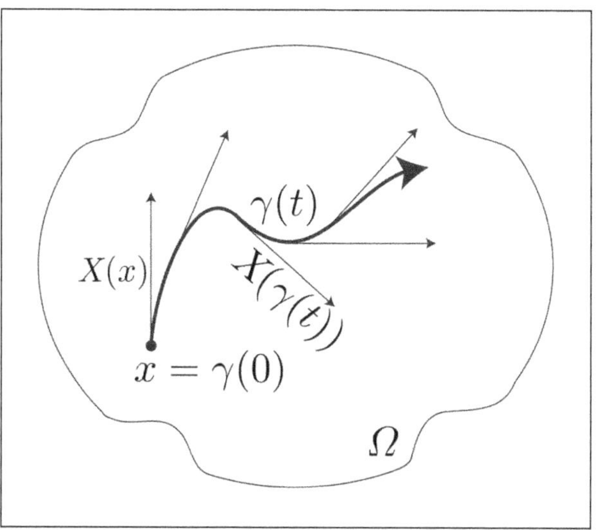

Fig. 1.1   An integral curve $\gamma(t)$ of X starting at x. The maximal solution $\gamma(t)$ will escape every compact set in Ω (i.e., it tends to the boundary of Ω when time tends to the endpoints of $\mathcal{D}(X, x)$; see App. B).

Observe that, fixing $x \in \Omega$, $\mathcal{D}(X, x)$ is the projection on the t-axis of a section of the open set $\mathcal{D}(X)$ in (1.8), this section being obtained by equating the space-variable with x (see also Fig. 1.2):
$$\mathcal{D}(X, x) = \{t \in \mathbb{R} : (t, x) \in \mathcal{D}(X)\}.$$
We remark that, unlike other open subsets in $\mathbb{R} \times \mathbb{R}^N$, $\mathcal{D}(X)$ consists of a family of intervals (all containing $0 \in \mathbb{R}$) placed along the t-axis:
$$\mathcal{D}(X) = \bigcup_{x \in \Omega} \mathcal{D}(X, x) \times \{x\}.$$

Remark 1.6 (Foliation into disjoint image sets of integral curves). If X is a C^1 v.f. on Ω, the image sets of two integral curves (*as subsets of* $\mathbb{R}^N$) can intersect; this is not in contrast with the uniqueness of (CP). Since the ODE defining an integral curve is autonomous we have the following simple fact (Exr. 1.9):

If $\gamma(t_1, X, x_1) = \gamma(t_2, X, x_2)$ then $\mathcal{D}(X, x_2) = t_2 - t_1 + \mathcal{D}(X, x_1)$ and
$$\gamma(t, X, x_1) = \gamma(t_2 - t_1 + t, X, x_2), \quad \text{for every } t \in \mathcal{D}(X, x_1).$$
As a consequence the image set of $\gamma(\cdot, X, x_1)$ coincides with that of $\gamma(\cdot, X, x_2)$.

As a result, Ω is foliated into the (disjoint) union of the image sets of all the integral curves that intersect each other. More precisely, we can partition the family
$$\left\{ \mathcal{D}(X, x) \ni t \mapsto \gamma(t, X, x) \right\}_{x \in \Omega}$$
into the disjoint union of the sub-families obtained by gathering together the integral curves with intersecting image sets. By passing from these sub-families to the image sets that they (uniquely) define, we obtain a disjoint union of sets foliating Ω. We leave the details to the reader.   ♯

Example 1.7. (a). If $X = \partial_{x_1} + 2x_2\,\partial_{x_3}$ in $\mathbb{R}^3$, since $X(x)$ is the vector function with entries $(1, 0, 2x_2)$, the integral curves of X are the solutions of

$$\dot{\gamma}_1(t) = 1, \quad \dot{\gamma}_2(t) = 0, \quad \dot{\gamma}_3(t) = 2\,\gamma_2(t).$$

They are of the form $\gamma(t) = (x_0 + t, y_0, z_0 + 2y_0\,t)$ for some $(x_0, y_0, z_0) \in \mathbb{R}^3$.

(b). The integral curves of the s.v.f. $x_2\,\partial_{x_1} - x_1\,\partial_{x_2}$ on $\mathbb{R}^2$ are solutions of

$$\dot{\gamma}_1(t) = \gamma_2(t), \quad \dot{\gamma}_2(t) = -\gamma_1(t).$$

This is a system of linear ODEs with constant coefficients, hence its solutions are of the form $\gamma(t) = \exp(A)\gamma_0$, where $A = \left(\begin{smallmatrix} 0 & 1 \\ -1 & 0 \end{smallmatrix}\right)$ and $\gamma_0 \in \mathbb{R}^2$. Thus one has

$$\gamma(t) = (x_0 \cos t + y_0 \sin t,\, y_0 \cos t - x_0 \sin t),$$

for some $(x_0, y_0) \in \mathbb{R}^2$. This simple example shows that computations with integral curves may be complicated, and only in extremely rare occasions can integral curves be explicitly determined.

(c). The (maximal) integral curve of $(1 + x^2)\,\partial_x$ on $\mathbb{R}^1$ starting at 0 is the map $\gamma(t) = \tan(t)$. It is defined only for $|t| < \pi/2$.

(d). The maximal solution of $X = x^2\,\partial_x$ on $\mathbb{R}^1$ starting at x is (see Exr. 1.1)

$$\gamma(t, X, x) = \frac{x}{1 - xt} \quad \text{and} \quad \mathcal{D}(X, x) = \begin{cases} (-\infty, 1/x), & \text{if } x > 0 \\ (-\infty, \infty), & \text{if } x = 0 \\ (1/x, \infty), & \text{if } x < 0. \end{cases}$$

Therefore, the associated set $\mathcal{D}(X)$ in (1.8) is $\mathcal{D}(X) := \big\{(t, x) \in \mathbb{R} \times \mathbb{R} : \ t\,x < 1\big\}$.

We next introduce yet another notation for the integral curves $\gamma(t, X, x)$. Here the focus is on the maximal integral curve *being a function of the starting point x*, hence the following notation privileging x.

Definition 1.8 (Flow of a vector field). Let X be a given C^1 vector field on Ω, and let $t \in \mathbb{R}$ be fixed. When this is defined, we set

$$\Psi_t^X(x) := \gamma(t, X, x). \tag{1.9}$$

This makes sense iff $x \in \Omega$ is such that $\mathcal{D}(X, x)$ contains t (in general, this may not be true of every $x \in \Omega$). When its domain is non-empty, the map $x \mapsto \Psi_t^X(x)$ is called the **flow of X at time** t. The notation Ψ_t will apply when X is understood.

Given $t \in \mathbb{R}$, we denote by Ω_t^X (possibly $\varnothing$) the subset of Ω of the points x such that the maximal integral curve of X starting at x survives for time t, i.e.,

$$\Omega_t^X := \big\{x \in \Omega : \ t \in \mathcal{D}(X, x)\big\}.$$

When this is not empty, Ω_t^X is exactly the domain of the flow map Ψ_t^X:

$$\Psi_t^X : \Omega_t^X \longrightarrow \Omega.$$

Convention. Throughout the book, we shall feel free to call Ψ_t^X **the flow of** X, forgetting to say 'at time t' at each occasion. However, we warn the reader that –in Differential Geometry– by 'flow' one usually means the map

$$\mathcal{D}(X) \ni (t,x) \mapsto \Psi_t^X(x).$$

This mild abuse will cause no confusion.

Remark 1.9. Note that, by Rem. 1.5, the set Ω_t^X is an open subset of Ω (possibly empty); indeed, $\Omega_t^X = \varnothing$ if no x in Ω is such that $\mathcal{D}(X,x)$ contains t; otherwise (if there exists $x \in \Omega$ such that $t \in \mathcal{D}(X,x)$), Ω_t^X is exactly the (non-empty) projection on the x-axis of a section of the open set $\mathcal{D}(X)$ in (1.8), this section being obtained by equating the time-variable with t (see also Fig. 1.2):

$$\Omega_t^X = \{x \in \Omega \,:\, (t,x) \in \mathcal{D}(X)\}.$$

We also have

$$\mathcal{D}(X) = \bigcup_{t\in\mathbb{R}} \{t\} \times \Omega_t^X, \quad \text{and} \quad \mathcal{D}(X) = \bigcup_{x\in\Omega} \mathcal{D}(X,x) \times \{x\}.$$

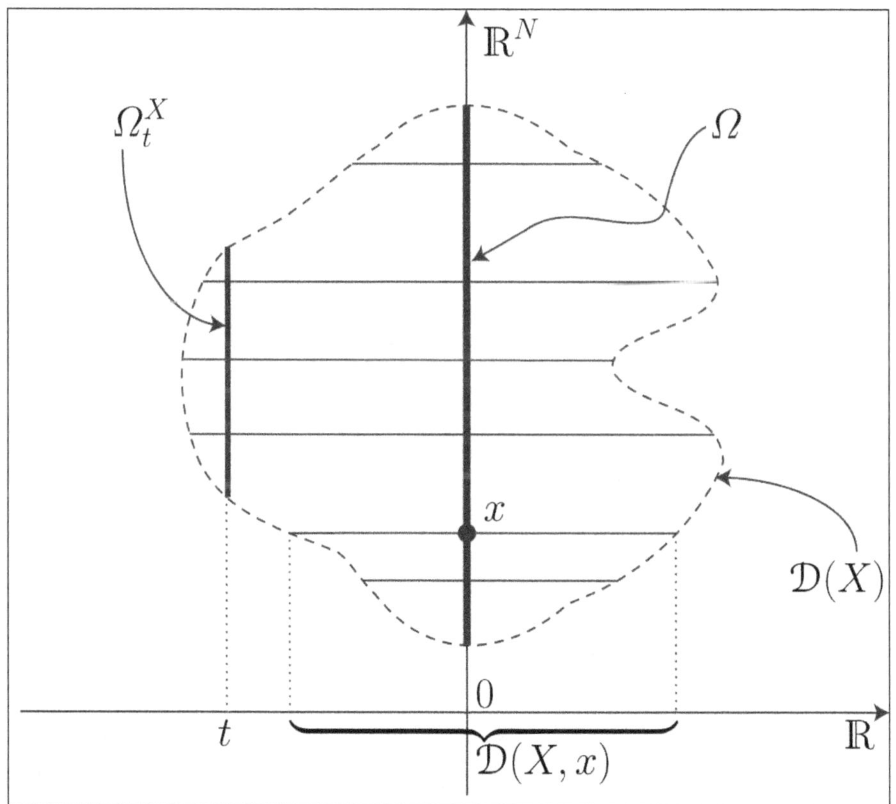

Fig. 1.2 The sets $\mathcal{D}(X)$, $\mathcal{D}(X,x)$, Ω_t^X. The figure contains a small abuse: the set Ω_t^X should be pictured on the space-axis $\mathbb{R}^N$ by projection on $\mathbb{R}^N$ of the set "Ω_t^X" depicted in the figure.

In the first identity, some of the summands Ω_t^X may be $\varnothing$, whereas in the second identity any summand $\mathcal{D}(X, x)$ is non-void. ♯

Remark 1.10. If X is C^1 on Ω, the following properties hold:

- $\Omega_0^X = \Omega$;
- if $t > 0$ and $\Omega_t^X \neq \varnothing$, then $\Omega_t^X \subseteq \Omega_s^X$ whenever $0 \leq s \leq t$;
- if $t < 0$ and $\Omega_t^X \neq \varnothing$, then $\Omega_t^X \subseteq \Omega_s^X$ whenever $t \leq s \leq 0$;
- $\Omega_t^X \uparrow \Omega$ as $t \downarrow 0$, and $\Omega_t^X \uparrow \Omega$ as $t \uparrow 0$;
- for every compact set $K \subset \Omega$ there exists $\varepsilon = \varepsilon(K, X, \Omega) > 0$ such that $K \subset \Omega_\varepsilon^X$ and $K \subset \Omega_{-\varepsilon}^X$ (see App. B). ♯

Example 1.11. (i). The flow at time t of X in Exm. 1.7-(a) is the affine map

$$\begin{pmatrix} x_1 \\ x_2 \\ x_3 \end{pmatrix} \mapsto \begin{pmatrix} t \\ 0 \\ 0 \end{pmatrix} + \begin{pmatrix} 1 & 0 & 0 \\ 0 & 1 & 0 \\ 0 & 2t & 1 \end{pmatrix} \cdot \begin{pmatrix} x_1 \\ x_2 \\ x_3 \end{pmatrix}.$$

Note that Ψ_t^X is invertible and its inverse is Ψ_{-t}^X.

(ii). The flow at time t of the v.f. $Y = x_2\,\partial_{x_1} - x_1\,\partial_{x_2}$ in Exm. 1.7-(b) is the clockwise rotation around the origin with angle t:

$$\begin{pmatrix} x_1 \\ x_2 \end{pmatrix} \mapsto \begin{pmatrix} \cos t & \sin t \\ -\sin t & \cos t \end{pmatrix} \cdot \begin{pmatrix} x_1 \\ x_2 \end{pmatrix}.$$

Again, note that Ψ_t^Y is invertible and its inverse is Ψ_{-t}^Y.

(iii). The flow at time t of the v.f. $X = x^2\,\partial_x$ in Exm. 1.7-(d) is

$$\Psi_t^X : \Omega_t^X \longrightarrow \mathbb{R}, \qquad \Psi_t^X(x) = \frac{x}{1 - x\,t},$$

and its domain is

$$\Omega_t^X = \begin{cases} (-\infty, 1/t), & \text{if } t > 0 \\ (-\infty, \infty), & \text{if } t = 0 \\ (1/t, \infty), & \text{if } t < 0. \end{cases}$$

The inverse map of Ψ_t^X is Ψ_{-t}^X, and this is defined on Ω_{-t}^X.

Obviously, the fact that in all the above examples we have $(\Psi_t^X)^{-1} = \Psi_{-t}^X$ and $\Psi_t^X(\Omega_t^X) = \Omega_{-t}^X$ is not sheer chance; see Rem. 1.15. ♯

We observe that, for a fixed $t \neq 0$, the map Ψ_t *may not be defined on the whole of* Ω, as Exm. 1.11-(iii) shows. Nonetheless, as anticipated in Rem. 1.10, for every compact subset K of Ω, there exists $\varepsilon > 0$ (depending on K) such that Ψ_t is defined on the entire K, for every $t \in [-\varepsilon, \varepsilon]$. This says that for every compact subset of Ω there exists $\varepsilon > 0$ such that any integral curve starting at a point of this compact set does survive for any small time in $[-\varepsilon, \varepsilon]$ (Cor. B.18 on page 364).

Remark 1.12 (Finer regularity of the flow). From general ODE Theory results concerning the regular dependence on the data (see App. B), if X is of class C^k on Ω (for $k = 1, 2, \ldots, \infty, \omega$), then the map $\mathcal{D}(X, x) \ni t \mapsto \Psi_t^X(x)$ is C^{k+1} (for any fixed $x \in \Omega$). Moreover, the map $\mathcal{D}(X) \ni (x, t) \mapsto \Psi_t^X(x)$ is C^k.

Furthermore, there also exist the partial derivatives of order $k+1$ of the following form (and the order of differentiation can be interchanged)

$$\frac{\partial^{k+1}}{\partial t\, \partial x_{i_1} \ldots \partial x_{i_k}} \Psi_t^X(x),$$

for any $i_1, \ldots, i_k \in \{1, \ldots, N\}$, and these functions are *continuous*. ♯

Example 1.13. For example, if $X = x\,|x|\,\partial_x$ on $\mathbb{R}^1$, one has (Exr. 1.14)

$$\Psi_t^X(x) = \begin{cases} (\text{if } x \neq 0) & \dfrac{x}{1 - |x|\,t} & \text{for } t < 1/|x| \\[2ex] (\text{if } x = 0) & 0 & \text{for } t \in \mathbb{R}. \end{cases}$$

It is easy to see that $(x, t) \mapsto \Psi_t^X(x)$ is C^1, that $\partial^2 \Psi_t^X(x)/\partial t\, \partial x$ exists and is continuous, but $\partial^2 \Psi_t^X(x)/\partial x^2$ does not exist at $x = 0$, when $t \neq 0$.

Note that X is C^1 but not C^2. ♯

1.2.1 *The semigroup property*

The following result is a consequence of the uniqueness of the solution of an ODE with C^1 coefficients.

Proposition 1.14 (Semigroup properties of flows). *Let X be a C^1 vector field on Ω and let $x \in \Omega$. We have the following facts:*

(1) *if $s, t + s \in \mathcal{D}(X, x)$ then $t \in \mathcal{D}(X, \gamma_X(s, x))$ and one has*

$$\gamma_X\big(t, \gamma_X(s, x)\big) = \gamma_X(t + s, x);$$

(2) *if $s \in \mathcal{D}(X, x)$ then $-s \in \mathcal{D}(X, \gamma_X(s, x))$ and one has*

$$\gamma_X\big(-s, \gamma_X(s, x)\big) = x;$$

(3) *if $\alpha, t \in \mathbb{R}$ are such that $\alpha\, t \in \mathcal{D}(X, x)$ then $t \in \mathcal{D}(\alpha\, X, x)$ and one has*

$$\gamma_X(\alpha\, t, x) = \gamma_{\alpha\, X}(t, x);$$

(4) *$\gamma(0, X, x) = x$ for every $x \in \Omega$.*

In terms of the flow of the vector field X, these can shortly be recast as follows:

(1′) *$\Psi_t^X \circ \Psi_s^X = \Psi_{t+s}^X$;*
(2′) *the maps Ψ_s^X and Ψ_{-s}^X are inverse to each other (see Rem. 1.15);*
(3′) *$\Psi_{\alpha\, t}^X = \Psi_t^{\alpha\, X}$; in particular $\Psi_{-t}^X = \Psi_t^{-X} = (\Psi_t^X)^{-1}$;*
(4′) *the map Ψ_0^X is the identity map of Ω.*

*Properties (1)-(1′) will be referred to as the **semigroup property** of the flow of X.*

Clearly, these statements can be rephrased in other ways; for example (see Exr. 1.3), the semigroup property (1) is equivalent to:

if $s \in \mathcal{D}(X, x)$ and $t \in \mathcal{D}(X, \gamma_X(s, x))$, then $t + s \in \mathcal{D}(X, x)$ (the latter being equal to $s + \mathcal{D}(X, \gamma_X(s, x)))$, and one has $\gamma_X(t, \gamma_X(s, x)) = \gamma_X(t + s, x)$.

Proof of Prop. 1.14. We restrict to prove property (1); we leave it to the reader to prove (3), whilst (2) follows from (1) by taking $t = -s$ and by observing that $\gamma_X(0, x) = x$ for every X and x.

Let $s, t + s \in \mathcal{D}(X, x)$. We show that $\gamma_X(t, \gamma_X(s, X))$ is defined and it coincides with $\gamma_X(t + s, x)$. For s fixed we set $\mu(r) := \gamma_X(r + s, x)$; note that μ is well posed on the interval $H := \mathcal{D}(X, x) - s$. For $r \in H$, we have

$$\dot{\mu}(r) = \frac{\mathrm{d}}{\mathrm{d}r}\big(\gamma_X(r + s, x)\big) = \frac{\mathrm{d}}{\mathrm{d}u}\Big|_{u = r + s} \gamma_X(u, x) = X(\gamma(r + s, x)) = X(\mu(r)).$$

Moreover, $\mu(0) = \gamma_X(s, x)$. This proves that $\mu : H \to \Omega$ solves

$$\dot{\mu} = X(\mu), \quad \mu(0) = \gamma_X(s, x).$$

On the other hand, by definition (see also Def. 1.4), this Cauchy problem is solved by $r \mapsto \gamma_X(r, \gamma_X(s, x))$ on the maximal domain $\mathcal{D}(X, \gamma_X(s, x))$. By uniqueness of the solution, this proves that the maximal interval $\mathcal{D}(X, \gamma_X(s, x))$ contains H, and in particular it contains $t = t + s - s \in \mathcal{D}(X, x) - s = H$ (as, by assumption, $t + s \in \mathcal{D}(X, x)$). Moreover,

$$\gamma_X(r + s, x) = \mu(r) = \gamma_X(r, \gamma_X(s, x)), \quad \forall r \in H.$$

By taking $r = t$ (note that $t \in H$) we get property (1). $\qquad\square$

Remark 1.15. For every $t \in \mathcal{D}(X, x)$ one has $\Psi_t^X(\Omega_t^X) = \Omega_{-t}^X$. This follows from (2) of Prop. 1.14 (see also Exr. 1.4 for a detailed guided proof of this fact). This shows that the functions

$$\Psi_t^X : \Omega_t^X \longrightarrow \Omega_{-t}^X \quad \text{and} \quad \Psi_{-t}^X : \Omega_{-t}^X \longrightarrow \Omega_t^X$$

are inverse to each other. $\qquad\qquad\sharp$

Remark 1.16. The semigroup property of the flow is so special that *it characterizes being a flow*. More precisely, suppose $F = F(t, x)$ is a C^1 function defined on an open set $A \subseteq \mathbb{R} \times \Omega$ and valued in Ω. Suppose that A has the following form: $A = \bigcup_{x \in \Omega} I_x \times \{x\}$, where, for any $x \in \Omega$, I_x is an open interval containing 0.

Assume the following two facts: $F(0, x) = x$ for any $x \in \Omega$; moreover, for any $x \in \Omega$ and any $t \in I_x$, one has the identity

$$F(s, F(t, x)) = F(s + t, x), \tag{1.10}$$

valid for any $s \in I_{F(t, x)}$ near 0. Then the C^1 vector field X defined by

$$\Omega \ni x \mapsto X(x) := \frac{\mathrm{d}}{\mathrm{d}t}\Big|_{t = 0} F(t, x)$$

is such that $F(t, x) = \Psi_t^X(x)$ for any $x \in \Omega$ and any $t \in I_x \subseteq \mathcal{D}(X, x)$. This follows immediately by differentiating (1.10) wrt s at $s = 0$ (Exr. 1.5). $\qquad\sharp$

1.2.2 *Global vector fields*

We give a definition which plays a crucial role when dealing with left-invariant vector fields on Lie groups.

Definition 1.17 (Global vector field). Let $\Omega \subseteq \mathbb{R}^N$ be open and let X be a C^1 vector field on Ω. We say that X is **global** if, for every $x \in \Omega$, the maximal domain $\mathcal{D}(X, x)$ of the (maximal) integral curve of X starting at x is the whole of $\mathbb{R}$.

We warn that some authors use the adjective 'complete' instead of 'global'.

Remark 1.18. Globality is a "fragile" property: two v.f.s can be global, but the same may not be true of their sum or of their commutator; indeed (Exr. 1.6):

(1) the vector fields in $\mathbb{R}^2$ defined by $X_1 = 6\,x_2^2\,\partial_{x_1}$, $X_2 = x_1\,\partial_{x_2}$ are global, but $X_1 + X_2$ is not global;

(2) the vector fields in $\mathbb{R}^2$ defined by $X_1 = (1+x_2^2)\,\partial_{x_1}$, $X_2 = x_1\,\partial_{x_2}$ are global, but $[X_1, X_2]$ is not global. ♯

General results on ODEs ensure that a sufficient condition for $X \in \mathfrak{X}(\mathbb{R}^N)$ to be global is that it grows at most linearly in x, that is

$$\exists\, C_1, C_2 > 0: \qquad \|X(x)\| \leq C_1 + C_2\,\|x\|, \quad \forall\, x \in \mathbb{R}^N. \tag{1.11}$$

Example 1.19. Let B be a real $N \times N$ matrix. We consider the vector field $\mathbf{B}$ whose coefficient vector at $x \in \mathbb{R}^N$ is Bx, i.e.,

$$\mathbf{B} := \sum_{j=1}^{N}\left(\sum_{k=1}^{N} b_{j,k}x_k\right)\frac{\partial}{\partial x_j}.$$

$\mathbf{B}$ clearly satisfies (1.11), so it is global. Since $\mathbf{B}(x) = Bx$, the Cauchy problem defining the flow of $\mathbf{B}$ is

$$\dot{\gamma}(t) = B\,\gamma(t), \quad \gamma(0) = x,$$

a constant-coefficient linear homogeneous problem. Consequently,

$$\Psi_t^{\mathbf{B}}(x) = e^{tB}x, \quad \text{for any } t \in \mathbb{R} \text{ and } x \in \mathbb{R}^N,$$

where $e^{tB} = \sum_{k=0}^{\infty}(tB)^k/k!$ is the usual matrix exponential. ♯

The sufficient condition (1.11) for globality is by no means a necessary one, as the following useful example shows.

Example 1.20 (Pyramid-shaped vector field). Suppose that X has the form

$$X(x) = \begin{pmatrix} a_1 \\ a_2(x_1) \\ a_3(x_1, x_2) \\ \vdots \\ a_N(x_1, \ldots, x_{N-1}) \end{pmatrix}, \qquad x \in \mathbb{R}^N,$$

for suitable C^1 functions $a_1, \ldots, a_N$ defined on $\mathbb{R}^N$ (here a_1 is constant). In this case we say that X is **pyramid-shaped**. Then X is global, and its integral curves can be determined by successive quadratures.

Indeed, the system defining $\gamma(t) = \gamma(t, X, x)$ is:

$$\begin{cases} \dot{\gamma}_1(t) = a_1, & \gamma_1(0) = x_1 \\ \dot{\gamma}_2(t) = a_2(\gamma_1(t)), & \gamma_2(0) = x_2 \\ \dot{\gamma}_3(t) = a_3(\gamma_1(t), \gamma_2(t)), & \gamma_3(0) = x_3 \\ \vdots & \vdots \end{cases}$$

This system can be solved starting from the first equation downwards:

$$\gamma_1(t) = x_1 + a_1\, t$$

$$\gamma_2(t) = x_2 + \int_0^t a_2(\gamma_1(s))\, \mathrm{d}s$$

$$\gamma_3(t) = x_3 + \int_0^t a_3(\gamma_1(s), \gamma_2(s))\, \mathrm{d}s, \qquad \ldots$$

Clearly the above functions $\gamma_i(t)$ are defined for every $t \in \mathbb{R}$.

For example, $X = \partial_{x_2} + x_1\, \partial_{x_2} + \frac{1}{2}\, x_1^2\, \partial_{x_4}$ has coefficients functions which do not grow linearly, nonetheless X is global since it is pyramid-shaped. $\qquad \sharp$

Remark 1.21. With the previous notations, if X is a C^1 global v.f. on Ω, then

- $\mathcal{D}(X, x) = \mathbb{R}$ for every $x \in \Omega$, and $\mathcal{D}(X) = \mathbb{R} \times \Omega$;
- $\Omega_t^X = \Omega$ for every $t \in \mathbb{R}$;
- for any $t \in \mathbb{R}$, $\Psi_t^X : \Omega \to \Omega$ is a C^1 diffeomorphism, with inverse Ψ_{-t}^X. $\qquad \sharp$

We leave as an exercise the following simple (but not very operative) fact.

Proposition 1.22. *Let X be a C^1 vector field on Ω. Then X is global if and only if there exists $\varepsilon > 0$ such that, for every $x \in \Omega$, $\mathcal{D}(X, x)$ contains $(-\varepsilon, \varepsilon)$.*

From Prop. 1.22 (and a simple compactness argument), it is easy to prove that any compactly supported vector field is global.

1.2.3 *Regular and singular points*

Let X be a v.f. on Ω; a point $x \in \Omega$ is said to be a **singular point** of X (or an **equilibrium point** of the flow of X) if $X(x) = 0$; otherwise, x is said to be a **regular point** of X. Clearly, if x is singular for X, then $\Psi_t^X(x) \equiv x$ for any $t \in \mathcal{D}(X, x) = \mathbb{R}$. The situation is quite different when x is a regular point:

Proposition 1.23. *Let X be a C^1 vector field on Ω, and suppose $x \in \Omega$ is a regular point of X. Then the map*

$$\mathcal{D}(X, x) \ni t \mapsto \gamma(t) := \gamma_X(t, x) \in \Omega$$

is a C^1 immersion, i.e., $\dot{\gamma}(t) \neq 0$ for every $t \in \mathcal{D}(X, x)$.

Proof. By contradiction, suppose x is a regular point of X and assume there exists $s \in \mathcal{D}(X, x)$ with $\dot{\gamma}(s) = 0$; then $X(\gamma(s)) = \dot{\gamma}(s) = 0$, so that $y := \gamma(s)$ is a singular point of X. Hence, by the argument preceding the statement of this proposition, $\gamma_X(t, y) = y$ for any $t \in \mathcal{D}(X, y) = \mathbb{R}$. As a consequence, by Prop. 1.14-(2) we get $x = \gamma_X(-s, \gamma_X(s, x)) = \gamma_X(-s, y) = y$; hence, $x = y$ and x would be at the same time a regular and a singular point of X, a contradiction. $\square$

The following result can be seen as a Frobenius-type theorem.

Theorem 1.24. *Suppose that X is a C^1 vector field on Ω, and that $x_0 \in \Omega$ is a regular point of X. Then there exists a C^1 change of variable $y = F(x)$ defined near x_0 such that the local expression of X in the new coordinates is $\partial/\partial y_1$.*

Once one knows the right F, the proof is a two-line argument. We indulge in the flow/geometrical idea behind the proof: the longer demonstration is worthwhile.

Proof. Suppose $x_0 = 0$ for notational simplicity. Let us construct F. Suppose $F : O \to \Omega'$ is a C^1 diffeomorphism as in the assertion, where $O \subseteq \Omega$ is an open neighborhood of 0 and $\Omega' \subseteq \mathbb{R}^N$ is open. We require that v.f. expressing X in the coordinates $y = F(x)$, say $\widetilde{X}$, be $\partial/\partial y_1$ on Ω'. See Fig. 1.3.

Taking into account how vector fields are transformed under a change of variable, we must have

$$J_F(x)\, X(x) = (1, 0, \ldots, 0)^T, \quad \forall\, x \in O. \tag{1.12}$$

If we write $F = (F_1, \ldots, F_N)$, this is equivalent to the system of PDEs

$$X F_1(x) = 1, \quad X F_2(x) = 0, \ldots, X F_N(x) = 0 \quad (x \in O). \tag{1.13}$$

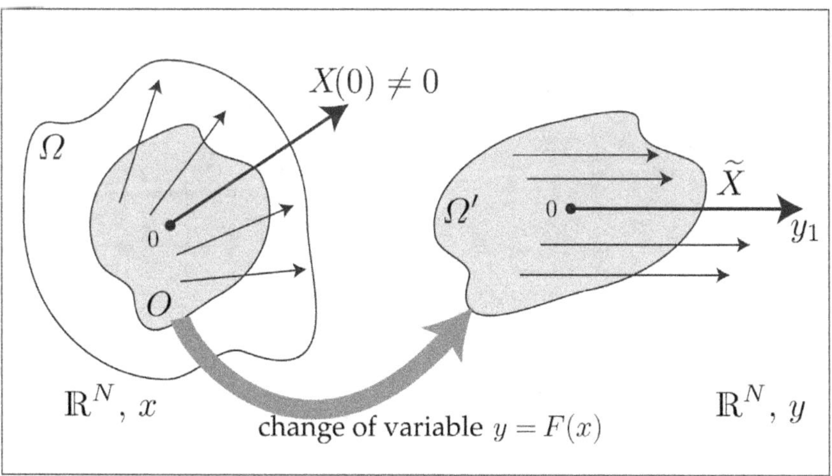

Fig. 1.3 The idea in the proof of Lem. 1.24, turning X into $\dfrac{\partial}{\partial y_1}$.

We remark that each equation is independent of the others, in the sense that each component function F_i appears only in the i-th row of the system. We can solve each equation by the so-called method of characteristics. Indeed, proceeding along the integral curves of X we can reduce each equation of (1.13) to an ODE. To this end, if $\gamma : I \to O$ is any integral curve of X, since $(XF_i)(\gamma(t)) = (F_i \circ \gamma)'(t)$, system (1.13) gives (when $x = \gamma(t)$)

$$(F_1 \circ \gamma)'(t) = 1, \quad (F_2 \circ \gamma)'(t) = 0, \ldots, (F_N \circ \gamma)'(t) = 0 \quad (t \in I).$$

Therefore we seek for $F_1, \ldots, F_N$ such that

$$F_1(\gamma(t)) = t + c_1(\gamma), \quad F_2(\gamma(t)) = c_2(\gamma), \ldots, F_N(\gamma(t)) = c_N(\gamma), \tag{1.14}$$

for every integral curve γ of X, with $c_1(\gamma), c_2(\gamma), \ldots, c_N(\gamma)$ constant functions along the curve γ (but possibly depending on the curve).

The natural strategy is to split the ambient space O into a family of disjoint integral curves of X and to define the maps F_i on each curve separately. A way to parameterize the integral curves in such a way that they are disjoint is to find a surface S in O which is transversal to each curve γ: this means that the coefficient vector of X (giving the direction of γ) must not lie on the tangent space of S (at least when γ intersects S).

Since near 0 the direction of X is approximately $X(0)$, we simply make the

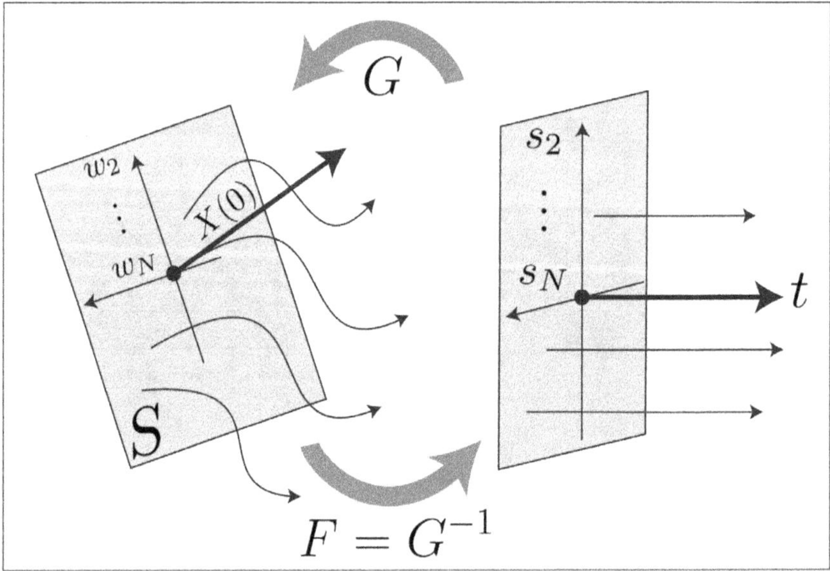

Fig. 1.4   The change of variable $F = G^{-1}$ in the proof of Thm. 1.24, and the geometric idea behind its construction: the integral curves of X (parameterized by the points of S and time t) are "flattened" to become lines, parallel to the t axis.

choice[2] $S := \mathrm{span}\{w_2, \ldots, w_N\}$, where

$$X(0), w_2, \ldots, w_N \quad \text{are linearly independent.} \tag{1.15}$$

This is possible because $X(0) \neq 0$ by hypothesis. It is reasonable to expect that the points of O near 0 are parameterized by the trajectories of the integral curves $\gamma_X(t, z)$ of X starting at points z of S near 0. Hence, we set

$$G(t; s_2, \ldots, s_N) := \gamma_X(t, s_2\, w_2 + \cdots + s_N\, w_N), \tag{1.16}$$

where G is defined on $[-\varepsilon, \varepsilon]^N$ and $\varepsilon > 0$ is so small that the rhs of (1.16) is defined for any $t, s_2, \ldots, s_N \in [-\varepsilon, \varepsilon]$. See Fig. 1.4. Owing to the Inverse Function Theorem, we can check if G defines a change of variable near $G(0) = 0$: this happens if $\mathcal{J}_G(0)$ is non-singular. Setting $s = (s_2, \ldots, s_N)$, we have

$$\left.\frac{\partial}{\partial t}\right|_{(0,0)} G(t; s) = \left.\frac{\mathrm{d}}{\mathrm{d}t}\right|_0 G(t; 0) = \left.\frac{\mathrm{d}}{\mathrm{d}t}\right|_0 \gamma_X(t, 0) = X(0).$$

Moreover, for every $i \in \{2, \ldots, N\}$ we have

$$\left.\frac{\partial}{\partial s_i}\right|_{(0,0)} G(t; s) = \left.\frac{\mathrm{d}}{\mathrm{d}s_i}\right|_0 G(0; 0, \ldots, s_i, \ldots, 0) = \left.\frac{\mathrm{d}}{\mathrm{d}s_i}\right|_0 \gamma_X(0, s_i\, w_i) = w_i.$$

This proves that the column-vectors of $\mathcal{J}_G(0)$ are $X(0), w_2, \ldots, w_N$, whence $\mathcal{J}_G(0)$ is non-singular in view of (1.15). This shows that $(t, s) \mapsto G(t; s)$ defines a system of coordinates near 0, by possibly shrinking ε. Returning to (1.14), it is natural to define the maps F_i in such a way that

$$\begin{cases} F_1\big(\gamma_X(t, s_2\, w_2 + \cdots + s_N\, w_N)\big) &= t + c_1(s) \\ F_2\big(\gamma_X(t, s_2\, w_2 + \cdots + s_N\, w_N)\big) &= c_2(s) \\ \qquad\qquad\qquad\qquad \vdots \\ F_N\big(\gamma_X(t, s_2\, w_2 + \cdots + s_N\, w_N)\big) &= c_N(s), \end{cases}$$

for suitable C^1 functions $c_i(s)$. Since we also need $F = (F_1, \ldots, F_N)$ to be a diffeomorphism, the simplest choice of the maps $c_i(s)$ is the following:

$$c_1(s) = 0, \qquad c_2(s) = s_2, \ldots, c_N(s) = s_N.$$

On account of (1.16), this is equivalent to $F(G(t; s)) = (t, s)$, that is, $F = G^{-1}$. Since G is a diffeomorphism of an open neighborhood of 0 onto an open neighborhood of 0, the definition $F := G^{-1}$ makes sense.

A posteriori, we check that the above defined F satisfies the requirements in the assertion of the lemma, that is, (1.12) holds near 0. We have

$$\mathcal{J}_F(x) = \mathcal{J}_{G^{-1}}(x) = \big(\mathcal{J}_G(G^{-1}(x))\big)^{-1};$$

hence (1.12) is equivalent to $X(x) = \mathcal{J}_G(G^{-1}(x))\, (1, 0, \ldots, 0)^T$; the latter is the first column of $\mathcal{J}_G(G^{-1}(x))$ and (see (1.16)) this column vector is given by

$$\left.\frac{\partial}{\partial t}\right|_{(t,s)=G^{-1}(x)} G(t; s) = \left.\frac{\partial}{\partial t}\right|_{(t,s)=G^{-1}(x)} \gamma_X(t, s_2\, w_2 + \cdots + s_N\, w_N)$$

$$= \big\{X\big(\gamma_X(t, s_2\, w_2 + \cdots + s_N\, w_N)\big)\big\}\big|_{(t,s)=G^{-1}(x)}$$

$$= \big\{X(G(t; s))\big\}\big|_{(t,s)=G^{-1}(x)} = X(x).$$

This demonstrates (1.12) and the proof is complete. $\qquad\qquad\qquad\square$

---

[2]Any other choice of the hypersurface S is possible, as long as $X(0)$ does not belong to the tangent space to S at 0. See [Lee (2013), Theorem 9.22].

1.3 Differentiation along a flow

The following very useful result will be used again and again in this book. Throughout the section, it is understood that $\Omega \subseteq \mathbb{R}^N$ is a non-void open set.

Theorem 1.25 (Differentiation along an integral curve). *Let X be a vector field on Ω. Let $\gamma_X(t)$ be any integral curve of X, defined on some open interval $I \subseteq \mathbb{R}$.*

(1). If X is of class C^1 and if $f \in C^1(\Omega)$ we have

$$\frac{\mathrm{d}}{\mathrm{d}t}(f(\gamma_X(t))) = (Xf)(\gamma_X(t)), \qquad t \in I. \tag{1.17}$$

(2). Let $k \geq 2$; if X of class C^{k-1} and if $f \in C^k(\Omega)$ we have

$$\frac{\mathrm{d}^k}{\mathrm{d}t^k}(f(\gamma_X(t))) = (X^k f)(\gamma_X(t)), \qquad t \in I. \tag{1.18}$$

(3). Let $n \in \mathbb{N}$ and suppose that X is of class C^n, and that $f \in C^{n+1}(\Omega)$; if $x \in \Omega$ and $\gamma_{X,x}(t)$ is the maximal integral curve of X starting at x, by Taylor's formula with an integral remainder we have the identity

$$f(\gamma_{X,x}(t)) = \sum_{k=0}^{n} \frac{(X^k f)(x)}{k!} t^k + \frac{1}{n!} \int_0^t (t-s)^n (X^{n+1} f)(\gamma_{X,x}(s))\,\mathrm{d}s, \tag{1.19}$$

holding true for every $t \in \mathcal{D}(X, x)$. In particular, if we replace f with any of the component functions of the identity map $I(x) = x$, we have the following identity (between vectors of $\mathbb{R}^N$)

$$\Psi_x^X(t) = \sum_{k=0}^{n} \frac{(X^k I)(x)}{k!} t^k + \frac{1}{n!} \int_0^t (t-s)^n (X^{n+1} I)\big(\Psi_s^X(x)\big)\,\mathrm{d}s, \tag{1.20}$$

valid for any $t \in \mathcal{D}(X, x)$.

(4). Let $n \in \mathbb{N}$ and suppose that X is of class C^n, and that $f \in C^{n+1}(\Omega)$; if $x \in \Omega$, we also get the following Taylor formulae with a Peano remainder of degree $n + 1$:

$$f(\gamma_{X,x}(t)) = \sum_{k=0}^{n+1} \frac{(X^k f)(x)}{k!} t^k + o(t^{n+1}), \quad \text{as } t \to 0, \tag{1.21}$$

$$\Psi_x^X(t) = \sum_{k=0}^{n+1} \frac{(X^k I)(x)}{k!} t^k + o(t^{n+1}), \quad \text{as } t \to 0. \tag{1.22}$$

(5). Consequently, if X is a smooth v.f. and if $f \in C^\infty(\Omega)$, the Maclaurin series of the function $t \mapsto f(\gamma_{X,x}(t))$ is

$$\sum_{k=0}^{\infty} \frac{(X^k f)(x)}{k!} t^k. \tag{1.23}$$

The simple proof is left as an exercise.

If $f \in C^1(\Omega, \mathbb{R}^m)$ (for some $m \geq 1$), the statement of Thm. 1.25 holds true unaltered, with our usual convention that Xf is the vector of $\mathbb{R}^m$ whose components are $Xf_1, \ldots, Xf_m$, where $f_1, \ldots, f_m$ are the component functions of f. For example, with the flow notation in (1.9), one has

$$\frac{\mathrm{d}}{\mathrm{d}t}(f(\Psi_t^X(x))) = (Xf)(\Psi_x^X(t)), \qquad \forall\, f \in C^1(\Omega, \mathbb{R}^m),$$

valid for $x \in \Omega$ and $t \in \mathcal{D}(X, x)$ and whenever X is of class C^1. More generally, for any $k \geq 2$, we have

$$\frac{\mathrm{d}^k}{\mathrm{d}t^k}(f(\Psi_t^X(x))) = (X^k f)(\Psi_x^X(t)), \qquad \forall\, f \in C^k(\Omega, \mathbb{R}^m),$$

valid for $x \in \Omega$ and $t \in \mathcal{D}(X, x)$, and whenever X is of class C^{k-1}.

For future reference, we explicitly write the following expansions:

$$\Psi_t^X(x) = x + t\, X(x) + o(t), \quad \text{as } t \to 0, \tag{1.24}$$

$$\Psi_t^X(x) = x + t\, X(x) + \frac{t^2}{2}\, X^2 I(x) + o(t^2), \quad \text{as } t \to 0. \tag{1.25}$$

Both are valid if X is of class C^1.

1.4 The equation of variation for the flow

Let X be a C^1 v.f. on Ω, and let $x \in \Omega$ be fixed. As discussed in Rem. 1.12, the map $t \mapsto \Psi_t^X(x)$ is C^2 on $\mathcal{D}(X, x)$, and (by the very definition of flow!) one has

$$\frac{\mathrm{d}}{\mathrm{d}t}\Psi_t^X(x) = X(\Psi_t^X(x)), \quad \Psi_0^X(x) = x. \tag{1.26}$$

The map $\mathcal{D}(X) \ni (t, x) \mapsto \Psi_t^X(x)$ is C^1 and it admits continuous mixed second-order partial derivatives

$$\frac{\partial^2}{\partial t \partial x_j}\Psi_t^X(x) \qquad (\text{for any } j = 1, \ldots, N),$$

and the order of differentiation can be interchanged. Thus we can take derivatives wrt x in (1.26) and interchange with $\frac{\mathrm{d}}{\mathrm{d}t}$, to obtain the notable fact

$$\frac{\mathrm{d}}{\mathrm{d}t}\mathcal{J}_{\Psi_t^X}(x) = \mathcal{J}_{XI}(\Psi_t^X(x)) \cdot \mathcal{J}_{\Psi_t^X}(x), \qquad \mathcal{J}_{\Psi_0^X}(x) = \mathbb{I}_N. \tag{1.27}$$

Here and throughout $\mathbb{I}_N$ denotes the $N \times N$ identity matrix. Incidentally, this shows that the matrix-valued map

$$\mathcal{D}(X, x) \ni t \mapsto W_x(t) := \mathcal{J}_{\Psi_t^X}(x)$$

solves the (matrix) *linear ODE system* (called **equation of variation** of (1.26))

$$\begin{cases} \frac{\mathrm{d}}{\mathrm{d}t}W_x(t) = A_x(t) \cdot W_x(t) \\[2mm] W_x(0) = \mathbb{I}_N, \end{cases} \qquad \text{where } A_x(t) := \mathcal{J}_{XI}(\Psi_t^X(x)). \tag{1.28}$$

Note that, since X is C^1, the matrix $A_x(t)$ has continuous entries. From (1.27) we also infer the integral identity

$$\mathcal{J}_{\Psi_t^X}(x) = \mathbb{I}_N + \int_0^t \mathcal{J}_{XI}(\Psi_\tau^X(x)) \cdot \mathcal{J}_{\Psi_\tau^X}(x)\, \mathrm{d}\tau, \qquad \forall\, t \in \mathcal{D}(X, x). \tag{1.29}$$

Remark 1.26. From general results on linear systems of ODEs (see Exr. 1.8), from the homogeneous system (1.28), we derive that, setting

$$w_x(t) := \det\left(\mathcal{J}_{\Psi_t^X}(x)\right), \qquad t \in \mathcal{D}(X, x),$$

the Wronskian function $t \mapsto w_x(t)$ satisfies the linear *scalar* ODE

$$w_x'(t) = \mathrm{trace}(A_x(t))\, w_x(t), \qquad t \in \mathcal{D}(X, x). \tag{1.30}$$

Solving this (separable) ODE, and on account of[3]

$$\mathrm{trace}(A_x(t)) \overset{(1.28)}{=} \mathrm{trace}\left(\mathcal{J}_{XI}(\Psi_t^X(x))\right) = \mathrm{div}(X)(\Psi_t^X(x)),$$

we obtain

$$w_x(t) = w_x(0)\, \exp\left(\int_0^t \mathrm{div}(X)(\Psi_\tau^X(x))\, \mathrm{d}\tau\right).$$

Since $w_x(0) = \det(\mathcal{J}_{\Psi_0^X}(x)) = \det(\mathbb{I}_N) = 1$, we finally derive

$$w_x(t) = \exp\left(\int_0^t \mathrm{div}(X)(\Psi_\tau^X(x))\, \mathrm{d}\tau\right).$$

1.4.1 *A Liouville Theorem for ODEs*

In the above computations we have proved the following notable result, also referred to as the Abel-Jacobi-Liouville identity.

Theorem 1.27 (Liouville identity). *Let X be a C^1 vector field on Ω, and let the point $x \in \Omega$ be fixed. Then, for any $t \in \mathcal{D}(X, x)$,*

$$\det\left(\mathcal{J}_{\Psi_t^X}(x)\right) = \exp\left(\int_0^t \mathrm{div}(X)(\Psi_\tau^X(x))\, \mathrm{d}\tau\right). \tag{1.31}$$

Incidentally, this shows that

$$\det\left(\mathcal{J}_{\Psi_t^X}(x)\right) > 0, \quad \forall\, t \in \mathcal{D}(X, x). \tag{1.32}$$

An important consequence of the equation of variation (1.28) is the following result, often called **Liouville's Theorem on flows**. In the sequel, given a Lebesgue measurable set $A \subseteq \mathbb{R}^N$, by $\mathrm{meas}(A)$ we denote the (Lebesgue) measure of A.

---

[3]For the sake of brevity, we denote by $\mathrm{div}(X)$ the function $\mathrm{div}(XI)$, where as usual XI denotes the coefficient vector of the v.f. X.

Theorem 1.28 (Variation of the measure under a flow). *Let X be a C^1 v.f. on the open set Ω. Let $t \in \mathbb{R}$ be such that $\Omega_t^X \neq \varnothing$. Then, for any measurable set $A \subseteq \Omega_t^X$,*

$$\mathrm{meas}(\Psi_t^X(A)) = \int_A \exp\left(\int_0^t \mathrm{div}(X)(\Psi_\tau^X(x))\,\mathrm{d}\tau\right)\mathrm{d}x. \tag{1.33}$$

If $t > 0$ and if $A \subset \Omega_t^X$ is compact, one has either of the formulae

$$\begin{aligned}
\frac{\mathrm{d}}{\mathrm{d}s}\Big(\mathrm{meas}(\Psi_s^X(A))\Big) &= \int_{\Psi_s^X(A)} \mathrm{div}(X)(x)\,\mathrm{d}x \\
&= \int_A \mathrm{div}(X)(\Psi_s^X(x))\,\det\left(\eth_{\Psi_s^X}(x)\right)\mathrm{d}x,
\end{aligned} \tag{1.34}$$

for any $s \in [0,t]$. On account of (1.31), this also gives

$$\frac{\mathrm{d}}{\mathrm{d}s}\Big(\mathrm{meas}(\Psi_s^X(A))\Big) = \int_A \mathrm{div}(X)(\Psi_s^X(x))\,\exp\left(\int_0^s \mathrm{div}(X)(\Psi_\tau^X(x))\,\mathrm{d}\tau\right)\mathrm{d}x.$$

Analogous formulae hold for $t < 0$ (and $s \in [t,0]$).

In particular, if $\mathrm{div}(X) \equiv 0$ on Ω then $\mathrm{meas}(\Psi_s^X(A))$ is constant wrt s.

Note that the first identity in (1.34) gives a meaningful interpretation of the divergence of a v.f. X and its role in the flow of X.

Proof. Let Ω, X, t be as in the assertion. If $A \subseteq \Omega_t^X$ is measurable we have

$$\mathrm{meas}(\Psi_t^X(A)) = \int_{\Psi_t^X(A)} \mathrm{d}y = \int_A \left|\det\left(\eth_{\Psi_t^X}(x)\right)\right|\mathrm{d}x,$$

where we used the change of variable $y = \Psi_t^X(x)$ (with $x \in A$). From (1.31) and (1.32) we get (1.33). Now, if $A \subset \Omega_t^X$ is compact and $0 \leq s \leq t$, we have (see Rem. 1.10) $A \subset \Omega_t^X \subseteq \Omega_s^X$. Thus we are entitled to put $t = s$ in (1.33) and we can differentiate wrt $s \in [0,t]$ by passing the derivative under the integral sign (remember that A is compact and that $\Psi_s^X(x)$ has continuous second-order derivatives of the form $\partial^2/\partial s\,\partial x_j$). From the ODE in (1.30),

$$\begin{aligned}
\frac{\mathrm{d}}{\mathrm{d}s}\Big(\mathrm{meas}(\Psi_s^X(A))\Big) &= \int_A \frac{\mathrm{d}}{\mathrm{d}s}\left\{\det\left(\eth_{\Psi_s^X}(x)\right)\right\}\mathrm{d}x \\
&= \int_A \mathrm{div}(X)(\Psi_s^X(x))\,\det\left(\eth_{\Psi_s^X}(x)\right)\mathrm{d}x.
\end{aligned}$$

This is the second identity in (1.34); by the substitution $y = \Psi_s^X(x)$ (with $x \in A$), we get the first one as well. $\qquad\square$

1.4.2 Further regularity of the flow

We end this section with some technical results that will be helpful in Sec. 1.5 and in Chap. 4. First we give a remark showing why the ensuing Lemmas 1.30 and 1.31 contain further regularities of the flow, not even contained in the fine regularity analysis in Rem. 1.12.

Remark 1.29. If X is C^1 on Ω, we know that the map $(t,x) \mapsto \Psi_t^X(x)$ is C^1 on $\mathcal{D}(X)$, and it admits continuous second-order partial derivatives provided that t appears at least once. As a consequence, the matrix-valued map

$$(t,x) \mapsto M(t,x) := \partial_{\Psi_t^X}(x)$$

is continuously differentiable in t, but it is only *continuous* wrt x. It is therefore not at all obvious if, given a function $f(t)$ (even C^1), the composite map

$$F(t) := M(t,f(t)) = \partial_{\Psi_t^X}(f(t))$$

is differentiable wrt t. When f is continuous we shall prove in Lem. 1.30 that $F(t)$ is differentiable at $t = 0$ via the equation of variation under its integral form (1.29); when $f(t) = \Psi_{-t}^X(x)$ we shall prove in Lem. 1.31 the differentiability of $F(t)$ for any t, via the semigroup property in Prop. 1.14.   ♯

Lemma 1.30. *Let X be C^1 vector field on Ω, and let $x \in \Omega$ be fixed. Suppose that (for some $\varepsilon > 0$) $f : [-\varepsilon, \varepsilon] \longrightarrow \Omega$ is a continuous function such that $f(0) = x$. Then the matrix-valued function $F(t) = \partial_{\Psi_t^X}(f(t))$ is differentiable at $t = 0$ and*

$$\frac{\mathrm{d}}{\mathrm{d}t}\Big|_{t=0} \partial_{\Psi_t^X}(f(t)) = \partial_{XI}(x). \tag{1.35}$$

Proof. First it is simple to observe that $F(t)$ is well posed on a neighborhood of $t = 0$. We use (1.29) with $x = f(t)$: for small t, this identity gives

$$F(t) = \partial_{\Psi_t^X}(f(t)) = \mathbb{I}_N + \int_0^t \partial_{XI}(\Psi_\tau^X(f(t))) \cdot \partial_{\Psi_\tau^X}(f(t))\, \mathrm{d}\tau.$$

The above integrand function is continuous, hence the Mean Value Theorem for integrals ensure the existence of some $\tau(t)$ in $[0,1]$ such that

$$F(t) = \mathbb{I}_N + t\, \partial_{XI}(\Psi_{t\tau(t)}^X(f(t))) \cdot \partial_{\Psi_{t\tau(t)}^X}(f(t)).$$

As a consequence (as $F(0) = \mathbb{I}_N$) we get

$$\frac{F(t) - F(0)}{t} = \partial_{XI}(\Psi_{t\tau(t)}^X(f(t))) \cdot \partial_{\Psi_{t\tau(t)}^X}(f(t)).$$

Since $\tau(t)$ is bounded in $[0,1]$, continuity arguments prove that the limit as $t \to 0$ in the above rhs is $\partial_{XI}(x) \cdot \partial_I(x) = \partial_{XI}(x)$. This ends the proof.   □

Lemma 1.31. *Let X be C^1 v.f. on Ω, and let $x \in \Omega$ be fixed. Then the matrix-valued map*

$$F(t) = \partial_{\Psi_{-t}^X}(\Psi_t^X(x))$$

is differentiable wrt $t \in \mathcal{D}(X,x)$, and either of the following identity holds

$$\frac{\mathrm{d}}{\mathrm{d}t}\Big(\partial_{\Psi_{-t}^X}(\Psi_t^X(x))\Big) = -\partial_{\Psi_{-t}^X}(\Psi_t^X(x)) \cdot \partial_{XI}(\Psi_t^X(x))$$

$$= -\Big(\partial_{\Psi_t^X}(x)\Big)^{-1} \cdot \partial_{XI}(\Psi_t^X(x)). \tag{1.36}$$

Note the similarity of (1.36) with the equation of variation (1.27).

Proof. Let $x \in \Omega$ and $t \in \mathcal{D}(X, x)$. Then $F(t)$ is well posed, since Ψ^X_{-t} is C^1. Owing to $\Psi^X_{-t}(\Psi^X_t(x)) = x$ (see Prop. 1.14), we recognize that $F(t)$ is nothing but

$$\left(\mathcal{J}_{\Psi^X_t}(x) \right)^{-1},$$

hence it is C^1 in t due to the fine regularity properties of the flow (see Rem. 1.12). Now (1.36) follows from the equation of variation (1.27) (and from Exr. 1.12; for another proof of this lemma, see Exr. 1.13). $\qquad\square$

1.5 Flowing through $X, Y, -X, -Y$: commutators

The aim of this section is to give a first important interpretation of the commutator $[X, Y]$. We prove the following crucial Thm. 1.32, of which we shall provide another proof (in the case of *smooth* v.f.s) in Chap. 3, after the *exponential formalism* has been developed. Another interpretation will be given in Sec. 4.3 of Chap. 4.

Although these topics will intervene in later chapters, we remark here that the study of the composition of flows of vector fields is of paramount importance in this book. In order to attack the investigation of the composition of flows of two *different* vector fields, the Theorem of Campbell, Baker, Hausdorff and Dynkin (which we shall briefly call *the Exponential Theorem*) will be our fundamental ally. We shall understand this fact in Chap. 2.

Theorem 1.32. *Let X, Y be C^1 vector fields on Ω and let $x \in \Omega$. Then, as $t \to 0$,*

$$\Psi^{-Y}_t \circ \Psi^{-X}_t \circ \Psi^Y_t \circ \Psi^X_t(x) = x + t^2 \, [X, Y](x) + o(t^2). \tag{1.37}$$

Thus, the t-function on the lhs of (1.37) admits the second-derivative at $t = 0$ given by $2\,[X,Y](x)$. Moreover, if X, Y are C^2 (so that $[X,Y]$ is C^1), as $t \to 0$ one has

$$\Psi^{-Y}_t \circ \Psi^{-X}_t \circ \Psi^Y_t \circ \Psi^X_t(x) = \Psi^{[X,Y]}_{t^2}(x) + o(t^2). \tag{1.38}$$

Formula (1.37) gives a first interpretation of the commutator $[X, Y]$ as a "weight" of how much the composition $\Psi^{-Y}_t \circ \Psi^{-X}_t \circ \Psi^Y_t \circ \Psi^X_t$ differs from the identity map (at least for $t \sim 0$) up to t^2-accuracy. See also Fig. 1.5. Indeed it can be rewritten as

$$\lim_{t \to 0} \frac{\Psi^{-Y}_t \circ \Psi^{-X}_t \circ \Psi^Y_t \circ \Psi^X_t(x) - x}{t^2} = [X, Y](x). \tag{1.39}$$

The proof of Thm. 1.32 in the case when X and Y are C^2 is simpler: see Exr. 1.17.

Proof of Thm. 1.32. We remark that (1.38) follows from (1.37) since (see formula (1.24) with X replaced by $[X, Y]$ and t replaced by t^2)

$$\Psi^{[X,Y]}_{t^2}(x) = x + t^2 \, [X, Y](x) + o(t^2), \quad \text{as } t \to 0.$$

Hence we attack the proof of (1.37) which is quite laborious. For $x \in \Omega$ fixed, we set $F(t) := \Psi^{-Y}_t \circ \Psi^{-X}_t \circ \Psi^Y_t \circ \Psi^X_t(x)$. It is easy to check that F is well posed on some neighborhood $[-\varepsilon, \varepsilon]$ of $t = 0$. Since X, Y are C^1, we infer that $\Psi_t^{\pm X}(x)$ and $\Psi_t^{\pm Y}(x)$ are C^1 maps of (x, t) on their domains, so that F is C^1 on $[-\varepsilon, \varepsilon]$. We need to show that $F'(t)$ is differentiable at $t = 0$ with $F'(0) = 0$ and $F''(0) = 2[X, Y](x)$; then we apply Taylor's formula with Peano's remainder (Exr. 1.10) and we get (1.37).

In order to compute $F'(t)$ we are entitled to use the formula

$$F'(t) = \left(\frac{\partial}{\partial t_1} + \frac{\partial}{\partial t_2} + \frac{\partial}{\partial t_3} + \frac{\partial}{\partial t_4} \right)\Bigg|_{t_1 = t_2 = t_3 = t_4 = t} \Psi_{t_1}^{-Y} \circ \Psi_{t_2}^{-X} \circ \Psi_{t_3}^{Y} \circ \Psi_{t_4}^{X}(x).$$

By the Chain Rule, this gives $F'(t) = G_1(t) + G_2(t) + G_3(t) + G_4(t)$, where

$$G_1'(t) = -Y(\Psi_t^{-Y} \circ \Psi_t^{-X} \circ \Psi_t^{Y} \circ \Psi_t^{X}(x))$$

$$G_2'(t) = -\mathcal{J}_{\Psi_t^{-Y}}(\Psi_t^{-X} \circ \Psi_t^{Y} \circ \Psi_t^{X}(x))\, X(\Psi_t^{-X} \circ \Psi_t^{Y} \circ \Psi_t^{X}(x))$$

$$G_3'(t) = \mathcal{J}_{\Psi_t^{-Y}}(\Psi_t^{-X} \circ \Psi_t^{Y} \circ \Psi_t^{X}(x))\, \mathcal{J}_{\Psi_t^{-X}}(\Psi_t^{Y} \circ \Psi_t^{X}(x))\, Y(\Psi_t^{Y} \circ \Psi_t^{X}(x))$$

$$G_4'(t) = \mathcal{J}_{\Psi_t^{-Y}}(\Psi_t^{-X} \circ \Psi_t^{Y} \circ \Psi_t^{X}(x))\, \mathcal{J}_{\Psi_t^{-X}}(\Psi_t^{Y} \circ \Psi_t^{X}(x))$$

$$\mathcal{J}_{\Psi_t^{Y}}(\Psi_t^{X}(x))\, X(\Psi_t^{X}(x)).$$

Since the Jacobian matrix of the identity map is $\mathbb{I}_N$, we get

$$F'(0) = -Y(x) - X(x) + Y(x) + X(x) = 0.$$

In order to get the existence and the value of $F''(0)$, we show that each $G_i(t)$ (for $i = 1, \ldots, 4$) is differentiable at $t = 0$. This is trivial for G_1 since Y is C^1 and

$$G_1''(0) = -\mathcal{J}_{YI}(x)\, F'(0) = 0.$$

In order to show that each $G_i(t)$ ($i = 2, 3, 4$) is differentiable at $t = 0$, we apply Leibniz's Rule and the Chain Rule, together with a repeated application of formula (1.35) in Lem. 1.30; after some computation, we get

$$G_2''(0) = \mathcal{J}_{YI}(x)\, X(x) - \mathcal{J}_{XI}(x)\, (-X(x) + Y(x) + X(x))$$

$$G_3''(0) = -\mathcal{J}_{YI}(x)\, Y(x) - \mathcal{J}_{XI}(x)\, Y(x) + \mathcal{J}_{YI}(x)\, (Y(x) + X(x))$$

$$G_4''(0) = -\mathcal{J}_{YI}(x)\, X(x) - \mathcal{J}_{XI}(x)\, X(x) + \mathcal{J}_{YI}(x)\, X(x) + \mathcal{J}_{XI}(x)\, X(x).$$

Gathering all together, we obtain

$$\begin{aligned} F''(0) &= G_1''(0) + G_2''(0) + G_3''(0) + G_4''(0) \\ &= X(YI)(x) + X(XI)(x) - Y(XI)(x) - X(XI)(x) + \\ &\quad - Y(YI)(x) - Y(XI)(x) + Y(YI)(x) + X(YI)(x) + \\ &\quad - X(YI)(x) - X(XI)(x) + X(YI)(x) + X(XI)(x) \\ &= 2\, X(YI)(x) - 2\, Y(XI)(x) = 2[X, Y](x). \end{aligned}$$

This ends the proof. $\qquad\square$

Due to its role, we introduce a selected notation for the path in Thm. 1.32.

Definition 1.33. Let X and Y be C^1 v.f.s on the open set $\Omega \subseteq \mathbb{R}^N$. We set

$$\Gamma_t^{X,Y}(x) := \Psi_t^{-Y} \circ \Psi_t^{-X} \circ \Psi_t^Y \circ \Psi_t^X(x), \tag{1.40}$$

for every $x \in \Omega$ and $t \in \mathbb{R}$ such that the rhs is defined.

The reader should not confuse the (image set of the) curve $t \mapsto \Gamma_t^{X,Y}(x)$ (this curve being of class C^k if X, Y are of class C^k) with the set obtained by the union of the image sets of the integral curves

$$\{\Psi_s^X(x_1) : s \in [0,t]\}, \qquad \{\Psi_s^Y(x_2) : s \in [0,t]\},$$
$$\{\Psi_s^{-X}(x_3) : s \in [0,t]\}, \qquad \{\Psi_s^{-Y}(x_4) : s \in [0,t]\},$$

with $x_1 = x$, $x_2 = \Psi_t^X(x_1)$, $x_3 = \Psi_t^Y(x_2)$, $x_4 = \Psi_t^{-X}(x_3)$. In general, this union of curves may not at all be a C^1 manifold. See also Fig. 1.5.

In Thm. 1.32 we have proved that $\Gamma_t^{X,Y}(x)$ admits the second derivative wrt t at 0, and that it holds that

$$\Gamma_t^{X,Y}(x) = x + t^2\,[X,Y](x) + o(t^2) \quad \text{as } t \to 0. \tag{1.41}$$

Remark 1.34. From general results of ODE Theory (see App. B) we know that, for any compact set $K \subset \Omega$, there exists $\varepsilon = \varepsilon(K, X, Y, \Omega) > 0$ such that $\Gamma_t^{X,Y}(x)$ is well defined for any $t \in [-\varepsilon, \varepsilon]$ and any $x \in K$. Moreover, if X and Y are regular of class C^k (for $k = 1, 2, \ldots, \infty, \omega$), then

$$t \mapsto \Gamma_t^{X,Y}(x) \text{ is } C^{k+1} \quad \text{and} \quad (x,t) \mapsto \Gamma_t^{X,Y}(x) \text{ is } C^k,$$

on their respective domains. Moreover, there also exist the partial derivatives of the following form (and the order of differentiation can be interchanged)

$$\frac{\partial^{k+1}}{\partial t\, \partial x_{i_1} \ldots \partial x_{i_k}} \Gamma_t^{X,Y}(x),$$

for any choice of $i_1, \ldots, i_k \in \{1, \ldots, N\}$. ♯

1.6 The product of exponentials: motivations

Let $\Omega \subseteq \mathbb{R}^N$ be open, let X be a smooth vector field on Ω, and let $x \in \Omega$. If we take $f = I$ (the identity map of $\mathbb{R}^N$) in (1.23) we see that the Maclaurin series of the function $t \mapsto \gamma_{X,x}(t)$ is

$$\gamma_{X,x}(t) \sim \sum_{k=0}^{\infty} \frac{(X^k I)(x)}{k!}\, t^k. \tag{1.42}$$

Here and in the sequel, the symbol '$\sim$' means that what comes to its right is the Maclaurin series of what comes to its left.

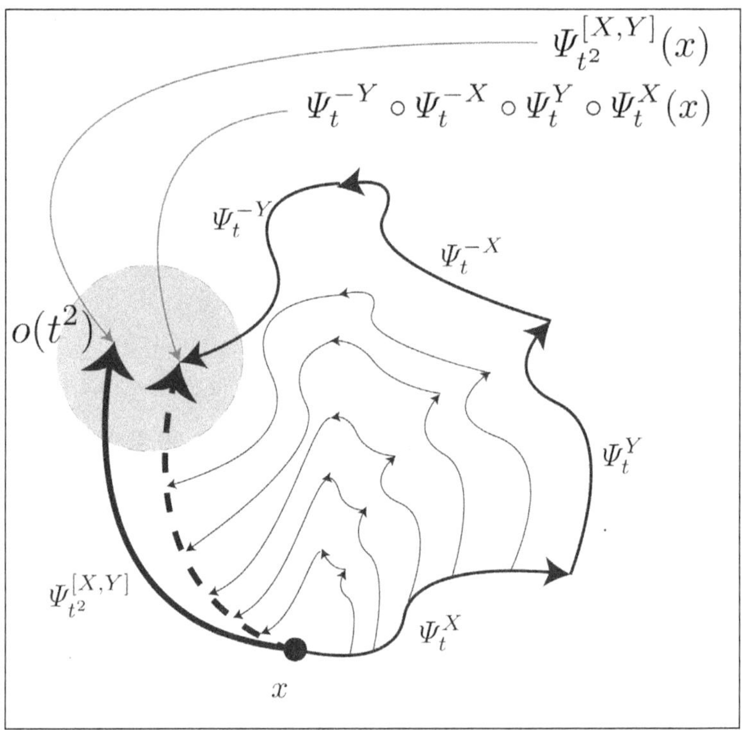

Fig. 1.5   The approximation of the flow of the commutator $[X, Y]$ (up to time t^2) by means of the flows of the vector fields $\pm X$, $\pm Y$. The dashed line represents the image set of the curve $t \mapsto \Gamma_t^{X,Y}(x)$ introduced in Def. 1.33, whereas the larger bold "square-shaped" path is the union of four image sets of integral curves of $\pm X$, $\pm Y$, and it shares with $\Gamma_t^{X,Y}(x)$ the starting and the ending points; the smaller "square-shaped" paths give values of $\Gamma_t^{X,Y}(x)$ for smaller times t. The grey set aims at picturing the error (of the order of $o(t^2)$) between $\Gamma_t^{X,Y}(x)$ and $\Psi_{t^2}^{[X,Y]}(x)$.

We observe that the formal series in (1.42) "looks like" an exponential (the "exponential" of tX evaluated at x, whatever this may mean). It is for this reason that we introduce a widely employed notation for the integral curve $\gamma(t, X, x)$.

Notation 1.35. Let X be a C^1 vector field on Ω; we use the notation

$$\exp(tX)(x) := \gamma(t, X, x),$$
$$\text{where } x \in \Omega \text{ and } t \in \mathcal{D}(X, x). \tag{1.43}$$

When time $t = 1$ is allowed, we call any map of the form $x \mapsto \exp(X)(x)$ an **exp-like map**, or simply the **exponential** of the vector field X.

The reader familiar with Lie group theory will not confuse this notion of exponential map with the map Exp on a Lie group $\mathbb{G}$, even though the latter is related to the former (here e is the identity of $\mathbb{G}$):

$$\mathrm{Exp}(X) = \exp(t\,X)(e)\Big|_{t=1}, \quad \text{for } X \in \mathrm{Lie}(\mathbb{G}).$$

According to Def. 1.8, the exp-like map $\exp(X)$ is nothing but the flow of X at time $t = 1$; in general (if time $t = 1$ is allowed at least for one point of Ω), this flow is defined only on the open subset Ω_1^X of Ω.

Remark 1.36. We warn the reader that the notation $\exp(tX)(x)$ is somewhat misleading (and it must be handled with care) since it does not involve t and X separately, like in $\gamma(t, X, x)$. For example, what does $\exp(2X)(x)$ mean? Is it $\gamma(2, X, x)$, or $\gamma(1, 2X, x)$, or $\gamma(4, \frac{1}{2}X, x)$?

In some sense this abuse of notation is justified by the homogeneity property in Prop. 1.14-(3). For instance, if $t \in \mathcal{D}(X, x)$, we can think of $\exp(tX)(x)$ as $\gamma(t, X, x)$ (this is its very definition) or as $\gamma(1, tX, x)$ (when $t \in \mathcal{D}(X, x)$, the well posedness of $\gamma(1, tX, x)$, i.e., $1 \in \mathcal{D}(tX, x)$, is contained in Prop. 1.14). ♮

Remark 1.37. If X is a vector field of class C^ω on Ω, then –by general results of ODEs– any integral curve of X is also of class C^ω on its maximal domain. This shows that for a suitable $\varepsilon > 0$ we have the *identity* of the function $\gamma_{X,x}(t)$ with its Maclaurin expansion on the interval $(-\varepsilon, \varepsilon)$ (see (1.23)):

$$\exp(tX)(x) = I(x) + t\,XI(x) + \frac{t^2}{2}\,X^2I(x) + \frac{t^3}{3!}\,X^3I(x) + \cdots, \qquad (1.44)$$

for every $|t| < \varepsilon$. Note that in computing X^nI, only the first order part of the higher order operator $X^n = X \circ \cdots \circ X$ (n times) matters. Hence, in simple cases (1.44) can be used to integrate explicitly the ODE defining $\gamma(t, x, X)$ (Exr. 1.20).

When X is C^ω, the well-known Cauchy-Hadamard Theorem, providing the radius of convergence ρ of a complex power series $\sum_k a_k(z - z_0)^k$ by the formula $1/\rho = \limsup_{k \to \infty} |a_k|^{1/k}$, ensures that the series in (1.44) converges for $|t| < \varepsilon$ if and only if $\rho \geq \varepsilon$, whence

$$\limsup_{k \to \infty} \left\| \frac{X^kI(x)}{k!} \right\|^{1/k} \leq 1/\varepsilon.$$

This gives the following bound for a C^ω vector field X (where M_x is a suitable positive constant depending on x)

$$\|X^kI(x)\| \leq (M_x)^k\,k!, \qquad k \in \mathbb{N}.$$

Since the same (small) ε can be chosen for every x in a fixed compact subset of Ω (see Sec. B.1 in App. B), we derive that, for any compact set $K \subset \Omega$ and any real analytic v.f. X on Ω, there exists $M = M(X, K) > 0$ such that

$$\|X^kI(x)\| \leq M^k\,k!, \qquad \text{for every } k \in \mathbb{N} \text{ and every } x \in K.$$

The basic result which will make a bridge with the algebraic investigations of the next Chap. 2 is the following one:

Theorem 1.38 (Expansion of the composition of two exponentials of v.f.s). *Let $X, Y \in \mathcal{X}(\Omega)$, $x \in \Omega$ and $f \in C^\infty(\Omega)$. Then, the Taylor series at $(0, 0)$ of*

$$(t, s) \mapsto f\big(\exp(tY)\big(\exp(sX)(x)\big)\big)$$

(which is well defined on an open neighborhood of $(0,0) \in \mathbb{R}^2$) is

$$\sum_{i,j\geq 0} \frac{(X^i Y^j f)(x)}{i!\, j!}\, s^i t^j. \tag{1.45}$$

Note the reverse order of X, Y in (1.45) with respect to their order of appearance in $\exp(tY)\big(\exp(sX)(x)\big)$ (when we read lexicographically).

Proof. Let x, X, Y be as in the assertion. The curve $\exp(sX)(x) = \gamma(s, X, x)$ is defined for $s \in \mathcal{D}(X, x)$ which is an open interval containing 0. Let $\varepsilon > 0$ be such that $[-\varepsilon, \varepsilon] \subset \mathcal{D}(X, x)$. Then the set $K := \{\gamma(s, X, x) : s \in [-\varepsilon, \varepsilon]\}$ is a compact subset of Ω. Hence, by general results of ODEs (see App. B), there exists $\delta > 0$ such that $\exp(tY)(k) = \gamma(t, Y, k)$ exists for every $t \in [-\delta, \delta]$ and *uniformly for every* $k \in K$. This shows that the function

$$F(s,t) := f\big(\exp(tY)\big(\exp(sX)(x)\big)\big) \tag{1.46}$$

is well posed for $(s, t) \in U := [-\varepsilon, \varepsilon] \times [-\delta, \delta]$, and it is smooth in the interior of U, an open neighborhood of $(0,0)$ in $\mathbb{R}^2$. See Fig. 1.6.

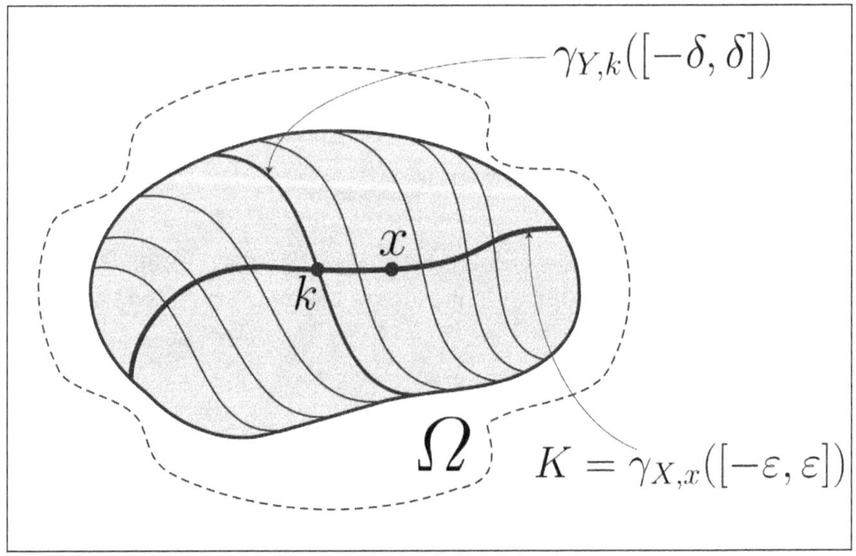

Fig. 1.6 The domain of the map $F(s, t)$ in (1.46).

For every $i, j \in \mathbb{N} \cup \{0\}$ we have the following calculation, based on (1.18):

$$\frac{\partial^{i+j} F}{\partial s^i \partial t^j}(0,0) = \frac{\partial^i}{\partial s^i}\Big|_{s=0}\left(\frac{\partial^j}{\partial t^j}\Big|_{t=0} F(s,t)\right)$$

$$= \frac{\partial^i}{\partial s^i}\Big|_{s=0}\left(\frac{\partial^j}{\partial t^j}\Big|_{t=0} f\big(\gamma_Y(t,z)\big)\right) \qquad \text{(with } z = \exp(sX)(x))$$

$$\stackrel{(1.18)}{=} \frac{\partial^i}{\partial s^i}\Big|_{s=0}\left((Y^j f)\big(\exp(sX)(x)\big)\right) = \frac{\partial^i}{\partial s^i}\Big|_{s=0}\left((Y^j f)(\gamma_X(s,x))\right)$$

$$\stackrel{(1.18)}{=} X^i(Y^j f)(x).$$

We have therefore proved that

$$\frac{\partial^{i+j} F}{\partial s^i \partial t^j}(0,0) = (X^i Y^j f)(x), \qquad \forall \, i, j \geq 0. \tag{1.47}$$

Since the Taylor series of $F(s,t)$ at $(0,0)$ is given by (see also Exr. 1.21),

$$\sum_{i,j \geq 0} \frac{\partial^{i+j} F(0,0)}{\partial s^i \partial t^j} \frac{s^i t^j}{i! \, j!},$$

then (1.45) follows from (1.47). $\qquad\qquad\square$

Now, *the series in (1.45) can formally be rewritten as the product of two formal power series of exponential type* (acting on f):

$$\text{Formally:} \qquad \sum_{i,j \geq 0} \frac{X^i Y^j}{i! \, j!} s^i t^j = \sum_{i \geq 0} \frac{(sX)^i}{i!} \sum_{j \geq 0} \frac{(tY)^j}{j!} =: e^{sX} e^{tY}.$$

If we are able to produce identities for the formal product $e^x e^y$ in the setting of the formal power series in two non-commuting indeterminates x and y, then we shall be able to improve our knowledge of the Taylor series (1.45), whence (by using exact expansions with remainders) we shall highly improve our knowledge of the composition of flows $\exp(tY)\big(\exp(sX)(x)\big)$.

The reader may still wonder why the study of the composition of exp-like maps is so important. The are many good answers (one of the most convincing ones will be given when dealing with the Connectivity Theorem in Chap. 6); presently, we confine ourselves to describing what happens to the composition of exp-like maps in the simplest of all *subelliptic* settings: the first Heisenberg group. Here we anticipate some terminology on Lie groups.

Example 1.39. Let us consider on $\mathbb{R}^3$ the vector fields

$$X_1 = \partial_{x_1} + 2x_2\,\partial_{x_3} \quad \text{and} \quad X_2 = \partial_{x_2} - 2x_1\,\partial_{x_3}.$$

These v.f.s are Lie-generators of the Lie algebra of the first Heisenberg group in $\mathbb{R}^3$, which is nilpotent of step 2. See Exm. C.28 (page 401).

Let us start from the origin $P_0 = (0,0,0)$ of $\mathbb{R}^3$ and let us run through the integral curve of X_1; after time t we get to the point $P_1 = \exp(tX_1)(P_0) = (t,0,0)$.

Then, if we proceed along the integral curve of X_2, after time t we get to the point $P_2 = \exp(tX_2)(P_1) = (t, t, -2t^2)$. Let us go two steps further: from P_2 we pass to $P_3 = \exp(-tX_1)(P_2) = (0, t, -4t^2)$ (along the integral curve of $-X_1$ at time t) and finally from P_3 we pass to $P_4 = \exp(-tX_2)(P_3) = (0, 0, -4t^2)$ (along the integral curve of $-X_2$ at time t). See Fig. 1.7.

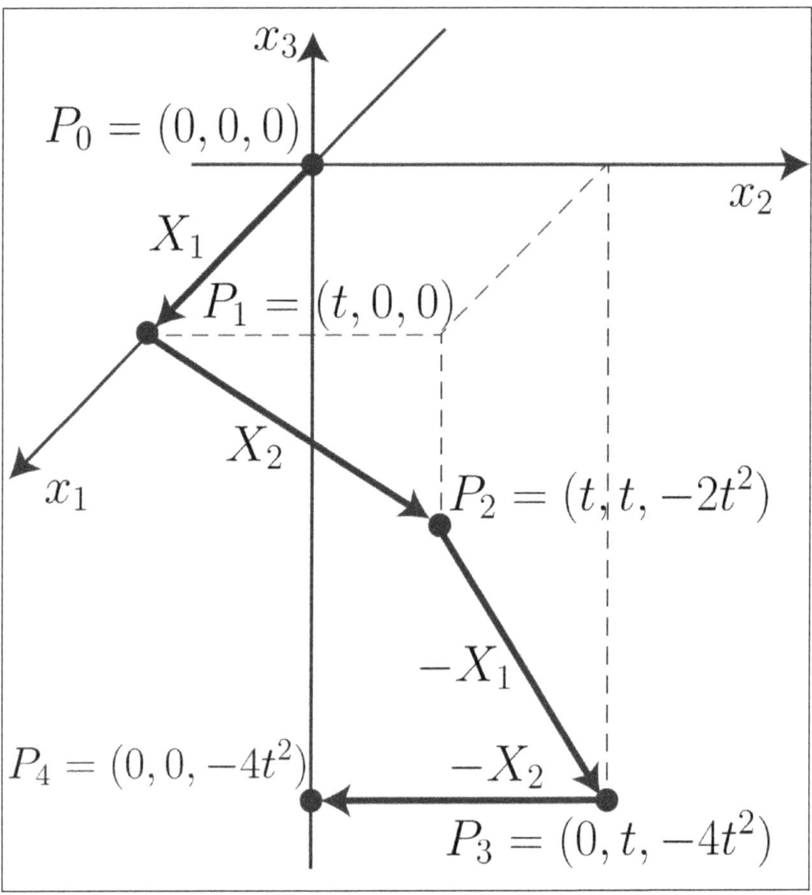

Fig. 1.7   How to obtain a (negative) variation along the x_3-coordinate, by flowing along the "horizontal" integral curves of the v.f.s $\pm X_1, \pm X_2$ in the Heisenberg group.

We have found a way to obtain a variation[4] in the third component of the starting point $(0, 0, 0)$, just proceeding along the (later to be called) "horizontal" directions of the vector fields $\pm X_1, \pm X_2$. This is a notable fact since, in terms of

---

[4]Actually we have obtained a negative variation, but we can get a positive one by reversing the roles of X_1 and X_2:

$$(0,0,0) \xrightarrow{\exp(tX_2)} Q_1 \xrightarrow{\exp(tX_1)} Q_2 \xrightarrow{\exp(-tX_2)} Q_3 \xrightarrow{\exp(-tX_1)} (0, 0, 4t^2).$$

coefficient vectors, $\partial_3 \equiv (0,0,1)$ *does not belong* to the span of $X_1 \equiv (1,0,2x_2)$ and $X_2 \equiv (0,1,-2x_1)$ for any choice of (x_1, x_2, x_3)!

In some sense, we have completely *regained* the direction of the vector field ∂_3 which, together with X_1 and X_2, gives a basis of $\mathbb{R}^3$ at any point:

$$\det \begin{pmatrix} 1 & 0 & 0 \\ 0 & 1 & 0 \\ 2x_2 & -2x_1 & 1 \end{pmatrix} \neq 0, \quad \forall \, (x_1, x_2, x_3) \in \mathbb{R}^3.$$

We plainly see that the fact that $[X_1, X_2]$ is proportional to X_3 is not mere chance. Indeed, as seen in Thm. 1.32, suitable compositions of exp-like maps of two v.f.s X_1 and X_2 can carry us (approximately) where the bracket $[X_1, X_2]$ leads us to.

This is a typical *subelliptic phenomenon*: a set of m vector fields, with m strictly smaller than the dimension N of the ambient space (here $m = 2$ and $N = 3$), may suffice to recover all the N directions of space. Note that

$$P_4 = \Big(\exp(-tX_2) \circ \exp(-tX_1) \circ \exp(tX_2) \circ \exp(tX_1) \Big)(0).$$

Why should we stop here? We may want to obtain approximations of commutators of lengths 3, 4, 5,...; this is not visible in the case of the Heisenberg group because all the commutators of lengths ≥ 3 of X_1, X_2 are vanishing, but in more general (e.g., non-nilpotent) frameworks the problem of the approximation of long commutators is crucial (we shall see this in Chap. 3). ♯

Now, this should be enough to convince the reader of the opportunity of studying more closely the composition of exp-like maps. Our starting point is the abstract algebraic setting, carried out in the next Chap. 2.

1.7 Exercises of Chap. 1

Exercise 1.1. Consider the smooth vector field $X = x^2\, \partial_x$ on $\mathbb{R}^1$. For every $x \in \mathbb{R}$, determine explicitly the integral curve $\gamma(t, X, x)$. Show that the maximal domain of this curve is

$$\mathcal{D}(X, x) = \begin{cases} (-\infty, 1/x), & \text{if } x > 0 \\ (-\infty, \infty), & \text{if } x = 0 \\ (1/x, \infty), & \text{if } x < 0. \end{cases}$$

Note that $\bigcap_{x \in \mathbb{R}} \mathcal{D}(X, x) = \{0\}$. Derive that, for every $t \neq 0$, the flow Ψ_t^X is not defined on the whole of $\mathbb{R}$ (determine the actual domain of Ψ_t^X).

Check the validity, in this explicit example, of the assertions in Rem. 1.10.

Exercise 1.2. Prove property (3) in Prop. 1.14 by imitating the proof of (1).

[*Hint:* Suppose $\alpha t \in \mathcal{D}(X, x)$ and consider the curve $\mu(r) := \gamma_X(\alpha r, x)$ defined for $r \in \frac{1}{\alpha} \mathcal{D}(X, x)$...]

Exercise 1.3. Let X be a C^1 v.f. Prove this restatement of the semigroup property:
if $s \in \mathcal{D}(X, x)$ and $t \in \mathcal{D}(X, \gamma_X(s, x))$, then $\mathcal{D}(X, x)$ coincides with the set $s + \mathcal{D}(X, \gamma_X(s, x))$, and one has $\gamma_X(t, \gamma_X(s, x)) = \gamma_X(t + s, x)$.

[*Hint:* Prove that $\nu(r) := \gamma_X(r + s, x)$ (defined on a suitable set...) solves the same Cauchy problem defining $\mu(r) := \gamma_X(r, \gamma_X(s, x))$.]

Exercise 1.4. Prove that, given a C^1 v.f. on Ω, for any $x \in \Omega$ one has

$$t \in \mathcal{D}(X, x) \qquad \Longrightarrow \qquad \Psi_t^X(\Omega_t^X) = \Omega_{-t}^X.$$

[*Hint:* Point (2) of Prop. 1.14 on page 11 gives

$$(\bigstar): \qquad t \in \mathcal{D}(X, x) \quad \Longrightarrow \quad \begin{pmatrix} -t \in \mathcal{D}(X, \Psi_t^X(x)) \\ \text{and } \Psi_{-t}^X(\Psi_t^X(x)) = x \end{pmatrix}.$$

Recognize that this gives $\Psi_t^X(\Omega_t^X) \subseteq \Omega_{-t}^X$. For the reverse inclusion, observe that, if $y \in \Omega_{-t}^X$ then $-t \in \mathcal{D}(X, y)$ so that (again from $(\bigstar)$)

$$t \in \mathcal{D}(X, \Psi_{-t}^X(y)),$$

which incidentally gives $\Psi_{-t}^X(y) \in \Omega_t^X$ and $y = \Psi_t^X(\Psi_{-t}^X(y)) \in \Psi_t^X(\Omega_t^X)$.]

Exercise 1.5. Prove the assertion in Rem. 1.16.

[*Hint:* Let X be as in Rem. 1.16 and set $\gamma(t) := F(t, x)$ (for $t \in I_x$); by differentiating (1.10) wrt s at $s = 0$, show that γ is an integral curve of X starting at x; thus $I_x \subseteq \mathcal{D}(X, x)$ and $F(t, x) = \Psi_t^X(x)$ for $(t, x) \in A$.]

Exercise 1.6. Prove the following facts.

(1) Consider the vector fields in $\mathbb{R}^2$ defined by $X_1 = 6\,x_2^2\,\partial_{x_1}$, $X_2 = x_1\,\partial_{x_2}$. Prove that X_1, X_2 are global but $X_1 + X_2$ is not.

[*Hint:* Denote the integral curve of $X_1 + X_2$ which starts at the point $(-2, 1)$ by $(\gamma_1(t), \gamma_2(t))$; show that $\ddot{\gamma}_2 = 6\,\gamma_2^2$ (plus suitable initial data on $\gamma_2(0), \dot{\gamma}_2(0)$), and this second order ODE is solved by $\gamma_2(t) = (1 + t)^{-2}$...]

(2) Consider the vector fields in $\mathbb{R}^2$ defined by $X_1 = (1 + x_2^2)\partial_{x_1}$, $X_2 = x_1\,\partial_{x_2}$. Prove that X_1, X_2 are global but $[X_1, X_2]$ is not global.

Exercise 1.7. Let $g : \Omega \subseteq \mathbb{R}^N \to O \subseteq \mathbb{R}^m$, $f : O \to \mathbb{R}^n$ be of class C^1, where Ω and O are open sets. Let also X be a vector field on Ω. Prove the formula

$$X(f \circ g)(x) = J_f(g(x)) \cdot Xg(x), \quad x \in \Omega.$$

[*Hint:* See formula (1.5) on page 4...]

Exercise 1.8. Provide all the details of the following classical argument of ODE Theory, used in this chapter.

(1) Consider a matrix-valued continuous function $I \ni t \mapsto A(t)$ (where $A(t)$ is an $n \times n$ real or complex matrix and $I \subseteq \mathbb{R}$ is an open interval). Let $\{u_1(t), \ldots, u_n(t)\}$ be linearly independent solutions of the ODE system $u'(t) = A(t)\, u(t)$ (here $u(t)$ is written as a column vector of $\mathbb{R}^n$). Let $U(t)$ be the matrix whose columns are $u_1(t), \ldots, u_n(t)$. Then recognize that $U'(t) = A(t)\, U(t)$, for $t \in I$.

(2) Consider the Wronskian function $w(t) = \det U(t)$. Write $U(t)$ in rows:
$$U(t) = \begin{pmatrix} R_1(t) \\ \vdots \\ R_n(t) \end{pmatrix}, \quad \text{where } R_i(t) \text{ is the } i\text{-th row of } U(t).$$
Prove that $w'(t)$ is equal to
$$\sum_{i=1}^{n} \det \begin{pmatrix} R_1(t) \\ \vdots \\ R_i'(t) \\ \vdots \\ R_n(t) \end{pmatrix}.$$

(3) From point (1) deduce that $R_i'(t) = \sum_{j=1}^{n} a_{i,j}(t)\, R_j(t)$ for any $i = 1, \ldots, n$.

(4) From points (2) and (3) infer that $w(t)$ satisfies the (linear, scalar) ODE $w'(t) = \mathrm{trace}(A(t))\, w(t)$, for $t \in I$. Deduce that
$$w(t) = c \exp \left(\int_0^t \mathrm{trace}(A(s))\, \mathrm{d}s \right), \qquad t \in I,$$
for some constant c (actually $c = w(0) = \det U(0)$).

Exercise 1.9. Prove the assertions made in Rem. 1.6.

[*Hint:* Setting $y := \gamma(t_1, X, x_1) = \gamma(t_2, X, x_2)$, prove that $t \mapsto \gamma(t, X, x_1)$ and $t \mapsto \gamma(t + t_2 - t_1, X, x_2)$ solve the same Cauchy problem: $\dot{\gamma} = X(\gamma),\, \gamma(t_1) = y$.]

Exercise 1.10. Prove the following Taylor formula with a Peano remainder under "minimal" regularity assumptions:

Let $f : [-\varepsilon, \varepsilon] \to \mathbb{R}$ (with $\varepsilon > 0$) be a differentiable function; assume that $f'(t)$ admits the first-order derivative at $t = 0$, say $f''(0)$; then
$$f(t) = f(0) + f'(0)\, t + \frac{f''(0)}{2}\, t^2 + o(t^2), \quad \text{as } t \to 0.$$

[*Hint:* Apply de l'Hôpital's Rule to $(f(t) - f(0) - f'(0)\, t - \frac{f''(0)}{2}\, t^2)/t^2 \ldots$]

Exercise 1.11. Suppose $I \subseteq \mathbb{R}$ is an open interval, $t_0 \in I$, and $\varphi \in C^{n+1}(I, \mathbb{R})$. The Fundamental Theorem of Calculus and the integration by parts give
$$\varphi(t) = \varphi(t_0) + \int_{t_0}^t \varphi'(s)\, \mathrm{d}s = \varphi(t_0) + (t - t_0)\, \varphi'(t_0) + \int_{t_0}^t (t - s)\, \varphi''(s)\, \mathrm{d}s.$$
By inductively integrating by parts, obtain Taylor's Formula with an integral remainder, that is (with the notation $\varphi^{(k)} = \frac{\mathrm{d}^k \varphi}{\mathrm{d}t^k}$):
$$\varphi(t) = \sum_{k=0}^{n} \frac{\varphi^{(k)}(t_0)}{k!} (t - t_0)^k + \frac{1}{n!} \int_{t_0}^t (t - s)^n\, \varphi^{(n+1)}(s)\, \mathrm{d}s, \qquad t \in I.$$

Exercise 1.12. Let $M(t)$ be a square matrix with differentiable entries for t in an interval I. Suppose that $M(t)$ is invertible for every $t \in I$. Prove that

$$\frac{\mathrm{d}}{\mathrm{d}t}\left\{\left(M(t)\right)^{-1}\right\} = -\left(M(t)\right)^{-1}\dot{M}(t)\left(M(t)\right)^{-1}, \qquad t \in I.$$

[*Hint:* Differentiate $M(t)\cdot\left(M(t)\right)^{-1}$ = identity matrix...]

Exercise 1.13. Give another proof of Lem. 1.31 by providing the details of the following calculation (here F is as in the statement of the cited lemma):

$$\lim_{\tau\to 0}\frac{F(t+\tau)-F(t)}{\tau}$$

$$= \lim_{\tau\to 0}\frac{1}{\tau}\left\{\mathcal{J}_{\Psi_{-t}^X}\left(\Psi_{-\tau}^X(\Psi_{t+\tau}^X(x)))\right)\cdot\mathcal{J}_{\Psi_{-\tau}^X}\left(\Psi_{t+\tau}^X(x)\right) - \mathcal{J}_{\Psi_{-t}^X}\left(\Psi_t^X(x)\right)\right\}$$

$$= \mathcal{J}_{\Psi_{-t}^X}(x')\cdot\frac{\mathrm{d}}{\mathrm{d}\tau}\Big|_{\tau=0}\left\{\mathcal{J}_{\Psi_{\tau}^{-X}}(\Psi_{\tau}^X(x'))\right\},$$

where $x' := \Psi_t^X(x)$. Then apply Lem. 1.30...

Exercise 1.14. Prove the assertions in Exm. 1.13; show that

$$\frac{\partial\Psi_t^X(x)}{\partial x} = \frac{1}{(1-|x|\,t)^2}, \qquad \frac{\partial^2\Psi_t^X(x)}{\partial t\,\partial x} = \frac{2\,|x|}{(1-|x|\,t)^3},$$

on $\mathcal{D}(X) = \{(x,t)\in\mathbb{R}^2 : |x|t < 1\}$. Finally, show that

$$\frac{\partial^2\Psi_t^X(x)}{\partial x^2} = \frac{2t}{(1-|x|\,t)^3}\,\frac{x}{|x|}, \qquad \text{if } x \neq 0.$$

Exercise 1.15. In $\mathbb{R}^1$ consider the v.f.s $X = x|x|\partial_x$ and $Y = \partial_x$. With the notation in (1.40) show that

$$\Gamma_t^{X,Y}(x) = \frac{t + \dfrac{x}{1-|x|t}}{1 + t\cdot\left|t + \dfrac{x}{1-|x|t}\right|} - t.$$

Show that, for $x = 0$, $t \mapsto \Gamma_t^{X,Y}(0)$ is C^2 (actually more than we expected, since X is C^1 but not C^2), but not C^3.

Exercise 1.16. Prove the following lemma, a degree-two expansion of the composition of two flows (to be studied in all generality in Chap. 3).

If X and Y are C^2 vector fields on Ω one has

$$\Psi_{t_2}^Y \circ \Psi_{t_1}^X(x) = x + t_1 X(x) + t_2 Y(x)$$

$$+ \frac{t_1^2}{2}X^2 I(x) + \frac{t_2^2}{2}Y^2 I(x) + t_1 t_2\,(X\circ Y)I(x)$$

$$+ o(t_1^2 + t_2^2), \qquad \text{as } (t_1,t_2)\to(0,0).$$

Give the details of the following argument:

(1) Set $F(t_1, t_2) := \Psi^Y_{t_2} \circ \Psi^X_{t_1}(x)$ and show that F is well posed and C^2 on some neighborhood of $(0, 0)$.

(2) Show that

$$F(0, 0) = x, \quad \frac{\partial F}{\partial t_1}(0, 0) = X(x), \quad \frac{\partial F}{\partial t_2}(0, 0) = Y(x),$$

$$\frac{\partial^2 F}{\partial t_1^2}(0, 0) = X^2 I(x), \quad \frac{\partial^2 F}{\partial t_2^2}(0, 0) = Y^2 I(x), \quad \frac{\partial^2 F}{\partial t_1 \partial t_2}(0, 0) = (X \circ Y) I(x).$$

(3) Apply Taylor's Formula (of order two) with a Peano remainder.

Exercise 1.17. Give a simpler proof of Thm. 1.32 in the case when X and Y are C^2, by providing the details of the following argument:

(1) Given X, Y of class C^2 on Ω, we let, for $s = (s_1, s_2, s_3, s_4)$ near $0 \in \mathbb{R}^4$, $F(s) := \Psi^{-Y}_{s_4} \circ \Psi^{-X}_{s_3} \circ \Psi^Y_{s_2} \circ \Psi^X_{s_1}(x)$. Show that F is C^2 near 0.

(2) Prove that the Maclaurin polynomial of degree 2 of $F(s)$ is

$$x + s_1\, X(x) + s_2\, Y(x) - s_3\, X(x) - s_4\, Y(x)$$

$$+ \frac{s_1^2}{2}\, X^2 I(x) + \frac{s_2^2}{2}\, Y^2 I(x) + \frac{s_3^2}{2}\, X^2 I(x) + \frac{s_4^2}{2}\, Y^2 I(x)$$

$$+ s_1 s_2\, (X \circ Y) I(x) - s_1 s_3\, X^2 I(x) - s_1 s_4\, (X \circ Y) I(x)$$

$$- s_2 s_3\, (Y \circ X) I(x) - s_2 s_4\, Y^2 I(x) + s_3 s_4\, (X \circ Y) I(x).$$

To this end, use Exr. 1.16, taking into account that when one computes the second-order derivatives of $F(s_1, s_2, s_3, s_4)$ at $(0, 0, 0, 0)$, then two out of the four variables can be taken to be null...

(3) Recognize that the Maclaurin expansion of degree 2 of $t \mapsto F(t, t, t, t)$ is obtained by taking $s_1 = \cdots = s_4 = t$; this will give (1.37).

Exercise 1.18. Consider the v.f.s $X := (1 + x^2)\partial_x$ and $Y := x\partial_x$ on $\mathbb{R}$. Prove the existence of a small $\varepsilon > 0$ such that

$$\Gamma^{X,Y}_t(0) = \tan\Big(-t + \arctan\big(e^t \tan t\big)\Big) e^{-t}, \quad \text{for all } |t| \leq \varepsilon,$$

and check the validity of formula (1.41): $\lim_{t \to 0} \frac{\Gamma^{X,Y}_t(0)}{t^2} = [X, Y](0) = 1$. Observe that, since X is *not* global, the map $\Gamma^{X,Y}_t(0)$ is defined only for $|t|$ small.

Exercise 1.19. Consider $X := \partial_{x_1} + e^{x_1}\partial_{x_2}$, $Y := x_1 \partial_{x_1}$ on $\mathbb{R}^2$. Write down the explicit expression of the path $t \mapsto \Gamma^{X,Y}_t(x)$, for any $x \in \mathbb{R}^2$, by completing the following argument:

(1) Prove that X and Y are global and show that, for all $t \in \mathbb{R}$,

$$\Psi^X_t(x) = \Big(x_1 + t, x_2 + e^{x_1}(e^t - 1)\Big), \quad \Psi^Y_t(x) = (x_1 e^t, x_2).$$

(2) Deduce from the above expressions of Ψ_t^X and Ψ_t^Y that, for all t,

$$\Gamma_t^{X,Y}(x) = \left(e^{-t}((x_1+t)e^t - t), x_2 + e^{x_1}(e^t - 1) + \exp\left(e^t(x_1+t)\right)(e^{-t} - 1) \right).$$

Check the validity of formula (1.41): $\lim_{t\to 0} \frac{\Gamma_t^{X,Y}(x) - x}{t^2} = [X, Y](x) = (1, -x_1 e^{x_1})$.

Exercise 1.20. Consider the vector fields $X = \partial_x$, $Y = x\,\partial_y$ and

$$Y + X + \frac{1}{2}[Y, X] = \partial_x + (x - \tfrac{1}{2})\partial_y.$$

Determine the integral curves (starting from a generic $(x_0, y_0) \in \mathbb{R}^2$) of these three vector fields, by solving the associated ODEs.

    Check your results by computing these same integral curves using only (1.44). Finally, verify the following identity (for every $(x_0, y_0) \in \mathbb{R}^2$)

$$\exp(X)(\exp(Y)(x_0, y_0)) = \exp\left(Y + X + \frac{1}{2}[Y, X]\right)(x_0, y_0),$$

which can be seen as a sort of "Exponential Theorem for smooth vector fields" (since here all higher order commutators vanish identically and the vector fields involved are real-analytic...), a topic which will be thoroughly studied in Chap. 3.

Exercise 1.21. Prove that the usual form of Taylor's series of a function $F(x)$ (defined and smooth on an open neighborhood of $0 \in \mathbb{R}^N$) at the point 0, i.e.,

$$\sum_{k=0}^{\infty} \sum_{|\alpha|=k} \frac{D_x^\alpha F(0)}{\alpha!} x^\alpha$$

can formally be rewritten (and reordered), when $N = 2$, as

$$\sum_{i,j\geq 0} \frac{\partial^{i+j} F(0,0)}{\partial x_1^i \partial x_2^j} \frac{x_1^i x_2^j}{i!\, j!}.$$

Here $|\alpha| = \alpha_1 + \cdots + \alpha_N$, $D_x^\alpha = \partial_{x_1}^{\alpha_1} \cdots \partial_{x_N}^{\alpha_N}$, $x^\alpha = x_1^{\alpha_1} \cdots x_N^{\alpha_N}$, and $\alpha! = \alpha_1! \cdots \alpha_N!$.

Exercise 1.22. Verify that the diagram

$$(0,0,0) \xrightarrow{\exp(tX_2)} Q_1 \xrightarrow{\exp(tX_1)} Q_2 \xrightarrow{\exp(-tX_2)} Q_3 \xrightarrow{\exp(-tX_1)} (0,0,4t^2)$$

described in the footnote on page 30 (see the notation therein) for the Heisenberg group $\mathbb{R}^3$ is correct.

Chapter 2

The Exponential Theorem

THE aim of this chapter is to give a self-contained algebraic proof of the so-called Theorem of Campbell, Baker, Hausdorff and Dynkin, which we shall call –for the sake of brevity– *the Exponential Theorem* or *the CBHD Theorem* (following the recent monograph by [Bonfiglioli and Fulci (2012)]). The motivation for an independent algebraic study of the composition of exponentials was given in Sec. 1.6: there we studied the composition $e^{-tY} \circ e^{-tX} \circ e^{tY} \circ e^{tX}$ in order to give a geometric interpretation of $[X, Y]$. With this motivation on the horizon, here we give the proof of the Exponential Theorem, hopefully the simplest out of many existing ones (see e.g., [Bonfiglioli and Fulci (2012)]).

A notable feature of the CBHD Theorem is that it profoundly rests on the validity (in an algebraic setting) of a suitable *ordinary differential equation*, which may be traced back to [Poincaré (1900)], and for this reason we name it '*the Poincaré ODE*' (three variants of this ODE are presented in this chapter). In the spirit, this chapter is constructed following this pattern:

(1) we establish the appropriate abstract algebraic setting;
(2) we prove the corresponding Exponential Theorem in that setting;
(3) we bring out an infinite family of identities by means of that theorem, that can profitably be used in many other different contexts.

We shall follow this pattern three times (only the first time details will be given, the other two cases following by analogy); the settings will be:

a. $\mathbb{K}\langle x, y\rangle[\![t]\!]$: the formal power series in t with coefficients in the algebra $\mathbb{K}\langle x, y\rangle$ of the polynomials in two non-commuting indeterminates x, y;
b. $\mathbb{K}\langle\!\langle x, y\rangle\!\rangle$: the formal power series in the non-commuting x and y;
c. $\mathbb{K}\langle x, y\rangle[\![s, t]\!]$: the formal power series in two commuting indeterminates s, t with coefficients in $\mathbb{K}\langle x, y\rangle$.

In Chap.s 3 and 13 we shall provide important applications of the Exponential Theorem in the study of the composition of vector fields, and in ODE theory. Applications to Lie groups will be given in Chap.s 14, 15, 17.

The only prerequisite for this chapter is some basic Algebra.

37

2.1   Main algebraic setting

From now on, we fix a field $\mathbb{K}$ of characteristic zero[1]; the cases $\mathbb{K} = \mathbb{Q}, \mathbb{R}, \mathbb{C}$ will be the most interesting for us. A self-contained exposition of the contents of this section can be found in [Bonfiglioli and Fulci (2012), Chapter 2].

We denote by $\mathbb{K}\langle x, y \rangle$ the unital associative algebra over $\mathbb{K}$ of the **polynomials in two *non-commuting* indeterminates**[2] x **and** y (these polynomials may have any degree, 0 included). The (associative but non-commutative) multiplication between polynomials is the obvious one.[3] For example

$$(2 + y + xyx^2)(x - yx) = 2x - yx - y^2x + xyx^3 - xyx^2yx.$$

As an associative algebra, $\mathbb{K}\langle x, y \rangle$ is naturally endowed with the structure of a Lie algebra by the related commutator (see Rem. A.2 on page 342). The Lie algebra $\mathbb{K}\langle x, y \rangle$ contains a selected subalgebra, the **Lie subalgebra generated by** x **and** y:

$$\mathcal{L}(x, y) := \bigcap_{\{x,y\} \subset \mathfrak{g} \subseteq \mathbb{K}\langle x,y \rangle} \mathfrak{g},$$

where the intersection runs through all the Lie subalgebras $\mathfrak{g}$ of $\mathbb{K}\langle x, y \rangle$ which contain x and y. Hence $\mathcal{L}(x, y)$ is the *smallest Lie subalgebra of* $\mathbb{K}\langle x, y \rangle$ *containing* x *and* y. Its elements are called **Lie polynomials** in x and y. For example

$$x + y + \tfrac{1}{2}\,[x,y], \quad [[x,y],[[x,y],y]], \quad [y,[y,[y,x]]]$$

are elements of $\mathcal{L}(x, y)$, whereas xy is not a Lie polynomial in x, y.

Notation 2.1. In the sequel, the letter $\mathscr{T}$ will also stand for $\mathbb{K}\langle x, y \rangle$, briefly.

Using the above Not. 2.1, we denote by $\mathscr{T}[\![t]\!]$ the unital associative algebra of the **formal power series** in the indeterminate t over the algebra $\mathscr{T} = \mathbb{K}\langle x, y \rangle$. The elements of $\mathscr{T}[\![t]\!]$ have the form

$$p = \sum_{k=0}^{\infty} a_k\, t^k, \quad \text{where } a_k \in \mathbb{K}\langle x, y \rangle \text{ for every } k \geq 0.$$

---

[1]This means that any integer multiple of the unit $1_{\mathbb{K}}$ of $\mathbb{K}$ is different form zero:

$$n\, 1_{\mathbb{K}} \neq 0 \text{ for all } n \in \mathbb{N}.$$

[2]Some authors prefer to use the name 'symbol' instead of 'indeterminate'. This will cause no confusion since we shall never treat polynomials in x and y as functions.

[3]This is quite an informal definition of $\mathbb{K}\langle x, y \rangle$. The formal definition is the following one: denoted by V the free $\mathbb{K}$-vector space generated by the set $\{x, y\}$ (i.e., V is the set of the formal linear combinations over $\mathbb{K}$ of x and y), then

$$\mathbb{K}\langle x, y \rangle := \mathscr{T}(V) = \bigoplus_{k=0}^{\infty} \mathscr{T}_k(V)$$

is the tensor algebra of V, where $\mathscr{T}_0(V) = \mathbb{K}$ and $\mathscr{T}_k(V) = V^{\otimes k}$ is the vector space of the tensors of degree k in the elements of V. See [Bonfiglioli and Fulci (2012), Theorem 2.40] for the canonical identification of $\mathscr{T}(V)$ with the free unital associative algebra over the set $\{x, y\}$ (namely, $\mathbb{K}\langle x, y \rangle$: the unital associative algebra over $\mathbb{K}$ of the polynomials in two non-commuting indeterminates x, y).

We say that a_0 is the zero-degree term of p and that, in general, $a_k \in \mathbb{K}\langle x, y \rangle$ is the k-th coefficient (or coefficient of degree k) of p.

The product in $\mathscr{T}[\![t]\!]$ is the usual Cauchy product of formal power series:

$$\left(\sum_{i=0}^{\infty} a_i\, t^i \right) \cdot \left(\sum_{j=0}^{\infty} b_j\, t^j \right) = \sum_{k=0}^{\infty} \left(\sum_{i+j=k} a_i\, b_j \right) t^k,$$

where $a_i, b_j \in \mathbb{K}\langle x, y \rangle$ for every $i, j \geq 0$.

In the sequel, we say that a non-vanishing formal power series

$$p = a_k t^k + a_{k+1} t^{k+1} + \cdots = \sum_{j=k}^{\infty} a_j t^j \quad \text{(with } a_k \neq 0\text{)}$$

has **minimum degree** k (denote by $\underline{\deg}(p)$). We also set $\underline{\deg}(0) := +\infty$.

Remark 2.2 (Metric on $\mathscr{T}[\![t]\!]$). $\mathscr{T}[\![t]\!]$ can be equipped with a *metric d* by setting

$$d(p, q) := \exp(-\underline{\deg}(p - q)),$$

with the convention that $e^{-\infty} = 0$. More explicitly, we have

$$d\left(\sum_{j \geq 0} a_j\, t^j, \sum_{j \geq 0} b_j\, t^j \right) = \begin{cases} 0, & \text{if } a_j = b_j \text{ for every } j \geq 0, \\ \exp\left(-\min\{j \geq 0 \,:\, a_j \neq b_j\} \right), & \text{otherwise.} \end{cases}$$

It is left to the reader to prove that $(\mathscr{T}[\![t]\!], d)$ is a *metric space* (see Exr. 2.1):

(1) $d(p, q) \geq 0$ for every $p, q \in \mathscr{T}[\![t]\!]$, and $d(p, q) = 0$ if and only if $p = q$;
(2) $d(p, q) = d(q, p)$ for every $p, q \in \mathscr{T}[\![t]\!]$;
(3) $d(p, q) \leq d(p, r) + d(r, q)$ for every $p, q, r \in \mathscr{T}[\![t]\!]$.

Actually, (3) holds in a stronger form (called *ultra-metric triangle inequality*):

$$d(p, q) \leq \max\{d(p, r), d(r, q)\}, \quad \text{for every } p, q, r \in \mathscr{T}[\![t]\!].$$

Moreover, given a sequence $\{p_n\}_n$ in $\mathscr{T}[\![t]\!]$, it holds that

$$p_n \xrightarrow{n \to \infty} 0 \text{ in } (\mathscr{T}[\![t]\!], d) \quad \Longleftrightarrow \quad \underline{\deg}(p_n) \xrightarrow{n \to \infty} \infty. \tag{2.1}$$

Remark 2.3 ($\mathscr{T}[\![t]\!]$ is a topological algebra). By means of (2.1) one easily proves that the algebraic operations

$$\mathscr{T}[\![t]\!] \times \mathscr{T}[\![t]\!] \ni (p, q) \mapsto p + q,\ p \cdot q \in \mathscr{T}[\![t]\!]$$

are continuous (the domain has the product topology). This means that $\mathscr{T}[\![t]\!]$ with the associated metric topology is a *topological algebra*. ♮

Remark 2.4 ($\mathscr{T}[\![t]\!]$ is complete). The most interesting fact about $(\mathscr{T}[\![t]\!], d)$ is that it is a *complete* metric space. We sketch the proof of this fact. Let $\{p_n\}_n$ be a Cauchy sequence in $\mathscr{T}[\![t]\!]$; the Cauchy condition means that we have

$$d(p_n, p_m) \longrightarrow 0, \quad \text{as } n, m \to \infty.$$

This means that the minimum degree of $p_n - p_m$ tends to ∞ as $n, m \to \infty$. This easily proves that, for every $k \in \mathbb{N}$, all the series p_n with n large enough (depending on k) share together the same polynomial part of degree $\leq k$. In a suitable sense, this ensures that the coefficients $a_{n,k} \in \mathbb{K}\langle x, y \rangle$ of

$$p_n = \textstyle\sum_{k=0}^{\infty} a_{n,k} \, t^k$$

eventually become constant. The proof that $\{p_n\}_n$ converges is simple.   ♯

Remark 2.5. Suppose that X is a vector space equipped with the structure (X, d) of a metric space satisfying the ultra-metric triangle inequality:

$$d(x, y) \leq \max\{d(x, z), d(z, y)\}, \quad \text{for every } x, y, z \in X.$$

Suppose also that (X, d) is complete and that d is translation-invariant:

$$d(x + z, y + z) = d(x, y), \quad \text{for every } x, y, z, \in X.$$

Then a series $\sum_{n \geq 0} x_n$ is convergent in X if and only if $x_n \to 0$ in X. This can be proved by means of the argument in Exr. 2.2.   ♯

Definition 2.6 (D_t on $\mathscr{T}[\![t]\!]$). We define the following endomorphism:

$$D_t : \mathscr{T}[\![t]\!] \longrightarrow \mathscr{T}[\![t]\!], \qquad \sum_{j \geq 0} a_j \, t^j \mapsto \sum_{j \geq 1} j \, a_j \, t^{j-1}.$$

Remark 2.7. Since $\mathbb{K}$ has characteristic zero, it is true that, for every $p \in \mathscr{T}[\![t]\!]$,

$$\underline{\deg}(D_t p) = \begin{cases} \underline{\deg}(p - a_0) - 1, & \text{if } \underline{\deg}(p) = 0, \\[2mm] \underline{\deg}(p) - 1, & \text{if } \underline{\deg}(p) \geq 1. \end{cases}$$

Here a_0 denotes the zero-degree term of $p = a_0 + a_1 t + \cdots$.   ♯

Remark 2.8. The hypothesis requiring that $\mathbb{K}$ has characteristic zero shows that $D_t p = D_t q$ if and only if $p - q$ has degree zero.

As a consequence, if we introduce the two sets consisting of the formal power series with zero-degree term equal to, respectively, 0 and 1

$$\mathscr{T}[\![t]\!]_+ := \Big\{ \textstyle\sum_{k=0}^{\infty} a_k t^k \in \mathscr{T}[\![t]\!] \, : \, a_0 = 0 \Big\}, \tag{2.2a}$$

$$1 + \mathscr{T}[\![t]\!]_+ := \Big\{ \textstyle\sum_{k=0}^{\infty} a_k t^k \in \mathscr{T}[\![t]\!] \, : \, a_0 = 1 \Big\}, \tag{2.2b}$$

then D_t is a bijection of $\mathscr{T}[\![t]\!]_+$ onto $\mathscr{T}[\![t]\!]$ and of $1 + \mathscr{T}[\![t]\!]_+$ onto $\mathscr{T}[\![t]\!]$.   ♯

Proposition 2.9 (D_t is a continuous derivation). *The operator D_t in Def. 2.6 is a derivation[4] of the associative algebra $\mathscr{T}[\![t]\!]$ (hence of the related Lie algebra too). Moreover, D_t is continuous wrt the topology of the metric space $(\mathscr{T}[\![t]\!], d)$.*

Proof. The first assertion amounts to $D_t(p \cdot q) = p \cdot D_t(q) + D_t(p) \cdot q$, a simple fact (Exr. 2.5). The second assertion is also left as an exercise (Exr. 2.6).   □

---

[4]See Def. A.4 on page 342.

We are ready to give two central definitions: those of Exp and Log.

Definition 2.10 (Exp **and** Log **on** $\mathscr{T}[\![t]\!]$ **).** With the notation in (2.2a) and (2.2b), we introduce the following maps:[5]

$$\mathrm{Exp} : \mathscr{T}[\![t]\!]_+ \longrightarrow 1 + \mathscr{T}[\![t]\!]_+ \qquad \mathrm{Exp}(p) := \sum_{k=0}^{\infty} \frac{1}{k!}\, p^k, \tag{2.3}$$

$$\mathrm{Log} : 1 + \mathscr{T}[\![t]\!]_+ \longrightarrow \mathscr{T}[\![t]\!]_+ \qquad \mathrm{Log}(1 + q) := \sum_{k=1}^{\infty} \frac{(-1)^{k+1}}{k}\, q^k. \tag{2.4}$$

In (2.3) and (2.4), the powers p^k and q^k are, obviously, powers in the associative algebra $\mathscr{T}[\![t]\!]$. It is left as an exercise to prove that Exp and Log are well posed (Exr. 2.7). It is also left as an exercise the next simple fact (Exr. 2.4).

Proposition 2.11. *For every $a \in \mathbb{K}\langle x, y\rangle$ one has*
$$D_t\big(\mathrm{Exp}(a\,t)\big) = a \cdot \mathrm{Exp}(a\,t) = \mathrm{Exp}(a\,t) \cdot a.$$

The following result is not surprising (but not so trivial, either).

Proposition 2.12. *The maps* Exp *and* Log *in Def. 2.10 are inverse to each other.*

Proof. We prove that $\mathrm{Log}(\mathrm{Exp}(p)) = p$ for every $p \in \mathscr{T}[\![t]\!]_+$; the proof that $\mathrm{Exp}(\mathrm{Log}(1 + q)) = 1 + q$ for every $q \in \mathscr{T}[\![t]\!]_+$ is analogous.

Since $\mathscr{T}[\![t]\!]$ is a topological algebra (Rem. 2.3), we have the computation:

$$\mathrm{Log}(\mathrm{Exp}(p)) = \mathrm{Log}\Big(1 + \sum_{k=1}^{\infty} \frac{p^k}{k!}\Big) = \sum_{h=1}^{\infty} \frac{(-1)^{h+1}}{h} \Big(\sum_{k=1}^{\infty} \frac{p^k}{k!}\Big)^h$$

$$= \sum_{h=1}^{\infty} \frac{(-1)^{h+1}}{h} \sum_{k_1=1}^{\infty} \frac{p^{k_1}}{k_1!} \cdots \sum_{k_h=1}^{\infty} \frac{p^{k_h}}{k_h!}$$

$$= \sum_{n=1}^{\infty} p^n \sum_{h=1}^{n} \sum_{k_1+\cdots+k_h=n} \frac{(-1)^{h+1}}{h} \frac{1}{k_1!\cdots k_h!}.$$

The very same computation holds when p is replaced with[6] a real $x < \ln 2$:

$$\ln(e^x) = \sum_{n=1}^{\infty} x^n \sum_{h=1}^{n} \sum_{k_1+\cdots+k_h=n} \frac{(-1)^{h+1}}{h} \frac{1}{k_1!\cdots k_h!}, \qquad \forall\, x < \ln 2.$$

Since we know from basic function theory that $\ln(e^x) = x$ for every real x (note that some results of elementary Analysis are required to prove this fact...), this means that the following identities in $\mathbb{Q}$ do hold true:

$$\sum_{h=1}^{n} \sum_{k_1+\cdots+k_h=n} \frac{(-1)^{h+1}}{h} \frac{1}{k_1!\cdots k_h!} = \begin{cases} 1 & \text{if } n = 1, \\ 0 & \text{if } n > 1. \end{cases}$$

Since any field with 0 characteristic has a sub-field which is isomorphic to $\mathbb{Q}$, analogous identities hold true in $\mathbb{K}$. Thus, going back to the former computation in $\mathscr{T}[\![t]\!]$, these same identities finally give $\mathrm{Exp}(\mathrm{Log}(p)) = p$. $\qquad\square$

---

[5]Note that $1/k!$ and $1/k$ make sense for $\mathbb{K}$ has characteristic 0.

[6]Note that we need to insert $q = e^x - 1$ in the Maclaurin expansion of $\log(1 + q)$; this requires that $q \in (-1, 1]$, that is $x \leq \ln 2$.

2.2   The Exponential Theorem for $\mathbb{K}\langle x, y\rangle[\![t]\!]$

Within the algebraic background introduced in the previous section, we can state and prove the following remarkable result, *the Exponential Theorem for $\mathscr{T}[\![t]\!]$*. Thm. 2.13 is a temporary version of what we shall call *the* Exponential Theorem, the latter being the analogue of Thm. 2.13 for the completion of $\mathbb{K}\langle x, y\rangle$ (Sec. 2.3).

In the literature, this theorem is often named after Campbell, Baker and Hausdorff. For other algebraic proofs, see [Bonfiglioli and Fulci (2012), Chapters 3, 4].

Theorem 2.13 (Exponential Theorem for $\mathscr{T}[\![t]\!]$). *Let $\mathscr{T} = \mathbb{K}\langle x, y\rangle$ denote the $\mathbb{K}$-algebra of the polynomials in two non-commuting indeterminates x and y (where $\mathbb{K}$ is a field of characteristic zero). Let us consider the formal power series in $\mathscr{T}[\![t]\!]$ defined by*

$$Z = \mathrm{Log}\big(\mathrm{Exp}(x\,t) \cdot \mathrm{Exp}(y\,t)\big),$$

and let us denote by $Z_n(x, y) \in \mathbb{K}\langle x, y\rangle$ its (uniquely defined) coefficients:

$$Z = \sum_{n=1}^{\infty} Z_n(x, y)\, t^n. \tag{2.5}$$

Then, for every $n \in \mathbb{N}$, $Z_n(x, y)$ is a Lie polynomial in x and y:

$$Z_n(x, y) \in \mathcal{L}(x, y), \quad \text{for every } n \in \mathbb{N}. \tag{2.6}$$

We shall refer to this theorem also as **the CBHD Theorem** (named after Campbell, Baker, Hausdorff and Dynkin, its major contributors, in a chronological order; actually, Ernesto Pascal should be equally mentioned, see [Achilles and Bonfiglioli (2012)]). For a more classical statement of the Exponential Theorem for formal power series in x, y, see Sec. 2.3.

Note that (2.6) is far from being obvious. Indeed, by the very definitions of Exp and Log, one obtains at once

$$Z = \mathrm{Log}(\mathrm{Exp}(x\,t) \cdot \mathrm{Exp}(y\,t)) = \mathrm{Log}\Big(1 + \sum_{i+j \geq 1} \frac{x^i\, y^j}{i!\, j!}\, t^{i+j}\Big)$$

$$= \sum_{k=1}^{\infty} \frac{(-1)^{k+1}}{k} \sum_{i_1+j_1,\ldots,i_k+j_k \geq 1} \frac{x^{i_1} y^{j_1} \cdots x^{i_k} y^{j_k}}{i_1!\, j_1! \, \cdots \, i_k!\, j_k!}\, t^{i_1+j_1+\cdots+i_k+j_k}$$

$$= \sum_{n=1}^{\infty} t^n \sum_{k=1}^{n} \frac{(-1)^{k+1}}{k} \sum_{\substack{(i_1,j_1),\ldots,(i_k,j_k) \neq (0,0) \\ i_1+j_1+\cdots+i_k+j_k=n}} \frac{x^{i_1} y^{j_1} \cdots x^{i_k} y^{j_k}}{i_1!\, j_1! \, \cdots \, i_k!\, j_k!}.$$

By the uniqueness of the decomposition $Z = \sum_n Z_n(x, y)\, t^n$ of Z as a power series in t, we have (for every $n \geq 1$)

$$Z_n(x, y) = \sum_{k=1}^{n} \frac{(-1)^{k+1}}{k} \sum_{\substack{(i_1,j_1),\ldots,(i_k,j_k) \neq (0,0) \\ i_1+j_1+\cdots+i_k+j_k=n}} \frac{x^{i_1} y^{j_1} \cdots x^{i_k} y^{j_k}}{i_1!\, j_1! \, \cdots \, i_k!\, j_k!}. \tag{2.7}$$

Thus the remarkable information (2.6) in the Exponential Theorem gives out the non-trivial fact that, for every $n \in \mathbb{N}$,

$$\sum_{k=1}^{n} \frac{(-1)^{k+1}}{k} \sum_{\substack{(i_1,j_1),\ldots,(i_k,j_k)\neq(0,0) \\ i_1+j_1+\cdots+i_k+j_k=n}} \frac{x^{i_1} y^{j_1} \cdots x^{i_k} y^{j_k}}{i_1! \, j_1! \, \cdots \, i_k! \, j_k!} \in \mathcal{L}(x,y).$$

2.2.1 *Two crucial lemmas of non-commutative algebra*

For the proof of the above theorem, we need some prerequisites of noncommutative Algebra. Following Def. A.5 on page 342, if $(A, *)$ is an associative algebra, in the sequel we use the notation 'ad' to denote the adjoint map in the Lie algebra associated with $(A, *)$.

Lemma 2.14. *Let $(A, *)$ be a unital associative algebra. Then one has*

$$(\operatorname{ad} b)^n(a) = \sum_{i=0}^{n} (-1)^i \binom{n}{i} b^{n-i} * a * b^i, \tag{2.8}$$

for every $n \in \mathbb{N} \cup \{0\}$ and every $a, b \in A$.

Note that the lhs of (2.8) is the higher-order commutator

$$\underbrace{[b, [b, \cdots [b, a] \cdots]]}_{n \text{ times}},$$

so that formula (2.8) gives the explicit expansion of this commutator (called *nested commutator*) in terms of summands of $*$-products.

Proof. Given $c \in A$, let us respectively denote by L_c and R_c the left and the right multiplication by c on the algebra A, i.e.,

$$L_c(a) := c * a, \quad R_c(a) := a * c, \qquad \forall \, a \in A.$$

With this notation we have $(\operatorname{ad} b)(a) = [b, a] = b * a + a * (-b) = (L_b + R_{-b})(a)$. Now, since L_b and R_{-b} commute (by the associativity of $*$), from the classical Newton's Binomial Formula we get

$$(\operatorname{ad} b)^n(a) = (L_b + R_{-b})^n(a) = \sum_{i=0}^{n} \binom{n}{i} (L_b)^{n-i} \circ (R_{-b})^i(a),$$

which is equivalent to (2.8). $\qquad\square$

Lemma 2.15. *Let $(A, *)$ be a unital associative algebra and let D be a derivation of A. Then, for every $n \in \mathbb{N}$ and every $u \in A$, one has*

$$D(u^n) = \begin{cases} \displaystyle\sum_{k=0}^{n-1} \binom{n}{k+1} u^{n-k-1} * (-\operatorname{ad} u)^k (Du) \\[2em] \displaystyle\sum_{k=0}^{n-1} \binom{n}{k+1} (\operatorname{ad} u)^k (Du) * u^{n-k-1}. \end{cases} \tag{2.9}$$

Since D is a derivation, one clearly has

$$D(u * \cdots * u) = (Du) * u * \cdots * u + u * (Du) * \cdots * u + \cdots + u * u * \cdots * (Du);$$

thus (2.9) means that we can write this sum in such a way that all the u's are shifted to the left or all to the right. For example

$$D(u^3) = \begin{cases} 3\, u^2 * Du + 3\, u * [-u, Du] + [-u, [-u, Du]] \\[2mm] 3\, Du * u^2 + 3\, [u, Du] * u + [u, [u, Du]]. \end{cases}$$

Proof. This is a simple exercise on the Induction Principle, left to the reader, based on the trivial trick reversing the order of a $*$-product:

$$a * b = b * a + [a, b], \qquad \text{for all } a, b \in A,$$

plus Pascal's Rule $\binom{n}{k+1} + \binom{n}{k} = \binom{n+1}{k+1}$. $\qquad\qquad\qquad\qquad\square$

Key consequences of Lemmas 2.14 and 2.15 are, respectively, Theorems 2.16 and 2.17 below. The first one can be referred to as **Hadamard's Lemma** in the setting of $\mathscr{T}[\![t]\!]$ (see also Exr. 2.14). It gives a compact expression for the conjugate by an exponential, that is the map

$$z \mapsto \mathrm{Exp}(u) \cdot z \cdot (\mathrm{Exp}(u))^{-1} \qquad (\text{with } u \in \mathscr{T}[\![t]\!]_+),$$

in terms of the exponential of the linear operator $\mathrm{ad}\, u$.

Theorem 2.16 (Conjugate by an Exponential). *For every $u \in \mathscr{T}[\![t]\!]_+$ one has*

$$\mathrm{Exp}(u) \cdot z \cdot \mathrm{Exp}(-u) = \sum_{n=0}^{\infty} \frac{1}{n!}\, (\mathrm{ad}\, u)^n (z), \qquad \text{for every } z \in \mathscr{T}[\![t]\!]. \qquad (2.10)$$

Equivalently, with a self-explaining compact notation:

$$e^u \cdot z \cdot e^{-u} = e^{\mathrm{ad}\, u}(z).$$

Proof. Since $\mathscr{T}[\![t]\!]$ is a topological algebra we have

$$\mathrm{Exp}(u) \cdot z \cdot \mathrm{Exp}(-u) = \sum_{i=0}^{\infty} \sum_{m=0}^{\infty} \frac{u^m \cdot z \cdot (-u)^i}{m!\, i!}$$

$$\left(\text{we reorder the sum as follows: } \sum_{i=0}^{\infty} \sum_{m=0}^{\infty} = \sum_{n=0}^{\infty} \sum_{i+m=n} \right)$$

$$= \sum_{n=0}^{\infty} \frac{1}{n!} \sum_{i=0}^{n} (-1)^i \binom{n}{i}\, u^{n-i} \cdot z \cdot u^i \overset{(2.8)}{=} \sum_{n=0}^{\infty} \frac{1}{n!}\, (\mathrm{ad}\, u)^n (z).$$

This ends the proof. $\qquad\qquad\qquad\qquad\qquad\qquad\qquad\qquad\qquad\qquad\qquad\square$

Without giving an independent meaning to the word "differential" in the present context, we name the following crucial result in a suggestive way:

Theorem 2.17 (Differential of the Exponential). *Let D be any continuous derivation of $\mathscr{T}[\![t]\!]$. Then the following formulae are satisfied*

$$D(\mathrm{Exp}(u)) = \begin{cases} \mathrm{Exp}(u) \cdot \displaystyle\sum_{k=1}^{\infty} \frac{1}{k!} (-\mathrm{ad}\, u)^{k-1}(Du) \\[2em] \displaystyle\sum_{k=1}^{\infty} \frac{1}{k!} (\mathrm{ad}\, u)^{k-1}(Du) \cdot \mathrm{Exp}(u) \end{cases} \tag{2.11}$$

for every $u \in \mathscr{T}[\![t]\!]_+$. With a compact self-explaining notation, these identities are

$$D(e^u) = e^u \cdot \frac{1 - e^{-\mathrm{ad}\, u}}{\mathrm{ad}\, u}(Du), \qquad D(e^u) = \frac{e^{\mathrm{ad}\, u} - 1}{\mathrm{ad}\, u}(Du) \cdot e^u.$$

Introducing the (complex entire) function φ

$$\varphi(z) := \frac{e^z - 1}{z} = \sum_{k=1}^{\infty} \frac{z^{k-1}}{k!}, \tag{2.12}$$

these formulae can be further rewritten in the following nice way:

$$D(e^u) = e^u \cdot \varphi(-\mathrm{ad}\, u)(Du), \qquad D(e^u) = \varphi(\mathrm{ad}\, u)(Du) \cdot e^u.$$

Proof. We prove the first identity in (2.11).

Since D is continuous, we can pass D under the series sign

$$D(\mathrm{Exp}(u)) = \sum_{n-1}^{\infty} \frac{1}{n!} D(u^n) \stackrel{(2.9)}{=} \sum_{n=1}^{\infty} \frac{1}{n!} \sum_{k=0}^{n-1} \binom{n}{k+1} u^{n-k-1} \cdot (-\mathrm{ad}\, u)^k (Du).$$

Now we use $\dfrac{1}{n!} \dbinom{n}{k+1} = \dfrac{1}{(k+1)!(n-k-1)!}$; then we interchange sums

$$\sum_{n=1}^{\infty} \sum_{k=0}^{n-1} = \sum_{k=0}^{\infty} \sum_{n=k+1}^{\infty},$$

and finally we rename the inner dummy index as $m = n - k - 1$. This gives

$$D(\mathrm{Exp}(u)) = \sum_{k=0}^{\infty} \sum_{m=0}^{\infty} \frac{u^m \cdot (-\mathrm{ad}\, u)^k (Du)}{(k+1)!\, m!}$$

(note that the two iterated series break up into a product)

$$= \mathrm{Exp}(u) \cdot \left(\sum_{k=0}^{\infty} \frac{(-\mathrm{ad}\, u)^k (Du)}{(k+1)!} \right) = \mathrm{Exp}(u) \cdot \sum_{k=1}^{\infty} \frac{1}{k!} (-\mathrm{ad}\, u)^{k-1}(Du).$$

The second identity in (2.11) can be proved analogously, by means of the second identity in (2.9). This ends the proof. $\qquad\square$

2.2.2 *Poincaré's ODE in the formal power series setting*

We are in a position to introduce the final argument for the CBHD Theorem. Since $x\,t, y\,t \in \mathbb{K}\langle x, y\rangle[\![t]\!]_+$, and since the product of two exponentials is a series with zero-degree term equal to 1, the following "path" is well defined:

$$Z(t) := \mathrm{Log}\big(\mathrm{Exp}(x\,t) \cdot \mathrm{Exp}(y\,t)\big). \tag{2.13}$$

(We use the functional notation $Z(t)$ despite no function of t is involved here; we use this abused notation because the present argument perfectly adapts to contexts where the above $Z(t)$ is actually a curve, as in Lie-group theory...)

By its very definition, $Z(t)$ satisfies the identity

$$\mathrm{Exp}(Z(t)) = \mathrm{Exp}(x\,t) \cdot \mathrm{Exp}(y\,t), \tag{2.14}$$

and, by Prop. 2.11, we have

$$D_t(\mathrm{Exp}(x\,t)) = x \cdot \mathrm{Exp}(x\,t), \qquad D_t(\mathrm{Exp}(y\,t)) = \mathrm{Exp}(y\,t) \cdot y. \tag{2.15}$$

Due to these facts, we have the following computation:

$$\frac{1 - e^{-\mathrm{ad}\,Z(t)}}{\mathrm{ad}\,Z(t)}(D_t Z(t)) \overset{(2.11)}{=} \mathrm{Exp}(-Z(t)) \cdot D_t(\mathrm{Exp}(Z(t)))$$

$$\overset{(2.14)}{=} \mathrm{Exp}(-Z(t)) \cdot D_t\big(\mathrm{Exp}(x\,t) \cdot \mathrm{Exp}(y\,t)\big)$$

$$\overset{(2.15)}{=} \mathrm{Exp}(-Z(t)) \cdot x \cdot \mathrm{Exp}(x\,t) \cdot \mathrm{Exp}(y\,t)+$$

$$+ \mathrm{Exp}(-Z(t)) \cdot \mathrm{Exp}(x\,t) \cdot \mathrm{Exp}(y\,t) \cdot y$$

$$\overset{(2.14)}{=} \mathrm{Exp}(-Z(t)) \cdot x \cdot \mathrm{Exp}(Z(t)) + y$$

$$\overset{(2.10)}{=} e^{-\mathrm{ad}\,Z(t)}(x) + y.$$

By equating the far left-hand and right-hand members we have

$$\frac{1 - e^{-\mathrm{ad}\,Z(t)}}{\mathrm{ad}\,Z(t)}(D_t Z(t)) = e^{-\mathrm{ad}\,Z(t)}(x) + y. \tag{2.16}$$

Remark 2.18. The (very remarkable!) formula (2.16) can be rewritten, with some obvious manipulation of formal power series of endomorphisms (Sec. 2.8), as an *autonomous non-linear ODE* in $Z(t)$:

$$D_t Z(t) = \frac{\mathrm{ad}\,Z(t)}{e^{\mathrm{ad}\,Z(t)} - 1}\big(x + e^{\mathrm{ad}\,Z(t)}(y)\big); \tag{2.17a}$$

in its turn the latter can also be rewritten in a more symmetric form:

$$D_t Z(t) = \frac{\mathrm{ad}\,Z(t)}{e^{\mathrm{ad}\,Z(t)} - 1}(x) + \frac{-\mathrm{ad}\,Z(t)}{e^{-\mathrm{ad}\,Z(t)} - 1}(y). \tag{2.17b}$$

Setting $\psi(z) := \dfrac{z}{1 - e^{-z}}$ (called **Todd's function**), this is equivalent to

$$D_t Z(t) = \psi(-\mathrm{ad}\,Z(t))(x) + \psi(\mathrm{ad}\,Z(t))(y). \tag{2.17c}$$

Yet another notation: if we use φ in (2.12), and if we observe that[7]

$$\mathbf{b}(z) := \frac{1}{\varphi(z)} = \psi(-z),$$

then (2.17c) becomes

$$D_t Z(t) = \mathbf{b}(\operatorname{ad} Z(t))(x) + \mathbf{b}(-\operatorname{ad} Z(t))(y). \tag{2.17d}$$

Remark 2.19. We want to explain what we mean by $\mathbf{b}(\pm\operatorname{ad} Z)$, when $Z = Z(t)$. First of all, since $\mathbf{b}$ is holomorphic on $B(0, 2\pi)$ (as the reader can verify), it can be expanded as a power series, that is,

$$\mathbf{b}(z) = \sum_{n=0}^{\infty} \frac{B_n}{n!} z^n, \quad \text{for all } z \in B(0, 2\pi), \tag{2.18}$$

and it is easy to recognize that the sequence $\{B_n\}_{n \geq 0}$ is in fact a sequence of *rational numbers* (referred to as the *Bernoulli numbers*). It is then natural to set (as $\mathbb{K}$ contains a subfield isomorphic to the field $\mathbb{Q}$)

$$\mathbf{b}(\pm\operatorname{ad} Z) : \mathscr{T}[\![t]\!] \to \mathscr{T}[\![t]\!], \quad \mathbf{b}(\pm\operatorname{ad} Z)(f) := \sum_{n=0}^{\infty} \frac{B_n}{n!} (\pm\operatorname{ad} Z)^n(f).$$

From $Z_0 = 0$ we have $\underline{\deg}(Z) \geq 1$ so that

$$\underline{\deg}((\operatorname{ad} Z)^n(f)) \geq n, \quad \forall\, n \in \mathbb{N}, \quad \forall\, f \in \mathscr{T}[\![t]\!].$$

This shows that the sequences $\{\frac{B_n}{n!}(\pm\operatorname{ad} Z)^n(f)\}_{n \in \mathbb{N}}$ converge to 0 in the ultra-metric space $(\mathscr{T}[\![t]\!], d)$ for all $f \in \mathscr{T}[\![t]\!]$; thus the above definition of $\mathbf{b}(\pm\operatorname{ad} Z)$ is well posed and $\mathbf{b}(\pm\operatorname{ad} Z)$ define two endomorphisms of $\mathscr{T}[\![t]\!]$. We remark that, since $\mathbf{b}(0) = 1 \neq 0$, the endomorphisms $\mathbf{b}(\pm\operatorname{ad} Z)$ are invertible (by arguing as in the Sec. 2.8), and we have

$$\mathbf{b}(\pm\operatorname{ad} Z) = \left(\frac{e^{\pm\operatorname{ad} Z} - 1}{\pm\operatorname{ad} Z} \right)^{-1}.$$

We have proved the following very important result

Theorem 2.20 (Poincaré's ODE in $\mathscr{T}[\![t]\!]$). *As usual, let $\mathscr{T} = \mathbb{K}\langle x, y \rangle$, and consider the formal power series in $\mathscr{T}[\![t]\!]$ defined by*

$$Z(t) = \operatorname{Log}(\operatorname{Exp}(x\, t) \cdot \operatorname{Exp}(y\, t)).$$

*Then the following **Poincaré ODE** is satisfied by $Z(t)$*

$$D_t Z(t) = \mathbf{b}(\operatorname{ad} Z(t))(x) + \mathbf{b}(-\operatorname{ad} Z(t))(y), \tag{2.19}$$

where $\mathbf{b}$ is the formal power series defined by $\mathbf{b}(z) = \dfrac{z}{e^z - 1}$, which is the generating function of the so-called Bernoulli numbers B_n:

$$\frac{z}{e^z - 1} = \sum_{n=0}^{\infty} \frac{B_n}{n!} z^n.$$

---

[7]This has to be read as an identity in the complex variable z and this identity holds true only for $|z| < 2\pi$; but we are not using here the topology and the convergence of the power series in $\mathbb{C}$: we are only considering formal series expansions and using them to rewrite power series in $\mathscr{T}[\![x, y]\!]$, where convergence matters are very easy to handle with.

We derived (2.19) as an identity in $\mathscr{T}[\![t]\!]$ involving the derivation D_t; later in this book, (2.19) will take life as a genuine ODE; for example, in the framework of finite-dimensional Lie algebras (Chap. 5) or on Lie groups (Chap. 14).

We finally show that (2.16) allows us to prove the Exponential Theorem on $\mathscr{T}[\![t]\!]$, without the need to invert the endomorphism

$$\frac{1 - e^{-\operatorname{ad} Z(t)}}{\operatorname{ad} Z(t)}$$

which acts on $D_t Z(t)$ in formula (2.16).

Proof of Thm. 2.13. Let us write $Z(t)$ (as an element of $\mathscr{T}[\![t]\!]$) as a series in t, say $Z(t) = \sum_{n=1}^{\infty} Z_n\, t^n$, where $Z_n \in \mathbb{K}\langle x, y \rangle$ for every $n \in \mathbb{N}$. Let us insert this series in both sides of (2.16), and let us highlight the powers of t. The lhs of (2.16) is:

$$\frac{1 - e^{-\operatorname{ad} Z(t)}}{\operatorname{ad} Z(t)}\big(D_t Z(t)\big) = \sum_{k=1}^{\infty} \frac{1}{k!}\,(-\operatorname{ad} Z(t))^{k-1}\Big(\sum_{n=1}^{\infty} n\, Z_n\, t^{n-1}\Big)$$

$$= \sum_{n=1}^{\infty} n\, Z_n\, t^{n-1} + \sum_{k=2}^{\infty} \frac{(-1)^{k-1}}{k!}\, \underbrace{\Big[Z(t) \cdots \Big[Z(t), \sum_{n=1}^{\infty} n\, Z_n\, t^{n-1} \Big] \cdots \Big]}_{k-1} =$$

(in the first sum, rename $n - 1$ as h; in the second sum, rename $k - 1$ as j)

$$\sum_{h=0}^{\infty} (h+1)\, Z_{h+1}\, t^h + \sum_{j=1}^{\infty} \sum_{n=1}^{\infty} \sum_{n_1,\ldots,n_j \geq 1} t^{n_1 + \cdots + n_j + n - 1} \frac{(-1)^j\, n}{(j+1)!}\, [Z_{n_1} \cdots [Z_{n_j}, Z_n] \cdots].$$

The rhs of (2.16) is:

$$e^{-\operatorname{ad} Z(t)}(x) + y = y + x + \sum_{k=1}^{\infty} \frac{(-1)^k}{k!}\,(\operatorname{ad} Z(t))^k(x)$$

$$= y + x + \sum_{k=1}^{\infty} \sum_{n_1,\ldots,n_k \geq 1} \frac{(-1)^k}{k!}\, [Z_{n_1} \cdots [Z_{n_k}, x] \cdots]\, t^{n_1 + \cdots + n_k}.$$

Next we turn our attention to equating, from the left and the right members of (2.16), the coefficients of $t^0, t, \ldots, t^h$: we obtain the family of identities

$$Z_1 = y + x$$

$$2\, Z_2 - \frac{1}{2}\, [Z_1, Z_1] = -[Z_1, x] \qquad \Big(\text{whence } Z_2 = \frac{1}{2}\, [x, y]\Big),$$

$$\cdots$$

$$(h+1) Z_{h+1} + \sum_{j=1}^{h} \sum_{\substack{n,n_1,\ldots,n_j \geq 1 \\ n_1 + \cdots + n_j + n - 1 = h}} \frac{(-1)^j n}{(j+1)!}\, [Z_{n_1} \cdots [Z_{n_j}, Z_n] \cdots]$$

$$= \sum_{k=1}^{h} \sum_{\substack{n_1,\ldots,n_k \geq 1 \\ n_1 + \cdots + n_k = h}} \frac{(-1)^k}{k!}\, [Z_{n_1} \cdots [Z_{n_k}, x] \cdots].$$

Now, the crucial remark is that Z_n, Z_{n_j}, Z_{n_k} in the above sums have subscripts n, n_j, n_k *less than or equal to h*. Consequently the above identities give a recursion formula which allows us to derive Z_{h+1} in terms of linear combinations of iterated brackets of $Z_1, \ldots, Z_h$.

Since $Z_1 = x + y$, $Z_2 = \frac{1}{2}[x, y]$ are Lie polynomials in x and y, we inductively infer that all the Z_h are Lie polynomials in x and y, and the Exponential Thm. 2.13 is finally proved by an inductive argument. $\qquad\qquad\qquad\qquad\qquad\qquad\qquad\Box$

2.3 The Exponential Theorem for $\mathbb{K}\langle\!\langle x, y \rangle\!\rangle$

Let x and y be non-commuting indeterminates. In the unital associative $\mathbb{K}$-algebra $\mathbb{K}\langle x, y \rangle$ of the polynomials in x and y, we denote by $\mathscr{T}_k(x, y)$ the vector space of the polynomials of homogenous degree k in x and y jointly. For example,

$$\mathscr{T}_0(x, y) = \mathbb{K}, \qquad \mathscr{T}_1(x, y) = \mathrm{span}\{x, y\}$$
$$\mathscr{T}_2(x, y) = \mathrm{span}\{x^2, xy, yx, y^2\}$$
$$\mathscr{T}_3(x, y) = \mathrm{span}\{x^3, x^2y, xyx, xy^2, y^3\}.$$

Clearly, $\mathbb{K}\langle x, y \rangle = \bigoplus_{k=0}^{\infty} \mathscr{T}_k(x, y)$. A **formal power series** in x and y is a formal sum of the following form

$$p = \sum_{k=0}^{\infty} p_k, \quad \text{where } p_k \in \mathscr{T}_k(x, y) \text{ for every } k \geq 0.$$

We denote the set of these formal power series by $\mathbb{K}\langle\!\langle x, y \rangle\!\rangle$ (or $\widehat{\mathscr{T}}$ for brevity). The latter can be endowed with the structure of a unital associative algebra by prolonging in the natural way the multiplication of $\mathbb{K}\langle x, y \rangle$. Moreover, $\widehat{\mathscr{T}}$ can also be endowed with the structure of a *complete ultra-metric space* ($\mathbb{K}\langle x, y \rangle$ being dense in $\widehat{\mathscr{T}}$ with this metric structure) with the metric

$$\widehat{d}(p, q) := \exp(-\underline{\deg}(p - q)),$$

where the minimum degree $\underline{\deg}(p)$ of $p = \sum_{k=0}^{\infty} p_k$ is the minimum k such that $p_k \neq 0$ if this exists, and $\underline{\deg}(0) := \infty$ (as usual we set $\exp(-\infty) := 0$).

A selected subspace of $\widehat{\mathscr{T}}$ is $\widehat{\mathcal{L}}(x, y)$ which, by definition, is the smallest *closed* Lie subalgebra of $\widehat{\mathscr{T}}$ containing $\{x, y\}$. Equivalently, $\widehat{\mathcal{L}}(x, y)$ is the closure of $\mathcal{L}(x, y)$ in the metric space $\widehat{\mathscr{T}}$. We say that $\widehat{\mathcal{L}}(x, y)$ is the **Lie algebra of the Lie series** in x and y. Any element of $\widehat{\mathcal{L}}(x, y)$ is of the form

$$\ell = \sum_{k=1}^{\infty} \ell_k, \quad \text{where } \ell_k \in \mathcal{L}(x, y) \cap \mathscr{T}_k(x, y) \text{ for every } k \geq 1,$$

and ℓ is called a **Lie series** in x and y. Clearly, $\mathcal{L}(x, y)$ is dense in $\widehat{\mathcal{L}}(x, y)$ (as topological subspaces of $(\widehat{\mathscr{T}}, \widehat{d}\,)$).

If $\widehat{\mathscr{T}}_+$ denotes the set of the formal power series $\sum_{k=0}^{\infty} p_k$ with vanishing term of degree zero p_0, one can define (inverse to each other) functions

$$\mathrm{Exp} : \widehat{\mathscr{T}}_+ \longrightarrow 1 + \widehat{\mathscr{T}}_+ \quad \text{and} \quad \mathrm{Log} : 1 + \widehat{\mathscr{T}}_+ \longrightarrow \widehat{\mathscr{T}}_+,$$

by means of the usual Maclaurin series of the exponential and logarithmic functions, as we did in Def. 2.10. Then *the Exponential Theorem for formal power series in x and y* can be stated as follows:

Theorem 2.21 (The Exponential Theorem for $\mathbb{K}\langle\langle x, y\rangle\rangle$). *Following all the above notations,* $\mathrm{Log}(\mathrm{Exp}(x) \cdot \mathrm{Exp}(y))$ *is a Lie series in x and y, that is*

$$\mathrm{Log}(\mathrm{Exp}(x) \cdot \mathrm{Exp}(y)) \in \widehat{\mathcal{L}}(x, y).$$

More precisely, one has

$$\mathrm{Log}(\mathrm{Exp}(x) \cdot \mathrm{Exp}(y)) = \sum_{n=1}^{\infty} Z_n(x, y), \tag{2.20}$$

where the homogeneous polynomials $Z_n(x, y)$ of degree n are the same as in (2.5), in the framework of the Exponential Theorem for $\mathbb{K}\langle x, y\rangle[\![t]\!]$.

Proof. This is a direct consequence of the Exponential Theorem for $\mathscr{T}[\![t]\!]$. Indeed, exactly as in the computation preceding (2.7), one has

$$\mathrm{Log}(\mathrm{Exp}(x) \cdot \mathrm{Exp}(y)) = \mathrm{Log}\left(1 + \sum_{i+j\geq 1} \frac{x^i y^j}{i!\, j!}\right)$$

$$= \sum_{n=1}^{\infty} \sum_{k=1}^{n} \frac{(-1)^{k+1}}{k} \sum_{\substack{(i_1,j_1),\dots,(i_k,j_k)\neq(0,0) \\ i_1+j_1+\cdots+i_k+j_k=n}} \frac{x^{i_1} y^{j_1} \cdots x^{i_k} y^{j_k}}{i_1!\, j_1! \cdots i_k!\, j_k!}.$$

Comparing to (2.7) this gives the decomposition (2.20) in $\mathbb{K}\langle\langle x, y\rangle\rangle$. Now, due to (2.6) we obtain that $\mathrm{Log}(\mathrm{Exp}(x) \cdot \mathrm{Exp}(y))$ is in fact a Lie series. $\quad\square$

2.4   Dynkin's Formula

In this section, we first provide a Dynkin-type formula (Sec. 2.4.1), then we prove Dynkin's original formula (Sec. 2.4.2).

2.4.1   *A Dynkin-type formula*

The aim of this section is to derive a *universal* formula for $Z_n(x, y)$ as a sum of Lie monomials in x and y.

Let us start from identity (2.17a) which is a sort of nonlinear ODE in $Z(t)$; the crucial tool is to rewrite it as a formal ODE "$D_t Z(t) = \cdots$" with a right-hand member which is *independent* of $Z(t)$: a direct "quadrature" argument will then give out $Z(t)$ explicitly. What is more important, this right-hand member will be a *Lie series* in $x\, t$, $y\, t$ so that the above quadrature argument will also prove that $Z(t)$ is a Lie series with a *universal* expression for $Z(t)$.

To begin with, we make the following claim:

$$\text{the rhs of (2.17a) can be rewritten } \textit{as a function of } e^{\mathrm{ad}\, Z(t)}. \tag{2.21}$$

The reason why this is so important is that

$$e^{\mathrm{ad}\, Z(t)} = e^{t\, \mathrm{ad}\, x} \circ e^{t\, \mathrm{ad}\, y}. \tag{2.22}$$

Hence, if we prove (2.21), then identity (2.22) will allow us to rewrite the mentioned rhs as a function of $\operatorname{ad} x, \operatorname{ad} y$ solely.

In its turn, identity (2.22) can be proved as follows:

$$
\begin{aligned}
e^{\operatorname{ad} Z(t)}(z) &\overset{(2.10)}{=} \operatorname{Exp}(Z(t)) \cdot z \cdot \operatorname{Exp}(-Z(t)) \\
&\overset{(2.14)}{=} \operatorname{Exp}(x\,t) \cdot \operatorname{Exp}(y\,t) \cdot z \cdot \operatorname{Exp}(-y\,t) \cdot \operatorname{Exp}(-x\,t) \\
&\overset{(2.10)}{=} \operatorname{Exp}(x\,t) \cdot e^{t\,\operatorname{ad} y}(z) \cdot \operatorname{Exp}(-x\,t) \\
&\overset{(2.10)}{=} e^{t\,\operatorname{ad} x}\left(e^{t\,\operatorname{ad} y}(z)\right) = e^{t\,\operatorname{ad} x} \circ e^{t\,\operatorname{ad} y}(z).
\end{aligned}
$$

Next, we turn our attention to the claimed (2.21). This comes from the following simple (but keystone!) identity between power series in one indeterminate z: from the formal power series

$$
\frac{\log(1+z)}{z} = \sum_{n=1}^{\infty} \frac{(-1)^{n+1}}{n} z^{n-1},
$$

we get[8]

$$
\frac{\omega}{e^{\omega}-1} = \frac{\log(1+(e^{\omega}-1))}{e^{\omega}-1} = \sum_{n=1}^{\infty} \frac{(-1)^{n+1}}{n}(e^{\omega}-1)^{n-1}. \tag{2.23}
$$

As the reader can see, *the crucial trick here is to write $\omega/(e^{\omega}-1)$ as a function of $e^{\omega}-1$.* A formal (but legitimate) substitution of $\operatorname{ad} Z(t)$ in place of ω yields

$$
\frac{\operatorname{ad} Z(t)}{e^{\operatorname{ad} Z(t)}-1} = \sum_{n=1}^{\infty} \frac{(-1)^{n+1}}{n}(e^{\operatorname{ad} Z(t)}-1)^{n-1}.
$$

This proves (2.21). Finally, the ODE (2.17a) becomes

$$
D_t Z(t) = \sum_{n=1}^{\infty} \frac{(-1)^{n+1}}{n}(e^{\operatorname{ad} Z(t)}-1)^{n-1}\left(x + e^{\operatorname{ad} Z(t)}(y)\right), \tag{2.24}
$$

and via (2.22) this can directly be integrated to obtain a formula for $Z(t)$ in terms of Lie polynomials: see Exr. 2.11. The resulting formula can be referred to as a Dynkin-type formula for $Z(t)$; it was indeed Dynkin who obtained a first *explicit* representation of $Z(t)$ in terms of Lie polynomials: see [Dynkin (1947, 2000)]. Dynkin's original formula is discussed in the next section.

---

[8]We are here manipulating formal power series; if –instead– we are interested in actual identities between complex functions and their series expansions, it is interesting to observe that, whereas the Maclaurin expansion of the lhs of (2.23) has radius of convergence 2π, the series in the rhs of (2.23) does not converge if $|e^{\omega}-1| > 1$, and the latter inequality describes a region of the complex plane much different from the disc of centre 0 and radius 2π; see Exr. 2.15.

2.4.2   *Dynkin's original formula*

In this section we present the original form of Dynkin's expression of the summands of the Lie series in $\mathbb{K}\langle\!\langle x, y\rangle\!\rangle$ defined by

$$\mathrm{Log}(\mathrm{Exp}(x) \cdot \mathrm{Exp}(y)) = \sum_{n=1}^{\infty} Z_n(x, y).$$

We use (without proving it) the following result, which is a particular case of
the so-called Lemma of Dynkin, Specht, Wever. For a proof of this useful lemma,
see e.g. [Bonfiglioli and Fulci (2012), Section 3.3.1]. As usual, we are here dealing
with a field $\mathbb{K}$ of characteristic zero.

Lemma 2.22. *Consider the unique linear map* $\varphi : \mathbb{K}\langle x, y\rangle \to \mathcal{L}(x, y)$ *such that* $\varphi(1) = 0$,
$\varphi(x) = x$, $\varphi(y) = y$ *and*

$$\varphi(x^{i_1} y^{j_1} \cdots x^{i_k} y^{j_k}) := \frac{(\mathrm{ad}\, x)^{i_1} (\mathrm{ad}\, y)^{j_1} \cdots (\mathrm{ad}\, x)^{i_k} (\mathrm{ad}\, y)^{j_k - 1}(y)}{i_1 + j_1 + \cdots + i_k + j_k},$$

for every choices of nonnegative integers $i_1, j_1, \ldots, i_k, j_k$ *and every* $k \in \mathbb{N}$.
     Then φ *is a projection, that is,* φ *is onto and* $\varphi(\ell) = \ell$ *for every* $\ell \in \mathcal{L}(x, y)$.

For example, since $xy - yx$ is a Lie polynomial, it is left unchanged by φ:

$$\varphi(xy - yx) = \frac{1}{2}[x, y] - \frac{1}{2}[y, x] = \frac{1}{2}[x, y] + \frac{1}{2}[x, y] = [x, y] = xy - yx.$$

By Thm. 2.13 we know that $Z_n(x, y)$ belongs to $\mathcal{L}(x, y)$ for every $n \in \mathbb{N}$. Thus, by
the previous lemma,

$$Z_n(x, y) = \varphi(Z_n(x, y)), \quad \forall\, n \in \mathbb{N}.$$

Now, by the explicit formula for $Z_n(x, y)$ in (2.7) and by the definition of φ, we
obtain the following explicit universal expression of $Z_n(x, y)$, valid for any n:

$$Z_n(x, y) = \frac{1}{n} \sum_{k=1}^{n} \frac{(-1)^{k+1}}{k} \sum_{\substack{(i_1,j_1),\ldots,(i_k,j_k)\neq(0,0) \\ i_1+j_1+\cdots+i_k+j_k=n}} \frac{(\mathrm{ad}\, x)^{i_1} (\mathrm{ad}\, y)^{j_1} \cdots (\mathrm{ad}\, x)^{i_k} (\mathrm{ad}\, y)^{j_k - 1}(y)}{i_1!\, j_1! \cdots i_k!\, j_k!}.$$

which is **Dynkin's Formula** for the summands of $\mathrm{Log}(\mathrm{Exp}(x) \cdot \mathrm{Exp}(y))$.
     Explicitly, we have just proved the following family of identities involving
polynomials in two non-commuting indeterminates x, y (valid for every $n \in \mathbb{N}$):

$$\sum_{k=1}^{n} \frac{(-1)^{k+1}}{k} \sum_{\substack{(i_1,j_1),\ldots,(i_k,j_k)\neq(0,0) \\ i_1+j_1+\cdots+i_k+j_k=n}} \frac{x^{i_1} y^{j_1} \cdots x^{i_k} y^{j_k}}{i_1!\, j_1! \cdots i_k!\, j_k!} \tag{2.25}$$

$$= \frac{1}{n} \sum_{k=1}^{n} \frac{(-1)^{k+1}}{k} \sum_{\substack{(i_1,j_1),\ldots,(i_k,j_k)\neq(0,0) \\ i_1+j_1+\cdots+i_k+j_k=n}} \frac{(\mathrm{ad}\, x)^{i_1} (\mathrm{ad}\, y)^{j_1} \cdots (\mathrm{ad}\, x)^{i_k} (\mathrm{ad}\, y)^{j_k - 1}(y)}{i_1!\, j_1! \cdots i_k!\, j_k!}.$$

Note that (*a priori*) the lhs of (2.25) is a polynomial in the *associative* algebra $\mathbb{K}\langle x, y\rangle$ whereas the rhs is a polynomial in the *Lie* algebra $\mathcal{L}(x, y)$. In this sense the above identity is highly non-trivial.

For example, when $n = 2$, (2.25) is equivalent to the identity

$$\frac{x^2}{2} + \frac{y^2}{2} + xy - \frac{1}{2}\left(x^2 + xy + yx + y^2\right) = \frac{1}{2}\left([x, y] - \frac{1}{2}([x, y] + [y, x])\right),$$

which, after some computation, boils down to $\frac{1}{2}\, xy - \frac{1}{2}\, yx = \frac{1}{2}\, [x, y]$.

Remark 2.23. By a universal property of $\mathbb{K}\langle x, y\rangle$ (as the tensor algebra of the free vector space over the set $\{x, y\}$; see [Bonfiglioli and Fulci (2012), Theorem 2.38]),

> *formula (2.25) holds true for every pair of elements X, Y*
> *of any unital associative algebra $(A, *)$, by replacing x with X, y with Y*
> *and the product in $\mathbb{K}\langle x, y\rangle$ with the $*$ product.*

For example, an analogue of (2.25) holds true for two smooth vector fields X and Y, since they belong to the unital associative algebra $(\mathcal{U}(\Omega), \circ)$ of the smooth linear differential operators on the open set $\Omega \subseteq \mathbb{R}^N$. We shall soon give an important application of this fact in Chap. 3 (see Sec. 3.2). ♯

For a later use, we fix once and for all the relevant notation concerning Dynkin's Formula, which will return again in this book.

Definition 2.24 (Dynkin polynomials). Let $(\mathfrak{g}, [\cdot, \cdot])$ be a Lie algebra (for example, an associative algebra with its commutator). For every $n \in \mathbb{N}$ we define

$$Z_n : \mathfrak{g} \times \mathfrak{g} \longrightarrow \mathfrak{g}, \quad (a, b) \mapsto Z_n(a, b), \qquad \text{where}$$

$$Z_n(a, b) := \tag{2.26}$$

$$\frac{1}{n} \sum_{k=1}^{n} \frac{(-1)^{k+1}}{k} \sum_{\substack{(i_1, j_1), \ldots, (i_k, j_k) \neq (0,0) \\ i_1 + j_1 + \cdots + i_k + j_k = n}} \frac{(\operatorname{ad} a)^{i_1} (\operatorname{ad} b)^{j_1} \cdots (\operatorname{ad} a)^{i_k} (\operatorname{ad} b)^{j_k - 1}(b)}{i_1!\, j_1! \,\cdots\, i_k!\, j_k!}.$$

Moreover, for every fixed $i, j \in \mathbb{N} \cup \{0\}$ with $(i, j) \neq (0, 0)$ we define

$$C_{i,j} : \mathfrak{g} \times \mathfrak{g} \longrightarrow \mathfrak{g}, \quad (a, b) \mapsto C_{i,j}(a, b), \qquad \text{where}$$

$$C_{i,j}(a, b) := \tag{2.27}$$

$$\frac{1}{i + j} \sum_{k=1}^{i+j} \frac{(-1)^{k+1}}{k} \sum_{\substack{(i_1, j_1), \ldots, (i_k, j_k) \neq (0,0) \\ i_1 + \cdots + i_k = i,\ j_1 + \cdots + j_k = j}} \frac{(\operatorname{ad} a)^{i_1} (\operatorname{ad} b)^{j_1} \cdots (\operatorname{ad} a)^{i_k} (\operatorname{ad} b)^{j_k - 1}(b)}{i_1!\, j_1! \,\cdots\, i_k!\, j_k!}.$$

Clearly, when $j_k = 0$ the associated summand in (2.26) and (2.27) is understood to end with $\cdots (\operatorname{ad} a)^{i_k - 1}(a)$. All adjoint maps are meant wrt the Lie bracket of $\mathfrak{g}$. We also set $Z_0(a,b) := 0$ and $C_{0,0}(a,b) := 0$ for every $a, b \in \mathfrak{g}$. If this makes sense (e.g., if the sum is convergent for some topology on $\mathfrak{g}$), we set

$$a \diamond b := \sum_{n=1}^{\infty} Z_n(a,b). \tag{2.28}$$

We say that the series in (2.28) is the **homogeneous CBHD series** and that (2.26) is **Dynkin's representation** of Z_n. We call the Z_n's the **homogeneous Dynkin polynomials**; we call the $C_{i,j}$'s the **non-homogeneous Dynkin polynomials**.

Remark 2.25. The only non-zero $C_{i,j}(x,y)$ with $i = 0$ or $j = 0$ are $C_{1,0}(x,y) = x$ and $C_{0,1}(x,y) = y$; otherwise $C_{i,0}(x,y) = C_{0,j}(x,y) = 0$ for every $i, j \geq 2$.   ♯

Remark 2.26. If we consider the formal sum $\sum_{n=1}^{\infty} Z_n(a,b)$, if we insert the explicit expression of $Z_n(a,b)$ from (2.26) and if we formally interchange the sums over n and over k, we obtain the following alternative rearrangement[9] for the series expressing $a \diamond b$:

$$a \diamond b = \sum_{n=1}^{\infty} Z_n(a,b) = \qquad \text{(after rearranging, when applicable)}$$

$$\sum_{k=1}^{\infty} \frac{(-1)^{k+1}}{k} \sum_{(i_1,j_1),\ldots,(i_k,j_k) \neq (0,0)} \frac{(\mathrm{ad}\, a)^{i_1} (\mathrm{ad}\, b)^{j_1} \cdots (\mathrm{ad}\, a)^{i_k} (\mathrm{ad}\, b)^{j_k - 1}(b)}{(i_1 + j_1 + \cdots + i_k + j_k) \cdot i_1! j_1! \cdots i_k! j_k!}.$$

For example, this is legitimate (not only on a formal level) when $\mathfrak{g}$ is a normed space over $\mathbb{R}$ or $\mathbb{C}$ (besides being a Lie algebra over the same field) and provided that absolute convergence is fulfilled (allowing us to interchange summations).   ♯

Remark 2.27. According to Rem. 2.26, it is evident that $C_{i,j}(x,y)$ *is the Lie polynomial obtained by summing all the monomials, taken from the Lie polynomial* $Z_{i+j}(x,y)$, *where x appears exactly i times and y appears exactly j times.*

As a consequence we can also formally represent $a \diamond b$ as the (double) series of the Lie polynomials $C_{i,j}(a,b)$:

$$\begin{array}{l} \text{(double series} \\ \text{rearrangement,} \\ \text{if applicable)} \end{array} \qquad a \diamond b = \sum_{n=1}^{\infty} Z_n(a,b) = \sum_{i,j=0}^{\infty} C_{i,j}(a,b), \tag{2.29}$$

or, more generally, we can represent each $Z_n(a,b)$ by means of the $C_{i,j}(a,b)$:

$$Z_n(a,b) = \sum_{i+j=n} C_{i,j}(a,b), \qquad n \in \mathbb{N}. \tag{2.30}$$

---

[9]In specialized topological contexts, attention must be paid in that rearranging a series may alter the sum or even the convergence/divergence of the series itself!

Example 2.28 (Some summands of the CBHD series). For example, some (tedious) explicit computations on the Lie summands expressing $Z_n(a, b)$ give

$$Z_1(a, b) = a + b$$

$$Z_2(a, b) = \frac{1}{2}\, [a, b]$$

$$Z_3(a, b) = \frac{1}{12}\, [a, [a, b]] + \frac{1}{12}\, [b, [b, a]]$$

$$Z_4(a, b) = -\frac{1}{24}\, [a, [b, [a, b]]]$$

$$Z_5(a, b) = -\frac{1}{120}\, \big([a[b[a[a, b]]]] + [b[a[b[b, a]]]]\big)+$$

$$+ \frac{1}{360}\, \big([b[a[a[a, b]]]] + [a[b[b[b, a]]]]\big)+$$

$$- \frac{1}{720}\, \big([a[a[a[a, b]]]] + [b[b[b[b, a]]]]\big)$$

$$Z_6(a, b) = \frac{1}{1440}\, [a[b[b[b[a, b]]]]] - \frac{1}{720}\, [a[a[b[b[a, b]]]]]+$$

$$+ \frac{1}{240}\, [a[b[b[a[a, b]]]]] - \frac{1}{1440}\, [b[a[a[a[b, a]]]]].$$

Note the *homogeneity* of each $Z_n(a, b)$ (which is a Lie polynomial homogeneous of degree n in a and b jointly). By Rem. 2.27 we can derive from the above formulae the expressions of the first few many $C_{i,j}(a, b)$:

$$C_{1,0}(a, b) = a, \qquad C_{0,1}(a, b) = b$$

$$C_{1,1}(a, b) = \frac{1}{2}\, [a, b]$$

$$C_{2,1}(a, b) = \frac{1}{12}\, [a, [a, b]], \qquad C_{1,2}(a, b) = \frac{1}{12}\, [b, [b, a]]$$

$$C_{2,2}(a, b) = -\frac{1}{24}\, [a, [b, [a, b]]], \qquad C_{3,1}(a, b) = C_{1,3}(a, b) = 0$$

$$C_{4,1}(a, b) = -\frac{1}{720}\, [a[a[a[a, b]]]], \qquad C_{1,4}(a, b) = -\frac{1}{720}\, [b[b[b[b, a]]]]$$

$$C_{3,2}(a, b) = -\frac{1}{120}\, [a[b[a[a, b]]]] + \frac{1}{360}\, [b[a[a[a, b]]]]$$

$$C_{2,3}(a, b) = -\frac{1}{120}\, [b[a[b[b, a]]]] + \frac{1}{360}\, [a[b[b[b, a]]]]$$

$$C_{5,1}(a, b) = C_{1,5}(a, b) = 0$$

$$C_{4,2}(a, b) = -\frac{1}{1440}\, [b[a[a[a[b, a]]]]], \qquad C_{2,4}(a, b) = \frac{1}{1440}\, [a[b[b[b[a, b]]]]]$$

$$C_{3,3}(a, b) = -\frac{1}{720}\, [a[a[b[b[a, b]]]]] + \frac{1}{240}\, [a[b[b[a[a, b]]]]].$$

2.5 Identities resulting from the Exponential Theorem

By using the symbols in Def. 2.24, the Exponential Thm. 2.13 states that

$$\mathrm{Exp}(x\, t) \cdot \mathrm{Exp}(y\, t) = \mathrm{Exp}\left(\textstyle\sum_{n=1}^{\infty} Z_n(x, y)\, t^n\right). \tag{2.31}$$

As is meaningfully shown in Sec. 2.3, we are allowed, roughly speaking, to take "$t = 1$" in (2.31). More precisely in the algebra $\mathbb{K}\langle\!\langle x, y\rangle\!\rangle$ of the formal power series in x, y we have the following equivalent identities:

$$\mathrm{Exp}(x) \cdot \mathrm{Exp}(y) = \mathrm{Exp}\big(\textstyle\sum_{n=1}^{\infty} Z_n(x, y)\big), \tag{2.32}$$

$$\mathrm{Log}\big(\mathrm{Exp}(x) \cdot \mathrm{Exp}(y)\big) = \textstyle\sum_{n=1}^{\infty} Z_n(x, y). \tag{2.33}$$

With the compact notation in (2.28) (the series $x \diamond y$ being convergent in the complete metric space $\mathbb{K}\langle\!\langle x, y\rangle\!\rangle$), these can be recast as

$$\mathrm{Exp}(x) \cdot \mathrm{Exp}(y) = \mathrm{Exp}(x \diamond y), \qquad \mathrm{Log}\big(\mathrm{Exp}(x) \cdot \mathrm{Exp}(y)\big) = x \diamond y. \tag{2.34}$$

Starting from (2.32) and (2.33), it is possible to derive infinitely many polynomial identities in the indeterminates x and y (and this also holds, *mutatis mutandis*, in any unital associative algebra!). To this end, we fix a notation: the monomial

$$c\, x^{i_1} y^{j_1} \cdots x^{i_n} y^{j_n} \qquad (c \in \mathbb{K} \setminus \{0\})$$

is said to have:

    (i) x-**degree** equal to $i_1 + \cdots + i_n$;
    (ii) y-**degree** equal to $j_1 + \cdots + j_n$;
    (iii) **joint-degree** equal to $i_1 + j_1 + \cdots + i_n + j_n$.

The degrees (x-, y-, or joint-) for a polynomial are defined, as usual, by taking maximum degrees (respectively, x-, y-, joint-) from its non-null summands.

Given a polynomial $p \in \mathbb{K}\langle x, y\rangle$ or a formal power series $p \in \mathbb{K}\langle\!\langle x, y\rangle\!\rangle$, for any $n, m \in \mathbb{N} \cup \{0\}$ we introduce projecting operators denoted by

$$p \mapsto \mathop{\mathrm{DEG}}_{n,\,m}(p) \quad \text{and} \quad p \mapsto \mathop{\mathrm{DEG}}_{n}(p), \tag{2.35}$$

by declaring, respectively, that:[10]

- $\mathop{\mathrm{DEG}}_{n,\,m}(p)$ is the sum of the monomials in p with x-degree $\leq n$, and (at the same time) having y-degree $\leq m$;
- $\mathop{\mathrm{DEG}}_{n}(p)$ is the sum of the monomials in p with joint-degree $\leq n$.

For example we have

$$\mathop{\mathrm{DEG}}_{n,\,m}\Big(\mathrm{Exp}(x) \cdot \mathrm{Exp}(y)\Big) = \sum_{0 \leq i \leq n,\ 0 \leq j \leq m} \frac{x^i\, y^j}{i!\, j!},$$

$$\mathop{\mathrm{DEG}}_{n}\Big(\mathrm{Exp}(x) \cdot \mathrm{Exp}(y)\Big) = \sum_{0 \leq i+j \leq n} \frac{x^i\, y^j}{i!\, j!}.$$

---

[10] These definitions are well posed due to the obvious linear independence between monomials containing x or y a *different* number of times.

As a consequence, from (2.32) we obtain the following identities, holding true for every $n, m \in \mathbb{N} \cup \{0\}$:

$$\sum_{0 \leq i \leq n,\ 0 \leq j \leq m} \frac{x^i\, y^j}{i!\, j!} = \underset{n,m}{\mathrm{DEG}} \left\{ \sum_{k=0}^{\infty} \frac{1}{k!} \left(\sum_{h=1}^{\infty} Z_h(x,y) \right)^k \right\},$$

and also

$$\sum_{0 \leq i+j \leq n} \frac{x^i\, y^j}{i!\, j!} = \underset{n}{\mathrm{DEG}} \left\{ \sum_{k=0}^{\infty} \frac{1}{k!} \left(\sum_{h=1}^{\infty} Z_h(x,y) \right)^k \right\}.$$

Let us now observe that, in the infinite sums on both rhs's of the above equalities, the result is unaltered if we bound summations up to some suitable finite degree: precisely, up to degree $\max\{n,m\}$ in the first identity, and up to degree n in the second one; furthermore, we can select out of $\sum_{h=1}^{\infty} Z_h$ only the summands $C_{i,j}$ which "really matter" in producing $\underset{n,m}{\mathrm{DEG}}$ or $\underset{n}{\mathrm{DEG}}$.

All these remarks prove the following result.

Theorem 2.29 (Identities from the Exponential Theorem). *Let* $\mathbb{K}\langle x, y \rangle$ *denote the algebra of the polynomials in two non-commuting indeterminates* x *and* y *(over a field of characteristic zero* $\mathbb{K}$*). Let* $n, m \in \mathbb{N} \cup \{0\}$.

Then we have the following identities:

$$\sum_{k=0}^{\max\{n,m\}} \frac{1}{k!} \left(\sum_{\substack{0 \leq i \leq n \\ 0 \leq j \leq m}} C_{i,j}(x,y) \right)^k = \sum_{\substack{0 \leq i \leq n \\ 0 \leq j \leq m}} \frac{x^i\, y^j}{i!\, j!} + \mathcal{R}_{n,m}(x,y), \qquad (2.36)$$

$$\sum_{k=0}^{n} \frac{1}{k!} \left(\sum_{h=1}^{n} Z_h(x,y) \right)^k = \sum_{0 \leq i+j \leq n} \frac{x^i\, y^j}{i!\, j!} + \mathcal{R}_n(x,y), \qquad (2.37)$$

where $\mathcal{R}_{n,m}(x,y) \in \mathbb{K}\langle x, y \rangle$ *is a finite sum of monomials all having* x*-degree* $\geq n+1$ *or* y*-degree* $\geq m+1$*, while* $\mathcal{R}_n(x,y) \in \mathbb{K}\langle x, y \rangle$ *is a finite sum of monomials all having joint-degree* $\geq n+1$.

Equivalently, $\mathcal{R}_{n,m}(x,y)$ *is a finite sum of monomials where* x *appears at least* $n+1$ *times* **or** y *appears at least* $m+1$ *times, while* $\mathcal{R}_n(x,y)$ *is a finite sum of monomials where the total number of* x **and** y *is at least* $n+1$.

Remark 2.30. Due to the universality of the non-commuting indeterminates x and y (see Rem. 2.23), the above identities specialize to identities on any unital associative algebra $(A, *)$, by replacing x, y with any pair of elements of A, and by the substitution of the product of $\mathbb{K}\langle x, y \rangle$ with $*$.

For instance we can consider $(A, *) = (\mathcal{U}(\Omega), \circ)$, the unital associative algebra of the smooth linear differential operators on an open set Ω, and choose $x = sX$, $y = tY$ where X, Y are smooth vector fields on Ω and with $s, t \in \mathbb{R}$. We shall do this in the next chapter. $\sharp$

2.6   The Exponential Theorem for $\mathbb{K}\langle x, y\rangle[\![s, t]\!]$

The aim of this section is to derive remarkable formal "PDEs" involving the CBHD series (see Thm. 2.35): this is the important *fil-rouge* connecting the present section to Sec. 2.2, where we obtained a formal ODE for the fundamental "curve" defined by $Z(t) = \mathrm{Log}(\mathrm{Exp}(xt) \cdot \mathrm{Exp}(yt))$ (see identities (2.17a)-to-(2.19)).

To this end, firstly (Sec. 2.6.1) we introduce the algebraic background on which all the computations will be performed: this is the unital associative algebra $\mathscr{T}[\![s, t]\!]$ of the formal power series in two *commuting* indeterminates s, t over the algebra $\mathscr{T} = \mathbb{K}\langle x, y\rangle$; secondly (Sec. 2.6.2) we derive the corresponding Exponential Theorem on (the ultra-metric complete space) $\mathscr{T}[\![s, t]\!]$, which boils down to a family of recursive identities usable in any Lie algebra.

2.6.1   *The algebra* $\mathbb{K}\langle x, y\rangle[\![s, t]\!]$

We introduce an appropriate algebraic setting (slightly different from those used so far: $\mathscr{T}[\![t]\!]$ or $\mathbb{K}\langle\!\langle x, y\rangle\!\rangle$) allowing us to prove a more general version of the Exponential Thm. 2.13. From now on, $\mathbb{K}$ denotes a fixed field of characteristic zero.

We consider the unital associative algebra $\mathscr{T}[\![s, t]\!]$ of the formal power series in **two *commuting* intedeterminates** s, t over the algebra $\mathscr{T} = \mathbb{K}\langle x, y\rangle$ (the latter is the unital associative algebra over $\mathbb{K}$ of the polynomials in two *non-commuting* indeterminates x, y). The elements of $\mathscr{T}[\![s, t]\!]$ have the form

$$f = \sum_{i,j=0}^{\infty} f_{i,j}\, s^i t^j, \quad \text{where } f_{i,j} \in \mathbb{K}\langle x, y\rangle \text{ for every } i, j \geq 0;$$

the sum and the product in $\mathscr{T}[\![s, t]\!]$ are defined in the (obvious) way:

$$\left(\sum_{i,j=0}^{\infty} f_{i,j}\, s^i t^j \right) + \left(\sum_{i,j=0}^{\infty} g_{i,j}\, s^i t^j \right) = \sum_{i,j=0}^{\infty} (f_{i,j} + g_{i,j}) s^i t^j;$$

$$\left(\sum_{i,j=0}^{\infty} f_{i,j}\, s^i t^j \right) \cdot \left(\sum_{i,j=0}^{\infty} g_{i,j}\, s^i t^j \right) = \sum_{i,j=0}^{\infty} \left(\sum_{\substack{h+h'=i \\ k+k'=j}} f_{h,k}\, g_{h',k'} \right) s^i t^j.$$

The scalar action of $\mathbb{K}$ on $\mathscr{T}[\![s, t]\!]$ is also defined component-wise.

Essentially in the same way as in $\mathscr{T}[\![t]\!]$, it is possible to introduce on $\mathscr{T}[\![s, t]\!]$ a notion of minimum degree and a translation-invariant metric d; we leave the details to the reader. It turns out that $(\mathscr{T}[\![s, t]\!], d)$ is a complete ultra-metric space, and $(\mathscr{T}[\![s, t]\!], +, \cdot)$ is a topological algebra.

The main motivation for $\mathscr{T}[\![s, t]\!]$ is the introduction of two *partial* differential operators ∂_s and ∂_t:

Definition 2.31 (Partial derivatives on $\mathscr{T}[\![s,t]\!]$). We define the endomorphisms $\partial_s, \partial_t : \mathscr{T}[\![s,t]\!] \to \mathscr{T}[\![s,t]\!]$ by setting

$$\partial_s\left(\sum_{i,j=0}^{\infty} f_{i,j}\, s^i t^j \right) := \sum_{i,j=0}^{\infty} (i+1)\, f_{i+1,j}\, s^i t^j,$$

$$\partial_t\left(\sum_{i,j=0}^{\infty} f_{i,j}\, s^i t^j \right) := \sum_{i,j=0}^{\infty} (j+1)\, f_{i,j+1}\, s^i t^j.$$

As for the map D_t on the associative algebra $\mathscr{T}[\![t]\!]$, it is easy to recognize that the two endomorphisms ∂_s and ∂_t of $\mathscr{T}[\![s,t]\!]$ are actually *derivations* of the associative algebra $(\mathscr{T}[\![s,t]\!], \cdot)$ (hence they are derivations of the related Lie algebra as well). Moreover, it is immediate to see that the maps ∂_s and ∂_t are *continuous* on the metric space $(\mathscr{T}[\![s,t]\!], d)$; we suggest that the reader provide a detailed proof.

One can define Exp and Log maps on $\mathscr{T}[\![s,t]\!]$ in the obvious way, enjoying analogous formulae as in the case of $\mathscr{T}[\![t]\!]$.

2.6.2 *The Exponential Theorem for $\mathbb{K}\langle x, y\rangle[\![s,t]\!]$*

In the algebraic context introduced in the previous section, we can state and prove the following version of the Exponential Theorem for the algebra $\mathscr{T}[\![s,t]\!]$. As we shall see in a moment, the proof of this theorem is crucially based on two formal "PDEs" holding true in $\mathscr{T}[\![s,t]\!]$, which we shall use in Chap. 18 in order to find a domain of convergence for the CBHD series.

Theorem 2.32 (The Exponential Theorem for $\mathscr{T}[\![s,t]\!]$). *Let $\mathscr{T} = \mathbb{K}\langle x, y\rangle$. Let us consider the power series in $\mathscr{T}[\![s,t]\!]$ defined as follows:*

$$C = \sum_{i,j=0}^{\infty} C_{i,j}\, s^i t^j := \mathrm{Log}\big(\mathrm{Exp}(x\,s) \cdot \mathrm{Exp}(y\,t)\big). \tag{2.38}$$

Then, the polynomials $C_{i,j} \in \mathbb{K}\langle x, y\rangle$ are Lie-polynomials in x and y:

$$C_{i,j} \in \mathcal{L}(x,y), \quad \text{for every } i,j \in \mathbb{N} \cup \{0\}. \tag{2.39}$$

As we did in proving the Exponential Theorem for $\mathscr{T}[\![t]\!]$, we begin by obtaining an *explicit* expression for the polynomials $C_{i,j}$, a priori only belonging to $\mathbb{K}\langle x, y\rangle$. First of all, since C is the logarithm of a series, then $C \in \mathscr{T}[\![s,t]\!]_+$, that is $C_{0,0} = 0$. Moreover, from the very definition of Exp and Log we get

$$C = \mathrm{Log}\big(\mathrm{Exp}(x\,s) \cdot \mathrm{Exp}(y\,t)\big) = \mathrm{Log}\Big(1 + \sum_{i+j=1}^{\infty} \frac{x^i y^j}{i!\,j!}\, s^i t^j\Big)$$

$$= \sum_{k=1}^{\infty} \frac{(-1)^{k+1}}{k} \sum_{(i_1,j_1),\dots,(i_k,j_k) \neq (0,0)} \frac{x^{i_1} y^{j_1} \cdots x^{i_k} y^{j_k}}{i_1!\,j_1!\dots i_k!\,j_k!}\, s^{i_1+\cdots+i_k}\, t^{j_1+\cdots+j_k}$$

$$= \sum_{i+j=1}^{\infty} \left(\sum_{k=1}^{i+j} \frac{(-1)^{k+1}}{k} \sum_{\substack{(i_1,j_1),\dots,(i_k,j_k) \neq (0,0) \\ i_1+\cdots+i_k = i,\; j_1+\cdots+j_k = j}} \frac{x^{i_1} y^{j_1} \cdots x^{i_k} y^{j_k}}{i_1!\,j_1!\dots i_k!\,j_k!} \right) s^i t^j,$$

and thus, for every $i, j \in \mathbb{N}$ with $i + j \geq 1$, we obtain

$$C_{i,j} = \sum_{k=1}^{i+j} \frac{(-1)^{k+1}}{k} \sum_{\substack{(i_1,j_1),\dots,(i_k,j_k) \neq (0,0) \\ i_1+\cdots+i_k = i,\ j_1+\cdots+j_k = j}} \frac{x^{i_1} y^{j_1} \cdots x^{i_k} y^{j_k}}{i_1! j_1! \dots i_k! j_k!}. \tag{2.40}$$

Identity (2.40) exhibits a deep connection between the polynomials $C_{i,j}$ and the Lie-polynomials $Z_n(x, y)$ introduced in Thm. 2.13 (see, precisely, formula (2.7)). Indeed, for every $n \in \mathbb{N}$, by summing up all the polynomials $C_{i,j}$ with $i+j = n$, from (2.40) we get

$$\sum_{i+j=n} C_{i,j} \overset{(2.40)}{=} \sum_{k=1}^{n} \frac{(-1)^{k+1}}{k} \sum_{\substack{(i_1,j_1),\dots,(i_k,j_k) \neq (0,0) \\ i_1+\cdots+i_k+j_1+\cdots+j_k = n}} \frac{x^{i_1} y^{j_1} \cdots x^{i_k} y^{j_k}}{i_1! j_1! \dots i_k! j_k!}$$

$$\overset{(2.7)}{=} Z_n(x, y).$$

This gives

$$\sum_{i+j=n} C_{i,j} = Z_n(x, y), \quad \text{for every } n \in \mathbb{N}. \tag{2.41}$$

We can rephrase identities (2.40) and (2.41) as follows: for every fixed natural n, the polynomial $C_{i,j}$ with $i + j = n$ is obtained from the Lie-polynomial $Z_n(x, y)$ by gathering together all the summands in $Z_n(x, y)$ having x-degree exactly equal to i and y-degree exactly equal to j (as $Z_n(x, y)$ is a homogeneous Lie-polynomial with joint degree n in x and y).

Remark 2.33. We remark that, due to (2.41), the Exponential Thm. 2.13 for $\mathscr{T}[\![t]\!]$ is a simple consequence of the Exponential Thm. 2.32 for $\mathscr{T}[\![s, t]\!]$.

Vice versa, one could prove Thm. 2.32 for $\mathscr{T}[\![s, t]\!]$ by starting from the Exponential Thm. 2.13 for $\mathscr{T}[\![t]\!]$. This is a consequence of Exr. 2.12, ensuring that, if p is a Lie-polynomial, then the sum of the summands of p having x-degree i and y-degree j is a Lie-polynomial as well (for any given $i, j \geq 0$). However, we shall give a proof of Thm. 2.32 independent of Thm. 2.13. ♯

Remark 2.34. Once we know (2.39), if φ is the Dynkin-Specht-Wever map in Lem. 2.22, we get $C_{i,j} = \varphi(C_{i,j})$, for any $i, j \geq 0$. This gives

$$C_{i,j} \overset{(2.40)}{=} \frac{1}{i+j} \sum_{k=1}^{i+j} \frac{(-1)^{k+1}}{k} \sum_{\substack{(i_1,j_1),\dots,(i_k,j_k) \neq (0,0) \\ i_1+\cdots+i_k = i,\ j_1+\cdots+j_k = j}} \frac{(\operatorname{ad} x)^{i_1} \cdots (\operatorname{ad} y)^{j_k-1}(y)}{i_1! j_1! \dots i_k! j_k!}$$

$$\overset{(2.27)}{=} C_{i,j}(x, y),$$

where Dynkin's $C_{i,j}$ have been introduced in Def. 2.24. The above identity

$$C_{i,j} = C_{i,j}(x, y) \quad (i, j \geq 0) \tag{2.42}$$

gives the non-trivial polynomial identity

$$\sum_{k=1}^{i+j} \frac{(-1)^{k+1}}{k} \sum_{\substack{(i_1,j_1),\ldots,(i_k,j_k) \neq (0,0) \\ i_1+\cdots+i_k=i,\ j_1+\cdots+j_k=j}} \frac{x^{i_1} y^{j_1} \cdots x^{i_k} y^{j_k}}{i_1! j_1! \ldots i_k! j_k!}$$

$$= \frac{1}{i+j} \sum_{k=1}^{i+j} \frac{(-1)^{k+1}}{k} \sum_{\substack{(i_1,j_1),\ldots,(i_k,j_k) \neq (0,0) \\ i_1+\cdots+i_k=i, \\ j_1+\cdots+j_k=j}} \frac{(\operatorname{ad} x)^{i_1} (\operatorname{ad} y)^{j_1} \cdots (\operatorname{ad} x)^{i_k} (\operatorname{ad} y)^{j_k-1}(y)}{i_1! j_1! \ldots i_k! j_k!}.$$

2.6.3 *Poincaré's PDEs on* $\mathbb{K}\langle x, y\rangle [\![s,t]\!]$

We are now ready to derive two formal "PDEs" previously anticipated for C, which is the real core of the proof of Thm. 2.32.

Theorem 2.35 (Poincaré's PDEs for C). *Let C be the series in $\mathscr{T}[\![s,t]\!]$ defined in (2.38) and let $\mathbf{b}$ be the complex holomorphic function defined as follows*

$$\mathbf{b} : B(0, 2\pi) \longrightarrow \mathbb{C}, \qquad \mathbf{b}(z) := \frac{z}{e^z - 1}, \tag{2.43}$$

where $B(0, 2\pi) := \{z \in \mathbb{C} : |z| < 2\pi\}$ (here $\mathbf{b}(0) = 1$).
   Then C satisfies the following identities (two "formal PDEs"):

$$\partial_s(C) = \mathbf{b}(\operatorname{ad} C)(x), \qquad \partial_t(C) = \mathbf{b}(-\operatorname{ad} C)(y). \tag{2.44}$$

Here $\mathbf{b}$ is used only under its formal power series expansion (see Rem. 2.19). We refer to the identities in (2.44) respectively as **the first and second Poincaré's PDEs on** $\mathbb{K}\langle x, y\rangle [\![s,t]\!]$. *With a compact notation we have*

$$\partial_s(C) = \frac{\operatorname{ad} C}{e^{\operatorname{ad} C} - 1}(x), \qquad \partial_t(C) = \frac{-\operatorname{ad} C}{e^{-\operatorname{ad} C} - 1}(y).$$

The second identity in (2.44) goes back to [Poincaré (1900)] (see equation (7) on p. 248); Poincaré wrote it (with a different formalism) as $\partial_t(C) = \psi(\operatorname{ad} C)(y)$, via what is now called Todd's function $\psi(z) = \mathbf{b}(-z)$.

Proof. We prove the first formal PDE in (2.44), since the second one can be proved analogously. First of all, from the very definition of C (and from the fact that Exp and Log are inverse to each other), one has $\operatorname{Exp}(C) = \operatorname{Exp}(x\,s) \cdot \operatorname{Exp}(y\,t)$. Our idea is to apply the endomorphism ∂_s on both sides of this identity: for the lhs, we have

$$\partial_s(\operatorname{Exp}(C)) = \frac{e^{\operatorname{ad} C} - 1}{\operatorname{ad} C}(\partial_s(C)) \cdot \operatorname{Exp}(C), \tag{2.45}$$

while for the rhs we have (remember that ∂_s is a derivation of $\mathscr{T}[\![s,t]\!]$)

$$\partial_s(\operatorname{Exp}(x\,s) \cdot \operatorname{Exp}(y\,t)) = \partial_s(\operatorname{Exp}(x\,s)) \cdot \operatorname{Exp}(y\,t)$$

$$= x \cdot \operatorname{Exp}(x\,s) \cdot \operatorname{Exp}(y\,t) = x \cdot \operatorname{Exp}(C). \tag{2.46}$$

By gathering together identities (2.45) and (2.46) we obtain

$$\frac{e^{\operatorname{ad}C} - 1}{\operatorname{ad}C}(\partial_s(C)) \cdot \operatorname{Exp}(C) = x \cdot \operatorname{Exp}(C),$$

and by right-multiplication on both sides by $\operatorname{Exp}(-C)$ we get

$$\frac{e^{\operatorname{ad}C} - 1}{\operatorname{ad}C}(\partial_s(C)) = x. \tag{2.47}$$

We are now ready to conclude: by inverting the endomorphism $\dfrac{e^{\operatorname{ad}C} - 1}{\operatorname{ad}C}$ acting on $\partial_s(C)$ in (2.47) (see Rem. 2.19) we obtain

$$\partial_s(C) = \left(\frac{e^{\operatorname{ad}C} - 1}{\operatorname{ad}C}\right)^{-1}(x) = \mathbf{b}(\operatorname{ad}C)(x),$$

and this is exactly the first formal PDE in (2.44). $\qquad\square$

2.7   More identities

Theorem 2.36 (Recursive identities for $\{C_{i,j}\}$). *Let C be the series in $\mathscr{T}[\![s,t]\!]$ defined in (2.38) and let $\{B_n\}_{n\geq 0}$ be the Bernoulli numbers (see the generating series in (2.18) relative to the function $\mathbf{b}$ defined in (2.43)). We have*

$$C_{0,0} = 0, \quad C_{1,0} = x, \quad C_{0,1} = y, \tag{2.48}$$

and, for all $i,j \in \mathbb{N} \cup \{0\}$ with $i+j \geq 1$, the following recursive identities hold:

$$
\begin{cases}
C_{i+1,j} = \dfrac{1}{i+1} \displaystyle\sum_{\substack{1 \leqslant h \leqslant i+j \\ (i_1,j_1),\ldots,(i_h,j_h) \neq (0,0) \\ i_1+\cdots+i_h = i,\; j_1+\cdots+j_h = j}} \dfrac{B_h}{h!}\,[C_{i_1,j_1}\ldots[C_{i_h,j_h},x]\ldots], \\[2.5em]
C_{i,j+1} = \dfrac{1}{j+1} \displaystyle\sum_{\substack{1 \leqslant h \leqslant i+j \\ (i_1,j_1),\ldots,(i_h,j_h) \neq (0,0) \\ i_1+\cdots+i_h = i,\; j_1+\cdots+j_h = j}} (-1)^h \dfrac{B_h}{h!}\,[C_{i_1,j_1}\ldots[C_{i_h,j_h},y]\ldots].
\end{cases}
\tag{2.49}
$$

Proof. First of all, since C is the logarithm of a series, then C belongs to $\mathscr{T}[\![s,t]\!]_+$, that is, $C_{0,0} = 0$. Moreover, by inserting into the first PDE of (2.44) the explicit expression of C we get (as $C_{0,0} = 0$)

$$\sum_{i,j=0}^{\infty} (i+1)\,C_{i+1,j}\,s^i t^j = \partial_s(C) = \mathbf{b}(\operatorname{ad}C)(x) = \sum_{h=0}^{\infty} \frac{B_h}{h!}\,(\operatorname{ad}C)^h(x)$$

$$= x + \sum_{h=1}^{\infty} \sum_{i_1+j_1,\ldots,i_h+j_h \geq 1} \frac{B_h}{h!}\,[C_{i_1,j_1}\ldots[C_{i_h,j_h},x]\ldots]\,s^{i_1+\cdots+i_h}\,t^{j_1+\cdots+j_h}$$

$$= x + \sum_{i+j=1}^{\infty} \left(\sum_{\substack{1 \leqslant h \leqslant i+j \\ (i_1,j_1),\ldots,(i_h,j_h) \neq (0,0) \\ i_1+\cdots+i_h = i,\; j_1+\cdots+j_h = j}} \frac{B_h}{h!}\,[C_{i_1,j_1}\ldots[C_{i_h,j_h},x]\ldots] \right) s^i t^j.$$

By equating the coefficients relative to the same powers of s and t in the above identity, we obtain $C_{1,0} = x$, and, for all $i, j \in \mathbb{N} \cup \{0\}$ with $i + j \geq 1$,

$$(i+1)\, C_{i+1,j} = \sum_{\substack{1 \leq h \leq i+j \\ (i_1,j_1),\dots,(i_h,j_h) \neq (0,0) \\ i_1+\cdots+i_h = i,\ j_1+\cdots+j_h = j}} \frac{B_h}{h!}\, [C_{i_1,j_1}, \dots, [C_{i_h,j_h}, x] \dots].$$

The proof of the third equality in (2.48) as well as the proof of the second identity in (2.49) are completely analogous. $\qquad\square$

We can finally give the awaited proof of the Exponential Thm. 2.32.

Proof of Thm. 2.32. We proceed by induction on $n := i + j$. If $n = 0$ or $n = 1$ there is nothing to prove, since $C_{0,0} = 0, C_{1,0} = x$ and $C_{0,1} = y$ belong to $\mathcal{L}(x,y)$.

Let $n \geq 2$ and let us assume the induction hypothesis:

$$C_{i,j} \in \mathcal{L}(x,y) \text{ for all } i, j \in \mathbb{N} \cup \{0\} \text{ with } i + j \leq n.$$

Let $i, j \in \mathbb{N} \cup \{0\}$ be such that $i + j = n + 1$; we have to show that $C_{i,j} \in \mathcal{L}(x,y)$.

- If $i \geq 1$ (that is, if $j < n + 1$), from the first identity in (2.49) we get

$$C_{i,j} = \frac{1}{i} \sum_{\substack{1 \leq h \leq i+j-1 \\ (i_1,j_1),\dots,(i_h,j_h) \neq (0,0) \\ i_1+\cdots+i_h = i-1,\ j_1+\cdots+j_h = j}} \frac{B_h}{h!}\, [C_{i_1,j_1}, \dots, [C_{i_h,j_h}, x] \dots],$$

and since the above $C_{i_1,j_1}, \dots, C_{i_h,j_h}$ are such that $i_1+j_1, \dots, i_h+j_h$ do not[11] exceed n, it follows from the inductive hypothesis that $C_{i,j} \in \mathcal{L}(x,y)$.

- If $i = 0$ (that is, if $j = n + 1 \geq 3$), we use the second identity in (2.49):

$$C_{0,j} = \frac{1}{j} \sum_{\substack{1 \leq h \leq n \\ (i_1,j_1),\dots,(i_h,j_h) \neq (0,0) \\ i_1+\cdots+i_h = 0,\ j_1+\cdots+j_h = j-1=n}} (-1)^h \frac{B_h}{h!}\, [C_{i_1,j_1}, \dots, [C_{i_h,j_h}, y] \dots].$$

Since $i_1 = \cdots = i_h = 0$ and $j_1,\dots,j_h$ do not exceed n, again by the inductive hypothesis we infer that $C_{0,j} \in \mathcal{L}(x,y)$, and this completes the proof. $\qquad\square$

We close the section with a useful corollary of Thm. 2.36, which will be fundamental in Chap. 18. Indeed, by gathering together Rem. 2.34 and the well-known *universal property* of the Lie algebra $\mathcal{L}(x,y)$ (see e.g., [Bonfiglioli and Fulci (2012), Sec. 2.2]) we immediately obtain the following:

Corollary 2.37 (Recursive identities for $\{C_{i,j}(a,b)\}$). *Let $(\mathfrak{g}, [\cdot,\cdot])$ be a Lie algebra over a field $\mathbb{K}$ of characteristic zero, and let $a, b \in \mathfrak{g}$ be arbitrarily fixed. Let finally $\{C_{i,j}(a,b)\}$ be Dynkin's non-homogeneous polynomials, defined in Def. 2.24, and let $\{B_n\}_{n\geq 0}$ be Bernoulli numbers in (2.18). We have*

$$C_{0,0}(a,b) = 0, \quad C_{1,0}(a,b) = a, \quad C_{0,1}(a,b) = b,$$

---

[11]Indeed $i_1 + j_1, \dots, i_h + j_h \leq i_1 + j_1 + \cdots + i_h + j_h = i + j - 1 = n + 1 - 1 = n$.

and, for all $i, j \in \mathbb{N} \cup \{0\}$ with $i + j \geq 1$, the following recursive identities hold:

$$C_{i+1,j}(a,b) = \tag{2.50}$$

$$\frac{1}{i+1} \sum_{\substack{1 \leq h \leq i+j \\ (i_1,j_1),\ldots,(i_h,j_h) \neq (0,0) \\ i_1+\cdots+i_h = i,\ j_1+\cdots+j_h = j}} \frac{B_h}{h!} \, [C_{i_1,j_1}(a,b),\ldots,[C_{i_h,j_h}(a,b),a]\ldots],$$

$$C_{i,j+1}(a,b) = \tag{2.51}$$

$$\frac{1}{j+1} \sum_{\substack{1 \leq h \leq i+j \\ (i_1,j_1),\ldots,(i_h,j_h) \neq (0,0) \\ i_1+\cdots+i_h = i,\ j_1+\cdots+j_h = j}} (-1)^h \frac{B_h}{h!} \, [C_{i_1,j_1}(a,b),\ldots,[C_{i_h,j_h}(a,b),b]\ldots].$$

Proof. This follows at once from identity (2.42), from (2.49) and from the mentioned universal property[12] of $\mathcal{L}(x,y)$. $\qquad\qquad\square$

Obviously, one can obtain analogous identities for the Dynkin polynomials $Z_n(x,y)$; due to the crucial role of these identities in deriving an ODE in finite-dimensional Lie algebras of vector fields, we shall do this in a selected section, see Sec. 5.3 on page 118.

2.8 Appendix: manipulations of formal series

In this short appendix we give some hint on how to formalize the manipulations of formal power series of endomorphisms used in the previous sections. The details are left to the interested reader.

To this end, let $\mathcal{A}$ be any of the ultra-metric topological spaces

$$\mathbb{K}\langle x,y\rangle[\![t]\!], \qquad \mathbb{K}\langle\!\langle x,y\rangle\!\rangle, \qquad \mathbb{K}\langle x,y\rangle[\![s,t]\!],$$

defined in the previous sections ($\mathbb{K}$ is a field of characteristic zero). On any of these algebras $\mathcal{A}$ we defined the notion of a minimum degree $\underline{\deg}$, deeply intertwined with the metric structure.

Let A be an endomorphism of $\mathcal{A}$ with the following property:

$$\underline{\deg}(A(p)) \geq \underline{\deg}(p) + 1, \quad \forall\, p \in \mathcal{A}. \tag{2.52}$$

This is the case, for example, if $A = \operatorname{ad} z$, with $z \in \mathcal{A}$ such that $\underline{\deg}(z) \geq 1$. Let $(a_n)_n$ be any sequence in $\mathbb{K}$. Then (2.52) allows to prove that the map

$$p \mapsto \sum_{n=0}^{\infty} a_n \, A^n(p)$$

---

[12] This universal property states that, if $(\mathfrak{g}, [\cdot,\cdot])$ is a Lie algebra and if $a, b \in \mathfrak{g}$ are arbitrary, then there exists a Lie-algebra morphism $\varphi_{a,b} : \mathcal{L}(x,y) \to \mathfrak{g}$ such that $\varphi_{a,b}(x) = a$ and $\varphi_{a,b}(y) = b$. Roughly speaking, it states that any identity in $\mathcal{L}(x,y)$ between commutators of any order of x and y produces an analogous identity in any Lie algebra $\mathfrak{g}$, if x and y are replaced by any pair of elements a and b of $\mathfrak{g}$, and if the commutator in $\mathcal{L}(x,y)$ is replaced by the Lie bracket of $\mathfrak{g}$. For a proof of this useful result, see [Bonfiglioli and Fulci (2012), Section 2.2].

is well posed, and it defines a continuous endomorphism of $\mathcal{A}$. We denote it by $\sum_n a_n A^n$. It is also easy to demonstrate that $\sum_n a_n A^n$ *is invertible if $a_0 \neq 0$.* Indeed one can construct A^{-1} under the form $A^{-1} = \sum_n b_n A^n$, for suitable b_n's defined recursively by requiring that $A^{-1}(A(p)) = p$ for every $p \in \mathcal{A}$.

We next suppose that $(a_n)_n$ and $(b_n)_n$ are arbitrary sequences in $\mathbb{K}$. If A is as above, it is quite simple to check that the *composition* of the endomorphisms $\sum_n a_n A^n$ and $\sum_n b_n A^n$ is $\sum_n c_n A^n$, where $\sum_n c_n z^n$ is the formal power series obtained from the *Cauchy product* of the formal power series $\sum_n a_n z^n$ and $\sum_n b_n z^n$ (this just means that $c_n = \sum_{i+j=n} a_i b_j$). Note that this gives another proof of the fact that, *whenever $\sum_n a_n z^n$ and $\sum_n b_n z^n$ are formal power series in $\mathbb{K}$ whose (Cauchy) product is $1_{\mathbb{K}}$, then $\sum_n a_n A^n$ and $\sum_n b_n A^n$ are inverse-to-each-other endomorphisms.*

We finally suppose that $\mathbb{K} = \mathbb{R}$ (the case $\mathbb{K} = \mathbb{C}$ goes along the same lines). If $\sum_n a_n z^n$ is the Maclaurin series of a smooth function $f : \mathbb{R} \to \mathbb{R}$ with $f(0) \neq 0$, then –by the previous facts– the inverse endomorphism of $\sum_n a_n A^n$ is $\sum_n b_n A^n$, where $\sum_n b_n z^n$ is the Maclaurin series of the *reciprocal* function $1/f$ of f. Note that the radius of convergence of $\sum_n b_n z^n$ may be small, even if that of $\sum_n a_n z^n$ is ∞; but this does not make any difference as long as series are considered in the metric space $\mathcal{A}$, where convergence issues are very easy to be handled.

To give an explicit example of what is stated above, we can take $\mathcal{A} = \mathscr{T}[\![t]\!]$ and $A = \mathrm{ad}\, Z(t)$, where $Z(t)$ is the element of $\mathscr{T}[\![t]\!]_+$ defined in (2.13). Then the endomorphism (introduced in the proof of Thm. 2.13)

$$\varphi(-\mathrm{ad}\, Z(t)) := \frac{1 - e^{-\mathrm{ad}\, Z(t)}}{\mathrm{ad}\, Z(t)}$$

is invertible and its inverse map is $\dfrac{\mathrm{ad}\, Z(t)}{1 - e^{-\mathrm{ad}\, Z(t)}}$; the latter is the endomorphism $\sum_n b_n (\mathrm{ad}\, Z(t))^n$ where $\sum_n b_n z^n$ is the Maclaurin series of $\psi(z) = z/(1 - e^{-z})$. As a consequence, the formal computations used to obtain formulae (2.17a)-to-(2.17c) or employed in the derivation of the formal PDEs (2.44) are perfectly legitimate.

2.9 Exercises of Chap. 2

Exercise 2.1. Prove the assertions in Rem. 2.2, including (2.1).

[*Hint:* For the proof of the ultra-metric triangle inequality, first prove the inequality $\underline{\deg}(p + q) \geq \min\{\underline{\deg}(p), \underline{\deg}(q)\}$ for every $p, q \in \mathscr{T}[\![t]\!]$.]

Furthermore let $\mathscr{T}[t]$ denote the subalgebra of $\mathscr{T}[\![t]\!]$ of the polynomials in t with coefficients in $\mathscr{T} = \mathbb{K}\langle x, y \rangle$. In other words, the element $\sum_{k=0}^{\infty} a_k t^k$ of $\mathscr{T}[\![t]\!]$ belongs to $\mathscr{T}[t]$ iff $a_k = 0$, for k large. Prove that $\mathscr{T}[t]$ is dense in $\mathscr{T}[\![t]\!]$ by showing

$$\sum_{k=0}^{N} a_k t^k \longrightarrow \sum_{k=0}^{\infty} a_k t^k, \quad \text{as } N \to \infty.$$

Together with the fact that $\mathscr{T}[\![t]\!]$ is complete, the above density property allows us to say that $\mathscr{T}[\![t]\!]$ *is an isometric completion of $\mathscr{T}[t]$.*

Exercise 2.2. Let (X, d) be a metric space with the ultra-metric triangle inequality:

$$d(x, y) \leq \max\{d(x, z), d(z, y)\}, \quad \text{for every } x, y, z \in X.$$

Prove that a sequence $\{x_n\}_n$ in X is a Cauchy sequence iff $d(x_n, x_{n+1}) \to 0$ as $n \to \infty$. Suppose furthermore that X is a vector space and that the ultra-metric d is translation-invariant, that is it satisfies

$$d(x, y) = d(x + z, y + z) \text{ for every } x, y, z \in X.$$

Deduce that a series $\sum_{n \geq 0} x_n$ in X satisfies the Cauchy condition if and only if $x_n \to 0$ as $n \to \infty$ in X.

This shows that, if (X, d) is an ultra-metric and complete metric vector space, and if d is translation-invariant (as happens in the case of $\mathscr{T}[\![t]\!]$), *a series* $\sum_{n \geq 0} x_n$ *is convergent in X if and only if $x_n \to 0$ in X.*

Exercise 2.3. Prove that the limit $\lim_{n \to \infty} \sum_{k=0}^{n} \frac{1}{k!}$ does not exist in the metric space $\mathscr{T}[\![t]\!]$, but the series $\sum_{k=0}^{\infty} k! \, t^k$ is convergent.

Exercise 2.4. Prove Prop. 2.11.   *Hint:* Show that

$$\partial_t\big(\mathrm{Exp}(a\,t)\big) = \begin{cases} a \cdot \Big(\sum_{k=1}^{\infty} \dfrac{a^{k-1}\,t^{k-1}}{(k-1)!}\Big) = a \cdot \mathrm{Exp}(a\,t) \\[2ex] \Big(\sum_{k=1}^{\infty} \dfrac{a^{k-1}\,t^{k-1}}{(k-1)!}\Big) \cdot a = \mathrm{Exp}(a\,t) \cdot a. \end{cases}$$

Exercise 2.5. Prove that ∂_t is a derivation of the associative algebra $\mathscr{T}[\![t]\!]$.

[*Hint:* If $p = \sum_i a_i t^i$ and $q = \sum_j b_j t^j$, prove that

$$\partial_t(p \cdot q) = \sum_{k \geq 0} \Big(\sum_{i+j=k+1} (i\,a_i)b_j\Big)t^k + \sum_{k \geq 0}\Big(\sum_{i+j=k+1} a_i(j\,b_j)\Big)t^k,$$

then rename $i - 1$ in the first sum as i' and $j - 1$ in the second sum as j'...]

Exercise 2.6. Show that ∂_t is a continuous.

[*Hint:* It suffices to prove continuity at 0; if $p_n \xrightarrow{n \to \infty} 0$ in $\mathscr{T}[\![t]\!]$, we have $\underline{\deg}(p_n) \xrightarrow{n \to \infty} \infty$; then apply Rem. 2.7 to show that $\partial_t p_n \xrightarrow{n \to \infty} 0$...]

Exercise 2.7. Prove that the series in (2.3) and (2.4) defining Exp and Log are convergent in the metric space $\mathscr{T}[\![t]\!]$.

[*Hint:* Using Rem. 2.5, it suffices to show that $p^k \xrightarrow{k \to \infty} 0$ for $p \in \mathscr{T}[\![t]\!]_+$; to this end, prove that $\underline{\deg}(p^k) \geq k$ and use (2.1)...]

Exercise 2.8. Prove the assertion made in Rem. 2.26.

[*Hint:* Note that if $i_1 + j_1 + \cdots + i_k + j_k = n$, where $(i_1, j_1), \ldots, (i_k, j_k) \neq (0, 0)$, then necessarily $k \leq n$...]

Exercise 2.9. Starting from the identity $\mathrm{Exp}(x \diamond y) = \mathrm{Exp}(x) \cdot \mathrm{Exp}(y)$ prove that $-(x \diamond y) = (-y) \diamond (-x)$, or equivalently $y \diamond x = -\big((-x) \diamond (-y)\big)$.

[*Hint:* Exploit the identity $\big(\mathrm{Exp}(z)\big)^{-1} = \mathrm{Exp}(-z)$ in the unital associative algebra $\mathbb{K}\langle\langle x, y\rangle\rangle$; moreover $\big(\mathrm{Exp}(x) \cdot \mathrm{Exp}(y)\big)^{-1} = \mathrm{Exp}(-y) \cdot \mathrm{Exp}(-x)$...]

Derive the following properties for the functions Z_n and $C_{i,j}$:

$$Z_n(b, a) = (-1)^{n+1} Z_n(a, b); \qquad C_{i,j}(b, a) = (-1)^{i+j+1} C_{j,i}(a, b).$$

Note that, for $n = 4$, the identity $Z_4(b, a) = -Z_4(a, b)$ together with the explicit expression of $Z_4(a, b)$ proves the commutator identity (which *must* be a consequence of the sole skew-symmetry plus the Jacobi identity)

$$[a, [b, [a, b]]] = -[b, [a, [b, a]]].$$

Prove it directly by unraveling both sides using $[a, b] = ab - ba$.

Exercise 2.10. Suppose that, in $\mathscr{T}[[t]]$, p satisfies the "ODE" $\partial_t p = \sum_{k=0}^{\infty} b_k\, t^k$, where $\sum_{k=0}^{\infty} b_k\, t^k \in \mathscr{T}[[t]]$ is assigned. Prove that, if we further have the "initial condition" stating that the zero-degree term of p is β_0, then $p = \beta_0 + \sum_{k=0}^{\infty} b_k \frac{t^{k+1}}{k+1}$.

Exercise 2.11. In this exercise we obtain a Dynkin-type formula for $Z(t)$ in terms of Lie polynomials. Provide the details of the following argument.

(1) Insert (2.22) in (2.24) and derive the ODE

$$D_t Z(t) = \sum_{n=1}^{\infty} \frac{(-1)^{n+1}}{n} \left(e^{t\,\mathrm{ad}\,x} e^{t\,\mathrm{ad}\,y} - 1\right)^{n-1}\left(x + y + \left(e^{t\,\mathrm{ad}\,x} e^{t\,\mathrm{ad}\,y} - 1\right)(y)\right).$$

(2) Prove the identity

$$e^{t\,\mathrm{ad}\,x} \diamond e^{t\,\mathrm{ad}\,y} - 1 = \sum_{(i,j) \neq (0,0)} \frac{t^{i+j}}{i!\,j!} (\mathrm{ad}\,x)^i (\mathrm{ad}\,y)^j,$$

and insert it in the above rhs.
(3) Expand the resulting rhs, then "integrate by series" (see Exr. 2.10), and obtain the formula

$$Z(t) = x + y + \sum_{(i,j) \neq (0,0)} \frac{t^{i+j+1}}{i+j+1} \frac{(\mathrm{ad}\,x)^i (\mathrm{ad}\,y)^j (y)}{i!j!} +$$

$$+ \sum_{n=2}^{\infty} \frac{(-1)^{n+1}}{n} \sum_{(i_1,j_1),\ldots,(i_{n-1},j_{n-1}) \neq (0,0)} \frac{t^{i_1+j_1+\cdots+i_{n-1}+j_{n-1}+1}}{i_1 + j_1 + \cdots + i_{n-1} + j_{n-1} + 1} \times$$

$$\times \frac{(\mathrm{ad}\,x)^{i_1} (\mathrm{ad}\,y)^{j_1} \cdots (\mathrm{ad}\,x)^{i_{n-1}} (\mathrm{ad}\,y)^{j_{n-1}} (x + y)}{i_1! j_1! \cdots i_{n-1}! j_{n-1}!} +$$

$$+ \sum_{n=1}^{\infty} \frac{(-1)^{n+1}}{n} \sum_{(i_1,j_1),\ldots,(i_n,j_n) \neq (0,0)} \frac{t^{i_1+j_1+\cdots+i_n+j_n+1}}{i_1 + j_1 + \cdots + i_n + j_n + 1} \times$$

$$\times \frac{(\mathrm{ad}\,x)^{i_1} (\mathrm{ad}\,y)^{j_1} \cdots (\mathrm{ad}\,x)^{i_n} (\mathrm{ad}\,y)^{j_n} (y)}{i_1! j_1! \cdots i_n! j_n!}.$$

Exercise 2.12. Let $i, j \in \mathbb{N} \cup \{0\}$ be fixed. With the notation of Section 2.5, let us consider the projecting map $\underset{i,j}{\mathrm{DEG}^*} : \mathbb{K}\langle x, y\rangle \longrightarrow \mathbb{K}\langle x, y\rangle$ which sends a polynomial to the sum of its summands having x-degree *exactly* i and having y-degree *exactly* j (it is understood that $\underset{i,j}{\mathrm{DEG}^*}(p) = 0$ if p has no such summand). Observe the difference with the operator $\underset{i,j}{\mathrm{DEG}}$ introduced in (2.35):

$$\underset{n,m}{\mathrm{DEG}} = \sum_{i \leq n,\, j \leq m} \underset{i,j}{\mathrm{DEG}^*}(p).$$

Prove that $\underset{i,j}{\mathrm{DEG}^*}(p)$ is a Lie polynomial whenever p is a Lie polynomial.

Hint: First prove the direct-sum decomposition

$$\mathbb{K}\langle x, y\rangle = \bigoplus_{i,j \geq 0} \mathscr{T}_{i,j}(x, y), \quad \text{where } \mathscr{T}_{i,j}(x, y) := \underset{i,j}{\mathrm{DEG}^*}(\mathbb{K}\langle x, y\rangle).$$

If φ is the Dynkin-Specht-Wever map in Lem. 2.22, show that

$$\varphi(\mathscr{T}_{i,j}(x, y)) \subseteq \mathscr{T}_{i,j}(x, y), \quad \forall\, i, j.$$

Finally, according to this direct-sum decomposition, given any $p \in \mathcal{L}(x, y)$, write $p = \sum_{i,j} p_{i,j}$ and apply φ to get $\sum_{i,j} p_{i,j} = p = \varphi(p) = \sum_{i,j} \varphi(p_{i,j})$...

Exercise 2.13. Determine the graphs of the (real version of the) functions met in this chapter (see Fig. 2.2 and 2.3):

(1) $\varphi : \mathbb{R} \to \mathbb{R}$, where $\varphi(x) = (e^x - 1)/x$ (setting $\varphi(0) = 1$);

(2) $g : \mathbb{R} \to \mathbb{R}$, where $\mathrm{b}(x) = \dfrac{x}{e^x - 1}$ (setting $\mathrm{b}(0) = 1$);

(3) Todd's function $\psi : \mathbb{R} \to \mathbb{R}$, where $\psi(x) = \dfrac{x}{1 - e^{-x}}$ (setting $\psi(0) = 1$).

Exercise 2.14. In the associative algebra $\mathbb{K}\langle\!\langle x, y\rangle\!\rangle$ of the formal power series in two non-commuting indeterminates x and y, prove the following formula, sometimes referred to as **Hadamard's Lemma**:

$$\mathrm{Exp}(x)\, y\, \mathrm{Exp}(-x) = \sum_{k=0}^{\infty} \frac{(\mathrm{ad}\, x)^k(y)}{k!} =: e^{\mathrm{ad}\, x}(y). \tag{2.53}$$

Here as usual, $(\mathrm{ad}\, x)(y) = [x, y] = xy - yx$. Clearly, this formula can be extended to many other contexts. For instance, prove that the same formula holds for two matrices x, y in the Banach space of the real square matrices (with the well-known meaning of the exponential of a matrix).

[*Hint:* Run over the proof of Thm. 2.16.]

Exercise 2.15. With the notation of Exr. 2.13 for b, (2.23) contains the identity:

$$\mathrm{b}(\omega) = \sum_{n=1}^{\infty} \frac{(-1)^{n+1}}{n} (e^\omega - 1)^{n-1}.$$

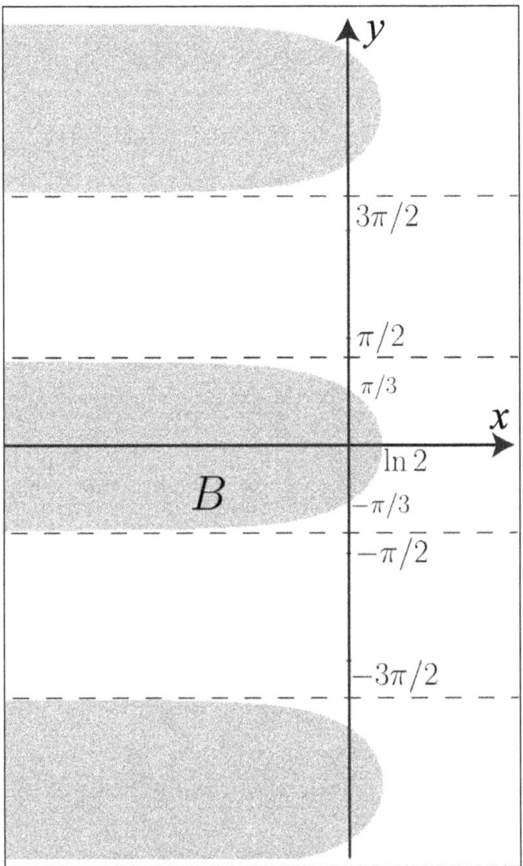

Fig. 2.1   The domain B in Exr. 2.15; the figure repeats periodically along the y-axis.

Show that the series in the rhs does not converge if $|e^\omega - 1| > 1$, whence this identity can hold true only in a subset of

$$B := \{\omega \in \mathbb{C} : |e^\omega - 1| \le 1\}.$$

Draw a picture of B (see Fig. 2.1) and compare it with the domain of **b** and with $B(0, 2\pi)$ (which is the disc of convergence of the Maclaurin series of **b**).

[*Hint:* Write $\omega = x + iy$; $|e^\omega - 1| \le 1$ is equivalent to $e^x < 2\cos y$...]

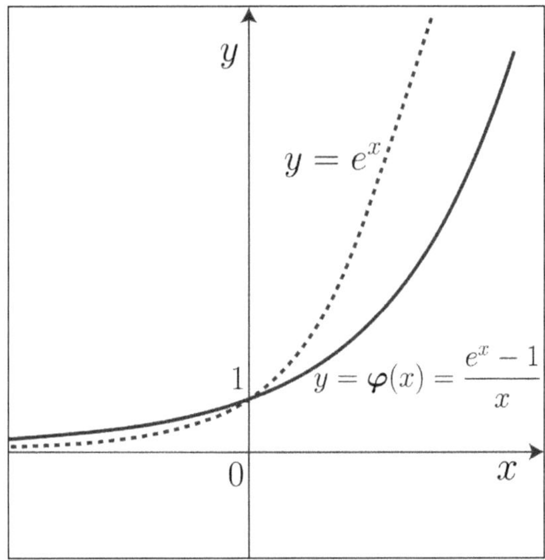

Fig. 2.2   Graphs of e^x (dashed line) and of $\varphi(x) = \dfrac{e^x - 1}{x}$.

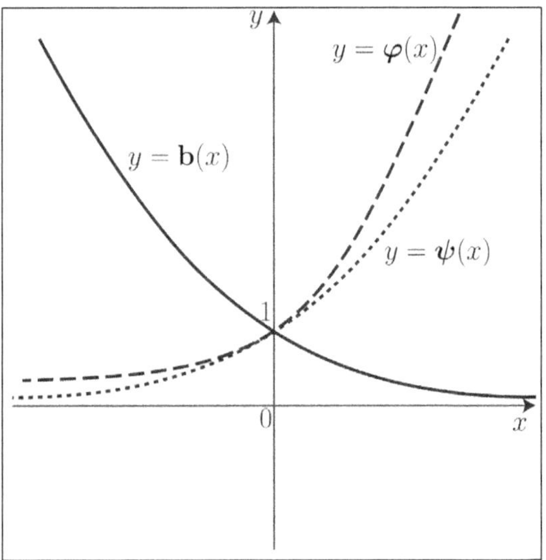

Fig. 2.3   Graphs of the functions $\varphi(x) = \dfrac{e^x - 1}{x}$, $\mathbf{b}(x) = \dfrac{x}{e^x - 1}$ and of the so-called Todd's function $\psi(x) = \dfrac{x}{1 - e^{-x}}$ (**b** and ψ are symmetric wrt the y axis).

Chapter 3

The Composition of Flows of Vector Fields

THE aim of this chapter is to apply the Exponential Theorem of Chap. 2 in order to approximate, with an arbitrary accuracy, the composition of the flows of two vector fields X, Y in terms of the flow of a third vector field $Z(X, Y)$. As is expected, $Z(X, Y)$ is intimately related to the Campbell, Baker, Hausdorff, Dynkin series associated with X and Y:

$$X + Y + \frac{1}{2}\,[X, Y] + \frac{1}{12}\,([X, [X, Y]] + [Y, [Y, X]]) - \frac{1}{24}\,[X, [Y, [X, Y]]] + \cdots .$$

Indeed, in order to approximate $\exp(Y) \circ \exp(X)$ via $\exp(Z(X, Y))$, the fundamental tool is furnished by the algebraic identities obtained from the Exponential Theorem in the previous chapter. As a natural consequence, we obtain a result which can legitimately be called *the Campbell, Baker, Hausdorff, Dynkin Theorem for flows of vector fields* (see Thm. 3.6).

Furthermore, like in Chap. 1 where we obtained an approximation of the flow of $t^2\,[X_1, X_2]$ by means of $\Psi_t^{-X_2} \circ \Psi_t^{-X_1} \circ \Psi_t^{X_2} \circ \Psi_t^{X_1}$, in the present chapter we shall *iterate this procedure* in order to approximate the flow of the long commutator

$$t^k\,[\cdots [[X_1, X_2], X_3], \cdots X_k],$$

by means of suitable compositions of the flows of $\pm X_1, \ldots, \pm X_k$ (thought of as 'horizontal' directions). This is the key ingredient for Chow's Connectivity Theorem, later to be derived in Chap. 6.

The proof of the approximation of the flow of $t^k\,[\cdots [[X_1, X_2], X_3], \cdots X_k]$ is very technical and lengthy, but the "spirit" behind it is simple:

- we first lift to an abstract level (formal power series in k indeterminates);
- we produce universal identities in this setting;
- we specialize these identities by putting v.f.s in place of the indeterminates.

This is the same approach that we followed in the previous chapter.

A prerequisite for this chapter is the CBHD Theorem of Chap. 2 (plus some elementary multivariate Analysis).

3.1   Again on commutators

Convention. Throughout this chapter, Ω will denote a fixed non-empty open subset of $\mathbb{R}^N$. This will be tacitly understood.

Unlike in Chap. 2, where we tried to keep the regularity of the v.f.s as low as possible, in this chapter we shall always deal with *smooth vector fields* on Ω. Throughout, we shall use the exponential formalism $\exp(tZ)$ for the flow map Ψ_t^Z of a smooth vector fields Z (see precisely (1.43), page 26). As pointed out in Rem. 1.36, this notation must be handled with some care. For example in the next result Thm. 3.2, we shall be dealing with the exp-like map

$$\exp(tX + tY + \tfrac{t^2}{2}[X, Y])(x);$$

since this is not exactly of the form $\exp(tZ)(x)$ (as Z here depends on t), we clarify once and for all the meaning of this notation in the following remark.

Remark 3.1. Let $X_1, \ldots, X_m$ be C^1 vector fields on Ω. If $\xi = (\xi_1, \ldots, \xi_m)$ belongs to $\mathbb{R}^m$, we consider the vector field $\xi \cdot X := \sum_{j=1}^m \xi_j X_j$ on Ω and the flow map $\Psi_t^{\xi \cdot X}(x)$, which is the solution to the Cauchy Problem

$$\dot{\gamma}(t) = (\xi \cdot X)(\gamma(t)), \quad \gamma(0) = x. \tag{3.1}$$

We are in the assumption of Thm. B.34 (page 373) in App. B: using the notation therein, we are considering a *parametric* ODE system associated with

$$f(t, x; \xi) = \sum_{j=1}^m \xi_j X_j(x),$$

which is C^1 on $\mathbb{R} \times \Omega \times \mathbb{R}^m$. With reference to the notation of Thm. B.34, due to the autonomous nature of the problem (3.1), we can take $t_0 = 0$ and drop the dependence on t_0 in the thesis of the theorem. Accordingly, we let

$$\mathcal{D}(x; \xi) = \big(\underline{\omega}(x; \xi), \overline{\omega}(x; \xi)\big)$$

denote the domain of the maximal solution of (3.1).

Then, due to Thm. B.34, the maps $\underline{\omega}, \overline{\omega}$ are, resp., upper semi-continuous and lower semi-continuous on $\Omega \times \mathbb{R}^m$. We have $\underline{\omega}(x; 0) = -\infty, \overline{\omega}(x; 0) = \infty$ for every $x \in \Omega$, since $\xi \cdot X$ is the null v.f. when $\xi = 0$. As a consequence, for every $x_0 \in \Omega$ there exists an open neighborhood $U(x_0)$ of x_0 in Ω and a positive $\delta(x_0)$ such that

$$\underline{\omega}(x; \xi) < -1 \quad \text{and} \quad \overline{\omega}(x; \xi) > 1, \quad \text{for every } x \in U(x_0) \text{ and } \|\xi\| \leq \delta(x_0).$$

By covering a compact $K \subset \Omega$ with a finite number of open sets $U(x_j)$ $(j \leq p)$ and by taking the minimum of $\delta(x_1), \ldots, \delta(x_p)$, say $\delta > 0$, we have

$$\underline{\omega}(x; \xi) < -1 \quad \text{and} \quad \overline{\omega}(x; \xi) > 1, \quad \text{for every } x \in K \text{ and } \|\xi\| \leq \delta.$$

In particular, the integral curve of $tX_1 + \cdots + t^m X_m$ starting at $x \in K$ survives for time in $[-1, 1]$ whenever t is small, say $t \in [-\varepsilon, \varepsilon]$. It is thus well posed

$$\Psi_1^{tX_1 + t^2 X_2 + \cdots + t^m X_m}(x), \quad \text{for every } x \in K \text{ and } |t| \leq \varepsilon.$$

This is what we mean, here and henceforth, for

$$\exp\big(tX_1 + t^2 X_2 + \cdots + t^m X_m\big)(x).$$

We begin with a crucial formula:

Theorem 3.2. *Let X, Y be smooth vector fields on Ω and let $x \in \Omega$. Then, as $t \to 0$,*

$$\exp(tY)(\exp(tX)(x)) = \exp\left(tX + tY + \frac{t^2}{2}[X, Y]\right)(x) + \mathcal{O}_x(t^3). \qquad (3.2)$$

More precisely, for every compact set $K \subset \Omega$ there exist $\varepsilon, C > 0$ such that

$$\left\| \exp(tY) \circ \exp(tX)(x) - \exp\left(tX + tY + \frac{t^2}{2}[X, Y]\right)(x) \right\| \leq C t^3, \qquad (3.3)$$

for every $x \in K$ and every $t \in \mathbb{R}$ satisfying $|t| \leq \varepsilon$.

Proof. Let $K \subset \Omega$ be a compact set, and let $x \in K$. The meaning of the exp-like map in the rhs of (3.2) is explained in Rem. 3.1: this map is well posed for any $t \in [-\varepsilon, \varepsilon]$, for some $\varepsilon > 0$ (depending on K). By Thm. 1.25-(3), we have

$$(\star) := \exp(tX + tY + \tfrac{t^2}{2}[X, Y])(x) =$$

$$x + tX(x) + tY(x) + \tfrac{t^2}{2}[X, Y](x) + \tfrac{1}{2}\left(tX + tY + \tfrac{t^2}{2}[X, Y]\right)^2 I(x) +$$

$$+ \frac{1}{2} \int_0^1 (1 - s)^2 \left(tX + tY + \tfrac{t^2}{2}[X, Y]\right)^3 I\left(\gamma(s, tX + tY + \tfrac{t^2}{2}[X, Y], x)\right) ds.$$

Since the trajectory of $s \mapsto \gamma(s, tX + tY + \tfrac{t^2}{2}[X, Y], x)$ remains in a compact subset of Ω (whenever $|s| \leq 1$, $|t| \leq \varepsilon$ and $x \in K$), the integral in the above formula is $\mathcal{O}(t^3)$ for every $x \in K$ and $|t| \leq \varepsilon$. The same is true of what we obtain by applying the following PDOs to the function I (and evaluating at x)

$$\tfrac{t^3}{4} X[X, Y], \quad \tfrac{t^3}{4} Y[X, Y], \quad \tfrac{t^3}{4}[X, Y]X, \quad \tfrac{t^3}{4}[X, Y]Y, \quad \tfrac{t^4}{8}[X, Y]^2$$

which is what one gets from the above second line, upon expansion of the square power. Therefore (as $[X, Y] + XY + YX = 2XY$),

$$(\star) = x + tX(x) + tY(x) + \tfrac{t^2}{2}(X^2 + Y^2 + 2XY)I(x) + \mathcal{O}_x(t^3),$$

where the remainder $\mathcal{O}_x(t^3)$ satisfies $\|\mathcal{O}_x(t^3)\| \leq C t^3$, uniformly for $x \in K$ and $|t| \leq \varepsilon$. The proof of (3.3) is complete if we show that

$$t \mapsto F_x(t) := \exp(tY)(\exp(tX)(x))$$

has the same type of expansion. We produce the Maclaurin expansion of $F_x(t)$ of degree 2 with an integral remainder:

$$F_x(t) = \sum_{k=0}^{2} \frac{F_x^{(k)}(0)}{k!} t^k + \frac{1}{2} \int_0^t (t - s)^2 F_x'''(s) \, ds.$$

From Thm. 1.38, one obtains the Maclaurin polynomial of $F_x(t)$ of degree 2:

$$\sum_{k=0}^{2} \frac{F_x^{(k)}(0)}{k!} t^k = \sum_{i+j \leq 2} \frac{(X^i Y^j I)(x)}{i!\, j!} t^{i+j}$$

$$= x + tX(x) + tY(x) + \tfrac{t^2}{2} X^2 I(x) + \tfrac{t^2}{2} Y^2 I(x) + t^2 (XY)I(x).$$

On the other hand, by continuity reasons (take into account that X and Y are smooth, hence the same is true of $\Psi_t^X(x)$ and $\Psi_t^Y(x)$ as functions of (x, t)), we easily infer that $\|F_x'''(s)\|$ is bounded by a constant, for any $|s| \leq |t|$, whenever $x \in K$ and $|t| \leq \varepsilon$. This ends the proof. $\qquad \square$

Notation 3.3. *In the sequel, given $X \in \mathcal{X}(\Omega)$, in any expression of the form $X + \mathcal{O}(t^n)$ (with $n \in \mathbb{N}$) we mean that $\mathcal{O}(t^n)$ is a smooth vector field Y_t on Ω, depending on a real parameter t in such a way that the component functions of $Y_t(x)$ are $\mathcal{O}(t^n)$ for x in some fixed compact subset of Ω and for t small enough, i.e., given a compact $K \subset \Omega$, there exist $C, \varepsilon > 0$ such that*

$$\|Y_t(x)\| \le C\,|t|^n, \qquad \text{for every } x \in K \text{ and } |t| \le \varepsilon, \tag{3.4}$$

where C and ε depend on n and K, but they are independent of x and t.

Roughly put, the ensuing identity (3.5) legitimates the passage of the big-oh $\mathcal{O}(t^{n+1})$ from the inside out of $\exp$; since this is only a technicality, the proof is sketched in Exr. 3.10

Lemma 3.4. *Let $n \in \mathbb{N}$. Let $X_1, \ldots, X_n \in \mathcal{X}(\Omega)$ (possibly identically zero) and let $x \in \Omega$. Let also $\mathcal{O}(t^{n+1})$ be a smooth vector field, in the sense of Not. 3.3.*
Then, there exists another vector field of the form $\widetilde{\mathcal{O}}(t^{n+1})$ such that

$$\exp\left(tX_1 + t^2 X_2 + \cdots + t^n X_n + \mathcal{O}(t^{n+1})\right)(x)$$
$$= \exp(tX_1 + t^2 X_2 + \cdots + t^n X_n)(x) + \widetilde{\mathcal{O}}(t^{n+1}). \tag{3.5}$$

Here, $\widetilde{\mathcal{O}}(t^{n+1})$ satisfies an estimate analogous to that in (3.4), uniformly for x in any compact set $K \subset \Omega$ and $t \in [-\varepsilon, \varepsilon]$.

The following result is important for the sequel. Under the exponential formalism that we have used so far, Cor. 3.5 below gives a refined version of Thm. 1.32 on page 23: note that now the remainder is qualified as a $\mathcal{O}(t^3)$ instead of a $o(t^2)$, and the proof is simpler (due to the smoothness of the v.f.s involved).

Corollary 3.5. *Let X, Y be smooth v.f.s on Ω and let $x \in \Omega$. Then, as $t \to 0$,*

$$\exp(-tY) \circ \exp(-tX) \circ \exp(tY) \circ \exp(tX)(x)$$
$$= \exp\left(t^2\,[X, Y]\right)(x) + \mathcal{O}(t^3). \tag{3.6}$$

More precisely, for every compact set $K \subset \Omega$ there exist $\varepsilon, C > 0$ such that

$$\left\| \exp(-tY) \circ \exp(-tX) \circ \exp(tY) \circ \exp(tX)(x) - \exp\left(t^2\,[X, Y]\right)(x) \right\| \le C\,t^3,$$

for every $x \in K$ and every $t \in \mathbb{R}$ satisfying $|t| \le \varepsilon$.

See also Fig. 3.3 to picture the meaning of this result. For an alternative proof of Cor. 3.5 independent of Lem. 3.4, see Exr. 3.8.

Proof. Both members of (3.6) are well posed for small t, say $t \in [-\varepsilon, \varepsilon]$. We repeatedly apply (3.2), plus the technical (3.5). In what follows, $\mathcal{O}(t^3)$ may denote any v.f.

as in Not. 3.3 (possibly different in any equality). We have

$$\exp(-tY) \circ \exp(-tX) \circ \exp(tY) \circ \exp(tX)(x)$$

$$= \exp(-tX - tY + \tfrac{t^2}{2}[X,Y])\big(\exp(tY)(\exp(tX)(x))\big) + \mathcal{O}(t^3)$$

(we now apply (3.2) with Y replaced by $-X - Y + \tfrac{t}{2}[X,Y]$,

with X replaced by Y and finally with x replaced by $z := \exp(tX)(x)$)

$$= \exp\big(tY - tX - tY + \tfrac{t^2}{2}[X,Y] + \tfrac{t^2}{2}[Y, -X - Y + \tfrac{t}{2}[X,Y]]\big)(z) + \mathcal{O}(t^3)$$

$$= \exp\big(-tX + t^2[X,Y] + \mathcal{O}(t^3)\big)(\exp(tX)(x)) + \mathcal{O}(t^3)$$

$$\stackrel{(3.5)}{=} \exp\big(-tX + t^2[X,Y]\big)(\exp(tX)(x)) + \mathcal{O}(t^3)$$

(another suitable application of (3.2))

$$= \exp\big(tX - tX + t^2[X,Y] + \tfrac{t^2}{2}[X, -X + t[X,Y]]\big)(x) + \mathcal{O}(t^3)$$

$$= \exp\big(t^2[X,Y] + \mathcal{O}(t^3)\big)(x) + \mathcal{O}(t^3) \stackrel{(3.5)}{=} \exp\big(t^2[X,Y]\big)(x) + \mathcal{O}(t^3).$$

This completes the proof. $\qquad\square$

3.2 Composition of flows of vector fields

Let X, Y be smooth v.f.s on Ω. Given a compact subset K of Ω, we intend to obtain an approximation (wrt s and t) of the composition of flows

$$(\Psi_t^Y \circ \Psi_s^X)(x), \quad x \in K,$$

by means of the flow of a third vector field, which can be obtained in a universal way from iterated brackets of tY and sX. To this end we will use the identities coming from the Exponential Theorem in Chap. 2.

Let $\varepsilon > 0$ be so small that $(\Psi_t^Y \circ \Psi_s^X)(x)$ is defined for every $x \in K$ and whenever $|s|, |t| \leq \varepsilon$. Let $n \in \mathbb{N}$ be arbitrarily fixed. We show that there exists a constant C_n and a smooth vector field $Z^{(n)}(s,t)$ (also depending on n but independent of x) such that (see also Fig. 3.1)

$$\left\| \Psi_t^Y(\Psi_s^X(x)) - \Psi_1^{Z^{(n)}(s,t)}(x) \right\| \leq C_n \big(|s| + |t|\big)^{n+1}, \tag{3.7}$$

for every $x \in K$ and every $t, s \in [-\varepsilon, \varepsilon]$. We can rewrite (3.7) as follows:

$$\exp(tY)(\exp(sX)(x)) = \exp\big(Z^{(n)}(s,t)\big)(x) + \mathcal{O}_{(s,t)\to(0,0)}\big(|s| + |t|\big)^{n+1}. \tag{3.8}$$

Note that the above symbol $\mathcal{O}$ is uniformly bounded, as long as $x \in K$ and as s, t are small. By general results on Taylor expansions of smooth functions, (3.8) will follow if we can define a s.v.f. $Z^{(n)}(s,t)$ (smoothly depending on (s,t)) such that the Taylor expansions at $(s,t) = (0,0)$ of the two functions

$$(s,t) \mapsto F(s,t) := \exp(tY)(\exp(sX)(x)),$$

$$(s,t) \mapsto G(s,t) := \exp(Z^{(n)}(s,t))(x)$$

coincide up to *joint-degree*[1] n in s, t. (See Exr. 3.4.)

For brevity, given a smooth function $H(s, t)$ near $(0, 0)$, we set

$$\mathrm{McL}_n(H) := \sum_{0 \le i+j \le n} \frac{\partial^{i+j} H}{\partial s^i \partial t^j}(0, 0) \frac{s^i t^j}{i! j!},$$

to denote the Maclaurin polynomial of joint-degree n in s, t of $H(s, t)$. Thus we need to prove that $\mathrm{McL}_n(F) = \mathrm{McL}_n(G)$. By Thm. 1.38, we have

$$\mathrm{McL}_n(F) = \sum_{0 \le i+j \le n} \frac{X^i Y^j I(x)}{i! \, j!} s^i t^j. \tag{3.9}$$

Thus we intend to prove that $\mathrm{McL}_n(G)$ is equal to (3.9) as well. We shall prove this with a "bridge" argument: we first show that $\mathrm{McL}_n(G) = \mathrm{McL}_n(H)$, where

$$H(s, t) := \sum_{k=0}^{n} \frac{1}{k!} \left(Z^{(n)}(s, t) \right)^k I(x),$$

and then we show that, in its turn, $\mathrm{McL}_n(H)$ is (3.9).

Due to above requirement on $Z^{(n)}(s, t)$, it is natural to define $Z^{(n)}(s, t)$ as the v.f. obtained from the formal sum $\sum_{h=1}^{\infty} Z_h(sX, tY)$ (see Def. 2.24 on page 53 for the meaning of Z_h), by preserving only the smallest numbers of terms, that is, by preserving the summands containing s, t with joint-degree $\le n$. This is precisely

$$Z^{(n)}(s, t) := \sum_{h=1}^{n} Z_h(sX, tY) = sX + tY + \frac{st}{2} [X, Y] + \cdots + Z_n(sX, tY). \tag{3.10}$$

By shrinking ε if necessary, we can suppose that $1 \in \mathcal{D}(Z^{(n)}(s, t), x)$ for every $x \in K$ and every $s, t \in [-\varepsilon, \varepsilon]$. By formula (1.19), we have the following identity

$$G(s, t) = \sum_{k=0}^{n} \frac{1}{k!} \left(Z^{(n)}(s, t) \right)^k I(x) +$$
$$+ \frac{1}{n!} \int_0^1 (1 - \rho)^n \left(Z^{(n)}(s, t) \right)^{n+1} I(\gamma(\rho, Z^{(n)}(s, t), x)) \, d\rho. \tag{3.11}$$

We observe that the integral remainder in (3.11) is $\mathcal{O}(|s| + |t|)^{n+1}$ as $(s, t) \to (0, 0)$ (in the precise meaning of the rhs of (3.7)), because the operator $(Z^{(n)}(s, t))^{n+1}$ can be expanded into a linear combination of differential operators whose coefficients depend polynomially on s and t and, precisely, they are polynomials in s, t with joint-degree $\ge n + 1$ (see (3.10)). This proves that, in looking for $\mathrm{McL}_n(G)$, we can ignore the integral remainder in formula (3.11). In other words, we have recognized that $\mathrm{McL}_n(G) = \mathrm{McL}_n(H)$.

---

[1]Here (see also page 56 where we defined the joint-degree in the setting of the polynomials in two non-commuting indeterminates x, y), we agree to say that a monomial $c\, s^i t^j$ (with $c \ne 0$) has joint-degree $i + j$ in s, t. Then one defines the joint-degree of a polynomial as the maximum of the joint-degrees of its monomials.

Hence we turn our attention to deriving $\mathrm{McL}_n(H)$. By (2.37) in Thm. 2.29 (via the specialization $x = sX$ and $y = tY$, see Rem. 2.30 for the legitimacy of this substitution argument) we have

$$\sum_{k=0}^{n} \frac{1}{k!} \left(\sum_{h=1}^{n} Z_h(sX, tY) \right)^k = \sum_{0 \le i+j \le n} \frac{X^i Y^j}{i!\, j!}\, s^i t^j + \mathcal{R}_n(sX, tY),$$

where $\mathcal{R}_n(sX, tY)$ is a linear combination of smooth differential operators, this linear combination having coefficients which depend polynomially on s, t and with joint-degrees $\ge n + 1$ in s, t. If we apply these differential operators to the identity function I and we evaluate at $x \in K$ we get

$$H(s, t) = \sum_{0 \le i+j \le n} \frac{X^i Y^j I(x)}{i!\, j!}\, s^i t^j + \mathcal{O}(|s| + |t|)^{n+1}.$$

The previous identity demonstrates that $\mathrm{McL}_n(H)$ is equal to (3.9). This ends our "bridge" argument. We have thus demonstrated the following very important:

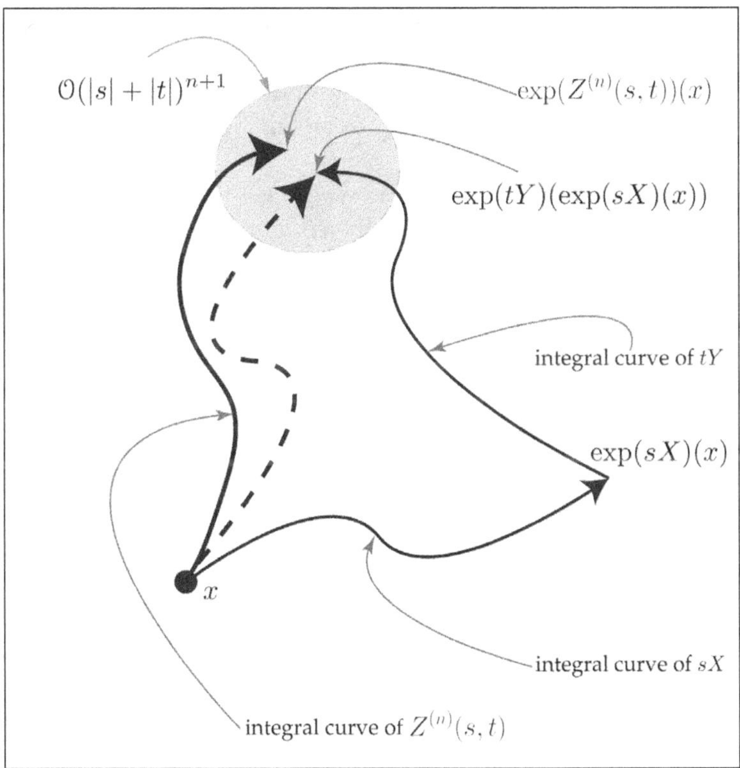

Fig. 3.1 Approximating the composition of the flows of X (with time s) and Y (with time t) by means of the flow of $Z^{(n)}(s, t)$ as in (3.12). The grey area depicts $\mathcal{O}(|s| + |t|)^{n+1}$.

Theorem 3.6 (CBHD Theorem for the approximation of flows of v.f.s). *Let* X, Y *be smooth vector fields on* Ω. *Let* $K \subset \Omega$ *be compact and let* $n \in \mathbb{N}$. *Consider the smooth vector field on* Ω *defined by*

$$Z^{(n)}(s,t) := \sum_{h=1}^{n} Z_h(sX, tY), \tag{3.12}$$

where Z_h *(the Dynkin polynomials of the Campbell-Baker-Hausdorff-Dynkin series) has been defined in Def. 2.24. Then there exists* $\varepsilon > 0$ *such that*

$$\exp(tY)(\exp(sX)(x)) = \exp\left(Z^{(n)}(s,t)\right)(x) + \mathcal{O}\left(|s| + |t|\right)^{n+1}, \tag{3.13}$$

for every $x \in K$ *and every* $s, t \in [-\varepsilon, \varepsilon]$.

More precisely, for any such s, t, *the remainder term can be bounded from above by* $M\left(|s| + |t|\right)^{n+1}$, *where the constant* M *depends on* K *but not on* $x \in K$ *(and* M *is independent of* $s, t \in [-\varepsilon, \varepsilon]$).

The case $n = 2$ and $s = t$ in the above theorem gives back Thm. 3.2.

3.3   Approximation for higher order commutators

Starting from (3.6), our next task is to show how to "generate" a higher order bracket $[\cdots [[X_1, X_2], X_3] \cdots X_k]$ starting from suitable compositions of the maps $\exp(\pm X_i)$ with $i = 1, \ldots, k$.

This is a crucial result "of subelliptic type": we learn how to approximate a movement along the integral curves of the commutator $[\cdots [[X_1, X_2], X_3] \cdots X_k]$ by means of movements along the integral curves of $\pm X_1, \ldots, \pm X_k$, often referred to as 'horizontal' directions. First we need some notation.

Notation 3.7. Given any pair of functions f and g, we set

$$\mathrm{comm}(f, g) := g^{-1} \circ f^{-1} \circ g \circ f, \quad \text{whenever this makes sense.}$$

Thus, given $X, Y \in \mathfrak{X}(\Omega)$ and $t \in \mathbb{R}$ we shall use the notation

$$\mathrm{comm}(\Psi_t^X, \Psi_t^Y) := (\Psi_t^Y)^{-1} \circ (\Psi_t^X)^{-1} \circ \Psi_t^Y \circ \Psi_t^X.$$

Equivalently, since $(\Psi_t^X)^{-1} = \Psi_t^{-X}$ (see Prop. 1.14-(2) on page 11), we get

$$\mathrm{comm}(\Psi_t^X, \Psi_t^Y) = \Psi_t^{-Y} \circ \Psi_t^{-X} \circ \Psi_t^Y \circ \Psi_t^X. \tag{3.14}$$

Remark 3.8. Note that (3.14) is certainly well posed for every $t \in \mathbb{R}$ if X and Y are global, and in this case it defines a function on the whole of Ω. In general, if $K \subset \Omega$ is compact, there exists $\varepsilon > 0$ such that $\mathrm{comm}(\Psi_t^X, \Psi_t^Y)$ is well posed as a function defined on K, for $|t| \leq \varepsilon$. (See App. B.)   ♮

Let $k \in \mathbb{N}$, $k \geq 2$, and suppose we are given a k-tuple of smooth vector fields $X_1, X_2 \ldots, X_k$ (not necessarily distinct) on the open set $\Omega \subseteq \mathbb{R}^N$. For any small $t \in \mathbb{R}$, we inductively define k functions that are compositions of maps of the form $\Psi_t^{\pm X_i}$ (with $i \in \{1, \ldots, k\}$) in the following recursive way:

$$\alpha_1(\Psi_t^{X_1}) := \Psi_t^{X_1}$$
$$\alpha_2(\Psi_t^{X_1}, \Psi_t^{X_2}) := \mathrm{comm}(\alpha_1(\Psi_t^{X_1}), \Psi_t^{X_2}) = \mathrm{comm}(\Psi_t^{X_1}, \Psi_t^{X_2})$$
$$\alpha_3(\Psi_t^{X_1}, \Psi_t^{X_2}, \Psi_t^{X_3}) := \mathrm{comm}(\alpha_2(\Psi_t^{X_1}, \Psi_t^{X_2}), \Psi_t^{X_3}) \tag{3.15}$$
$$\vdots$$
$$\alpha_k(\Psi_t^{X_1}, \ldots, \Psi_t^{X_k}) := \mathrm{comm}(\alpha_{k-1}(\Psi_t^{X_1}, \ldots, \Psi_t^{X_{k-1}}), \Psi_t^{X_k}).$$

For example, the first few terms are

$$\alpha_1(\Psi_t^{X_1}) = \Psi_t^{X_1}$$
$$\alpha_2(\Psi_t^{X_1}, \Psi_t^{X_2}) = \Psi_t^{-X_2} \circ \Psi_t^{-X_1} \circ \Psi_t^{X_2} \circ \Psi_t^{X_1}$$
$$\alpha_3(\Psi_t^{X_1}, \Psi_t^{X_2}, \Psi_t^{X_3}) =$$
$$\Psi_t^{-X_3} \circ \left(\Psi_t^{-X_2} \circ \Psi_t^{-X_1} \circ \Psi_t^{X_2} \circ \Psi_t^{X_1}\right)^{-1} \circ \Psi_t^{X_3} \circ \left(\Psi_t^{-X_2} \circ \Psi_t^{-X_1} \circ \Psi_t^{X_2} \circ \Psi_t^{X_1}\right)$$
$$= \Psi_t^{-X_3} \circ \Psi_t^{-X_1} \circ \Psi_t^{-X_2} \circ \Psi_t^{X_1} \circ \Psi_t^{X_2} \circ \Psi_t^{X_3} \circ \Psi_t^{-X_2} \circ \Psi_t^{-X_1} \circ \Psi_t^{X_2} \circ \Psi_t^{X_1}.$$

Remark 3.9. It is not difficult to show that $\alpha_k(\Psi_t^{X_1}, \ldots, \Psi_t^{X_k})$ is the composition of $3 \cdot 2^{k-1} - 2$ flows of the form $\Psi_t^{\pm X_i}$, with $i \in \{1, \ldots, k\}$ (Exr. 3.5).

k	1	2	3	4	5	6	7	8	9	10
$3 \cdot 2^{k-1} - 2$	1	4	10	22	46	94	190	382	766	1534

It is very important to observe that in defining α_k we are implicitly defining, in a unique way, an (ordered) h-tuple of vector fields

$$(\varepsilon_1 X_{i_1}, \varepsilon_2 X_{i_2}, \ldots, \varepsilon_h X_{i_h}), \qquad \text{with } h = 3 \cdot 2^{k-1} - 2,$$

such that $\varepsilon_1, \ldots, \varepsilon_h \in \{-1, +1\}$ and $i_1, \ldots, i_h \in \{1, \ldots, k\}$, this h-tuple being characterized by the fact that

$$\alpha_k(\Psi_t^{X_1}, \ldots, \Psi_t^{X_k}) = \Psi_t^{\varepsilon_1 X_{i_1}} \circ \Psi_t^{\varepsilon_2 X_{i_2}} \circ \cdots \circ \Psi_t^{\varepsilon_h X_{i_h}}.$$

For example, when $k = 2$ this 4-tuple is $(-X_2, -X_1, X_2, X_1)$, whilst when $k = 3$ the associated 10-tuple is

$$\left(-X_3, -X_1, -X_2, X_1, X_2, X_3, -X_2, -X_1, X_2, X_1\right). \tag{3.16}$$

See also Fig. 3.2. We also remark that, with the above notation, for any $k \in \mathbb{N}$ with $k \geq 2$ (setting again $h = 3 \cdot 2^{k-1} - 2$), one has

$$\varepsilon_1 X_{i_1} + \varepsilon_2 X_{i_2} + \cdots + \varepsilon_h X_{i_h} = 0, \quad \varepsilon_1 + \varepsilon_2 + \cdots + \varepsilon_h = 0,$$

as can easily be proved by induction on $k \in \mathbb{N}$. This means that one has the alternative representation for $\alpha_k(\Psi_t^{X_1}, \dots, \Psi_t^{X_k})$ as

$$\Psi_1^{\varepsilon_1 t X_{i_1}} \circ \Psi_1^{\varepsilon_2 t X_{i_2}} \circ \cdots \circ \Psi_1^{\varepsilon_h t X_{i_h}} =: \Psi_1^{Y_1} \circ \Psi_1^{Y_2} \circ \cdots \circ \Psi_1^{Y_h},$$

where $Y_1 + \cdots + Y_h = 0$. This simple fact has a meaning in Control Theory.   ♯

We are ready for our main result on the maps α_k. See Fig. 3.2 for the case $k = 3$.

Theorem 3.10. *Let $k \geq 2$, and suppose we are given a k-tuple of (not necessarily distinct) smooth v.f.s $X_1, X_2 \dots, X_k$ on Ω. Let $\alpha_k(\Psi_t^{X_1}, \dots, \Psi_t^{X_k})$ be as in (3.15).*
Then, for every compact set $K \subset \Omega$ there exist constants $\varepsilon, C > 0$ such that

$$\begin{aligned}
\alpha_k(\Psi_t^{X_1}, \dots, \Psi_t^{X_k})(x) &= \exp\left(t^k[\cdots[X_1, X_2]\cdots X_k]\right)(x) + \mathcal{O}_x(t^{k+1}) \\
&= \Psi_{t^k}^{[\cdots[X_1, X_2]\cdots X_k]}(x) + \mathcal{O}_x(t^{k+1}),
\end{aligned}$$

(3.17)

for every $x \in K$ and every $t \in \mathbb{R}$ satisfying $|t| \leq \varepsilon$; here $\|\mathcal{O}_x(t^{k+1})\| \leq C\, t^{k+1}$, uniformly for $x \in K$ and $t \in [-\varepsilon, \varepsilon]$.

Proof. Identity (3.17) will follow if we prove that the smooth functions

$$f(t) := \alpha_k(\Psi_t^{X_1}, \dots, \Psi_t^{X_k})(x), \qquad g(t) := \exp\left(t^k[\cdots[X_1, X_2]\cdots X_k]\right)(x)$$

have the same Maclaurin polynomial of degree k. For brevity, given a smooth function $h(t)$ near 0, we set

$$\mathrm{McL}_k(h) := \sum_{i=0}^{k} \frac{\mathrm{d}^i h(0)}{\mathrm{d}t^i} \frac{t^i}{i!},$$

to denote the Maclaurin polynomial of degree k of $h(t)$. Thus we need to prove that $\mathrm{McL}_k(f) = \mathrm{McL}_k(g)$. By Thm. 1.25 on page 18, we have, as $s \to 0$,

$$\exp\left(s[\cdots[X_1, X_2]\cdots X_k]\right)(x) = x + s[\cdots[X_1, X_2]\cdots X_k](x) + \mathcal{O}(s^2).$$

By replacing s with t^k we get

$$g(t) = x + t^k[\cdots[X_1, X_2]\cdots X_k](x) + \mathcal{O}(t^{2k}), \quad \text{as } t \to 0.$$

Thus $\mathrm{McL}_k(g)$ is equal to

$$x + t^k[\cdots[X_1, X_2]\cdots X_k](x). \tag{3.18}$$

We have to prove that the same polynomial is $\mathrm{McL}_k(f)$.
    Using the notations introduced in Rem. 3.9 we have

$$f(t) = \exp(t\varepsilon_1 X_{i_1}) \circ \exp(t\varepsilon_2 X_{i_2}) \circ \cdots \circ \exp(t\varepsilon_h X_{i_h})(x).$$

By arguing as in the proof of Thm 1.38 (page 27), one can prove that the Taylor series at the point $0 \in \mathbb{R}^h$ of the function

$$F(s_1, \dots, s_h) := \exp(s_1\varepsilon_1 X_{i_1}) \circ \exp(s_2\varepsilon_2 X_{i_2}) \circ \cdots \circ \exp(s_h\varepsilon_h X_{i_h})(x)$$

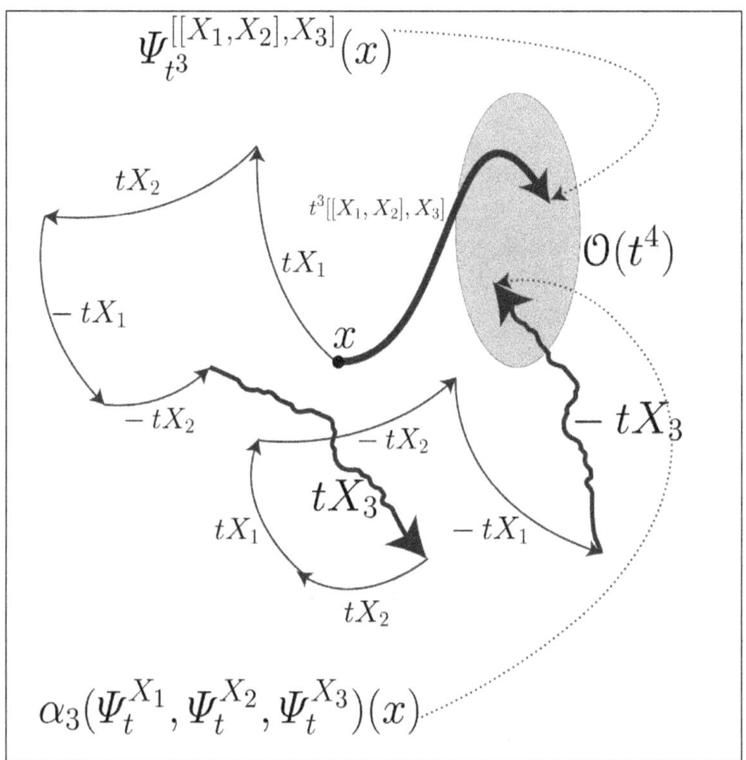

Fig. 3.2　The case $k = 3$ in Thm. 3.10: approximating the flow of the commutator $t^3[[X_1, X_2], X_3]$ of length 3, by means of 10 'horizontal' integral curves of $\pm tX_1, \pm tX_2, \pm tX_3$. The shaded set pictures the error term $\mathcal{O}(t^4)$.

is the following one (note the reverse order of the vector fields):

$$\sum_{n_1,n_2,\dots,n_h \geq 0} \frac{(\varepsilon_h X_{i_h})^{n_h} \cdots (\varepsilon_2 X_{i_2})^{n_2}(\varepsilon_1 X_{i_1})^{n_1} I(x)}{n_h! \cdots n_2! \, n_1!} \, s_h^{n_h} \cdots s_2^{n_2} s_1^{n_1}.$$

This is left as an exercise, Exr. 3.6. Since $f(t) = F(t, t, \dots, t)$, we infer that $\mathrm{McL}_k(f)$ is given by (see Exr. 3.7)

$$\sum_{\substack{n_1,n_2,\dots,n_h \geq 0 \\ n_1+n_2+\cdots+n_h \leq k}} \frac{(\varepsilon_h X_{i_h})^{n_h} \cdots (\varepsilon_2 X_{i_2})^{n_2}(\varepsilon_1 X_{i_1})^{n_1} I(x)}{n_h! \cdots n_2! \, n_1!} \, t^{n_1+n_2+\cdots+n_h}. \tag{3.19}$$

Now we get to the crucial part of the proof. If we intend to attack the sum in (3.19) and we want to show that it actually coincides with (3.18), we are naturally led to rely on polynomial identities coming from formal power-series computation. Indeed, there is no reason why the identity '(3.18)=(3.19)' should be true only for smooth vector fields $X_1, \dots, X_k$ rather than being valid, broadly speaking, at a...*higher level of truth*; in other words, the cited identity must be the consequence of the next identity between non-commuting indeterminates $x_1, \ldots, x_k$:

$$\sum_{\substack{n_1, n_2, \ldots, n_h \geq 0 \\ n_1 + n_2 + \cdots + n_h \leq k}} \frac{(\varepsilon_h x_{i_h})^{n_h} \cdots (\varepsilon_2 x_{i_2})^{n_2} (\varepsilon_1 x_{i_1})^{n_1}}{n_h! \cdots n_2! \, n_1!} = 1 + [[x_1, x_2] \cdots x_k]. \qquad (3.20)$$

The proof of the tricky identity (3.20) is quite technical and we postpone it to an appendix reading in Sec. 3.4.

By the universality of the non-commuting indeterminates $x_1, \ldots, x_k$, we are then entitled to make the substitutions $x_i := t X_i$ (for $i = 1, \ldots, k$). If id denotes the identical differential operator, we deduce that

$$\sum_{\substack{n_1, n_2, \ldots, n_h \geq 0 \\ n_1 + n_2 + \cdots + n_h \leq k}} \frac{(\varepsilon_h X_{i_h})^{n_h} \cdots (\varepsilon_2 X_{i_2})^{n_2} (\varepsilon_1 X_{i_1})^{n_1}}{n_h! \cdots n_2! \, n_1!} \, t^{n_1 + n_2 + \cdots + n_h} =$$

$$= \mathrm{id} + t^k [[X_1, X_2] \cdots X_k].$$

By evaluating both of these differential operators on $I(x)$, we finally discover that $\mathrm{McL}_k(f)$ (which is given by (3.19)) is equal to $x + t^k [[X_1, X_2] \cdots X_k](x)$, which, in its turn, is $\mathrm{McL}_k(g)$ (see (3.18)). This completes the proof. $\qquad \square$

3.4   Appendix: another identity between formal power series

Here we prove (3.20). All that we have to do is reproducing the recursive sequence α_k in the setting of formal power series.

To this aim, let $x_1, \ldots, x_k$ be non-commuting indeterminates and let us use the symbol $\mathbb{R}\langle\!\langle x_1, \ldots, x_k \rangle\!\rangle$ to denote the real algebra of the formal power series in $x_1, \ldots, x_k$. This can be constructed as in Sec. 2.3 for $\mathbb{K}\langle\!\langle x, y \rangle\!\rangle$.

Our task is to consider the series

$$\sum_{n_1, n_2, \ldots, n_h \geq 0} \frac{(\varepsilon_h x_{i_h})^{n_h} \cdots (\varepsilon_2 x_{i_2})^{n_2} (\varepsilon_1 x_{i_1})^{n_1}}{n_h! \cdots n_2! \, n_1!},$$

and to rewrite it (obviously with the aid of the Exponential Theorem) as the exponential of a Lie series. Note that the last series is equal to

$$\mathrm{Exp}(\varepsilon_h x_{i_h}) \cdots \mathrm{Exp}(\varepsilon_2 x_{i_2}) \, \mathrm{Exp}(\varepsilon_1 x_{i_1}). \qquad (3.21)$$

The h-tuple (where $h = 3 \cdot 2^{k-1} - 2$)

$$\left(\varepsilon_h x_{i_h}, \ldots, \varepsilon_2 x_{i_2}, \varepsilon_1 x_{i_1} \right)$$

is obtained by reversing the original h-tuple associated with α_k

$$\left(\varepsilon_1 X_{i_1}, \varepsilon_2 X_{i_2}, \ldots, \varepsilon_h X_{i_h} \right),$$

and by also replacing capitals with lower cases. As a consequence, the law of formation of the iterated product (3.21) is analogous to that of α_k, with the only difference that we replace the former "commutation" of functions

$$\mathrm{comm}(f, g) := g^{-1} \circ f^{-1} \circ g \circ f,$$

with the "commutation" of exponential series in the reverse order:

$$\mathrm{Exp}(u)\,\mathrm{Exp}(v)\,(\mathrm{Exp}(u))^{-1}\,(\mathrm{Exp}(v))^{-1}.$$

It is therefore convenient to give the following definition

$$\mathrm{Comm}(A,B) := A\,B\,A^{-1}\,B^{-1},$$

whenever $A, B \in \mathbb{R}\langle\!\langle x_1, \dots, x_k \rangle\!\rangle$ possess their multiplicative inverses A^{-1} and B^{-1}. This is always true, for example, if A and B are products of elements of the form $\mathrm{Exp}(\pm x_i)$, $i \in \{1, \dots, k\}$.

With this notation at hand, we set (note the analogy with (3.15)):

$$\beta_1(x_1) := \mathrm{Exp}(x_1)$$
$$\beta_2(x_1, x_2) := \mathrm{Comm}(\beta_1(x_1), \mathrm{Exp}(x_2))$$
$$= \mathrm{Exp}(x_1)\mathrm{Exp}(x_2)\mathrm{Exp}(-x_1)\mathrm{Exp}(-x_2)$$
$$\vdots$$
$$\beta_k(x_1, \dots, x_k) := \mathrm{Comm}\big(\beta_{k-1}(x_1, \dots, x_{k-1}), \mathrm{Exp}(x_k)\big).$$

All $\beta_1, \dots, \beta_k$ are products of formal power series $\mathrm{Exp}(\pm x_i)$ with i in $\{1, \dots, k\}$. We note that $\beta_k(x_1, \dots, x_k)$ is completely determined by an ordered h-tuple (with $h = 3 \cdot 2^{k-1} - 2$) of the form $(\delta_1 x_{j_1}, \delta_2 x_{j_2}, \dots, \delta_h x_{j_h})$, such that $\delta_1, \dots, \delta_h \in \{-1, +1\}$ and $j_1, \dots, j_h \in \{1, \dots, k\}$, this h-tuple being characterized by the fact that

$$\beta_k(x_1, \dots, x_k) = \mathrm{Exp}(\delta_1 x_{j_1})\,\mathrm{Exp}(\delta_2 x_{j_2}) \cdots \mathrm{Exp}(\delta_h x_{j_h}).$$

For example, the 4-tuple associated with β_2 is $(x_1, x_2, -x_1, -x_2)$, whilst the 10-tuple related to β_3 is

$$(x_1, x_2, -x_1, -x_2, x_3, x_2, x_1, -x_2, -x_1, -x_3);$$

indeed, by the definition of β_1, β_2 one has (we write $e^{\cdots}$ instead of $\mathrm{Exp}(\dots)$)

$$\beta_3(x_1, x_2, x_3) = \mathrm{Comm}\big(\beta_2(x_1, x_2), e^{x_3}\big) = \mathrm{Comm}\left(e^{x_1}e^{x_2}e^{-x_1}e^{-x_2}, e^{x_3}\right)$$
$$= e^{x_1}e^{x_2}e^{-x_1}e^{-x_2}e^{x_3}\left(e^{x_1}e^{x_2}e^{-x_1}e^{-x_2}\right)^{-1}e^{-x_3}$$
$$= e^{x_1}e^{x_2}e^{-x_1}e^{-x_2}e^{x_3}e^{x_2}e^{x_1}e^{-x_2}e^{-x_1}e^{-x_3}.$$

Note that the 10-tuple associated with β_3 is the reverse of the 10-tuple associated with α_3 (see (3.16)).

In general, if $h = 3 \cdot 2^{k-1} - 2$, *the h-tuple associated with $\beta_k(x_1, \dots, x_k)$ is the reverse one of the h-tuple associated with $\alpha_k(\Psi_t^{X_1}, \dots, \Psi_t^{X_k})$, obtained by replacing $X_1, \dots, X_k$ with $x_1, \dots, x_k$ respectively.*

As a corollary of this fact, we deduce that

$$(\delta_1, \delta_2, \dots, \delta_h) = (\varepsilon_h, \varepsilon_{h-1}, \dots, \varepsilon_1), \quad (j_1, j_2, \dots, j_h) = (i_h, i_{h-1}, \dots, i_1),$$

or equivalently

$$\mathrm{Exp}(\varepsilon_h x_{i_h}) \cdots \mathrm{Exp}(\varepsilon_2 x_{i_2})\,\mathrm{Exp}(\varepsilon_1 x_{i_1}) = \beta_k(x_1, \dots, x_k), \tag{3.22}$$

where $\varepsilon_h, \ldots, \varepsilon_1$ and $i_h, \ldots, i_1$ are the very same as those in Rem. 3.9.

Now, by means of the Exponential Theorem (see e.g. (2.34) and the notation in (2.29)), we can infer that, for every formal power series A, B with vanishing zero-degree term, it holds

$$\mathrm{Exp}(A) \cdot \mathrm{Exp}(B) = \mathrm{Exp}(A \diamond B) = \mathrm{Exp}\Big(A + B + \frac{1}{2}[A, B] + \{\cdots\}\Big),$$

where $\{\cdots\}$ denotes a Lie series whose summands are commutators of lengths ≥ 3 in A, B. This easily proves the following identity

$$\mathrm{Comm}(\mathrm{Exp}(A), \mathrm{Exp}(B)) = \mathrm{Exp}\Big([A, B] + \Big\{ \begin{array}{c} \text{series of brackets} \\ \text{of lengths} \geq 3 \text{ in } A, B \end{array} \Big\}\Big). \qquad (3.23)$$

By an inductive argument we can therefore prove that

$$\beta_k(x_1, \ldots, x_k) = \mathrm{Exp}\Big([[x_1, x_2]\cdots x_k] + \Big\{ \begin{array}{c} \text{series of brackets of lengths} \\ \geq k+1 \text{ in } x_1, \ldots, x_k \end{array} \Big\}\Big). \qquad (3.24)$$

For example, the case $k = 2$ is (3.23) with $A = x_1$ and $B = x_2$. We prove the case $k = 3$, leaving the general case to the interested reader:

$$\beta_3(x_1, x_2, x_3) = \mathrm{Comm}\big(\beta_2(x_1, x_2), \mathrm{Exp}(x_3)\big)$$

$$\overset{(3.23)}{=} \mathrm{Comm}\Big(\mathrm{Exp}\big([x_1, x_2] + \{\text{lengths} \geq 3 \text{ in } x_1, x_2\}\big), \mathrm{Exp}(x_3)\Big)$$

$$\overset{(3.23)}{=} \mathrm{Exp}\Big([[x_1, x_2], x_3] + \{\text{lengths} \geq 3 \text{ in } [x_1, x_2], x_3\}\Big)$$

$$= \mathrm{Exp}\Big([[x_1, x_2], x_3] + \{\text{lengths} \geq 4 \text{ in } x_1, x_2, x_3\}\Big).$$

Taking into account (3.22) and (3.24) we deduce that

$$\mathrm{Exp}(\varepsilon_h x_{i_h}) \cdots \mathrm{Exp}(\varepsilon_2 x_{i_2}) \mathrm{Exp}(\varepsilon_1 x_{i_1})$$

$$= \mathrm{Exp}\Big([[x_1, x_2]\cdots x_k] + \Big\{ \begin{array}{c} \text{series of brackets} \\ \text{of lengths} \geq k+1 \text{ in } x_1, \ldots, x_k \end{array} \Big\}\Big).$$

This is an identity in $\mathbb{R}\langle\langle x_1, \ldots, x_k \rangle\rangle$. We can make a projection of this identity on the space of the polynomials in $x_1, \ldots, x_k$ having degree $\leq k$ (jointly in all the indeterminates). We note that the term in braces does not contribute at all to this projection; furthermore, the same is true of the powers of $[[x_1, x_2]\cdots x_k]$ with exponent ≥ 2. This gives, precisely, the needed (3.20).

3.5   Exercises of Chap. 3

In this chapter, we repeatedly used the following fact, also generalized to vector-valued functions f, and possibly replacing t with the 2-dimensional (s, t).

Exercise 3.1. Let $n \in \mathbb{N}$, let $\Omega \subseteq \mathbb{R}^{1+N}$ and let $f = f(t, x)$ be in $C^{n+1}(\Omega, \mathbb{R})$. More-over, let I be a compact interval in $\mathbb{R}$ and let K be a compact subset of $\mathbb{R}^N$ such that $I \times K \subset \Omega$. Fixing $x \in K$ and $t_0 \in I$, write down the Taylor expansion of

$t \mapsto f(t, x)$ at $t = t_0$ with a Lagrange remainder. Derive that there exists a constant C (depending on f, n, I, K) such that

$$\left| f(t, x) - \sum_{k=0}^{n} \frac{1}{k!} \frac{\partial^k}{\partial t^k}\Big|_{t=t_0} f(t, x)\,(t - t_0)^k \right| \leq C\,(t - t_0)^{n+1},$$

uniformly for $t \in I$ and $x \in K$.

Exercise 3.2. Prove the following facts:

(a) Given $n, m \in \mathbb{N} \cup \{0\}$, prove the existence of $C_{n,m} > 0$ such that

$$|s|^n |t|^m \leq C_{n,m}(|s|^{n+m} + |t|^{n+m}), \quad \forall\, s, t \in \mathbb{R}. \tag{3.25}$$

[*Hint:* Use a homogeneity argument: given $(s, t) \in \mathbb{R}^2 \setminus \{(0,0)\}$ let us set

$$\sigma = \frac{s}{\sqrt[n+m]{|s|^{n+m} + |t|^{n+m}}}, \qquad \tau = \frac{t}{\sqrt[n+m]{|s|^{n+m} + |t|^{n+m}}}.$$

Show that the choice

$$C_{n,m} := \max_{K_{n,m}} |\sigma|^n |\tau|^m \quad (\text{with } K_{n,m} := \{(\sigma, \tau) : |\sigma|^{n+m} + |\tau|^{n+m} = 1\})$$

does the job... Alternatively, use polar coordinates...]

(b) Let $n \in \mathbb{N}$. Prove the existence of a constant $C_n > 0$ such that

$$|s|^n + |t|^n \leq (|s| + |t|)^n \leq C_n(|s|^n + |t|^n), \quad \forall\, s, t \in \mathbb{R}.$$

[*Hint:* Apply Newton's binomial to $(|s| + |t|)^n$ and use (3.25) on the summands.]

(c) Let $n, m \in \mathbb{N} \cup \{0\}$. Prove that, for when $|s|, |t| \leq 1$, both

$$(|s| + |t|)^{n+m+1} \quad \text{and} \quad (|s| + |t|)^{n \vee m+1}$$

are of the form $\mathcal{O}(|s|^{n+1} + |t|^{m+1})$. [*Hint:* Use the preceding results...]

Exercise 3.3. Let $n, m \in \mathbb{N} \cup \{0\}$. Prove that the maximum $k \geq 0$ such that

$$s^n t^m = \mathcal{O}(|s| + |t|)^k, \quad \text{as } (s, t) \to (0, 0)$$

is $k = n + m$.   [*Hint:* Consider the ratio $s^n t^m / (|s| + |t|)^k$ and use a homogeneity argument, replacing s and t respectively with $(|s| + |t|) \cdot \sigma$ and $(|s| + |t|) \cdot \tau$, where (σ, τ) belongs to the compact set where $|\sigma| + |\tau| = 1$...]

Exercise 3.4. Review Taylor's formula with Lagrange's remainder (see the notation of Exr. 1.21, page 36): if $\Omega \subseteq \mathbb{R}^N$ is convex, if $k \in \mathbb{N}$, $f \in C^{k+1}(\Omega, \mathbb{R})$ and if $x_0 \in \Omega$, then for every $x \in \Omega$ there exists $\xi_x \in \Omega$ such that

$$f(x) = \sum_{|\alpha| \leq k} \frac{D_x^\alpha f(x_0)}{\alpha!}(x - x_0)^\alpha + \sum_{|\alpha| = k+1} \frac{D_x^\alpha f(\xi_x)}{\alpha!}(x - x_0)^\alpha.$$

Take $N = 2$ and denote by (s, t) the points of $\mathbb{R}^2$. Let Ω be an open neighborhood of $(0,0)$ containing the square $[-\varepsilon, \varepsilon] \times [-\varepsilon, \varepsilon]$. Let $n, m \in \mathbb{N}$. Deduce from the previous formula that if $\frac{\partial^{\alpha_1+\alpha_2} f(0,0)}{\partial s^{\alpha_1} \partial t^{\alpha_2}} = 0$ for every (α_1, α_2) such that $0 \leq \alpha_1 \leq n$ and $0 \leq \alpha_2 \leq m$, then it holds that

$$|f(s,t)| \leq C\left(|s|^{n+1} + |t|^{m+1}\right), \quad \text{whenever } |s|, |t| \leq \varepsilon,$$

for some $C > 0$ (depending on f, ε, n, m). Analogously, if $\frac{\partial^{\alpha_1+\alpha_2} f(0,0)}{\partial s^{\alpha_1} \partial t^{\alpha_2}} = 0$, for every $(\alpha_1, \alpha_2) \in (\mathbb{N} \cup \{0\})^2$ such that $\alpha_1 + \alpha_2 \leq n$, then it holds that

$$|f(s,t)| \leq C\left(|s| + |t|\right)^{n+1}, \quad \text{whenever } |s|, |t| \leq \varepsilon.$$

[*Hint:* Using Exr. 3.3, the second result is a consequence of the recalled Taylor's formula with Lagrange's remainder; for the first result, choose $k = n + m$...]

Exercise 3.5. Prove that $\alpha_k(\Psi_t^{X_1}, \ldots, \Psi_t^{X_k})$ defined in (3.15) is the composition of $3 \cdot 2^{k-1} - 2$ flows of the form $\Psi_t^{\pm X_i}$, with $i \in \{1, \ldots, k\}$.

Exercise 3.6. Prove the following generalization of Thm. 1.38 on page 27:

Let $X_1, \ldots, X_h \in X(\Omega)$, $x \in \Omega$ and $f \in C^\infty(\Omega, \mathbb{R})$. Then, the Taylor series at the point $0 \in \mathbb{R}^h$ of the function

$$F(s_1, \ldots, s_h) := f\left(\exp(s_h X_h) \circ \cdots \circ \exp(s_2 X_2) \circ \exp(s_1 X_1)(x) \right)$$

(which is smooth and well defined on an open neighborhood of $0 \in \mathbb{R}^h$) is equal to

$$\sum_{n_1, n_2, \ldots, n_h \geq 0} \frac{X_1^{n_1} X_2^{n_2} \cdots X_h^{n_h} f(x)}{n_1! \, n_2! \cdots n_h!} \, s_1^{n_1} s_2^{n_2} \cdots s_h^{n_h}.$$

Exercise 3.7. Suppose $F(s_1, \ldots, s_h)$ is smooth in a neighborhood of $0 \in \mathbb{R}^h$ and that its Taylor expansion at 0 is given by

$$\sum_{n_1, \ldots, n_h \geq 0} a_{n_1, \ldots, n_h} \, s_1^{n_1} s_2^{n_2} \cdots s_h^{n_h}.$$

Prove that the Maclaurin expansion of $f(t) := F(t, \ldots, t)$ is

$$\sum_{n=0}^{\infty} \left(\sum_{\substack{n_1, \ldots, n_h \geq 0 \\ n_1 + n_2 + \cdots + n_h = n}} a_{n_1, \ldots, n_h} \right) t^n.$$

Exercise 3.8. Here is an alternative proof of Cor. 3.5. Provide details:

(1) use Exr.s 3.6 and 3.7 to show that

$$\sum_{n_1, \ldots, n_4 \geq 0} \frac{X^{n_1} Y^{n_2} (-X)^{n_3} (-Y)^{n_4}}{n_1! \cdots n_4!} I(x) \, t^{n_1 + \cdots + n_4}$$

    is the Maclaurin series of the lhs of (3.6);

(2) derive that the Maclaurin polynomial of degree two of the lhs of (3.6) is $x + t^2[X, Y](x)$;

(3) by (1.20), prove that $x + t^2[X, Y](x)$ is the degree-two Maclaurin polynomial of $\exp\left(t^2 [X, Y]\right)(x)$ as well, and deduce that (3.6) holds true.

Exercise 3.9. Check the validity of identity (3.20) when $k = 2$.
  [*Hint:* First compute $\varepsilon_1, \ldots, \varepsilon_4$ and $i_1, \ldots, i_4$.]

Exercise 3.10. Prove Lem. 3.4 by giving the details of this argument.

(1). Both members of (3.5) are well-posed for small t (see Rem. 3.1...).
(2). Let $X_t := tX_1 + t^2 X_2 + \cdots + t^n X_n$ and let Y_t be the big-oh $\mathcal{O}(t^{n+1})$ in the statement of the lemma; deduce from (1.20) that

$$\exp(X_t + Y_t)(x) = \sum_{k=0}^{n} \frac{1}{k!} (X_t + Y_t)^k I(x) + \tag{3.26}$$

$$+ \frac{1}{n!} \int_0^1 (1-s)^n (X_t + Y_t)^{n+1} I(\gamma(s, X_t + Y_t, x)) \, ds.$$

(3). Observe that for every $k \in \mathbb{N}$ we have

$$(X_t + Y_t)^k = (X_t)^k + \left\{ \begin{array}{c} \text{sum of products where the} \\ \text{factor } Y_t \text{ appears at least once} \end{array} \right\} = (X_t)^k + \mathcal{O}_1(t^{n+1}),$$

where $\mathcal{O}_1(t^{n+1})$ is a higher-order PDO whose coefficients are $\mathcal{O}(t^{n+1})$.
(4). Deduce that, when $k = n + 1$, $(X_t + Y_t)^{n+1} = \mathcal{O}_2(t^{n+1})$.
(5). Derive from the expansion (3.26) that

$$\exp(X_t + Y_t)(x) = \sum_{k=0}^{n} \frac{1}{k!} (X_t)^k I(x) + \mathcal{O}_3(t^{n+1}).$$

(6). The proof of (3.5) is complete if one shows that $\exp(X_t)(x)$ has the same expansion. Prove this as a consequence of (3.26) by observing that

$$\frac{1}{n!} \int_0^1 (1-s)^n (X_t)^{n+1} I(\gamma(s, X_t, x)) \, ds = \mathcal{O}_4(t^{n+1}).$$

Exercise 3.11. Prove the following alternative versions of Thm. 3.6 (you may benefit from the previous exercises); in (B), we set $n \vee m := \max\{n, m\}$.

(A). *Let X, Y be C^∞ v.f.s on Ω. Let $K \subset \Omega$ be compact and let $n, m \geq 0$. Consider*

$$Z^{(n,m)}(s,t) := \sum_{0 \leq i \leq n, \, 0 \leq j \leq m} C_{i,j}(sX, tY),$$

where $C_{i,j}$ is defined in Def. 2.24. Then there exists $\varepsilon > 0$ such that

$$\exp(tY)(\exp(sX)(x)) = \exp\left(Z^{(n,m)}(s,t)\right)(x) + \mathcal{O}\left(|s|^{n+1} + |t|^{m+1}\right), \tag{3.27}$$

for every $x \in K$ and every $s, t \in [-\varepsilon, \varepsilon]$.

(B). *In the above assumptions, using the notation in (3.12), consider the smooth v.f.*

$$Z^{(n \vee m)}(s,t) = \sum_{k \leq n \vee m} Z_k(sX, tY) = \sum_{i+j \leq n \vee m} C_{i,j}(sX, tY),$$

where Z_k and $C_{i,j}$ are defined in Def. 2.24. Then there exists $\varepsilon > 0$ such that

$$\exp(tY)(\exp(sX)(x)) = \exp\left(Z^{(n \vee m)}(s,t)\right)(x) + \mathcal{O}\left(|s|^{n+1} + |t|^{m+1}\right), \tag{3.28}$$

for every $x \in K$ and every $s, t \in [-\varepsilon, \varepsilon]$.

If from formulae (3.13) or (3.28) we want to generate a remainder which is $\mathcal{O}(|s|^3 + |t|^2)$, the smallest n we can take is $n = 2 = \max\{2, 1\}$, since[2] $\mathcal{O}(|s| + |t|)^3$ is also $\mathcal{O}(|s|^3 + |t|^2)$; the approximating v.f. is then $Z_1(sX, tY) + Z_2(sX, tY)$, which, in terms of the summands $C_{i,j}$, is equal to $C_{1,0} + C_{0,1} + C_{1,1}$, all evaluated at (sX, tY). Instead, if we want to generate the same remainder $\mathcal{O}(|s|^3 + |t|^2)$ this time by formula (3.27), we must take $n = 2$ and $m = 1$; the approximating v.f. is then $C_{1,0} + C_{0,1} + C_{1,1} + C_{2,1}$, having one more summand if compared to the previous approximating v.f.

Note that, for arbitrary n and m, the summands $C_{i,j}$ composing $Z^{(n,m)}$ and those in $Z^{(n \vee m)}$ cannot be compared. For example, when $n = m$, $Z^{(n \vee m)} = Z^{(n)}$ contains less summands $C_{i,j}$ than $Z^{(n,n)}$. Instead, if we take, for example, $n = 6$ and $m = 2$, then $Z^{(n \vee m)} = Z^{(6)}$ contains 17 summands, whereas $Z^{(6,2)}$ contains 14 summands.[3] The advantage of using $Z^{(6)}$ lies in that it produces a remainder of the form $\mathcal{O}(|s| + |t|)^7 = \mathcal{O}(|s|^7 + |t|^7)$, whereas $Z^{(6,2)}$ produces a remainder of the form $\mathcal{O}(|s|^7 + |t|^3)$.

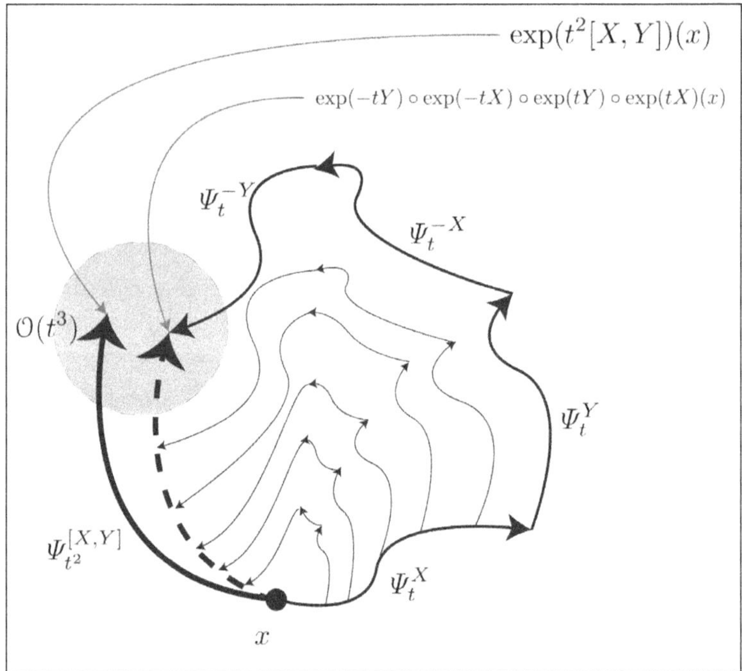

Fig. 3.3 The approximation of the flow of the commutator $[X, Y]$ (up to time t^2) by means of the flows of the vector fields $\pm X$, $\pm Y$ in Cor. 3.5. See also Fig. 1.5 on page 26.

---

[2] Indeed, Exr. 3.2 ensures that $(|s| + |t|)^3 = \mathcal{O}(|s|^3 + |t|^3) = \mathcal{O}(|s|^3 + |t|^2)$.

[3] Prove this, also showing that not every summand in $Z^{(6,2)}$ is a summand of $Z^{(6)}$.

Chapter 4

Hadamard's Theorem for Flows

THE aim of this chapter is to study how a vector field Y changes under the action of the flow of another vector field X: this fundamental topic is studied here under the name of *Hadamard's Theorem for flows*. The reason for this (non-standard) naming is the analogy with the so-called Hadamard's formula for formal power series in the indeterminates x, y

$$e^x\, y\, e^{-x} = e^{\operatorname{ad} x} y = y + [x, y] + \tfrac{1}{2}\,[x, [x, y]] + \tfrac{1}{3!}\,[x, [x, [x, y]]] + \cdots.$$

Indeed, our central target in this chapter is $\mathrm{d}\Psi^X_{-t}Y$, the pushforward of the v.f. Y under the diffeomorphism Ψ^X_{-t}, the latter being the flow of X running backward in time. If we interpret $\mathrm{d}\Psi^X_{-t}Y$ as the result of a change of coordinates (via Ψ^X_{-t}) applied to Y, the analogy with $e^x\, y\, e^{-x}$ becomes more apparent. This analogy will eventually be crystal clear, if we consider the formula

$$\mathrm{d}\Psi^X_{-t}Y = e^{\operatorname{ad} tX}Y,$$

which we obtain in Thm. 4.27. Here, the higher order commutators of X and Y must satisfy some growth assumption: for example, if X and Y belong to a finite-dimensional Lie algebra of v.f.s this growth assumption is automatically fulfilled.

As important as Poincaré's ODE (Chap. 2), another ODE-like formula plays the leading role here, namely *Hadamard's Formula*

$$\frac{\mathrm{d}}{\mathrm{d}t}\left(\mathrm{d}\Psi^X_{-t}Y\right) = \mathrm{d}\Psi^X_{-t}[X, Y],$$

an identity of time-dependent vector fields on the set $\mathcal{D}(X)$. When $t = 0$, the previous identity gives a remarkable interpretation of the commutator $[X, Y]$, via the notion of the so-called *Lie derivative*.

A prerequisite for this chapter is the knowledge of some elementary theory of smooth manifolds, for which the reader is referred to, e.g., [Lee (2013)].

4.1 Preliminaries on derivations and differentials

In order to preserve as much as possible the elegant approach of Differential Geometry, in this chapter we consider vector fields on an n-dimensional manifold M. All the basic results of Chap. 1 hold true in this setting as well: the reader is referred to [Lee (2013), Chapter 9].

We adopt the notation of Lee's book (occasionally with some minor modification, for example for the differential of F at p, which we denote $d_p F$ instead of dF_p), except for the notations –concerning flows– already introduced in Chap. 1, namely $\Psi_t^X(x)$, $\mathcal{D}(X,x)$, $\mathcal{D}(X)$, Ω_t^X. In Lee's book these are denoted with very similar notations: respectively by $\theta_t(x)$, $\mathcal{D}^{(x)}$, $\mathcal{D}$, M_t.

Sometimes we review the definitions from (basic) smooth manifold theory that we need to use, sometimes we do not. For instance, among the notions that we do *not* review, we have: the very definition of a smooth manifold; the smooth manifold structure of the finite dimensional vector spaces; charts and local coordinates; coordinate basis vectors; smooth maps; the tangent bundle and its smooth structure; the identification of smooth vector fields with derivations of C^∞. The reader unfamiliar with the manifold setting is –at any moment– free to think of M as an open subset of $\mathbb{R}^N$ with its standard differentiable structure.

Let M be a smooth manifold.[1] A tangent vector $\mathbf{v} \in T_x M$ at $x \in M$ is thought of as a derivation at x, i.e., $\mathbf{v} : C^\infty(M) \to \mathbb{R}$ is linear and it satisfies

$$\mathbf{v}(f\,g) = f(x)\,\mathbf{v}(g) + g(x)\,\mathbf{v}(f), \quad \forall\, f, g \in C^\infty(M).$$

Sometimes we may want to write $\mathbf{v}_x$ for $\mathbf{v} \in T_x M$, to emphasise the role of x. If M, N are smooth manifolds, $F : M \to N$ is a smooth map and $x \in M$, the differential of F at x, denoted by $d_x F$ is the linear map $d_x F : T_x M \to T_{F(x)} N$ sending $\mathbf{v} \in T_x M$ to the derivation at $F(x)$ acting as follows:

$$d_x F(\mathbf{v})g := \mathbf{v}(g \circ F), \quad \forall g \in C^\infty(N).$$

Thus, $d_x F(\mathbf{v})$ belongs to $T_{F(x)} N$ for any $\mathbf{v} \in T_x M$.

Example 4.1 (Velocity vector). Let $\gamma : I \to M$ be a smooth map, where $I \subseteq \mathbb{R}$ is an open interval: we say that γ is a curve in M. For any $t_0 \in I$, we set

$$\dot{\gamma}(t_0) := d_{t_0}\gamma\left(\left.\frac{d}{dt}\right|_{t_0}\right) \in T_{\gamma(t_0)}M,$$

where $d/dt|_{t_0}$ is the standard coordinate basis vector of $T_{t_0}\mathbb{R}$. The notations

$$\left.\frac{d}{dt}\right|_{t_0}\gamma(t) \quad \text{or} \quad \left.\frac{d}{dt}\right|_{t=t_0}\gamma(t)$$

will also stand for $\dot{\gamma}(t_0)$. Hence, $\dot{\gamma}(t_0)$ acts on $f \in C^\infty(M)$ as follows:

$$\dot{\gamma}(t_0)f = \left.\frac{d}{dt}\right|_{t_0} f(\gamma(t)).$$

---

[1]We only consider manifolds *without boundary*, in the sense of [Lee (2013), p. 26].

Example 4.2 (The tangent space to a vector space). Let V be a finite-dimensional real vector space. We assume that the reader is familiar with the natural smooth manifold structure of V. Given $x \in V$, we identify $T_x V$ with V by associating to any $v \in V$ the derivation at x defined by

$$C^\infty(V) \ni f \mapsto \frac{\partial f}{\partial v}(x) := \frac{\mathrm{d}}{\mathrm{d}t}\Big|_{t=0} f(x + tv). \tag{4.1}$$

We leave it as an exercise to show that $v \mapsto \frac{\partial}{\partial v}|_x$ is indeed a linear bijection from V to $T_x V$, for any $x \in V$. $\qquad\qquad\qquad\qquad\qquad\qquad\qquad\qquad\qquad\qquad\qquad\qquad\sharp$

As usual, if $TM = \coprod_{x \in M} T_x M$ is the tangent bundle[2] of M, for any $x \in M$ we are free to identify $T_x M$ with its image in TM under the map

$$T_x M \to TM, \quad \mathbf{v}_x \mapsto (x, \mathbf{v}_x).$$

By putting together the maps $\mathrm{d}_x F$ as the points x vary in M, we obtain the so-called global differential $\mathrm{d}F : TM \to TN$, which is the smooth map whose restriction to each $T_x M \subseteq TM$ is $\mathrm{d}_x F$. This simply means that

$$\mathrm{d}F(x, \mathbf{v}) = (F(x), \mathrm{d}_x F(\mathbf{v})), \quad \forall\, x \in M,\ \mathbf{v} \in T_x M.$$

A smooth vector field on M is a smooth section of the canonical projection map $\pi : TM \to M$, $\pi(x, \mathbf{v}) = x$. We sometimes abridge 'vector field' with v.f., and 'smooth vector field' with s.v.f.

We denote by $\mathcal{X}(M)$ the set of the smooth vector fields on M. Strictly speaking, $X : M \to TM$ is a smooth map such that

$$x \mapsto X(x) = (x, X_x), \quad \text{with } X_x \in T_x M \text{ for all } x \in M.$$

We straightaway renounce the notation $X(x)$, and *we identify the C^∞ v.f.s on M with the derivations of $C^\infty(M)$*; in other words $X \in \mathcal{X}(M)$ iff $X : C^\infty(M) \to C^\infty(M)$ is a linear map such that

$$X(f\,g) = f\,Xg + g\,Xf, \quad \forall\, f, g \in C^\infty(M).$$

Throughout Xf denotes the smooth function on M defined by $(Xf)(x) := X_x(f)$ for any $x \in M$, where it is understood that

$$X_x \in T_x M, \quad \text{for every } X \in \mathcal{X}(M) \text{ and every } x \in M.$$

Now, a natural question arises: *do differentials act in some way on vector fields?* If M, N are smooth manifolds, if $F : M \to N$ is smooth and if $X \in \mathcal{X}(M)$, since X_x is a derivation at any point $x \in M$, then $\mathrm{d}_x F(X_x)$ is meaningful as a derivation at

---

[2]From a strictly set-theoretic point of view, the disjoint union defining TM is simply

$$\bigcup_{x \in M} \{x\} \times T_x M = \big\{(x, \mathbf{v}) : x \in M,\ \mathbf{v} \in T_x M\big\}.$$

In comparison to the usual notation for disjoint unions, we take the liberty of interchanging the position of the index x and the element $\mathbf{v}$ of the indexed set $T_x M$.

$F(x) \in N$. Essentially, if we go back to the definition of X as a map $X : M \to TM$, we can compose it with the global differential $dF : TM \to TN$, obtaining
$$dF \circ X : M \to TN, \quad x \mapsto (F(x), d_x F(X_x)).$$
This is however much less interesting than asking for the existence of *a vector field defined on N*, say $\widetilde{X}$, such that the derivation $\widetilde{X}_{F(x)}$ coincides with $d_x F(X_x)$ for any $x \in M$. One immediately recognizes that, if F is not surjective, the definition of $\widetilde{X}_y$ for y outside the image set of F is quite arbitrary, and if F is not injective (say $F(x_1) = F(x_2)$ with $x_1 \neq x_2$), the derivation $\widetilde{X}_{F(x_1)} = \widetilde{X}_{F(x_2)}$ should coincide with both $d_{x_1} F(X_{x_1})$ and $d_{x_2} F(X_{x_2})$, and this may be quite hopeless.

A positive answer to the above question can be given when we deal with a *smooth diffeomorphism $F : M \to N$*. Indeed the v.f. $\widetilde{X}$ on N defined by
$$\widetilde{X}_y := d_{F^{-1}(y)} F(X_{F^{-1}(y)}) \qquad (\text{for } y \in N)$$
is well-posed on $N = F(M)$, and it is the *unique* v.f. on N that satisfies
$$\widetilde{X}_{F(x)} = d_x F(X_x), \qquad \forall\, x \in M. \tag{4.2}$$
We therefore introduce another (more interesting) notion of dF:

Definition 4.3 (Push-forward of a v.f. by a diffeomorphism). Let M and N be smooth manifolds. If $F : M \to N$ is a smooth diffeomorphism and if X is any vector field on M, we denote by dFX the vector field on N defined by
$$(dFX)_y := d_{F^{-1}(y)} F(X_{F^{-1}(y)}), \quad y \in N. \tag{4.3}$$
We also write $dF(X)$ when further mathematical objects may create notational ambiguity. We say that dFX is the **pushforward** of the v.f. X by the diffeomorphism F; we indifferently denote dF also by F_*.

We also say that dF is the **differential** of the diffeomorphism F, as a map from $\mathfrak{X}(M)$ to $\mathfrak{X}(N)$ (not to be confused with the global differential dF from TM to TN).

Remark 4.4 (Characterizations of F_*). When $F : M \to N$ is a diffeomorphism, the defining identity (4.3) for F_* is equivalent to the following facts:

(1) $(F_*X)_{F(x)} = d_x F(X_x)$ for every $x \in M$;

(2) $X(g \circ F)(x) = (F_*X\, g)(F(x))$, for all $g \in C^\infty(N)$ and all $x \in M$;

(3) $(F_*X\, g)(y) = X(g \circ F)(F^{-1}(y))$, for all $g \in C^\infty(N)$ and all $y \in N$;

(4) going back to the definition of a v.f. as a map from the manifold to its tangent bundle, and using the notion of the global differential between tangent bundles, F_* is characterized by the commutativity of

$$
\begin{array}{ccc}
TM & \xrightarrow{\;\;dF\;\;} & TN \\[2pt]
{\scriptstyle X}\big\uparrow & & \big\uparrow{\scriptstyle F_*X} \\[2pt]
M & \xrightarrow{\;\;F\;\;} & N
\end{array}
$$

Indeed, the image of x under $dF \circ X$ is $(F(x), d_x F(X_x))$, and the image of x under $(F_*X) \circ F$ is the same: $(F(x), (F_*X)_{F(x)})$.

Roughly put, we can picture F_*X (as a v.f. on N) as obtained from X (as a v.f. on M) via the *change of variable*

$$M \ni x \mapsto F(x) = y \in N.$$

The relationship between X and $\widetilde{X}$ described in (4.2) is called *F-relatedness*, and is characterized –in terms of flows– in Sec. 4.2.

4.1.1 *Time-dependent vector fields*

In Secs. 4.3 and 4.4 we shall make use of time-dependent vector fields, a notion that we now review.

Let M be a manifold and let Ω be an open subset of $\mathbb{R} \times M$ (with the product-manifold structure); a smooth **time-dependent vector field** on Ω is a smooth map $V : \Omega \to TM$ such that $V(t,x) \in T_xM$ for any $(t,x) \in \Omega$ (here we are using the routine identification $T_xM \equiv \{x\} \times T_xM \subseteq TM$). Given t, if we denote by Ω_t the t-section of Ω, i.e.,

$$\Omega_t := \Big\{ x \in M \,\Big|\, (t,x) \in \Omega \Big\},$$

then $V^t := V(t,\cdot)$ is a smooth v.f. on Ω_t when the latter (which is always an open subset of M) is not empty. Given $(t,x) \in \Omega$, we can define the t-derivative of the time-dependent v.f. $V(t,x)$ as the v.f. on Ω_t acting on functions as follows:

$$\left(\frac{\mathrm{d}}{\mathrm{d}t} V(t,x) \right) f(x) := \frac{\mathrm{d}}{\mathrm{d}s}\bigg|_{s=t} \Big(s \mapsto V(s,x)f \Big), \quad \forall\, f \in C^\infty(M),\ x \in \Omega_t. \tag{4.4}$$

This is well posed since Ω is open, so that, if $(t,x) \in \Omega$, there exists an open interval J containing t such that $(t,x) \in \Omega$ for any $t \in J$. It is easy to check that $\frac{\mathrm{d}}{\mathrm{d}t}V(t,x)$ defines a time-dependent vector field on Ω; indeed (Exr. 4.5)

$$\frac{\mathrm{d}}{\mathrm{d}t}V(t,x) \in T_xM \quad \text{for every } (t,x) \in \Omega.$$

An integral curve of a time-dependent v.f. $V(t,x)$ on Ω is (similar to the autonomous case) a C^∞ curve $\gamma : I \to M$ ($I \subseteq \mathbb{R}$ being an interval) such that

- $(t,\gamma(t)) \in \Omega$ for every $t \in I$;
- $\gamma'(t) = V(t,\gamma(t))$ for every $t \in I$.

4.2 Relatedness of vector fields and flows

Definition 4.5 (Relatedness of vector fields). Let M, N be smooth manifolds and let $f : M \to N$ be a smooth map. We say that two vector fields $X \in \mathcal{X}(M)$ and $Y \in \mathcal{X}(N)$ are f-**related** if one has

$$\mathrm{d}_xf(X_x) = Y_{f(x)} \quad \text{for every } x \in M. \tag{4.5}$$

By unraveling the definition, X and Y are f-related iff

$$X(g \circ f)(x) = (Yg)(f(x)), \qquad \forall\, g \in C^\infty(N), \quad \forall\, x \in M. \tag{4.6}$$

Proposition 4.6 (Consistency of f-relatedness). *Let $f : M \to N$ be a smooth map. Let $X_1, X_2 \in \mathfrak{X}(M)$ and $Y_1, Y_2 \in \mathfrak{X}(N)$. Suppose that, for $i = 1, 2$, X_i is f-related to Y_i. Then $[X_1, X_2]$ is f-related to $[Y_1, Y_2]$.*

The simple proof is left as an exercise.

Example 4.7 (X and F_*X are F-related). Suppose that $F : M \to N$ is a C^∞-diffeomorphism and X is any s.v.f. on M; then X is F-related to the s.v.f. F_*X on N introduced in Def. 4.3. This follows from Rem. 4.4-(1).

As a consequence of Prop. 4.6, we deduce that $F_* : \mathfrak{X}(M) \to \mathfrak{X}(N)$ is a Lie algebra homomorphism, i.e.,

$$F_*[X_1, X_2] = [F_*X_1, F_*X_2], \quad \forall\, X_1, X_2 \in \mathfrak{X}(M). \tag{4.7}$$

This is also a consequence of the fact that, when $F : M \to N$ is a diffeomorphism and $X \in \mathfrak{X}(M)$, there exists one and only one v.f. on N which is F-related to X, and this is F_*X. Formula (4.7) is referred to as the **naturality of the Lie bracket under push-forwards.**   ♮

We need a result on flows of f-related v.f.s, of independent interest. Given $X \in \mathfrak{X}(M)$, we exploit our usual notations: $t \mapsto \Psi_t^X(x)$ is the integral curve of X starting at $x \in M$; its maximal domain is denoted by $\mathcal{D}(X, x)$; given $t \in \mathbb{R}$, M_t^X is either empty or is the largest set of points $x \in M$ such that $x \mapsto \Psi_t^X(x)$ is defined; finally $\mathcal{D}(X)$ is the full domain of $(t, x) \mapsto \Psi_t^X(x)$, i.e.,

$$\mathcal{D}(X) = \Big\{ (t, x) \in \mathbb{R} \times M \,:\, x \in M,\, t \in \mathcal{D}(X, x) \Big\}.$$

By the very definition of an integral curve, we have

$$\frac{\mathrm{d}}{\mathrm{d}t}\Big\{ g(\Psi_t^X(x)) \Big\} = \mathrm{d}_{\Psi_t^X(x)} g\Big(X_{\Psi_t^X(x)} \Big), \quad \forall\, x \in M,\, t \in \mathcal{D}(X, x), \tag{4.8}$$

whatever the smooth function $g : M \to N$ (N being any smooth manifold).

Lemma 4.8 (Relatedness and flows). *Let $f : M \to N$ be a smooth map of smooth manifolds M, N. Suppose that $X \in \mathfrak{X}(M)$ is f-related to $Y \in \mathfrak{X}(N)$.*

Then, for every $x \in M$, one has $\mathcal{D}(X, x) \subseteq \mathcal{D}(Y, f(x))$; moreover

$$f(\Psi_t^X(x)) = \Psi_t^Y(f(x)) \quad \text{for every } x \in M \text{ and every } t \in \mathcal{D}(X, x).$$

In particular, if X is complete, then Y is complete as well.

Proof. We prove that $t \mapsto \gamma(t) := f(\Psi_t^X(x))$ solves the same Cauchy problem solved by $t \mapsto \Psi_t^Y(f(x))$, at least when t belongs to $\mathcal{D}(X, x)$; this will give both the assertions of the lemma. We have:

- $\gamma(t)|_{t=0} = f(x)$: this is trivial since $\Psi_t^X(x)|_{t=0} = x$;
- $\dot{\gamma}(t) = Y_{\gamma(t)}$: this is a consequence of the computation

$$\frac{\mathrm{d}}{\mathrm{d}t}\{f(\Psi_t^X(x))\} \overset{(4.8)}{=} \mathrm{d}_{\Psi_t^X(x)}f\left(X_{\Psi_t^X(x)}\right) \overset{(4.5)}{=} Y_{f(\Psi_t^X(x))} = Y_{\gamma(t)}.$$

This ends the proof. $\qquad\qquad\qquad\qquad\qquad\qquad\qquad\qquad\qquad\qquad\qquad\square$

Summing up what we know so far, we have the following remarkable characterization of f-relatedness in terms of flows. See Fig. 4.1.

Proposition 4.9 (Characterization of relatedness). *Let M, N be smooth manifolds and let $f : M \to N$ be a smooth map. Let also $X \in \mathcal{X}(M)$ and $Y \in \mathcal{X}(N)$.*

Then, the following facts are equivalent:

(1) X and Y are f-related;

(2) $\mathrm{d}_x f(X_x) = Y_{f(x)}$ for every $x \in M$;

(3) $X(g \circ f)(x) = (Yg)(f(x))$ for every $g \in C^\infty(N)$ and every $x \in M$;

(4) $f(\Psi_t^X(x)) = \Psi_t^Y(f(x))$ for every $x \in M$ and $t \in \mathcal{D}(X, x) \subseteq \mathcal{D}(Y, f(x))$;

(5) $f(\Psi_t^X(x)) = \Psi_t^Y(f(x))$ for every $x \in M$ and every t in some open neighborhood of $t = 0$ (possibly depending on x).

Proof. (2) is the very definition of (1), and (3) is a restatement of (2); Lem. 4.8 shows that (1) implies (4), while (4) trivially implies (5), since $\mathcal{D}(X, x)$ is an open neighborhood of $t = 0$. We are left to prove that (5) implies (1).

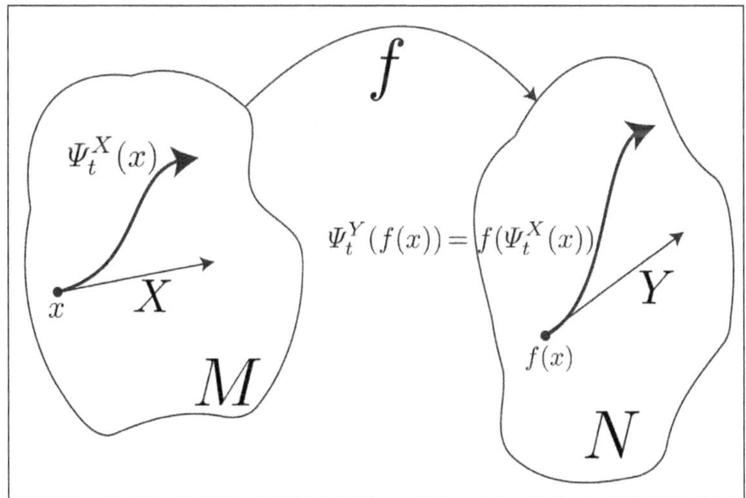

Fig. 4.1 f-related vector fields X, Y, and the behaviour of their flows (Prop. 4.9).

Let $x \in M$ be fixed and let $I(x)$ be the open neighborhood of $t = 0$ mentioned in (5); if we differentiate both sides of the identity in (5) (and if we use (4.5) and the definition of flow), we get

$$\mathrm{d}_{\Psi_t^X(x)} f\left(X_{\Psi_t^X(x)}\right) = Y_{\Psi_t^Y(f(x))}, \qquad t \in I(x).$$

Taking $t = 0 \in I(x)$ we get (2), which means that X and Y are f-related. $\qquad\square$

As a direct corollary of Prop. 4.9 we have the following result:

Corollary 4.10. *Let $f : M \to N$ be a C^∞ map, and let $X \in \mathfrak{X}(M)$ and $Y \in \mathfrak{X}(N)$ be f-related. Then, for any $t \in \mathbb{R}$ one has $f(M_t^X) \subseteq N_t^Y$, and $\Psi_t^Y \circ f = f \circ \Psi_t^X$ on M_t^X, i.e., the following diagram is commutative:*

$$
\begin{array}{ccc}
M_t^X & \xrightarrow{\ \ f\ \ } & N_t^Y \\
\Psi_t^X \downarrow & & \downarrow \Psi_t^Y \\
M_{-t}^X & \xrightarrow{\ \ f\ \ } & N_{-t}^Y
\end{array}
$$

From Prop. 4.9 and Def. 4.3 we obtain the following result:

Corollary 4.11. *Let $F : M \to N$ be a smooth diffeomorphism of smooth manifolds. For every $X \in \mathfrak{X}(M)$ and every $x \in M$, we have*

$$F(\Psi_t^X(x)) = \Psi_t^{F_*X}(F(x)),$$

for every $t \in \mathcal{D}(X, x)$. Moreover, $\mathcal{D}(X, x) = \mathcal{D}(F_*X, F(x))$. Equivalently, if $M_t^X \neq \varnothing$, the flow of F_*X is given by

$$\Psi_t^{F_*X} = F \circ \Psi_t^X \circ F^{-1}, \quad \text{defined on } N_t^{F_*X} = F(M_t^X). \tag{4.9}$$

In particular, if X is global, the same is true of F_*X and, for every $t \in \mathbb{R}$,

$$F \circ \Psi_t^X = \Psi_t^{F_*X} \circ F \quad \text{on } M.$$

4.2.1   *Invariance of a vector field under a map*

In order to introduce the notion of invariance of a vector field under a map, we first give an example.

Example 4.12 (Invariance of a vector field under its flows). Let X be a smooth v.f. on the manifold M. Let $t \in \mathbb{R}$ be such that $M_t^X \neq \varnothing$. For any fixed such t, we know that $\Psi_t^X(M_t^X) = M_{-t}^X$; moreover

$$\Psi_t^X : M_t^X \longrightarrow M_{-t}^X$$

is well-posed and it defines a C^∞-diffeomorphism. We can think of $M_{\pm t}^X$ as smooth manifolds in themselves, since they are open subsets of M.

We claim that, *for every t such that $M_t^X \neq \varnothing$, the vector field X (restricted to M_t^X) is Ψ_t^X-related to itself (restricted to M_{-t}^X)*; written in terms of differentials:
$$\mathrm{d}_x \Psi_t^X(X_x) = X_{\Psi_t^X(x)}, \quad \text{for every } x \in M_t^X. \tag{4.10}$$

We can rewrite this as an identity between v.f.s on the open set M_{-t}^X:
$$\mathrm{d}\Psi_t^X\left(X\big|_{M_t^X}\right) = X\big|_{M_{-t}^X} \quad \text{on } M_{-t}^X.$$

Here we are using the pushforward $\mathrm{d}\Psi_t^X \equiv (\Psi_t^X)_*$ instead of the pointwise differential used in (4.10); this is legitimate since Ψ_t^X is a C^∞-diffeomorphism. See Fig. 4.2. Clearly, (4.10) is equivalent to
$$X(f \circ \Psi_t^X)(x) = (Xf)(\Psi_t^X(x)), \quad \text{for } x \in M_t^X \text{ and } f \in C^\infty(M_{-t}^X).$$

Proof of (4.10). We prove the claimed (4.10). Let t be fixed in such a way that $M_t^X \neq \varnothing$, and let $x \in M_t^X$; since M_t^X is an open set, and as $\Psi_s^X(x) = x$ when $s = 0$, we infer that the semigroup property
$$\Psi_t^X(\Psi_s^X(x)) = \Psi_{t+s}^X(x)$$
is valid for any s in a small neighborhood, say J, of $s = 0$ (J depends on t and x). If we take derivatives at both sides wrt s, we get (for $s \in J$)
$$\mathrm{d}_{\Psi_s^X(x)} \Psi_t^X\left(\frac{\mathrm{d}}{\mathrm{d}s}\Psi_s^X(x)\right) = \left(\frac{\mathrm{d}}{\mathrm{d}r}\Psi_r^X(x)\right)\Big|_{r=t+s}.$$
Since $t \mapsto \Psi_t^X(x)$ solves the ODE $\frac{\mathrm{d}}{\mathrm{d}t}\Psi_t^X(x) = X_{\Psi_t^X(x)}$, the above identity gives
$$\mathrm{d}_{\Psi_s^X(x)} \Psi_t^X\left(X_{\Psi_s^X(x)}\right) = X_{\Psi_{t+s}^X(x)}, \quad s \in J.$$
If we take $s = 0$ (and we remember that $\Psi_0^X(x) = x$), we get (4.10). $\qquad\square$

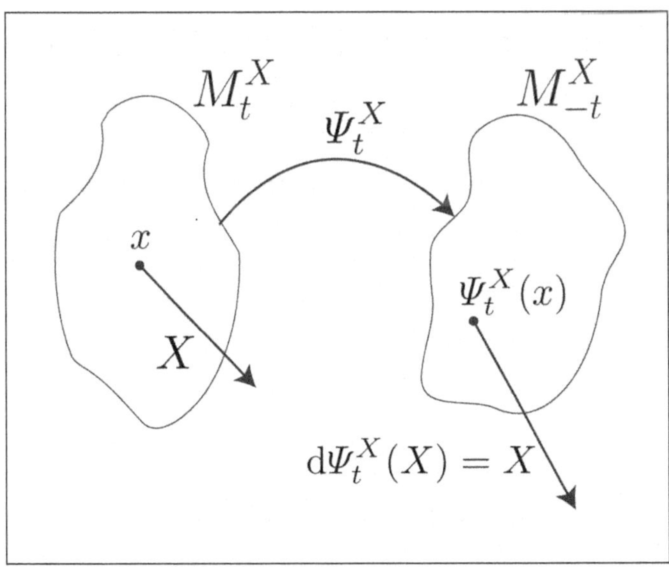

Fig. 4.2 The invariance of a vector field X under its flow map Ψ_t^X.

Remark 4.13. By taking into account the definition of flow, we infer that for any $X \in \mathfrak{X}(M)$ we have the identity of the following three objects:

$$\frac{\mathrm{d}}{\mathrm{d}t}(\Psi_t^X(x)) = X_{\Psi_t^X(x)} = \mathrm{d}_x \Psi_t^X(X_x),$$

for every $x \in M_t^X$ (or, equivalently, for every $(t,x) \in \mathcal{D}(X)$).   ♯

Definition 4.14 (f-invariant vector field). Let $f : M \to M$ be a smooth map. A smooth vector field X on M is said to be f**-invariant** (or **invariant under** f), if X is f-related to itself, i.e., $\mathrm{d}_x f(X_x) = X_{f(x)}$ for every $x \in M$.

A nice example of invariance is contained in Exm. 4.12 where it is shown that (disregarding for a moment the problem of the domain of definition of Ψ_t^X) *any X is invariant under any of its flow maps* Ψ_t^X. If, for example, X is global on M, the latter assertion is correct for any $t \in \mathbb{R}$, since $\Psi_t^X : M \to M$ is a smooth diffeomorphism for any real time t.

As is well known, another meaningful example of invariance occurs for the left-invariant vector fields on Lie groups.

Remark 4.15. Taking into account Prop. 4.9, for any $X \in \mathfrak{X}(M)$ and any function $f \in C^\infty(M, M)$, the following facts are equivalent:

(1) X is f-invariant,
(2) $X(g \circ f)(x) = (Xg)(f(x))$, for every $x \in M$ and every $g \in C^\infty(M)$,
(3) $f(\Psi_t^X(x)) = \Psi_t^X(f(x))$ for every $x \in M$ and $t \in \mathcal{D}(X, x) \subseteq \mathcal{D}(X, f(x))$,
(4) $f(\Psi_t^X(x)) = \Psi_t^X(f(x))$ for every $x \in M$ and every t in some open neighborhood of $t = 0$ (possibly depending on x).

Property (3) states that *X is f-invariant iff f commutes with the flows of X.*   ♯

Remark 4.16 (Invariant v.f.s wrt a family of functions). Let M be a smooth manifold, and let $\mathfrak{F}$ be a family of smooth maps from M to M. Let

$$I(\mathfrak{F}) := \Big\{ X \in \mathfrak{X}(M) : \ X \text{ is } f\text{-invariant, for every } f \in \mathfrak{F} \Big\}.$$

Clearly, $I(\mathfrak{F})$ is a vector space. If $X, Y \in I(\mathfrak{F})$, then Prop. 4.6 ensures that $[X, Y]$ belongs to $I(\mathfrak{F})$. This shows that the set $I(\mathfrak{F})$ is a Lie-subalgebra of $\mathfrak{X}(M)$, called *the Lie algebra of the $\mathfrak{F}$-invariant smooth vector fields.*

For example, if G is a Lie group and $\mathfrak{F}$ is the family of the left translations on G, then $I(\mathfrak{F})$ is nothing but the Lie algebra of G.   ♯

4.3   Commutators and Lie-derivatives

We now give another interpretation of the commutator $[X, Y]$ as a weight of the change of Y under the action of (the integral curves of) X. To this aim, we need to introduce the notion of *Lie derivative*. First we point our attention to the way that vector fields are transformed under the flows of other v.f.s.

Convention. *Throughout what follows, given $X, Y \in \mathfrak{X}(M)$, if $t \in \mathbb{R}$ is such that $M_t^X \neq \varnothing$, when we write $\mathrm{d}\Psi_{-t}^X Y$ it is understood that Y denotes the restriction of Y to M_{-t}^X, so that $\mathrm{d}\Psi_{-t}^X Y$ is well posed and –in its turn– the latter is implicitly understood as a smooth v.f. on M_t^X. See Fig. 4.3.*

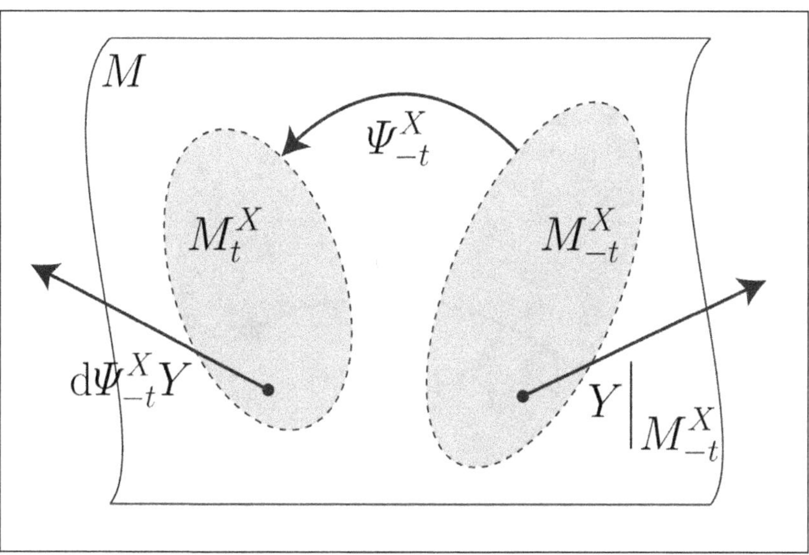

Fig. 4.3 Convention: when we write $\mathrm{d}\Psi_{-t}^X Y$, it is understood that Y denotes $Y|_{M_{-t}^X}$, the restriction of Y to M_{-t}^X, so that $\mathrm{d}\Psi_{-t}^X Y$ is a smooth v.f. on M_t^X.

Under the above convention on notations, in this section we study $\mathrm{d}\Psi_{-t}^X Y$. It is therefore worthwhile spending a few remarks on how we can picture this interesting object. It can be thought of as:

- the result of a *change of coordinate* argument (Rem. 4.17);
- a curve in $T_x M$, if we consider $(\mathrm{d}\Psi_{-t}^X Y)_x$ for $x \in M_t^X$ generic (Rem. 4.18).

Remark 4.17 ("Push-backward" of a vector field along the flows of another s.v.f.). Let X, Y be smooth v.f.s on M. Let us fix $t \in \mathbb{R}$ such that $M_t^X \neq \varnothing$; as argued in Exm. 4.12, we know that
$$\Psi_t^X : M_t^X \longrightarrow M_{-t}^X, \quad x \mapsto y = \Psi_t^X(x)$$
is a C^∞ diffeomorphism. Its inverse map is another flow map:
$$\Psi_{-t}^X : M_{-t}^X \longrightarrow M_t^X, \quad y \mapsto x = \Psi_{-t}^X(y), \tag{4.11}$$
the "backward flow" of X. Since Ψ_{-t}^X is also a diffeomorphism, we can picture it as a *change of coordinates*: in particular, every v.f. on M_{-t}^X can be pushed-forward via Ψ_{-t}^X to a related v.f. on M_t^X. Loosely speaking, since Ψ_{-t}^X appears to run backward in time, we may say that the change of coordinates (4.11) operates a *"push-backward"* (the term pull-back not being appropriate here, since it has another meaning in Differential Geometry).

We are interested in discovering how the change of coordinates Ψ^X_{-t} transforms any other[3] vector field Y. To this end, it is sufficient to unravel the definition of a change of coordinates: (4.11) transforms Y (which operates in the y variable on M^X_{-t}) into a vector field, $\widetilde{Y}^t$ say, operating in the x variable on M^X_t as follows: given any smooth function $f = f(x)$ on M^X_t we have

$$\widetilde{Y}^t f(x) = Y(f \circ \Psi^X_{-t})(\Psi^X_t(x)).$$

With the notation of differentials in Def. 4.3, we recognize that $\widetilde{Y}^t = \mathrm{d}\Psi^X_{-t}Y$ on M^X_t. Roughly put, the pushforward of Y via $\mathrm{d}\Psi^X_{-t}$ may be referred to as a *pushbackward of Y along the flow of X*. See also Fig. 4.4.   ♮

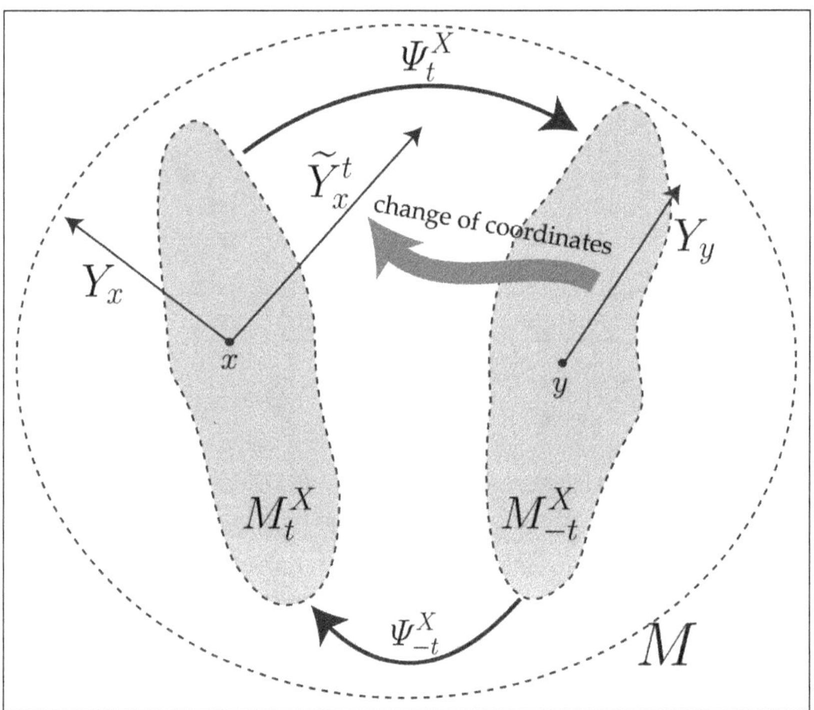

Fig. 4.4   An interpretation of the vector field $\widetilde{Y}^t$ resulting from the change-of-coordinate argument in Rem. 4.17. In the figure, the vector field $\widetilde{Y}^t$ is obtained from Y (on M^X_{-t}) via the change of coordinates $y \mapsto x = \Psi^X_{-t}(y)$. In terms of differentials, we know that $\widetilde{Y}^t$ is nothing but $\mathrm{d}\Psi^X_{-t}Y$.

Remark 4.18 (The curve $t \mapsto (\mathrm{d}\Psi^X_{-t}Y)_x$ in T_xM). Let us fix $x \in M$. Since in writing $\mathrm{d}\Psi^X_{-t}Y$ we mean $Y = Y|_{M^X_{-t}}$, we see that

$$\text{the curve } t \mapsto (\mathrm{d}\Psi^X_{-t}Y)_x \qquad \begin{array}{l} \text{is defined for } t \in \mathcal{D}(X, x), \\ \text{and is valued in } T_xM. \end{array}$$

---

[3]Another interesting question is to look for the transformed of X: in Rem. 4.12 we have seen that X is transformed into itself by any of its flows!

See Fig. 4.5 on page 105. Following Sec. 4.1.1, this shows that $(t, x) \mapsto (\mathrm{d}\Psi^X_{-t}Y)_x$ *is a smooth time-dependent vector field on the set* $\mathcal{D}(X)$*. Its t-derivative is therefore well posed, in the sense of (4.4).* ♮

For future reference, we review how $\mathrm{d}\Psi^X_{-t}Y$ acts on functions:

$$\left(\mathrm{d}\Psi^X_{-t}Y\right)f(x) = Y(f \circ \Psi^X_{-t})(\Psi^X_t(x)), \tag{4.12}$$

valid for every $x \in M^X_t$ and any $f \in C^\infty(M^X_t)$ (or equivalently for $(t, x) \in \mathcal{D}(X)$ and $f \in C^\infty(M)$).

Definition 4.19 (Lie derivative). Let X and Y be smooth vector fields on M. For any $x \in M$ we define a derivation at x by setting

$$(\mathcal{L}_X(Y))_x := \lim_{t \to 0} \frac{(\mathrm{d}\Psi^X_{-t}Y)_x - Y_x}{t}.$$

Equivalently, in the sense of derivatives of time-dependent v.f.s (see Rem. 4.18)

$$(\mathcal{L}_X(Y))_x = \frac{\mathrm{d}}{\mathrm{d}t}\Big|_{t=0}\left\{(\mathrm{d}\Psi^X_{-t}Y)_x\right\}.$$

We say that $(\mathcal{L}_X(Y))_x$ is the **Lie derivative of Y along X at the point** x, and, for short, that $\mathcal{L}_X(Y)$ is the **Lie derivative of Y along** X.

By the very definition (4.3) of the pushforward, we know that

$$(\mathrm{d}\Psi^X_{-t}Y)_x = \mathrm{d}_{\Psi^X_t(x)}\Psi^X_{-t}\left(Y_{\Psi^X_t(x)}\right), \quad \text{for every } x \in M^X_t,$$

and this is an equality of tangent vectors in T_xM, since $\Psi^X_{-t}(\Psi^X_t(x)) = x$ for every $t \in \mathcal{D}(X, x)$. Thus, $(\mathcal{L}_X(Y))_x$ is indeed a derivation at x. It is worth mentioning that, as we have to perform a limit as $t \to 0$, the well posedness of $(\mathcal{L}_X(Y))_x$ is also a consequence of the fact that any $x \in M$ belongs to M^X_t for small $|t|$.

We shall see in a moment that the map $M \ni x \mapsto (\mathcal{L}_X(Y))_x$ defines a smooth vector field. This is contained in the following theorem.

Theorem 4.20 (Lie derivatives and brackets). *Let X and Y be smooth vector fields on M. Then $\mathcal{L}_X(Y) = [X, Y]$; more precisely,*

$$(\mathcal{L}_X(Y))_x = [X, Y]_x \quad \text{for every } x \in M, \tag{4.13}$$

an equality of tangent vectors at the point x. Consequently we have

$$\frac{\mathrm{d}}{\mathrm{d}t}\Big|_{t=0}\left\{(\mathrm{d}\Psi^X_{-t}Y)f(x)\right\} = [X, Y]f(x), \tag{4.14}$$

for any $x \in M$ and any $f \in C^\infty(M)$.

When $M = \Omega$ is an open subset of $\mathbb{R}^N$, the same formula holds true for C^1 vector fields X, Y on Ω, and f of class C^1 on Ω (real or vector valued).

For the sake of future reference, if we further unravel the definitions, identity (4.14) is equivalent to (see (4.12))

$$\frac{\mathrm{d}}{\mathrm{d}t}\Big|_{t=0}\Big\{Y(f\circ\Psi^X_{-t})(\Psi^X_t(x))\Big\} = [X,Y]f(x), \quad \forall\, x\in M,\ f\in C^\infty(M). \tag{4.15}$$

Formula (4.15) is a special case of a more general formula (given in Thm. 4.22) providing the derivative wrt t of the function in curly braces *at any t*.

Remark 4.21. From Rem. 4.18 we know that $\mathrm{d}\Psi^X_{-t}(Y)$ is a C^∞ *time-dependent v.f.* defined on M^X_t. The arbitrariness of $x\in M$ and $f\in C^\infty(M)$ in (4.14) suggests to write

$$\frac{\mathrm{d}}{\mathrm{d}t}\Big|_{t=0}\Big\{\mathrm{d}\Psi^X_{-t}Y\Big\} = [X,Y] \quad \text{on } M,$$

as an equality of v.f.s on M. We warn the reader that this is an abuse of notation, because $\mathrm{d}\Psi^X_{-t}Y$ is defined only on M^X_t. The fact that M^X_t invades M as $t\downarrow 0$ is the redeeming feature that makes this abuse more tolerable.   ♯

Proof of Thm. 4.20. We verify (4.13) in a local chart around x. This boils down to assuming that X,Y are C^1 v.f.s on some open set $\Omega\subseteq\mathbb{R}^N$ and $x\in\Omega$. Note that we are strongly lowering the regularity assumption on X,Y (from C^∞ to C^1) aiming at the result stated in the last sentence of the theorem.

We need to prove the identity of the two derivations at x in (4.13). Since two derivations *in space* are coincident iff they act in the same way when applied to the identity map $I:\mathbb{R}^N\to\mathbb{R}^N$, we need to prove that, setting $F(t):=(\mathrm{d}\Psi^X_{-t}Y)I$ (which is a curve in $\mathbb{R}^N$), the function $F(t)$ is differentiable at $t=0$, and it holds that $F'(0)=[X,Y](x)$.

To this aim, we have the following fact (we apply (4.12) to the component functions of I; we can benefit of the fact that v.f.s in $\mathbb{R}^N$ are allowed to act on vector valued functions, see e.g., (1.4) on page 3):

$$F(t) = (Y\Psi^X_{-t})(\Psi^X_t(x)) = \mathfrak{J}_{\Psi^X_{-t}}(\Psi^X_t(x))\cdot Y(\Psi^X_t(x)). \tag{4.16}$$

(See also Fig. 4.6 on page 112.) We apply Leibniz rule to compute $F'(0)$; the second factor in the far rhs of (4.16) is C^1 in t and

$$\frac{\mathrm{d}}{\mathrm{d}t}\Big|_{t=0}\Big\{Y(\Psi^X_t(x))\Big\} = X(YI)(x). \tag{4.17}$$

The first factor in the far rhs of (4.16) is differentiable at $t=0$ due to Lem. 1.30 on page 22; putting together (1.35) and (4.17) we obtain that $F'(0)$ is equal to

$$-\mathfrak{J}_{XI}(x)Y(x) + \mathfrak{J}_I(x)X(YI)(x) = -Y(XI)(x) + X(YI)(x) = [X,Y](x).$$

This is what we wanted to prove.   □

4.4 Hadamard's Theorem for flows

The next important result, Thm. 4.22, provides an improvement of Thm. 4.20. Despite this is non-standard, we decided to name this theorem *Hadamard's Formula*, in view of the analogy with the so-called Hadamard's formula for formal power series in the indeterminates x, y (see (2.53) on page 68)

$$e^x \, y \, e^{-x} = e^{\operatorname{ad} x} y. \tag{4.18}$$

The following formula (4.21) provides a notable improvement of (4.15), since Ψ_0^X is the identity map.

Theorem 4.22 (Hadamard's Formula). *Let X, Y be smooth vector fields on the smooth manifold M. Then we have*

$$\frac{\mathrm{d}}{\mathrm{d}t}\left(\mathrm{d}\Psi_{-t}^X Y\right) = \mathrm{d}\Psi_{-t}^X[X, Y], \tag{4.19}$$

as an identity of time-dependent vector fields on the set $\mathcal{D}(X) \subseteq \mathbb{R} \times M$, or equivalently as an identity of smooth vector fields on the set $M_t^X \subseteq M$.

More precisely, for every $(t, x) \in \mathcal{D}(X)$ we have the identity in $T_x M$

$$\frac{\mathrm{d}}{\mathrm{d}t}\left\{\mathrm{d}_{\Psi_t^X(x)}\Psi_{-t}^X\left(Y_{\Psi_t^X(x)}\right)\right\} = \mathrm{d}_{\Psi_t^X(x)}\Psi_{-t}^X\left([X, Y]_{\Psi_t^X(x)}\right). \tag{4.20}$$

Consequently, for any $f \in C^\infty(M)$, any $x \in M$ and any $t \in \mathcal{D}(X, x)$, one has

$$\frac{\mathrm{d}}{\mathrm{d}t}\left\{Y(f \circ \Psi_{-t}^X)(\Psi_t^X(x))\right\} = [X, Y](f \circ \Psi_{-t}^X)(\Psi_t^X(x)). \tag{4.21}$$

When $M = \Omega$ is an open subset of $\mathbb{R}^N$, the same formula holds true for C^1 vector fields X, Y on Ω, and f of class C^1 on Ω (real or vector valued).

With the terminology introduced in Rem. 4.17, formula (4.19) states that the derivative of the time-dependent vector field obtained by "pushing-backward" Y along the flow of X is equal to the vector field obtained by "pushing-backward" $[X, Y]$ under the same flow.

Proof. Since in this book we shall be interested not only in C^∞ v.f.s, let us prove the theorem under the lower regularity formulation in the last sentence of its statement. The *smooth* manifold case is simpler, since it is based on the semigroup property in Prop. 1.14 and (4.14) (see e.g., [Lee (2013), Prop. 9.41]).

To this end, it is sufficient to prove (4.21) when f is replaced by the identity map of $\mathbb{R}^N$, since any v.f. X in space is identified by its coefficient vector XI. Therefore, we need to prove that, if X, Y are C^1 v.f.s on Ω,

$$\frac{\mathrm{d}}{\mathrm{d}t}\left\{Y(\Psi_{-t}^X)(\Psi_t^X(x))\right\} = ([X, Y]\Psi_{-t}^X)(\Psi_t^X(x)), \tag{4.22}$$

valid for $x \in \Omega$ and $t \in \mathcal{D}(X, x)$; as per usual, we allow v.f.s operate componentwise on vector valued functions.

If $(x,t) \in \mathcal{D}(X)$, then $\Omega_t^X \neq \varnothing$, and we know that $\Psi_t^X : \Omega_t^X \to \Omega_{-t}^X$ is a C^1-diffeomorphism with inverse Ψ_{-t}^X. Hence, for any $x \in \Omega_t^X$, the map

$$\mathcal{D}(X,x) \ni t \mapsto F(t) := (Y\Psi_{-t}^X)(\Psi_t^X(x))$$

is well posed. More explicitly we have

$$F(t) = \partial_{\Psi_{-t}^X}(\Psi_t^X(x))\, Y(\Psi_t^X(x)).$$

The second factor in the above rhs is obviously C^1 in t; the same is true of the first factor, due to Lem. 1.31 on page 22. If we apply Leibniz rule and we take into account formula (1.36) in the mentioned Lem. 1.31, we get

$$F'(t) = -\partial_{\Psi_{-t}^X}(\Psi_t^X(x))\, \partial_{XI}(\Psi_t^X(x))\, Y(\Psi_t^X(x)) + \partial_{\Psi_{-t}^X}(\Psi_t^X(x))\, X(YI)(\Psi_t^X(x))$$

$$= \partial_{\Psi_{-t}^X}(\Psi_t^X(x)) \left(-Y(XI)(\Psi_t^X(x)) + X(YI)(\Psi_t^X(x)) \right)$$

$$= \partial_{\Psi_{-t}^X}(\Psi_t^X(x))\, [X,Y](\Psi_t^X(x)) = ([X,Y]\Psi_{-t}^X)(\Psi_t^X(x)).$$

This is precisely (4.22), and (4.21) follows directly from it. $\qquad\square$

Corollary 4.23. *In the hypotheses of Thm. 4.22, for $(t,x) \in \mathcal{D}(X)$ we have*

$$\frac{\mathrm{d}}{\mathrm{d}t}\left(\mathrm{d}\Psi_{-t}^X Y\right)_x = \left[\mathrm{d}\Psi_{-t}^X X, \mathrm{d}\Psi_{-t}^X Y\right]_x \quad \text{in } T_x M.$$

Thereby, since X is invariant under Ψ_{-t}^X, this is equivalent to

$$\frac{\mathrm{d}}{\mathrm{d}t}\left(\mathrm{d}\Psi_{-t}^X(Y)\right)_x = \left[X, \mathrm{d}\Psi_{-t}^X(Y)\right]_x \quad \text{in } T_x M, \qquad (4.23)$$

valid for $t \in \mathcal{D}(X,x)$ (i.e., for $(t,x) \in \mathcal{D}(X)$).

Proof. It is a restatement of (4.19), taking into account (4.7) and (4.10). $\qquad\square$

As a matter of fact, (4.23) looks like an ODE solved by $t \mapsto \mathrm{d}\Psi_{-t}^X(Y)$; for the time being, it is however not so clear the setting where this ODE can be made precise. Is this an ODE in the finite-dimensional $T_x M$, or rather in the (infinite dimensional!) vector space of the smooth vector fields on M_t^X?

A positive answer to the first question is hindered by the fact that the rhs of (4.23) requires the knowledge of $\mathrm{d}\Psi_{-t}^X(Y)$ outside $T_x M$; on the other hand the infinite-dimensionality of $\mathfrak{X}(M_t^X)$ in the second question (and the t-dependence of the domain) is rather unpleasant.

A very convenient hypothesis making things easier to handle is when X and Y generate a finite-dimensional Lie sub-algebra of $\mathfrak{X}(M)$: (4.23) then becomes a genuine ODE (on a finite-dimensional vector space); we shall return to this topic in Chap. 13 (see Rem. 13.3 on page 269).

The previous Thm. 4.22 can be generalized to obtain the following result, of paramount importance. Here, as usual, we set $(\mathrm{ad}\, X)Y = [X,Y]$, so that

$$(\mathrm{ad}\, X)^k(Y) = \underbrace{[X,[X\cdots[X,Y]\cdots]]}_{k\ \text{times}}.$$

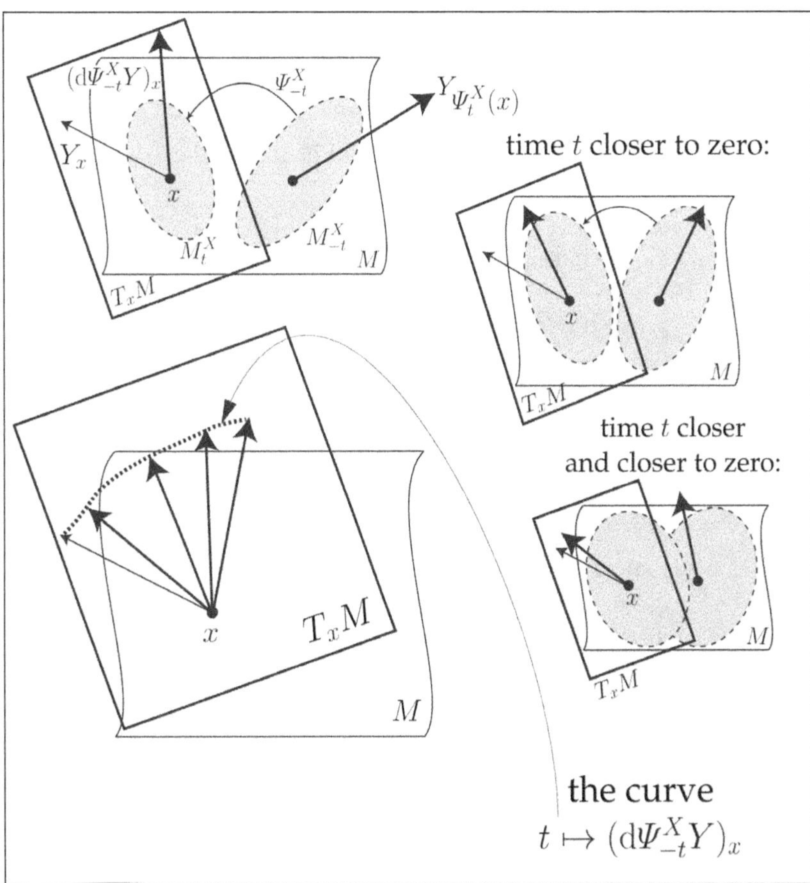

Fig. 4.5 Rem 4.18 gives another interpretation of $\mathrm{d}\Psi^X_{-t}Y$, as the curve $t \mapsto (\mathrm{d}\Psi^X_{-t}Y)_x$ (the dotted curve in the largest figure), defined for $t \in \mathcal{D}(X,x)$ and valued in $T_x M$ for any t. In the smaller figures on the right, we see the dynamics of the picture for times closer to 0: the sets $M^X_{\pm t}$ become larger and larger, and $(\mathrm{d}\Psi^X_{-t}Y)_x$ tends to Y_x.

Theorem 4.24 (Higher order Hadamard's Formula). *Let X, Y be smooth vector fields on M and let $k \in \mathbb{N}$. Then we have*

$$\left(\frac{\mathrm{d}}{\mathrm{d}t}\right)^k \left(\mathrm{d}\Psi^X_{-t}Y\right) = \mathrm{d}\Psi^X_{-t}((\mathrm{ad}\,X)^k Y), \tag{4.24}$$

as an identity of time-dependent vector fields on $\mathcal{D}(X)$.

Consequently, for any $f \in C^\infty(M)$ and any $(t,x) \in \mathcal{D}(X)$, one has

$$\left(\frac{\mathrm{d}}{\mathrm{d}t}\right)^k \left\{Y(f \circ \Psi^X_{-t})(\Psi^X_t(x))\right\} = ((\mathrm{ad}\,X)^k Y)(f \circ \Psi^X_{-t})(\Psi^X_t(x)).$$

When $M = \Omega$ is an open subset of $\mathbb{R}^N$, the same formula holds true for C^k vector fields X, Y on Ω, and f of class C^{k+1} on Ω (real or vector valued).

The proof is a simple induction argument which we omit.

Taking into account the invariance of X under Ψ^X_{-t} and the naturality of the Lie bracket (see (4.7)), formula (4.24) is equivalent to

$$\left(\frac{\mathrm{d}}{\mathrm{d}t}\right)^k \left(\mathrm{d}\Psi^X_{-t}Y\right) = (\mathrm{ad}\,X)^k (\mathrm{d}\Psi^X_{-t}Y) \quad \text{in } T_x M,$$

holding true for any $(t, x) \in \mathcal{D}(X)$. As usual, in this formula X is considered as a v.f. on M^X_t whereas Y is viewed as a v.f. on M^X_{-t}.

4.5   Commuting vector fields

In this section we prove the following theorem:

Theorem 4.25 (Commuting vector fields). *Let X, Y be smooth vector fields on M. Then X and Y commute if and only if their flows commute; more precisely, $[X, Y] \equiv 0$ on M if and only if, for every $x \in M$, there exists $\varepsilon(x) > 0$ such that*

$$\Psi^X_t \circ \Psi^Y_s(x) = \Psi^Y_s \circ \Psi^X_t(x), \quad \text{whenever } |t|, |s| < \varepsilon(x).$$

When $M = \Omega \subseteq \mathbb{R}^N$ is open, the same holds true if X, Y are simply of class C^1.

Proof. First, let us suppose that $[X, Y] \equiv 0$ on M. Let O be any open subset of M such that $\overline{O}$ is compact. Let $\varepsilon = \varepsilon(K) > 0$ be so small that $\Psi^Y_s \circ \Psi^X_t(x)$ and $\Psi^X_t \circ \Psi^Y_s(x)$ are defined for $|s|, |t| \leq \varepsilon$ uniformly for $x \in O$. We aim to prove that

$$\Psi^Y_s \circ \Psi^X_t(x) = \Psi^X_t \circ \Psi^Y_s(x) \qquad (x \in O, \ |s|, |t| \leq \varepsilon). \tag{4.25}$$

The arbitrariness of O will then give the proof. By Prop.4.9-(5), (4.25) is equivalent to requiring that the v.f. $X|_O$ be Ψ^Y_s-related to the vector field

$$X\big|_{\Psi^Y_s(O)} \qquad \text{(for any fixed } s \text{ such that } |s| \leq \varepsilon\text{).}$$

In its turn, by the very definition of relatedness, the latter statement is equivalent to the identity $\mathrm{d}_x \Psi^Y_s(X_x) = X_{\Psi^Y_s(x)}$ for every $x \in O$ and every $s \in [-\varepsilon, \varepsilon]$. If we set $y = \Psi^Y_s(x)$, we prove this identity under the equivalent form

$$\mathrm{d}_{\Psi^Y_{-s}(y)} \Psi^Y_s\left(X_{\Psi^Y_{-s}(y)}\right) = X_y \quad \text{for every } s \in [-\varepsilon, \varepsilon] \text{ and } y \in \Psi^Y_s(O).$$

Let O' be a subset of M containing every set of the form $\Psi^Y_s(O)$ for all s in $[-\varepsilon, \varepsilon]$. We are left to prove that, for any given $y \in O'$, the function

$$[-\varepsilon, \varepsilon] \ni s \mapsto F(s) := \mathrm{d}_{\Psi^Y_{-s}(y)} \Psi^Y_s\left(X_{\Psi^Y_{-s}(y)}\right)$$

is constant (bear in mind that Ψ^Y_0 is the identity). This is a consequence of

$$F'(s) = -\mathrm{d}_{\Psi^Y_{-s}(x)} \Psi^Y_s\left([Y, X]_{\Psi^Y_{-s}(x)}\right),$$

which derives directly from formula (4.20). The assumption $[X, Y] \equiv 0$ gives $F'(s) = 0$ for every s, as desired.

Vice versa, let us suppose that (4.25) is fulfilled. We make a computation in local coordinates, so that we are allowed to apply the results of Chap. 1 on flow calculus in space. By taking $s = t$ in (4.25) we recognize that

$$\Gamma^{X,Y}_t(x) = \Psi^Y_{-t} \circ \Psi^X_{-t} \circ \Psi^Y_t \circ \Psi^X_t(x)$$

is constant, for t in a neighborhood of $t = 0$. Since $\frac{\mathrm{d}^2}{\mathrm{d}t^2}\big|_{t=0}\Gamma^{X,Y}_t(x) = 2[X, Y]_x$, as a consequence of (1.41) (page 25), we derive that $[X, Y] = 0$ on M. $\qquad\square$

4.6 Hadamard's Theorem for flows in space

When $M = \Omega$ is an open subset of $\mathbb{R}^N$, we can provide an integral version of Thm. 4.22, a straightforward consequence of (4.22) and of the Fundamental Theorem of Calculus; moreover, Thm. 4.24 and Taylor's expansion with an integral remainder give at once the following result. The results contained in this section can also be used as local results in any coordinate chart for a smooth manifold M.

Corollary 4.26. *Let X, Y be C^1 vector fields on the open set $\Omega \subseteq \mathbb{R}^N$. Then, for any $x \in \Omega$ and any $t \in \mathcal{D}(X, x)$, we have*

$$(\mathrm{d}\Psi^X_{-t} Y)(x) = Y(x) + \int_0^t \mathrm{d}\Psi^X_{-\tau}[X, Y](x)\, \mathrm{d}\tau.$$

By unraveling the definition of the pushforward, this formula states that

$$Y(f \circ \Psi^X_{-t})(\Psi^X_t(x)) = Yf(x) + \int_0^t [X, Y](f \circ \Psi^X_{-\tau})(\Psi^X_\tau(x))\, \mathrm{d}\tau,$$

for any $x \in \Omega$, any $t \in \mathcal{D}(X, x)$ and any f of class C^1 on Ω (real or vector valued).

If $k \in \mathbb{N}$ and if X, Y are C^{k+1} vector fields on Ω, for any x, t as above we have

$$(\mathrm{d}\Psi^X_{-t} Y)(x) = \sum_{j=0}^{k} ((\operatorname{ad} X)^j Y)(x)\, \frac{t^j}{j!} +$$
$$+ \frac{1}{k!} \int_0^t (t - \tau)^k \mathrm{d}\Psi^X_{-\tau}((\operatorname{ad} X)^{k+1} Y)(x)\, \mathrm{d}\tau, \tag{4.26a}$$

and, more generally, for any (real or vector valued) f of class C^{k+1} on Ω,

$$Y(f \circ \Psi^X_{-t})(\Psi^X_t(x)) = \sum_{j=0}^{k} ((\operatorname{ad} X)^j Y) f(x)\, \frac{t^j}{j!} +$$
$$+ \frac{1}{k!} \int_0^t (t - \tau)^k ((\operatorname{ad} X)^{k+1} Y)(f \circ \Psi^X_{-\tau})(\Psi^X_\tau(x))\, \mathrm{d}\tau. \tag{4.26b}$$

4.6.1 *Series expansibility*

A natural question arises: is it legitimate to let $k \to \infty$ in (4.26a) and (4.26b)? Be this legitimate, it would depend on suitable growth estimates of $(\operatorname{ad} X)^k Y$. For example, if X and Y belong to a finite dimensional Lie algebra (e.g., the Lie algebra of a Lie group) this is always the case. This fact, and much more, is contained in the following result.

The next theorem contains a strikingly surprising result of *real-analyticity*, even if the vector fields involved may not be of class C^ω.

Theorem 4.27. *Let X, Y be C^∞ vector fields on the open set $\Omega \subseteq \mathbb{R}^N$. Suppose that, for any compact set $K \subset \Omega$, there exists $\mathbf{c} = \mathbf{c}(K, X, Y) > 0$ such that*

$$\sup_{x \in K} \left\| ((\operatorname{ad} X)^j Y)(x) \right\| \leq \mathbf{c}^j, \quad \text{for every } j \in \mathbb{N}. \tag{4.27}$$

Then, for any fixed $x \in \Omega$, the function $t \mapsto (\mathrm{d}\Psi^X_{-t}Y)(x)$ is real analytic on $\mathcal{D}(X,x)$ (the maximal domain of $t \mapsto \Psi^X_t(x)$), and we have the expansion

$$(\mathrm{d}\Psi^X_{-t}Y)(x) = \sum_{j=0}^{\infty} \frac{(\operatorname{ad} tX)^j Y}{j!}(x), \qquad t \in \mathcal{D}(X,x). \qquad (4.28a)$$

More generally, for any $x \in \Omega$ and any (real or vector valued) $f \in C^{\infty}(\Omega)$, we have

$$Y(f \circ \Psi^X_{-t})(\Psi^X_t(x)) = \sum_{j=0}^{\infty} \frac{(\operatorname{ad} tX)^j Y}{j!} f(x), \qquad t \in \mathcal{D}(X,x), \qquad (4.28b)$$

the power series in t in the rhs being convergent and, therefore, defining a real-analytic function of t on $\mathcal{D}(X,x)$.

We remark the following notable fact contained in this theorem: *a priori*, the function $t \mapsto \mathrm{d}\Psi^X_{-t}(Y)(x)$ is of class C^{∞} (since this is true of X and Y); the above result ensures that this function is actually real-analytic on its domain!

In a compact (self-explained) notation, formula (4.28a) becomes

$$\mathrm{d}\Psi^X_{-t}Y = e^{\operatorname{ad} tX}Y \quad \text{on } \Omega^X_t. \qquad (4.29)$$

The latter can also be seen as an identity of time-dependent vector fields on $\mathcal{D}(X)$. It is worth observing the splendid analogy of (4.29) with Hadamard's formula (for formal power series) in (4.18).

In Chap. 13, we shall show that a geometrically meaningful sufficient condition for (4.27) to hold is when X and Y belong to a finite dimensional Lie-algebra of smooth v.f.s (see Thm. 13.1 on page 268).

Remark 4.28. If, in addition to the assumptions of Thm. 4.27, X is a global vector field, then $\Omega^X_t = \Omega$ for every $t \in \mathbb{R}$, and $\Psi^X_{-t} \equiv \Psi^{-tX}_1$ so that (4.29) is equivalent to $\mathrm{d}\Psi^{-tX}_1(Y) = e^{\operatorname{ad} tX}Y$ on the whole of Ω, for any $t \in \mathbb{R}$. In this formula it is implicitly contained the fact that the series defining $e^{\operatorname{ad} tX}Y$ in the rhs is convergent *for every* $t \in \mathbb{R}$, hence the rhs is a vector field on Ω whose coefficient functions (depending on $t \in \mathbb{R}$) are real-analytic functions of t on the entire real line $\mathbb{R}$.   ♯

Proof. Clearly, (4.28b) follows from (4.28a), since

$$Y(f \circ \Psi^X_{-t})(\Psi^X_t(x)) = \nabla(f \circ \Psi^X_{-t})(\Psi^X_t(x)) \cdot Y(\Psi^X_t(x))$$

$$= \nabla f(x) \cdot \mathcal{J}_{\Psi^X_{-t}}(\Psi^X_t(x)) \cdot Y(\Psi^X_t(x)) = \nabla f(x) \cdot Y(\Psi^X_{-t})(\Psi^X_t(x))$$

$$\overset{(4.28a)}{=} \sum_{j=0}^{\infty} \nabla f(x) \cdot ((\operatorname{ad} X)^j Y)(x) \frac{t^j}{j!} = \sum_{j=0}^{\infty} ((\operatorname{ad} X)^j Y) f(x) \frac{t^j}{j!}.$$

Since the last series is the Taylor series of $t \mapsto Y(f \circ \Psi^X_{-t})(\Psi^X_t(x))$ at $t = 0$ (see (4.26b)), this demonstrates, provided that (4.28a) is valid, the analyticity of the latter function and the expansion (4.28b) on the whole of $\mathcal{D}(X,x)$.

We are left with (4.28a). Let $x \in \Omega$ be fixed and let us set $F(t) := (\mathrm{d}\Psi^X_{-t}Y)(x)$, for $t \in \mathcal{D}(X, x)$. In order to prove the real-analyticity of F on $\mathcal{D}(X, x)$, from a well-known result on the expansibility of Taylor series (see Exr. 4.1), it suffices to prove the existence, for any fixed $T \in \mathcal{D}(X, x)$, of a constant $C_T > 0$ such that

$$\sup_{t \in [0,T]} \|F^{(j)}(t)\| \le (C_T)^j, \quad \text{for every } j \in \mathbb{N} \cup \{0\}. \tag{4.30}$$

If $T < 0$, it is understood that the interval $[0, T]$ has to be replaced (here and in the sequel) by $[T, 0]$. From (4.24) we know that

$$F^{(j)}(t) = \mathrm{d}\Psi^X_{-t}((\mathrm{ad}\, X)^j Y)(x), \quad \forall\, j \ge 0.$$

The set $K_T := \{\Psi^X_t(x) : t \in [0, T]\}$ is a compact subset of Ω, hence, by hypothesis (4.27), there exists $\mathbf{c}_T > 0$ such that

$$\left\|((\mathrm{ad}\, X)^j Y)(\Psi^X_t(x))\right\| \le (\mathbf{c}_T)^j, \quad \text{for every } j \in \mathbb{N} \text{ and } t \in [0, T]. \tag{4.31}$$

We therefore have the following computation:[4]

$$\sup_{t \in [0,T]} \|F^{(j)}(t)\| = \sup_{t \in [0,T]} \left\|((\mathrm{ad}\, X)^j Y)\Psi^X_{-t}(\Psi^X_t(x))\right\|$$

$$= \sup_{t \in [0,T]} \left\|\mathfrak{J}_{\Psi^X_{-t}}(\Psi^X_t(x)) \cdot ((\mathrm{ad}\, X)^j Y)(\Psi^X_t(x))\right\|$$

$$\overset{(4.31)}{\le} \sup_{t \in [0,T]} \sup_{z \in K_T} \left\|\left|\mathfrak{J}_{\Psi^X_{-t}}(z)\right|\right\| \cdot (\mathbf{c}_T)^j =: C \cdot (\mathbf{c}_T)^j.$$

If we show that $C < \infty$, then (4.30) will follow with the choice $C_T = \max\{1, C\} \cdot \mathbf{c}_T$, thus ending the proof. Now, the finiteness of C straightforwardly follows from Exr. 4.4. The proof is complete. $\qquad\square$

4.6.2 *Conjugation of flows*

Given two v.f.s $X, Y \in \mathfrak{X}(\Omega)$, we consider the "conjugation-like" flow composition $\Psi^X_{-t} \circ \Psi^Y_s \circ \Psi^X_t$. This is defined on a suitable open subset of Ω (depending on X, Y and on s, t), say $\Omega^{X,Y}_{s,t}$; when X and Y are global v.f.s, this set is simply Ω and the composition is defined for any $s, t \in \mathbb{R}$. From formula (4.9), we have

$$\Psi^X_{-t} \circ \Psi^Y_s \circ \Psi^X_t = \Psi^{\mathrm{d}\Psi^X_{-t}Y}_s \quad \text{on } \Omega^{X,Y}_{s,t}.$$

If X, Y satisfy the assumptions of Thm 4.27 (for example, if X, Y belong to a finite-dimensional Lie sub-algebra of $\mathfrak{X}(\Omega)$), this formula gives another declination of the Hadamard-type formulae in this chapter (see (4.18)), namely

$$\Psi^X_{-t} \circ \Psi^Y_s \circ \Psi^X_t = \Psi^{e^{\mathrm{ad}\, tX}Y}_s \quad \text{on } \Omega^{X,Y}_{s,t}. \tag{4.32}$$

For future reference, we explicitly state (4.32) under simplified assumptions:

---

[4]We follow the notation in (4.33) (page 111) for the operator norm $\||\cdot\||$.

Proposition 4.29. *Let X, Y be smooth vector fields on Ω; suppose that X and Y belong[5] to a finite-dimensional Lie sub-algebra of Ω. Suppose furthermore that X and Y are global. Then $e^{\operatorname{ad} tX} Y = \sum_{j=0}^{\infty} \frac{(\operatorname{t\,ad} X)^j}{j!} Y$ is well-posed, and*

$$\Psi_{-t}^X \circ \Psi_s^Y \circ \Psi_t^X(x) = \Psi_s^{e^{\operatorname{ad} tX} Y}(x),$$

for every $s, t \in \mathbb{R}$ and every $x \in \Omega$.

4.7 Exercises of Chap. 4

Exercise 4.1. (The Expansibility Theorem). Let $I \subseteq \mathbb{R}$ be an open interval, $t_0 \in I$ and $f \in C^\infty(I, \mathbb{R})$. Suppose that $t \in I$ is such that there exists $M_t > 0$ for which the following estimate holds:

$$\sup_{\tau \in [t_0, t]} |f^{(k)}(\tau)| \leq (M_t)^k, \quad \text{for every } k \in \mathbb{N} \cup \{0\}.$$

(Here $[t_0, t]$ must be replaced with $[t, t_0]$ when $t < t_0$.)

Then we have the Taylor expansion

$$f(t) = \sum_{k=0}^{\infty} \frac{f^{(k)}(t_0)}{k!} (t - t_0)^k.$$

[*Hint:* Apply Taylor's Formula with Lagrange's remainder...]

Exercise 4.2. Consider the vector fields $X = (1 + x_1^2)\partial_{x_1} + \partial_{x_2}$, $Y = x_2^2 \partial_{x_2}$ on the open set $\Omega = \mathbb{R}^2$. Show that

$$\Omega_t^X = \left\{ (x_1, x_2) \in \mathbb{R}^2 : -\frac{\pi}{2} - t < \arctan(x_1) < \frac{\pi}{2} - t \right\},$$

and prove that, for every $x \in \Omega_t^X$, one has $(Y \Psi_{-t}^X)(\Psi_t^X(x)) = (0, (x_2 + t)^2)$. Check the validity of formula (4.22) by showing that

$$\frac{\mathrm{d}}{\mathrm{d}t}(Y \Psi_{-t}^X)(\Psi_t^X(x)) = ([X, Y]\Psi_{-t}^X)(\Psi_t^X(x)) = (0, 2(x_2 + t)),$$

for every $x \in \mathbb{R}^2$ and for every $t \in \mathcal{D}(X, x)$.

Exercise 4.3. Prove the following facts:

(1) Let $V : [a, b] \to \mathbb{R}^N$, $V(t) = (v_1(t), \dots, v_N(t))$, where the functions v_i are real-valued and integrable on $[a, b]$; setting

$$\int_a^b V(t)\,\mathrm{d}t := \left(\int_a^b v_1(t)\,\mathrm{d}t, \dots, \int_a^b v_1(t)\,\mathrm{d}t \right),$$

prove that $\left\| \int_a^b V(t)\,\mathrm{d}t \right\| \leq \int_a^b \|V(t)\|\,\mathrm{d}t$.

Hint: Let $\xi := \int_a^b V(t)\,\mathrm{d}t$ and argue as follows:

$$\|\xi\|^2 = \left\langle \int_a^b V(t)\,\mathrm{d}t, \xi \right\rangle = \int_a^b \langle V(t), \xi \rangle\,\mathrm{d}t \leq \int_a^b \|V(t)\| \cdot \|\xi\|\,\mathrm{d}t.$$

---

[5]This is equivalent to saying that $\mathrm{Lie}\{X, Y\}$ is finite-dimensional, where $\mathrm{Lie}\{X, Y\}$ is the smallest Lie sub-algebra of $\mathfrak{X}(\Omega)$; see Def. A.9.

(2) Given any real matrix A, we use the operator norm

$$|||A||| := \sup_{\xi \in \mathbb{R}^N : \|\xi\| \leq 1} \|A\xi\|. \tag{4.33}$$

Here, $\|\cdot\|$ indifferently denotes the Euclidean norm on $\mathbb{R}^m$ for any $m \in \mathbb{N}$.

Let $A(t) = (a_{i,j}(t))_{i,j \leq N}$, where the functions $a_{i,j}$ are real-valued and integrable on $[a,b]$; setting

$$\int_a^b A(t)\,\mathrm{d}t := \left(\int_a^b a_{i,j}(t)\,\mathrm{d}t \right)_{i,j \leq N},$$

deduce from (1) above that, if $|||\cdot|||$ is the operator norm in (4.33), one has

$$\left\|\left\| \int_a^b A(t)\,\mathrm{d}t \right\|\right\| \leq \int_a^b |||A(t)|||\,\mathrm{d}t.$$

Exercise 4.4. Prove the following *a priori* bound of the derivatives of $x \mapsto \Psi_t^X(x)$ in terms of the flow itself and of the derivatives of XI:

Let X be a C^1 v.f. on Ω. Let $x \in \Omega$ and let $[a,b]$ be any compact subinterval of the maximal domain $\mathcal{D}(X,x)$. Then

$$\left\|\left\| \partial_{\Psi_t^X}(x) \right\|\right\| \leq \exp\left(\max_{\tau \in [a,b]} |||\partial_{XI}(\Psi_\tau^X(x))||| \right), \tag{4.34}$$

for every $t \in [a,b]$.

[*Hint:* From the integral equation of variation (1.29) (page 20), derive the following estimate (exploit Exr. 4.3 as well)

$$\left\|\left\| \partial_{\Psi_t^X}(x) \right\|\right\| \leq 1 + \int_0^t |||A_x(\tau)|||\cdot\left\|\left\| \partial_{\Psi_\tau^X}(x) \right\|\right\|\,\mathrm{d}\tau.$$

Observe that $K_x := \{\Psi_\tau^X(x) : \tau \in [a,b]\}$ is a compact subset of Ω, so that

$$c(x) := \max_{\tau \in [a,b]} |||A_x(\tau)||| < \infty.$$

Apply Gronwall's Lemma (see Lem. B.19 on page 365) to obtain (4.34).]

Exercise 4.5. Given a time-dependent vector field $V(t,x)$ on Ω, show that, for any $(t,x) \in \Omega$, $\frac{\mathrm{d}}{\mathrm{d}t} V(t,x)$ is a derivation at x.

[*Hint:* given $f,g \in C^\infty(M)$, prove that

$$\left(\frac{\mathrm{d}}{\mathrm{d}t} V(t,x)\right) fg = g(x)\left(\frac{\mathrm{d}}{\mathrm{d}t} V(t,x)\right) f + f(x)\left(\frac{\mathrm{d}}{\mathrm{d}t} V(t,x)\right) g,$$

recalling that $V(t,x) \in T_x M$ for any $(t,x) \in \Omega$.]

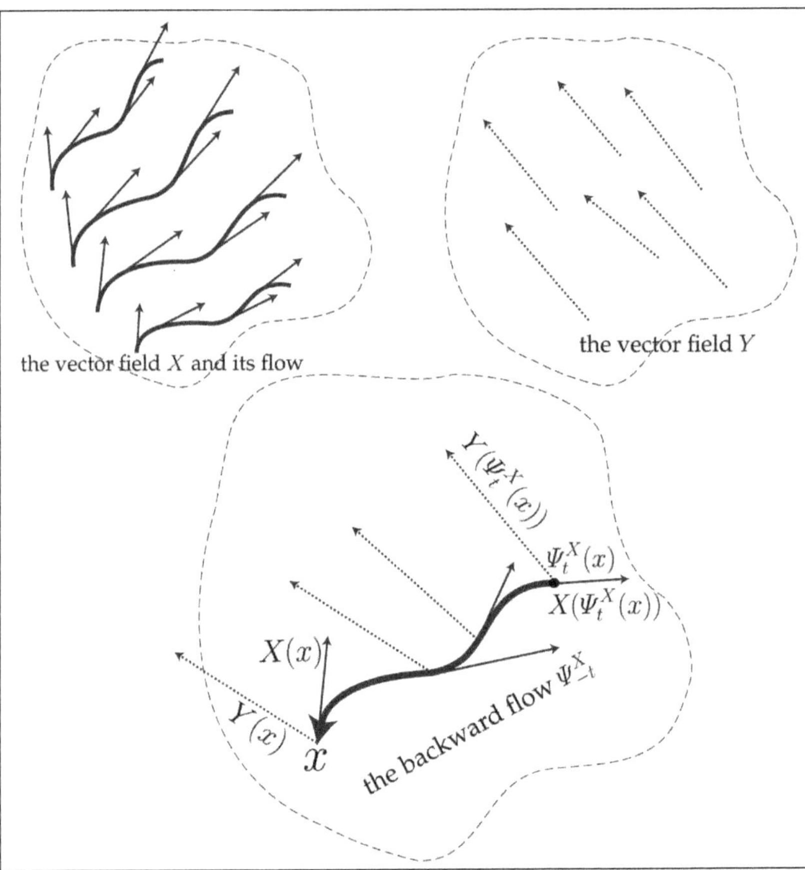

Fig. 4.6 *The function $(Y\Psi_{-t}^X)(\Psi_t^X(x))$ in the proof of Thm. 4.20.* This is well posed in local coordinates in $\mathbb{R}^N$. We know that, in general, for any $\mathbb{R}^N$-valued function f which is C^1 around z, $Yf(z)$ is the directional derivative of f at z along the vector $Y(z)$. Hence, $(Y\Psi_{-t}^X)(\Psi_t^X(x))$ can be interpreted as the directional derivative of the backward flow map Ψ_{-t}^X (the bold arrowed curve in the lowest figure) at the point $\Psi_t^X(x)$ along the vector $Y(\Psi_t^X(x))$. Thm. 4.20 states that the derivative of this function at $t = 0$ is equal to $[X,Y](x)$. See also Rem. 4.17 for a change-of-coordinate interpretation of this function.

Chapter 5

The CBHD Operation on Finite Dimensional Lie Algebras

THE aim of this chapter is to consider, *in the realm of finite-dimensional Lie algebras* $\mathfrak{g}$, the Campbell-Baker-Hausdorff-Dynkin series (CBHD series, for short), previously investigated in Chap. 2. The best result one can hope for is a *local* convergence of the CBHD series in $\mathfrak{g}$, since global convergence does not hold in general.

For example, if $A = \left(\begin{smallmatrix} 0 & -\alpha \\ \alpha & 0 \end{smallmatrix}\right)$ (with $\alpha = 5\pi/4$) and $B = \left(\begin{smallmatrix} 0 & 1 \\ 0 & 0 \end{smallmatrix}\right)$, it is not difficult to prove that there does not exist any real matrix C such that

$$e^A \, e^B = e^C.$$

See [Wei (1963)] (or Exr. 5.11). This interdicts the convergence of the homogenous CBHD series $\sum_{n=1}^{\infty} Z_n(A, B)$ defined in Chap. 2, namely

$$A \diamond B = A + B + \tfrac{1}{2}[A, B] + \tfrac{1}{12}[A, [A, B]] + \tfrac{1}{12}[B, [B, A]] + \cdots,$$

otherwise one should mandatorily[1] have $e^{A \diamond B} = e^A \, e^B$.

This chapter also contains counterparts of the *Poincaré's ODEs/PDEs* encountered in Chap. 2, here adapted to the differentiable framework of finite-dimensional Lie algebras. This is a setting where a differentiable structure is perfectly meaningful, so that these differential equations are genuine ODEs/PDEs, of a great relevance for the Lie group theory.

If x and y are close to 0, we shall prove that

$$\gamma(t) = \sum_{n=1}^{\infty} Z_n(x, y) \, t^n$$

solves in $\mathfrak{g}$ (at least when $t \in [-1, 1]$) the Poincaré ODE

$$\gamma' = \mathbf{b}(\operatorname{ad}\gamma)(x) + \mathbf{b}(-\operatorname{ad}\gamma)(y),$$

where $\mathbf{b}(z) = \frac{z}{e^z - 1}$ is the generating function of the *Bernoulli numbers*. This ODE has applications in Lie group theory (see Chaps. 14 and 17). Finally, Sec. 5.4 will show that the CBHD operation $\diamond$ has a "germ of associativity" on a small neighborhood of $0 \in \mathfrak{g}$; this important result will serve as a crucial point in solving Lie's Third Theorem in its local form (Chap. 15).

Prerequisites for this chapter are the notion and the properties of the CBHD series (Chap. 2), plus some basic theory of real-analytic functions of one real variable.

---

[1] A proof can be found in [Bonfiglioli and Fulci (2012), Sec. 5.6].

5.1 Local convergence of the CBHD series

In this section we provide an important result of local convergence of the CBHD series (see also Sec. 18.1 for an enlarged domain of convergence).

Notation 5.1. Throughout this chapter, $\mathfrak{g}$ will denote a finite-dimensional Lie algebra over $\mathbb{K}$, where $\mathbb{K}$ is $\mathbb{R}$ or $\mathbb{C}$.

If $N := \dim_{\mathbb{K}}(\mathfrak{g})$, we can consider some norm $\|\cdot\|_\mathcal{V}$ on $\mathfrak{g}$ by fixing some basis $\mathcal{V} = \{v_1, \dots, v_N\}$ of $\mathfrak{g}$ and by setting

$$\|x_1 v_1 + \cdots + x_N v_N\|_\mathcal{V} := \sqrt{|x_1|^2 + \cdots + |x_N|^2}, \quad (x_1, \dots, x_N) \in \mathbb{K}^N.$$

In an obvious way (see Exr. 5.1) we obtain an isomorphism of vector spaces and of metric spaces between $(\mathfrak{g}, \|\cdot\|_\mathcal{V})$ and the usual normed space $\mathbb{K}^N$. On account of Exr. 5.4, since the map $\mathfrak{g} \times \mathfrak{g} \ni (a, b) \mapsto [a, b] \in \mathfrak{g}$ is bilinear, in view of the finite-dimensionality of $\mathfrak{g}$ we can deduce that it is also continuous (on $\mathfrak{g}$ we fix the topology of the metric space $(\mathfrak{g}, \|\cdot\|_\mathcal{V})$, whereas on $\mathfrak{g} \times \mathfrak{g}$ we fix the associated product topology).

By the characterization of the continuous multilinear functions in Exr. 5.2, we derive the existence of a constant $M > 0$ such that

$$\big\|[a, b]\big\|_\mathcal{V} \leq M \, \|a\|_\mathcal{V} \cdot \|b\|_\mathcal{V}, \quad \text{for every } a, b \in \mathfrak{g}.$$

If we multiply both sides of this inequality by M, and if we set $\|x\| := M \, \|x\|_\mathcal{V}$ (for $x \in \mathfrak{g}$), we obtain that $\|\cdot\|$ satisfies the **Lie-sub-multiplicative inequality**

$$\big\|[a, b]\big\| \leq \|a\| \cdot \|b\|, \quad \text{for every } a, b \in \mathfrak{g}. \tag{5.1}$$

This gives the following inequality for higher order commutators:

$$\Big\|[a_1 [a_2 \cdots [a_{k-1}, a_k] \cdots]]\Big\| \leq \|a_1\| \cdot \|a_2\| \cdots \|a_k\|, \tag{5.2}$$

valid for every $k \geq 2$ and every $a_1, \dots, a_k \in \mathfrak{g}$.

Due to the equivalence of all norms on finite-dimensional vector spaces (see Exr. 5.3), the topologies of $(\mathfrak{g}, \|\cdot\|_\mathcal{V})$ and $(\mathfrak{g}, \|\cdot\|)$ are the same, so that both metric spaces have the same convergent sequences/series. In the sequel, we tacitly understand that $\|\cdot\|$ satisfies the Lie-sub-multiplicative inequality (5.1). Incidentally, we have proved the following fact:

Remark 5.2. Let $(\mathfrak{g}, [\cdot, \cdot])$ be a finite-dimensional Lie algebra over $\mathbb{K}$; then there exists a norm $\|\cdot\|$ on $\mathfrak{g}$ satisfying (5.1).   ♯

In the following statement, we remind that the homogeneous CBHD series has been introduced in Def. 2.24 on page 53.

Theorem 5.3. *If $\|\cdot\|$ is a Lie-sub-multiplicative norm on the finite-dimensional Lie algebra $\mathfrak{g}$, we consider the ball around $0 \in \mathfrak{g}$ defined by*

$$U := \Big\{ a \in \mathfrak{g} : \|a\| \leq h \Big\}, \quad \text{with } h = \frac{\log 2}{3} \approx 0.231. \tag{5.3}$$

Then the homogeneous Campbell-Baker-Hausdorff-Dynkin series $a \diamond b = \sum_{n=1}^{\infty} Z_n(a,b)$ is uniformly convergent on $U \times U$; it is also absolutely convergent[2] for any $a, b \in U$.

Furthermore, for any $i, j \geq 0$, let $C_{i,j}(a,b)$ denote the Dynkin polynomials in (2.27); the following double series are also uniformly convergent on $U \times U$

$$\sum_{i=0}^{\infty} \Big(\sum_{j=0}^{\infty} C_{i,j}(a,b) \Big), \quad \sum_{j=0}^{\infty} \Big(\sum_{i=0}^{\infty} C_{i,j}(a,b) \Big),$$

and their sum is equal to $a \diamond b$. The same is true of

$$\sum_{k=1}^{\infty} \frac{(-1)^{k+1}}{k} \sum_{(i_1,j_1),\ldots,(i_k,j_k) \neq (0,0)} \frac{(\operatorname{ad} a)^{i_1} (\operatorname{ad} b)^{j_1} \cdots (\operatorname{ad} a)^{i_k} (\operatorname{ad} b)^{j_k - 1}(b)}{i_1! j_1! \cdots i_k! j_k! (i_1 + j_1 + \cdots + i_k + j_k)},$$

where this series of infinite sums has to be interpreted in the sense of Lem. 5.10.

It is worth observing that the following proof is crucially based on the explicit Dynkin representation of $Z_n(a,b)$ and $C_{i,j}(a,b)$.

Proof. By a general result on the commutativity/associativity of absolutely convergent series (Lem. 5.10; see also Exr. 5.5), and due to the fact that

$$Z_n(a,b) = \sum_{i+j=n} C_{i,j}(a,b) \quad \text{(for any } n \in \mathbb{N}\text{)},$$

it is sufficient to prove that the following series of nonnegative real numbers

$$\sum_{k=1}^{\infty} \frac{1}{k} \sum_{\mathcal{S}_k} \frac{\sup_{a,b \in U} \big\| (\operatorname{ad} a)^{i_1} (\operatorname{ad} b)^{j_1} \cdots (\operatorname{ad} a)^{i_k} (\operatorname{ad} b)^{j_k - 1}(b) \big\|}{i_1! j_1! \cdots i_k! j_k! (i_1 + j_1 + \cdots + i_k + j_k)} \tag{5.4}$$

is finite. Here we have used the notation

$$\mathcal{S}_k := \Big\{ (i_1, j_1, \ldots, i_k, j_k) \in (\mathbb{N} \cup \{0\})^{2k} : (i_1, j_1), \ldots, (i_k, j_k) \neq (0,0) \Big\}.$$

It is important here to observe that the sufficiency of this for the uniform convergence of all the series mentioned in the theorem depends on the fact that $(\mathfrak{g}, \| \cdot \|)$ *is a complete normed space* (i.e., a Banach space), the completeness of $\mathfrak{g}$ being a consequence of its finite-dimensionality.

If $a, b \in U$, by applying (5.2) we see that (5.4) is bounded from above by

$$\sum_{k=1}^{\infty} \frac{1}{k} \sum_{(i_1,j_1),\ldots,(i_k,j_k) \neq (0,0)} \frac{\|a\|^{i_1 + \cdots + i_k} \cdot \|b\|^{j_1 + \cdots + j_k}}{i_1! j_1! \cdots i_k! j_k!}$$

$$= \sum_{k=1}^{\infty} \frac{1}{k} \Big(\sum_{(i,j) \neq (0,0)} \frac{\|a\|^i}{i!} \cdot \frac{\|b\|^j}{j!} \Big)^k = \sum_{k=1}^{\infty} \frac{1}{k} \big(e^{\|a\|} e^{\|b\|} - 1 \big)^k =: (\star).$$

---

[2] By this we mean that $\sum_n \|Z_n(a,b)\| < \infty$.

Observe that we have replaced $1/(i_1 + j_1 + \cdots + i_k + j_k)$ with 1. Now we use the well-known Maclaurin expansion (see Exr. 5.6)

$$\sum_{k=1}^{\infty} \frac{1}{k} q^k = -\log(1-q), \qquad |q| < 1. \tag{5.5}$$

We can make use of this formula for our $(\star)$, since $|e^{\|a\|} e^{\|b\|} - 1| = e^{\|a\|+\|b\|} - 1 < 1$; indeed, this is equivalent to $\|a\|+\|b\| < \log 2$, which, in its turn, is certainly ensured by $a, b \in U$ (see (5.3)). Therefore

$$(\star) = -\log\left(2 - e^{\|a\|+\|b\|}\right) = \log\left(\frac{1}{2 - e^{\|a\|+\|b\|}}\right) < \infty.$$

This gives the following interesting estimate:

$$\|a \diamond b\| \leq \log\left(\frac{1}{2 - e^{\|a\|+\|b\|}}\right), \quad \text{whenever } \|a\| + \|b\| < \ln 2. \tag{5.6}$$

Finally, if we want to pass to the sup over $a, b \in U$, we run over the above computations replacing $\|a\|, \|b\|$ with their common upper bound $h = \frac{\log 2}{3}$; this proves that the infinite sum in (5.4) is bounded from above by

$$\sum_{k=1}^{\infty} \frac{1}{k} \left(e^{2h} - 1\right)^k = \log\left(\frac{1}{2 - e^{2h}}\right) \approx 0.885.$$

We are indeed entitled to apply (5.5) since $q := e^{2h} - 1 = \sqrt[3]{4} - 1 \approx 0.587$ is such that $|q| < 1$. This ends the proof. $\qquad\square$

Remark 5.4. A closer inspection to the previous proof shows that the only crucial ingredients are:

    (1) $(\mathfrak{g}, \|\cdot\|)$ is a Banach space,
    (2) the map $\mathfrak{g} \times \mathfrak{g} \ni (a, b) \mapsto [a, b]$ is continuous.

When (1) and (2) are satisfied, one says that the triple $(\mathfrak{g}, [\cdot, \cdot], \|\cdot\|)$ is a *Banach-Lie algebra*. As a consequence, we have also proved that *the homogeneous CBHD series converges (uniformly) on a neighborhood of the origin of every Banach-Lie algebra (not necessarily finite-dimensional)*. $\qquad\sharp$

5.2   Recursive identities for Dynkin's polynomials

Let us consider the entire function $\varphi(z) := \frac{e^z - 1}{z}$ and let $\mathbf{b}$ be the reciprocal of φ, defined and holomorphic on the disk $B(0, 2\pi) \subseteq \mathbb{C}$. From Exr. 5.7 we know that

$$\mathbf{b}(z) := \frac{z}{e^z - 1} = \sum_{n=0}^{\infty} \frac{B_n}{n!} z^n \qquad (|z| < 2\pi), \tag{5.7}$$

where $\{B_n\}_{n \geq 0}$ is the sequence of the **Bernoulli numbers** (see also Exr. 5.8). Actually, since the only non-zero B_n with an odd n is $B_1 = -1/2$, one has

$$\mathbf{b}(z) = \sum_{n=0}^{\infty} \frac{B_n}{n!} z^n = 1 - \frac{z}{2} + \sum_{n=1}^{\infty} \frac{B_{2n}}{(2n)!} z^{2n}.$$

Moreover, we highlight that

$$\mathbf{b}(-z) := \frac{-z}{e^{-z} - 1} = 1 + \frac{z}{2} + \sum_{n=1}^{\infty} \frac{B_{2n}}{(2n)!} z^{2n} \qquad (|z| < 2\pi),$$

and the function $\psi(z) := \mathbf{b}(-z)$ is the so-called **Todd's function**.

Let us now consider the formal ODE in the algebra $\mathbb{K}\langle x, y\rangle[\![t]\!]$ obtained in formula (2.19), page 47:

$$D_t Z(t) = \mathbf{b}(\operatorname{ad} Z(t))(x) + \mathbf{b}(-\operatorname{ad} Z(t))(y), \tag{5.8}$$

where $Z(t) = \sum_{n=1}^{\infty} Z_n(x, y)\, t^n$ is the homogeneous CBHD series.

We want to obtain some useful recursive identities between the Lie polynomials $Z_n(x, y)$ by inserting the series expansion of $Z(t)$ in the above ODE, exactly like we did for the $C_{i,j}(x, y)$ in Thm. 2.36 on page 62.

For the lhs of (5.8) we immediately have

$$\partial_t Z(t) = \sum_{n=0}^{\infty} (n + 1) Z_{n+1}(x, y)\, t^n, \tag{5.9}$$

while for the rhs of (5.8) we get (by taking into account the Maclaurin expansion of b recalled above)

$$\mathbf{b}(\operatorname{ad} Z(t))(x) + \mathbf{b}(-\operatorname{ad} Z(t))(y) =$$

$$= x - \frac{1}{2}(\operatorname{ad} Z(t))(x) + y + \frac{1}{2}(\operatorname{ad} Z(t))(y) + \sum_{n=1}^{\infty} \frac{B_{2n}}{(2n)!}(\operatorname{ad} Z(t))^{2n}(x + y)$$

$$= x + y + \frac{1}{2} \sum_{n=1}^{\infty} [Z_n(x, y), y - x]\, t^n \tag{5.10}$$

$$+ \sum_{n=2}^{\infty} \Bigg(\sum_{\substack{2 \leqslant 2i \leqslant n \\ k_1, \ldots, k_{2i} \geqslant 1 \\ k_1 + \cdots + k_{2i} = n}} \frac{B_{2i}}{(2i)!}\, [Z_{k_1}(x, y) \cdots [Z_{k_{2i}}(x, y), x + y] \cdots] \Bigg) t^n.$$

By equating the coefficients of the same power of t appearing in the power series (5.9) and (5.10), we derive the following result.

Theorem 5.5 (The recursive identities for Dynkin's polynomials $Z_n(a, b)$). *Let $(\mathfrak{g}, [\cdot, \cdot])$ be a Lie algebra over a field $\mathbb{K}$ of characteristic zero, and let $a, b \in \mathfrak{g}$ be arbitrarily fixed. Let us denote, as usual, by $Z_n(a, b)$ the Dynkin polynomials of the homogeneous CBHD series (see Def. 2.24). Finally, let $\{B_n\}_{n\geq 0}$ be the Bernoulli numbers, obtained by the generating function $\mathbf{b}$ in (5.7).*

With all these notations we have the system of identities

$$\begin{cases} Z_1(a,b) = a+b, \\ Z_2(a,b) = \frac{1}{4}[a-b, Z_1(a,b)] \\ \\ Z_{n+1}(a,b) = \dfrac{1}{2(n+1)}\,[a-b, Z_n(a,b)] + \\ \qquad + \dfrac{1}{n+1} \displaystyle\sum_{\substack{2 \leqslant 2i \leqslant n \\ k_1,\ldots,k_{2i} \geqslant 1 \\ k_1 + \cdots + k_{2i} = n}} \dfrac{B_{2i}}{(2i)!}\,[Z_{k_1}(a,b)\cdots[Z_{k_{2i}}(a,b), a+b]\cdots]. \end{cases} \tag{5.11}$$

As we obtained the above system of identities starting from Poincaré's ODE in $\mathscr{T}[\![t]\!]$ (Thm. 2.20), we may refer to (5.11) as the **Poincaré's system for the Dynkin polynomials** in the Liea algebra $\mathfrak{g}$.

5.3 Poincaré's ODE on Lie algebras

In this section, we consider a finite-dimensional Lie algebra $(\mathfrak{g}, [\cdot, \cdot])$ over $\mathbb{R}$ equipped with a Lie-sub-multiplicative norm as in (5.1); with a little more effort, all the results of this section can be extended to any real or complex *Banach-Lie algebra* (see Rem. 5.4).

Let us fix any $z \in \mathfrak{g}$ satisfying $\|z\| < 2\pi$; we can define on the normed space $(\mathfrak{g}, \|\cdot\|)$ a linear map in the following way (see (5.7) for the B_k)

$$\begin{aligned} \mathbf{b}(\mathrm{ad}\,z) : \mathfrak{g} &\longrightarrow \mathfrak{g} \\ x &\mapsto \mathbf{b}(\mathrm{ad}\,z)(x) := \textstyle\sum_{k=0}^{\infty} \frac{B_k}{k!}\,(\mathrm{ad}\,z)^k(x). \end{aligned} \tag{5.12}$$

This definition is well-posed since the above series is absolutely convergent:

$$\sum_{k=0}^{\infty} \left\| \frac{B_k}{k!}\,(\mathrm{ad}\,z)^k(x) \right\| \;\leq\; \sum_{k=0}^{\infty} \left| \frac{B_k}{k!} \right| \cdot \|(\mathrm{ad}\,z)^k(x)\|$$

$$\overset{(5.1)}{\leq} \sum_{k=0}^{\infty} \left| \frac{B_k}{k!} \right| \cdot \|z\|^k \, \|x\| = \|x\| \sum_{k=0}^{\infty} \left| \frac{B_k}{k!} \right| \cdot \|z\|^k < \infty.$$

The last inequality is a consequence of $\|z\| < 2\pi$ and the fact that the complex function $\mathbf{b}$ in (5.7) is analytic of the complex disc or radius 2π and centre 0.

Since $(\mathfrak{g}, \|\cdot\|)$ is finite-dimensional,[3] we can consider ODEs and Cauchy problems on $\mathfrak{g}$. We fix once and for all $x, y \in \mathfrak{g}$ and, with the notation in (5.12), we consider the following Cauchy problem on $\mathfrak{g}$:

$$(\mathrm{CP}): \qquad \gamma' = \mathbf{b}(\mathrm{ad}\,\gamma)(x) + \mathbf{b}(-\mathrm{ad}\,\gamma)(y), \quad \gamma(0) = 0. \tag{5.13}$$

---

[3]Actually it is sufficient that $(\mathfrak{g}, \|\cdot\|)$ be a Banach space, provided that one knows some ODE Theory on Banach spaces.

We call the ODE in (5.13) **the Poincaré ODE on** $\mathfrak{g}$ (relative to x and y). As is explained in [Bonfiglioli and Fulci (2012), Sec. 1.1.2.2, pag.12], it was Poincaré who first introduced a similar Cauchy problem in studying what -eventually- was to be named the Baker-Campbell-Hausdorff series; actually he introduced (in the context of continuous groups of transformations) the Cauchy problem
$$\gamma' = \mathbf{b}(-\operatorname{ad}\gamma)(y), \quad \gamma(0) = x.$$
This is the reason why we name after Poincaré the results in this section, as we did in Secs. 2.2.2 and 2.6.3.

Remark 5.6. Fixing $x, y \in \mathfrak{g}$, the function of $g \in \mathfrak{g}$
$$f(g) := \mathbf{b}(\operatorname{ad} g)(x) + \mathbf{b}(-\operatorname{ad} g)(y)$$
is well posed for g in $B(0, 2\pi) := \{g \in \mathfrak{g} : \|g\| < 2\pi\}$; hence (5.13) is an *autonomous and non-linear Cauchy problem on* $\mathfrak{g}$. Moreover, $g \mapsto f(g)$ is real analytic on $B(0, 2\pi)$; as a consequence, from general results on C^ω dependence for ODEs, it follows that the maximal solution $\gamma(t)$ of (5.13) is real-analytic on its maximal domain. ♮

We have the following result.

Theorem 5.7 (Poincaré ODE on finite-dimensional Lie algebras). *Let* $\mathfrak{g}$ *be a real finite-dimensional Lie algebra. Let us set*
$$\varepsilon := \frac{1}{2} \log(2 - e^{-2\pi}) \approx 0.3461. \tag{5.14}$$
Let $\| \cdot \|$ *be any Lie-sub-multiplicative norm on* $\mathfrak{g}$*. For every* $x, y \in \mathfrak{g}$ *with* $\|x\|, \|y\| < \varepsilon$, *the Cauchy problem associated with Poincaré's ODE*
$$(\mathrm{CP}): \quad \gamma' = \mathbf{b}(\operatorname{ad}\gamma)(x) + \mathbf{b}(-\operatorname{ad}\gamma)(y), \quad \gamma(0) = 0$$
admits a (unique) maximal solution $\gamma(t)$ *whose maximal domain I contains* $[-1, 1]$*. Moreover* $\gamma(t)$ *is real-analytic in I and its Maclaurin expansion converges on* $[-1, 1]$*. Finally,* $\gamma(t)$ *is explicitly given, at least for* $t \in [-1, 1]$*, by the formula*
$$\gamma(t) = \sum_{n=1}^{\infty} Z_n(x, y)\, t^n, \tag{5.15}$$
where $Z_n(x, y)$ are the Dynkin polynomials of the homogeneous CBHD series.

Proof. We claim that, if $\{a_n\}_{n\in\mathbb{N}}$ is a sequence in $\mathfrak{g}$ such that
$$\sum_{n=1}^{\infty} \|a_n\| < 2\pi,$$
then the analytic function $\gamma : [-1, 1] \longrightarrow \mathfrak{g}$ defined by $\gamma(t) = \sum_{n=1}^{\infty} a_n\, t^n$ solves (CP) *if and only if* the coefficients a_n satisfy the following system:
$$\begin{cases} a_1 = x + y, \\[2mm] a_2 = \frac{1}{4}[x - y, a_1] \\[4mm] \quad\vdots \\[2mm] a_{n+1} = \dfrac{1}{2(n+1)}\,[x - y, a_n] + \\[4mm] \quad + \dfrac{1}{n+1} \displaystyle\sum_{\substack{2 \leq 2i \leqslant n \\ k_1,\ldots,k_{2i} \geqslant 1 \\ k_1+\cdots+k_{2i}=n}} \dfrac{B_{2i}}{(2i)!}\,[a_{k_1}\cdots[a_{k_{2i}}, x + y]\cdots]. \end{cases} \tag{5.16}$$

This is the typical *Ansatz* technique for solving an ODE, and (if need be) the details can be found in Exr. 5.13.

We now claim that it is possible to choose $\varepsilon > 0$ so small that

$$\sum_{n=1}^{\infty} \|Z_n(x,y)\| < 2\pi, \quad \text{for every } x,y \in \mathfrak{g} \text{ satisfying } \|x\|, \|y\| < \varepsilon. \tag{5.17}$$

Indeed, the estimate (5.6) can be applied, so that a sufficient condition for (5.17) to hold is that

$$\log\left(\frac{1}{2 - e^{\|x\|+\|y\|}}\right) < 2\pi, \quad \text{i.e., } \|x\| + \|y\| < \log(2 - e^{-2\pi}).$$

Hence it suffices to take ε as in (5.14). By applying the above argument for the sequence $a_n = Z_n(x,y)$ (with $\|x\|, \|y\| < \varepsilon$) we infer that the function

$$\gamma : [-1,1] \longrightarrow \mathfrak{g} \quad \gamma(t) := \sum_{n=1}^{\infty} Z_n(x,y)\, t^n$$

solves (CP) if and only if (5.16) is satisfied with $Z_n(x,y)$ replacing a_n, and the latter is true, due to Thm. 5.5. $\qquad\qquad\square$

5.3.1 *More Poincaré-type ODEs*

Arguing exactly as in the above proof, via another power-series *Ansatz*, one can prove the following result, this time by using the identities contained in Cor. 2.37 (page 63) concerning the series

$$Z(tx,y) = \sum_{i=0}^{\infty}\left(\sum_{j=0}^{\infty} C_{i,j}(x,y)\right)t^i, \quad Z(x,ty) = \sum_{j=0}^{\infty}\left(\sum_{i=0}^{\infty} C_{i,j}(x,y)\right)t^j.$$

Theorem 5.8. *Let $\mathfrak{g}$ be a real finite-dimensional Lie algebra; let $\|\cdot\|$ be any Lie-sub-multiplicative norm on $\mathfrak{g}$. Let the $C_{i,j}$ be the non-homogeneous Dynkin polynomials introduced in Def. 2.24 (page 53). Let ε be as in (5.14).*

Then, for every $x,y \in \mathfrak{g}$ satisfying $\|x\|, \|y\| < \varepsilon$, the following facts hold true.

(1) *The Cauchy problem*

$$\gamma' = \mathbf{b}(\mathrm{ad}\,\gamma)(x), \quad \gamma(0) = y \tag{5.18}$$

has a maximal solution coinciding for $t \in [-1,1]$ with the real-analytic function

$$\gamma(t) := Z(tx,y) = tx + y + \sum_{i=1}^{\infty}\left(\sum_{j=1}^{\infty} C_{i,j}(x,y)\right)t^i. \tag{5.19}$$

(2) *The Cauchy problem*

$$\mu' = \mathbf{b}(-\mathrm{ad}\,\mu)(y), \quad \mu(0) = x \tag{5.20}$$

has a maximal solution coinciding for $t \in [-1,1]$ with the real-analytic function

$$\mu(t) := Z(x,ty) = x + ty + \sum_{j=1}^{\infty}\left(\sum_{i=1}^{\infty} C_{i,j}(x,y)\right)t^j. \tag{5.21}$$

*We refer to the ODEs in (5.18) and in (5.20) respectively as **the first and second Poincaré's ODEs on** $\mathfrak{g}$.*

The well-posedness (for small $\|x\|, \|y\|$) of the inner series in (5.21) and (5.19) is a consequence of Thm. 5.3.

5.4 The local associativity of the CBHD series

The aim of this section is to prove the following theorem, which will play a central role in establishing the *local* Third Theorem of Lie, in Chap. 15.

Theorem 5.9 (Local associativity of the CBHD operation). *Let $\mathfrak{g}$ be a real finite-dimensional Lie algebra. Let $\| \cdot \|$ be a Lie-sub-multiplicative norm on $\mathfrak{g}$. Let*

$$\Omega := \{a \in \mathfrak{g} : \|a\| < \varepsilon\}, \quad \text{with } \varepsilon := \frac{1}{2} \log \left(2 - \frac{1}{\sqrt[3]{2}} \right) \approx 0.0937. \tag{5.22}$$

As usual, $a \diamond b = \sum_n Z_n(a, b)$ is the sum of the homogeneous CBHD series, when this is defined. Then, for any $a, b \in \Omega$, we have $a \diamond b \in U$, where U is as in (5.3). Thus, $a \diamond (b \diamond c)$ and $(a \diamond b) \diamond c$ are both well posed for $a, b, c, \in \Omega$.

Most importantly, the following local associativity of $\diamond$ holds true:

$$a \diamond (b \diamond c) = (a \diamond b) \diamond c, \quad \text{for every } a, b, c \in \Omega. \tag{5.23}$$

The ball Ω as it may seem small, we shall derive great results from it!

Proof. Since ε in (5.22) is less than h in (5.3), we have $\Omega \subset U$. Thus, $a \diamond b$ is well posed for every $a, b \in \Omega$. Moreover, for any such a and b we have[4]

$$\|a \diamond b\| \overset{(5.6)}{\leq} \log \left(\frac{1}{2 - e^{\|a\| + \|b\|}} \right) \leq \log \left(\frac{1}{2 - e^{2\varepsilon}} \right) \leq \frac{\log 2}{3},$$

with the best ε chosen as in (5.22). This proves that $a \diamond b \in U$ for any $a, b \in \Omega$.

Next we turn to the local associativity. This must be a consequence of some "inborn" properties of the summands in the CBHD series, so an algebraic digression is needed; we split the rest of the proof in three steps.

STEP I: We first prove the associative property of $\diamond$ in an appropriate abstract setting. To this end, we consider $\mathbb{Q}\langle\langle x, y, z \rangle\rangle$, the unital associative algebra (over $\mathbb{Q}$) of the formal series in three non-commuting indeterminates x, y, z; we shortly denote this algebra by A. One can define Exp and Log maps in the obvious ways, respectively defined on A_+ and $1 + A_+$, where A_+ is the set of the elements of A lacking the zero-degree term.

Exactly as in Chap. 2, one can prove the Exponential Theorem on A:

$$\mathrm{Log}(\mathrm{Exp}(p) \mathrm{Exp}(q)) = \sum_{n=1}^{\infty} Z_n(p, q), \quad \forall \, p, q \in A_+, \tag{5.24}$$

---

[4]Note that $x \mapsto \log \left(\dfrac{1}{2 - e^x} \right)$ is monotone non-decreasing on $(-\infty, \ln 2)$.

where Z_n are the Dynkin polynomials (associated with the Lie algebra of A). If we shift to the notation '$p \diamond q$' for the two (equal) sides of (5.24), we see that (due to the multiplicative nature of the lhs) $\diamond$ is associative on A_+. Indeed,

$$p \diamond (q \diamond r) = \mathrm{Log}\big(\mathrm{Exp}(p)\,\mathrm{Exp}(q \diamond r)\big) = \mathrm{Log}\big(\mathrm{Exp}(p)\,(\mathrm{Exp}(q)\,\mathrm{Exp}(r))\big)$$
$$= \mathrm{Log}\big((\mathrm{Exp}(p)\,\mathrm{Exp}(q))\,\mathrm{Exp}(r)\big) = (p \diamond q) \diamond r.$$

Since x, y, z belong to A_+, we get in particular

$$x \diamond (y \diamond z) = (x \diamond y) \diamond z. \tag{5.25}$$

We can project this identity on the vector space $A_{i,j,k}$, which is the subset of A of the polynomials with degree i wrt x, j wrt y, k wrt z; this follows from the obvious decomposition $A = \prod_{i,j,k \geq 0} A_{i,j,k}$.

The Lie-nature of $\diamond$ easily proves the existence of uniquely defined *Lie-polynomials* $U_{i,j,k}, V_{i,j,k} \in A_{i,j,k}$ such that

$$x \diamond (y \diamond z) = \sum_{i,j,k \geq 0} U_{i,j,k}(x, y, z) \quad \text{and}$$
$$(x \diamond y) \diamond z = \sum_{i,j,k \geq 0} V_{i,j,k}(x, y, z). \tag{5.26}$$

In its turn, (5.25) ensures that

$$U_{i,j,k}(x, y, z) = V_{i,j,k}(x, y, z), \quad \text{for every } i, j, k \geq 0. \tag{5.27}$$

From the universal property of the free Lie algebra $\mathcal{L}(x, y, z)$ (the Lie polynomials in the indeterminates x, y, z; see e.g., [Bonfiglioli and Fulci (2012), Sec. 2.2]) we see that, for any real Lie algebra $\mathfrak{g}$, there exist well defined maps

$$U_{i,j,k}, V_{i,j,k} : \mathfrak{g} \times \mathfrak{g} \times \mathfrak{g} \longrightarrow \mathfrak{g}, \qquad (a, b, c) \mapsto U_{i,j,k}(a, b, c), V_{i,j,k}(a, b, c),$$

obtained by replacing x, y, z with any $a, b, c \in \mathfrak{g}$ (and inserting the Lie bracket of $\mathfrak{g}$ in place of the trivial bracket between polynomials). Identity (5.27) yields

$$U_{i,j,k} \equiv V_{i,j,k} \quad \text{on } \mathfrak{g} \times \mathfrak{g} \times \mathfrak{g}, \quad \text{for every } i, j, k \geq 0. \tag{5.28}$$

If we prove that

$$\sum_{i,j,k \geq 0} \|U_{i,j,k}(a, b, c)\| < \infty \quad \text{for every } a, b, c \in \Omega, \tag{5.29}$$

then (due to (5.28)) the same is true with $V_{i,j,k}$ replacing $U_{i,j,k}$; thus (owing to Lem. 5.10), the triple series $\sum_{i,j,k} U_{i,j,k}(a, b, c)$ and $\sum_{i,j,k} V_{i,j,k}(a, b, c)$ will support any kind of commutation/association of their summands, so that (5.26) will straightforwardly imply our thesis (5.23).

STEP II: We are left to prove (5.29); to this end we need a representation of $U_{i,j,k}(x, y, z)$ *à la* Dynkin, which can be obtained with some patience starting from a representation of $x \diamond (y \diamond z)$ in terms of Dynkin polynomials. It is not difficult to verify that the following (awkward!) formula holds; here we avoid multiple subindices, and we allow ourselves the liberty to use repeatedly the index k even if in every line the notation should be changed:

$$x \diamond (y \diamond z) =$$

$$\sum_{k \geq 1} \frac{(-1)^{k+1}}{k} \sum_{(i_1,j_1),\ldots,(i_k,j_k) \neq (0,0)} \frac{1}{i_1! j_1! \cdots i_k! j_k! (i_1 + j_1 + \cdots + i_k + j_k)}$$

$$\sum_{k \geq 1} \frac{(-1)^{k+1}}{k} \sum_{(a_1,\alpha_1),\ldots,(a_k,\alpha_k) \neq (0,0)} \frac{1}{a_1! \alpha_1! \cdots a_k! \alpha_k! (a_1 + \alpha_1 + \cdots + a_k + \alpha_k)}$$

$$\ddots \sum_{k \geq 1} \frac{(-1)^{k+1}}{k} \sum_{(b_1,\beta_1),\ldots,(b_k,\beta_k) \neq (0,0)} \frac{1}{b_1! \beta_1! \cdots b_k! \beta_k! (b_1 + \beta_1 + \cdots + b_k + \beta_k)}$$

$$\vdots \qquad \vdots \qquad \vdots$$

$$\sum_{k \geq 1} \frac{(-1)^{k+1}}{k} \sum_{(c_1,\gamma_1),\ldots,(c_k,\gamma_k) \neq (0,0)} \frac{1}{c_1! \gamma_1! \cdots c_k! \gamma_k! (c_1 + \gamma_1 + \cdots + c_k + \gamma_k)}$$

$$\ddots \sum_{k \geq 1} \frac{(-1)^{k+1}}{k} \sum_{(d_1,\delta_1),\ldots,(d_k,\delta_k) \neq (0,0)} \frac{1}{d_1! \delta_1! \cdots d_k! \delta_k! (d_1 + \delta_1 + \cdots + d_k + \delta_k)}$$

$$\underbrace{[x \cdots [x}_{i_1 \text{ times}}$$

$$\left. \begin{array}{c} \Big[\underbrace{y \cdots [y}_{a_1} \underbrace{[z \cdots [z}_{\alpha_1} \ldots \ldots \underbrace{[y \cdots [y}_{a_k} \underbrace{[z \cdots [z, z]}_{\alpha_k} \cdots \Big] \\[1em] \ddots \Big[\underbrace{y \cdots [y}_{b_1} \underbrace{[z \cdots [z}_{\beta_1} \ldots \ldots \underbrace{[y \cdots [y}_{b_k} \underbrace{[z \cdots [z, z]}_{\beta_k} \cdots \Big] \end{array} \right\} j_1 \text{ times}$$

$$\vdots \qquad \vdots \qquad \vdots$$

$$\underbrace{[x \cdots [x}_{i_k \text{ times}}$$

$$\left. \begin{array}{c} \Big[\underbrace{y \cdots [y}_{c_1} \underbrace{[z \cdots [z}_{\gamma_1} \ldots \ldots \underbrace{[y \cdots [y}_{c_k} \underbrace{[z \cdots [z, z]}_{\gamma_k} \cdots \Big] \\[1em] \ddots \Big[\underbrace{y \cdots [y}_{d_1} \underbrace{[z \cdots [z}_{\delta_1} \ldots \ldots \underbrace{[y \cdots [y}_{d_k} \underbrace{[z \cdots [z, z]}_{\delta_k} \cdots \Big] \end{array} \right\} j_k \text{ times.}$$

Then, $U_{i,j,k}(x,y,z)$ is the polynomial obtained by selecting, from the above series, the summands where x, y, z appear (resp.) i, j, k times.

STEP III: Suppose we have written $U_{i,j,k}(x,y,z)$, and we have faithfully replaced x, y, z with a, b, c and we have accordingly inserted the Lie bracketing of $\mathfrak{g}$.

We are finally ready to obtain an upper estimate for the series in (5.29): to this end, we obtain a majorant series if we substitute $1/k$ for $(-1)^{k+1}/k$, and if we further replace with a 1 all the fractions

$$\frac{1}{i_1 + j_1 + \cdots + i_k + j_k}, \quad \frac{1}{a_1 + \alpha_1 + \cdots + a_k + \alpha_k}, \ldots;$$

finally, we use the sub-multiplicative inequality (5.2) of $\|\cdot\|$. This all leads to the majorant series

$$\sum_{k \geq 1} \frac{1}{k} \sum_{(i_1,j_1),\ldots,(i_k,j_k) \neq (0,0)} \frac{\|a\|^{i_1+\cdots+i_k}}{i_1! j_1! \cdots i_k! j_k!}$$

$$\left(\sum_{h \geq 1} \frac{1}{h} \sum_{(r_1,s_1),\ldots,(r_h,s_h) \neq (0,0)} \frac{\|b\|^{r_1+\cdots+r_h} \|c\|^{s_1+\cdots+s_h}}{r_1! s_1! \cdots r_h! s_h!} \right)^{j_1+\cdots+j_k}.$$

Arguing as in the proof of Thm. 5.3, the series in parentheses is equal to

$$\log \left(\frac{1}{2 - e^{\|b\|+\|c\|}} \right), \quad \text{since } \|b\| + \|c\| < \ln 2.$$

As a consequence, our majorant series is

$$\sum_{k \geq 1} \frac{1}{k} \sum_{(i_1,j_1),\ldots,(i_k,j_k) \neq (0,0)} \frac{\|a\|^{i_1+\cdots+i_k}}{i_1! j_1! \cdots i_k! j_k!} \left(\log \left(\frac{1}{2 - e^{\|b\|+\|c\|}} \right) \right)^{j_1+\cdots+j_k}.$$

For the same reason, once we know that

$$\|a\| + \log \left(\frac{1}{2 - e^{\|b\|+\|c\|}} \right) < \ln 2, \tag{5.30}$$

the majorant series becomes

$$\log \left(\frac{1}{2 - \dfrac{e^{\|a\|}}{2 - e^{\|b\|+\|c\|}}} \right) < \infty.$$

It is a simple exercise, left to the interested reader, to check that (5.30) holds true with our choice $\|a\|, \|b\|, \|c\| < \varepsilon$, with ε as in (5.22). $\qquad\square$

5.5   Appendix: multiple series in Banach spaces

In this chapter, we have used of the following Real Analysis lemma, on the commutativity/associativity/reordering of absolutely series in a Banach space. Roughly put, the following result is an analog of Fubini-Tonelli's theorem for multiple integrals of absolutely integrable functions (and the counting measure).

Lemma 5.10. *Let $(X, \|\cdot\|)$ be a Banach space. Let $\mathcal{S} \subset X$ be a denumerable set, and let $\mathcal{S}$ be indexed over a (denumerable) set of indices $\mathcal{A}$: $\mathcal{S} = \{s_a : a \in \mathcal{A}\}$. Suppose that, for some enumeration $\{a_n : n \in \mathbb{N}\}$ of $\mathcal{S}$ we have*

$$\sum_{n=1}^{\infty} \|a_n\| < \infty. \tag{5.31}$$

*In this case we say that $\sum_{s \in S} s$ (or the series $\sum_n a_n$) is **absolutely convergent**.*

Then there exists $\ell \in X$, indifferently denoted by

$$\ell = \sum_{s \in S} s \quad or \quad \ell = \sum_{a \in A} s_a, \tag{5.32}$$

*satisfying the **strong commutativity** property: for any bijection $\pi : \mathbb{N} \to S$, it holds that*

$$\ell = \sum_{n=1}^{\infty} \pi(n).$$

*Moreover, the following **strong associativity** property holds as well: for any denumerable family $\{S_n\}_{n \in \mathbb{N}}$ of (finite or denumerable) subsets of S partitioning S (i.e., they are pairwise disjoint and their union is S), one has*

$$\ell = \sum_{n \in \mathbb{N}} \left(\sum_{s \in S_n} s \right).$$

Here, for any n, $\sum_{s \in S_n} s$ is unambiguously defined according to the notation in (5.32)

Lem. 5.10 allows to manipulate in a very simple way double series, triple series, etc., as is shown in Exm.s 5.11 and 5.12.

Proof. Condition (5.31) ensures that the sequence of the partial sums of the series $\sum_n a_n$ is a Cauchy sequence in X:

$$\left\| \sum_{n=m}^{m+p} a_n \right\| \leq \sum_{n=m}^{m+p} \|a_n\| \longrightarrow 0, \quad \text{as } m \to \infty, \text{ for any } p.$$

Hence $\sum_n a_n$ is convergent, to ℓ say, since X is Banach.

To prove the strong commutativity, we slightly change the notation: let us consider a bijection $\pi : \mathbb{N} \to \mathbb{N}$, and let us prove that $\sum_n a_{\pi(n)}$ converges to ℓ. Due to (5.31) and $\ell = \sum_n a_n$, given $\varepsilon > 0$ there exists $n(\varepsilon)$ such that

$$\left\| \ell - \sum_{n \leq n(\varepsilon)} a_n \right\| < \varepsilon \quad \text{and} \quad \sum_{n > n(\varepsilon)} \|a_n\| < \varepsilon. \tag{5.33}$$

There clearly[5] exists $m(\varepsilon) \gg 1$ such that

$$\{a_n : n \leq n(\varepsilon)\} \subseteq \{a_{\pi(m)} : m \leq m(\varepsilon)\}. \tag{5.34}$$

As a consequence of (5.33) and (5.34), for any $p \in \mathbb{N}$ we get

$$\left\| \ell - \sum_{m=1}^{m(\varepsilon)+p} a_{\pi(m)} \right\| \leq \left\| \ell - \sum_{n \leq n(\varepsilon)} a_n \right\| + \sum_{n > n(\varepsilon)} \|a_n\| < 2\varepsilon.$$

This proves that $\ell = \sum_{m=1}^{\infty} a_{\pi(m)}$. We next turn to the strong associativity, and we take a partition $\{S_n\}_n$ of S. Clearly, for any fixed $n \in \mathbb{N}$, we have

$$\sum_{s \in S_n} \|s\| \leq \sum_{n=1}^{\infty} \|a_n\| < \infty,$$

---

[5]It obviously suffices to take $m(\varepsilon) := \max\{\pi^{-1}(1), \pi^{-1}(2), \ldots, \pi^{-1}(n(\varepsilon))\}$.

so that any (possibly) infinite sum $\sum_{s \in \mathcal{S}_n} s$ (adopting the notation in (5.32)) converges in X, to ℓ_n say, whatever the chosen enumeration of $\mathcal{S}_n$; this follows from what we have already proved. We need to show that $\ell = \sum_{n=1}^{\infty} \ell_n$.

Let us use a double-index notation, and we enumerate $\mathcal{S}_n$ as follows (when some $\mathcal{S}_n$ is finite, we artificially add zero summands)

$$\mathcal{S}_n = \{a_{n,j} : j \in \mathbb{N}\}, \quad n \in \mathbb{N}.$$

With this notation we have $\ell_n = \sum_{j=1}^{\infty} a_{n,j}$ for any n. Since $\{\mathcal{S}_n\}_n$ is a partition of $\mathcal{S}$, the doubly infinite matrix

$$
\begin{array}{lllll}
\mathcal{S}_1 : & a_{1,1} & a_{1,2} & a_{1,3} & \cdots \\
\mathcal{S}_2 : & a_{2,1} & a_{2,2} & a_{2,3} & \cdots \\
\mathcal{S}_3 : & a_{3,1} & a_{3,2} & a_{3,3} & \cdots \\
& \vdots & \vdots & \vdots & \ddots
\end{array}
$$

exhibits –once– all the elements of $\mathcal{S}$; we enumerate the elements of this matrix in a square-shaped way:

$$a_{1,1}, \qquad a_{2,1}, a_{2,2}, a_{1,2}, \qquad a_{3,1}, a_{3,2}, a_{3,3}, a_{2,3}, a_{1,3}, \qquad \cdots$$

We denote by $(a_n)_n$ the resulting sequence; for example, note that $a_1, \ldots, a_{n^2}$ contain precisely the first n elements of the first n rows of the matrix (i.e., the first n elements of $\mathcal{S}_1, \ldots, \mathcal{S}_n$).

From the first part of the proof, we have $\ell = \sum_{n=1}^{\infty} a_n$ and $\sum_n \|a_n\| < \infty$; as a consequence, if $\varepsilon > 0$ is fixed, there exists $n(\varepsilon)$ such that

$$\forall\, N \geq n(\varepsilon) \quad \text{one has} \quad \left\| \ell - \sum_{n=1}^{N} a_n \right\| < \varepsilon \quad \text{and} \quad \sum_{n=N+1}^{\infty} \|a_n\| < \varepsilon. \tag{5.35}$$

On the other hand, from $\ell_n = \sum_{j=1}^{\infty} a_{n,j}$ and from $\sum_{j=1}^{\infty} \|a_{n,j}\| < \infty$, we deduce that, for any fixed $n \in \mathbb{N}$, there exists $j(n, \varepsilon)$ such that

$$\left\| \ell_n - \sum_{j=1}^{j(n,\varepsilon)} a_{n,j} \right\| < \frac{\varepsilon}{2^n} \quad \text{and} \quad \sum_{j=j(n,\varepsilon)+1}^{\infty} \|a_{n,j}\| < \frac{\varepsilon}{2^n}.$$

From this, if we choose $J(n, \varepsilon) := \max\{j(1,\varepsilon), j(2,\varepsilon), \ldots, j(n,\varepsilon), n\}$, we have

$$\left\| \ell_1 + \cdots + \ell_n - \left(\sum_{j=1}^{J(n,\varepsilon)} a_{1,j} + \cdots + \sum_{j=1}^{J(n,\varepsilon)} a_{n,j} \right) \right\| < \frac{2\varepsilon}{2} + \cdots + \frac{2\varepsilon}{2^n} < 2\varepsilon.$$

In its turn, the distance of the sum (of sums) in the above parentheses from

$$a_1 + \cdots + a_{n^2} \quad \text{is less than} \quad \|a_{n^2+1}\| + \|a_{n^2+2}\| + \cdots < \varepsilon,$$

the last inequality following from (5.35), provided that $n^2 \geq n(\varepsilon)$. Again from (5.35), the distance from ℓ and $a_1 + \cdots + a_{n^2}$ is less than ε, with the same choice $n^2 \geq n(\varepsilon)$. Summing up, we have proved that, whenever $n^2 \geq n(\varepsilon)$, we have

$$\|\ell_1 + \cdots + \ell_n - \ell\| < 4\varepsilon,$$

and this ends the proof. $\qquad\qquad\qquad\qquad\qquad\qquad\qquad\qquad\qquad\qquad \square$

Essentially, the above proof hides an argument based on the interchange of two limits, with some kind of "uniformity" of one limit wrt the other (here uniformity is played by (5.35)); see Exr. 5.9.

Example 5.11 (Triple series). Let $\{c_{i,j,k} : i,j,k \geq 0\}$ is a family in the Banach space X satisfying $\sum_{i,j,k \geq 0} \|c_{i,j,k}\| < \infty$. Since $\mathcal{S} := \mathbb{N} \cup \{0\})^3$ is denumerable, we are allowed to use, in an unambiguous way, the notation $\ell = \sum_{i,j,k \geq 0} c_{i,j,k}$, where ℓ is the unique element of X with the meaning in Lem. 5.10.

With the partition

$$(\mathbb{N} \cup \{0\})^3 = \bigcup_{i \in \mathbb{N} \cup \{0\}} \mathcal{S}_i, \quad \text{with } \mathcal{S}_i = \{(i,j,k) : j,k \in \mathbb{N} \cup \{0\}\},$$

and since any $\mathcal{S}_i$ can be further partitioned as

$$\mathcal{S}_i = \bigcup_{j \in \mathbb{N} \cup \{0\}} \mathcal{S}_{i,j}, \quad \text{with } \mathcal{S}_{i,j} = \{(i,j,k) : k \in \mathbb{N} \cup \{0\}\},$$

if we apply twice the strong associativity in Lem. 5.10 we have

$$\sum_{i,j,k \geq 0} c_{i,j,k} = \sum_{i \geq 0} \left(\sum_{j,k \geq 0} c_{i,j,k} \right) = \sum_{i \geq 0} \left(\sum_{j \geq 0} \left(\sum_{k \geq 0} c_{i,j,k} \right) \right).$$

Here we also tacitly apply the strong commutativity in Lem. 5.10: indeed, when we write for example $\sum_{k \geq 0} c_{i,j,k}$, we mean that (for fixed $i,j \in \mathbb{N} \cup \{0\}$) we are allowed to enumerate $\mathcal{S}_{i,j}$ at will: we can use $\pi : \mathbb{N} \cup \{0\} \to \mathcal{S}_{i,j}$ defined by $\pi(k) := (i,j,k)$, and $\sum_{(i,j,k) \in \mathcal{S}_{i,j}} c_{i,j,k}$ equals the sum of $\sum_{k=0}^{\infty} c_{\pi(k)} = \sum_{k=0}^{\infty} c_{i,j,k}$. ♯

Example 5.12 (Double series). Let $\{c_{i,j} : i,j \geq 0\}$ be a family in the Banach space X satisfying $\sum_{i,j \geq 0} \|c_{i,j}\| < \infty$. As in Exm. 5.11, we make the partition

$$(\mathbb{N} \cup \{0\})^2 = \bigcup_{i \in \mathbb{N} \cup \{0\}} \mathcal{S}_i, \quad \text{with } \mathcal{S}_i = \{(i,j) : j \in \mathbb{N} \cup \{0\}\},$$

and, by the associativity in Lem. 5.10, there exists a unique $\ell \in X$ such that

$$\ell = \sum_{i,j \geq 0} c_{i,j} = \sum_{i \geq 0} \left(\sum_{j \geq 0} c_{i,j} \right).$$

Again we are tacitly applying the strong commutativity property: indeed, by $\sum_{j \geq 0} c_{i,j}$, we mean that (for a fixed $i \in \mathbb{N} \cup \{0\}$) we can freely enumerate $\mathcal{S}_i$ via the bijection $\pi : \mathbb{N} \cup \{0\} \to \mathcal{S}_i$ defined by $\pi(j) := (i,j)$, and $\sum_{(i,j) \in \mathcal{S}_i} c_{i,j}$ is equal to the sum of the series $\sum_{j=0}^{\infty} c_{\pi(j)} = \sum_{j=0}^{\infty} c_{i,j}$. Owing to Lem. 5.10, we can perform summation in other ways, for example $\ell = \sum_{n=0}^{\infty} (\sum_{i+j=n} c_{i,j})$. Indeed, the last is obtained by strong associativity with the following choice

$$\mathcal{S}_n = \{c_{n,0}, c_{n-1,1}, c_{n-2,2}, \ldots, c_{0,n}\}, \quad n \in \mathbb{N}.$$

5.6   Exercises of Chap. 5

Exercise 5.1. Let V be a finite-dimensional vector space over $\mathbb{K}$ (where $\mathbb{K}$ is $\mathbb{R}$ or $\mathbb{C}$). Let $\mathcal{V} = \{v_1, \ldots, v_n\}$ be a basis of V. Prove that

$$\|x_1 v_1 + \cdots + x_n v_n\|_{\mathcal{V}} := \sqrt{|x_1|^2 + \cdots + |x_n|^2} \qquad (x_1, \ldots, x_n \in \mathbb{K})$$

defines a norm on V. Observe that, with this choice, the map

$$\mathbb{K}^n \ni (x_1, \ldots, x_n) \mapsto x_1 v_1 + \cdots + x_n v_n \in V$$

is both an isomorphism of vector spaces and an isomorphism of the metric spaces $(V, \|\cdot\|_{\mathcal{V}})$ and $\mathbb{K}^n$ (with the ordinary Euclidean or Hermitian metric).

Exercise 5.2. Let $(V, \|\cdot\|)$ be a normed vector space (over $\mathbb{R}$ or $\mathbb{C}$). Let $k \in \mathbb{N}$ and suppose that the map

$$F : \underbrace{V \times \cdots \times V}_{k\ \text{times}} \longrightarrow V$$

is a k-linear map. Prove that F is continuous (wrt the obvious topologies) if and only if there exists $M > 0$ such that

$$\|F(v_1, \ldots, v_k)\| \leq M \, \|v_1\| \cdots \|v_k\|, \qquad \forall \, v_1, \ldots, v_k \in V.$$

[*Hint:* For the 'if' part of the proof, it may be helpful to use the identity

$$F(v_1, \ldots, v_k) - F(w_1, \ldots, w_k) = \sum_i F(w_1, \ldots, w_{i-1}, v_i - w_i, v_{i+1}, \ldots, v_k).$$

(Prove this identity!) For the 'only if' part of the proof, first prove the existence of $\varepsilon > 0$ such that

$$F\left(\underbrace{B(0, \varepsilon) \times \cdots \times B(0, \varepsilon)}_{k\ \text{times}} \right) \subseteq B(0, 1).$$

(Here $B(0, r)$ is the metric ball in $(V, \|\cdot\|)$ with radius r and centre 0.)
     Then prove and use the following identity

$$\|F(v_1, \ldots, v_k)\| = \left(\frac{2}{\varepsilon}\right)^k \|v_1\| \cdots \|v_k\| \cdot \left\| F\left(\frac{\varepsilon}{2} \frac{v_1}{\|v_1\|}, \ldots, \frac{\varepsilon}{2} \frac{v_k}{\|v_k\|} \right) \right\|,$$

valid for $v_1, \ldots, v_k \neq 0$...]

Exercise 5.3. Let V be a finite-dimensional vector space over $\mathbb{K}$ (where $\mathbb{K}$ is $\mathbb{R}$ or $\mathbb{C}$). Prove that any two norms on V are equivalent; that is, if $\|\cdot\|_i$ $(i = 1, 2)$ are two norms on V, there exist positive constants C_i $(i = 1, 2)$ such that

$$C_1 \|v\|_1 \leq \|v\|_2 \leq C_2 \|v\|_1, \qquad \forall \, v \in V.$$

For the proof, follow the steps below:

     (1) It suffices to assume that $\|\cdot\|_1$ is the norm $\|\cdot\|_{\mathcal{V}}$ in Exr. 5.1, associated with some basis $\mathcal{V} = \{v_1, \ldots, v_n\}$ of V (why?).

(2) Prove that the function
$$\mathbb{K}^n \ni (x_1, \ldots, x_n) \mapsto \|x_1 v_1 + \ldots + x_n v_n\|_2$$
is continuous (with respect to the usual topology on $\mathbb{K}^n$).

(3) Let $K := \{\xi \in V : \|\xi\|_V = 1\}$. From the continuity of the function in the previous point, derive the existence of strictly positive constants C_i ($i = 1, 2$) such that $C_1 \leq \|\xi\|_2 \leq C_2$ for every $\xi \in K$.

(4) Deduce that $C_1 \|v\|_V \leq \|v\|_2 \leq C_2 \|v\|_V$ for every $v \in V$.

Exercise 5.4. Let $(V, \|\cdot\|)$ be a finite-dimensional normed vector space (over $\mathbb{R}$ or $\mathbb{C}$). Let $F : V \times V \to V$ be bilinear. Prove that F is continuous. (An analogous result is valid for any k-linear map F.)

[*Hint:* Use the result in Exr. 5.2. In order to apply that result, first show that, if $\{v_1, \ldots, v_n\}$ is a basis of V, one has
$$\left\| F(x_1 v_1 + \cdots + x_n v_n, y_1 v_1 + \cdots + y_n v_n) \right\| \leq M \sum_{i,j} |x_i| \cdot |y_j|,$$
where $M = \max_{i,j} \|F(v_i, v_j)\|$. In order to end the proof it may be useful to use the equivalence of all norms, see Exr. 5.3 (and maybe you may benefit of the Cauchy-Schwarz inequality...).]

Exercise 5.5. Let $(X, \|\cdot\|)$ be a Banach space (i.e., a complete normed vector space) over $\mathbb{R}$ or $\mathbb{C}$. Let A be any set and suppose we are given a sequence of functions $f_n : A \to X$ (with $n \in \mathbb{N}$). Suppose that $\sum_{n=1}^{\infty} \sup_{a \in A} \|f_n(a)\| < \infty$. Prove that the series $\sum_{n=1}^{\infty} f_n(a)$ is convergent in X for every $a \in A$. Finally show that the series of functions $\sum_{n=1}^{\infty} f_n$ is uniformly convergent on A to the (well-given) function
$$f : A \longrightarrow X, \qquad f(a) := \sum_{n=1}^{\infty} f_n(a), \quad a \in A.$$

Exercise 5.6. Prove (5.5) by means of the well-known expansion of $\log(1 + q)$, or by the following alternative argument:

starting from the sum of the geometric series $\sum_{k=0}^{\infty} x^k$ (with $|x| < 1$), integrate by series, after showing that this is possible in view of the uniform convergence of this series on intervals of the form $[0, q]$ or $[q, 0]$, with $|q| < 1$...

Exercise 5.7. (Bernoulli numbers). Consider $\varphi(z) = \frac{e^z - 1}{z}$ with $z \in \mathbb{C}$. Denote, as usual, by $\mathbf{b}(z)$ the reciprocal of this function on its natural domain.

(1) Prove that the largest disc about 0 where $\mathbf{b}$ is holomorphic has radius 2π.

(2) Show that for $|z| < 2\pi$ one has $\mathbf{b}(z) + \frac{1}{2} z = \cosh(z/2) \frac{z/2}{\sinh(z/2)}$. Deduce that $\mathbf{b}(z) + \frac{z}{2}$ is an even function.

(3) For every $n \geq 0$, set $B_n := \frac{\mathrm{d}^n}{\mathrm{d}z^n}\big|_{z=0} \mathbf{b}(z)$ (the **Bernoulli numbers**). From point (1) derive that the power series $\sum_{n=0}^{\infty} \frac{B_n}{n!} z^n$ has radius of convergence 2π and that $\mathbf{b}(z)$ coincides with the sum of this series on the disc $\{|z| < 2\pi\}$. From point (2) deduce that $B_0 = 1$, $B_1 = -\frac{1}{2}$, and $B_{2k+1} = 0$ for every $k \in \mathbb{N}$. More properties of the B_n are contained in Exr. 5.8.

Exercise 5.8. (More on Bernoulli numbers). Consider $\mathbf{b}(z) = \frac{z}{e^z - 1}$ as a complex function, and let $\{b_n\}_{n \geq 0}$ be the sequence of the coefficients of its Maclaurin series $\sum_n b_n z^n$. By taking into account that $B_n = b_n\, n!$ one can derive, from the ensuing (a)-to-(g), analogous properties of the B_n (see also [Wang and Guo (1989), §1.1]):

(a). $b_n \in \mathbb{Q}$ for all $n \geq 0$ and $b_0 = 1$.

    [*Hint:* Use the fact that $\frac{e^z - 1}{z}\, \mathbf{b}(z) = 1$ for all $|z| < 2\pi$...]

(b). $b_1 = -\frac{1}{2}$ and $b_{2n+1} = 0$ for all $n \geq 1$.

    [*Hint:* Prove that the function $\mathbf{h}(z) := \mathbf{b}(z) + \frac{z}{2}$ is an *even* function and deduce that $\mathbf{h}^{(2n+1)}(0) = 0$ for all $n \geq 0$...]

(c). Deduce from the explicit expression of $\mathbf{b}$ the identity

$$\Big(\mathbf{b}(z) - \mathbf{b}(2z)\Big)\mathbf{b}(z) = \frac{z}{2}\,\mathbf{b}(2z), \quad \text{for all } |z| < \pi.$$

(d). For all $n > 2$, one has $\sum_{\substack{i+j=n \\ i,j \neq 1}} b_i\, b_j\, (2^i - 1) = 0$.

    [*Hint:* Use part (c) and take derivatives at $z = 0$...]

(e). For all $n \geq 2$, the following recursive identity holds true:

$$b_{2n} = -\frac{1}{2^{2n} - 1} \sum_{j=1}^{n-1} (2^{2j} - 1)\, b_{2j}\, b_{2(n-j)}. \tag{5.36}$$

    [*Hint:* A restatement of (d)...]

(f). $(-1)^{n-1} b_{2n} > 0$ for all $n \geq 1$ (i.e., $b_{2n} = (-1)^{n-1} |b_{2n}|$).

    [*Hint:* Proceed by induction on n with the aid of identity (5.36).]

(g). The sequence $\{b_n\}_n$ is characterized by the recursive definition

$$b_0 = 1, \qquad b_n = -\sum_{k=0}^{n-1} \frac{b_k}{(n+1-k)!} \qquad (n \geq 1).$$

    [*Hint:* Take derivatives at $z = 0$ of $\dfrac{e^z - 1}{z}\, \mathbf{b}(z) = 1$...]

Exercise 5.9. (Double limits). Let (X, d) be a complete metric space. Let $\{\ell_n^k\}_{k,n \in \mathbb{N}}$ be a double sequence in X. Suppose that:

    (i) the limit $\ell^k := \lim_{n \to \infty} \ell_n^k$ exists, uniformly with respect to k;
    (ii) for every fixed $n \in \mathbb{N}$, the limit $\ell_n := \lim_{k \to \infty} \ell_n^k$ exists.

More precisely, the first assumption means that

$$\forall\,\varepsilon > 0 \ \exists\, n(\varepsilon) \in \mathbb{N}: \quad d(\ell^k, \ell_n^k) < \varepsilon \qquad \forall\, n \geq n(\varepsilon),\ \forall\, k \in \mathbb{N}.$$

This must not be confused with the non-uniform information in (ii), namely

$$\forall\, n \in \mathbb{N},\ \forall\, \varepsilon > 0 \ \exists\, k(n,\varepsilon) \in \mathbb{N}: \quad d(\ell_n, \ell_n^k) < \varepsilon \qquad \forall\, k \geq k(n,\varepsilon).$$

Then all the following limits exist and all are equal

$$\lim_{k\to\infty} \ell^k = \lim_{n\to\infty} \ell_n = \lim_{n,k\to\infty} \ell_n^k.$$

[*Hint:* First prove that $\{\ell^k\}_k$ is a Cauchy sequence in (X,d); to this aim take, with the above notation, $h, k \geq k(n(\varepsilon), \varepsilon)$ and make a triangle-inequality estimate of $d(\ell^h, \ell^k)$...; then set $\ell := \lim_k \ell^k$ and use the inequality

$$d(\ell, \ell_n) \leq d(\ell, \ell^\nu) + d(\ell^\nu, \ell_n^\nu) + d(\ell_n^\nu, \ell_n) < 3\,\varepsilon,$$

where $\nu = \nu(n,\varepsilon)$ is chosen in such a way that $\nu \geq \max\{k(n,\varepsilon), k(\varepsilon)\}$ where $k(\varepsilon)$ is such that $d(\ell, \ell^k) < \varepsilon$ for any $k \geq k(\varepsilon)$...; for the last assertion use

$$d(\ell, \ell_n^k) \leq d(\ell, \ell^k) + d(\ell^k, \ell_n^k) < 2\,\varepsilon,$$

and take $k, n \geq \max\{k(\varepsilon), n(\varepsilon)\}$...]

Exercise 5.10. (Double series). Let $(X, \|\cdot\|)$ be a Banach space over $\mathbb{R}$. Let $(a_{i,j})_{i,j\geq 0}$ be a double sequence in X such that (as a double series of nonnegative real numbers) $\sum_{i,j\geq 0} \|a_{i,j}\| < \infty$. Prove the following facts:

(1) for any fixed $i \geq 0$, the series $A_i := \sum_{j=0}^{\infty} a_{i,j}$ is convergent in X;
(2) for any fixed $j \geq 0$, the series $B_j := \sum_{i=0}^{\infty} a_{i,j}$ is convergent in X;
(3) $\sum_{i=0}^{\infty} A_i, \sum_{j=0}^{\infty} B_j$ are convergent in X to the same element, say ℓ;
(4) with the above notation it holds that $\ell = \sum_{n=0}^{\infty} C_n$ where we have set $C_n := \sum_{i+j=n} a_{i,j}$, the series $\sum_n C_n$ being absolutely convergent in X.

[*Hint:* Just observe that all the series

$$\sum_j \|a_{i,j}\|, \quad \sum_i \|a_{i,j}\|, \quad \sum_i \|A_i\|, \quad \sum_j \|B_j\|, \quad \sum_n \|C_n\|$$

are bounded from above by the majorant series $\sum_{i,j\geq 0} \|a_{i,j}\|$.

For the claimed equalities $\sum_i A_i = \sum_j B_j = \sum_n C_n$ one may want to use Exr. 5.9 for the double sequence $s_{N,M} := \sum_{i=0}^{N} \sum_{j=0}^{M} a_{i,j}$, or Lem. 5.10.]

Exercise 5.11. Consider the matrices A, B in the *incipit* of this chapter. Prove that there does not exist any real 2×2 matrix C such that $e^A e^B = e^C$ (here e^A denotes the exponential of a square matrix). Complete this argument:

(1) Observe that B is nilpotent of degree 2 and diagonalize A; infer that[6]

$$e^A = \begin{pmatrix} -1/\sqrt{2} & 1/\sqrt{2} \\ -1/\sqrt{2} & -1/\sqrt{2} \end{pmatrix}, \qquad e^B = \begin{pmatrix} 1 & 1 \\ 0 & 1 \end{pmatrix}.$$

---

[6] Alternatively, use the following fact (holding for any $\alpha, \beta \in \mathbb{R}$):

$$\text{if } R(\alpha, \beta) = \begin{pmatrix} \alpha & -\beta \\ \beta & \alpha \end{pmatrix} \quad \text{then} \quad e^{R(\alpha,\beta)} = \begin{pmatrix} e^\alpha \cos\beta & -e^\alpha \sin\beta \\ e^\alpha \sin\beta & e^\alpha \cos\beta \end{pmatrix},$$

which can be obtained by observing that the matrix $R(\alpha, \beta)$ is the matrix of the endomorphism of $\mathbb{R}^2 \equiv \mathbb{C}$ obtained by multiplication by $\alpha + i\beta \in \mathbb{C}$.

(2) Deduce that $e^A \, e^B = \begin{pmatrix} -1/\sqrt{2} & 0 \\ -1/\sqrt{2} & -\sqrt{2} \end{pmatrix} =: D.$

(3) Show that there does not exist any real 2×2 matrix whose square matrix equals D; otherwise

$$\begin{pmatrix} -1/\sqrt{2} & 0 \\ -1/\sqrt{2} & -\sqrt{2} \end{pmatrix} = \begin{pmatrix} a^2 + bc & b(a+d) \\ c(a+d) & d^2 + bc \end{pmatrix},$$

for some real numbers a, b, c, d; show that this is absurd in the real field.[7]

(4) Finally, if there existed C such that $\exp(C) = D$, then the square of $\exp(C/2)$ would be equal to D, but this is excluded by point (3).

Exercise 5.12. Prove Thm. 5.8 by imitating the proof of Thm. 5.7.

Exercise 5.13. Motivate the *Ansatz* in the proof of Thm. 5.7 as follows:

(1) γ solves (CP) on $[-1, 1]$ if and only if

$$\dot{\gamma}(t) = \mathbf{b}(\operatorname{ad}\gamma(t))(x) + \mathbf{b}(-\operatorname{ad}\gamma(t))(y), \quad \text{for all } t \in [-1, 1]. \tag{5.37}$$

(2) The above rhs makes sense owing to the assumption on $\sum_{n=1}^{\infty} \|a_n\|$.

[*Hint:* observe that $\|\gamma(t)\| \leq \sum_{n=1}^{\infty} \|a_n\| \cdot |t^n| \leq \sum_{n=1}^{\infty} \|a_n\| < 2\pi$.]

(3) γ solves (CP) if and only if the coefficients of the Maclaurin expansions of the map $t \mapsto \dot{\gamma}(t)$ equal those of the map

$$t \mapsto h(t) := \mathbf{b}(\operatorname{ad}\gamma(t))(x) + \mathbf{b}(-\operatorname{ad}\gamma(t))(y).$$

[*Hint:* both sides of (5.37) are g-valued analytic functions of $t \in [-1, 1]$...]

(4) The Maclaurin expansion of $\dot{\gamma}(t)$ is

$$\sum_{n=0}^{\infty} (n+1)a_{n+1}t^n.$$

(5) The Maclaurin expansion of $h(t)$ is

$$x + y + \frac{1}{2}\sum_{n=1}^{\infty} [a_n, y - x]t^n + \sum_{n=2}^{\infty} t^n \sum_{\substack{2 \leq 2i \leq n \\ k_1, \ldots, k_{2i} \geq 1 \\ k_1 + \cdots + k_{2i} = n}} \frac{B_{2i}}{(2i)!}[a_{k_1} \cdots [a_{k_{2i}}, x + y] \cdots].$$

[*Hint:* $\sum_{n=1}^{\infty} \|a_n\| < 2\pi$ allows for series-reordering arguments...]

(6) Compare the expansions in (4) and (5) and derive that γ solves (CP) if and only if the coefficients a_n satisfy (5.16).

---

[7]Note that here the underlying field plays a major role. Indeed, as can be seen by using the Jordan normal form, in the complex case the map $A \mapsto e^A$ is onto the set of the invertible matrices.

Chapter 6

The Connectivity Theorem

THE aim of this chapter is twofold: firstly, we introduce the notion of *Hörmander system* of smooth vector fields; secondly, we prove the Connectivity Theorem attributed to [Carathéodory (1909); Chow (1939); Hermann (1968); Rashevskiĭ (1938)]. The latter is one of the most important subelliptic-type results of this book: it states that, given a Hörmander system of smooth vector fields

$$X = \{X_1, \ldots, X_m\}$$

on a connected open set $\Omega \subseteq \mathbb{R}^N$, any pair of points $x, y \in \Omega$ can be connected by a continuous curve $\gamma_{x,y}$ in Ω, which is piecewise an integral curve of

$$\pm X_1, \ldots, \pm X_m.$$

By definition, Hörmander vector fields satisfy the so-called *rank condition*, also known as the *bracket-generating condition* in Control Theory, where it plays a remarkable role in reachability theory. Hörmander vector fields are also fundamental in the study of hypoellipticity for linear PDEs, due to the celebrated result by [Hörmander (1967)]. See the recent monograph by [Bramanti (2014)] for an introduction to the multiple aspects of the theory of Hörmander v.f.s.

The proof of the Connectivity Theorem will be based on Thm. 3.10 in Chap. 3, where we showed how to approximate the flow of $[[X_1, X_2] \cdots X_k]$ through elementary flows of $\pm X_1, \ldots, \pm X_k$. Bearing this in mind, it can be understood why the linear independence of a set of iterated commutators of $X_1, \ldots, X_m$ can reword (via the Inverse Function Theorem) into the possibility of connecting two points x and y by a curve $\gamma_{x,y}$ as above.

In this chapter we set the basis for the definition of the so-called *control distance* associated with the family X, more closely investigated in Chap. 7. To this end, we introduce and characterize the *X-subunit paths*: the idea of modeling an appropriate geometry attached to X by means of the X-subunit paths proved to be one of the most fruitful in the theory of sub-elliptic PDEs.

A prerequisite for this chapter is Chap. 3; we shall also make use of some basic Linear Algebra in App. A.

6.1 Hörmander systems of vector fields

Throughout, it is understood that Ω is a non-empty open subset of $\mathbb{R}^N$.

Convention. In the sequel, given a set of smooth vector fields

$$X = \{X_1, \ldots, X_m\}$$

on Ω, when we refer to **the Lie algebra generated by** X, we mean that this is $\mathrm{Lie}\{X\}$ as the Lie subalgebra generated by X *with respect to the Lie algebra* $\mathfrak{X}(\Omega)$ *of the smooth vector fields on* Ω *endowed with the usual commutator of v.f.s.*

Furthermore, the vector space structure of $\mathfrak{X}(\Omega)$ is the one inherited by the vector space of the endomorphisms of $C^\infty(\Omega)$: with this vector space structure, $X_1, \ldots, X_m \in \mathfrak{X}(\Omega)$ are linearly dependent if and only if there exist $c_1, \ldots, c_m \in \mathbb{R}$ such that $\sum_{i=1}^m c_i X_i$ is the null v.f., that is its components are *identically vanishing*.

Remark 6.1. Taking into account the above convention, given a set of smooth vector fields $X_1, \ldots, X_m$ on Ω, the linear independence of $X_1, \ldots, X_m$ *as linear differential operators* (i.e., as elements of the vector space $\mathfrak{X}(\Omega)$) must not be confused with the linear independence of the *vectors of* $\mathbb{R}^N$ given by $X_1(x), \ldots, X_m(x)$ (for $x \in \Omega$). See the next example. ♯

Example 6.2. The above assertion is motivated by the following examples:

(1) The vector fields in $\mathbb{R}^2$ defined by $X_1 = \partial_{x_1}$ and $X_2 = x_1 \, \partial_{x_2}$ are linearly independent (in $\mathfrak{X}(\mathbb{R}^2)$), since the only constants c_1, c_2 turning

$$c_1 X_1 + c_2 X_2 = c_1 \, \partial_{x_1} + c_2 x_1 \, \partial_{x_2}$$

into the null vector field, must make the two functions c_1 and $c_2 x_1$ *identically vanishing*, whence $c_1 = c_2 = 0$.

Nonetheless, $X_1(0) = (1,0)$ and $X_2(0) = (0,0)$ are not linearly independent as vectors of $\mathbb{R}^2$. Note that in this case $\dim(\mathrm{span}\{X_1(x), X_2(x)\})$ is not constant with $x \in \mathbb{R}^2$.

(2) The vector fields in $\mathbb{R}^1$ defined by $X_1 = \partial_{x_1}$ and $X_2 = x_1 \, \partial_{x_1}$ are such that the set of vectors $\{X_1(x), X_2(x)\}$ is linearly dependent for every $x \in \mathbb{R}^1$, but X_1, X_2 are not linearly dependent in the vector space $\mathfrak{X}(\mathbb{R}^1)$. ♯

It is known that none of the above situations can occur for left-invariant v.f.s on a Lie group (see Prop. C.5). In the absence of left-invariance, there is a partial result relating linear independence of $\{X_1, \ldots, X_m\}$ in $\mathfrak{X}(\Omega)$ to that of the vectors $\{X_1(x), \ldots, X_m(x)\}$ in $\mathbb{R}^N$ (the proof is in Exr. 6.8):

Proposition 6.3. *Let $X \subseteq \mathfrak{X}(\Omega)$. For every fixed $x \in \Omega$ one has*

$$\dim\big(\mathrm{span}\{Y(x) \in \mathbb{R}^N \,:\, Y \in X\}\big) \leq \dim(\mathrm{span}\{X\}).$$

In the above lhs, the dimension is meant for vector subspaces of $\mathbb{R}^N$, whereas, in the rhs, for vector subspaces of $\mathfrak{X}(\Omega)$. As a consequence, given $X_1, \ldots, X_m \in \mathfrak{X}(\Omega)$, if there exists $x \in \Omega$ such that $X_1(x), \ldots, X_m(x)$ are linearly independent in $\mathbb{R}^N$, then $X_1, \ldots, X_m$ are linearly independent in $\mathfrak{X}(\Omega)$.

Remark 6.4. If $x \in \Omega$ and $X \subseteq \mathfrak{X}(\Omega)$, we have (by Prop. 6.3)

$$\dim\Big(\{Y(x) : Y \in \mathrm{Lie}\{X\}\}\Big) \leq \min\{N, r\}, \quad \text{where } r = \dim(\mathrm{Lie}\{X\}).$$

Here $r \in [0, \infty]$ is understood as the dimension of $\mathrm{Lie}\{X\}$ as a vector subspace of the infinite-dimensional vector space $\mathfrak{X}(\Omega)$. $\qquad\qquad\qquad\qquad\qquad\qquad\sharp$

We now introduce a key definition for this chapter.

Definition 6.5 (Hörmander system of vector fields). Let X be any set of smooth vector fields on Ω. We say that X is a **Hörmander system (of vector fields)** on Ω if the following condition is satisfied

$$\dim\Big(\{Y(x) \in \mathbb{R}^N : Y \in \mathrm{Lie}\{X\}\}\Big) = N, \quad \text{for every } x \in \Omega. \tag{6.1}$$

As usual, $\mathrm{Lie}\{X\}$ denotes the Lie subalgebra generated by X in $\mathfrak{X}(\Omega)$ (see Def. A.9).

Alternatively, we shall also say that X is a set of **Hörmander vector fields** or a **Hörmander family of vector fields**, if (6.1) is satisfied. Hypothesis (6.1) is also known as **Hörmander's rank condition** for X, or **bracket generating property**. Note that condition (6.1) is well posed since $\{Y(x) : Y \in \mathrm{Lie}\{X\}\}$ is a vector subspace of $\mathbb{R}^N$ (Exr. 6.1), and (6.1) can be rewritten as

$$\{Y(x) : Y \in \mathrm{Lie}\{X\}\} = \mathbb{R}^N, \quad \text{for every } x \in \Omega.$$

Remark 6.6. Let $X \subseteq \mathfrak{X}(\Omega)$. Since the nested brackets of X span $\mathrm{Lie}\{X\}$ (see Thm. A.11), the family X is a Hörmander system on Ω iff, for every $x \in \Omega$, there exist N nested brackets $Y_1, \ldots, Y_N$ of X (depending on x) such that $Y_1(x), \ldots, Y_N(x)$ are linearly independent (i.e., they are a basis of $\mathbb{R}^N$). $\qquad\qquad\qquad\sharp$

Definition 6.7 (Hörmander operator). Let $X = \{X_1, \ldots, X_m, Y\}$ be a Hörmander system of vector fields on Ω; then the second order PDO

$$X_1^2 + \cdots + X_m^2 + Y$$

is called a **Hörmander operator** on Ω, also addressed to as a **Hörmander sum of squares plus a drift** (here, Y is the drift). If $Y \equiv 0$, the PDO $X_1^2 + \cdots + X_m^2$ is said to be a **Hörmander sum of squares** on Ω.

Example 6.8. The vector fields on $\mathbb{R}^2$ defined by $X_1 = \partial_{x_1}$, $X_2 = x_1 \partial_{x_2}$ form a Hörmander system, as $\mathrm{Lie}\{X_1, X_2\} = \mathrm{span}\{\partial_{x_1}, x_1 \partial_{x_2}, \partial_{x_2}\}$, so that, for every $x \in \mathbb{R}^2$, the vector space $\{Y(x) : Y \in \mathrm{Lie}\{X_1, X_2\}\}$ contains $(1, 0)$ and $(0, 1)$ and it is therefore 2-dimensional. $\qquad\qquad\qquad\qquad\qquad\qquad\qquad\qquad\qquad\sharp$

Very important results where Hörmander's rank condition plays a key role are given in the following sections.

6.2 A useful Linear Algebra lemma

In order to understand the meaning of the so-called X-subunit curves (see Def. 6.13), we first need a lemma of Linear Algebra which will be useful not only in the present chapter. First a definition.

Definition 6.9. Let $\mathcal{V} = \{v_1, \ldots, v_m\}$ be a family of vectors of $\mathbb{R}^N$. We say that $v \in \mathbb{R}^N$ is $\mathcal{V}$-**subunit** if

$$\langle v, \xi \rangle^2 \leq \sum_{j=1}^{m} \langle v_j, \xi \rangle^2, \qquad \text{for every } \xi \in \mathbb{R}^N. \tag{6.2}$$

We denote by $\mathrm{Sub}(\mathcal{V})$ the set of all the $\mathcal{V}$-subunit vectors in $\mathbb{R}^N$.

Remark 6.10. If we test (6.2) with a vector ξ which is orthogonal to every v_j, we get $v \perp \xi$. This shows that $\mathrm{Sub}(\mathcal{V}) \perp \{v_1, \ldots, v_m\}^{\perp}$, whence (Exr. 6.5)

$$\mathrm{Sub}(\mathcal{V}) \subseteq \mathrm{span}\{v_1, \ldots, v_m\}. \tag{6.3}$$

It is also straightforward to recognize that $\mathrm{Sub}(\mathcal{V})$ is a convex subset of $\mathbb{R}^N$ since (6.2) is equivalent to

$$|\langle v, \xi \rangle| \leq \sqrt{\sum_{j=1}^{m} \langle v_j, \xi \rangle^2}, \qquad \text{for every } \xi \in \mathbb{R}^N.$$

When $\mathcal{V}$ is a family of linearly independent vectors, it is quite simple to characterize $\mathrm{Sub}(\mathcal{V})$ (see Exr. 6.6). More generally, we have the following remark.

Remark 6.11. Let $\mathcal{V} = \{v_1, \ldots, v_m\}$ be a family of vectors of $\mathbb{R}^N$. We denote by $S = S(\mathcal{V})$ the $N \times m$ matrix whose column vectors are, orderly, $v_1, \ldots, v_m$. The matrix $S\,S^T$ is (symmetric and) positive semidefinite and

$$\langle S\,S^T \xi, \xi \rangle^2 = \|S^T \xi\|^2 = \sum_{j=1}^{m} \langle v_j, \xi \rangle^2.$$

This shows that the rhs of (6.2) is the quadratic form associated with $S\,S^T$. Moreover, if we consider the linear map $T : \mathbb{R}^m \to \mathbb{R}^N$ defined by

$$T(x_1, \ldots, x_m) = \sum_{j=1}^{m} x_j\, v_j, \tag{6.4}$$

the matrix representing T (wrt the canonical bases) is S, since

$$\sum_{j=1}^{m} x_j\, v_j = S\,x, \quad \text{where } x = \begin{pmatrix} x_1 \\ \vdots \\ x_m \end{pmatrix}.$$

Due to (6.3), any $v \in \mathrm{Sub}(\mathcal{V})$ can be represented as $v = S\,x$ with many x in $\mathbb{R}^m$; this representation becomes unique, say $v = S\,c$, if we also require

$$c \in (\ker S)^{\perp} = \mathrm{range}(S^T).$$

This unique c gives the so-called **canonical coordinates** of v wrt $\mathcal{V}$, and it minimizes the norm $\|x\|$ of all the x such that $v = S\,x$. (See Sec. A.3.)   ♯

Thus, we have the following general characterization of $\mathrm{Sub}(\mathcal{V})$. For the proof of the following result, and for the concept of Moore-Penrose pseudo-inverse of a linear map, the reader is referred to Sec. A.3 in App. A.[1]

Proposition 6.12. *Let* $\mathcal{V} = \{v_1, \ldots, v_m\}$ *be a family of vectors of* $\mathbb{R}^N$, *and let* $\mathrm{Sub}(\mathcal{V})$ *be the set of the* $\mathcal{V}$-*subunit vectors, according to Def. 6.9.*

Then $v \in \mathrm{Sub}(\mathcal{V})$ *if and only if* v *is of the form* $v = \sum_{j=1}^{m} c_j\, v_j$, *with* $c_1, \ldots, c_m \in \mathbb{R}$ *satisfying* $\sum_{j=1}^{m} c_j^2 \leq 1$.

Equivalently, if T *is as in (6.4) and if* $D = \{x \in \mathbb{R}^m : \|x\| \leq 1\}$ *is the unit ball of* $\mathbb{R}^m$, *then* $\mathrm{Sub}(\mathcal{V}) = T(D)$. *More precisely, for any* $v \in \mathrm{Sub}(\mathcal{V})$ *one has* $v = \sum_{j=1}^{m} c_j\, v_j$ *with* $c = T^{\dagger}(v)$, *and the latter belongs to* D. *The entries of* c *are the canonical coordinates of* v *wrt* $\mathcal{V}$, *in the sense of Rem. 6.11. Here,* $T^{\dagger}$ *is the Moore-Penrose pseudo-inverse of* T.

6.3 The Connectivity Theorem

The aim of this section is to prove the Connectivity Theorem, Thm. 6.22. First we need the definitions of X-subunit curve and of X-connectedness.

6.3.1 X-subunit curves and X-connectedness

A function $f : [a, b] \to \mathbb{R}$ is called absolutely continuous if for every $\varepsilon > 0$ there exists $\delta_\varepsilon > 0$ such that, whenever we have disjoint subintervals $(a_1, b_1), \ldots, (a_n, b_n)$ of $[a, b]$ satisfying

$$\sum_{j=1}^{n} |b_j - a_j| < \delta_\varepsilon,$$

then $\sum_{j=1}^{n} |f(b_j) - f(a_j)| < \varepsilon$. It is known from Real Analysis that the absolutely continuous functions can be characterized in the following way: $f : [a, b] \to \mathbb{R}$ is absolutely continuous if and only if (simultaneously)

(1) f is differentiable almost everywhere on $[a, b]$;
(2) f' is Lebesgue-integrable on $[a, b]$;
(3) $f(x) = f(a) + \int_a^x f'(t)\, \mathrm{d}t$ for every $x \in [a, b]$.

The well-known formula of integration by parts holds true for absolutely continuous functions. In the sequel, given $\gamma : [a, b] \to \mathbb{R}^N$, we say that γ is absolutely continuous if this is true of any of its component functions.

Definition 6.13 (X-subunit curve). Let $X = \{X_1, \ldots, X_m\}$ be a family of vector fields on Ω. An absolutely continuous[2] curve $\gamma : [a, b] \to \Omega$ is said to be X-**subunit (in Ω)** if, for almost every $t \in [a, b]$,

$$\langle \dot{\gamma}(t), \xi \rangle^2 \leq \sum_{j=1}^{m} \langle X_j(\gamma(t)), \xi \rangle^2, \qquad \text{for every } \xi \in \mathbb{R}^N. \tag{6.5}$$

---

[1]A different proof of Prop. 6.12 is given in [Bonfiglioli *et al.* (2007), Lemma 5.16.10, page 330].

[2]Some authors require that X-subunit curves be Lipschitz continuous (or continuous and piecewise C^1) instead of absolutely continuous.

We denote by $\mathcal{S}_\Omega(X)$ the set of all the X-subunit curves (in Ω). Whenever Ω is understood, we simply write $\mathcal{S}(X)$.

Note that, in the above definition, we did not require the smoothness of the vector fields $X_1, \ldots, X_m$. Actually, many interesting examples already arise for locally Lipschitz continuous vector fields.

Remark 6.14. In view of Def. 6.9, we recognize that (6.5) is equivalent to the requirement that $\dot\gamma(t)$ be $\mathcal{V}(t)$-subunit, where

$$\mathcal{V}(t) = \Big\{ X_1(\gamma(t)), \ldots, X_m(\gamma(t)) \Big\}.$$

Remark 6.15. If $\gamma \in \mathcal{S}(X)$, by choosing $\xi = \dot\gamma(t)/\|\dot\gamma(t)\|$ in (6.5) (in the non-trivial case $\dot\gamma(t) \neq 0$), and by applying the Cauchy-Schwarz inequality in the rhs, we get $\|\dot\gamma(t)\|^2 \le \sum_{j=1}^{m} \big\| X_j(\gamma(t)) \big\|^2$, whence[3]

$$\|\dot\gamma(t)\| \le \sum_{j=1}^{m} \big\| X_j(\gamma(t)) \big\|. \tag{6.6}$$

Example 6.16. Let $X = \{X_1, \ldots, X_m\}$ be locally Lipschitz continuous v.f.s on Ω and $j \in \{1, \ldots, m\}$. We highlight that, if $\gamma(t)$ is any integral curve of $\pm X_j$, then the restriction of γ to any compact subinterval of its maximal domain is X-subunit: indeed (6.5) is trivially satisfied since $\dot\gamma(t) = \pm X_j(\gamma(t))$.   ♮

More generally, we have the following result.

Proposition 6.17 (Characterization of X-subunit curves).
Suppose that we are given a family $X = \{X_1, \ldots, X_m\}$ of vector fields on Ω. Let also $\gamma : [a, b] \to \Omega$ be an absolutely continuous curve. A necessary and sufficient condition for γ to be X-subunit is that it satisfies, for almost every $t \in [a, b]$,

$$\dot\gamma(t) = \sum_{j=1}^{m} \alpha_j(t)\, X_j(\gamma(t)), \tag{6.7a}$$

where $\alpha_1, \ldots, \alpha_m$ are real-valued functions on $[a, b]$ such that

$$\sum_{j=1}^{m} |\alpha_j(t)|^2 \le 1. \tag{6.7b}$$

Proof of Prop. 6.17. This follows from Rem. 6.14 and the characterization of subunit vectors in Prop. 6.12.   □

Remark 6.18. Actually, the function $\alpha = (\alpha_1, \ldots, \alpha_m)$ in (6.7a) can be chosen to be *measurable* on $[a, b]$; we shall not give the proof of this last fact,[4] but we limit ourselves to mentioning that one can replace α by the function

$$c(t) := (c_1(t), \ldots, c_m(t)) = (T_t)^{\dagger}(\dot\gamma(t)),$$

---

[3]Here we apply $\sqrt{\sum_j a_j^2} \le \sum_j |a_j|$, holding true for any $a_1, \ldots, a_m \in \mathbb{R}$.

[4]For measurable selection theorems, see e.g., [Agrachev *et al.* (2016); Castaing and Valadier (1977); Hermes and LaSalle (1969)].

where $T_t : \mathbb{R}^m \to \mathbb{R}^N$ is the linear map defined by
$$T_t(x) = \sum_{j=1}^m x_j \, X_j(\gamma(t)) = S(\gamma(t)) \, x.$$
Here S denotes the $N \times m$-matrix valued map defined on Ω whose columns are given by the coefficients of $X_1, \ldots, X_m$.

With the definition introduced in Rem. 6.11, $c(t)$ gives the canonical coordinates of $\dot\gamma(t)$ wrt $\{X_1(\gamma(t)), \ldots, X_m(\gamma(t))\}$, and it holds that
$$c(t) \in \big(\ker S(\gamma(t))\big)^{\perp} \quad \text{for every } t.$$

Definition 6.19 (X-connectedness). Let $\Omega \subseteq \mathbb{R}^N$ be an open and connected set. Let $X = \{X_1, \ldots, X_m\}$ be a family of vector fields on Ω.

We say that Ω is X-**connected** if for every pair of points x, y in Ω there exists
$$\gamma \in \mathcal{S}_\Omega(X), \ \ \gamma : [a, b] \to \Omega \quad \text{such that } \gamma(a) = x \text{ and } \gamma(b) = y.$$
As usual, $\mathcal{S}_\Omega(X)$ denotes the set of the X-subunit curves in Ω (see Def. 6.13).

In the sequel, given two curves $\gamma : [a, b] \to \mathbb{R}^N$ and $\mu : [c, d] \to \mathbb{R}^N$ with $\gamma(b) = \mu(c)$, we say that the curve
$$\Gamma : [a, b + d - c] \to \mathbb{R}^N, \quad \Gamma(t) := \begin{cases} \gamma(t) & \text{if } t \in [a, b] \\ \mu(c + t - b) & \text{if } t \in (b, b + d - c] \end{cases}$$
is the **gluing of γ and μ**; we can also use the names 'link' or 'sum' as synonyms for 'gluing'. It is obvious that, if $\gamma, \mu \in \mathcal{S}(X)$ and if the gluing is possible (i.e., the ending point of γ is equal to the starting point of μ), then their gluing still belongs to $\mathcal{S}(X)$ as well (Exr. 7.3). Furthermore, X-subunit curves are "reversible" (Exr. 7.2).

Example 6.20. Let $X = \{X_1, \ldots, X_m\}$ be locally Lipschitz continuous v.f.s on Ω. Every curve obtained by successively gluing any finite number of integral curves of vector fields chosen amongst $\pm X_1, \ldots, \pm X_m$ is X-subunit. $\sharp$

Remark 6.21. (1). If X and Y are C^1 v.f.s on Ω and $x \in \Omega$ is fixed, any point of the image set of the curve $\Gamma(t) = \Psi_t^Y \circ \Psi_t^X(x)$ can be reached by a continuous curve which is piecewise an integral curve of X and of Y (this is –in particular– an absolutely continuous curve).

Indeed, we first run through the integral curve $s \mapsto \Psi_s^X(x)$ up to time $s = t$, then we proceed along the integral curve $s \mapsto \Psi_s^Y(z)$ up to time $s = t$, where $z = \Psi_t^X(x)$. *This must not be confused with the trajectory of* $\Gamma(t)$, which is a smooth curve, whereas the curve obtained by gluing the cited integral curves of X and Y is not smooth, in general. (See also Fig. 3.1 on page 77.)

(2). In particular, if X_1 and X_2 are C^1 on Ω and $x \in \Omega$, it is not in general true that $t \mapsto \Gamma(t) = \Psi_t^{X_2} \circ \Psi_t^{X_1}(x)$ is X-subunit, where $X = \{X_1, X_2\}$ (see Exr. 6.9). This is not in contradiction to the fact that $t \mapsto \Psi_t^{X_1}(x_1)$ and $t \mapsto \Psi_t^{X_2}(x_2)$ are X-subunit for any $x_1, x_2 \in \Omega$ (so that any of their gluing is X-subunit as well). Once again this shows that the process of gluing must not be confused with the above Γ.

(3). If X_1 and X_2 are C^2 on Ω, it is not in general true that an integral curve of $[X_1, X_2]$ is $\{X_1, X_2\}$-subunit (see e.g., Exr. 6.10). $\sharp$

6.3.2 *The Connectivity Theorem for Hörmander vector fields*

We have the following spectacular theorem, usually attributed to [Carathéodory (1909); Chow (1939); Hermann (1968); Rashevskiĭ (1938)].

Theorem 6.22 (The Connectivity Theorem). *Let $\Omega \subseteq \mathbb{R}^N$ be an open and connected set. Let $X = \{X_1, \ldots, X_m\}$ be a Hörmander system of smooth vector fields on Ω. Then Ω is X-connected, according to Def. 6.19.*

Furthermore, any pair of points in Ω can be connected by an X-subunit curve in Ω which is piecewise an integral curve of the vector fields $\pm X_1, \ldots, \pm X_m$.

Proof of Thm. 6.22. The proof, divided in three steps, is (almost) completely based on Thm. 3.10 on page 80.

STEP I. Let us denote by $\mathcal{S}^*(X)$ the subset of $\mathcal{S}(X)$ of all the continuous curves supported in Ω which are piecewise integral curves of $\pm X_1, \ldots, \pm X_m$. We shall prove, in Step III below, that the following property is satisfied:

$$\text{For every } x \in \Omega \text{ there exists a neighborhood } U_x \subseteq \Omega \text{ of } x$$
$$\text{such that every point of } U_x \text{ can be joined to } x \text{ by a curve in } \mathcal{S}^*(X). \tag{6.8}$$

Once (6.8) has been proved, the proof of the theorem is a simple argument on connected sets. Indeed, fixing any $x_0 \in \Omega$, we denote by Ω_0 the set of the points of Ω joined to x_0 by curves in $\mathcal{S}^*(X)$. Clearly $\Omega_0 \neq \varnothing$ since $x_0 \in \Omega_0$. We show in Exr. 6.12 that Ω_0 is open and closed relative to Ω; this gives $\Omega_0 = \Omega$ (since Ω is connected), which proves the second statement of the Connectivity Theorem, which clearly implies the first statement as well.

We are left with the proof of (6.8). This is done in the following Steps II, III.

STEP II. Let $x_0 \in \Omega$. Since $X = \{X_1, \ldots, X_m\}$ is a Hörmander system of vector fields, the vector space $\{Y(x_0) : Y \in \mathrm{Lie}\{X\}\}$ is N-dimensional. Since left-nested brackets span $\mathrm{Lie}\{X\}$ (see Thm. A.11), there exists a set of left-nested commutators of $X_1, \ldots, X_m$, say $\{Y_1, \ldots, Y_N\}$, such that

$$Y_1(x_0), \ldots, Y_N(x_0) \text{ are linearly independent.} \tag{6.9}$$

Roughly, the idea of the proof is the following one: if $Y_1, \ldots, Y_N$ are as above, the function

$$(s_1, \ldots, s_N) \mapsto \Psi_{s_1}^{Y_1} \circ \Psi_{s_2}^{Y_2} \circ \cdots \circ \Psi_{s_N}^{Y_N}(x_0)$$

is a C^∞ diffeomorphism of an open neighborhood of 0 in $\mathbb{R}^N$ onto an open neighborhood of x_0 in $\mathbb{R}^N$ (Exr. 6.7). This gives local Y-connectivity, where $Y = \{Y_1, \ldots, Y_N\}$. Since any Y_i is left-nested, by means of Thm. 3.10, we can approximate each $\Psi_{s_i}^{Y_i}$ (for $i = 1, \ldots, N$) by means of a composition of flows of $\pm X_1, \ldots, \pm X_m$, and this will produce X-connectivity.

We now proceed with the proof of the theorem. Each Y_i has the form

$$[\cdots [X_{j_1}, X_{j_2}] \cdots X_{j_k}],$$

where $(j_1, \ldots, j_k)$ is a k-tuple (for some $k \in \mathbb{N}$) of elements of $\{1, \ldots, m\}$. (We remark that $j_1, \ldots, j_k$ and k all depend on i, but we temporarily avoid to make explicit this dependence for the simplicity of notation).

By Thm. 3.10, there exists a *suitable composition of $3 \cdot 2^{k-1} - 2$ flows (at time t) of* $\pm X_1, \ldots, \pm X_m$, which we denoted by

$$\alpha_k(\Psi_t^{X_{j_1}}, \Psi_t^{X_{j_2}}, \ldots, \Psi_t^{X_{j_k}})$$

(using the notation in (3.15) on page 79), such that

$$\alpha_k(\Psi_t^{X_{j_1}}, \ldots, \Psi_t^{X_{j_k}})(x) = \exp\left(t^k[\cdots[X_{j_1}, X_{j_2}]\cdots X_{j_k}]\right)(x) + \mathcal{O}(t^{k+1}),$$

for x in any compact subset K of Ω and for small t (the "smallness" of t depending on K). Choosing $t = s^{1/k}$ (for s positive and small), we have[5]

$$\alpha_k(\Psi_{s^{1/k}}^{X_{j_1}}, \ldots, \Psi_{s^{1/k}}^{X_{j_k}})(x) = \exp\left(s[\cdots[X_{j_1}, X_{j_2}]\cdots X_{j_k}]\right)(x) + \mathcal{O}\left(s^{\frac{k+1}{k}}\right)$$

$$= x + s[\cdots[X_{j_1}, X_{j_2}]\cdots X_{j_k}](x) + \mathcal{O}\left(s^{\frac{k+1}{k}}\right). \tag{6.10}$$

By Thm. 3.10, the same kind of formula is valid for every choice of k vector fields replacing $X_{j_1}, X_{j_2}, \ldots, X_{j_k}$; for example we can interchange some of these vector fields and then obtain an analogue of (6.10).

For the sake of simplicity, we set, for $x \in K$ and $s > 0$ small,

$$E(s, x) := \alpha_k(\Psi_{s^{1/k}}^{X_{j_1}}, \Psi_{s^{1/k}}^{X_{j_2}}, \ldots, \Psi_{s^{1/k}}^{X_{j_k}})(x). \tag{6.11}$$

Note that $E(s, x)$ is well defined (and smooth) for x in any open set compactly contained in Ω and for $s > 0$ sufficiently small.

We next turn to prolong $E(s, x)$ in order to allow small *negative* values of s, in such a way that the resulting prolongation $E^*(s, x)$ is C^1 wrt (s, x).

To this end it is convenient to set

$$E^*(s, x) := \begin{cases} E(s, x) & \text{if } s \geq 0, \\ \alpha_k\left(\Psi_{|s|^{1/k}}^{X_{j_2}}, \Psi_{|s|^{1/k}}^{X_{j_1}}, \Psi_{|s|^{1/k}}^{X_{j_3}} \cdots, \Psi_{|s|^{1/k}}^{X_{j_k}}\right)(x) & \text{if } s < 0, \end{cases}$$

for s sufficiently small (so that all terms make sense). Observe that, with respect to the definition of $E(s, x)$ in (6.11), in the case $s < 0$ we have simply interchanged X_{j_1} and X_{j_2}. By means of the trivial fact that

$$[\cdots[[X_{j_2}, X_{j_1}], X_{j_3}]\cdots X_{j_k}] = -[\cdots[[X_{j_1}, X_{j_2}], X_{j_3}]\cdots X_{j_k}],$$

and owing to (6.10) (see also the emphasized comment immediately below formula (6.10)), it follows that

$$E^*(s, x) = \begin{cases} x + s[\cdots[[X_{j_1}, X_{j_2}], X_{j_3}]\cdots X_{j_k}](x) + \mathcal{O}\left(s^{\frac{k+1}{k}}\right) & (s \geq 0), \\ x - |s|[\cdots[[X_{j_1}, X_{j_2}], X_{j_3}]\cdots X_{j_k}](x) + \mathcal{O}\left(|s|^{\frac{k+1}{k}}\right) & (s < 0). \end{cases}$$

---

[5]We use formula (1.20) on page 18, noticing that $s^2 = \mathcal{O}\left(s^{\frac{k+1}{k}}\right)$ since $2 \geq \frac{k+1}{k}$.

Since $(k+1)/k > 1$ and $-|s| = s$ when $s < 0$, it is a simple exercise to recognize that $E^*(s,x)$ is indeed C^1 wrt (s,x). Moreover, the above formula gives

$$\frac{\partial}{\partial s}\Big|_{s=0} E^*(s,x) = [\cdots[[X_{j_1}, X_{j_2}], X_{j_3}]\cdots X_{j_k}](x) = Y_i(x). \qquad (6.12)$$

STEP III. Let x_0 and $Y_1 \ldots, Y_N$ be as in the *incipit* of Step II. Since all the v.f.s $Y_1 \ldots, Y_N$ are left-nested brackets, we can construct the associated functions $E_1^*(s,x), \ldots, E_N^*(s,x)$ as in the previous Step, relative to $Y_1 \ldots, Y_N$ respectively. We then define the function which maps $s = (s_1, \ldots, s_N)$ into

$$\Theta(s) := E_1^*\Big(s_1, E_2^*\big(s_2, \cdots E_{N-1}^*\big(s_{N-1}, E_N^*(s_N, x_0)\big)\cdots\big)\Big).$$

Θ is well defined and C^1 for s in a small neighborhood of $0 \in \mathbb{R}^N$; moreover Θ is valued in Ω. It is important to observe that, since each $E_i^*(s_i, x)$ is a composition of flows of $\pm X_1, \ldots, \pm X_m$ starting from x, it follows inductively that $\Theta(s_1, \ldots, s_N)$ *is a composition of flows of* $\pm X_1, \ldots, \pm X_m$ *passing through* x_0 *when* $s = 0 \in \mathbb{R}^N$.

Therefore, by repeatedly applying Rem. 6.21-(1), we see that any point $\Theta(s)$ can be reached by a continuous curve which is piecewise an integral curve of $\pm X_1, \ldots, \pm X_m$, that is, by a curve in $\mathcal{S}^*(X)$. This curve is well defined in Ω if s is sufficiently small: how small s must be chosen, it depends on the finite (although very long!) sequence of $3 \cdot 2^{k_i} - 2$ vector fields used to decompose any Y_i (which is a left-nested bracket of length k_i, say) into flows of $\pm X_1, \ldots, \pm X_m$. Ultimately, the "smallness" of s depends only on x_0, on Ω and on the system X.

We claim that Θ is a C^1-diffeomorphism of a neighborhood of 0 in $\mathbb{R}^N$ onto a neighborhood of x_0 in Ω. To prove this claim, owing to the Inverse Function Theorem, it is enough to prove that $\mathcal{J}_\Theta(0)$ is invertible. The columns of $\mathcal{J}_\Theta(0)$ are

$$\frac{\partial}{\partial s_i}\Big|_{s=0} \Theta(s), \qquad i = 1, \ldots, N.$$

Since $E_N^*(0, \cdot), E_{N-1}^*(0, \cdot), \ldots, E_1^*(0, \cdot)$ are all equal to the identity map, one has

$$\frac{\partial}{\partial s_i}\Big|_{s=0} \Theta(s) = \frac{\mathrm{d}}{\mathrm{d}s_i}\Big|_{s_i=0}\Big\{\Theta(s_1, \ldots, s_i, \ldots, s_N)\Big\}\Big|_{s_1,\ldots,s_{i-1},s_{i+1},\ldots,s_N=0}$$

$$= \frac{\mathrm{d}}{\mathrm{d}s_i}\Big|_{s_i=0} \Theta(0, \ldots, s_i, \ldots, 0) = \frac{\mathrm{d}}{\mathrm{d}s_i}\Big|_{s_i=0} E_i^*(s_i, x_0) = Y_i(x_0).$$

In the last equality we have used (6.12) (remember that, by construction, $E_i^*(s_i, x)$ is as in Step II relative to the nested bracket Y_i).

The above computation demonstrates that $\mathcal{J}_\Theta(0)$ is the matrix whose columns are given by the vectors $Y_1(x_0), \ldots, Y_N(x_0)$. By (6.9) we infer that $\mathcal{J}_\Theta(0)$ is invertible, as we wanted to prove.

By the Inverse Function Theorem, we deduce that there exist an open neighborhood Q of 0 in $\mathbb{R}^N$ and an open neighborhood U_{x_0} of x_0 in Ω such that

$$\Theta|_Q : Q \longrightarrow U_{x_0}$$

is a C^1-diffeomorphism. Since $U_{x_0} = \Theta(Q)$, it follows that any point of U_{x_0} is of the form $\Theta(s)$ for $s \in Q$. Because of the fact that $\Theta(s)$ is a composition of flows of $\pm X_1, \ldots, \pm X_m$ starting from x_0, statement (6.8) is completely proved. $\qquad \square$

Remark 6.23. As shown by Thm. 6.22, in order to have X-connectivity the hypothesis that X is a Hörmander system is sufficient, *but it is by no means a necessary condition*: for example, the vector fields in $\mathbb{R}^2$ $X_1 = \partial_{x_1}$, $X_2 = \max\{0, x_1\}\,\partial_{x_2}$ are such that $\mathbb{R}^2$ is $\{X_1, X_2\}$-connected, but they do not form a Hörmander system, as the reader can easily check. ♮

6.4 Exercises of Chap. 6

Exercise 6.1. Let $X \subseteq \mathcal{X}(\Omega)$. Prove that $\{X(x) :\ X \in \mathrm{Lie}\{X\}\}$ is a vector subspace of $\mathbb{R}^N$, where $\mathrm{Lie}\{X\}$ is the Lie algebra generated by X in $\mathcal{X}(\Omega)$.

Exercise 6.2. Consider the system $X = \{X_1, X_2\}$ on $\mathbb{R}^2$ given by $X_1 = x_1^2\,\partial_{x_1}$ and $X_2 = x_1\,\partial_{x_2}$. Prove that ($X_1$ is repeated n times)

$$[X_1 \cdots [X_1, X_2]\cdots] = n!\, x_1^{n+1}\,\partial_{x_2}, \quad \forall\, n \in \mathbb{N}.$$

Deduce that $\mathrm{Lie}\{X\}$ is infinite-dimensional. Prove that X is not a Hörmander system on any open set containing points of the x_2-axis. Prove that X is a Hörmander system on the complement of the x_2-axis.

Exercise 6.3. Let $X = \{X_1, \ldots, X_m\} \subset \mathcal{X}(\Omega)$ be a system of Hörmander vector fields on Ω. We introduce a notation for left-nested commutators: given $k \in \mathbb{N}$ and $J = (j_1, \ldots, j_k) \in \{1, \ldots, m\}^k$ (the latter being the k-fold Cartesian product of $\{1, \ldots, m\}$ with itself), we set $X_J := [\cdots [X_{j_1}, X_{j_2}] \cdots X_{j_k}]$.

(1). Prove that the following function r is well posed: we define $r : \Omega \to \mathbb{N}$ by requiring that $r(x)$ is the minimum $n \in \mathbb{N}$ such that the dimension of

$$W_n(x) := \mathrm{span}\{X_J(x) : J \in \{1, \ldots, m\}^k,\ 1 \le k \le n\}$$

is equal to N. (Show in particular that $r(x)$ cannot be ∞.)

(2). Prove that for every $x \in \Omega$ there exists an open neighborhood U_x of x such that $r(y) \le r(x)$ for every $y \in U_x$. In particular, $\limsup_{y \to x} r(y) \le r(x)$ for every $x \in \Omega$, that is r is upper semi-continuous.

[*Hint:* If a square matrix $(a_{i,j}(x))$, whose entries are continuous functions on Ω, is invertible for some x, then it is invertible for y in a neighborhood of x...]

(3). Note that the above function r may not be continuous. Indeed, find out the function $r : \mathbb{R}^2 \to \mathbb{N}$ related to $X_1 = \partial_{x_1}$, $X_2 = x_1\,\partial_{x_2}$. (Note that, if x is on the x_2-axis, the above inequality $r(y) \le r(x)$ may be strict for infinitely many points.)

(4). Finally, suppose that $K \subset \Omega$ is compact. Prove that there exists n_0 such that $W_{n_0}(x)$ has dimension N for every $x \in K$; roughly, this means that the left-nested commutators of $X_1, \ldots, X_m$ of length $\le n_0$ suffice to span $\mathbb{R}^N$ at every point of K.

Exercise 6.4. Prove the following facts:

(1) Let $v_1, \ldots, v_m \in \mathbb{R}^N$ and $c_1, \ldots, c_m \in \mathbb{R}$. Show that
$$\left\langle \sum_{j=1}^m c_j\, v_j, \xi \right\rangle^2 \le \sum_{i=1}^m c_i^2 \cdot \sum_{j=1}^m \langle v_j, \xi \rangle^2, \qquad \forall\, \xi \in \mathbb{R}^N.$$

(2) Let $X = \{X_1, \ldots, X_m\}$ be a set of v.f.s on the open set $\Omega \subseteq \mathbb{R}^N$. Let $\gamma : [0, T] \to \Omega$ be an absolutely continuous curve such that
$$\dot\gamma(t) = \sum_{j=1}^m \alpha_j(t)\, X_j(\gamma(t)),$$
for almost every $t \in [0, T]$, where $\alpha_1, \ldots, \alpha_m$ are real-valued functions on $[0, T]$. Suppose that there exists $M > 0$ such that
$$\sqrt{\sum_{j=1}^m |\alpha_j(t)|^2} \le M \quad \text{for almost every } t \in [0, T].$$
Prove that $\Gamma : [0, M\,T] \to \Omega$ defined by $\Gamma(t) := \gamma(t/M)$ is X-subunit.

(3) If γ is as in the previous point and $M = 1$, deduce that γ is X-subunit.

Exercise 6.5. Let $v, v_1, \ldots, v_m$ be vectors of $\mathbb{R}^N$ and suppose that
$$\langle v, \xi \rangle^2 \le \sum_{j=1}^m \langle v_j, \xi \rangle^2, \quad \text{for every } \xi \in \mathbb{R}^N.$$
Prove that $v \in \mathrm{span}\{v_1, \ldots, v_m\}$. [*Hint:* If ξ is orthogonal to $v_1, \ldots, v_m$ then...]

Exercise 6.6. Let $\{v_1, \ldots, v_m\}$ be a set of vectors of $\mathbb{R}^N$. We say that the matrix $G := (\langle v_i, v_j \rangle)_{i,j \le m}$ is the *Gram matrix* of (the ordered m-tuple) $(v_1, \ldots, v_m)$.

(1). Prove that G is invertible iff $\{v_1, \ldots, v_m\}$ are linearly independent.

[*Hint:* Write G as $S^T \cdot S$ where S is the $N \times m$ matrix whose columns are given by the coordinates of the vectors $v_1, \ldots, v_m$. If the latter are dependent, then what can you say of the rank of S (and of G consequently)? Vice versa, suppose the given vectors are independent, and prove that G is positive definite...]

(2). Let $v_1, \ldots, v_m \in \mathbb{R}^N$ be linearly independent, and let $c_1, \ldots, c_m \in \mathbb{R}$. Prove
$$(\bigstar) \qquad \left\langle \sum_{j=1}^m c_j\, v_j, \xi \right\rangle^2 \le \sum_{j=1}^m \langle v_j, \xi \rangle^2, \qquad \forall\, \xi \in \mathbb{R}^N$$
if and only if $\sum_{j=1}^m c_j^2 \le 1$.

[*Hint:* One of the implications is contained in Exr. 6.4-1. For the other implication, provide the details of the following argument:

(a) Observe that if ξ is orthogonal to $v_1, \ldots, v_m$ then $(\bigstar)$ gives no information. It then suffices to consider $\xi = \xi_1 v_1 + \cdots + \xi_m v_m$ with $\xi_1, \ldots, \xi_m \in \mathbb{R}$.
(b) If $x = (\xi_1, \ldots, \xi_m)^T$ and $c = (c_1, \ldots, c_m)^T$, recognize that $(\bigstar)$, owing to (a), is equivalent to $\langle G\,x, c \rangle^2 \le \|G\,x\|^2$, where $G = (\langle v_i, v_j \rangle)_{i,j \le m}$.
(c) Use part (1) of the exercise choosing a suitable x in the step (b) above...]

Exercise 6.7. Let $Y_1, \ldots, Y_N \in \mathfrak{X}(\Omega)$. Let also $x_0 \in \Omega$ and suppose that $Y_1(x_0), \ldots, Y_N(x_0)$ are linearly independent. Prove that the function
$$s = (s_1, \ldots, s_N) \mapsto \Theta(s) := \Psi_{s_1}^{Y_1} \circ \Psi_{s_2}^{Y_2} \circ \cdots \circ \Psi_{s_N}^{Y_N}(x_0)$$
is a C^∞ diffeomorphism between open neighborhoods of 0 and x_0.

[*Hint:* Prove that the columns of $J_\Theta(0)$ are $Y_1(x_0), \ldots, Y_N(x_0)$ (see Exr. 3.6 on page 86), then use the Inverse Function Theorem.]

Exercise 6.8. Prove Prop. 6.3 by completing the following argument:

(1) Let $V := \mathrm{span}\{Y(x) \in \mathbb{R}^N : Y \in X\}$. One can select $X_1, \ldots, X_m \in X$ (why?) such that $X_1(x), \ldots, X_m(x)$ is a basis of V. Prove that $X_1, \ldots, X_m$ are linearly independent in $\mathcal{X}(\Omega)$, by contradiction:

(2) if there exists $(\alpha_1, \ldots, \alpha_m) \in \mathbb{R}^m \setminus \{0\}$ such that $\sum_{i=1}^m \alpha_i X_i$ is the null vector field, then $\sum_{i=1}^m \alpha_i X_i(x) = 0$, against (1);

(3) since $\mathrm{span}\{X_1, \ldots, X_m\} \subseteq \mathrm{span}\{X\}$, derive that

$$\dim(V) \overset{(1)}{=} m = \dim(\mathrm{span}\{X_1, \ldots, X_m\}) \leq \dim(\mathrm{span}\{X\}).$$

Exercise 6.9. (a). Consider $X_1 = \partial_{x_1}$ and $X_2 = \partial_{x_2}$ in $\mathbb{R}^2$. Let $X = \{X_1, X_2\}$. Given any $x \in \mathbb{R}^2$, prove that $\Psi_t^{X_1}(x) = (x_1 + t, x_2)$ and $\Psi_t^{X_2}(x) = (x_1, x_2 + t)$ are X-subunit, but this is not true of $\Gamma(t) := \Psi_t^{X_2} \circ \Psi_t^{X_1}(x) = (x_1 + t, x_2 + t)$.

[*Hint:* Show that (6.5) for Γ reads like $(\xi_1 + \xi_2)^2 \leq \xi_1^2 + \xi_2^2$, which is false!]

However a rescaling in the velocity of Γ turns it into the X-subunit curve

$$t \mapsto \widetilde{\Gamma}(t) := \Gamma(t/\sqrt{2}) = (x_1 + t/\sqrt{2}, x_2 + t/\sqrt{2}).$$

[*Hint:* Now (6.5) for $\widetilde{\Gamma}$ reads like $(\xi_1 + \xi_2)^2 \leq 2(\xi_1^2 + \xi_2^2)$, which is true!]

(b). Consider the v.f.s $X_1 = \partial_{x_1} + 2 x_2 \partial_{x_3}$ and $X_2 = \partial_{x_2} - 2 x_1 \partial_{x_3}$ in $\mathbb{R}^3$. Let $X = \{X_1, X_2\}$. Prove that the following curve is not X-subunit

$$t \mapsto \Gamma(t) := \Psi_t^{X_2} \circ \Psi_t^{X_1}(0) = (t, t, -2 t^2).$$

[*Hint:* Use the characterization in Prop. 6.17.]

Exercise 6.10. Consider $X_1 = \partial_{x_1} + 2 x_2 \partial_{x_3}$ and $X_2 = \partial_{x_2} - 2 x_1 \partial_{x_3}$ in $\mathbb{R}^3$. Let $X = \{X_1, X_2\}$. Prove that the following path is not X-subunit:

$$t \mapsto \Gamma(t) := \Psi_t^{[X_1, X_2]}(0) = (0, 0, -4 t).$$

[*Hint:* Show that (6.5) for Γ reads like $16 \xi_3^2 \leq \xi_1^2 + \xi_2^2$, which is false!]

Exercise 6.11. Prove that the operator $L = (x_1 \partial_{x_2})^2 + \partial_{x_1}$ is a Hörmander operator in $\mathbb{R}^2$, but it is not true that the matrix $A(x)$ of the principal part of L satisfies $\mathrm{trace}(A(x)) > 0$ for every $x \in \mathbb{R}^2$.

Note the fundamental role of the drift vector field ∂_{x_1} in making L satisfy Hörmander's Rank Condition.

Exercise 6.12. With the notation of Step I of the proof of Thm. 6.22, prove that Ω_0 is connected by taking for granted (6.8), and by the following argument.

(1) $\Omega_0 \neq \varnothing$. [*Hint:* $x_0 \in \Omega_0$!]

(2) Ω_0 *is open.* [*Hint:* If $x \in \Omega_0$, let $\gamma : [a_1, b_1] \to \Omega$ be in $\mathcal{S}^*(X)$ such that $\gamma(a_1) = x_0$ and $\gamma(b_1) = x$. For every $y \in U_x$ there exists $\mu : [a_2, b_2] \to \Omega$ in $\mathcal{S}^*(X)$ such that $\mu(a_2) = x$ and $\mu(b_2) = y$. By gluing γ and μ, it follows that x_0 and y are joined by a curve in $\mathcal{S}^*(X)$; thus $U_x \subseteq \Omega_0$. See Fig. 6.1.]

(3) Ω_0 *is closed (in Ω).* [*Hint:* Let $x \in \Omega$ be such that there exists a sequence $x_j \in \Omega_0$ with $\lim_{j \to \infty} x_j = x$. Let U_x be as in (6.8) and let j be so large that $x_j \in U_x$. Then x_j is joined to x by a curve in $\mathcal{S}^*(X)$ by (6.8); moreover x_j is joined to x_0 by another curve in $\mathcal{S}^*(X)$ since $x_j \in \Omega_0$. Infer that x and x_0 are joined by a curve in $\mathcal{S}^*(X)$, so that $x \in \Omega_0$. See Fig. 6.2.]

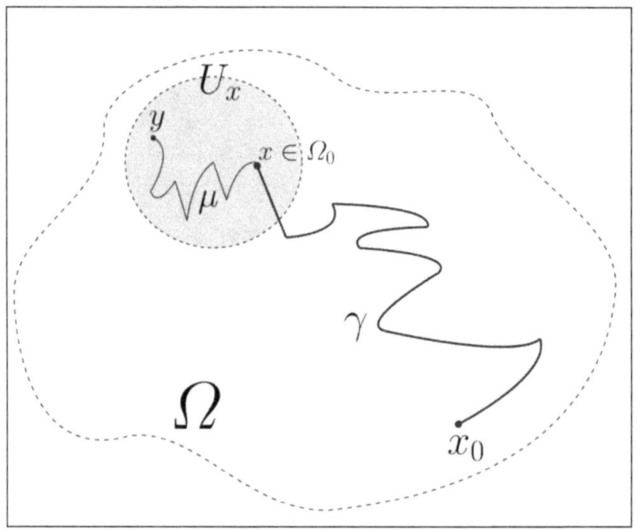

Fig. 6.1   Proof of the Connectivity Theorem: Ω_0 is open.

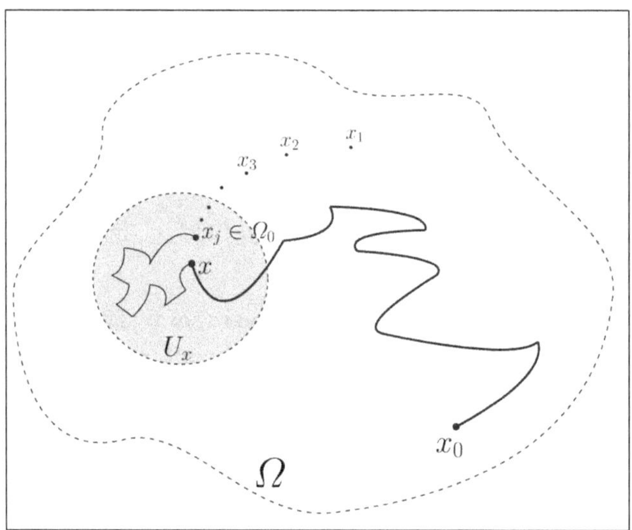

Fig. 6.2   Proof of the Connectivity Theorem: Ω_0 is closed.

Chapter 7

The Carnot-Carathéodory distance

THE aim of this chapter is to provide a very short introduction to the notion of X-*control distance* d_X (also referred to as the *Carnot-Carathéodory distance*) related to a family

$$X = \{X_1, \ldots, X_m\}$$

of vector fields on an open set Ω in space $\mathbb{R}^N$. Broadly, the d_X distance of $x, y \in \Omega$ is obtained by minimizing the life-time of the X-subunit curves connecting x and y. Different definitions of d_X will be given, and their equivalence will be proved. As already remarked in dealing with the Connectivity Theorem, the modeling of a "more intrinsic" geometry attached to X by means of X-subunit curves proves a powerful tool in the theory of sub-elliptic PDEs, in Control Theory and in Geometric Measure Theory.

We shall overview some basic topological properties of (Ω, d_X), such as the comparison with the underlying Euclidean topology, local-compactness, and length-space properties. Some pathological aspects of d_X are not so uncommon as one may expect: for example, boundedness wrt d_X may have little to share with boundedness in the Euclidean metric, even for Hörmander systems X.

Nonetheless, the Hörmander assumption is much convenient for a satisfactory connection between the metric topology of (Ω, d_X) and the underlying Euclidean topology; thus, we shall benefit of the work done in the proof of the Connectivity Thm. 6.22 when demonstrating that

$$\mathbf{c}^{-1}\|x - y\| \le d_X(x, y) \le \mathbf{c}\,\|x - y\|^{1/k} \quad \text{for every } x, y \in K,$$

where K is compact, $\mathbf{c} = \mathbf{c}(K) > 0$ and the commutators of lengths $\le k$ span the tangent space at any $x \in K$.

A more exhaustive theory of CC spaces would comprise the study of geodesics, and advanced topics of Geometric Measure Theory, such as the study of boundaries and perimeters, or differentiability issues and the Sobolev theory. Unfortunately all these topics are not suited for our introductory slant, and we cannot help but recommend advanced references.

Prerequisites for this chapter are the investigations in Chap. 6.

147

7.1 The X-control distance

Throughout this chapter, the following conventions are understood:

- Ω is a non-empty open subset of $\mathbb{R}^N$;
- $X = \{X_1, \ldots, X_m\}$ is a set of locally Lipschitz continuous v.f.s on Ω;
- when dealing with an L^∞ function $f : A \to [-\infty, \infty]$, if we write $\sup_A |f|$, then we mean the *essential* sup of f on A.

These tacit assumptions will not be restated in the sequel.

Moreover, we use the notions of X-subunit curve and X-connectedness introduced in Def.s 6.13 and 6.19 in the previous chapter; accordingly, $\mathcal{S}_\Omega(X)$ denotes the set of the X-subunit curves in Ω. We shall use the following definitions:[1]

Definition 7.1 (X-horizontal curve, X-trajectory). We say that an absolutely continuous curve $\gamma : [a, b] \to \Omega$ (with $a \leq b$) is:

- an X**-horizontal** curve (in Ω), if

$$\dot{\gamma}(t) = \sum_{j=1}^{m} \alpha_j(t)\, X_j(\gamma(t)) \quad \text{for almost every } t \in [a, b], \tag{7.1}$$

for some real-valued measurable functions $\alpha_1, \ldots, \alpha_m$ on $[a, b]$ such that

$$\|\alpha\|_\infty := \sup_{t \in [a,b]} \left(\sum_{j=1}^{m} |\alpha_j(t)|^2 \right)^{1/2} < \infty; \tag{7.2}$$

- an X**-trajectory** (in Ω), if it is an X**-horizontal** curve defined on $[0, 1]$.

By means of Prop. 6.17, we recognize that $\gamma : [a, b] \to \Omega$ is X-subunit iff γ is X-horizontal and (7.1) holds with $\|\alpha\|_\infty \leq 1$.

We shall use the following notation: when $x \in \Omega$, we denote by $S_X(x)$ the $N \times m$ matrix whose columns are[2] $X_1(x), \cdots, X_m(x)$ (in this order). We write $S(x)$ instead of $S_X(x)$ when X is understood. From the trivial identity

$$\sum_{j=1}^{m} \alpha_j\, X_j(x) = S_X(x)\, \alpha, \quad \text{where} \quad \alpha = \begin{pmatrix} \alpha_1 \\ \vdots \\ \alpha_m \end{pmatrix},$$

we get that (7.1) can (and always will) be written as

$$\dot{\gamma}(t) = S_X(\gamma(t))\, \alpha(t) \qquad (\text{briefly,} \quad \dot{\gamma} = S(\gamma)\alpha).$$

---

[1] It is necessary to warn the reader that this terminology is not generally fixed in the literature: some authors call an X-admissible curve what we name X-horizontal; some others call X-horizontal what we named X-subunit. It is clearly only a matter of convention, and no issues will arise once the definitions are clearly given.

[2] As per usual, $X(x)$ denotes the column vector of the coefficient functions of the v.f. X at x.

Remark 7.2. We know that the decomposition (7.1) may hold true for many functions $\alpha(t)$; however, $\alpha(t)$ is unique if we require that it belong to the orthogonal of $\ker\left(S(\gamma(t))\right)$, which is also equivalent to requiring that $\alpha(t)$ belong to the range of $S^T(\gamma(t))$. This unique $\alpha(t)$ gives, in the terminology of Rem. 6.11, the canonical coordinates of $\dot{\gamma}(t)$ wrt $\{X_1(\gamma(t)), \ldots, X_m(\gamma(t))\}$, and we call it the vector of the X**-canonical coordinates** of $\dot{\gamma}$. ♯

Remark 7.3. *Any X-horizontal curve* (hence any X-trajectory or any X-subunit curve as well) *is Lipschitz continuous wrt Euclidean metric.*

Indeed, let $\gamma : [a, b] \to \Omega$ satisfy $\dot{\gamma} = S(\gamma)\,\alpha$ (a.e. on $[a, b]$) with a finite $\|\alpha\|_\infty$; then, for any $s, t \in [a, b]$ we have (as γ is absolutely continuous)

$$\|\gamma(t) - \gamma(s)\| = \left\| \int_s^t \dot{\gamma}(\tau)\,\mathrm{d}\tau \right\| = \left\| \int_s^t S(\gamma(\tau))\,\alpha(\tau)\,\mathrm{d}\tau \right\|$$

$$\leq \sup_{x \in \gamma([a,b])} |||S(x)||| \cdot \|\alpha\|_\infty\, |t - s|.$$

The sup in the far rhs is finite since the entries of S are continuous (and $\gamma([a, b])$ is compact). This is why *it is not restrictive to replace the assumption of absolute continuity in Def. 7.1 with that of Lipschitz continuity.* ♯

We fix another definition: let $x, y \in \mathbb{R}^N$ and suppose that $\gamma : [a, b] \to \mathbb{R}^N$ is a continuous curve such that $\gamma(a) = x$ and $\gamma(b) = y$; then we say that γ **connects** x **and** y. Clearly, the order of x and y is immaterial: if $\gamma : [a, b] \to \mathbb{R}^N$ connects x and y, then $\tilde{\gamma} : [a, b] \to \mathbb{R}^N$ defined by $\tilde{\gamma}(t) := \gamma(b + a - t)$ connects $y = \tilde{\gamma}(a)$ and $x = \tilde{\gamma}(b)$; moreover, $\tilde{\gamma}$ has the same image set of γ.

We are ready for the main definition of this chapter.

Definition 7.4 (X-control distance). If $\gamma : [a, b] \to \Omega$ is an X-subunit curve, we set $\ell(\gamma) := b - a$. Let $x, y \in \Omega$; we give the following definition:

- if there is no γ in $\mathcal{S}_\Omega(X)$ connecting x and y, we set $d_X(x, y) := \infty$;
- otherwise, we set

$$d_X(x, y) := \inf \left\{ \ell(\gamma) \,\middle|\, \gamma \in \mathcal{S}_\Omega(X),\ \gamma \text{ connects } x \text{ and } y \right\}. \qquad (7.3)$$

Clearly, d_X is finite on $\Omega \times \Omega$ if and only if Ω is X-connected (Def. 6.19).

The map d_X is called the X**-control distance** or the **Carnot-Carathéodory distance** (also abbreviated as **CC distance**) related to X (and to Ω).

Remark 7.5. If $\gamma : [a, b] \to \Omega$ is X-horizontal, and $\dot{\gamma} = S(\gamma)\alpha$ (a.e.), then

$$d_X(\gamma(s), \gamma(t)) \leq \|\alpha\|_\infty\, |t - s|, \qquad \forall\, t, s \in [a, b]. \qquad (7.4)$$

Indeed, we can suppose $\|\alpha\|_\infty > 0$ (otherwise γ is trivially constant), and we can take $a \leq s < t \leq b$; then we consider the rescaling

$$\mu : [A, B] \to \Omega \quad \text{defined by } \mu(\tau) = \gamma(\tau/\|\alpha\|_\infty),$$

with $A = \|\alpha\|_\infty\, s$ and $B = \|\alpha\|_\infty\, t$. This μ is obviously X-subunit, so that

$$d_X(\gamma(s), \gamma(t)) = d_X(\mu(A), \mu(B)) \leq \ell(\mu) = B - A = \|\alpha\|_\infty\, (t - s).$$

As a particular case of (7.4) (or directly from the definition of d_X), we deduce that, if $\gamma : [a, b] \to \Omega$ is X-subunit, then

$$d_X(\gamma(s), \gamma(t)) \leq |t - s|, \qquad \forall\, t, s \in [a, b]. \tag{7.5}$$

Example 7.6. Let $E = \{\partial_1, \ldots, \partial_N\}$ in $\mathbb{R}^N$. It is not difficult to prove (e.g., by the characterization of the X-subunit curves in (6.7a) and (6.7b)) that the associated E-control distance d_E is nothing but the Euclidean metric on $\mathbb{R}^N$.

Beware that, in general, d_X *also depends on the understood set* Ω, despite we avoid any notation in d_X that keeps track of Ω. For example, if $E = \{\partial_1, \partial_2\}$ in $\mathbb{R}^2$ and if Ω is the set by depriving $Q = (-2, 2) \times (-2, 2)$ of the closed segment on the x_2-axis joining $(0, 0)$ and $(0, 2)$, the d_E distance of $(-1, 1)$ and $(1, 1)$ (relative to Ω) is greater than their d_E distance relative to Q.   ♯

Remark 7.7. When it is not X-connected, Ω splits into a family of metric spaces $\Omega = \bigcup_i \Omega_i$, where $x, y \in \Omega_i$ if and only if x and y are connectable by an X-subunit curve in Ω_i. In this case, (Ω_i, d_X) is a metric space, and the distance between Ω_i and Ω_j is infinite iff $i \neq j$.

An example is given by $X = \{\partial_{x_1}\}$ and $\Omega = \mathbb{R}^2$: the associated Ω_i are the lines parallel to the x_1-axis, and d_X restricted to any such line is Euclidean distance; any two points x, y with $x_2 \neq y_2$ have infinite d_X-distance.   ♯

The name 'distance' in Def. 7.4 is justified by the following remarkable result.

Theorem 7.8. *If Ω is X-connected, then (Ω, d_X) is a metric space.*

Example 7.9. Since a Hörmander system X always provides X-connectivity on any connected open set (see Thm. 6.22), we infer that *the X-control distance associated with a system of Hörmander vector fields is a metric.*

Beware that, even if in the Hörmander case the X-connectivity can be achieved by integral curves of $\pm X_1, \ldots, \pm X_m$ which connect x and y (in the proof of Thm. 6.22 we denoted the set of these curves by $\mathcal{S}^*(X)$), this does not mean that in (7.3) we can replace $\mathcal{S}(X)$ with $\mathcal{S}^*(X)$. If we replaced $\mathcal{S}(X)$ with $\mathcal{S}^*(X)$, we would obtain

$$d_X^*(x, y) := \inf \left\{ \ell(\gamma) \,\middle|\, \gamma \in \mathcal{S}^*(X), \ \gamma \text{ connects } x \text{ and } y \right\},$$

which is greater than or equal to d_X, since $\mathcal{S}^*(X) \subseteq \mathcal{S}(X)$.

For example, in the simple Euclidean case when $E = \{\partial_1, \partial_2\}$ in $\mathbb{R}^2$, it may be easily proved that $d_E^*(x, y) = \sum_{j=1}^{2} |x_j - y_j|$, the so-called distance associated with the "taxicab norm" $\| \cdot \|_1$, where $\|(x_1, x_2)\|_1 = |x_1| + |x_2|$. Note that $\|x\|_1$ may be strictly greater than the Euclidean norm of x.   ♯

For the proof of Thm. 7.8 we need a preliminary result, of independent interest. In the sequel, we denote by $B_E(x, r)$ the Euclidean ball of centre $x \in \mathbb{R}^N$ and radius $r > 0$, and by $\| \cdot \|$ we denote the Euclidean norm on $\mathbb{R}^N$. The notation $A \Subset B$ means that the closure of A is contained in B.

Proposition 7.10. *Let Ω be X-connected. Suppose that $B_E(x,r) \Subset \Omega$. Then, there exists a constant $\mathbf{c} > 0$ (depending on Ω, X, x, r) such that*

$$d_X(x,y) \geq \mathbf{c}\,\|x-y\|, \quad \text{for every } y \in B_E(x,r). \tag{7.6}$$

If $K \subset \Omega$ is compact, then there exists $\mathbf{c} > 0$ (only depending on Ω, X, K) such that $d_X(x,y) \geq \mathbf{c}\,\|x-y\|$ for every $x, y \in K$.

Proof. We can suppose $x \neq y$ otherwise (7.6) is a trivial consequence of the non-negativity of d_X. For any x, r such that $B_E(x,r) \Subset \Omega$, we set

$$M(x,r) := \sup_{y \in B_E(x,r)} \sum_{j=1}^{m} \|X_j(y)\|. \tag{7.7}$$

Note that $M(x,r)$ is finite since $X_1, \ldots, X_m$ are continuous and $B_E(x,r) \Subset \Omega$. We show the following inequality

$$M(x, \|x-y\|)\, d_X(x,y) \geq \|x-y\|, \quad \text{whenever } B_E(x, \|x-y\|) \Subset \Omega. \tag{7.8}$$

This gives at once (7.6) by choosing $\mathbf{c} = (\max\{1, M(x,r)\})^{-1}$ (note that we have $M(x, \|x-y\|) \leq M(x,r)$ if $y \in B_E(x,r)$).

By contradiction, assume (7.8) is false for some x, y in Ω such that

$$B_E(x, \|x-y\|) \Subset \Omega.$$

Then, as Ω is X-connected, there exists a curve $\gamma \in \mathcal{S}(X)$, $\gamma : [0,T] \to \Omega$ with $\gamma(0) = x$, $\gamma(T) = y$ and such that the following inequality is fulfilled (here we exploit the "inf-definition" of d_X)

$$M(x, \|x-y\|)\,T < \|x-y\|.$$

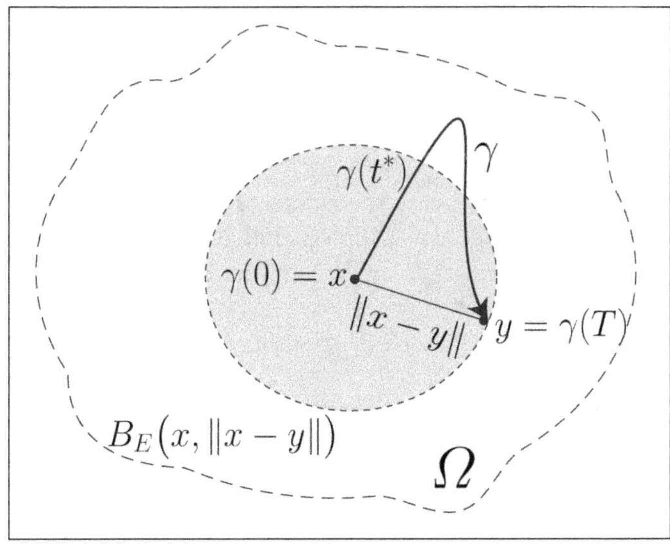

Fig. 7.1 The figure in the proof of Prop. 7.10.

As a consequence, if we consider the so-called first exit-time t^* of γ with respect to the set $B_E(x, \|x - y\|)$ (see Fig. 7.1), that is

$$t^* := \sup\left\{t \in [0, T] \ : \ \gamma(s) \in B_E(x, \|x - y\|) \text{ for } 0 \le s \le t\right\},$$

we have the following computation (we use the absolute continuity of γ)

$$\|\gamma(t^*) - x\| = \left\|\int_0^{t^*} \dot{\gamma}(s)\,\mathrm{d}s\right\| \le \int_0^{t^*} \|\dot{\gamma}(s)\|\,\mathrm{d}s \overset{(6.6)}{\le} \int_0^{t^*} \sum_{j=1}^m \|X_j(\gamma(s))\|\,\mathrm{d}s$$

$$\left(\text{note that } \gamma(s) \in B_E(x, \|x - y\|) \text{ for every } t \in [0, t^*)\right)$$

$$\le M(x, \|x - y\|)\,t^* \le M(x, \|x - y\|)\,T \lneqq \|x - y\|.$$

Since we obviously have $\|\gamma(t^*) - x\| = \|x - y\|$, the above chain of inequalities yields a contradiction, hence (7.8) holds true. The last statement of the proposition is an adaptation of the previous arguments (see Exr. 7.6). $\qquad\square$

We are ready for the following proof.

Proof of Thm. 7.8. The theorem follows from the properties of d_X below.

(1). We trivially have $d_X(x, y) \ge 0$ for every $x, y \in \Omega$ and $d_X(x, x) = 0$.

(2). For $x, y \in \Omega$ it holds that $d_X(x, y) = d_X(y, x)$ (Exr. 7.2).

(3). For $x, y, z \in \Omega$, one has $d_X(x, z) \le d_X(x, y) + d_X(y, z)$. Indeed, given $\varepsilon > 0$ there exist X-subunit curves $\gamma : [a_1, b_1] \to \Omega$, $\mu : [a_2, b_2] \to \Omega$ with

$$\gamma(a_1) = x, \quad \gamma(b_1) = y, \qquad \mu(a_2) = y, \quad \mu(b_2) = z,$$

and such that $T_1 := b_1 - a_1 \le d_X(x, y) + \varepsilon$ and $T_2 := b_2 - a_2 \le d_X(y, z) + \varepsilon$. By gluing γ and μ, it is easy to construct an X-subunit curve Γ defined on the interval $[0, T_1 + T_2]$ connecting x and z (see Exr. 7.3). As a consequence,

$$d_X(x, z) \le \ell(\Gamma) = T_1 + T_2 \le d_X(x, y) + d_X(y, z) + 2\varepsilon.$$

Letting $\varepsilon \to 0$, the triangle inequality for d_X follows.

(4). The only non-trivial fact is the axiom of positivity, that is $d_X(x, y) = 0$ implies $x = y$. To this end, let $x, y \in \Omega$ be such that $d_X(x, y) = 0$. Let $r > 0$ be so small that $B_E(x, r) \Subset \Omega$. We set $B := B_E(x, r)$ for brevity. If $y \in B$, by (7.6) we derive that (for some positive $\mathbf{c} = \mathbf{c}(x, r)$) $\mathbf{c}\,\|x - y\| \le d_X(x, y) = 0$, whence $x = y$.

We can therefore suppose that $y \notin B$. Since $d_X(x, y) = 0$ and $r > 0$ there exists $\gamma_n \in \mathcal{S}(X)$, $\gamma_n : [0, T_n] \to \Omega$ with $\gamma_n(0) = x$, $\gamma_n(T_n) = y$ and such that $T_n \to 0$ as $n \to \infty$. Let us consider the first exit time t^* of γ_n out of B:

$$t^* := \sup\left\{t \in [0, T_n] \ : \ \gamma_n(s) \in B \text{ for } 0 \le s \le t\right\}.$$

Clearly, $t^* > 0$ and $\gamma_n(s) \in B$ for every $s \in [0, t^*)$. For any such s we have (again in view of (7.6))

$$\mathbf{c}\,\|x - \gamma_n(s)\| \le d_X(x, \gamma_n(s)) \le s \le t^* \le T_n.$$

Thus $\mathbf{c}\,\|x - \gamma_n(s)\| \leq T_n$ for any $s \in [0, t^*)$. Letting $s \to t^*$ from below, we get $\mathbf{c}\,\|x - \gamma_n(t^*)\| \leq T_n$. As $\gamma_n(t^*) \in \partial B$ we infer that $\mathbf{c}\,r \leq T_n$; letting $n \to \infty$ this gives the absurd $\mathbf{c}\,r = 0$. Thus, it is not possible that $y \notin B$, and the proof is complete with the previous case $y \in B$. $\qquad\square$

Remark 7.11. Let $x_0 \in \Omega$ be fixed, and let $r_0 > 0$ be such that $B_E(x_0, r_0) \Subset \Omega$. From Exr. 7.4 we infer that, if $M_0 := M(x_0, r_0)$ is as in (7.7), then any X-subunit curve $\gamma : [0, T] \to \Omega$ starting at x_0 and such that $T < r_0/M_0$ also satisfies

$$\gamma([0, T]) \subseteq B_E(x_0, r_0).$$

This implies (Exr. 7.5) that *the d_X-ball of centre x_0 and radius $\rho_0 < r_0/M_0$ is contained in $B_E(x_0, r_0)$*. This fact has nice topological consequences (Sec. 7.3). See Fig. 7.2. $\quad\natural$

7.2 Some equivalent definitions of d_X

In the literature, there are many equivalent ways of defining the X-control distance (see e.g., [Jerison and Sánchez-Calle (1987); Nagel *et al.* (1985)]). For example, it is immaterial (Exr. 7.1) if in the definition of an X-subunit curve $\gamma : [a, b] \to \Omega$ (see Def. 6.13) we take $[0, T]$ (with $T \geq 0$) instead of $[a, b]$.

Another equivalent definition is the following one.

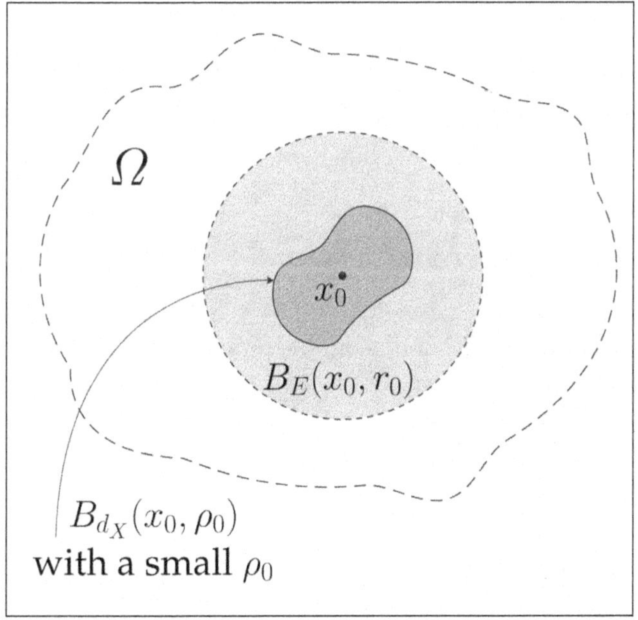

Fig. 7.2   Rem. 7.11: any d_X-ball $B_{d_X}(x_0, \rho_0)$ with a small radius $\rho_0 < r_0/M_0$ (where we have set $M_0 = \sup\limits_{B_E(x_0, r_0)} \sum_{j=1}^m \|X_j\|$) cannot escape the Euclidean ball $B_E(x_0, r_0)$.

Proposition 7.12. *The points $x, y \in \Omega$ can be connected by an X-trajectory if and only if they can be connected by an X-subunit curve, and (see (7.2))*

$$d_X(x,y) = \inf \left\{ \|\alpha\|_\infty \; : \; \begin{array}{l} \gamma : [0,1] \to \Omega \text{ is an } X\text{-trajectory, it} \\ \text{connects } x, y, \text{ and } \dot\gamma = S(\gamma)\alpha \text{ (a.e.)} \end{array} \right\}. \tag{7.9}$$

Thus, the rhs of (7.9) can serve as an equivalent definition of the CC distance.

Proof. If $\gamma : [0,T] \to \Omega$ is X-subunit, then the rescaling $\mu : [0,1] \to \Omega$ defined by $\mu(t) := \gamma(Tt)$ is an X-trajectory, with $\dot\mu = S(\mu)\alpha$ and $\|\alpha\|_\infty \leq T$ (see also Prop. 6.17). This shows that the rhs of (7.9) is $\leq d_X(x,y)$.

Vice versa, the fact that the rhs of (7.9) is also $\geq d_X(x,y)$ follows from (7.4) (since the domain of an X-trajectory is $[0,1]$). $\qquad\square$

In geodesic problems, the following definition is very useful.

Let $p \in [1,\infty]$ be arbitrarily fixed. Let $\gamma : [0,1] \to \Omega$ be an X-trajectory and let $c(t) = (c_1(t), \ldots, c_m(t))$ be the canonical coordinates of $\dot\gamma(t)$ (see Rem. 6.18). By definition of X-trajectory, we have $c_j \in L^\infty([0,1])$ for $j \leq m$. We define

$$\ell_p(\gamma) := \Big\| |c(t)| \Big\|_{L^p} = \begin{cases} \left(\displaystyle\int_0^1 |c(t)|^p \, \mathrm{d}t \right)^{1/p} & \text{if } p \in [1,\infty), \\[1em] \displaystyle\sup_{t \in [0,1]} |c(t)| & \text{if } p = \infty. \end{cases}$$

Here and throughout, if $c \in \mathbb{R}^m$ we denote by $|c|$ the Euclidean norm of c.

Accordingly, one defines

$$d_X^{(p)}(x,y) := \inf \left\{ \ell_p(\gamma) \;\Big|\; \gamma \text{ is an } X\text{-trajectory connecting } x \text{ and } y \right\}.$$

We recognize that $\ell_\infty(\gamma) = \|c\|_\infty$ (compare to (7.2)), whence $d_X \equiv d_X^{(\infty)}$ on account of (7.9). Luckily, we also have the following fact:

Proposition 7.13. *For any $x, y \in \Omega$ one has*

$$d_X^{(p)}(x,y) = d_X(x,y) \quad \forall \, p \in [1,\infty].$$

Proof. On the one hand we have $d_X^{(1)} \leq d_X^{(p)} \leq d_X^{(\infty)} = d_X$, as a consequence of Hölder's inequality for the L^p norms on the (probability) space $[0,1]$.

We are then left to show that $d_X^{(\infty)} \leq d_X^{(1)}$. We only give a sketch of the proof. Let γ be an X-trajectory connecting x and y, and let $c(t)$ provide the canonical coordinates of $\dot\gamma(t)$. We can clearly assume $\|c\|_{L^1} > 0$. Let us consider the function $\varphi : [0,1] \to [0,1]$ defined by

$$\varphi(t) := \frac{1}{\|c\|_{L^1}} \int_0^t |c(s)| \, \mathrm{d}s.$$

φ is absolutely continuous and non-decreasing. For simplicity, we assume that φ is everywhere differentiable and strictly monotonic, with

$$\varphi'(t) = \frac{|c(t)|}{\|c\|_{L^1}} > 0, \quad \text{for every } t \in [0,1].$$

(The proof in the general case goes along the same lines, taking some care in differentiation/inversion matters.) Let $\psi : [0,1] \to [0,1]$ be the inverse function of φ and consider the re-parameterized curve $\Gamma(s) := \gamma(\psi(s))$. Obviously, $\Gamma : [0,1] \to \Omega$ is absolutely continuous and it connects x and y; it is also an X-trajectory, since (for almost every s) one has

$$\dot{\Gamma}(s) = \psi'(s)\,\dot{\gamma}(\psi(s)) = \sum_{j=1}^{m} \alpha_j(s)\,X_j(\Gamma(s)), \quad \text{with}$$

$$\alpha_j(s) = \frac{\|c\|_{L^1}}{|c(\psi(s))|}\, c_j(\psi(s)), \quad \text{and} \quad \ell_\infty(\Gamma) = \sup_{s \in [0,1]} |\alpha(s)| = \|c\|_{L^1} < \infty.$$

As a consequence,[3] $d_X^{(\infty)}(x,y)$ does not exceed $\ell_\infty(\Gamma) = \|c\|_{L^1} = \ell_1(\gamma)$, whence $d_X^{(\infty)}(x,y) \le \ell_1(\gamma)$; taking the inf over the X-trajectories γ connecting x and y, we get $d_X^{(\infty)}(x,y) \le d_X^{(1)}(x,y)$ as desired. $\qquad\qquad\square$

Next we establish a metric characterization of X-trajectories. First, we say that a curve $\gamma : [a,b] \to \Omega$ is L-**Lipschitz** (wrt d_X), if $L \ge 0$ and

$$d_X(\gamma(t),\gamma(s)) \le L\,|t-s|, \quad \text{for any } t,s \in [a,b].$$

Remark 7.14. If $\gamma : [a,b] \to \Omega$ is L-Lipschitz wrt d_X, then it is Lipschitz continuous wrt Euclidean metric as well. Indeed, on account of Prop. 7.10, since $K := \gamma([a,b])$ is compact, for any $t,s \in [a,b]$ we have

$$\|\gamma(t) - \gamma(s)\| \le \mathbf{c}^{-1}\,d_X(\gamma(t),\gamma(s)) \le \mathbf{c}^{-1}\,L\,|t-s|,$$

so that γ is Lipschitz continuous wrt Euclidean metric. $\qquad\qquad\sharp$

Proposition 7.15. *Let Ω be X-connected. A curve $\gamma : [0,1] \to \Omega$ is L-Lipschitz (wrt d_X) if and only if it is an X-trajectory and $\dot{\gamma} = S(\gamma)\alpha$, with $\|\alpha\|_\infty \le L$.*
The choice $L = 1$ provides a characterization of the X-subunit curves.

Proof. If γ is an X-trajectory, then (7.4) shows that γ is $\|a\|_\infty$-Lipschitz.

Vice versa, suppose $\gamma : [0,1] \to \Omega$ is L-Lipschitz. By a simple rescaling, we can suppose that $L = 1$, so that we need to prove that a 1-Lipschitz curve $\gamma : [0,T] \to \Omega$ is X-subunit. By Rem. 7.14, γ is Lipschitz continuous (wrt Euclidean metric), hence it is a.e. differentiable; let $t \in [0,T]$ be a differentiability point of γ and let us prove that $\dot{\gamma}(t) = S(\gamma(t))a(t)$, with $\|a(t)\|_\infty \le 1$. For simplicity we consider $t = 0$.

We henceforth fix $k \in \mathbb{N}$ and we let $\varepsilon_k := (k+1)/k^2$. Let us consider the points $\gamma(0)$ and $\gamma(1/k)$; since γ is 1-Lipschitz, we have $d_X(\gamma(0),\gamma(1/k)) \le 1/k < \varepsilon_k$. Since Ω is X-connected, there exists an X-subunit curve γ_k in Ω connecting $\gamma(0)$ and $\gamma(1/k)$ with $\ell(\gamma_k)$ close to $d_X(\gamma(0),\gamma(1/k))$. Since the latter is $< \varepsilon_k$, we can suppose (up to a constant prolongation of γ_k) that γ_k is defined on $[0,\varepsilon_k]$. We have $\dot{\gamma}_k = S(\gamma_k)\alpha_k$ a.e. on $[0,\varepsilon_k]$, with $\|\alpha_k\|_\infty \le 1$.

---

[3] Here we are also using the minimality property in Rem. A.24 on page 350.

Since $\varepsilon_k \to 0$, we can appeal to the assertions in Rem. 7.11 and infer that the trajectories of all the curves γ_k are contained in a fixed compact subset of Ω. As a result (Prop. 7.10), there exists a constant $\mathbf{c} > 0$ (independent of k) such that, for any $a, b \in [0, \varepsilon_k]$ one has

$$\|\gamma_k(a) - \gamma_k(b)\| \le \mathbf{c}^{-1} d_X(\gamma_k(a), \gamma_k(b)) \le \mathbf{c}^{-1}|a - b|. \tag{7.10}$$

In the last inequality we used the fact that γ_k is X-subunit, together with (7.5). Consider the following computations:

$$\frac{\gamma(1/k) - \gamma(0)}{1/k} = \frac{\gamma(\varepsilon_k) - \gamma(0)}{1/k} = k \int_0^{\varepsilon_k} S(\gamma_k(t))\, \alpha_k(t)\, \mathrm{d}t$$

$$= k \int_0^{\varepsilon_k} \Big(S(\gamma_k(t)) - S(\gamma(0)) \Big)\, \alpha_k(t)\, \mathrm{d}t + k \int_0^{\varepsilon_k} S(\gamma(0))\, \alpha_k(t)\, \mathrm{d}t =: \mathrm{I}_k + \mathrm{II}_k.$$

Since the entries of S are locally Lipschitz continuous, and as $\|\alpha_k\|_\infty \le 1$, there exists a constant $C > 0$ such that

$$\|\mathrm{I}_k\| \le k \int_0^{\varepsilon_k} C\, \|\gamma_k(t) - \gamma(0)\|\, \mathrm{d}t \overset{(7.10)}{\le} C\, k \int_0^{\varepsilon_k} t\, \mathrm{d}t = \frac{C}{2}\, k\, \varepsilon_k^2.$$

Since the above rhs vanishes as $k \to 0$, we infer that $\lim_k \mathrm{I}_k = 0$. This gives

$$\dot\gamma(0) = \lim_{k \to \infty} \frac{\gamma(1/k) - \gamma(0)}{1/k} = \lim_{k \to \infty} \mathrm{II}_k = S(\gamma(0)) \cdot \lim_{k \to \infty} k \int_0^{\varepsilon_k} \alpha_k(t)\, \mathrm{d}t.$$

In particular, the limit in the far rhs exists and we denote it by $\alpha(0)$, so that the above identity reads $\dot\gamma(0) = S(\gamma(0))\, \alpha(0)$. Clearly we have

$$\|\alpha(0)\| \le \liminf_{k \to \infty} k \int_0^{\varepsilon_k} \|\alpha_k(t)\|\, \mathrm{d}t \le \liminf_{k \to \infty} k\, \varepsilon_k = 1.$$

The case of a general $t \ne 0$ is analogous, only complicated by notations, and we leave it to the interested reader. This completes the proof. $\qquad\square$

7.3   Basic topological properties of the CC-distance

Throughout this section, we suppose that Ω is X-connected, and d_X denotes its X-control distance. We also employ the following notation for the metric open ball of centre $x \in \Omega$ and radius $r > 0$:

$$B_X(x, r) := \{y \in \Omega : d_X(x, y) < r\}.$$

From Prop. 7.10 one gets the following fact (see also Exr. 7.15):

Proposition 7.16. *The Euclidean open sets in Ω are open sets in the metric space (Ω, d_X) as well. Equivalently,* id $: (\Omega, d_X) \longrightarrow (\Omega, \|\cdot\|)$ *is a continuous map.*

Hence, the d_X-topology coincides with the Euclidean topology of Ω if and only if d_X is continuous on $\Omega \times \Omega$ (the latter is equipped with the Euclidean product topology).

Dealing with metric spaces (which are first countable), continuity and sequential continuity are equivalent: hence Prop. 7.16 reads

$$\left(x_n \xrightarrow{n\to\infty} x \quad \text{wrt } d_X\right) \quad \Longrightarrow \quad \left(x_n \xrightarrow{n\to\infty} x \quad \text{wrt } \|\cdot\|\right).$$

Proof. Given any $x_0 \in \Omega$, due to Rem. 7.11 we know that for any $r_0 > 0$ there exists $\rho_0 > 0$ such that $B_X(x_0, \rho_0) \subseteq B_E(x_0, r_0)$. This shows that any Euclidean open set is also open wrt d_X. $\qquad\square$

Remark 7.17. Since the d_X-metric topology is finer than the Euclidean topology of Ω, any Euclidean closed set in Ω is closed wrt d_X. This shows that

$$\mathrm{CL}_E(A) \supseteq \mathrm{CL}_X(A), \quad \text{for every } A \subseteq \Omega;$$

here, CL_E, CL_X denote closures wrt Euclidean metric and wrt d_X, respectively. Analogously, any compact set in the d_X topology is automatically a compact set in the Euclidean topology. $\qquad\sharp$

Unfortunately, id $: (\Omega, \|\cdot\|) \longrightarrow (\Omega, d_X)$ is, in general, *not necessarily continuous*, as the next example shows.

Example 7.18. Consider the vector fields in $\mathbb{R}^2$ defined by

$$X_1 = \partial_{x_1}, \quad X_2 = \max\{0, x_1\}\,\partial_{x_2},$$

and let $X = \{X_1, X_2\}$. It is easy to recognize that $\mathbb{R}^2$ is X-connected. Surprisingly (or not), the sequence $(-1, 1/n)$, converging to $x_0 = (-1, 0)$ as $n \to \infty$ in the Euclidean metric, does not converge to x_0 in the d_X metric (Exr. 7.8). It is easy to prove that the d_X-ball of centre $(-1, -1)$ and radius $1/2$ is

$$B_X((-1, -1), 1/2) = \left\{(x_1, -1) \in \mathbb{R}^2 : -3/2 < x_1 < -1/2\right\},$$

which is not open in the Euclidean metric. See Fig. 7.3. This pathology is not due to the lack of smoothness of X_2: we can obtain an analogous example by replacing X_2, for instance, with the smooth vector field $\psi(x_1)\,\partial_{x_2}$, where

$$\psi(x_1) = \begin{cases} 0 & \text{if } x_1 \leq 0 \\ \exp(-1/x_1) & \text{if } x_1 > 0. \end{cases}$$

The following result shows that, if X is a Hörmander system on Ω, the topology on Ω associated with the X-control distance coincides with the Euclidean topology. Hence, from a mere topological point of view, we cannot distinguish d_X from the Euclidean distance.

Theorem 7.19. *Let $X = \{X_1, \ldots, X_m\}$ be a set of Hörmander v.f.s on Ω. Then, for every compact set $K \subset \Omega$, there exist $k \in \mathbb{N}$ and $\mathbf{c} > 0$ (both depending on X, K, Ω) such that*

$$d_X(x, y) \leq \mathbf{c}\, \|x - y\|^{1/k}, \quad \text{for every } x, y \in K. \tag{7.11}$$

More precisely, one can take k as the smallest integer such that the span of the iterated commutators of $X_1, \ldots, X_m$ of length $\leq k$ is $\mathbb{R}^N$ at every point of K.

Moreover, $d_X : \Omega \times \Omega \to \mathbb{R}$ is continuous (in the usual Euclidean topology), that is id $: (\Omega, \|\cdot\|) \to (\Omega, d_X)$ is a continuous map (in fact, a homeomorphism), and the Euclidean topology of Ω coincides with the d_X-metric topology.

Putting together Prop. 7.10 and (7.11), we infer that, if X is a Hörmander system, for every compact $K \subset \Omega$, there exist $\mathbf{c} > 0$ and $k \in \mathbb{N}$ such that

$$\mathbf{c}^{-1}\|x - y\| \le d_X(x, y) \le \mathbf{c}\,\|x - y\|^{1/k}, \quad \text{for every } x, y \in K. \tag{7.12}$$

Proof. We prove (7.11), since the last statement follows from it and Prop. 7.16.

In order to prove (7.11), it suffices to run over the proof of the Connectivity Thm. 6.22: we inherit the notation of that proof. We fix any $x_0 \in K$ and we choose $Y_1, \ldots, Y_N$ as in (6.9) in such a way that they are nested commutators of the v.f.s in X of length $\le k$. Since the associated map Θ (page 142) is a C^1-diffeomorphism of a neighborhood of 0 in $\mathbb{R}^N$ onto a neighborhood of x_0, there exist $\rho, \varepsilon, M > 0$ such that $B_E(x_0, \varepsilon)$ is contained in $\Theta(B_E(0, \rho))$, and

$$\|\Theta(s_1) - \Theta(s_2)\| \ge M\,\|s_1 - s_2\|, \quad \forall\, s_1, s_2 \in B_E(0, \rho). \tag{7.13}$$

Let us fix $x \in B_E(x_0, \varepsilon)$ and let us take $\|s\| < \rho$ such that $x = \Theta(s)$. With reference to the form of Θ, namely

$$\Theta(s) = E_1^*\Big(s_1, E_2^*\big(s_2, \cdots E_{N-1}^*\big(s_{N-1}, E_N^*(s_N, x_0)\big) \cdots\big)\Big),$$

one has that x is the last term of the finite sequence $x_1, x_2, \ldots, x_N$, where

$$x_1 = E_N^*(s_N, x_0), \quad x_2 = E_{N-1}^*(s_{N-1}, x_1), \quad x_3 = E_{N-2}^*(s_{N-2}, x_2), \quad \text{etc.}$$

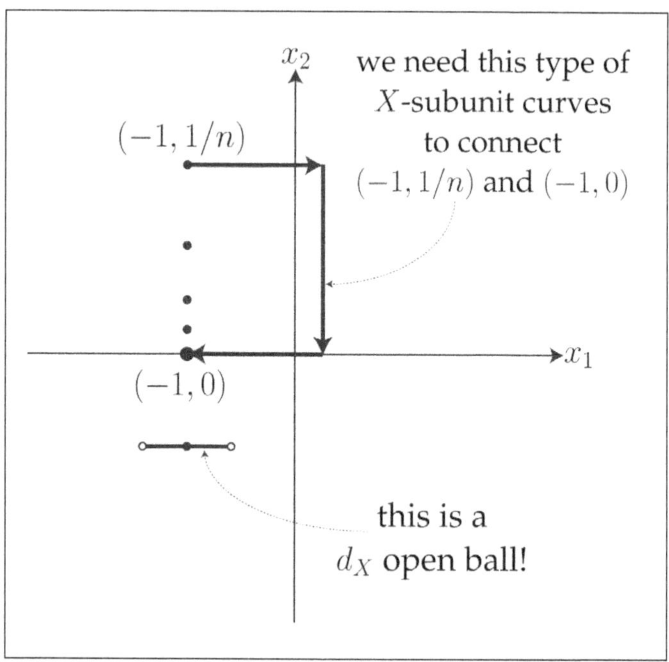

Fig. 7.3 Exm. 7.18: an example of a CC metric whose topology is strictly finer than the Euclidean topology; the segment $(-3/2, -1/2) \times \{-1\}$ is an open ball wrt d_X. "Long" X-subunit curves (if measured wrt Euclidean metric) are needed to connect $(-1, 1/n)$ and $(-1, 0)$.

It is very easy to recognize that

$$d_X\left(x, \alpha_k(\Psi_t^{X_{j_1}}, \Psi_t^{X_{j_2}}, \ldots, \Psi_t^{X_{j_k}})(x)\right) \leq |t|\,(3 \cdot 2^{k-1} - 2).$$

Since the maps E_i^* are modeled on objects of the form

$$\alpha_k(\Psi_t^{X_{j_1}}, \Psi_t^{X_{j_2}}, \ldots, \Psi_t^{X_{j_k}})(x), \quad \text{with } t = s^{1/k},$$

we deduce that $d_X(x_{i-1}, x_i) \leq C\,|s_{N-i+1}|^{1/k}$, for $i = 1, \ldots, N$ (where $C > 0$ depends on K and k). Summing up, using also (7.13), we have

$$d_X(x_0, x) \leq \sum_{i=1}^{N} d_X(x_{i-1}, x_i) \leq C \sum_{i=1}^{N} |s_{N-i+1}|^{1/k} \leq C\,N\,\|s\|^{1/k}$$

$$\leq \frac{C\,N}{M^{1/k}}\,\|\Theta(s) - \Theta(0)\|^{1/k} = \frac{C\,N}{M^{1/k}}\,\|x - x_0\|^{1/k}.$$

A covering argument of the compact set K with finitely many balls of the form $B_E(x_0, \varepsilon)$ proves (7.11). $\qquad\square$

7.3.1 *Euclidean boundedness of the d_X balls*

Boundedness wrt d_X can be very different from boundedness in the Euclidean metric of Ω, even in the case of Hörmander systems:

Example 7.20. In $\mathbb{R}^2$, let us set $X = \{X_1, X_2\}$, where $X_1 = \partial_{x_1}$, $X_2 = (1 + x_2^2)\,\partial_{x_2}$. Clearly, X is a Hörmander system so that $\mathbb{R}^2$ is X-connected. As is shown in Exr. 7.7, the sequence of points $P_n = \left(0, \tan(\frac{\pi}{2} - \frac{1}{n})\right)$ satisfies

$$d_X(0, P_n) < \pi/2, \quad \text{but} \quad \|P_n\| \xrightarrow{n \to \infty} \infty,$$

where $\|\cdot\|$ is Euclidean norm on $\mathbb{R}^2$. Hence $B_X(0, \pi/2)$ is bounded wrt d_X, but unbounded in the Euclidean metric. Thus, $D := \{x \in \mathbb{R}^2 : d_X(0, x) \leq \pi/2\}$ is not compact in the Euclidean topology. Note that the Euclidean topology and the d_X-topology of $\mathbb{R}^2$ coincide, because X is Hörmander (Thm. 7.19), so that the closed d_X-ball D is not compact even in the d_X-topology. We shall see in Exm. 7.29 that this implies that $(\mathbb{R}^2, d_X)$ is not complete.

The fact that X_2 *is not a global vector field* plays a major role in this example: indeed, the curve $\gamma : (-\pi/2, \pi/2) \to \mathbb{R}^2$, $\gamma(t) = (0, \tan(t))$ is the maximal integral curve of X_2 starting at $(0, 0)$, hence the restriction of γ to any interval $[a, b]$ in $(-\pi/2, \pi/2)$ is X-subunit (Exm. 6.16). $\qquad\sharp$

A partial positive result is the following one.

Proposition 7.21. *For any point $x \in \Omega$ there exists $\rho > 0$ such that $B_X(x, \rho)$ is bounded in the Euclidean metric.*

More precisely, if $r > 0$ is such that $B_E(x, r) \Subset \Omega$ and if $M := M(x, r)$ is as in (7.7), then $B_X(x, \rho) \subseteq B_E(x, r)$ whenever $\rho\,M < r$.

Proof. This follows directly from Rem. 7.11. $\qquad\square$

Another pertinent result is the following one.

Proposition 7.22. *If d_X is continuous, then (Ω, d_X) is locally compact. Hence, this is true if X is a Hörmander set of vector fields (see Thm. 7.19).*

More precisely, the closure (wrt d_X or wrt Euclidean toplogy) of any ball $B_X(x, \rho)$ as in Prop. 7.21 is a compact set in (Ω, d_X).

Proof. Let $x \in \Omega$ be arbitrary, and let $\rho > 0$ be so small that $B_X(x, \rho)$ is bounded in the Euclidean metric (Prop. 7.21). Since d is continuous, by Prop. 7.16 we infer that the d_X-topology coincides with the Euclidean topology, so that (using the notation of Rem. 7.17) $\mathrm{CL}_X(B_X(x, \rho))$ coincides with the Euclidean closure, say $\overline{B_X(x, \rho)}$, of $B_X(x, \rho)$. Therefore, $\overline{B_X(x, \rho)}$ is a compact set in the Euclidean sense, hence in (Ω, d_X) as well. This ends the proof. $\qquad\square$

Remark 7.23. It is not difficult to prove (e.g., by using Gronwall's Lemma) that, if the v.f.s in X are *globally* Lipschitz continuous, then the d_X-balls are bounded in the Euclidean metric. $\qquad\qquad\sharp$

Example 7.24. The control distance d_X on $\mathbb{R}^2$ in Exm. 7.18 provides an example of a non-locally compact CC space. To see this, we argue by contradiction: suppose that 0 has a d_X-compact neighborhood; this implies the existence of a d_X-ball B centred at 0 whose d_X-closure, say $D := \mathrm{CL}_X(B)$, is d_X-compact. It can be proved that B contains a square $[-\varepsilon, \varepsilon] \times [-\varepsilon, \varepsilon]$ (Exr. 7.14). The following family is a d_X-open cover of $\mathbb{R}^2$, hence of D:

$$\mathfrak{U} := \{U, \ U_y\}_{y \in \mathbb{R}},$$

where U is the half-plane $\{x \ : \ x_1 > -\varepsilon/2\}$ (this is Euclidean-open, hence d_X-open), while U_y is the half-line $\{x \ : \ x_1 < 0, \ x_2 = y\}$ (this is d_X-open as observed in Exm. 7.18). Now, U is not enough to cover D since it leaves out the segment $S := \{-\varepsilon\} \times [-\varepsilon, \varepsilon]$; on the other hand, again because of S, no finite sub-family of $\{U_y\}_y$ can cover D. This shows that $\mathfrak{U}$ is an open cover of D not allowing any finite sub-cover; hence D is not compact. $\qquad\qquad\sharp$

A useful improvement of Prop. 7.21 is the following result. In what follows, an over-line denotes the closure of a set wrt Euclidean topology.

Lemma 7.25. *Let D be bounded wrt Euclidean metric, and assume that $\overline{D} \subseteq \Omega$. Then, there exists a positive $R_0 \ll 1$ and a (Euclidean) compact set $K \subset \Omega$ such that, for every $x \in D$ and for every $r \leq R_0$, one has $B_X(x, r) \subseteq K$.*

Proof. Let $\varepsilon > 0$ be so small that the closure K of $\bigcup_{x \in \overline{D}} B_E(x, \varepsilon)$ is a compact subset of Ω. With the notation in (7.7), we let

$$\mathbf{M} := \sup\{1 + M(x, \varepsilon) \ : \ x \in \overline{D}\}.$$

Clearly, $0 < \mathbf{M} < \infty$. We set $R_0 := \varepsilon/(2\,\mathbf{M})$. Let $x \in D$ and $r \le R_0$; let also $y \in B_X(x, r)$. Then, there exists an X-subunit curve $\gamma : [0, T] \to \Omega$ connecting x and y, and satisfying

$$T < d_X(x, y) + R_0 < r + R_0 \le 2\,R_0 = \varepsilon/\mathbf{M} \le \varepsilon/M(x, \varepsilon).$$

Then, due to Rem. 7.11, we derive that $\gamma([0, T]) \subseteq B_E(x, \varepsilon)$. In particular, we have $y = \gamma(T) \in B_E(x, \varepsilon)$. This ends the proof. $\qquad\square$

7.3.2 Length space property

As in Riemann integration theory, given a compact real interval $[a, b]$, we say that $\sigma = \{t_0, t_1, \ldots, t_n\}$ is a partition of $[a, b]$ if

$$a = t_0 \le t_1 \le \cdots \le t_n = b.$$

The set of the partitions of $[a, b]$ will be denoted by $\Omega[a, b]$.

Let (M, d) be a metric space; $\gamma : [a, b] \to M$ is said to be d-**rectifiable** if

$$\mathrm{Var}(\gamma) := \sup_{\sigma \in \Omega[a,b]} \left\{ \sum_{j=1}^{n} d(\gamma(t_{j-1}), \gamma(t_j)) \;\middle|\; \sigma = \{t_0, \ldots, t_n\} \right\} < \infty.$$

$\mathrm{Var}(\gamma)$ is referred to as the **total variation** of γ. Moreover, (M, d) is said to be a **length space** if, for every $x, y \in M$ one has

$$d(x, y) = \inf_{\gamma} \left\{ \mathrm{Var}(\gamma) \right\},$$

where the inf runs over the continuous d-rectifiable curves $\gamma : [a, b] \to M$ connecting x and y. It is part of the definition of a length space to require that this set is always non-void, for every $x, y \in M$. We have the following fact.

Proposition 7.26. *(Ω, d_X) is a length space, i.e., for any $x, y \in \Omega$ it holds that*

$$d_X(x, y) = \inf \left\{ \mathrm{Var}(\gamma) \;\middle|\; \begin{array}{l} \gamma : [a, b] \to \Omega \text{ is continuous,} \\ d_X\text{-rectifiable, and connects } x, y \end{array} \right\}. \tag{7.14}$$

Moreover, any X-subunit curve γ is d_X-rectifiable, and $\mathrm{Var}(\gamma) \le \ell(\gamma)$.

Proof. The inequality $d_X(x, y) \le \mathrm{Var}(\gamma)$ (if γ is as in the rhs of (7.14)) is a consequence of the triangle inequality for d_X; vice versa, if $\gamma : [0, T] \to \Omega$ is X-subunit and connects x and y, one uses the inequality (see Exr. 7.9)

$$\sum_{j=1}^{n} d_X(\gamma(t_{j-1}), \gamma(t_j)) \le \sum_{j=1}^{n}(t_j - t_{j-1}) = T.$$

The latter (besides showing that an X-subunit curve is d_X-rectifiable) implies that the inf in the rhs of (7.14) is less than or equal to T; we conclude by passing to the inf over the X-subunit curves connecting x and y. $\qquad\square$

Corollary 7.27. *For any $x \in \Omega$ and any $r > 0$ one has that*

$$\text{the } d_X\text{-closure of } B_X(x,r) \text{ is } \left\{ y \in \Omega : d_X(x,y) \leq r \right\}, \text{ and}$$

$$\text{the } d_X\text{-boundary f } B_X(x,r) \text{ is } \left\{ y \in \Omega : d_X(x,y) = r \right\}.$$

Proof. This is true in any length space (Exr. 7.11), and so is (Ω, d_X) (Prop. 7.26). $\quad\square$

We have the following notable result, a partial statement of a Hopf-Rinow type theorem in our setting:[4]

Theorem 7.28. *Suppose that* id $: (\Omega, \|\cdot\|) \longrightarrow (\Omega, d_X)$ *is continuous. Then the following facts are equivalent:*

    (a) *for every $x \in \Omega$ and $r > 0$, the closed d_X-ball $\{y : d(x,y) \leq r\}$ is compact;*
    (b) *the metric space (Ω, d_X) is complete.*

Proof. (a) $\Rightarrow$ (b): Let $(x_n)_n$ be a Cauchy sequence in (Ω, d_X); it is trivial to show that $(x_n)_n$ is contained in a closed d_X-ball. Since the latter is compact by assumption, we can extract a convergent sub-sequence; hence $(x_n)_n$ is convergent, as it is a Cauchy sequence with a convergent sub-sequence.

(b) $\Rightarrow$ (a): Let $x \in \Omega$ be fixed, and denote by $D(r)$ the closed d_X-ball of centre x and radius r. Let also

$$I := \{r \geq 0 : D(r) \text{ is compact}\}.$$

It is obvious that I is an interval, and the continuity of d_X implies that I contains some interval $[0, \rho]$, with $\rho > 0$ (Prop. 7.22). We need to show that $I = [0, \infty)$; to this end, we show that I is open and closed wrt $[0, \infty)$.

We show that I is open: given $r \in I$, we use the local compactness of (Ω, d_X) to cover the (compact) $D(r)$ with finitely many balls $B_i = B_X(x_i, r_i)$ with compact closure in Ω. Then it is easy to show that $\bigcup_i B_i$ is compact and it contains a set of the form $D(r + \delta)$, for some $\delta > 0$; thus $r + \delta \in I$.

The proof that I is closed is (only) sketched in Exr. 7.16. $\quad\square$

Example 7.29. Consider the CC space $(\mathbb{R}^2, d_X)$ in Exm. 7.20. We have seen that it contains a closed d_X-ball which is not d_X-compact. Since d_X is continuous, as X is a Hörmander system (Thm. 7.19), in view of Thm 7.28 we deduce that $(\mathbb{R}^2, d_X)$

---

[4]In a locally compact length spaces (M, d), the following facts are equivalent (see, e.g., [Cohn-Vossen (1936)] or [Busemann (1955)]):

- closed balls are compact;

- M is complete;

- every geodesic $\gamma : [0, \varepsilon) \to M$ (with $\varepsilon > 0$) can be completed.

Bear in mind that our CC spaces are always length spaces (Prop. 7.26), and, if d_X is continuous, they are also locally compact (Prop. 7.22).

is *not complete.* Observe that this is not in contradiction to the fact that the d_X-topology of $\mathbb{R}^2$ is equivalent to the Euclidean topology (and with the fact that $\mathbb{R}^2$ with the Euclidean metric is complete).

This example also shows that the double inequality in (7.12) (holding true in the present example since X is Hörmander) does not guarantee that $(\mathbb{R}^2, d_X)$ has the same Cauchy sequences of the metric space $(\mathbb{R}^2, \| \cdot \|)$.

7.4   Exercises of Chap. 7

Exercise 7.1. Suppose that $\gamma : [a, b] \to \Omega$ is X-subunit. Prove that

$$\Gamma : [0, b - a] \to \Omega, \quad \Gamma(t) := \gamma(a + t)$$

is X-subunit and has the same image set of γ (preserving starting/ending points).

Exercise 7.2. Let $\gamma : [0, T] \to \Omega$ be X-subunit. Prove that

$$\Gamma : [0, T] \to \Omega, \quad \Gamma(t) := \gamma(T - t)$$

is X-subunit. Note that $\Gamma(0) = \gamma(T)$ and $\Gamma(T) = \gamma(0)$.

[This easily proves the symmetry of the X-control distance.]

Exercise 7.3. Let $\gamma : [0, T_1] \to \Omega$ and $\mu : [0, T_2] \to \Omega$ be X-subunit curves satisfying $\gamma(T_1) = \mu(0)$. Prove that the following gluing curve is X-subunit:

$$\Gamma : [0, T_1 + T_2] \to \Omega, \quad \Gamma(t) := \begin{cases} \gamma(t), & \text{if } t \in [0, T_1] \\ \mu(t - T_1), & \text{if } t \in (T_1, T_1 + T_2]. \end{cases}$$

[This is used to prove the triangle inequality for the X-control distance.]

Exercise 7.4. Prove the following fact: if $B_E(x, r) \Subset \Omega$ and $M(x, r)$ is as in (7.7), then any X-subunit curve $\gamma : [0, T] \to \Omega$ starting at x with $T < r/M(x, r)$ also satisfies $\gamma([0, T]) \subseteq B_E(x, r)$.

Hint: By contradiction let $t^* \in (0, T]$ be the first exit time of γ wrt $B_E(x, r)$; then $\gamma(t^*) \in \partial B_E(x, r)$ so that

$$r = \left\| \int_0^{t^*} \dot{\gamma}(t)\, dt \right\| \leq \int_0^{t^*} \sum_{j=1}^m \|X_j(\gamma(t))\|\, dt \leq M(x, r)\, T < r.$$

Exercise 7.5. Let $x_0 \in \Omega$ be fixed, and let $r_0 > 0$ be such that $B_E(x_0, r_0) \Subset \Omega$. With the notation of the previous exercise, prove that the d_X-ball of centre x_0 and radius $\rho_0 < r_0/M_0$ is contained in $B_E(x_0, r_0)$.

[*Hint:* If $d(x_0, y) < \rho_0$, by the definition of d_X there exists an X-subunit curve $\gamma : [0, T] \to \Omega$ connecting x_0 and y with $T < \rho_0$; then use Exr. 7.4 to infer that $y = \gamma(T) \in B_E(x_0, r_0)$...]

Exercise 7.6. Prove the last statement of Prop. 7.10 by the following argument:

(1) It is possible to find a small $\varepsilon = \varepsilon(K, \Omega) > 0$ such that the closure of the set $K_\varepsilon := \bigcup_{x \in K} B_E(x, \varepsilon)$ is contained in Ω.

(2) The number $M = M(K, \Omega, X) := \sup \left\{ \sum_{j=1}^m \|X_j(y)\| \ : \ y \in K_\varepsilon \right\}$ is finite. Fixing $x, y \in K$, let us take $\rho := \min\{\varepsilon, \|x - y\|\}$; recognize that

- $B_E(x, \rho) \Subset \Omega$,
- $M(x, \rho) \leq M$ (see the notation in (7.7)),
- $y \notin B_E(x, \rho)$

(3) Take any X-subunit $\gamma : [0, T] \to \Omega$ connecting x and y. Recognize that $\gamma(T)$ does not belong to $B_E(x, \rho)$ so that, by Exr. 7.4, $T \geq \rho/M(x, \rho)$.

(4) Infer that $T \geq \rho/M$; due to the definition of ρ this leads to at least one of the following situations:

- if $\rho = \|x - y\|$, then $T \geq \|x - y\|/M =: \mathbf{c}\,\|x - y\|$,
- if $\rho = \varepsilon$, then

$$T \geq \frac{\varepsilon}{M} \geq \frac{\varepsilon}{M}\,\frac{\|x - y\|}{\operatorname{diam}(K)} =: \mathbf{c}\,\|x - y\|.$$

(5) In either cases, $T \geq \mathbf{c}\,\|x - y\|$ and by passing to the inf over γ one gets $d_X(x, y) \geq \mathbf{c}\,\|x - y\|$.

Exercise 7.7. Consider the v.f.s in $\mathbb{R}^2$ in Exm. 7.20. Check that $X = \{X_1, X_2\}$ is a Hörmander system in $\mathbb{R}^2$. Consider the associated X-control distance d_X. Why is $\mathbb{R}^2$ X-connected? Next, provide the details of the following argument:

(1) The integral curve of X_2 starting from (x_0, y_0) is

$$t \mapsto \big(x_0, \tan(t + \arctan(y_0))\big).$$

(2) Deduce that, for every $n \in \mathbb{N}$, one has

$$d_X\Big((0, 0), \big(0, \tan(\tfrac{\pi}{2} - \tfrac{1}{n})\big)\Big) \leq \pi/2 - 1/n < \pi/2.$$

(3) However we have $\big\|\big(0, \tan(\tfrac{\pi}{2} - \tfrac{1}{n})\big)\big\| \to \infty$ as $n \to \infty$.

Derive the existence of a subset of $\mathbb{R}^2$ which is bounded wrt d_X but unbounded in the Euclidean metric.

Exercise 7.8. Let $X = \{X_1, X_2\}$ in $\mathbb{R}^2$, where

$$X_1 = \partial_{x_1}, \quad X_2 = \max\{0, x_1\}\,\partial_{x_2}.$$

(1) Prove that the integral curves of X_1 are the lines parallel to the x_1 axis; prove also that the integral curve of X_2 starting at (x_0, y_0) is

$$t \mapsto \begin{cases} (x_0, y_0) & \text{if } x_0 \leq 0, \\ (x_0, y_0 + x_0\, t) & \text{if } x_0 > 0, \end{cases}$$

i.e., the integral curves of X_2 are given by constant points (x_0, y_0) if $x_0 \leq 0$, or by the lines parallel to the x_2 axis if $x_0 > 0$.

(2) Deduce that $\mathbb{R}^2$ is X-connected (by means of curves which are integral curves of $\pm X_1, \pm X_2$).

(3) Consider $z = (-1, 0)$ and the sequence $z_n = (-1, 1/n)$. Prove that there exists a constant $\varepsilon > 0$ such that $d_X(z, z_n) > \varepsilon$ for every n.

[*Hint:* A subunit curve $\gamma : [0, T] \to \mathbb{R}^2$ connecting z and z_n must satisfy (6.7a). Write explicitly this ODE in the present case, and observe that $\gamma_2(t) = 0$ for $0 \leq t \leq \varepsilon$ for some $\varepsilon > 0$. This implies that $T \geq \varepsilon$...]

(4) Derive that the identity map id $: (\mathbb{R}^2, d_E) \to (\mathbb{R}^2, d_X)$ is not continuous, where d_E is the Euclidean metric.

Exercise 7.9. Prove that, if $\gamma : [a, b] \to \Omega$ is X-subunit, then

$$d_X(\gamma(t_1), \gamma(t_2)) \leq t_2 - t_1, \quad \text{whenever } a \leq t_1 \leq t_2 \leq b.$$

[*Hint:* Consider the restriction of γ to $[t_1, t_2]$...]

Exercise 7.10. In this exercise, (M, d) is a length space, and it is understood that M is equipped with the metric topology. Moreover, for $x \in M$ and $r > 0$ we set $B_d(x, r) := \{y \in M : d(x, y) < r\}$. Prove the following facts.

(1). Using the very definition of $\mathrm{Var}(\gamma)$, show the *additivity property of* Var: if γ is d-rectifiable, for any partition $\{t_0, \ldots, t_n\}$ of $[a, b]$, we have

$$\mathrm{Var}(\gamma) = \sum_{i=1}^{n} \mathrm{Var}\left(\gamma|_{[t_{i-1}, t_i]}\right). \tag{7.15}$$

(2). Prove this *lower semi-continuity property of* Var: if $\gamma, \gamma_n : [a, b] \to M$ are curves such that γ_n point-wise converges to γ, then

$$\liminf_{n \to \infty} \mathrm{Var}(\gamma_n) \geq \mathrm{Var}(\gamma).$$

(3). Prove the following *mesh property of* Var: if $\gamma : [a, b] \to M$ is continuous and d-rectifiable, for every $\varepsilon > 0$ there exists $\delta(\varepsilon) > 0$ such that, for any partition $\{t_0, \ldots, t_n\}$ of $[a, b]$ with $\sup_{1 \leq j \leq n} |t_j - t_{j-1}| \leq \delta(\varepsilon)$, then

$$0 \leq \mathrm{Var}(\gamma) - \sum_{j=1}^{n} d(\gamma(t_{j-1}), \gamma(t_j)) < \varepsilon. \tag{7.16}$$

[*Hint:* Use the definition of a length space and the Heine-Borel theorem...]

Exercise 7.11. Whereas in an arbitrary metric space this is not always the case, in a length space we have

$$\overline{B_d(x, r)} = \{y \in M : d(x, y) \leq r\}, \qquad \partial B_d(x, r) = \{y \in M : d(x, y) = r\}.$$

Prove this, by completing the following argument.

(1) Recognize that both the above identities are consequences of the following claim: if $d(x, y) = r$, then there exist $y_n \in B_d(x, r)$ such that $y_n \to y$. Then prove the claim as below.

(2) There exist d-rectifiable curves $\gamma_n : [a_n, b_n] \to M$ connecting x and y, and such that $\lim_n \mathrm{Var}(\gamma_n) = d(x, y) = r$;

(3) From the Intermediate Value Theorem, there exists $\tau_n \in (a_n, b_n)$ such that

$$d(x, \gamma_n(\tau_n)) = \frac{n-1}{n}\, r.$$

The choice $y_n := \gamma_n(\tau_n)$ does the required job in the claim, as

$$r \le d(x, y_n) + d(y_n, y) \le \mathrm{Var}\big(\gamma_n|_{[a_n, \tau_n]}\big) + \mathrm{Var}\big(\gamma_n|_{[\tau_n, b_n]}\big) \stackrel{(7.15)}{=} \mathrm{Var}(\gamma_n) \to r.$$

Exercise 7.12. We provide a review on *arc-length parameterizations* in a length space (M, d). Let $\gamma : [\alpha, \beta] \to M$ be a continuous d-rectifiable curve with $\mathrm{Var}(\gamma) > 0$. Let us consider the map

$$[\alpha, \beta] \ni t \mapsto f(t) := \mathrm{Var}\big(\gamma|_{[\alpha, t]}\big).$$

Prove the following property of f (use the additivity property (7.15)):

$$f(t_2) - f(t_1) = \mathrm{Var}\big(\gamma|_{[t_1, t_2]}\big), \qquad \text{for } \alpha \le t_1 \le t_2 \le \beta.$$

Then deduce the following facts:

- f is non-decreasing;
- if $f(t_1) = f(t_2)$ then $\gamma(t_1) = \gamma(t_2)$;
- f is continuous (for the proof of this fact, one may benefit of (7.16)...).

All these properties entitle us to set the following definition:

$$\Gamma : [0, \mathrm{Var}(\gamma)] \longrightarrow M, \qquad s \mapsto \Gamma(s) := \gamma(t(s)),$$

where, for any $s \in [0, \mathrm{Var}(\gamma)]$, $t(s) \in [\alpha, \beta]$ has been chosen[5] in some way so that $f(t(s)) = s$, i.e., $\mathrm{Var}\big(\gamma|_{[\alpha, t(s)]}\big) = s$. We can also assume that $s \mapsto t(s)$ is non-decreasing. We say that Γ is the **arc-length parameterization** of γ. Prove that:

(1) it holds that

$$d(\Gamma(s_2), \Gamma(s_1)) \le s_2 - s_1, \quad \text{for } 0 \le s_1 \le s_2 \le \mathrm{Var}(\gamma); \tag{7.17}$$

(2) deduce that Γ is continuous and d-rectifiable;

(3) the property (7.15) also ensures that (whenever $0 \le s_1 \le s_2 \le \mathrm{Var}(\gamma)$)

$$\mathrm{Var}\big(\Gamma|_{[t_1, t_2]}\big) = \mathrm{Var}\Big(\gamma|_{[t(s_1), t(s_2)]}\Big) = s_2 - s_1.$$

For the first equality see [Busemann (1955), eq.(5.13) p. 22].

Exercise 7.13. Let (Ω, d) be a length space (with Ω an open subset of $\mathbb{R}^N$). Suppose $\gamma_n : [\alpha, \beta] \to \Omega$ are continuous curves such that:

- there exists $M > 0$ such that $\mathrm{Var}(\gamma_n) \le M$, for every $n \in \mathbb{N}$;
- there exists a compact subset of Ω containing $\gamma_n([\alpha, \beta])$, for every $n \in \mathbb{N}$.

---

[5] The way $t(s)$ is chosen does not affect the definition of $\Gamma(s)$.

Then there exists a subsequence $(n_k)_k$ and re-parameterizations $\widetilde{\gamma}_{n_k}$ of γ_{n_k}, all defined on $[0,1]$, such that, as $k \to \infty$, the sequence $\widetilde{\gamma}_{n_k}$ uniformly converges on $[0,1]$ to a continuous d-rectifiable curve $\widetilde{\gamma}$.

[*Hint:* Extract a subsequence (still denoted by γ_k) such that $\gamma_k(\alpha)$ converges in Ω; then consider the arc-length parameterization Γ_k of γ_k, and re-scale it by setting
$$\widetilde{\gamma}_k(s) := \Gamma_k(s\,\mathrm{Var}(\gamma_k)), \quad s \in [0,1].$$
Show that the family $\{\widetilde{\gamma}_k\}_k$ is equi-bounded and equi-continuous (the latter follows from (7.17)); finally apply Arzelà-Ascoli Theorem...]

Exercise 7.14. Consider d_X in Exm. 7.18; let $r > 0$ and prove that $B_X((0,0),r)$ is a neighborhood of $(0,0)$ in the Euclidean topology by the following argument.

(1) Let $\varepsilon > 0$ and $y > 0$, and run through the integral curve of X_1 starting at $(-\varepsilon, y)$ up to time 2ε; the endpoint is (ε, y). Starting from the latter, run through the integral curve of $-X_2$ up to time $t = y/\varepsilon$; the endpoint is $(\varepsilon, 0)$. From the latter, run through the integral curve of $-X_1$ up to time $t = \varepsilon$; the endpoint is $(0,0)$.

(2) Recognize that the gluing γ of the above paths is an X-subunit curve of total length $3\varepsilon + y/\varepsilon$, connecting $(-\varepsilon, y)$ and $(0,0)$; thus, the d_X-distance between any point touched by γ and $(0,0)$ is $\leq 3\varepsilon + y/\varepsilon$.

(3) Prove that, if $0 < y < r^2/12$, there is some $\varepsilon > 0$ for which $3\varepsilon + y/\varepsilon < r$. Show that some $\varepsilon > 0$ can be chosen independently of $y \in [0, r^2/12]$.

(4) By repeating the same argument for $y < 0$, show that there is a rectangle centred at $(0,0)$ contained in $B_X((0,0),r)$.

Exercise 7.15. Let Ω be X-connected. Prove that $d_X : \Omega \times \Omega \to [0, \infty)$ is continuous (on the domain, we consider the Euclidean product topology) if and only if the map $\mathrm{id} : (\Omega, \|\cdot\|) \longrightarrow (\Omega, d_X)$ is continuous.

Hint: One implication is trivial. For the other, bear in mind that the triangle inequality for $d = d_X$ gives $|d(a,b) - d(b,c)| \leq d(a,c)$ so that
$$|d(x_n, y_n) - d(x_0, y_0)| \leq |d(x_n, y_n) - d(x_n, y_0)| + |d(x_n, y_0) - d(x_0, y_0)|$$
$$\leq d(y_n, y_0) + d(x_n, x_0).$$

Exercise 7.16. We sketch the rest of the proof of '(b) $\Rightarrow$ (a)' in Thm. 7.28. We tacitly resume the therein notations. Let $[0, R) \subseteq I$ and let $(z_n)_n$ be a sequence in $D(R)$; let also $\varepsilon_k = R/k$. Using the length space properties of d_X, one can construct a double sequence $x_n^k \in D(R - \varepsilon_k/2)$ such that $d(x_n^k, z_n) \leq \varepsilon_k$. The sequence $(x_n^1)_n$ has a convergent subsequence, say $\left(x_{n(1,j)}^1\right)_j$; accordingly we can extract from $\left(x_{n(1,j)}^2\right)_j$ a further converging subsequence $\left(x_{n(2,j)}^2\right)_j$. Proceeding like-wise, we infer that $\left(x_{n(j,j)}^k\right)_j$ converges for every fixed k.

Prove that $\left(z_{n(j,j)}\right)_j$ is Cauchy. From the completeness of (Ω, d_X), the latter sequence is convergent; since $(z_n)_n$ was arbitrary in $D(R)$, we have proved that $D(R)$ is sequentially compact, hence compact. Therefore $R \in I$, and I is closed.

Chapter 8

The Weak Maximum Principle

THE aim of this chapter is to introduce an important class of linear second order PDOs: the *semielliptic operators* of the second order L; this simply means that the second-order matrix of L is everywhere positive semidefinite. Then we shall establish one of the most fundamental tools used for the study of these operators, of an undisputed independent interest in the PDE literature: *the Weak Maximum Principle* (WMP, for short). For example, the WMP is one of the corner-stones of a satisfactory Potential Theory for L.

Broadly put, we say that L satisfies the WMP on the open set Ω if every u in $C^2(\Omega)$ satisfying $Lu \geq 0$ is forced to be non-positive on Ω whenever this is true on $\partial\Omega$ (in a suitable weak sense). Applications of the Weak Maximum Principle will be provided in the next Chap. 9, such as uniqueness (or comparison principles) for the solution of the Dirichlet problem.

As we shall see in Sec. 8.5, the semielliptic assumption on L is the natural hypothesis for the validity of the WMP on bounded open sets. It is however not sufficient as it stands, and some assumptions on the sign of the zero-order term c of L are needed. For example, $c < 0$ is a sufficient condition, or $c \leq 0$ together with the existence of a so-called L-barrier for L.

On the face of it, a seemingly aside result presented in this chapter will take an unexpected major role in Maximum Propagation issues (Chap. 10): when $c \equiv 0$, no function $u \in C^2(\Omega)$ satisfying $Lu \gneq 0$ can possess a local maximum point in Ω.

This chapter can be read independently of the previous ones, in that we do not make use of former results; however, it serves as a link for the investigation of Chap. 10 regarding the Maximum Propagation Principle and the Strong Maximum Principle. The latter topics are, instead, deeply connected with former topics and techniques of the book, such as the Connectivity Theorem of Chap. 6, and with the use of the integral curves of vector fields.

The only prerequisite for this chapter is the knowledge of some basic Linear Algebra and elementary Calculus.

169

8.1 Main definitions

Throughout what follows, D will denote a fixed non-empty open subset of $\mathbb{R}^N$, without the need to repeat it.

For $i, j \in \{1, \ldots, N\}$, let $a_{i,j}(x)$, $b_i(x)$, $c(x)$ be fixed real-valued functions defined on D. We consider the linear second-order partial differential operator (PDO, for short) defined by

$$L := \sum_{i,j=1}^{N} a_{i,j} \frac{\partial^2}{\partial x_i\, \partial x_j} + \sum_{i=1}^{N} b_i \frac{\partial}{\partial x_i} + c. \tag{8.1}$$

This means that L operates on functions $u \in C^2(D)$ in the following way:

$$u \mapsto Lu = \sum_{i,j=1}^{N} a_{i,j}\, \partial_{i,j} u + \sum_{i=1}^{N} b_i\, \partial_i u + c\, u.$$

The term $c = L(1)$ is called the **zero-order term** of L. We say that L is **homogeneous** if its zero-order term is identically vanishing. Finally we set

$$A(x) := \big(a_{i,j}(x)\big)_{i,j \leq N}, \qquad x \in D. \tag{8.2}$$

Since we are interested in evaluating L on C^2 functions only, for which the Schwarz Theorem applies, taking into account that[1]

$$\sum_{i,j=1}^{N} a_{i,j}\partial_{i,j} u = \sum_{i,j=1}^{N} \frac{a_{i,j} + a_{j,i}}{2}\, \partial_{i,j} u \qquad \forall\, u \in C^2,$$

replacing $a_{i,j}$ with $\dfrac{a_{i,j} + a_{j,i}}{2}$ if necessary, it is not restrictive to suppose that

the matrix $A(x)$ is symmetric for every $x \in D$.

The above hypothesis will be tacitly understood in the sequel.

If L is as in (8.1), the PDO $L_0 := \sum_{i,j=1}^{N} a_{i,j} \frac{\partial^2}{\partial x_i\, \partial x_j}$ is called the **principal part** (or the **second-order part**) of L. Moreover, the matrix $A(x)$ in (8.2) is referred to as the matrix of the principal part of L, or shortly as the **principal matrix** of L.

If L and A are as in (8.1) and (8.2) respectively, for a fixed $x \in D$ we set[2]

$$q_L(x, \cdot) : \mathbb{R}^N \longrightarrow \mathbb{R}, \quad q_L(x, \xi) := \langle A(x)\xi, \xi \rangle = \sum_{i,j=1}^{N} a_{i,j}(x)\, \xi_i \xi_j.$$

We say that $q_L(x, \cdot)$ is the **characteristic (quadratic) form** of L at $x \in D$.

We are ready for our main definition.

Definition 8.1 (Semielliptic operator). Let L be the differential operator in (8.1), and let $A(x)$ be the associated (symmetric) matrix in (8.2). We say that L is a:

---

[1]See also Exr. 8.1.

[2]Here and throughout $\langle \cdot, \cdot \rangle$ denotes the usual inner product of $\mathbb{R}^N$, whilst $\| \cdot \|$ denotes the Euclidean norm on $\mathbb{R}^N$ (or on any other space $\mathbb{R}^m$, $m \geq 1$). Also, when vectors occur in matrix computations, like in "$A(x)\xi$", we understand that $\xi \in \mathbb{R}^N$ is thought of as an $N \times 1$ column matrix.

- **semielliptic operator** if $A(x)$ is positive semidefinite for every $x \in D$;
- **elliptic operator** if $A(x)$ is positive definite for every $x \in D$;
- **uniformly elliptic operator** if there exists a positive constant λ_0 such that, for any $x \in D$, one has $\langle A(x)\xi, \xi \rangle \geq \lambda_0 \|\xi\|^2$ for every $\xi \in \mathbb{R}^N$.

The notion of uniformly elliptic operator will not be used in the sequel, and it is provided here for the sake of completeness only.

Frequently, a semielliptic operator which is not elliptic is called *degenerate elliptic*. We shall avoid the adjective 'degenerate' since important classes of PDOs (e.g., the sub-Laplacians on the Lie groups studied in [Bonfiglioli *et al.* (2007)]) are semielliptic without being elliptic *at any point*, yet they are well-behaved under so many points of view that calling them 'degenerate' would be...definitely unfair!

Remark 8.2. For any fixed $x \in D$, we denote by $\lambda_1(x), \ldots, \lambda_N(x)$ (counted with multiplicities) the eigenvalues of $A(x)$. Note that, by the Spectral Theorem for symmetric matrices, any $\lambda_i(x)$ is real since $A(x) = (A(x))^T$.

Obviously, the above definitions in Def. 8.1 can be rephrased in terms of the eigenvalues of A in the following way:

(1) L is semielliptic if and only if $\lambda_1(x), \ldots, \lambda_N(x) \geq 0$ for every $x \in D$;
(2) L is elliptic if and only if $\lambda_1(x), \ldots, \lambda_N(x) > 0$ for every $x \in D$;
(3) L is uniformly elliptic if and only if there exists $\lambda_0 > 0$ such that, for every $x \in D$, one has $\lambda_1(x), \ldots, \lambda_N(x) \geq \lambda_0$. $\qquad\qquad\sharp$

The following is a very important example for the purposes of this book.

Example 8.3 (Sum of squares of smooth vector fields). Let $X_0, X_1, \ldots, X_m$ be smooth vector fields on D. Let us consider the PDO $L = \sum_{i=1}^m X_i^2 + X_0$. We say that X_0 is the **drift** of L and that L is a **sum of squares (of smooth vector fields) plus a drift**. If $X_0 \equiv 0$, we simply say that L is a **sum of squares** (of smooth vector fields). Actually, the following results also hold if $X_1, \ldots, X_m$ are merely C^1 on D (so that X_i^2 makes sense) and if X_0 is any v.f. on D.

As the reader will easily prove in Exr. 8.4, the quadratic form of L is associated with the symmetric matrix $A = S \cdot S^T$, where S is the $N \times m$ matrix whose columns are obtained by transposing the vectors of the coefficients of $X_1, \ldots, X_m$ in this order. Note that A is positive semidefinite since, for every ξ and x, one has

$$\langle A(x)\xi, \xi \rangle = \langle S(x)^T \xi, S(x)^T \xi \rangle = \|S(x)^T \xi\|^2 = \sum_{i=1}^m \langle X_i(x), \xi \rangle^2 \geq 0. \qquad (8.3)$$

Hence a sum of squares plus a drift is a *homogeneous semielliptic operator*. Since

$$\mathrm{rank}(A) = \mathrm{rank}(S \cdot S^T) \leq \mathrm{rank}(S) \leq \min\{N, m\},$$

we deduce that *when $m < N$ the PDO L cannot be elliptic at any point*. This can also be derived from formula (8.3), which clearly shows that the set

$$\mathrm{Isotr}(A(x)) := \{\xi \in \mathbb{R}^N : q_L(x, \xi) = 0\} = \{\xi \in \mathbb{R}^N : \langle A(x)\xi, \xi \rangle = 0\}$$

(the set of the isotropic vectors for the quadratic form associated with the matrix $A(x)$) is the vector space orthogonal to $X_1(x), \ldots, X_m(x)$. Note that the vector space $\{X_1(x), \ldots, X_m(x)\}^\perp$ has dimension $\geq N - m$, and it collapses to $\{0\}$ iff $X_1(x), \ldots, X_m(x)$ generate $\mathbb{R}^N$. It is interesting to observe that the above computations also show that (Fig. 8.1)

$$q_L(x, \xi) = 0 \quad \text{iff} \quad \xi \in \ker(A(x)) = \ker(S(x)^T) = \{X_1(x), \ldots, X_m(x)\}^\perp.$$

8.2 Picone's Weak Maximum Principle

We begin with an important definition:

Definition 8.4 (Picone's Weak Maximum Principle on an open set). Let L be the PDO in (8.1), defined on D, and let Ω be an open subset of D. We say that L satisfies the **(Picone) Weak Maximum Principle on** Ω if, for every function $u \in C^2(\Omega)$ satisfying the conditions

$$\begin{cases} Lu \geq 0 & \text{on } \Omega \\ \limsup_{x \to y} u(x) \leq 0 & \text{for every } y \in \partial\Omega, \end{cases} \tag{8.4}$$

it necessarily holds that $u \leq 0$ on the whole of Ω.

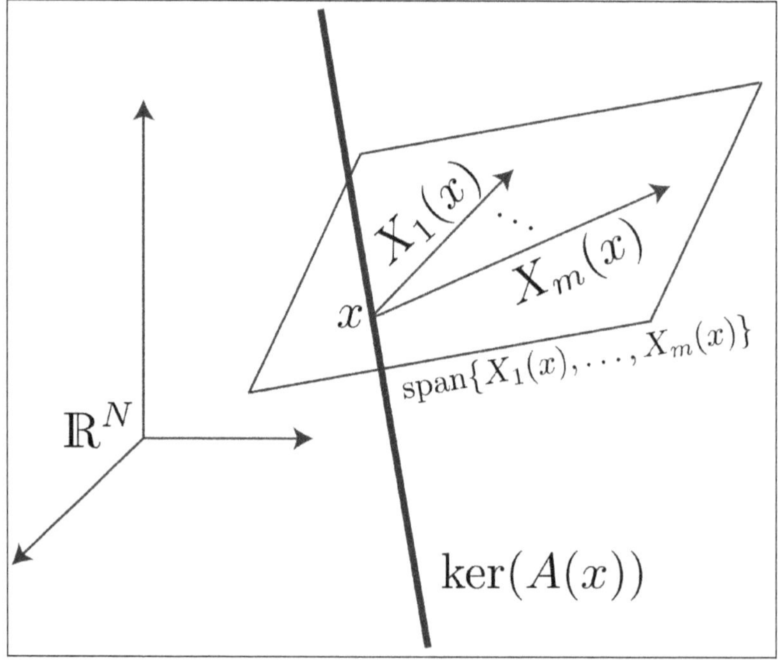

Fig. 8.1 The geometry of the isotropic set $\ker(A(x))$ associated with the matrix $A(x)$ of the quadratic form of a sum of squares $X_0 + \sum_{i=1}^m X_i^2$: we have that $\ker(A(x))$ is the orthogonal space of $\mathrm{span}\{X_1(x), \ldots, X_m(x)\}$.

We shall often abbreviate 'Weak Maximum Principle' by **WMP**.

For example, the second condition in (8.4) is certainly satisfied if
$$u \in C(\overline{\Omega}), \qquad u|_{\Omega} \in C^2(\Omega), \qquad u|_{\partial\Omega} \leq 0.$$
Throughout, the first two conditions in the above displayed line will be abbreviated by simply writing $u \in C(\overline{\Omega}) \cap C^2(\Omega)$.

Convention. In the sequel, Ω *will always denote a non-empty open subset of* $\mathbb{R}^N$; *we* shall often tacitly assume this fact. Moreover, L will always tacitly denote a second order differential operator as in (8.1) defined on the open set D.

Remark 8.5. Certainly, the reader is familiar with a very simple form of the WMP: indeed, if $L = \frac{d^2}{dx^2}$ on $D = \mathbb{R}$, the C^2 functions u satisfying $Lu = u'' \geq 0$ on an interval $\Omega = (a, b)$ are nothing but the C^2 *convex functions* on (a, b). Since convex functions admit limits (in the extended sense) at boundary points of their interval of definition, the second condition in (8.4) can be rewritten as
$$\lim_{x \to a^+} u(x) \leq 0 \quad \text{and} \quad \lim_{x \to b^-} u(x) \leq 0. \tag{8.5}$$
Because of the fact that the graph of a convex function lies below any of its secants, (8.5) easily implies that $u \leq 0$ on (a, b). Thus the (uniformly elliptic) PDO $\frac{d^2}{dx^2}$ satisfies the WMP on any bounded open interval of $\mathbb{R}$. ‡

The task of this section is:

> *to find sufficient conditions for a* semielliptic *operator L* > *to satisfy the WMP on any* bounded *open set Ω contained in D.*

First we provide a few remarks on the properness of the assumptions that Ω be bounded and that L be semielliptic.

Remark 8.6 (Convenient assumptions in order to have a WMP). Let us consider the following explicit examples:

(1). The necessity of the hypothesis that Ω be bounded will be justified in Rem. 8.19, taking into account the counterexample provided by
$$u(x_1, x_2) = e^{x_1} \sin x_2, \qquad \Omega = \{(x_1, x_2) \in \mathbb{R}^2 : x_2 > 0\},$$
which violates the WMP for the *classical Laplace operator* in $\mathbb{R}^2$
$$L = \Delta := (\partial_{x_1})^2 + (\partial_{x_2})^2.$$
Note that Δ is uniformly elliptic in $\mathbb{R}^2$ and Ω is unbounded.

(2). It is also easy to produce an example of a non-semielliptic L violating the WMP: for instance, the *d'Alembert (wave) operator* in $\mathbb{R}^2$
$$L = \square := (\partial_{x_1})^2 - (\partial_{x_2})^2$$
(which is not semielliptic at any point) does not satisfy the WMP in (8.4) for the following choice of u and Ω (see Exr. 8.6):
$$u(x_1, x_2) = \sin x_1 \sin x_2, \qquad \Omega = (0, \pi) \times (0, \pi).$$

(3). Finally, the importance of the semielliptic assumption on L is discussed in Sec. 8.5, where we prove that a *necessary* condition for the WMP to be satisfied by a homogeneous L (with continuous coefficients) on every bounded subset of D is that L *be semielliptic on D*.

(4). Later on, we shall realize that, in order to have a satisfactory WMP, some assumptions are necessary on the zero-order term of L as well: the most natural requirement is that *the zero-order term of L be ≤ 0*. Indeed, even the simple one-dimensional example of the operator $Lu = u'' + u$ in $\mathbb{R}$, which violates the WMP in (8.4) with the choice $u(x) = \sin x$, $\Omega = (0, \pi)$ (see Exr. 8.7), shows that one cannot expect the WMP without some assumption on the sign of $c = L(1)$ (which is $\equiv 1$ in this example).

(5). The failure of the second condition in (8.4) at just a single point of $\partial\Omega$ may cause the failure of the thesis $u \leq 0$, even for PDOs for which the WMP holds on every bounded open set Ω. For example, the function

$$u(x_1, x_2) = \frac{1 - x_1^2 - x_2^2}{(1 - x_1)^2 + x_2^2}$$

is Δ-harmonic in $D = \{(x_1, x_2) \in \mathbb{R}^2 : x_1^2 + x_2^2 < 1\}$, where Δ is the usual Laplace operator, a PDO satisfying the WMP on every bounded open set. Note that $u \equiv 0$ on $\partial D \setminus \{(1, 0)\}$, but $u > 0$ on D (Exr. 8.12).   ♯

For our investigation of the WMP, we begin with a very general Lemma.

Lemma 8.7 (Weierstrass). *Let $u : A \longrightarrow [-\infty, \infty]$ be a function defined on a bounded set $A \subset \mathbb{R}^N$. There exists $x_0 \in \overline{A}$ such that*

$$\sup_{A} u = \sup_{A \cap N} u, \quad \textit{for every neighborhood } N \textit{ of } x_0. \tag{8.6}$$

When (8.6) holds, we say that u **peaks at** x_0 (note that x_0 does not necessarily belong to A, and that, if $x_0 \in A$, x_0 may not be a maximum point of u; Fig. 8.2).

Proof. We argue by contradiction by supposing that for every $x \in \overline{A}$ there exists an open neighborhood N_x of x such that

$$\sup_{A \cap N_x} u < \sup_{A} u. \tag{8.7}$$

From the open cover $\{N_x\}_{x \in \overline{A}}$ of the compact set $\overline{A}$, we can extract a finite subcover $\{N_{x_1}, \ldots, N_{x_p}\}$. Since $A = \bigcup_{i=1}^{p} A \cap N_{x_i}$, we obtain

$$\sup_{A} u = \max_{i=1}^{p} \sup_{A \cap N_{x_i}} u \overset{(8.7)}{<} \max_{i=1}^{p} \sup_{A} u = \sup_{A} u,$$

which is clearly a contradiction. This ends the proof.   □

Remark 8.8. When A is an open set, any point of $\overline{A}$ is an accumulation point of A, so that condition (8.6) is equivalent to

$$
\begin{cases}
\sup_A u = \max\left\{ u(x_0), \limsup_{A \ni x \to x_0} u(x) \right\} & \text{if } x_0 \in A, \\
\sup_A u = \limsup_{A \ni x \to x_0} u(x) & \text{if } x_0 \in \partial A.
\end{cases}
$$

See also Fig. 8.2. This follows from the very definition of $\limsup$:

$$
\limsup_{A \ni x \to x_0} u(x) = \lim_{r \to 0^+} \sup_{\substack{x \in A: \\ 0 \neq \|x - x_0\| < r}} u(x).
$$

In view of this remark, Lem. 8.7 immediately proves the following fact.

Corollary 8.9. *If Ω is a bounded open set and $u : \Omega \to \mathbb{R}$ is continuous, there exists $x_0 \in \overline{\Omega}$ such that $\sup_\Omega u = \limsup_{\Omega \ni x \to x_0} u(x)$.*

We are ready for our first version of the WMP: we consider a *semielliptic* operator with a *negative zero-order term* (see also Rem. 8.6-(4)).

Theorem 8.10 (WMP for a semielliptic PDO with negative zero-order term).
Let L be semielliptic on D. Suppose that the zero-order term of L is strictly negative. Then L satisfies the Weak Maximum Principle on any bounded open set $\Omega \subseteq D$.

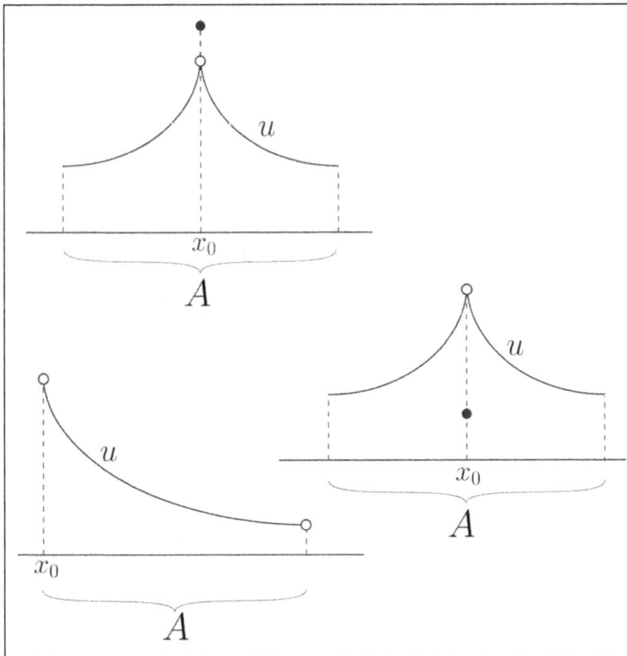

Fig. 8.2 Some different ways a function u on A can peak at x_0. Following Rem. 8.8, note that in the first figure one has $\sup_A u = u(x_0)$, whereas in the second and third it holds $\sup_A u = \limsup_{x \to x_0} u(x)$.

Proof. Suppose that Ω is bounded. Let $u \in C^2(\Omega)$ satisfy (8.4); we need to prove that $u \leq 0$ on Ω. By Cor. 8.9, let $x_0 \in \overline{\Omega}$ be such that

$$\sup_{\Omega} u = \limsup_{x \to x_0} u(x). \tag{8.8}$$

If $x_0 \in \partial\Omega$, by the second condition of (8.4) we infer $\limsup_{x \to x_0} u(x) \leq 0$, so that (8.8) gives at once $\sup_{\Omega} u \leq 0$, that is $u \leq 0$ on Ω.

We are left with the case when $x_0 \in \Omega$. In this case, as $u \in C(\Omega)$, we have

$$\sup_{\Omega} u \overset{(8.8)}{=} \limsup_{x \to x_0} u(x) = u(x_0),$$

so that x_0 is an *interior maximum point* of u. By elementary multivariate Calculus we derive that $\nabla u(x_0) = 0$ and that the Hessian matrix of u at x_0 (which we denote by $H(x_0)$) is negative semidefinite. Thus

$$Lu(x_0) = \sum_{i,j=1}^{N} a_{i,j}(x_0)\, \partial_{i,j} u(x_0) + \sum_{i=1}^{N} b_i(x_0)\, \partial_i u(x_0) + c(x_0)\, u(x_0)$$

$$= \mathrm{trace}\big(A(x_0)\, H(x_0)\big) + c(x_0)\, u(x_0) \leq c(x_0)\, u(x_0)$$

(remember that $A(x_0) \geq 0$ and $H(x_0) \leq 0$ and apply Féjer's Lem. A.23).

This proves that $Lu(x_0) \leq c(x_0)\, u(x_0)$. By the first condition in (8.4) we thus have

$$0 \leq Lu(x_0) \leq c(x_0)\, u(x_0),$$

whence $c(x_0)\, u(x_0) \geq 0$. As $c < 0$, we derive $u(x_0) \leq 0$. Recalling that we have $u(x_0) = \sup_{\Omega} u$, we deduce once again that $u \leq 0$ on Ω. This ends the proof. $\qquad\square$

Remark 8.11. In the above argument we have proved the following fact (interesting in its own right), a direct consequence of Féjer's Lem. A.23:

If L is a semielliptic operator (with no assumption on its zero-order term), and if u is in $C^2(\Omega)$ has a local interior maximum point at $x_0 \in \Omega$, then

$$Lu(x_0) \leq c(x_0)\, u(x_0).$$

From this simple fact we derive the following useful result. ♮

Corollary 8.12. *Let L be a semielliptic operator on D with zero-order term $c \leq 0$. Suppose that $\Omega \subseteq D$. Then a function $u \in C^2(\Omega)$ satisfying $Lu \gneqq 0$ cannot possess a local maximum point x_0 in Ω such that $u(x_0) \geq 0$.*

When $c \equiv 0$ (i.e., if L is homogeneous), no function $u \in C^2(\Omega)$ satisfying $Lu \gneqq 0$ can possess a local maximum point in Ω.

The last statement of this corollary is (partially) related to the so-called Strong Maximum Principle for L, to which we shall extensively return later on (Chap. 10).

Proof. Let $c \leq 0$ and assume that $u \in C^2(\Omega)$ satisfies $Lu > 0$. By contradiction, suppose that there exists a local maximum point $x_0 \in \Omega$ of u such that $u(x_0) \geq 0$. Arguing as in Rem. 8.11 we get the absurd

$$0 < Lu(x_0) \leq \underbrace{c(x_0)}_{\leq 0} \cdot \underbrace{u(x_0)}_{\geq 0} \leq 0.$$

The same argument applies if $c \equiv 0$ with no hypothesis on $u(x_0)$. $\qquad\square$

Due to the importance of homogeneous PDOs, we need to overcome the restrictive assumption $c < 0$ in Thm. 8.10. Since positive zero-order terms are not suited for WMPs (Exr. 8.7), the most natural assumption is $c \leq 0$. In Thm. 8.16 below we establish a WMP in this situation. In comparison with the hypotheses of Thm. 8.10, we shall make a (seemingly) stronger assumption: we shall require the existence of a so-called *barrier function for L*. Later on in Thm. 8.17, we shall also establish a WMP with no restriction on the sign of c, on condition that we strengthen the hypothesis on our barrier function. Here are the relevant definitions:

Definition 8.13 (Barrier function). A function $h \in C^2(\Omega)$ will be called:

- an *L-barrier for Ω*, if it satisfies

$$Lh < 0 \quad \text{and} \quad h \geq 0 \quad \text{on } \Omega; \tag{8.9a}$$

- a **strong L-barrier for Ω**, if it satisfies

$$Lh < 0 \quad \text{on } \Omega \quad \text{and} \quad \inf_{\Omega} h > 0. \tag{8.9b}$$

Equivalently, we shall say that Ω possesses an *L-barrier* (or a *strong L-barrier*, resp.) if there exists h as in (8.9a) (or as in (8.9b), resp.).

See also Fig. 8.3. We observe that if Ω possesses an L-barrier (or a strong L-barrier) then every open subset of Ω does the same.

Remark 8.14. Note that, even if the existence of an L-barrier is not required in Thm. 8.10, *this is implicitly satisfied when the zero-order term is negative*, for the function $h \equiv 1$ is (strictly) positive and it satisfies $Lh = c < 0$, in other words $h \equiv 1$ is an L-barrier (actually, strong) for any open set. $\qquad\sharp$

Remark 8.15. Suppose that L is semielliptic and satisfies $L(1) \leq 0$, and suppose that h is an L-barrier for Ω; *then the second condition in (8.9a) gives $h > 0$ on Ω.*

Indeed, if $Lh < 0$ and $h \geq 0$ on Ω, one cannot have $h(x_0) = 0$ at some $x_0 \in \Omega$; otherwise, if such an x_0 existed, the function $u := -h$ would satisfy $Lu = -Lh > 0$ and it would possess a nonnegative (actually, null) maximum on Ω (notice that $u(x_0) = -h(x_0) = 0$ and $u = -h \leq 0$ on Ω), but this is contradicts the result in Cor. 8.12 (remember that $L(1) \leq 0$ by assumption). $\qquad\sharp$

Theorem 8.16 (WMP for a semielliptic PDO with $c \leq 0$ and with barriers). *Let L be semielliptic on D. Suppose that its zero-order term $c = L(1)$ satisfies $c \leq 0$ on D.*

Then L satisfies the Weak Maximum Principle on every bounded open subset of D which possesses an L-barrier.

Proof. Let $\Omega \subseteq D$ be a bounded open set and let $u \in C^2(\Omega)$ satisfy (8.4); we need to prove that $u \leq 0$ on Ω. We shall demonstrate this in the following way: let h be an L-barrier for Ω (which exists by hypothesis) and, for every $\varepsilon > 0$, we set $u_\varepsilon := u - \varepsilon\, h$; we shall prove that $u_\varepsilon \leq 0$ on Ω. Letting $\varepsilon \to 0^+$ we shall obtain the desired inequality $u \leq 0$ on Ω.

By Cor. 8.9 (relative to the function u_ε), there exists $x_0 \in \overline{\Omega}$ such that

$$\sup_{\Omega} u_\varepsilon = \limsup_{x \to x_0} u_\varepsilon(x). \tag{8.10}$$

We distinguish two cases:

- $x_0 \in \partial\Omega$. In this case, by the second condition of (8.4) and by the fact that $u_\varepsilon = u - \varepsilon\, h \leq u$ (as $\varepsilon > 0$ and $h \geq 0$), we infer that

$$\limsup_{x \to x_0} u_\varepsilon(x) \leq \limsup_{x \to x_0} u(x) \leq 0.$$

Thus, (8.10) gives $\sup_\Omega u_\varepsilon \leq 0$, that is $u_\varepsilon \leq 0$ on Ω as desired.

- $x_0 \in \Omega$. In this case, as u_ε is continuous on Ω, from (8.10) it easily follows that x_0 is an interior maximum point of u_ε on Ω:

$$\max_{\Omega} u_\varepsilon = u_\varepsilon(x_0). \tag{8.11}$$

By Rem. 8.11 we have

$$L u_\varepsilon(x_0) \leq c(x_0)\, u_\varepsilon(x_0). \tag{8.12}$$

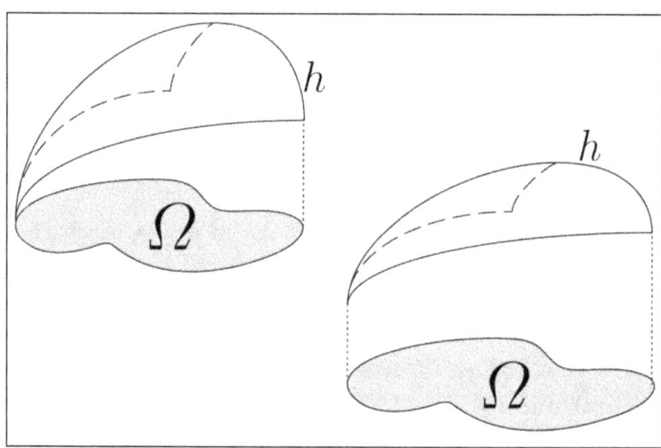

Fig. 8.3   Roughly put, if one consider as a prototype the (trivial) example of $L = \frac{\mathrm{d}^2}{\mathrm{d}x^2}$, we can picture a function h satisfying $Lh < 0$ as a (strictly) "concave" function; hence the picture draws (top to bottom) an L-barrier h for Ω and a strong L-barrier h for Ω.

On the other hand, on the whole of Ω it holds that

$$Lu_\varepsilon = Lu - \varepsilon\, Lh \overset{(8.4)}{\geq} -\varepsilon\, Lh \overset{(8.9a)}{>} 0.$$

Thus (8.12) implies that

$$0 < c(x_0)\, u_\varepsilon(x_0). \tag{8.13}$$

This implies that $c(x_0) \neq 0$ so that (since $c \leq 0$) we have the strict inequality $c(x_0) < 0$. Consequently, by dividing both sides of (8.13) by $c(x_0)$, we derive that $u_\varepsilon(x_0) < 0$. This fact, in view of (8.11), proves that $u_\varepsilon < 0$ on Ω. $\qquad\square$

We are ready for removing any condition on the sign of c. This only requires a stronger assumption on the barrier.

Theorem 8.17 (WMP for a semielliptic PDO possessing strong barriers). *Let L be semielliptic on D. No assumptions are made on the sign of $L(1)$. Then L satisfies the Weak Maximum Principle on every bounded open set possessing a strong L-barrier.*

For an example of a semielliptic PDO L with a *positive* zero-order term, possessing a strong L-barrier on a bounded open set Ω (hence L satisfies the WMP on Ω, in force of the above theorem), see Exm. 9.7.

Proof. Suppose that Ω be bounded. Let $u \in C^2(\Omega)$ satisfy (8.4); we need to prove that $u \leq 0$ on Ω. We shall prove this by showing that $v := \frac{u}{h} \leq 0$ on Ω, where h is a strong L-barrier for Ω (whose existence is granted by hypothesis). Being $h > 0$, this will obviously give $u \leq 0$ on Ω, as desired.

Our task is to prove that $v \leq 0$ on Ω. *We shall do this by showing that v is a solution to a system of inequalities as in* (8.4), *with L replaced by a semielliptic PDO L^* which is known to satisfy the WMP on Ω: indeed L^* will have a negative zero-order term and we shall be entitled to apply Thm. 8.10.*

To begin with, notice that by (8.9b) h satisfies

$$Lh < 0 \quad \text{on } \Omega \qquad \text{and} \quad h \geq m := \inf_\Omega h > 0. \tag{8.14}$$

In particular, for every $x \in \Omega$ one has

$$h(x) \geq m > 0 \quad \Rightarrow \quad 0 < \frac{1}{h(x)} \leq \frac{1}{m} < \infty.$$

Thus, the boundedness-from-above (and the nonnegativity) of $1/h$ and the second condition in (8.4) ensure that (Exr. 8.10)

$$\limsup_{x \to y} v(x) = \limsup_{x \to y} \frac{u(x)}{h(x)} \leq 0, \qquad \text{for any } y \in \partial\Omega.$$

Next we turn to the main computation; from $u = v \cdot h$ we get

$$Lu = h \cdot \left(\underbrace{\sum_{i,j} a_{i,j}\partial_{i,j}v}_{} + \sum_j \underbrace{\left(b_j + 2\sum_i \frac{a_{i,j}}{h}\,\partial_i h\right)}_{=:\, b_j^*}\partial_j v + \underbrace{\frac{Lh}{h}}_{=:\, c^*}\,v \right);$$

hence, if we set $L^* := \sum_{i,j} a_{i,j} \partial_{i,j} + \sum_j b_j^* \partial_j + c^*$, we have proved that

$$Lu = h \cdot L^* v, \quad \text{that is} \quad L^* v = \frac{Lu}{h}.$$

From hypotheses $Lu \geq 0$ and $h > 0$ (see the first inequality in (8.4) and the second one in (8.14)), we derive $L^* v \geq 0$. Now L^* is a semielliptic operator (because the characteristic forms of L^* and L coincide!) with zero-order term $c^* = Lh/h$ strictly negative (see (8.14)). Also, we have proved so far that

$$\begin{cases} L^* v \geq 0 & \text{on } \Omega, \\ \limsup_{x \to y} v(x) \leq 0 & \text{for every } y \in \partial\Omega. \end{cases}$$

By the WMP in Thm. 8.10 (applied to L^*), we derive that $v \leq 0$ on Ω. $\qquad\square$

We conclude this section with an example of a WMP on *unbounded* sets. This type of results may be very complicated (depending on how much they are to be weak the hypotheses imposed on the open set destined to satisfy the WMP), and we restrict to simple corollaries of the WMP on bounded sets.

In the next proposition we meet again the assumption $L(1) \leq 0$; as we shall see in Sec. 8.3, this condition has the much convenient consequence that, if $Lu \geq 0$ (such a function will soon be called **L-subharmonic**) and if M is a nonnegative constant, then $v := u - M$ also satisfies $Lv \geq 0$: indeed

$$L(u - M) = \underbrace{Lu}_{\geq 0} - \underbrace{M}_{\geq 0} \cdot \underbrace{L(1)}_{\leq 0} \geq 0.$$

Hence, *the assumption $L(1) \leq 0$ allows for downwards-translation arguments of the functions involved, preserving L-subharmonicity.*

Proposition 8.18 (WMP on unbounded sets). *Let D be unbounded and suppose that L satisfies $L(1) \leq 0$ on D. Suppose furthermore that L satisfies the Weak Maximum Principle on every bounded open subset of D.*
Then, if $\Omega \subseteq D$ is unbounded and $u \in C^2(\Omega)$ satisfies

$$\begin{cases} Lu \geq 0 & \text{on } \Omega, \\[2mm] \limsup_{x \to y} u(x) \leq 0 & \text{for every } y \in \partial\Omega, \\[2mm] \limsup_{\|x\| \to \infty} u(x) \leq 0, \end{cases} \tag{8.15}$$

it necessarily holds that $u \leq 0$ on the whole of Ω.
The second condition in (8.15) may be empty if $\partial\Omega = \varnothing$, i.e. if $\Omega = \mathbb{R}^N$.

Proof. The third condition in (8.15) ensures that, for every $\varepsilon > 0$, there exists $R(\varepsilon) > 0$ such that $u(x) \leq \varepsilon$ for every $x \in \Omega$ such that $\|x\| \geq R(\varepsilon)$. For $\varepsilon = 1/n$

(with $n \in \mathbb{N}$) we consider $\rho_n := \max\{n, R(1/n)\}$. Note that $\rho_n \to \infty$ as $n \to \infty$ and $u(x) \leq 1/n$ for every $x \in \Omega$ with $\|x\| \geq \rho_n$.

Let us consider $\Omega_n := \Omega \cap B(0, \rho_n)$, where $B(0, \rho_n) = \{x \in \mathbb{R}^N : \|x\| < \rho_n\}$ (see Fig. 8.4). We observe that Ω_n is a bounded open subset of D and that the function $u_n := u - 1/n$ satisfies

$$Lu_n = Lu - L(1)/n \geq -L(1)/n \geq 0 \quad (\text{since } L(1) \leq 0);$$

we also claim that $\limsup\limits_{\Omega_n \ni x \to y} u_n(x) \leq 0$ for every $y \in \partial\Omega_n$. Indeed, one has

$$\partial\Omega_n \subseteq \left(\Omega \cap \partial B(0, \rho_n)\right) \cup \left(\overline{B(0, \rho_n)} \cap \partial\Omega\right).$$

We therefore have two cases:

- for every $y \in \Omega \cap \partial B(0, \rho_n)$, u is continuous at y, whence

$$\limsup\limits_{\Omega_n \ni x \to y} u(x) = u(y) \leq 1/n \qquad (\text{since } \|y\| = \rho_n);$$

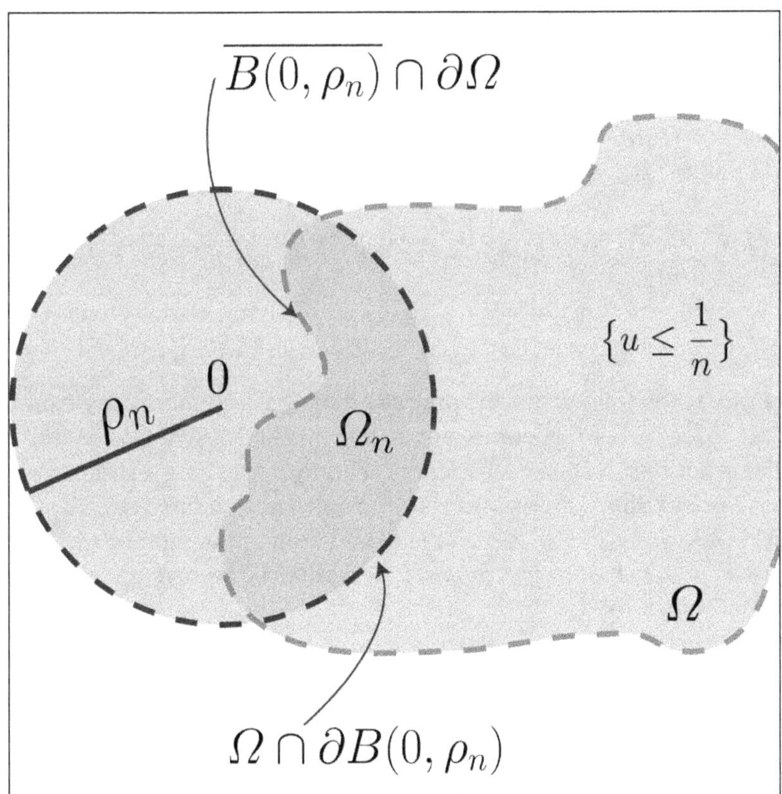

Fig. 8.4 The set Ω_n in the proof of Prop. 8.18.

- for every $y \in \overline{B(0, \rho_n)} \cap \partial\Omega$, we can exploit the second condition of (8.15), deducing that

$$\limsup_{\Omega_n \ni x \to y} u(x) \leq \limsup_{\Omega \ni x \to y} u(x) \leq 0 \leq 1/n \qquad (\text{since } y \in \partial\Omega).$$

Since $u_n = u - \frac{1}{n}$, these facts imply that $\limsup_{\Omega_n \ni x \to y} u_n(x) \leq 0$ for every $y \in \partial\Omega_n$.

Due to the fact that, by hypothesis, L satisfies the WMP on Ω_n, we infer that $u_n \leq 0$ on Ω_n, that is $u(x) \leq 1/n$ for every $x \in \Omega$ such that $\|x\| < \rho_n$. Letting $n \to \infty$ and exploiting the fact that $\rho_n \to \infty$ as $n \to \infty$ we immediately derive that $u \leq 0$ on the whole of Ω. $\qquad\square$

Remark 8.19. We highlight the fact that the third condition in (8.15) cannot be erased, without possibly loosing the WMP, as the example of the function $u(x_1, x_2) = e^{x_1} \sin x_2$ shows, when $L = \Delta$ in $\mathbb{R}^2$ and if we consider the upper half-plane $\Omega = \{(x_1, x_2) \in \mathbb{R}^2 : x_2 > 0\}$ (Exr. 8.5). $\qquad\sharp$

8.3　Existence of L-barriers

Since the existence of L-barriers is, after all, crucial for the WMP to hold for a semielliptic operator L, we begin with a very promising example in this direction.

Example 8.20 (Parabolic-type operators). Suppose that, after we have renamed and (possibly) reordered the coordinates x of $\mathbb{R}^N$ under the form (y, t) (with $y \in \mathbb{R}^{N-1}$ and $t \in \mathbb{R}$), the PDO L can be put in the following form

$$L = \sum_{i,j=1}^{N-1} a_{i,j}(y, t) \frac{\partial^2}{\partial y_i \partial y_j} + \sum_{i=1}^{N-1} b_i(y, t) \frac{\partial}{\partial y_i} + \tau(y, t) \frac{\partial}{\partial t} + c(y, t),$$

where the coefficient functions of L satisfy the following assumptions:

$$c \leq 0 \quad \text{and} \quad \tau < 0 \quad \text{on } D \qquad (\text{or } c \leq 0 \text{ and } \tau > 0 \text{ on } D).$$

Roughly put, when this occurs one may say that L is of **parabolic type** (notice that the second order part of L operates only in the y variables, whereas ∂_t only appears in the first order part, with a coefficient function τ possessing a strict sign).

Then we claim that L possesses a strong L-barrier (defined on $\mathbb{R}^N$). Hence, if the matrix $(a_{i,j}(y, t))_{i,j \leq N-1}$ is positive semidefinite at any point $(y, t) \in D$, *any semielliptic PDO L of the previous form satisfies the WMP on every bounded open subset of D* (due to Thm. 8.16).

Indeed, the claimed strong L-barrier h can be chosen as follows:

$$h(y, t) := \begin{cases} e^{+t} + 1, & \text{if } \tau < 0 \text{ on } D, \\ e^{-t} + 1, & \text{if } \tau > 0 \text{ on } D. \end{cases}$$

This follows from the fact that, if $h_\pm := e^{\pm t} + 1$, one has $h_\pm \geq 1$ on $\mathbb{R}^N$ and

$$L(h_\pm)(y, t) = \pm\tau(y, t)\, e^{\pm t} + \underbrace{c(y, t)}_{\leq 0}\, \underbrace{h_\pm(y, t)}_{> 0} \leq \pm\tau(y, t)\, e^{\pm t},$$

and the latter function is strictly negative precisely if one chooses h_+ when $\tau < 0$, and if one chooses h_- when $\tau > 0$. $\sharp$

Example 8.21. By using Exm. 8.20, we deduce that the PDO on $\mathbb{R}^2$ defined by

$$L = x_1^2 \frac{\partial^2}{\partial x_1^2} - \frac{\partial}{\partial x_2}$$

satisfies the WMP on every bounded open subset of $\mathbb{R}^2$. Note that L is semielliptic (but not elliptic) on $\mathbb{R}^2$ and the principal part of L is identically 0 on $\{x_1 = 0\}$. $\sharp$

Hitherto we have not assumed any regularity on the functions $a_{i,j}$, b_i, c of L. We shall do it now, by establishing sufficient conditions for the existence of L-barriers on bounded sets, when more general PDOs L are concerned than the parabolic-type operators in Exm. 8.20.

Throughout, we write $\Omega \Subset D$ if the closure of Ω is compact (i.e., if Ω is bounded) and $\overline{\Omega} \subset D$; in this case we say that Ω is *compactly contained in D*. Note that if $\Omega \Subset D$ then no boundary point of Ω can fall on the boundary points of D.

Proposition 8.22 (WMP for semielliptic PDOs with ellipticity direction). *Let L be semielliptic on D, and assume that $c = L(1) \leq 0$ on D. Let $\Omega \subseteq D$ be bounded, and suppose that the coefficients $a_{i,j}, b_j$ fulfil the following conditions:*
there exists $i \in \{1, \ldots, N\}$ such that

$$\inf_{\Omega} a_{i,i} > 0 \quad and \quad \inf_{\Omega} b_i > -\infty. \tag{8.16}$$

Then Ω admits a (smooth and strong) L-barrier. Hence, (8.16) guarantees that L satisfies the WMP on Ω (see Thm. 8.16).

For example, (8.16) is satisfied for every $\Omega \Subset D$, whenever L has continuous coefficient-functions $a_{i,j}, b_j, c$ such that $c \leq 0$ and if

$$a_{1,1} > 0 \quad on\ D. \tag{8.17}$$

This last condition holds true for any elliptic operator,[3] so that an elliptic operator L, with non-positive zero-order term $L(1)$, satisfies the WMP on every open set compactly contained in D. Conditions (8.16) are sufficient for the WMP to hold on Ω, but they are not necessary (see Exm. 8.23).

When the first condition in (8.16) is satisfied, we say that the x_i-axis is an **ellipticity direction** for L on Ω.

Apparently, since we shall prove the existence of a *strong* L-barrier for Ω, one may think that the hypothesis $c \leq 0$ can be removed (as Thm. 8.17 would suggest); unfortunately this hypothesis is used in proving the existence of the strong L-barrier, whence it cannot be dropped (see Exr. 8.7, involving $L = \partial_{x,x} + 1$ on $\mathbb{R}$ which violates the WMP on $\Omega = (0, \pi)$, yet L being elliptic).

---

[3]Indeed, if L is elliptic one has $a_{i,i}(x) = \langle A(x)e_i, e_i \rangle > 0$ for every $i = 1, \ldots, N$. Here e_i denotes the i-th element of the canonical basis of $\mathbb{R}^N$.

Proof. Let $\Omega \subseteq D$ be bounded and let $i \leq N$ be such that (8.16) is verified. Let us fix $\lambda \in \mathbb{R}$, $\lambda \gg 1$ such that

$$\lambda > -\inf_{\Omega} b_i / \inf_{\Omega} a_{i,i}. \tag{8.18a}$$

Due to (8.16), λ is finite. For this fixed λ, we next choose $M \gg 1$ such that

$$M > \max_{x \in \overline{\Omega}}\{\exp(\lambda\, x_i)\}. \tag{8.18b}$$

Such an M exists due to the hypothesis that Ω is bounded. We claim that the function $h(x) := M - \exp(\lambda\, x_i)$ is a (strong) L-barrier for Ω (see Fig. 8.5).

The inequality $\inf_\Omega h > 0$ is trivially satisfied owing to (8.18b). On the other hand, the inequality $Lh(x) < 0$ for $x \in \Omega$ comes from

$$Lh(x) = -a_{i,i}(x)\,\lambda^2\, e^{\lambda\, x_i} - b_i(x)\,\lambda\, e^{\lambda\, x_i} + \underbrace{c(x)\, h(x)}_{\leq 0}$$

$$\leq -\lambda^2\, e^{\lambda\, x_i} \underbrace{\left(a_{i,i}(x) + \frac{b_i(x)}{\lambda} \right)}_{>0} < 0.$$

In the first inequality we used the hypothesis $c \leq 0$ (and the nonnegativity of h); in the last inequality we used (8.18a). This ends the proof. $\qquad\square$

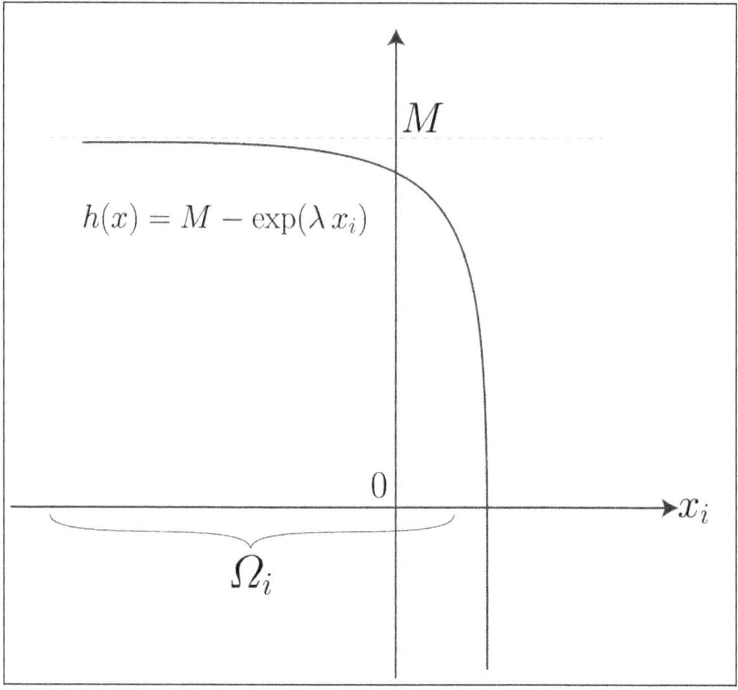

Fig. 8.5 The barrier function in the proof of Prop. 8.22. Here Ω_i is the projection of Ω on the i-th coordinate axis of $\mathbb{R}^N$, which is an ellipticity direction for L on Ω.

Example 8.23. Note that the operator

$$L = x_1^2 \frac{\partial^2}{\partial x_1^2} - \frac{\partial}{\partial x_2} \quad \text{in } \mathbb{R}^2$$

considered in Exm. 8.21 does not satisfy the assumptions of Prop. 8.22 on any open set Ω intersecting the axis $\{x_1 = 0\}$. However, we know from Exm. 8.20 that such an L possesses the strong L-barrier $e^{x_2} + 1$, whence it satisfies the WMP on every bounded open subset of $\mathbb{R}^2$. This shows that *conditions* (8.16) *and* (8.17) *are sufficient but not necessary for the WMP to hold.*   ♯

Example 8.24 (Laplace's operator). The Laplace operator $\Delta = \frac{\partial^2}{\partial x_1^2} + \cdots + \frac{\partial^2}{\partial x_N^2}$ on $\mathbb{R}^N$ is a (uniformly) elliptic operator. The associated matrix $A(x)$ is the identity matrix of order N, whence $a_{1,1} \equiv 1 > 0$. By Prop. 8.22, we derive that Δ satisfies the WMP on every bounded open subset of $\mathbb{R}^N$.

Other examples of remarkable operators satisfying the WMP (see Exr. 8.11) on the bounded open sets of $\mathbb{R}^3$ are given by the following PDOs:

(1) *The Heat operator in* $\mathbb{R}^3$: $(\partial_{x_1})^2 + (\partial_{x_2})^2 - \partial_{x_3}$;
(2) *Kolmogorov's operator in* $\mathbb{R}^3$: $(\partial_{x_1})^2 + x_1\, \partial_{x_2} - \partial_{x_3}$;
(3) *Mumford's operator:* $(\partial_{x_3})^2 - (\cos x_3\, \partial_{x_1} + \sin x_3\, \partial_{x_2})$;
(4) *The Kohn-Laplacian on the Heisenberg group in* $\mathbb{R}^3$:

$$(\partial_{x_1} + 2x_2\, \partial_{x_3})^2 + (\partial_{x_2} - 2x_1\, \partial_{x_3})^2.$$

By Thm. 8.17, the validity of the WMP for a semielliptic operator L is ensured by the existence of strong L-barriers. In its turn, the existence of strong L-barriers *on "small" sets compactly contained in* D is guaranteed by very mild conditions on the coefficients of L, as is shown in the following result, establishing a *local* WMP.

Proposition 8.25 (Local WMP for semielliptic operators). *Let L be semielliptic on D; suppose its coefficients $a_{i,j}, b_j, c$ are continuous and that* $\mathrm{trace}(A(x))$ *is strictly positive on D. No assumption is made on the sign of c.*

Then, for every $x \in D$ there exists a bounded open neighborhood $D_x \subseteq D$ of x such that L satisfies the WMP on any open subset of D_x.

Actually, the hypothesis of continuity of c and of the b_j's can be replaced with their being locally bounded, as the following proof shows. Also, the continuity of the entries of A can be relaxed by assuming only that the $a_{i,i}$'s are continuous.

We highlight that the "local nature" of the above result cannot be improved without further assumptions on the sign of c. For instance, the operator $\partial_{x,x} + 1$ in $\mathbb{R}$ does not satisfy the WMP on the open set $(0, \pi)$ (consider $u(x) = \sin x$; Exr. 8.7).

Proof. In view of Thm. 8.17, we turn our attention to the construction of a strong L-barrier in some neighborhood of each point of D.

Let $\xi \in D$ be fixed. The hypothesis $\mathrm{trace}(A(\xi)) > 0$ ensures that there exists $i \in \{1, \ldots, N\}$ such that $a_{i,i}(\xi) > 0$. The continuity of $a_{i,i}$ implies that there exists a neighborhood U of ξ (compactly contained in D) such that $\min_{\overline{U}} a_{i,i} > 0$. For the simplicity of notation we suppose $i = 1$. Let us set

$$\alpha := \min_{\overline{U}} a_{1,1}, \quad \beta := \min_{\overline{U}} b_1, \quad \gamma := \max_{\overline{U}} |c|.$$

We have $\alpha > 0$ and $\beta, \gamma \in \mathbb{R}$. Next we choose $\lambda > 0$ large enough so that

$$\alpha + \frac{\beta}{\lambda} - \frac{5\gamma}{\lambda^2} > 0. \tag{8.19}$$

Accordingly, we choose a small $\delta > 0$ in such a way that

$$e^{\lambda\delta} \leq 2. \tag{8.20}$$

Let now D_ξ be an open neighborhood of ξ of diameter smaller than the above δ and compactly contained in U. We claim that the following function

$$h(x) := 3 - \exp(\lambda(x_1 - \xi_1))$$

is a strong L-barrier for D_ξ (hence h is a strong L-barrier for every subset of D_ξ). First we note that, since the diameter of D_ξ is $\leq \delta$,

$$\inf_{x \in D_\xi} h(x) \geq 3 - \exp\left(\lambda \sup_{x \in D_\xi} \|x - \xi\|\right) \geq 3 - e^{\lambda\delta} \overset{(8.20)}{\geq} 1.$$

Finally, for every $x \in D_\xi$ we have (since $D_\xi \Subset U$ and $h > 0$ on D_ξ)

$$Lh(x) = -a_{1,1}(x)\,\lambda^2\,e^{\lambda(x_1-\xi_1)} - b_1(x)\,\lambda\,e^{\lambda(x_1-\xi_1)} + c(x)\left(3 - e^{\lambda(x_1-\xi_1)}\right)$$

$$\leq -\lambda^2\,e^{\lambda(x_1-\xi_1)} \min_{\overline{U}} a_{1,1} - \lambda\,e^{\lambda(x_1-\xi_1)} \min_{\overline{U}} b_1 + \left(3 - e^{\lambda(x_1-\xi_1)}\right) \cdot \max_{\overline{U}} |c|$$

$$= -e^{\lambda(x_1-\xi_1)}\lambda^2\left(\alpha + \frac{\beta}{\lambda} - \frac{\gamma}{\lambda^2}\left(3\,e^{-\lambda(x_1-\xi_1)} - 1\right)\right).$$

Now, for $x \in D_\xi$ it holds that

$$3\,e^{-\lambda(x_1-\xi_1)} - 1 \leq 3\,e^{\lambda|x_1-\xi_1|} - 1 \leq 3\,e^{\lambda\delta} - 1 \overset{(8.20)}{\leq} 5.$$

We have therefore proved that

$$Lh(x) \leq -e^{\lambda(x_1-\xi_1)}\lambda^2\left(\alpha + \frac{\beta}{\lambda} - \frac{5\gamma}{\lambda^2}\right).$$

Owing to (8.19), this shows that $Lh < 0$ on D_ξ, whence the proof that h is a strong L-barrier for D_ξ is complete. $\qquad\square$

The hypothesis $\mathrm{trace}(A(x)) > 0$ contained in the previous proposition is very natural for the following reason.

Definition 8.26 (Non-totally degenerate operator). Let L be the PDO in (8.1), defined on D. We say that L is **non-totally degenerate** on D (**NTD**, for short) if the matrix $A(x)$ is not identically zero for every $x \in D$.

Hence, for a non-totally degenerate PDO of the second-order, we are sure that (around every point x) this is a genuine *second-order* PDO.

Remark 8.27. By simple results of Linear Algebra, if L is semielliptic (i.e., if $A(x)$ is positive semidefinite for every $x \in D$) then *L is non-totally degenerate if and only if* $\mathrm{trace}(A(x)) > 0$ *on D.*

Indeed, suppose $A = (a_{i,j})$ is a symmetric positive semidefinite matrix. It clearly holds that $A \neq 0$ if and only if $\mathrm{trace}(A) > 0$. Indeed, A has only non-negative real eigenvalues, so that $\mathrm{trace}(A) > 0$ iff one at least of the eigenvalues is > 0, and this prevents A from being the zero matrix, since A is diagonalizable. ♯

Example 8.28. The operator $L = x_1^2(\partial_{x_1})^2 - \partial_{x_2}$ in Exm. 8.23 is *not* a non-totally degenerate PDO on $\mathbb{R}^2$ since the associated matrix

$$A(x) = \begin{pmatrix} x_1^2 & 0 \\ 0 & 0 \end{pmatrix} \qquad \text{vanishes on the } x_2\text{-axis.}$$

However, due to its parabolic-type nature, we know from Exm. 8.20 that such an L satisfies the WMP on every open subset of $\mathbb{R}^2$. ♯

Taking into account Rem. 8.27, we can rephrase Prop. 8.25 as follows:

Proposition 8.29 (Local WMP for NTD semielliptic PDOs). *Let L be semielliptic on $D \subseteq \mathbb{R}^N$. Suppose its coefficients $a_{i,j}, b_j, c$ are continuous and that L is non-totally degenerate on D. Then, for every $x \in D$ there exists a bounded open neighborhood $D_x \subseteq D$ of x such that L satisfies the WMP on any open subset of D_x.*

Even in the case $L(1) \leq 0$, the non-totally degenerate assumption is not sufficient for the WMP to hold on all bounded open sets $\Omega \Subset D$ (see Exm. 8.30).

Example 8.30. Consider the v.f. on $\mathbb{R}^2$ $X = x_2 \partial_{x_1} - x_1 \partial_{x_2}$. The PDO $L = X^2$ is semielliptic on $\mathbb{R}^2$, for it is a sum of squares (see Exm. 8.3). Besides, L is NTD on $D := \mathbb{R}^2 \setminus \{0\}$, since its principal matrix is

$$A(x_1, x_2) = \begin{pmatrix} x_2^2 & -x_1 x_2 \\ -x_1 x_2 & x_1^2 \end{pmatrix}.$$

Consider the annulus $\Omega = \{x \in \mathbb{R}^2 : \pi < x_1^2 + x_2^2 < 3\pi\}$, which is compactly contained in D. The function $u(x) = \sin(x_1^2 + x_2^2)$ satisfies $Lu \equiv 0$ on Ω (one has $Xu \equiv 0$, due to the radial nature of u) and $u \equiv 0$ on $\partial\Omega$. Since u changes it sign on Ω, the WMP is clearly violated. This shows that the condition on the non-total degeneracy is not sufficient for the WMP to hold, not even for homogeneous PDOs.

We remark that the fact that the above L loses the NTD property at 0 can easily be circumvented, to produce an example of a sum of squares which is NTD on the whole of $\mathbb{R}^2$ but it violates the WMP on a bounded open set (Exr. 8.13). ♯

Before passing to the corollaries of the WMP, the reader may find it convenient to visualize the sufficient conditions for the WMP obtained so far in this chapter, all summarized in Fig. 8.8 on page 196.

8.4 The parabolic Weak Maximum Principle

Let $D \subseteq \mathbb{R}^{N+1}$ be an open set, and suppose that the PDO L has the following "parabolic" form:

$$L := \sum_{i,j=1}^{N} a_{i,j}(x,t)\,\frac{\partial^2}{\partial x_i\,\partial x_j} + \sum_{i=1}^{N} b_i(x,t)\,\frac{\partial}{\partial x_i} + c(x,t) - \frac{\partial}{\partial t}.$$

In this very short section we denote by $z = (x,t)$ the points of $\mathbb{R}^{N+1}$, with $x \in \mathbb{R}^N$ and $t \in \mathbb{R}$. We suppose that L is semielliptic on D (i.e., $(a_{i,j}(x,t))_{i,j}$ is positive semidefinite for any $(x,t) \in D$) and that $c = L(1) \leq 0$.

Remark 8.31. Since, as already shown in Exm. 8.20, L possesses the strong L barrier $e^t + 1$ on D, we see that L satisfies the WMP on every bounded open set in D. We establish a stronger form of the WMP in Thm. 8.32. ♮

Given an open set $\Omega \subseteq \mathbb{R}^{N+1}$ and a $T \in \mathbb{R}$, we use the notation

$$\Omega_T := \Omega \cap \{(x,t) \in \mathbb{R}^{N+1} : t < T\}.$$

To avoid trivialities, in the sequel we assume that $\Omega_T \neq \varnothing$. We also set

$$\partial_T \Omega := \overline{(\partial\Omega) \cap \{(x,t) \in \mathbb{R}^{N+1} : t < T\}}, \tag{8.21}$$

and we call $\partial_T \Omega$ the **parabolic boundary** of Ω_T. See also Fig. 8.6.

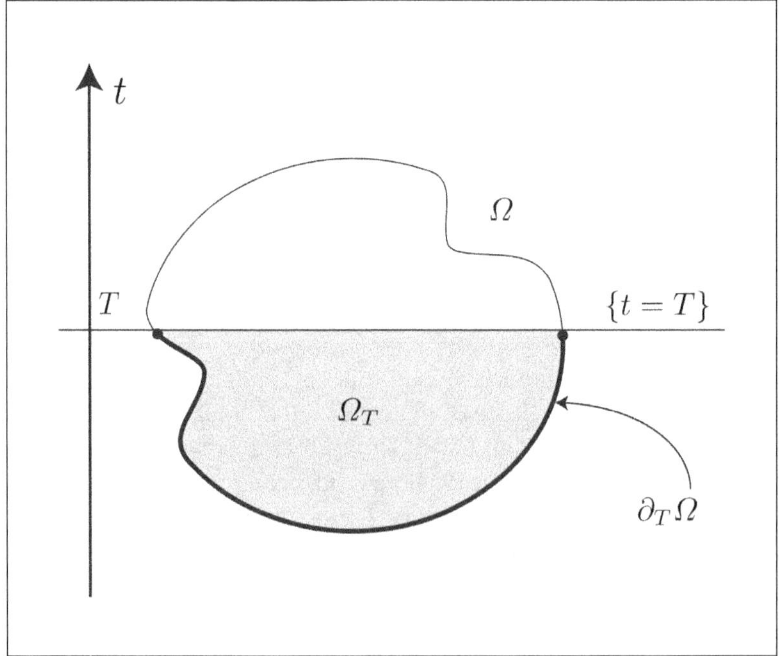

Fig. 8.6   The parabolic boundary $\partial_T \Omega$.

Theorem 8.32 (Parabolic Weak Maximum Principle). *Let the above assumptions on L apply. Let $\Omega \subseteq D$ be a bounded open set, and let $T \in \mathbb{R}$. Then the following stronger form of the WMP for L holds true.*

If $u \in C^2(\Omega_T)$ satisfies the conditions

$$\begin{cases} Lu \geq 0 & on\ \Omega_T \\ \displaystyle\limsup_{z \to \zeta} u(z) \leq 0 & for\ every\ \zeta \in \partial_T\Omega, \end{cases} \tag{8.22}$$

then $u \leq 0$ on the whole of Ω_T.

*We call this the **Parabolic Weak Maximum Principle** for L.*

Note that, wrt the usual WMP, no assumption is made on the boundary behavior of u on $\partial\Omega \setminus \partial_T\Omega$; actually, *a posteriori*, the thesis $u \leq 0$ on Ω_T will give $\limsup_{x \to y} u(x) \leq 0$ also at any $y \in \partial\Omega \setminus \partial_T\Omega$.

Proof. It suffices to prove that, for every $\tau < T$, u is ≤ 0 on Ω_τ. To prove this, we fix once and for all $\tau < T$, and we first claim that u is bounded from above on Ω_τ. By Cor. 8.9, there exists $\zeta_0 \in \overline{\Omega_\tau}$ such that

$$M_\tau := \sup_{\Omega_\tau} u = \limsup_{z \to \zeta_0,\ z \in \Omega_\tau} u(z).$$

If $\zeta_0 \in \Omega_\tau$, then $\zeta_0 \in \Omega_T$, whence (for continuity reasons) $M_\tau = u(\zeta_0) < \infty$. If, instead, $\zeta_0 \in \partial(\Omega_\tau)$, then two cases can occur, since[4] it holds that $\partial(\Omega_\tau)$ is contained in $(\Omega \cap \{t = \tau\}) \cup \partial_T\Omega$:

- $\zeta_0 \in \Omega \cap \{t = \tau\}$; in this case, arguing as above, $M_\tau = u(\zeta_0) < \infty$;
- $\zeta_0 \in \partial_T\Omega$; then, from the second condition in (8.22),

$$M_\tau = \limsup_{z \to \zeta_0,\ z \in \Omega_\tau} u(z) \leq \limsup_{z \to \zeta_0} u(z) \leq 0.$$

This shows that M_τ is finite. Let us now consider the function on Ω_τ

$$w_\tau(x,t) := \frac{1}{\tau - t}.$$

Then w_τ is an L-barrier for Ω_τ, since $w_\tau > 0$ for $t < \tau$ and (since $c \leq 0$)

$$Lw_\tau = \frac{c(x,t)}{\tau - t} - \frac{1}{(\tau - t)^2} \leq -\frac{1}{(\tau - t)^2} < 0, \qquad \text{for } t < \tau. \tag{8.23}$$

We let u be as in the statement and we show that, for any $\varepsilon > 0$, it holds that $u_\varepsilon := u - \varepsilon\, w_\tau \leq 0$ on Ω_τ; letting $\varepsilon \to 0^+$ we then get $u \leq 0$ on Ω_τ. We claim that the non-positivity of u_ε on Ω_τ is a consequence of the usual WMP (Rem. 8.31) on Ω_τ (the latter set is bounded, as Ω is bounded). Indeed:

$$Lu_\varepsilon = Lu - \varepsilon\, Lw_\tau \overset{(8.22)}{\geq} -\varepsilon\, Lw_\tau \overset{(8.23)}{>} 0 \qquad \text{on } \Omega_\tau.$$

Moreover, if $\zeta \in \partial(\Omega_\tau)$, arguing as above we have only two cases:

---

[4] We are using the following facts: from $\partial(A \cap B) \subseteq (\partial A \cap \overline{B}) \cup (\overline{A} \cap \partial B)$, one has

$$\partial(\Omega_\tau) = \partial(\Omega \cap \{t < \tau\}) \subseteq (\partial\Omega \cap \{t \leq \tau\}) \cup (\overline{\Omega} \cap \{t = \tau\}) =: F_1 \cup F_2.$$

Clearly, $F_1 \subseteq \partial_T\Omega$; moreover $F_2 = (\Omega \cap \{t = \tau\}) \cup (\partial\Omega \cap \{t = \tau\})$ and $\partial\Omega \cap \{t = \tau\} \subseteq \partial_T\Omega$.

- if $\zeta \in \partial_T \Omega$, from the second condition in (8.22) we infer

$$\limsup_{\Omega_\tau \ni z \to \zeta} u_\varepsilon(z) = \limsup_{\Omega_\tau \ni z \to \zeta} \big(u(z) - \underbrace{\varepsilon\, w(z)}_{>0} \big) \le \limsup_{\Omega_T \ni z \to \zeta} u(z) \le 0;$$

- if $\zeta \in \Omega \cap \{t = \tau\}$, then, since u is bounded from above on Ω_τ by M_τ,

$$\limsup_{\Omega_\tau \ni z \to \zeta} u_\varepsilon(z) \le \limsup_{t \to \tau^-} \Big(M_\tau - \frac{\varepsilon}{\tau - t} \Big) = -\infty \le 0.$$

From these facts and the WMP on Ω_τ, we get $u_\varepsilon \le 0$ on Ω_τ, as desired. $\qquad\square$

8.5 Appendix: the role of semiellipticity in the WMP

The aim of this section is to show the role of the semielliptic assumption in establishing the WMP; for a homogeneous operator L we shall show that semiellipticity is *necessary* for the WMP to hold:

Theorem 8.33. *Let L be a PDO as in (8.1) with vanishing zero-order term c and with continuous coefficients $a_{i,j}, b_i$ on the open set $D \subseteq \mathbb{R}^N$. Suppose that L satisfies the WMP on every bounded open set $\Omega \Subset D$. Then L is semielliptic.*

Proof. We argue by contradiction, supposing L is not semielliptic. The proof is complete if we can produce an open set $\Omega \Subset D$ and $u \in C^2(\Omega)$ such that

$$\begin{cases} Lu \ge 0 & \text{on } \Omega, \\ \limsup_{x \to y} u(x) \le 0 & \text{for every } y \in \partial\Omega, \end{cases} \tag{8.24}$$

with u positive in at least one point of Ω.

To begin with, since L is not semielliptic, there exist $y \in D$ and a vector ν in $\mathbb{R}^N \setminus \{0\}$ such that $\langle A(y)\nu, \nu \rangle < 0$. By a suitable linear change of coordinates,[5] it is not restrictive to suppose that $\nu = (1, 0, \dots, 0)$, so that

$$a_{1,1}(y) < 0. \tag{8.25}$$

We choose the function u under the very simple form

$$u(x) = \beta - \Big(\alpha\, (x_1 - y_1)^2 + (x_2 - y_2)^2 + \cdots + (x_N - y_N)^2 \Big),$$

where $\alpha, \beta > 0$ will be determined in the sequel. We observe that u is smooth on $\mathbb{R}^N$ and that (see also Fig. 8.7) one has

$$\Omega := \{u > 0\} = \Big\{ x \in \mathbb{R}^N : \alpha\, (x_1 - y_1)^2 + (x_2 - y_2)^2 + \cdots + (x_N - y_N)^2 < \beta \Big\}.$$

The second condition of (8.24) is a consequence of the continuity of u together with the fact that $u = 0$ on $\partial\Omega$. Also, u is positive on the whole of Ω. The proof is complete if we can choose $\alpha, \beta > 0$ such that $Lu \ge 0$ on Ω and $\Omega \Subset D$.

---

[5]By taking into account the contents of Exr. 9.4, it follows that an effective linear change of coordinates doing the job required here is the linear map $x \mapsto M\, x$, where M is a non-singular matrix whose first row is ν.

To simplify the notation, we set $\gamma_1 := \alpha$ and $\gamma_2 = \cdots = \gamma_N = 1$. A simple computation gives

$$Lu(x) = -2\sum_i a_{i,i}(x)\,\gamma_i - 2\sum_i b_i(x)\,\gamma_i\,(x_i - y_i).$$

One has $Lu(y) = -2\sum_i a_{i,i}(y)\,\gamma_i = -2\alpha\,a_{1,1}(y) - 2\sum_{i=2}^{N} a_{i,i}(y)$. Hence, owing to (8.25), we can first choose $\alpha \gg 1$ such that $Lu(y) > 0$. By the continuity of the coefficients of L, there exists an open neighborhood of y, say W (depending on α), which is compactly contained in D and such that $Lu > 0$ on W. We now only need to choose $\beta > 0$ so small that the ellipsoid Ω is contained in W. This gives $Lu > 0$ on $\Omega \Subset D$, as we desired. $\qquad\square$

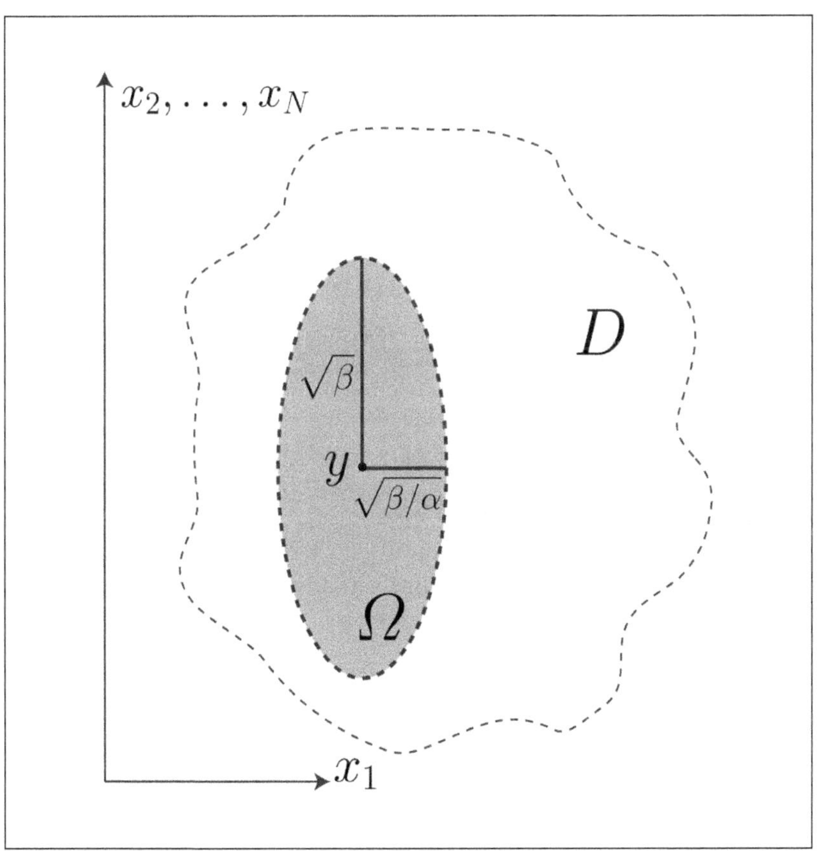

Fig. 8.7   The ellipsoid Ω in the proof of Thm. 8.33.

8.6 Exercises of Chap. 8

Exercise 8.1. For $i, j \in \{1, \ldots, N\}$, let $\widetilde{a}_{i,j}$ be functions on the open set D. Set

$$a_{i,j} := \frac{\widetilde{a}_{i,j} + \widetilde{a}_{j,i}}{2}, \quad i, j \in \{1, \ldots, N\}.$$

Then, for every $u \in C^2(D)$ and $x \in D$, we have

$$\sum_{i,j=1}^{N} \widetilde{a}_{i,j}(x)\, \partial_{i,j} u(x) = \sum_{i,j=1}^{N} a_{i,j}(x)\, \partial_{i,j} u(x).$$

Note that $a_{i,j} = a_{j,i}$ for every $i, j \leq N$. This is the reason why we are entitled to assume that $A(x)$ in (8.2) is symmetric.[6]

Exercise 8.2. Prove that the operator in $\mathbb{R}^2$

$$L = (1 - x^2)\left(\frac{\partial}{\partial x}\right)^2 + 2xy\, \frac{\partial^2}{\partial y \partial x} + (1 - y^2)\left(\frac{\partial}{\partial y}\right)^2$$

is elliptic in $D = \{(x, y) \in \mathbb{R}^2 : x^2 + y^2 < 1\}$, but not uniformly elliptic in D.

Exercise 8.3. Provide a detailed proof of the assertions in Rem. 8.5.

Exercise 8.4. Let $X_0, X_1, \ldots, X_m$ be smooth v.f.s on D and consider the differential operator $L = \sum_{k=1}^{m} X_k^2 + X_0$. Let also

$$X_i = \sum_{h=1}^{N} \alpha_{h,i}\, \partial_h, \quad i = 0, 1, \ldots, m.$$

Prove that L has the coordinate expression (8.1), with

$$a_{i,j} = \sum_{k=1}^{m} \alpha_{i,k}\, \alpha_{j,k}, \quad b_i(x) = \alpha_{i,0} + \sum_{k=1}^{m}\sum_{h=1}^{N} \alpha_{h,k}\, \partial_h \alpha_{i,k}, \quad c \equiv 0,$$

for any $i, j \in \{1, \ldots, N\}$. Deduce that one also has

$$A(x) = S(x) \cdot S(x)^T, \quad \text{where} \quad S(x) = \big(\alpha_{i,k}(x)\big)_{\substack{i=1,\ldots,N \\ k=1,\ldots,m}},$$

i.e., $S(x)$ is the $N \times m$ matrix whose columns are obtained by transposing the vectors of the coefficients of $X_1, \ldots, X_m$ at x. Derive that L is a semielliptic homogenous operator, but L cannot be elliptic at any point *whenever* $m < N$. Prove that the quadratic form of L at $x \in D$ is given by

$$q_L(x, \xi) = \|S(x)^T \xi\|^2 = \sum_{k=1}^{m} \langle X_k(x), \xi \rangle^2, \quad \xi \in \mathbb{R}^N.$$

Derive that, for any given $x \in D$, the set of the vectors ξ annihilating the quadratic form $q_L(x, \cdot)$ is given by (prove the equalities!)

$$\ker(A(x)) = \ker(S(x)^T) = \big\{X_1(x), \ldots, X_m(x)\big\}^{\perp}.$$

---

[6]Obviously, this is a consequence of simple results of Linear Algebra:

$$\operatorname{trace}(R) = \operatorname{trace}(R^T), \quad \operatorname{trace}(RS) = \operatorname{trace}(SR), \quad \sum_{i,j} a_{i,j} b_{i,j} = \operatorname{trace}(AB^T).$$

Exercise 8.5. Consider Laplace's Δ on $\mathbb{R}^2$. Let also $\Omega = \{(x, y) \in \mathbb{R}^2 : y > 0\}$. The function $u(x, y) = e^x \sin y$ is smooth on $\mathbb{R}^2$, it satisfies (actually with equalities!)

$$\Delta u \geq 0 \quad \text{on } \Omega, \qquad u \leq 0 \quad \text{on } \partial\Omega,$$

but it is not true that $u \leq 0$ on Ω. Hence Δ does not satisfy the WMP on Ω. Note that Ω is unbounded.

Exercise 8.6. Consider the wave operator $L = (\partial_x)^2 - (\partial_y)^2$ in $\mathbb{R}^2$. Observe that L is not semielliptic. Let $\Omega = (0, \pi) \times (0, \pi)$. Prove that the function $u(x, y) = \sin x \sin y$ satisfies (actually with equalities!)

$$Lu \geq 0 \quad \text{on } \Omega, \qquad u \leq 0 \quad \text{on } \partial\Omega,$$

but it is not true that $u \leq 0$ on Ω. Indeed, u attains its maximum on Ω at the interior point $(\pi/2, \pi/2)$ where u equals 1. Note that Ω is bounded, but L is not semielliptic.

In the sequel, we use the definitions of L-harmonic/L-subharmonic functions, for which the reader is referred to Def. 9.1 on page 198.

Exercise 8.7. (a). Consider the operator $L = (\partial_x)^2 + 1$ in $\mathbb{R}$. Observe that the zero-order term of L is positive. Let $\Omega = (0, \pi)$. Show that the function $u(x) = \sin x$ satisfies (actually with equalities!)

$$Lu = u'' + u \geq 0 \quad \text{on } \Omega, \qquad u \leq 0 \quad \text{on } \partial\Omega,$$

but it is not true that $u \leq 0$ on Ω. Indeed, u attains its maximum on Ω at the interior point $\pi/2$ where u equals 1.

(b). Consider the operator $L = (\partial_x)^2 + (\partial_y)^2 + 2$ in $\mathbb{R}^2$. Note that it is uniformly elliptic in $\mathbb{R}^2$. Prove that $u(x, y) = \sin x \sin y$ is L-harmonic, it is identically null on $\partial\Omega$, where $\Omega = (0, \pi) \times (0, \pi)$, but $u > 0$ on Ω (hence u attains its maximum at an interior point of Ω). Does this contradict any of the WMPs given in this chapter? (Compare this example with the result in Prop.8.25.)

(c). More generally, given any constant $C > 0$, prove that $L = (\partial_x)^2 + (\partial_y)^2 + C$ in $\mathbb{R}^2$ violates the WMP.

Hint: Consider $u(x, y) = \sin(x\sqrt{C/2}) \sin(y\sqrt{C/2})$ on the open set
$$\Omega = (0, \pi\sqrt{2/C}) \times (0, \pi\sqrt{2/C}).$$

Exercise 8.8. Prove the following "restricted" version (applying only to *strictly L-subharmonic functions*) of the WMP for PDOs with *non-positive c*:

Let L be a semielliptic operator on the open set $D \subseteq \mathbb{R}^N$. Suppose that the zero-order term c of L satisfies $c \leq 0$ on D.

Then L satisfies the following property: if Ω is a bounded open subset of D and if $u \in C^2(\Omega)$ is such that

$$\begin{cases} Lu \gneq 0 & \text{on } \Omega, \\[2mm] \limsup_{x \to y} u(x) \leq 0 & \text{for every } y \in \partial\Omega, \end{cases}$$

then $u \leq 0$ on Ω.

Observe the analogy with the WMP: the only difference lies in the stronger assumption $Lu > 0$ on Ω (playing the role of an L-barrier!).

[*Hint:* Argue as in the proof of Thm. 8.10, up to the point where it is shown that $Lu(x_0) \leq c(x_0)u(x_0)$, where x_0 is an interior maximum point of u. The hypotheses $Lu > 0$ and $c \leq 0$ give $c(x_0) < 0$...]

Exercise 8.9. Prove the WMP for *homogeneous* semielliptic operators possessing L-barriers on bounded open sets $\Omega \subseteq D$.

[*Hint:* Run over the proof of Thm. 8.16 showing that (with the notation of that proof) the case $x_0 \in \Omega$ *cannot occur*. The proof is then much simpler.]

Exercise 8.10. Observe that, if $\limsup_{x \to x_0} u(x) \leq 0$ and $w > 0$, one cannot conclude that $\limsup_{x \to x_0} w(x)\,u(x) \leq 0$.

[*Hint:* Consider the case $u = x\sin(1/x)$, $w(x) = 1/x^2$ and $x_0 = 0$.]

If, in addition to the above hypotheses, one knows that w is bounded from above, show that $\limsup_{x \to x_0} w(x)\,u(x) \leq 0$ holds true.

Exercise 8.11. Consider the following operators L in $\mathbb{R}^3$ (whose points are denote by (x, y, t)). Write down explicitly the matrix A associated with their characteristic form. Deduce that they are semielliptic homogeneous PDOs. Finally, prove that they satisfy the WMP on every bounded open set.

(1) *The Heat operator in $\mathbb{R}^3$:*
$$L = (\partial_x)^2 + (\partial_y)^2 - \partial_t.$$

(2) *The Kohn-Laplacian on the Heisenberg group in $\mathbb{R}^3$: $L = X^2 + Y^2$,*
$$\text{where } X = \partial_x + 2y\,\partial_t \text{ and } Y = \partial_y - 2x\,\partial_t.$$

(3) *Kolmogorov's operator in $\mathbb{R}^3$: $L = X^2 + X_0$,*
$$\text{where } X = \partial_x,\ X_0 = x\,\partial_y - \partial_t.$$

(4) *Mumford's operator: $L = X^2 - X_0$,*
$$\text{where } X = \partial_t,\ X_0 = \cos t\,\partial_x + \sin t\,\partial_y.$$

Exercise 8.12. Prove that the function
$$u = \frac{1 - x^2 - y^2}{(1 - x)^2 + y^2}$$
is Δ-harmonic in $D = \{(x, y) \in \mathbb{R}^2 : x^2 + y^2 < 1\}$, where $\Delta = (\partial_x)^2 + (\partial_y)^2$ is the usual Laplace operator. Note that $u > 0$ on D and $u \equiv 0$ on $\partial D \setminus \{(1, 0)\}$. Explain why this does not contradict the WMP.

Hint: It may help calculating
$$\limsup_{D \ni (x,y) \to (1,0)} u(x, y).$$

Exercise 8.13. Let $\varphi \in C_0^\infty(\mathbb{R}^2)$ be a nonnegative cut-off function such that $\varphi \equiv 1$ on $\{x_1^2 + x_2^2 < 1\}$ and $\varphi \equiv 0$ on $\{x_1^2 + x_2^2 > 2\}$. Consider the v.f.s on $\mathbb{R}^2$ defined by

$$X = x_2 \partial_{x_1} - x_1 \partial_{x_2}, \quad X_1 = \varphi(x)\,\partial_{x_1}, \quad X_2 = \varphi(x)\,\partial_{x_2}.$$

Prove that (see also Exm. 8.30):

- $L_\varphi := X^2 + X_1^2 + X_2^2$ is semielliptic on $\mathbb{R}^2$; observe that the eigenvalues of the principal matrix of L_φ at $x \in \mathbb{R}^2$ are $\varphi(x)^2 \geq 0$ and $x_1^2 + x_2^2 + \varphi^2(x) \geq 1$ so that L_φ is clearly semielliptic but not elliptic;
- L_φ is NTD on $\mathbb{R}^2$ since the trace of its principal matrix is

$$x_1^2 + x_2^2 + 2\,\varphi^2(x) \geq 1;$$

Nonetheless, observe that, if

$$\Omega = \{x \in \mathbb{R}^2 : \pi < x_1^2 + x_2^2 < 3\pi\},$$

the function $u(x) = \sin(x_1^2 + x_2^2)$ satisfies $L_\varphi u \equiv 0$ on Ω and $u \equiv 0$ on $\partial\Omega$. Since u changes it sign on Ω, the WMP is clearly violated for L_φ on Ω.

$$L = \sum a_{i,j}\partial_{i,j} + \sum b_j\partial_j + c \quad \text{on } D$$

with $(a_{i,j}) \geq 0$ and continuous $a_{i,j}, b_j, c$

$c < 0$	WMP on every bounded $\Omega \subseteq D$	Thm. 8.10
$c \leq 0$	WMP on every bounded $\Omega \subseteq D$ with an L-barrier: $\begin{cases} Lh < 0 \\ h \geq 0 \end{cases}$	Thm. 8.16
	for example: L of parabolic type as in Exm. 8.20	

$\left.\begin{array}{c} c \leq 0 \\[1em] \exists\, i:\ a_{i,i} > 0 \end{array}\right\}$	WMP on any bounded $\Omega \subset \overline{\Omega} \subset D$	Prop. 8.22
	for example: there exists i such that $a_{i,i} \equiv 1$	

any c	WMP on every bounded $\Omega \subseteq D$ with a strong L-barrier: $\begin{cases} Lh < 0 \\ \inf h > 0 \end{cases}$	Thm. 8.17
$\left.\begin{array}{c} \text{any } c \\[1em] \text{trace}(A) > 0 \end{array}\right\}$	local WMP	Prop. 8.25
	Beware the counterexample in Exm. 8.30!	

Fig. 8.8   A précis of the Weak Maximum Principles in this Chapter.

Chapter 9

Corollaries of the Weak Maximum Principle

THE aim of this chapter is to collect many consequences of the Weak Maximum Principles established in Chap. 8: after the introduction of the C^2 subharmonic functions wrt to the semielliptic operator L, we shall prove some *comparison principles* and *a priori estimates*, the *uniqueness* of the classical solution of the Dirichlet problem, the *Maximum-Modulus Principle* and the *Maximum Principle*. The latter is of paramount importance and it establishes that, when L has no zero-order term, any L-subharmonic function takes its maximum at the boundary of any bounded set where L satisfies the WMP.

As their very names recall, some of these results have counterparts in the theory of holomorphic functions; not by chance, here we prove the Fundamental Theorem of Algebra as a trivial corollary of the WMP for the classical harmonic functions vanishing at infinity.

As an application of these results, in Sec. 9.4 we briefly investigate the so-called Green and Poisson operators related to L, denoted by G_Ω and P_Ω: when L satisfies the WMP on the bounded open set Ω, and when the Dirichlet problem

$$(\text{DP}) \qquad \begin{cases} Lu = -f \text{ on } \Omega \\ u = \varphi \quad \text{ on } \partial\Omega \end{cases}$$

has a classical solution for every $f \in C(\overline{\Omega})$ and every $\varphi \in C(\partial\Omega)$, we denote by $G_\Omega(f)$ and $P_\Omega(\varphi)$, respectively, the solutions of (DP) associated with the choices $\varphi = 0$ and $f = 0$, respectively. As a consequence of the WMP and of the *a priori* estimates, we shall see that G_Ω and P_Ω are linear, positive and continuous operators (defined on suitable function spaces).

Natural prosecutions of this chapter are the next Chaps. 10 and 11, where the so-called *Strong Maximum Principle* is investigated.

A prerequisite for this chapter is the knowledge of the contents relating the WMP in the previous Chap. 8.

9.1 Comparison principles

We begin with some definitions. Throughout, L is a second order PDO as in (8.1) defined on the open set $D \subseteq \mathbb{R}^N$ and Ω is an open subset of D. We tacitly inherit all the notations and definitions of Chap. 8.

Definition 9.1 (L-harmonic and L-subharmonic/superharmonic C^2 functions).
We say that a function $u \in C^2(\Omega)$ is:

L**-harmonic** in Ω, if $Lu(x) = 0$ for every $x \in \Omega$;
L**-subharmonic** in Ω, if $Lu(x) \geq 0$ for every $x \in \Omega$;
L**-superharmonic** in Ω, if $Lu(x) \leq 0$ for every $x \in \Omega$;

strictly L-subharmonic in Ω, if $Lu(x) > 0$ for every $x \in \Omega$;
strictly L-superharmonic in Ω, if $Lu(x) < 0$ for every $x \in \Omega$.

We shall denote the set of the L-harmonic functions in Ω by $\mathcal{H}(\Omega)$ (or $\mathcal{H}_L(\Omega)$); the set of the L-subharmonic (resp., L-superharmonic) functions in Ω will be denoted by $\underline{\mathcal{S}}(\Omega)$ (resp., by $\overline{\mathcal{S}}(\Omega)$). The notations $\underline{\mathcal{S}}_L, \overline{\mathcal{S}}_L$ will also occasionally occur.

Actually, the natural definition of L-subharmonic (and L-superharmonic) function is more general, and it can be given for semi-continuous functions (Exr. 9.1). Nonetheless, in all the interesting cases, for functions of class C^2 this more general definition is equivalent to the above one.

In order to remedy the apparent discrepancy between the name *sub*-harmonic and the inequality $Lu \geq 0$ (it seems more natural to call 'subharmonic' a function which satisfies an inequality with a '$\leq$' sign...), some authors prefer to consider PDOs of the form $-L$, where L is semielliptic. In this case, however, one has to deal with PDOs whose principal matrix is *negative* semidefinite; this unpleasant fact justifies our choice in Def. 9.1 to tolerate the cited unaesthetic discrepancy.

As Exr. 9.1 shows, the name *subharmonic* perfectly reflects the fact that the graph of a subharmonic function on a domain *lies below* the graph of a harmonic function satisfying certain boundary conditions.

Remark 9.2. Obviously $\mathcal{H}(\Omega)$ is a real vector space, whereas $\underline{\mathcal{S}}(\Omega)$ and $\overline{\mathcal{S}}(\Omega)$ are convex cones (i.e., they are closed under linear combinations with nonnegative scalar coefficients). Also, it trivially[1] holds that

$$\underline{\mathcal{S}}(\Omega) \cap \overline{\mathcal{S}}(\Omega) = \mathcal{H}(\Omega), \qquad \overline{\mathcal{S}}(\Omega) = -\underline{\mathcal{S}}(\Omega).$$

We remark that $\mathcal{H}(\Omega)$ *contains the constant functions if and only if L is homogenous* (i.e., $L(1) = 0$). Analogously, *the constant functions are L-superharmonic functions if and only if $L(1) \equiv 0$.*

---

[1]The triviality of this result is due to the fact that we have defined (for the time being) L-subharmonic/superharmonic functions in the framework *of C^2 functions*. For semi-continuous functions, this fact is less obvious.

We also have the following simple fact: *the positive constant functions are L-superharmonic if and only if $L(1) \leq 0$; analogously, the negative constant functions are L-subharmonic if and only if $L(1) \leq 0$.* ♮

The previous WMPs can obviously be restated in terms of L-subharmonic functions: namely, L satisfies the WMP on Ω if and only if, for any L-subharmonic u in Ω, it holds that $u \leq 0$ on Ω whenever $\limsup_{x \to y} u \leq 0$ for every $y \in \partial\Omega$.

We are ready for two of the main applications of the WMP.

Theorem 9.3 (Comparison Principle). *Suppose that L satisfies the Weak Maximum Principle on the bounded open set Ω. Let u and v be, respectively, an L-subharmonic and an L-superharmonic function in Ω.*
Then, the following conditions

$$\begin{cases} \limsup_{x \to y} u(x) \leq \liminf_{x \to y} v(x) & \text{for every } y \in \partial\Omega \\[2mm] \text{for every } y \in \partial\Omega \text{ one at least of the above } \limsup, \liminf \text{ is finite} \end{cases} \tag{9.1}$$

imply that $u \leq v$ on Ω.

For example, the Comparison Principle holds true for functions u and v satisfying, along with (9.1), one of the following three conditions:

$$\big(u \in \underline{\mathcal{S}}(\Omega), \; v \in \mathcal{H}(\Omega)\big); \quad \big(u \in \mathcal{H}(\Omega), \; v \in \overline{\mathcal{S}}(\Omega)\big); \quad \big(u \in \mathcal{H}(\Omega), \; v \in \mathcal{H}(\Omega)\big).$$

Proof. We let $w := u - v$. Then $w \in \underline{\mathcal{S}}(\Omega)$ since $Lw = Lu - Lv \geq 0$ (since $Lu \geq 0$ and $Lv \leq 0$ by hypothesis). Moreover, for every $y \in \partial\Omega$ we have

$$\limsup_{x \to y} w(x) \leq \limsup_{x \to y} u(x) - \liminf_{x \to y} v(x) \overset{(9.1)}{\leq} 0.$$

Note that in the last inequality we have also used the second assumption in (9.1) that rules out any indeterminate form $\pm\infty \mp \infty$.

By the WMP for L on Ω we infer that $w \leq 0$, that is $u \leq v$ on Ω. $\square$

Let now Ω be a bounded open subset of D. Let us consider two functions $f : \Omega \to \mathbb{R}$ and $\varphi : \partial\Omega \to \mathbb{R}$. We say that a function u is a **(classical) solution of the Dirichlet problem**

$$\text{(DP)} \qquad \begin{cases} Lu = f & \text{on } \Omega \\ u = \varphi & \text{on } \partial\Omega, \end{cases} \tag{9.2}$$

if the following conditions are satisfied:

(1) $u \in C(\overline{\Omega})$ and $u|_\Omega \in C^2(\Omega)$,
(2) $Lu(x) = f(x)$ for every $x \in \Omega$,
(3) $u(y) = \varphi(y)$ for every $y \in \partial\Omega$.

When (1) is satisfied, we shall simply write, with a small abuse of notation,

$$u \in C(\overline{\Omega}) \cap C^2(\Omega).$$

Theorem 9.4 (Uniqueness of the solution of the Dirichlet problem). *Suppose that L satisfies the Weak Maximum Principle on the bounded open set Ω.*

Then, for any choice of f and φ, there exists at most one classical solution of the Dirichlet problem (9.2).

Proof. Let u_1, u_2 be two classical solutions of (DP) and let $u = u_1 - u_2$. Then $Lu = Lu_1 - Lu_2 = f - f = 0$ on Ω; moreover

$$u|_{\partial\Omega} = u_1|_{\partial\Omega} - u_2|_{\partial\Omega} = \varphi - \varphi = 0.$$

The WMP for L ensures that $u \leq 0$ on Ω, that is $u_1 \leq u_2$. By reversing the roles of u_1, u_2 we obtain the reverse inequality $u_2 \leq u_1$. Thus $u_1 \equiv u_2$. $\qquad\square$

Remark 9.5. In the previous proof we implicitly proved the following fact:

If L satisfies the WMP on Ω, the Dirichlet problem on Ω with $f \equiv 0$ and $\varphi \equiv 0$ is uniquely solved by $u \equiv 0$.

Equivalently, if L satisfies the WMP on Ω, the unique $u \in C(\overline{\Omega}) \cap C^2(\Omega)$ which is L-harmonic in Ω and null on $\partial\Omega$ is the null function. $\qquad\sharp$

9.2   Maximum-modulus and Maximum Principle

In the next result we are interested in a comparison of the values attained by a non-negative function on $\overline{\Omega}$ with the values it attains on $\partial\Omega$. To this aim it is convenient that the positive constant functions be L-superharmonic. This is the reason why we assume that $L(1) \leq 0$ (see Rem. 9.2).

Theorem 9.6 (Maximum-Modulus Principle). *Let L satisfy $L(1) \leq 0$ on D. Suppose that $\Omega \subseteq D$ is bounded and that L satisfies the WMP on Ω.*

Let $u \in C(\overline{\Omega}) \cap C^2(\Omega)$ be L-harmonic in Ω. Then we have

$$\max_{\overline{\Omega}} |u| = \max_{\partial\Omega} |u|. \tag{9.3}$$

Proof. Let $M := \max_{\partial\Omega} |u|$. This number is ≥ 0 and finite for $\partial\Omega \subset \overline{\Omega}$ is compact and $u \in C(\overline{\Omega})$. By definition of M one has $\pm u \leq M$ on $\partial\Omega$. Since u is in $\mathcal{H}(\Omega)$, one has $\pm u \in \mathcal{H}(\Omega)$ as well; moreover since $L(1) \leq 0$ and $M \geq 0$, the constant function equal to M belongs to $\overline{\mathcal{S}}(\Omega)$; indeed

$$L(M) = \underbrace{M}_{\geq 0} \cdot \underbrace{L(1)}_{\leq 0} \leq 0.$$

The Comparison Principle in Thm. 9.3 ensures that $\pm u \leq M$ in Ω which proves that $\max_{\overline{\Omega}} |u| \leq M$. On the other hand one has

$$\max_{\overline{\Omega}} u \leq M = \max_{\partial\Omega} |u| \leq \max_{\overline{\Omega}} |u|.$$

This gives the desired (9.3). $\qquad\qquad\square$

Remark 9.7. *The condition $L(1) \leq 0$ in Thm. 9.6 cannot be dropped.* For example, consider the operator $L = \Delta + 1/2$ (where Δ is the Laplace operator in $\mathbb{R}^2$); since the function $h(x) = \sin(x_1)\sin(x_2)$ satisfies $\Delta h = -2h$ in $\mathbb{R}^2$, it turns out that $Lh = -\frac{3}{2}h$. Note that h is an L-barrier for $\Omega = (0,\pi) \times (0,\pi)$ (since $h > 0$ there), but it is not a strong L-barrier on Ω since $\inf_{\Omega} h = 0$.

If we take $\varepsilon > 0$ small, and if we set $\Omega_\varepsilon := (\varepsilon, \pi - \varepsilon) \times (\varepsilon, \pi - \varepsilon)$, then $\inf_{\Omega_\varepsilon} h > 0$ so that h is a strong L-barrier on Ω_ε; due to Thm. 8.17, L satisfies the WMP on Ω_ε.

Let us consider the function $u(x) = \sin(x_1/\sqrt{2})\sin(x_2/\sqrt{2})$; then $Lu = 0$ in $\mathbb{R}^2$. If ε is so small that $\pi/\sqrt{2} \in (\varepsilon, \pi - \varepsilon)$, then

$$\max_{\overline{\Omega_\varepsilon}} |u| \geq u(\pi/\sqrt{2}, \pi/\sqrt{2}) = 1.$$

On the other hand, since $|u| \leq 1$, this gives $\max_{\overline{\Omega_\varepsilon}} |u| = 1$ (if ε is as above). It is not difficult to recognize that

$$\max_{\partial\Omega_\varepsilon} |u| = \max\left\{ \sin\left(\frac{\varepsilon}{\sqrt{2}}\right), \sin\left(\frac{\pi - \varepsilon}{\sqrt{2}}\right) \right\}.$$

For small ε, the above rhs is small; for such an ε, one then deduces that

$$\max_{\overline{\Omega_\varepsilon}} |u| > \max_{\partial\Omega_\varepsilon} |u|,$$

and the thesis of Thm. 9.6 does not hold true. $\qquad\qquad\sharp$

Theorem 9.8 (L-harmonic functions vanishing at ∞). *Let L be defined on $\mathbb{R}^N$ and suppose its zero-order term $L(1)$ is ≤ 0; assume furthermore that L satisfies the Weak Maximum Principle on every bounded open subset of $\mathbb{R}^N$.*

Let u be an L-harmonic function vanishing at infinity. Then $u \equiv 0$.

Proof. By hypothesis $Lu = 0$ and $\lim_{\|x\|\to\infty} u(x) = 0$. This gives

$$L(\pm u) = 0 \geq 0 \quad \text{and} \quad \limsup_{\|x\|\to\infty}(\pm u(x)) = 0 \leq 0.$$

By Prop. 8.18 it follows that $\pm u \leq 0$, that is $u \equiv 0$. $\qquad\qquad\square$

There are plenty of other corollaries and variants of the WMP: some of them can be found in Exr. 9.2 (which we strongly recommend to the reader's attention). For future reference (and due to its paramount importance), here we explicitly provide the following one, concerning L-subharmonic functions and their maxima.

We remark that this is the only example where we assume homogeneity of L (i.e., $L(1) \equiv 0$) rather than the usual condition $L(1) \leq 0$: this is due to our need for all constant functions to be L-superharmonic (in order to use them, roughly put, as rhs for a comparison argument).

Theorem 9.9 (The Maximum Principle). *Let L be a homogeneous PDO on $D \subseteq \mathbb{R}^N$ (that is, $L(1) \equiv 0$). Suppose that $\Omega \subseteq D$ is bounded and that L satisfies the WMP on Ω. Let $u \in C(\overline{\Omega}) \cap C^2(\Omega)$ be L-subharmonic in Ω. Then*

$$\sup_{\Omega} u = \sup_{\partial\Omega} u. \tag{9.4}$$

Furthermore, if u is only in $C^2(\Omega)$ and it is L-subharmonic in Ω, one has

$$\sup_{\Omega} u = \sup \left\{ \limsup_{x \to y} u(x) \; : \; y \in \partial\Omega \right\}. \tag{9.5}$$

Remark 9.10. We point out that the hypothesis of homogeneity of L cannot be removed from Thm. 9.9 (even by possibly replacing it with $L(1) \leq 0$), as the case of $L = \Delta - 1$ shows; indeed, one can take $\Omega = \{x \in \mathbb{R}^N : \|x\| < 1\}$ and the (strictly) L-subharmonic function $u(x) = -3\,N - \|x\|^2$ which satisfies (Exr. 9.6)

$$\sup_{\Omega} u > \sup_{\partial\Omega} u.$$

Note that $\Delta - 1$ satisfies the WMP on every bounded open set (Thm. 8.10). ♯

Proof. Let $u \in C(\overline{\Omega}) \cap C^2(\Omega)$ be L-subharmonic in Ω. Let $M := \sup_{\partial\Omega} u$. Then $M \in \mathbb{R}$, arguing as in the proof of Thm. 9.6. *Note that we do not know the sign of M* (this distinguishes the present situation to what happened in the proof of the Maximum-Modulus Principle).

As L is homogeneous, the constant function M is L-superharmonic in Ω (actually it is in $\mathcal{H}(\Omega)$); moreover $u \leq M$ on $\partial\Omega$ by the choice of M. From the Comparison Principle in Thm. 9.3, we get $u \leq M$ on Ω, which implies $\sup_{\Omega} u \leq M$. Since u is continuous on $\overline{\Omega}$, we have $\sup_{\overline{\Omega}} u = \sup_{\Omega} u$, whence

$$\sup_{\Omega} u \leq M = \sup_{\partial\Omega} u \leq \sup_{\overline{\Omega}} u = \sup_{\Omega} u,$$

and the proof of (9.4) is complete.

We now prove (9.5), dropping the assumption $u \in C(\overline{\Omega})$. Let us denote by M the rhs of (9.5). It is easy to recognize that $M \leq \sup_{\Omega} u$. Therefore, if $M = \infty$, there is nothing to prove. We can thus suppose that $M \in \mathbb{R}$. By applying the WMP[2] to the L-subharmonic function $u - M$, one infers that $\sup_{\Omega} u \leq M$. This ends the proof, since $M \leq \sup_{\Omega} u$. $\square$

For another Maximum Principle-type result for PDOs L which are not necessarily homogeneous, see Thm. 9.16 in Sec. 8.5.

9.2.1 *The parabolic case*

All the previous results have parabolic counterparts, taking into account the parabolic WMP in Thm 8.32. We summarize it in the following result.

---

[2]Note that $\limsup_{x \to y}(u(x) - M) \leq 0$ for any $y \in \partial\Omega$ owing to the definition of M.

Theorem 9.11 (Parabolic Maximum Principle). *Let $D \subseteq \mathbb{R}^{N+1}$ be an open set, and let L be a semielliptic PDO of the form*

$$L := \sum_{i,j=1}^{N} a_{i,j}(x,t)\, \frac{\partial^2}{\partial x_i\, \partial x_j} + \sum_{i=1}^{N} b_i(x,t)\, \frac{\partial}{\partial x_i} + c(x,t) - \frac{\partial}{\partial t},$$

with $c = L(1) \leq 0$ on D. Suppose that $\Omega \subseteq D$ is a bounded open set and, for $T \in \mathbb{R}$, let us set $\Omega_T := \Omega \cap \{(x,t) \in \mathbb{R}^{N+1} : t < T\}$, which we suppose to be non-void. As in (8.21), we denote by $\partial_T \Omega$ the parabolic boundary of Ω_T. Then the following facts hold.

1. *(**Parabolic Comparison Principle**). Let u and v be, respectively, an L-subharmonic and an L-superharmonic function in Ω_T. Then, the following conditions*

$$\begin{cases} \limsup\limits_{z \to \zeta} u(z) \leq \liminf\limits_{z \to \zeta} v(z) \quad \text{for every } \zeta \in \partial_T \Omega \\[2mm] \text{for every } \zeta \in \partial_T \Omega \text{ one at least of the above } \limsup, \liminf \text{ is finite} \end{cases}$$

   imply that $u \leq v$ on Ω_T.

2. *(**Uniqueness of the parabolic (DP) solution**). For any $f : \Omega_T \to \mathbb{R}$ and any $\varphi : \partial_T \Omega \to \mathbb{R}$, there exists at most one solution of the Dirichlet problem*

$$\text{(DP)} \qquad \begin{cases} Lu = f \text{ on } \Omega_T \\ u = \varphi \quad \text{on } \partial_T \Omega, \end{cases}$$

   that is a function u satisfying:

   (a) $u \in C(\Omega_T \cup \partial_T \Omega)$ *and* $u|_{\Omega_T} \in C^2(\Omega_T)$,
   (b) $Lu(z) = f(z)$ *for every* $z \in \Omega_T$,
   (c) $u(\zeta) = \varphi(\zeta)$ *for every* $\zeta \in \partial_T \Omega$.

3. *(**Parabolic Maximum-Modulus Principle**). Let $u \in C(\Omega_T \cup \partial_T \Omega) \cap C^2(\Omega_T)$ be an L-harmonic function in Ω_T. Then*

$$\sup_{\Omega_T \cup \partial_T \Omega} |u| = \sup_{\partial_T \Omega} |u|.$$

4. *(**The Parabolic Maximum Principle**). Let L be homogeneous (i.e., $L(1) \equiv 0$) and suppose that $u \in C(\Omega_T \cup \partial_T \Omega) \cap C^2(\Omega_T)$ is L-subharmonic in Ω_T. Then*

$$\sup_{\Omega_T} u = \sup_{\partial_T \Omega} u.$$

The proof is left as an exercise (Exr. 9.7), based on Thm 8.32.

We end the chapter with a remarkable corollary of the WMP.

Theorem 9.12 (The Fundamental Theorem of Algebra). *Let $p(z)$ be a complex polynomial of positive degree. Then $p(z)$ has at least one zero in $\mathbb{C}$.*

Proof. Suppose that $p(z) \neq 0$ for every $z \in \mathbb{C}$. Hence $q(z) := 1/p(z)$ is an entire function. We let $q(z) = u(x, y) + i\, v(x, y)$, with uniquely determined real-valued C^2 (actually C^ω!) functions u, v defined on $\mathbb{R}^2 \equiv \mathbb{C}$.

By exploiting the Cauchy-Riemann equation

$$0 = \frac{\partial}{\partial \bar{z}} q(z) = \frac{1}{2} \left(\frac{\partial}{\partial x} + i\, \frac{\partial}{\partial y} \right) (u(x, y) + i\, v(x, y)),$$

it is easy to see that u and v are Δ-harmonic on $\mathbb{R}^2$, where $\Delta = (\partial_x)^2 + (\partial_y)^2$. Since p has positive degree, then $q(z)$ vanishes as $|z| \to \infty$, as the following computation shows (here $n \geq 1$ and $a_n \neq 0$):

$$q(z) = \frac{1}{a_n\, z^n + a_{n-1}\, z^{n-1} + \cdots + a_0} = \frac{1/z^n}{a_n + a_{n-1}/z + \cdots + a_0/z^n} \xrightarrow{|z| \to \infty} 0.$$

This implies that u and v vanish as $\|(x, y)\| \to \infty$. Since Δ satisfies the WMP on every bounded open subset of $\mathbb{R}^2$ (Exm. 8.24), then Thm. 9.8 ensures that $u, v \equiv 0$. This gives $1/p = q \equiv 0$, which is absurd. $\qquad\square$

9.3 An a priori estimate

In this section we prove an alternative version of the Maximum-Modulus Principle in Thm. 9.6, under the assumption that Ω possesses a strong L-barrier h satisfying the strong bound $\sup_\Omega Lh \leq -1$. Instead of dealing with L-harmonic functions u (and with a zero-order term $c \leq 0$), we make no assumption on the nullity of Lu (nor on the sign of c), and this will produce an upper bound for $\max_{\overline{\Omega}} |u|$ also in terms of $\sup_\Omega |Lu|$ (and in terms of $\max\{0, c\}$). This gives a very useful example of an *a priori* estimate of u.

Theorem 9.13. *Let L be semielliptic on D and let $\Omega \subseteq D$ be bounded. Suppose that Ω admits a strong L-barrier (Def. 8.13) with the additional assumptions*

$$M := \sup_\Omega h < \infty \quad \text{and} \quad -\alpha := \sup_\Omega Lh < 0. \tag{9.6}$$

We make no assumption on the zero-order term $c = L(1)$ of L except that

$$C := \sup_{x \in \Omega} \max\{0, c(x)\} < \infty. \tag{9.7}$$

(Condition (9.7) is satisfied if, for instance, $\Omega \Subset D$ and c is continuous, or whenever $c \leq 0$; in the latter case $C = 0$.)

Then, for any $u \in C(\overline{\Omega}) \cap C^2(\Omega)$, we have the a priori estimate

$$\max_{\overline{\Omega}} |u| \leq \left(1 + \frac{CM}{\alpha} \right) \cdot \max_{\partial \Omega} |u| + \frac{M}{\alpha} \sup_\Omega |Lu|. \tag{9.8}$$

More explicitly:

$$\max_{\overline{\Omega}} |u| \leq \left(1 + \|c^+\|_\infty \frac{\|h\|_\infty}{\inf_\Omega(-Lh)} \right) \cdot \max_{\partial \Omega} |u| + \frac{\|h\|_\infty}{\inf_\Omega(-Lh)} \sup_\Omega |Lu|,$$

where $c^+ = \max\{0, c\}$ is the positive part of c, and $\|c^+\|_\infty, \|h\|_\infty$ denote the sup over Ω of the (nonnegative) functions c^+ and h.

Before giving the proof of this result, some remarks are in order.

Remark 9.14. (a). If $L(1) \leq 0$ and $Lu = 0$, (9.8) gives out

$$\max_{\overline{\Omega}} |u| \leq \max_{\partial\Omega} |u|,$$

a slightly weaker version of (9.3) in the Maximum-Modulus Principle.

(b). Suppose that L is homogenous and it satisfies the WMP on Ω. Then a sufficient condition for Ω to admit a strong L-barrier h as in Thm. 9.13 is that the following Dirichlet Problem admit a classical solution $h \in C(\overline{\Omega}) \cap C^2(\Omega)$:

$$\begin{cases} Lh = -1 & \text{on } \Omega \\ h = 1 & \text{on } \partial\Omega. \end{cases}$$

Indeed, in this case $L(-h) = 1 \geq 0$ on Ω so that $-h$ is L-subharmonic; as a consequence, the Maximum Principle in Thm. 9.9 ensures that

$$\inf_{\Omega} h = -\sup_{\Omega}(-h) = -\sup_{\partial\Omega}(-h) = 1.$$

Proof of Thm. 9.13. Let the notations of the statement of the theorem apply. We set $\|Lu\|_\infty := \sup_\Omega |Lu|$. If $\|Lu\|_\infty = \infty$, (9.8) is trivially true (since $h > 0$ on Ω, so that $M > 0$). Otherwise, if $\|Lu\|_\infty < \infty$, consider the two functions on Ω

$$v_\pm(x) := \pm u(x) - \max_{\partial\Omega} |u| - \left(C \max_{\partial\Omega} |u| + \|Lu\|_\infty\right)\frac{h(x)}{\alpha}.$$

On Ω we have

$$\begin{aligned}
L(v_\pm) &= \pm Lu - c\max_{\partial\Omega}|u| - \left(C\max_{\partial\Omega}|u| + \|Lu\|_\infty\right)\frac{Lh}{\alpha} \\
&\geq \pm Lu - c\max_{\partial\Omega}|u| + C\max_{\partial\Omega}|u| + \|Lu\|_\infty \\
&\geq \|Lu\|_\infty \pm Lu + (C - c)\max_{\partial\Omega}|u|.
\end{aligned}$$

Here we have used $c \leq \max\{0, c\} \leq C < \infty$ (due to (9.7)), together with the fact that $Lh \leq \sup_\Omega Lh = -\alpha < 0$ (whence $-Lh/\alpha \geq 1$, see (9.6)). Since both $\|Lu\|_\infty \pm Lu$ and $C - c$ are ≥ 0, we deduce that $L(v_\pm) \geq 0$ on Ω. Moreover, on $\partial\Omega$ we have (bearing in mind that $h \geq 0$ since h is an L-barrier, and $\alpha > 0$)

$$v_\pm|_{\partial\Omega} \leq \pm u|_{\partial\Omega} - \max_{\partial\Omega} |u| \leq 0.$$

We can apply the WMP for L on Ω (in Thm. 8.17), due to the very existence of a strong L-barrier for Ω; therefore $v_\pm \leq 0$ on Ω. This gives

$$\begin{aligned}
|u(x)| &\leq \max_{\partial\Omega} |u| + \left(C \max_{\partial\Omega} |u| + \|Lu\|_\infty\right)\frac{h(x)}{\alpha} \\
&\leq \max_{\partial\Omega} |u| + \left(C \max_{\partial\Omega} |u| + \|Lu\|_\infty\right)\frac{M}{\alpha}
\end{aligned}$$

and this implies (9.8). $\qquad\square$

9.4   Application: Green and Poisson operators

This short section provides examples of the applications of the Maximum Principles of this chapter; furthermore, we seize the opportunity to glance at some crucial topics: integral representation formulae and hypoelliptic operators.

Let L be a semielliptic homogeneous operator on $D \subseteq \mathbb{R}^N$, and let $\Omega \Subset D$ be a bounded open set. Suppose that L satisfies the WMP on Ω. Let us further assume that, for every $f \in C(\overline{\Omega})$ and every $\varphi \in C(\partial\Omega)$, the Dirichlet problem

$$\begin{cases} Lu = -f \text{ on } \Omega \\ u = \varphi \quad \text{ on } \partial\Omega \end{cases} \tag{9.9}$$

admits a classical solution $u \in C^2(\Omega) \cap C(\overline{\Omega})$. By Thm. 9.4, this solution is unique, and we denote it by

$$\Omega \ni x \mapsto u(x) =: D_\Omega(f, \varphi)(x).$$

If $u \in C^2(\overline{\Omega})$ (this means that u is the restriction to $\overline{\Omega}$ of a C^2 function defined on an open neighborhood of $\overline{\Omega}$), then u is the (unique) solution to (9.9), when $f = -Lu$ and $\varphi = u|_{\partial\Omega}$. This means that

$$u = D_\Omega(-Lu, u|_{\partial\Omega}), \qquad \forall \, u \in C^2(\overline{\Omega}). \tag{9.10}$$

As observed in Rem. 9.14-(b), the function $h = D_\Omega(1,1)$ is a strong L-barrier satisfying the requirements of Thm. 9.13; the estimate in (9.8) can be rewritten as follows (with reference to the notation of the statement of Thm. 9.13, here we have $C = 0$ and $\alpha = 1$):

$$\sup_{\overline{\Omega}} |D_\Omega(f, \varphi)| \le \max_{\partial\Omega} |\varphi| + M \max_{\overline{\Omega}} |f|, \tag{9.11}$$

where $M = \sup_\Omega h$. We denote $D_\Omega(f, 0)$ and $D_\Omega(0, \varphi)$ respectively by $G_\Omega(f)$ and $P_\Omega(\varphi)$. Again by the uniqueness result in Thm. 9.4, one has

$$D_\Omega(f, \varphi) = P_\Omega(\varphi) + G_\Omega(f), \tag{9.12}$$

for any $\varphi \in C(\partial\Omega)$ and $f \in C(\overline{\Omega})$. Gathering together (9.10) and (9.12) one gets the **representation formula**

$$u = P_\Omega(u|_{\partial\Omega}) - G_\Omega(Lu), \qquad \forall \, u \in C^2(\overline{\Omega}). \tag{9.13}$$

We now equip G_Ω and P_Ω with a precise functional framework. Indeed, let

$$C_{(0)}(\overline{\Omega}) := \{u \in C(\overline{\Omega}) \, : \, u = 0 \text{ on } \partial\Omega\}.$$

Then the operator $G_\Omega : C(\overline{\Omega}) \longrightarrow C_{(0)}(\overline{\Omega})$ mapping any function $f \in C(\overline{\Omega})$ to the classical solution $u = G_\Omega(f)$ of

$$\begin{cases} Lu = -f \quad \text{on } \Omega \\ u = 0 \qquad \text{on } \partial\Omega \end{cases}$$

will be referred to as the **Green operator** (relative to Ω and L).

Analogously, the operator $P_\Omega : C(\partial\Omega) \longrightarrow C(\overline{\Omega})$ mapping $\varphi \in C(\partial\Omega)$ to the classical solution $v = P_\Omega(\varphi)$ of

$$\begin{cases} Lv = 0 & \text{on } \Omega \\ v = \varphi & \text{on } \partial\Omega \end{cases}$$

will be referred to as the **Poisson operator**, or **harmonic operator** (relative to Ω and L). Clearly, both G_Ω and P_Ω are linear operators. Most importantly, *they are nonnegative operators*, in the sense that

$$G_\Omega(f) \geq 0 \quad \text{for every } f \geq 0,$$

$$P_\Omega(\varphi) \geq 0 \quad \text{for every } \varphi \geq 0.$$

This is a *consequence of the Weak Maximum Principle* applied to the L-subharmonic functions $-G_\Omega(f), -P_\Omega(\varphi)$ having non-positive boundary values on $\partial\Omega$. What is also most remarkable is that G_Ω and P_Ω are *continuous operators*: this is a consequence of Thm. 9.13. Indeed, with a small abuse of notation, for any compact set K and any $u \in C(K)$ we set

$$\|u\|_\infty := \max_{x \in K} |u(x)|.$$

With this notation, from (9.11) we obtain

$$\|G_\Omega(f)\|_\infty \leq M \, \|f\|_\infty, \qquad\qquad \forall \, f \in C(\overline{\Omega}), \qquad\qquad (9.14)$$

$$\|P_\Omega(\varphi)\|_\infty \leq \|\varphi\|_\infty, \qquad\qquad \forall \, \varphi \in C(\partial\Omega). \qquad\qquad (9.15)$$

Since $C(\overline{\Omega})$, $C(\partial\Omega)$ and $C_{(0)}(\overline{\Omega})$ equipped with the norm $\|\cdot\|_\infty$ are Banach spaces, (9.14) and (9.15) show that G_Ω and P_Ω are continuous operators.

Here, we only anticipate that real-valued linear positive operators (defined on suitable spaces of continuous functions) can be represented[3] as integral operators with respect to appropriate Radon measures. For example, this is the point of view of Potential Theory (in representing the subharmonic functions). As a by-product, identity (9.13) can be rewritten as an *integral representation formula* for C^2 functions u on $\overline{\Omega}$ in terms of the values of u on $\partial\Omega$ and in terms of Lu on Ω.

Remark 9.15. In most cases, it may be simpler to prove the solvability of the Dirichlet problem (9.9) for suitable subclasses of f's and φ's. For instance, when dealing with C^∞-*hypoelliptic operators*, it is expectable that (9.9) be solvable when $f \in C^\infty(\Omega) \cap C(\overline{\Omega})$.

If this is the case, Thm. 9.13 can be used to continue G_Ω, in a unique way, to a linear continuous operator defined on $C(\overline{\Omega})$. Indeed, given $f \in C(\overline{\Omega})$, by mollification one can construct a sequence $f_n \in C^\infty(\overline{\Omega})$ uniformly convergent to f on $\overline{\Omega}$. Then one defines $G_\Omega(f)$ as the uniform limit of $G_\Omega(f_n)$; the latter limit exists owing to (9.14), since one has

$$\|G_\Omega(f_n) - G_\Omega(f_m)\|_\infty = \|G_\Omega(f_n - f_m)\|_\infty \leq M \, \|f_n - f_m\|_\infty.$$

We leave the details to the reader, who should be now convinced that the *a priori* estimate in Thm. 9.13, which is a *summa* of the Maximum Principles considered in this chapter, is by all means a fundamental tool. ♮

---

[3]See the so-called Riesz(-Markov-Kakutani) Representation Theorem in [Rudin (1987)].

9.5　Appendix: Another Maximum Principle

The aim of this section is to provide a variant of the Maximum Principle in Thm. 9.9 when not necessarily homogeneous PDOs are involved.

Theorem 9.16. *Let L be a PDO as in (8.1), with $c = L(1) \leq 0$ on D. Suppose that the homogeneous operator $L - c$ satisfies the WMP on every bounded open subset of D.*
　If $\Omega \subseteq D$ is bounded, and $u \in C(\overline{\Omega}) \cap C^2(\Omega)$ is L-subharmonic in Ω, then

$$\sup_{\Omega} u \leq \sup_{\partial\Omega} u^+, \tag{9.16}$$

where $u^+ = \max\{u, 0\}$ is the nonnegative part of u.

As anticipated in Rem. 9.10, we remark that we are not allowed, in general, to replace u^+ with u in (9.16), due to the presence of the zero-order term c (even when this is ≤ 0); see Exr. 9.6. Moreover, it is trivial to get an example of the strict inequality in (9.16), by taking $L = \Delta$ in $\mathbb{R}^N$, $u \equiv -1$ and $\Omega \in \mathbb{R}^N$.

Proof. Let us set $\Omega^+ := \{x \in \Omega : u(x) > 0\}$. Obviously, Ω^+ is an open subset of Ω, possibly empty (the latter happening precisely when $u \leq 0$ on Ω).
　If $\Omega^+ = \varnothing$, we have $\sup_{\Omega} u \leq 0 = \sup_{\partial\Omega} u^+$ and the theorem is proved. Hence we suppose that $\Omega^+ \neq \varnothing$. Note that this ensures the following fact

$$\sup_{\Omega} u = \sup_{\Omega^+} u. \tag{9.17}$$

Let us consider the homogeneous operator $L_h := L - c$; on Ω^+ we have

$$L_h u = L u - c\,u \geq -c\,u \geq 0.$$

Hence u is L_h-subharmonic in Ω^+. Since L_h satisfies the WMP on Ω^+ (due to the hypothesis), we can apply the Comparison Principle in Thm. 9.9 for the PDO L_h and for the open set Ω^+, obtaining

$$\sup_{\Omega^+} u = \sup_{\partial(\Omega^+)} u. \tag{9.18}$$

Let $x_0 \in \partial(\Omega^+)$ be such that $u(x_0) = \sup_{\partial(\Omega^+)} u$. This x_0 exists as u belongs to $C(\overline{\Omega})$. From (9.18) we get $u(x_0) = \sup_{\Omega^+} u > 0$ since $\Omega^+ \neq \varnothing$. This prevents x_0 from belonging to Ω (otherwise we would have $x_0 \in \Omega^+$); therefore $x_0 \in \partial\Omega$. This proves the last inequality of the following chain:

$$\sup_{\Omega} u \overset{(9.17)}{=} \sup_{\Omega^+} u \overset{(9.18)}{=} \sup_{\partial(\Omega^+)} u = u(x_0) = u^+(x_0) \leq \sup_{\partial\Omega} u^+.$$

This ends the proof.　　□

9.6 Exercises of Chap. 9

Exercise 9.1. (The general definition of a subharmonic function). Suppose that L satisfies the WMP on every open set which is compactly contained in D. Let $u \in \underline{\mathcal{S}}(\Omega)$. Prove the following property:

(S). *For every $V \Subset \Omega$ and every $h \in C^2(V) \cap C(\overline{V})$ which is L-harmonic in V, the condition $u \leq h$ on ∂V implies $u \leq h$ on V.*

Note that in stating condition (S) there is no need to suppose that u is C^2. Actually, for upper semi-continuous functions $u : \Omega \to [-\infty, \infty)$ which are not identically $-\infty$ on any connected component of Ω, property (S) is traditionally assumed as the very definition of an L-subharmonic function.

Exercise 9.2. Let $\Omega \subseteq D$ be bounded and suppose that L is homogeneous and it satisfies the Weak Maximum Principle on Ω. Prove the following facts.

- If u is L-subharmonic in Ω and $\limsup_{x \to y} u(x) \leq M$ for every $y \in \partial\Omega$, then $u \leq M$ in Ω.
- If $u \in C(\overline{\Omega}) \cap C^2(\Omega)$ is L-harmonic in Ω and $u \equiv 0$ on $\partial\Omega$, then $u \equiv 0$.
- If $u \in C(\overline{\Omega}) \cap C^2(\Omega)$ is L-harmonic in Ω and $m \leq u \leq M$ on $\partial\Omega$, then $m \leq u \leq M$ in Ω.
- Let u be a classical solution of the Dirichlet problem (9.2) with $f = 0$ (hence u is L-harmonic in Ω). Prove the following bounds for u on Ω:

$$\min_{\partial\Omega} \varphi \leq u(x) \leq \max_{\partial\Omega} \varphi, \qquad \forall\, x \in \Omega.$$

Exercise 9.3. (Harmonic functions touching at infinity). Let L be defined on $\mathbb{R}^N$ and such that $L(1) \leq 0$; suppose also that L satisfies the Weak Maximum Principle on every bounded open subset of $\mathbb{R}^N$. Prove that if u, v are L-harmonic in $\mathbb{R}^N$ and $\lim_{\|x\| \to \infty}(u(x) - v(x)) = 0$, then $u \equiv v$ on $\mathbb{R}^N$.

Exercise 9.4. Let L be a PDO as in (8.1) on the non-empty open set $D \subseteq \mathbb{R}^N$. Let $\psi : D \to \widetilde{D}$ be a C^2-diffeomorphism onto the open set $\widetilde{D} \subseteq \mathbb{R}^N$, and let $\mathcal{J}_\psi(x)$ be the Jacobian matrix of ψ at $x \in D$. We denote the coordinates on $\widetilde{D}$ by y. Let us define the operator $\widetilde{L}$ on $\widetilde{D}$ (which is, roughly put, the operator L written in the new y-coordinates) by setting

$$\widetilde{L}u(y) := L(u \circ \psi)(\psi^{-1}(y)), \qquad y \in \widetilde{D},$$

for every $u = u(y)$ of class C^2 in $\widetilde{D}$. Prove that

$$\widetilde{L} = \sum_{i,j=1}^{N} \widetilde{a}_{i,j}(y)\, \frac{\partial^2}{\partial y_i\, \partial y_j} + \sum_{i=1}^{N} \widetilde{b}_i(y)\, \frac{\partial}{\partial y_i} + \widetilde{c}(y),$$

where, for any $y \in \widetilde{D}$,

$$\widetilde{A}(y) = \big(\widetilde{a}_{i,j}(y)\big)_{i,j} = \mathfrak{J}_\psi(\psi^{-1}(y)) \cdot A(\psi^{-1}(y)) \cdot \big(\mathfrak{J}_\psi(\psi^{-1}(y))\big)^T,$$

$$\widetilde{b}(y) = \big(\widetilde{b}_1(y) \cdots \widetilde{b}_N(y)\big)^T = \mathfrak{J}_\psi(\psi^{-1}(y)) \cdot b(\psi^{-1}(y)) + (L_0\psi)(\psi^{-1}(y)),$$

$$\widetilde{c}(y) = (c \circ \psi^{-1})(y).$$

Here $A = \big(a_{i,j}\big)_{i,j}$, $b = \big(b_1 \cdots b_N\big)^T$, L_0 is the principal part of L, and by $L_0\psi$ we mean (with a small abuse of notation) the column-vector function whose components are obtained by evaluating L_0 on the N components of ψ.

Deduce that L is semielliptic on D if and only if $\widetilde{L}$ is semielliptic on $\widetilde{D}$ (hence, semiellipticity is an invariant notion under change of coordinates).

In particular, if $\psi(x) = Mx$ is the linear change of coordinates associated with the $N \times N$ non-singular matrix M, one obtains

$$\widetilde{A}(y) = M \cdot A(M^{-1}y) \cdot M^T, \quad \widetilde{b}(y) \quad = M \cdot b(M^{-1}y), \quad \widetilde{c}(y) = c(M^{-1}y).$$

Exercise 9.5. Prove the following fact as a corollary of Thm. 9.16:

Let L be a PDO as in (8.1), with $c = L(1) \leq 0$ on the open set D. Suppose that the homogeneous operator $L - c$ satisfies the Weak Maximum Principle on every bounded open subset of D. Let Ω be a bounded open subset of D, and let $u \in C(\overline{\Omega}) \cap C^2(\Omega)$ be L-harmonic in Ω. Then,

$$\inf_{\partial\Omega} u^- \leq \sup_\Omega u \leq \sup_{\partial\Omega} u^+,$$

where $u^+ = \max\{u, 0\}$ and $u^- = \min\{u, 0\}$.

[*Hint:* Note that $(-u)^+ = -u^-$...]

Exercise 9.6. Consider the operator $L = \Delta - 1$ in $\mathbb{R}^N$, where Δ is the Laplace operator in $\mathbb{R}^N$. Let $\Omega = \{x \in \mathbb{R}^N : \|x\| < 1\}$ and let $u(x) = -3\,N - \|x\|^2$.

Prove that u is (strictly) L-subharmonic on $\mathbb{R}^N$, L satisfies the WMP on every bounded open subset of $\mathbb{R}^N$, and $\sup_\Omega u > \sup_{\partial\Omega} u$. This shows that u^+ cannot be replaced with u in (9.16), and that the hypothesis of homogeneity of L in Thm. 9.9 cannot be dropped.

Exercise 9.7. Prove the assertions in Thm. 9.11.

Chapter 10

The Maximum Propagation Principle

T HE aim of this chapter is to prove the so-called *Maximum Propagation Principle* for semielliptic operators L, and to derive from it the *Strong Maximum Principle* (SMP, for short) for selected classes of PDOs.

Actually, we shall prove the Strong Maximum Principle for a sub-class of the Hörmander operators: those of the form $L = \sum_{j=1}^{m} X_j^2 + X_0$, where $X_1, \ldots, X_m$ are Hörmander vector fields (see Chap. 6).

In studying the SMP, we are interested in how large the set $F(u)$ where an L-subharmonic function attains it maximum may be; in particular, we aim to investigate those vector fields whose integral curves "propagate" $F(u)$. The richer the class of these vector fields, the larger $F(u)$. For the SMP, we first need the Maximum Propagation Principle: for its proof, we follow the classical approach in the celebrated paper by [Bony (1969)]:

(1) Given a set F, we are interested in the class of v.f.s X whose integral curves γ are compelled to remain in F: we say that F is X-*invariant*.

(2) The remarkable Nagumo and Bony Theorem will tell us that F is X-invariant if and only if X is *tangent to* F: due to the variety of sets F involved, this notion of tangentiality is not the one of Differential Geometry, but it widely generalizes it.

(3) The *Hopf Lemma* for L will show that any X which is a so-called *principal v.f. for* L is automatically tangent to $F(u)$: hence $F(u)$ is X-invariant, so that $F(u)$ propagates along the integral curves of X. This is the Maximum Propagation Principle along the principal v.f.s for L.

Finally, for the sub-class of the Hörmander PDOs L described in the *incipit* of this introduction, we shall discover (by the Connectivity Theorem of Chap. 6, or by an independent study on the algebraic properties of the set of the tangent v.f.s) that we have at disposal so many tangent v.f.s to $F(u)$ that the latter propagates in every space direction. This is the SMP for L.

Prerequisites for this chapter are some results on semiellipticity established in Chap. 8, and the approximation of the flow of $[X, Y]$ via the flows of $\pm X$ and $\pm Y$, proved in Chap. 3.

211

10.1   Assumptions on the operators

In this chapter, we shall be dealing with the following type of PDOs.

Let $\varnothing \neq D \subseteq \mathbb{R}^N$ be an open set. For every $i, j \in \{1, \ldots, N\}$, we assume that $a_{i,j} = a_{j,i}$ and b_i are fixed real-valued continuous functions on D. We consider the second order linear homogeneous (i.e., $L(1) \equiv 0$) PDO on D

$$L := \sum_{i,j=1}^{N} a_{i,j}(x)\, \frac{\partial^2}{\partial x_i\, \partial x_j} + \sum_{i=1}^{N} b_i(x)\, \frac{\partial}{\partial x_i}. \tag{10.1}$$

The investigation of the non-homogeneous case, under the assumption that L has a *non-positive* zero-order term $c = L(1)$, could be easily obtained with some modifications, and we leave this to the reader.

As we did in Chap. 8, we introduce the notation

$$A(x) := \big(a_{i,j}(x)\big)_{i,j \leq N}, \qquad x \in D,$$

for the matrix of the principal part of L. The map $\xi \mapsto q_L(x, \xi) := \langle A(x)\xi, \xi \rangle$ denotes, as usual, the characteristic form of L (at $x \in D$).

Convention. *In the sequel, we understand the following assumption:*

$$L \text{ is semielliptic,}$$

that is, $A(x)$ is positive semidefinite for every $x \in D$. Throughout the chapter, the above notations and assumptions on L are tacitly assumed.

Example 10.1. We have the following examples.

(1) Any sum of squares of C^1 vector fields $L = \sum_{j=1}^{m} X_j^2$ on D is of the form required above. Indeed, by Exm. 8.3 we know that L is semielliptic, since

$$q_L(x, \xi) = \sum_{j=1}^{m} \langle X_j(x), \xi \rangle^2, \qquad x \in D, \quad \xi \in \mathbb{R}^N. \tag{10.2}$$

(2) Suppose $L = \sum_{j=1}^{m} X_j^2 + X_0$ is a sum of squares of C^1 v.f.s $X_1, \ldots, X_m$ plus a C^0 drift X_0 on D. We claim that L is semielliptic. Indeed this property only depends on the principal part of L, which is completely determined by $\sum_{j=1}^{m} X_j^2$; hence our claim follows from (1).

(3) If $a_{i,j} = a_{j,i}$ are C^1 functions on D $(i, j \leq N)$, any divergence form PDO

$$L = \sum_{i=1}^{N} \frac{\partial}{\partial x_i} \left(\sum_{j=1}^{N} a_{i,j}(x)\, \frac{\partial}{\partial x_j} \right) = \mathrm{div}\left(A(x) \cdot \nabla^T \right)$$

satisfies the above assumptions, provided that $A(x)$ is positive semidefinite at any $x \in D$. Under this assumption on A, the same is true of

$$L = \frac{1}{V(x)} \sum_{i=1}^{N} \frac{\partial}{\partial x_i} \left(V(x) \sum_{j=1}^{N} a_{i,j}(x)\, \frac{\partial}{\partial x_j} \right),$$

with $V \in C^1$ and $V > 0$ on D. The latter is the typical form of the Laplace-Beltrami operator written in coordinates, and of many meaningful PDOs on Lie groups, where V is the density of the Haar measure (Sec. C.3).   ♯

10.2 Principal vector fields

The primary aim of this chapter is to prove the so-called *Maximum Propagation Principle,* which we now introduce. We first need the concept of a principal vector field wrt the PDO L, the latter being assumed to be as in Sec. 10.1.

Definition 10.2 (Principal vector field for L). Let X be a vector field on D. We say that X is a **principal vector field for L (on D)** if for every $x \in D$ there exists a real number $\lambda(x) > 0$ such that

$$\langle X(x), \xi \rangle^2 \leq \lambda(x) \langle A(x)\xi, \xi \rangle, \qquad \text{for every } \xi \in \mathbb{R}^N. \tag{10.3}$$

If, as usual, $\mathfrak{X}(D)$ denotes the set of the smooth vector fields on D, we define

$$\mathrm{Pr}(L) := \{X \in \mathfrak{X}(D) \,:\, X \text{ is a principal v.f. for } L\}. \tag{10.4}$$

Observe that (10.3) can be rewritten as

$$\langle X(x), \xi \rangle^2 \leq \lambda(x)\, q_L(x, \xi), \qquad \text{for every } \xi \in \mathbb{R}^N.$$

Note that $\xi \mapsto \langle X(x), \xi \rangle^2$ is the quadratic form associated with $X(x)\,(X(x))^T$; if $X = \sum_{j=1}^N \alpha_j(x)\partial_j$, this matrix is simply $\left(\alpha_i(x)\,\alpha_j(x)\right)_{i,j}$.

As we did for the definition of X-subunit paths (see Prop. 6.17), with the aid of some elementary Linear Algebra we can provide a simple characterization of the principality of a vector field.

Proposition 10.3 (Characterizations of principality). *The v.f. X is principal for L (on D) if and only if one of the following equivalent conditions is satisfied for every $x \in D$:*

(1) the vector $X(x)$ belongs to $(\ker A(x))^\perp$;
(2) the vector $X(x)$ is a linear combination of the columns (or of the rows) of $A(x)$;
(3) every isotropic vector for $A(x)$ is orthogonal to $X(x)$:

$$\mathrm{Isotr}(A(x)) \subseteq (X(x))^\perp, \quad \forall\, x \in D.$$

Proof. See Sec. A.2 in App. A. $\qquad\qquad\qquad\qquad\qquad\qquad\qquad\qquad\qquad\square$

Example 10.4. The v.f. $X = \partial_t$ is not principal for the Heat operator $L = \Delta - \partial_t$ in $\mathbb{R}^{N+1}$; indeed $X(x)$ is the $(N+1)$-th vector in the canonical basis of $\mathbb{R}^{N+1}$, whereas the columns of the associated matrix $A(x)$ are the other first N vectors of the canonical basis of $\mathbb{R}^{N+1}$. $\qquad\qquad\qquad\qquad\qquad\sharp$

Example 10.5. Consider the PDO $L = \sum_{j=1}^m X_j^2 + X_0$, where $X_0, \ldots, X_m$ are C^1 vector fields on D. Then, *any vector field of the form*

$$X = \sum_{j=1}^m g_j(x)\, X_j,$$

where $g_1, \ldots, g_m$ are real-valued functions on D, is principal for L. Indeed, owing to the Cauchy-Schwarz inequality in $\mathbb{R}^m$, we have

$$\langle X(x), \xi \rangle^2 = \left(\sum_{j=1}^m g_j(x) \cdot \langle X_j(x), \xi \rangle \right)^2 \le \sum_{j=1}^m |g_j(x)|^2 \cdot \sum_{j=1}^m \langle X_j(x), \xi \rangle^2$$

$$\overset{(10.2)}{=} \widetilde{\lambda}(x)\, q_L(x, \xi), \qquad \text{where } \widetilde{\lambda}(x) := \sum_{j=1}^m |g_j(x)|^2.$$

This gives (10.3) by taking $\lambda(x) := \max\{\widetilde{\lambda}(x), 1\}$. By means of the characterization in Prop. 10.3-(2), this means that, for every $x \in D$, $\sum_{j=1}^m g_j(x)\, X_j(x)$ is a linear combination of the column-vectors (or of the row-vectors) of

$$A(x) = S(x) \cdot S(x)^T, \quad \text{where } S(x) := \big(X_1(x) \, \cdots \, X_m(x) \big).$$

Indeed this $A(x)$ is the matrix of the principal part of $\sum_{j=1}^m X_j^2 + X_0$. This fact is not surprising, since, for any $N \times m$ matrix S one has (Exr. A.11)

$$\mathrm{range}(S) = \mathrm{range}(S\, S^T),$$

where $\mathrm{range}(C)$ denotes, in general, the span of the column vectors of the matrix C. As a consequence

$$\mathrm{range}(A(x)) = \mathrm{range}(S(x)) = \mathrm{span}\{X_1(x), \ldots, X_m(x)\}, \quad \forall\, x \in D.$$

Hence *a v.f. X is principal for $L = \sum_{j=1}^m X_j^2 + X_0$ if and only if*

$$X(x) \in \mathrm{span}\{X_1(x), \ldots, X_m(x)\} \quad \forall\, x \in D.$$

Example 10.6. In particular, if the functions $g_1, \ldots, g_m$ in Exm. 10.5 are all chosen to be 0 except for one, which is identically $+1$ or -1, we see that

$$\pm X_1, \ldots, \pm X_m \text{ are principal vector fields for } L = \sum_{j=1}^m X_j^2 + X_0. \qquad \sharp$$

Remark 10.7. By arguing as in Exm. 10.5 one recognizes that $\mathrm{Pr}(L)$ *is a module over* $C^\infty(D)$, that is

$$X, Y \in \mathrm{Pr}(L), \quad f, g \in C^\infty(D) \quad \Longrightarrow \quad f\, X + g\, Y \in \mathrm{Pr}(L).$$

The next example shows that $\mathrm{Pr}(L)$ need not be a Lie-subalgebra of $\mathcal{X}(D)$. $\qquad \sharp$

Example 10.8. Consider the Kohn-Laplacian on the Heisenberg group in $\mathbb{R}^3$:

$$L = X_1^2 + X_2^2, \qquad \text{where } X_1 = \partial_1 + 2\, x_2\, \partial_3, \ X_2 = \partial_2 - 2\, x_1\, \partial_3.$$

Since L is a sum of squares, the associated second order matrix $A(x)$ is

$$A(x) = S(x) \cdot S(x)^T, \qquad \text{where } S(x) = \begin{pmatrix} 1 & 0 \\ 0 & 1 \\ 2\, x_2 & -2\, x_1 \end{pmatrix}.$$

Taking into account Exm. 10.5, we have $\mathrm{range}(A(x)) = \mathrm{range}(S(x))$ and X is principal for L iff it is of the form

$$X = g_1(x)\, (\partial_1 + 2\, x_2\, \partial_3) + g_2(x)\, (\partial_2 - 2\, x_1\, \partial_3),$$

for some real-valued functions g_1, g_2. Note that, even if X_1, X_2 are principal for L, their commutator $[X_1, X_2] = -4\, \partial_3$ is not principal for L, because $(0, 0, -4)$ does not belong to $\mathrm{range}(S(x))$, for any $x \in \mathbb{R}^3$.

    Thus, $\mathrm{Pr}(L)$ may fail to be a Lie algebra of vector fields. $\qquad \sharp$

10.3 Maximum Propagation and Strong Maximum Principle

We are ready to state the main result of this chapter. The notations and hypotheses of Sec. 10.1 apply.

Theorem 10.9 (Maximum Propagation Principle). *Let L be semielliptic on D. Let $\Omega \subseteq D$ be an open set. For every function $u \in C^2(\Omega)$ satisfying $Lu \geq 0$ and $u \leq 0$ on Ω, the set $F(u) = \{x \in \Omega : u(x) = 0\}$ contains the trajectories, starting at points of F, of the integral curves of any C^1 principal vector field for L.*

In this case we say that the set $F(u)$, which is the set of the maximum points of u (when non-void), **propagates** along the trajectories of the integral curves of the C^1 principal vector fields for L, whence the name of the theorem, the *Maximum Propagation Principle*.

We shall prove Thm. 10.9 in the next sections. We observe that, as will be clear in our proof, we can also consider locally Lipschitz continuous principal vector fields for L instead of C^1 ones, which is assumed for simplicity.

Remark 10.10. Since L is homogeneous, the hypothesis $u \leq 0$ in Thm. 10.9 is irrelevant: it suffices to remove it and to replace $F(u)$ by $F = \{x \in \Omega : u(x) = M\}$, where $M = \sup_\Omega u$. If $M = \infty$ or, more generally, if F is empty, then there is nothing to prove (since no integral curve can start at a point of an empty set!). Otherwise, if there exists $x \in \Omega$ such that $u(x) = \sup_\Omega u$, then x is a maximum point of u and (since L is homogeneous) we can apply Thm. 10.9 to $v = u - M$ (which satisfies $Lv = Lu \geq 0$ and $v \leq 0$).

We therefore obtain that, *if it exists, the maximum of an L-subharmonic function $u \in C^2(\Omega)$ propagates along the trajectories of the C^1 principal vector fields for L starting at the points where this maximum is attained.* ♯

We are ready to introduce the second main topic of this chapter, the so-called *Strong Maximum Principle*. First we give the relevant definition.

Definition 10.11 (Strong Maximum Principle). We say that L satisfies the **Strong Maximum Principle** (SMP, for short) on the connected open set Ω if it satisfies the following condition: for every function $u \in C^2(\Omega)$ such that

$$Lu \geq 0 \quad \text{and} \quad u \leq 0 \quad \text{on } \Omega, \tag{10.5}$$

the existence of $x_0 \in \Omega$ such that $u(x_0) = 0$ implies that $u \equiv 0$ on the whole of Ω.

More generally, if L satisfies the SMP on the connected open set Ω, and if $u \in C^2(\Omega)$ is such that $Lu \geq 0$ and u attains its maximum in Ω, then u is constant (Rem. 10.10).

The strict relationship between the Maximum Propagation Theorem and the SMP is clear: indeed, roughly put, if L admits sufficiently many principal v.f.s running throughout Ω, then the maximum of an L-subharmonic function propagates everywhere and the SMP holds.

Therefore, once the Maximum Propagation Principle is established, from the Connectivity Theorem we straightforwardly derive the following fundamental result, establishing the SMP for a class of Hörmander PDOs.

Theorem 10.12 (The SMP for $\sum_{j=1}^{m} X_j^2 + Y$, with $\{X_j\}_{j=1}^{m}$ Hörmander v.f.s).
Let $\{X_1, \ldots, X_m\}$ be a Hörmander system of vector fields on D. Moreover, let Y be a continuous vector field on D.

Then the operator $L = \sum_{j=1}^{m} X_j^2 + Y$ (a special case of a Hörmander operator) satisfies the Strong Maximum Principle on every connected open subset of D.

Proof. Let $u \in C^2(\Omega)$ be as in (10.5), and assume that ξ belongs to the set

$$F(u) := \{x \in \Omega : u(x) = 0\}.$$

By Exm. 10.1-(2) we know that L is semielliptic on D, for it is a sum of squares of v.f.s plus a drift. By Exm. 10.6, $\pm X_1, \ldots, \pm X_m$ are principal vector fields for L.

We are therefore entitled to apply the Maximum Propagation Thm. 10.9, and deduce that the set $F(u)$ contains the trajectories of any integral curve of the vector fields $\pm X_1, \ldots, \pm X_m$ which starts at a point of $F(u)$. By crucially exploiting the fact that $\{X_1, \ldots, X_m\}$ is a Hörmander system, by means of the Connectivity Theorem (Thm. 6.22 on page 140) we derive that any point η of Ω can be connected to ξ by a path which is piecewise an integral curve of $\pm X_1, \ldots, \pm X_m$. Thus, every point of Ω belongs to $F(u)$, i.e., $u \equiv 0$. See Fig. 10.1. $\qquad\square$

For an alternative proof of Thm. 10.12 *independent of the Connectivity Theorem,* see Rem. 10.39. More generally, arguing as in the above proof, one can demonstrate

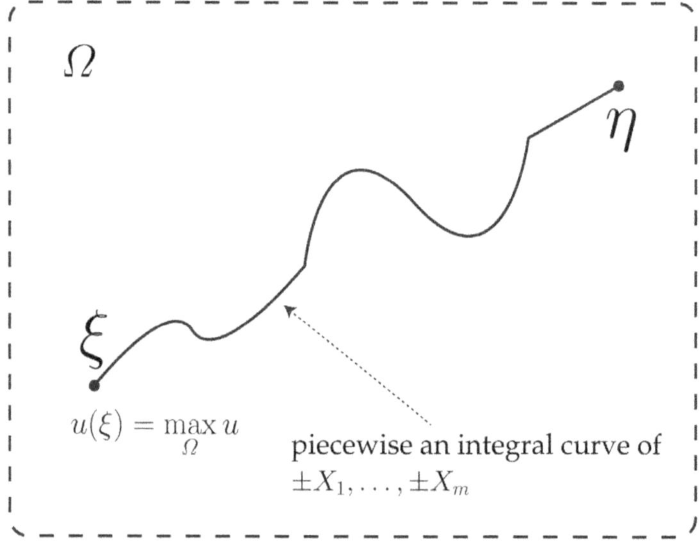

Fig. 10.1   The Strong Maximum Principle (with an argument based on the Connectivity Theorem) for $L = \sum_{j=1}^{m} X_j^2 + Y$, where $\{X_1, \ldots, X_m\}$ are Hörmander v.f.s.

the following theorem, the SMP for operators whose principal vector fields satisfy a connectivity property:

Theorem 10.13 (SMP for operators with connectivity along principal v.f.s). *Let Ω be a connected open set. If every pair of points of Ω can be connected by a continuous curve in Ω which is piecewise an integral curve of some principal vector fields for L, then L satisfies the Strong Maximum Principle on Ω.*

Analogously, if Ω is as above, then L satisfies the Strong Maximum Principle on every connected open subset of Ω.

Remark 10.14. It is important to observe that the Strong Maximum Principle may fail to hold true for a Hörmander operator $L = \sum_{j=1}^{m} X_j^2 + Y$. Indeed L is a Hörmander operator iff $\{Y, X_1, \dots, X_m\}$ is a Hörmander system, but this does not necessarily imply that $\{X_1, \dots, X_m\}$ is a Hörmander system. $\sharp$

Example 10.15. For example, if $L = \sum_{j=1}^{N} (\partial_{x_j})^2 - \partial_t$ is the Heat operator in $\mathbb{R}^{N+1}$, then L *does not satisfy the SMP!* Indeed, let us consider the so-called *fundamental solution* (with pole at the origin) of $\Delta - \partial_t$, namely the function

$$\Gamma(x,t) := \begin{cases} \dfrac{1}{(4\pi t)^{N/2}} \exp\left(-\dfrac{\|x\|^2}{4t}\right), & \text{if } t > 0 \\[2ex] 0, & \text{if } t \le 0 \text{ and } (x,t) \ne (0,0). \end{cases}$$

It can easily be proved that $-\Gamma(x,t)$ is L-harmonic (hence L-subharmonic) on the connected open set $\Omega := \mathbb{R}^{N+1} \setminus \{(0,0)\}$; it is ≤ 0 on Ω, but it vanishes at infinitely many points of Ω without being identically 0. Note that in this case $\{\partial_t, \partial_{x_1}, \dots, \partial_{x_N}\}$ is Hörmander on Ω, but this is false of $\{\partial_{x_1}, \dots, \partial_{x_N}\}$. $\sharp$

As the name itself suggests, the Strong Maximum Principle implies the Weak Maximum Principle, as the following result proves.

Proposition 10.16 (SMP implies WMP). *Suppose that L satisfies the Strong Maximum Principle on every bounded connected open set $\Omega \subseteq D$. Then L satisfies the Weak Maximum Principle on every bounded open set Ω contained in D.*

As the example of the Heat operator shows (Exm. 10.15), an operator may satisfy the Weak Maximum Principle, yet violating the Strong Maximum Principle.

Proof. Let $\Omega \subset D$ be a bounded open set and let $u \in C^2(\Omega)$ be such that $Lu \ge 0$ on Ω and $\limsup_{x \to y} u(y) \le 0$ for every $y \in \partial\Omega$. We want to prove that $u \le 0$ on Ω. Since Ω is bounded, by Cor. 8.9, there exists $x_0 \in \overline{\Omega}$ such that

$$\limsup_{x \to x_0} u(x) = \sup_{\Omega} u.$$

If $x_0 \in \partial\Omega$, we have $\sup_{\Omega} u = \limsup_{x \to x_0} u(x) \le 0$ (by the hypotheses on u), whence $u \le 0$ on Ω.

If $x_0 \in \Omega$, from $u(x_0) = \limsup_{x \to x_0} u(x) = \sup_{\Omega} u$ we deduce that x_0 is a maximum point of u on Ω. Since L satisfies the SMP, this implies that $u \equiv u(x_0)$

on Ω_0, the connected component of Ω containing x_0. Since $\partial\Omega_0 \subseteq \partial\Omega$, taking any $y \in \partial\Omega_0$ we obtain that

$$\sup_{\Omega} u = u(x_0) = \limsup_{\Omega_0 \ni x \to y} u(x) \leq \limsup_{\Omega \ni x \to y} u(x) \leq 0.$$

(In the second equality we used the fact that $u \equiv u(x_0)$ on Ω_0.) This again implies that $u \leq 0$ on Ω and the proof is complete. $\qquad\square$

Remark 10.17. As a consequence of Thm. 10.12 and Prop. 10.16 we deduce that any operator of the form $L = \sum_{j=1}^{m} X_j^2 + Y$, where $\{X_1, \ldots, X_m\}$ is a Hörmander system of vector fields on D (and Y is continuous on D), satisfies the Weak Maximum Principle on every bounded open set $\Omega \subseteq D$.

The same is true of any operator L satisfying the connectivity assumption along principal v.f.s as in Thm. 10.13. $\qquad\sharp$

10.4   Invariant sets and the Nagumo-Bony Theorem

The statement of the Maximum Propagation Thm. 10.9 suggests an independent study of the *invariance* of a set wrt the trajectories of a vector field; more precisely, given a closed set F, we aim to give a characterization of the vector fields whose integral curves are constrained to remain in F once they touch F at a point at least. We then begin with the relevant definition.

For the rest of the section, $\Omega \subseteq \mathbb{R}^N$ is a non-empty open set.

Definition 10.18 (Invariant set wrt a vector field). Let X be a vector field on Ω, and let F be a subset of Ω.

We say that F is **positively X-invariant** (or **positively invariant wrt X**) if, for every integral curve γ of X, $\gamma : [0, T] \to \Omega$ satisfying $\gamma(0) \in F$, we have $\gamma(t) \in F$ for every $t \in [0, T]$.

We say that F is **X-invariant** (or **invariant wrt X**) if it is positively invariant with respect to X and to $-X$.

Remark 10.19. It is easy to recognize that F is X-invariant if and only if, for every integral curve $\gamma : [a, b] \to \Omega$ of X (with $a < 0 < b$) such that $\gamma(0) \in F$, one has $\gamma(t) \in F$ for every $t \in [a, b]$.

The role of 0 is immaterial and, by re-parametrization, one can check that F *is X-invariant if and only if, for every integral curve* $\gamma : [a, b] \to \Omega$ *of X such that* $\gamma([a, b]) \cap F \neq \varnothing$, *one has* $\gamma(t) \in F$ *for every* $t \in [a, b]$. $\qquad\sharp$

Remark 10.20. By means of the notion of invariant set wrt a v.f., we can restate the thesis of the Maximum Propagation Principle in Thm. 10.9 as follows:

If L is semielliptic on Ω and if $u \in C^2(\Omega)$ is L-subharmonic and nonnegative, then the set $F(u) = \{x \in \Omega : u(x) = 0\}$ (when non-void) is X-invariant, for every C^1 principal vector field X for L. $\qquad\sharp$

Due to its importance in maxima propagation, it is now of our concern to find an effective characterization of X-invariance: this will be given in Thm. 10.28 (where –more generally– positive invariance is studied). Roughly put, if we try to picture a set F which captures the integral curves of a v.f. X, we spontaneously pass through the idea, coming from Differential Geometry, that X is somehow "tangent" to F.

Unfortunately, since we want to deal with sets $F(u)$ (see Rem. 10.20) which are made of maximum points $x \in \Omega$ of u, we have $\nabla u(x) = 0$; thus, we cannot expect $F(u)$ to be a submanifold of Ω. For this reason we have to consider a milder notion of "tangentiality", which we now introduce.

In what follows, we shall denote by $\| \cdot \|$ the usual Euclidean norm and by $B(z, r)$ the Euclidean ball of centre z and radius r:

$$B(z, r) := \{x \in \mathbb{R}^N \ : \ \|x - z\| < r\}.$$

Definition 10.21. Let F be a relatively closed subset of Ω and let $y \in \Omega \cap \partial F$. We say that a non-null vector $\nu \in \mathbb{R}^N$ is **externally orthogonal to F at y** if

$$\overline{B\big(y + \nu, \|\nu\|\big)} \subseteq (\Omega \setminus F) \cup \{y\}.$$

In this case we shall write $\nu \perp F$ at y. (See Fig. 10.2.) We also let

$$F^* := \Big\{y \in \Omega \cap \partial F \,\big|\, \text{there exists } \nu \text{ externally orthogonal to } F \text{ at } y\Big\}.$$

Since we are mainly interested in vectors which are *externally* orthogonal to F, we shall briefly say that ν is *orthogonal to F at y*, without reference to 'externality': this is the reason for the brief notation '$\nu \perp F$ at y'.

Remark 10.22. With the above notation, we remark that, provided F is a relatively closed *proper subset* of Ω and if Ω is connected, then (Exr. 10.6) $F^* \neq \varnothing$. ♮

Since "tangentiality" seems a good notion when X-invariance is concerned, it is convenient to give the following definition.

Definition 10.23 (Tangent vector field to a set). Let F be a relatively closed subset of Ω. Suppose that X is a vector field on Ω.
We say that X is **tangent to F** if

$$\langle X(y), \nu \rangle = 0, \quad \forall y \in F^*, \ \forall \nu \perp F \text{ at } y. \tag{10.6}$$

We tacitly mean that condition (10.6) is fulfilled whenever $F^* = \varnothing$. We set

$$\mathrm{Tg}(F) := \{X \in \mathcal{X}(\Omega) \ : \ X \text{ is tangent to } F\}. \tag{10.7}$$

Remark 10.24. It is a simple exercise to recognize that $\mathrm{Tg}(F)$ is a vector subspace of $\mathcal{X}(\Omega)$ and, more generally, *it is a module over $C^\infty(\Omega)$*, that is

$$X, Y \in \mathrm{Tg}(F), \quad f, g \in C^\infty(\Omega) \quad \Longrightarrow \quad f X + g Y \in \mathrm{Tg}(F).$$

Later on (Prop. 10.38), we shall see that $\mathrm{Tg}(F)$ is a Lie-subalgebra of $\mathcal{X}(\Omega)$. ♮

Next we turn to the characterization of positive X-invariance of a set F.

Remark 10.25. If F is relatively closed in Ω, it is not difficult to verify that

$$\langle X(y), \nu \rangle \leq 0, \quad \forall\, y \in F^*, \ \forall\, \nu \perp F \text{ at } y \tag{10.8}$$

is *necessary* for the positive X-invariance of F (see Fig. 10.3).

Indeed, let $y \in F^*$, $\nu \perp F$ at y and let $\gamma : [0, T] \to \Omega$ be an integral curve of X such that $\gamma(0) = y$. By definition of orthogonality of ν at y we have

$$\overline{B(y + \nu, \|\nu\|)} \subseteq (\Omega \setminus F) \cup \{y\}. \tag{10.9}$$

Then, if F is positively X-invariant (i.e., $\gamma([0, T]) \subseteq F$), from (10.9) we have

$$\|\gamma(t) - (y + \nu)\|^2 \geq \|\nu\|^2 \quad \text{and} \quad \|\gamma(0) - (y + \nu)\|^2 = \|\nu\|^2,$$

for every $t \in [0, T]$. This means that the C^1 real-valued function

$$t \mapsto \|\gamma(t) - (y + \nu)\|^2$$

has a minimum point at $t = 0$. As a consequence

$$0 \leq \frac{\mathrm{d}}{\mathrm{d}t}\bigg|_{t=0}\left(\|\gamma(t) - (y + \nu)\|^2\right) = 2\,\langle \dot{\gamma}(0), \gamma(0) - (y + \nu)\rangle = \langle X(y), -\nu\rangle.$$

Hence (10.8) is satisfied.   ♮

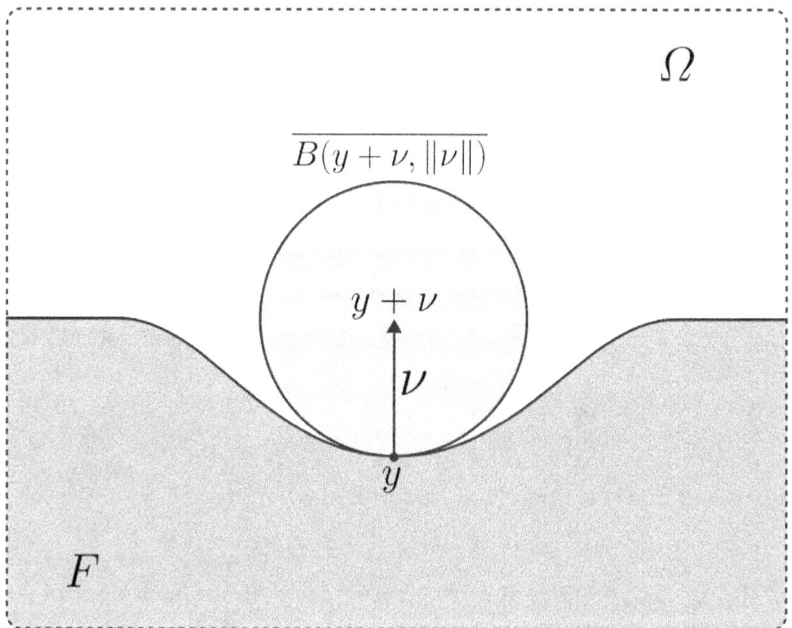

Fig. 10.2   The typical geometry for the study of invariant sets: the vector ν is externally orthogonal to the set F at the point y.

We are about to show that (10.8) is also *sufficient* for the positive X-invariance of F. To his end, we need the following Real Analysis lemmas.

Lemma 10.26. *Let $g : [0, T] \to \mathbb{R}$ be a continuous function such that*

$$\limsup_{h \to 0^-} \frac{g(t + h) - g(t)}{h} \leq M \quad \forall t \in (0, T], \tag{10.10}$$

for a suitable $M \in \mathbb{R}$. Then $g(t) \leq g(0) + M\, t$, for all $t \in [0, T]$.

The proof is left as a guided exercise, Exr. 10.4.

The following result, when g is differentiable, is a consequence of the well-known Gronwall's Lemma for ODEs (see Exr. 10.5).

Lemma 10.27. *Let $g : [0, T] \to \mathbb{R}$ be a continuous and nonnegative function such that $g(0) = 0$ and*

$$\limsup_{h \to 0^-} \frac{g(t + h) - g(t)}{h} \leq L\, g(t) \quad \forall t \in (0, T], \tag{10.11}$$

for a suitable constant $L \geq 0$. Then $g \equiv 0$ on $[0, T]$.

Proof. Let $\varepsilon > 0$ be so small that $\varepsilon < T$ and $L\varepsilon < 1$. We show that $g \equiv 0$ on $[0, \varepsilon]$; by repeating the same argument finitely many times we derive that $g \equiv 0$

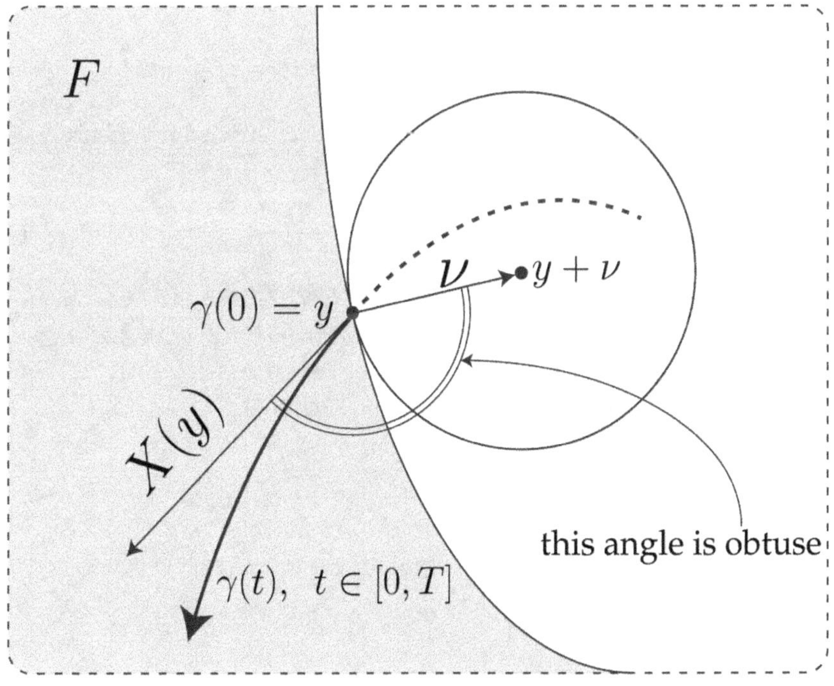

Fig. 10.3　Condition (10.8) for the positive X-invariance of F.

on $[0, T]$. From (10.11) we see that g satisfies (10.10) on $[0, \varepsilon]$, with the constant $M = L \sup_{[0,\varepsilon]} g$. Thus, from Lem. 10.26 (and $g(0) = 0$), we get

$$g(t) \leq M\,t \leq M\,\varepsilon = L\,\varepsilon \, \sup_{[0,\varepsilon]} g, \qquad \forall\, t \in (0, \varepsilon].$$

By taking the supremum over $[0, \varepsilon]$ we get

$$\sup_{[0,\varepsilon]} g \leq L\,\varepsilon \, \sup_{[0,\varepsilon]} g \quad \Longrightarrow \quad (1 - L\varepsilon) \sup_{[0,\varepsilon]} g \leq 0.$$

Since $L\varepsilon < 1$ and $g \geq 0$, this is possible only if $g \equiv 0$ on $[0, \varepsilon]$. $\square$

Theorem 10.28 (Nagumo, Bony). *Let X be a C^1 vector field on Ω, and suppose that F is a relatively closed subset of Ω.*

   Then F is positively X-invariant if and only if

$$\langle X(y), \nu \rangle \leq 0, \quad \text{for every } y \in F^* \text{ and every } \nu \perp F \text{ at } y. \tag{10.12}$$

Proof. Due to Rem. 10.25, we only need to show the "if" part of the assertion. To this end, let $\gamma : [0, T] \to \Omega$ be an integral curve of X such that $x_0 := \gamma(0)$ belongs to F. For $t \in [0, T]$ we define

$$\delta(t) = \mathrm{dist}(\gamma(t), F) := \inf \left\{ \|\gamma(t) - z\| \, : \, z \in F \right\}.$$

Note that $\delta(0) = 0$ and δ is continuous and nonnegative. We need to prove that, under condition (10.12), we have $\delta(t) = 0$ for every $t \in [0, T]$. Due to Lem. 10.27, it is enough to prove that

$$\Psi(t) := \limsup_{h \to 0^-} \frac{\delta(t + h) - \delta(t)}{h} \leq L\,\delta(t), \quad \forall\, t \in (0, T], \tag{10.13}$$

for some constant $L > 0$. To this aim, let $V \Subset \Omega$ be a bounded neighborhood of x_0 containing $\gamma([0, T])$, and let

$$L := \sup_{x, z \in V,\ x \neq z} \frac{\|X(x) - X(z)\|}{\|x - z\|} \tag{10.14}$$

be the Lipschitz constant of X on V. By shrinking T if necessary,[1] we can suppose that $V = B(x_0, r)$ with $B(x_0, 2r) \subseteq \Omega$. We shall prove (10.13) with this choice of L.

   If $\delta(t) = 0$, inequality (10.13) is trivial, since $h < 0$ and $\delta(t + h) \geq 0$. Suppose $\delta(t) > 0$ and choose a sequence $h_n < 0$, $h_n \to 0$ such that

$$\Psi(t) = \lim_{n \to \infty} \frac{\delta(t + h_n) - \delta(t)}{h_n}.$$

Let us denote $x := \gamma(t)$ and $x_n := \gamma(t + h_n)$. Since $\gamma([0, T]) \subset B(x_0, r)$ and $B(x_0, 2r) \subseteq \Omega$, for every n there exists a point $z_n \in F \cap B(x_0, r)$ such that

$$\|x_n - z_n\| = \mathrm{dist}(x_n, F) = \mathrm{dist}(\gamma(t + h_n), F) = \delta(t + h_n).$$

---

[1] Otherwise we apply this same argument on a partition of $[0, T]$ into small segments, say $[0, T_1]$, $[T_1, T_2]$, $\ldots, [T_n, T]$, proving that $\delta \equiv 0$ on $[0, T_1]$, then on $[T_1, T_2]$, and so forth.

Obviously, by choosing a subsequence if necessary, we may suppose that z_n converges to some $z \in F \cap \overline{B(x_0, r)}$. As a consequence, since $x_n \to x$, one has

$$\|x - z\| = \lim_{n \to \infty} \|x_n - z_n\| = \lim_{n \to \infty} \mathrm{dist}(x_n, F)$$
$$= \mathrm{dist}(x, F) = \mathrm{dist}(\gamma(t), F) = \delta(t). \tag{10.15}$$

Moreover, from $\|x - z\| = \mathrm{dist}(x, F)$ (and the minimizing property of the distance from a set) it is easy to check that

$$\nu := \tfrac{1}{2}(x - z) \perp F \text{ at } z. \tag{10.16}$$

See Fig. 10.4. Then

$$\delta(t + h_n) - \delta(t) = \|x_n - z_n\| - \|x - z\| \geq \|x_n - z_n\| - \|x - z_n\|$$
$$\geq -\frac{\langle x - z_n, x - x_n \rangle}{\|x - z_n\|} = \frac{\langle x_n - x, x - z_n \rangle}{\|x - z_n\|}.$$

In the first inequality we used the fact that $\|x - z\| = \mathrm{dist}(x, F)$ and $z_n \in F$; the second inequality is a consequence of the following estimate (obtained by taking $a = x$, $b = z_n$, $c = x_n$ in Exr. 10.3)

$$\|x - z_n\| \leq \|x_n - z_n\| + \frac{\langle z_n - x, x_n - x \rangle}{\|z_n - x\| \cdot \|x_n - x\|} \|x - x_n\|.$$

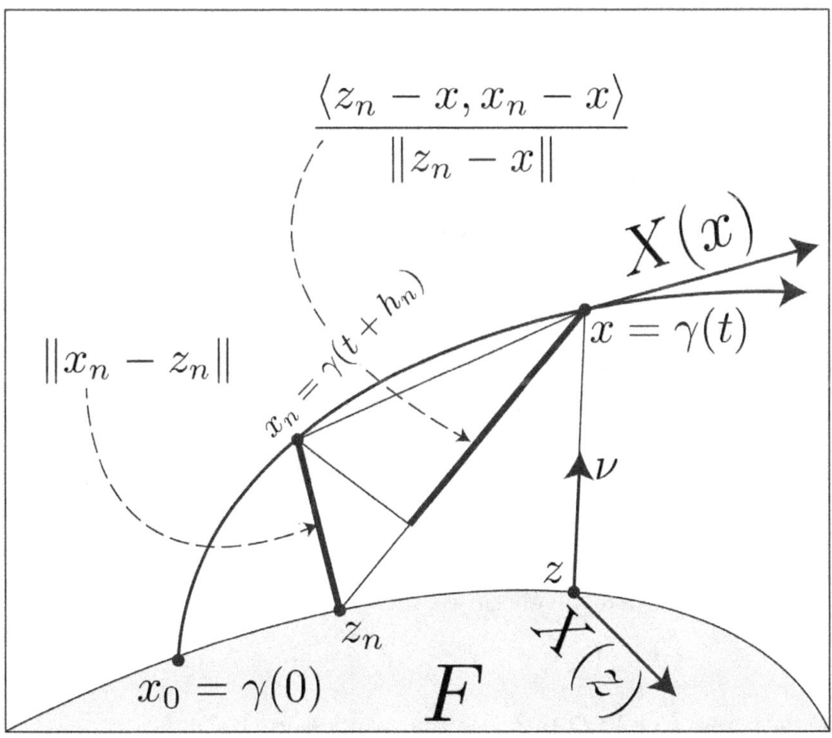

Fig. 10.4 The geometric construction in the proof of the Nagumo-Bony Thm. 10.28.

Hence (taking into account that $h_n < 0$), we have the following calculation:

$$\Psi(t) = \lim_{n \to \infty} \frac{\delta(t + h_n) - \delta(t)}{h_n} \le \lim_{n \to \infty} \left\langle \frac{x_n - x}{h_n}, \frac{x - z_n}{\|x - z_n\|} \right\rangle$$

$$= \lim_{n \to \infty} \left\langle \frac{\gamma(t + h_n) - \gamma(t)}{h_n}, \frac{x - z_n}{\|x - z_n\|} \right\rangle = \left\langle \dot{\gamma}(t), \frac{x - z}{\|x - z\|} \right\rangle$$

$$= \frac{2}{\|x - z\|} \langle X(\gamma(t)), \nu \rangle = \frac{2}{\|x - z\|} \langle X(x), \nu \rangle$$

$$= \frac{2}{\|x - z\|} \left\{ \langle X(x) - X(z), \nu \rangle + \langle X(z), \nu \rangle \right\} \le 2 \frac{\langle X(x) - X(z), \nu \rangle}{\|x - z\|} \, .$$

In the last inequality we used (10.16) and the hypothesis (10.12) applied when y is equal to z (giving $\langle X(z), \nu \rangle \le 0$). This produces the estimate

$$\Psi(t) \le 2 \frac{\langle X(x) - X(z), \nu \rangle}{\|x - z\|} \, .$$

From the Cauchy-Schwarz inequality, together with $\|\nu\| = \|x - z\|/2$, we get that $\Psi(t) \le \|X(x) - X(z)\|$. From the definition (10.14) of L, we finally get

$$\Psi(t) \le L \, \|x - z\| \overset{(10.15)}{=} L \, \delta(t).$$

This completes the proof of (10.13).   □

With Def. 10.23 at hand, the Nagumo-Bony Thm. 10.28 provides us with the following crucial result (see Fig. 10.5).

Corollary 10.29 (Equivalence of invariance and tangentiality). *Let X be a C^1 v.f. on Ω. Suppose that F is a relatively closed subset of Ω.*
    Then, F is X-invariant if and only if X is tangent to F.

Proof. By Def. 10.18, F is X-invariant iff F is positively invariant wrt X and $-X$. By the Nagumo-Bony Thm. 10.28, this is equivalent to

$$\langle \pm X(y), \nu \rangle \le 0, \quad \forall y \in F^*, \ \forall \nu \perp F \text{ at } y.$$

Hence (10.6) holds true, which means that X is tangent to F.   □

Remark 10.30. If F is a relatively closed subset of Ω, we let

$$\mathfrak{a}(F) := \{X \in \mathfrak{X}(\Omega) \, : \, F \text{ is } X\text{-invariant}\}.$$

We deduce from Cor. 10.29 that (see the notation in (10.7)) $\mathrm{Tg}(F) = \mathfrak{a}(F)$. In Sec. 10.6 (Prop. 10.38), we shall discover that $\mathrm{Tg}(F) = \mathfrak{a}(F)$ is not only a module over $C^\infty(\Omega)$ (Rem. 10.24), but *it is also a Lie-subalgebra of* $\mathfrak{X}(\Omega)$.   ♯

10.5 The Hopf Lemma

The proof of Thm. 10.9 requires a last preliminary result, a version of a classical result of elliptic PDO theory, the so-called *Hopf Lemma*, generalized to an operator L as in the hypothesis of Thm. 10.9.

Lemma 10.31 (Hopf-type Lemma). *Let L be semielliptic on D. Let $\Omega \subseteq D$ be a connected open set. Let us assume that $u \in C^2(\Omega, \mathbb{R})$ satisfies $Lu \geq 0$ and $u \leq 0$ on Ω. We set $F(u) := \{x \in \Omega : u(x) = 0\}$. Assume finally that $F(u)$ is a proper subset of Ω.*
Then, for every $y \in F(u)^*$ and every $\nu \perp F(u)$ at y, we have

$$\langle A(y)\,\nu, \nu \rangle = 0, \tag{10.17}$$

that is, the characteristic form of L at y is null at ν. As a consequence, we have

$$\langle X(y), \nu \rangle = 0, \quad \text{for every } y \in F(u)^* \text{ and every } \nu \perp F(u) \text{ at } y, \tag{10.18}$$

whenever X is a principal vector field for L.
In terms of Def. 10.23, this says that any principal v.f. for L is tangent to $F(u)$.

Before giving the proof of Hopf's Lem. 10.31 (where we use a remarkable technique based on the WMP), we collect a remark and two corollaries.

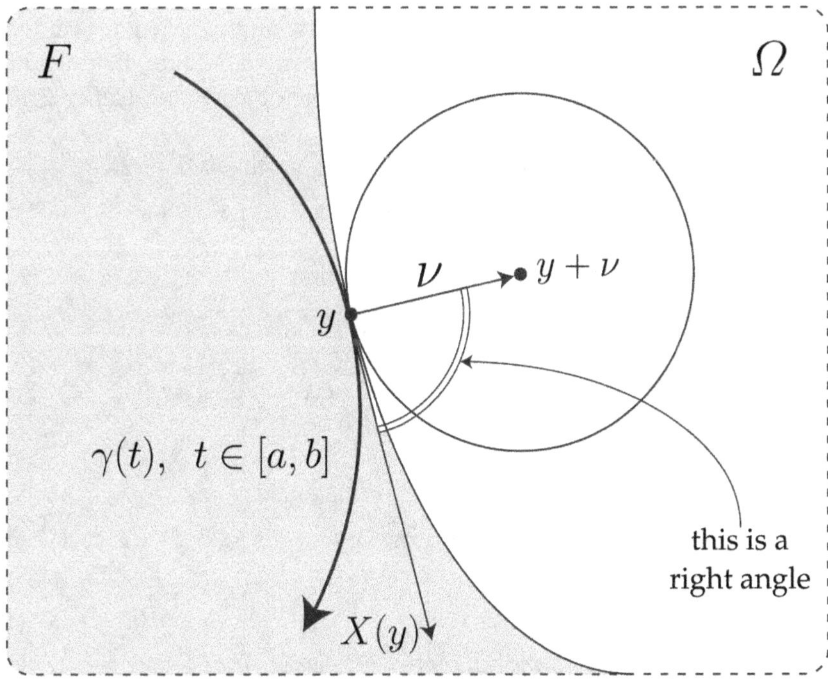

Fig. 10.5 The characterization of the X-invariance of F in terms of the tangentiality of X to F. The curve $\gamma : [a, b] \to \Omega$ is an integral curve of X such that $y \in \gamma([a, b])$.

Remark 10.32. Note that (10.17) says that ν is an isotropic vector for the quadratic form associated with $A(y)$. The characterization of $\mathrm{Isotr}(A(y))$ contained in Exr. A.9, namely $\mathrm{Isotr}(A(y)) = \ker(A(y))$, ensures that $\nu \in \ker(A(y))$. On the other hand, the characterization of principality contained in Prop. 10.3 implies that, if X is principal for L, then $X(y)$ is orthogonal to $\ker(A(y))$, hence to ν.

This confirms (10.18), which can also be seen as a consequence of the definition of the principality of X, by (10.17). ♮

Corollary 10.33. *Consider an operator L of the form $L = \sum_{j=1}^{m} X_j^2 + Y$, where $X_1, \dots, X_m$ are C^1 v.f.s and Y is a continuous v.f. on D. Let u and $F(u)$ be as in Hopf's Lem. 10.31. Then $X_1, \dots, X_m$ are tangent to $F(u)$.*

This follows from Lem. 10.31 and the fact that $X_1, \dots, X_m$ are principal vector fields for L, see Exm. 10.6.

The above Hopf-type Lem. 10.31 for L gives at once the Strong Maximum Principle for elliptic operators:

Corollary 10.34 (SMP for elliptic PDOs). *Suppose that L is elliptic in D. Then L satisfies the Strong Maximum Principle on every connected open set $\Omega \subseteq D$.*

Proof. Indeed, if L is elliptic then it is obviously semielliptic. With the notations in the statement of Lem. 10.31 for u and $F(u)$, we deduce that, if $F(u) \neq \varnothing$ then $F(u) = \Omega$. Otherwise there would exist $y \in F(u)^*$ and $\nu \perp F(u)$ at y so that, by (10.17), $\langle A(y)\nu, \nu \rangle = 0$. Since $A(y)$ is (strictly!) positive definite, this would give $\nu = 0$, contrary to the definition of an orthogonal vector at a point. Hence $F(u) = \Omega$ whenever $F(u) \neq \varnothing$.

This is just a restatement of the fact that L satisfies the SMP on Ω. $\square$

For the proof of Lem. 10.31 we first need a preliminary result:

Lemma 10.35. *Let P be a second order PDO of the form*

$$P = \sum_{i,j=1}^{N} \alpha_{i,j}\,\partial_{i,j} + \sum_{j=1}^{N} \beta_j\,\partial_j + \gamma,$$

where $\alpha_{i,j} = \alpha_{j,i}$, β_j and γ are continuous functions on an open set $\Omega \subseteq \mathbb{R}^N$. Suppose that there exist $y \in \Omega$ and $\nu \in \mathbb{R}^N \setminus \{0\}$ such that

$$\sum_{i,j=1}^{N} \alpha_{i,j}(y)\,\nu_i\,\nu_j > 0. \tag{10.19}$$

Then, setting $B = B(y + \nu, \|\nu\|)$, there exists a function $h \in C^\infty(\mathbb{R}^N, \mathbb{R})$ with the following properties:

$$h > 0 \text{ on } B, \qquad h \equiv 0 \text{ on } \partial B, \qquad h < 0 \text{ outside } \overline{B}; \tag{10.20}$$

$$Ph > 0 \quad \text{on some ball centred at } y, \text{ say } B(y, \delta), \tag{10.21}$$

where $\delta > 0$ depends on ν and on the coefficients of P near y; we can take

$$h(x) := \exp(-\lambda \|x - (y + \nu)\|^2) - \exp(-\lambda \|\nu\|^2), \tag{10.22}$$

with $\lambda > 0$ sufficiently large (depending on ν and $\alpha_{i,j}(y)$ and $\beta_j(y)$, for $i, j \leq N$).

*For future reference, we say that h in (10.22) is a **Hopf function** relative to the operator P, the point y, and the vector ν. (See Fig. 10.6.)*

Proof. Let h be as in (10.22). For brevity, we set $z := y + \nu$ and $r := \|\nu\|$. The constant $\lambda \gg 1$ will be chosen in a moment. Properties (10.20) are obvious by the definition of h. Furthermore, a direct computation shows that

$$Ph(y) = 4\lambda^2 e^{-\lambda r^2} \cdot \left\{ \langle H(y)\nu, \nu \rangle + \frac{1}{2\lambda}\Big(\langle \beta(y), \nu \rangle - \text{trace}(H(y)) \Big) \right\},$$

where we have set $H(y) := (\alpha_{i,j}(y))_{i,j}$ and $\beta(y) := (\beta_1(y), \ldots, \beta_N(y))$. Thus, assumption (10.19) ensures[2] the existence of $\lambda \gg 1$ such that $Ph(y) > 0$; by the continuity of the coefficients of P we infer the existence of a (small) open ball $B(y, \delta)$, compactly contained in Ω, such that (10.21) holds true. Clearly, δ also depends on λ. See also Fig. 10.7. This ends the proof. $\square$

We are ready for the following proof.

Proof of the Hopf-type Lem. 10.31. Obviously, (10.18) is a consequence of (10.17) (and of the definition of a principal v.f.), since for any principal v.f. X for L we have

$$0 \leq \langle X(y), \nu \rangle^2 \overset{(10.3)}{\leq} \lambda(y) \langle A(y)\nu, \nu \rangle \overset{(10.17)}{=} 0.$$

From $\lambda(y) > 0$ we obtain $\langle X(y), \nu \rangle = 0$.

We then prove (10.17). We denote $F(u)$ shortly by F. Let $y \in F^*$ and let $\nu \perp F$ at y; then, by definition, $\overline{B(y + \nu, \|\nu\|)} \subseteq (\Omega \setminus F) \cup \{y\}$. We now argue by contradiction

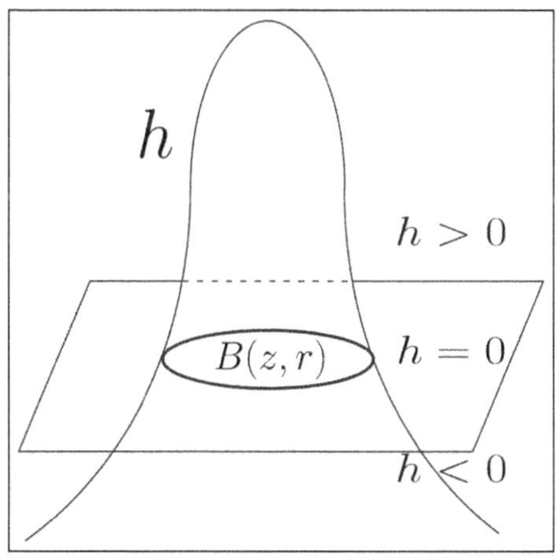

Fig. 10.6   The sign of the Hopf function h in Lem. 10.35; here $z = y + \nu$ and $r = \|\nu\|$.

[2]It suffices to take $\lambda > \dfrac{\big| \langle \beta(y), \nu \rangle - \text{trace}(H(y)) \big|}{2 \langle H(y)\nu, \nu \rangle}.$

assuming that (10.17) is false. Hence (taking into account that $A(y)$ is positive semidefinite), we have

$$\langle A(y)\,\nu,\nu\rangle > 0. \tag{10.23}$$

Thus, we are in a position to construct a Hopf function h as in Lem. 10.35, relative to the operator L (on Ω) and to the point y and the direction ν: let h be as in (10.22), and let $\delta > 0$ be as in the statement of Lem. 10.35. For brevity, we set $z = y + \nu$, $r = \|\nu\|$ and $V = B(y,\delta)$.

Let us split the boundary of V into the sets (see Fig. 10.8)

$$\Gamma_1 := \partial V \setminus \overline{B(z,r)} \quad \text{and} \quad \Gamma_2 := \partial V \cap \overline{B(z,r)}. \tag{10.24}$$

Since Γ_2 is a compact subset of $\Omega \setminus F$ and $u < 0$ in $\Omega \setminus F$, there exists $\varepsilon > 0$ such that $u + \varepsilon\,h < 0$ on Γ_2. On the other hand, as $h < 0$ outside $\overline{B(z,r)}$ (see (10.20)) and $u \le 0$ in Ω (by the hypotheses on u), we have $u + \varepsilon\,h < 0$ on Γ_1 too. Therefore, setting $u_\varepsilon := u + \varepsilon\,h$, we infer

$$u_\varepsilon < 0 \quad \text{on } \partial V = \Gamma_1 \cup \Gamma_2. \tag{10.25a}$$

Bearing in mind that $Lu \ge 0$ on Ω by hypothesis, and that $Lh > 0$ on V (by construction of V), we get

$$L(u_\varepsilon) = Lu + \varepsilon\,Lh \ge \varepsilon\,Lh > 0 \quad \text{in } V. \tag{10.25b}$$

This proves that u_ε is *strictly* L-subharmonic on V. Finally, since $u(y) = 0$ (because $y \in F^* \subseteq F$ and u vanishes on F) and since $h(y) = 0$ (see (10.20), and remember that $y \in \partial B(z,r) = \partial B(y+\nu, \|\nu\|)$), we get

$$u_\varepsilon(y) = 0. \tag{10.25c}$$

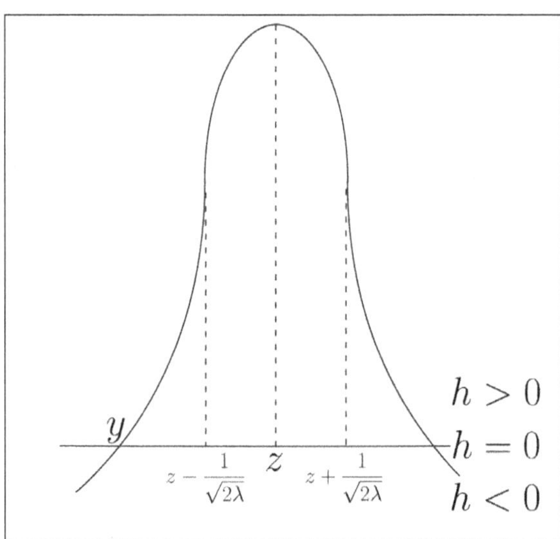

Fig. 10.7 The graph of h in (10.22) when $x \in \mathbb{R}^1$ (here $z = y + \nu$ and $r = \|\nu\|$). Note that the two inflexion points $z \pm 1/\sqrt{2\lambda}$ of h tend to z as $\lambda \to \infty$; this ensures that h is convex in a neighborhood of $y = z - r$. Roughly put, this convexity property reflects the "sub-harmonic" condition $Ph(y) > 0$.

Now, the facts contained in (10.25a), (10.25b) and (10.25c) do not get along together! Indeed, setting $\alpha = \max_{\overline{V}} u_\varepsilon$, we have $\alpha \geq u_\varepsilon(y) = 0$, by (10.25c). Hence this nonnegative maximum is not attained on ∂V, for u_ε is negative on ∂V by (10.25a). As a consequence, u_ε is a strictly L-subharmonic function on V (by (10.25b)) attaining its maximum at an interior point of V. Since L is a semielliptic homogeneous operator, this is impossible, in view of Cor. 8.12 on page 176. Therefore, our assumption (10.23) is not legitimate, and this proves (10.17) by contradiction. This ends the proof. $\qquad\square$

For a summary of the construction of the tricky perturbation function $u + \varepsilon\,h$ in the proof of Hopf's Lem. 10.31, see Fig. 10.11 on page 236.

Remark 10.36. In the notation and assumptions of Lem. 10.31, assume that $F(u)$ is a proper subset of Ω.

By gathering together the notation of (10.4) for the smooth principal vector fields for L (relative to Ω) and the notation of (10.7) for the smooth tangent vector fields to a set, we deduce from the Hopf-type Lem. 10.31 that

$$\mathrm{Pr}(F(u)) \subseteq \mathrm{Tg}(F(u)),$$

for any function u as above. In general, the set $\mathrm{Tg}(F(u))$ will be larger than $\mathrm{Pr}(F(u))$ (see Prop. 10.38 in the next section). $\qquad\sharp$

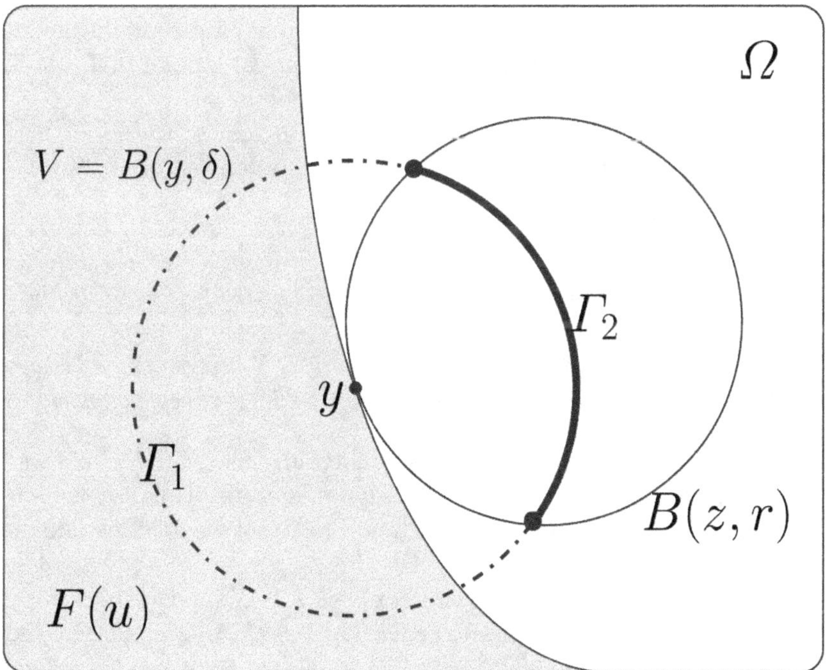

Fig. 10.8   The sets Γ_1, Γ_2 in (10.24), a tricky geometric construction in the proof of Lem. 10.31.

10.6 The proof of the Maximum Propagation Principle

Nagumo-Bony's Cor. 10.29, together with the Hopf-type Lem. 10.33, has the following keystone consequence, which is a restatement of the Maximum Propagation Principle (see Rem. 10.20):

Corollary 10.37. *Let $\Omega \subseteq \mathbb{R}^N$ be a connected open set. Suppose that $u \in C^2(\Omega, \mathbb{R})$ satisfies $Lu \geq 0$ and $u \leq 0$ on Ω.*

If $F(u) = \{x \in \Omega : u(x) = 0\}$ is not empty, then $F(u)$ is invariant wrt every C^1 principal vector field X for L.

Proof. If $F(u) = \Omega$ there is nothing to prove. Suppose that $F(u) \neq \Omega$ (so that $F(u)^* \neq \varnothing$ by Rem. 10.22). Let X be a principal v.f. for L. Due to the Hopf-type Lem. 10.31 we know that X is tangent to $F(u)$. By the equivalence of tangentiality and invariance in Cor. 10.29, this is equivalent to the X-invariance of $F(u)$.   □

We are ready to give the proof of the Maximum Propagation Principle:

Proof of Thm. 10.9. Let $\Omega \subseteq D$ be open and let $u \in C^2(\Omega, \mathbb{R})$ be such that $Lu \geq 0$ and $u \leq 0$ on Ω; let us suppose that $F(u) = \{x \in \Omega : u(x) = 0\}$ is not empty. By replacing Ω with any of its connected components, we suppose Ω is connected.

By Cor. 10.37, $F(u)$ is invariant wrt every C^1 principal vector field X for L. By the very definition of X-invariance, this means that $F(u)$ contains the integral curves of any C^1 principal vector field for L starting at a point of $F(u)$. This is exactly the thesis of the Maximum Propagation Theorem.   □

The following is another consequence of Nagumo-Bony's Cor. 10.29.

Proposition 10.38. *Let F be a relatively closed and proper subset of Ω. Let X, Y be C^2 vector fields in Ω, and suppose that F is X- and Y-invariant (or, equivalently, suppose that X and Y are tangent to F). Then F is also $[X, Y]$-invariant (or, equivalently, $[X, Y]$ is tangent to F).*

As a consequence, the set $\mathfrak{a}(F) := \{X \in \mathfrak{X}(\Omega) : F \text{ is } X\text{-invariant}\}$ is a Lie algebra of vector fields and a C^∞-module, and it coincides with $\mathrm{Tg}(F)$ (see the notation of (10.7)).

Proof. As for the last statement, the equality $\mathfrak{a}(F) = \mathrm{Tg}(F)$ follows from Rem. 10.30; besides, we know that $\mathrm{Tg}(F)$ is a C^∞-module. Thus, we are left with the proof that $\mathfrak{a}(F)$ is a Lie algebra, which will follow if we demonstrate the first statement of the proposition.

To this end, let X, Y be C^2 v.f.s, and suppose F is X- and Y-invariant. Since $[X, Y]$ is C^1, we can use Cor. 10.29 and prove that F is $[X, Y]$-invariant by showing that $[X, Y]$ is tangent to F. Let $y \in F^*$ and let $\nu \perp F$ at y. For every $t > 0$ define

$$\Gamma(t) := \left(\Psi^Y_{-\sqrt{t}} \circ \Psi^X_{-\sqrt{t}} \circ \Psi^Y_{\sqrt{t}} \circ \Psi^X_{\sqrt{t}} \right)(y).$$

Let $T > 0$ be such that $\Gamma(t) \in \Omega$ for $0 \le t \le T$. By (1.39) (page 23) we have

$$\lim_{t \to 0^+} \frac{\Gamma(t) - y}{t} = [X, Y](y). \tag{10.26}$$

On the other hand, since F is X and Y invariant, then $\Gamma(t) \in F$ for every $t \in [0, T]$. (See Fig. 1.5 on page 26 and its caption for a thorough explanation of $\Gamma([0, T]) \subseteq F$.)

As a consequence, since $\overline{B(y + \nu, \|\nu\|)} \subseteq (\Omega \setminus F) \cup \{y\}$ and $\Gamma(0) = y$,

$$\|\Gamma(t) - (y + \nu)\|^2 \ge \|\nu\|^2 = \|\Gamma(0) - (y + \nu)\|^2.$$

Then, by using (10.26), we get

$$0 \le \frac{\mathrm{d}}{\mathrm{d}t}\Big|_{t=0} \left(\|\Gamma(t) - (y + \nu)\|^2 \right) = 2 \left\langle \dot{\Gamma}(0), \Gamma(0) - (y + \nu) \right\rangle$$
$$= -2 \left\langle [X, Y](y), \nu \right\rangle.$$

Hence

$$\left\langle [X, Y](y), \nu \right\rangle \le 0, \quad \forall y \in F^*, \ \forall \nu \perp F \text{ at } y.$$

By interchanging X and Y, we analogously get $\left\langle [Y, X](y), \nu \right\rangle \le 0$. By skew-symmetry of the bracket we derive $\left\langle [X, Y](y), \nu \right\rangle = -\left\langle [Y, X](y), \nu \right\rangle$, so that the previous inequalities can hold simultaneously iff

$$\left\langle [X, Y](y), \nu \right\rangle = 0, \quad \forall y \in F^*, \ \forall \nu \perp F \text{ at } y.$$

Thus $[X, Y]$ is tangent to F as needed. $\qquad\qquad\qquad\qquad\qquad\qquad\square$

Remark 10.39. Prop. 10.38 leads to an alternative proof of the SMP in Thm. 10.12, *a proof which is completely independent of the Connectivity Thm. 6.22.*

Indeed let $X_1, \ldots, X_m$ be Hörmander v.f.s on D, and suppose that Y is a C^0 v.f. on D. Let $L = \sum_{j=1}^m X_j^2 + Y$ and let us take any connected open set $\Omega \subseteq D$. Suppose also that $u \in C^2(\Omega, \mathbb{R})$ satisfies $Lu \ge 0$ and $u \le 0$ on Ω.

By Cor. 10.33, we know that $X_j \in \mathrm{Tg}(F(u))$ for any $j \le m$. By Prop. 10.38 we know that $\mathrm{Tg}(F(u))$ is a Lie algebra of vector fields, whence

$$\mathrm{Lie}\{X_1, \ldots, X_m\} \subseteq \mathrm{Tg}(F(u)).$$

If $F(u)$ were a proper subset of Ω, there would exist $y \in F(u)^*$ and $\nu \perp F(u)$ at y. Since $\{X_1, \ldots, X_m\}$ is a Hörmander system, there exist $Y_1, \ldots, Y_N$ in the Lie algebra $\mathrm{Lie}\{X_1, \ldots, X_m\}$ such that $\{Y_1(y), \ldots, Y_N(y)\}$ is a basis of $\mathbb{R}^N$. Since

$$\{Y_1, \ldots, Y_N\} \subset \mathrm{Lie}\{X_1, \ldots, X_m\} \subseteq \mathrm{Tg}(F(u)),$$

then Y_i is tangent to $F(u)$ for every $i = 1, \ldots, N$. By the definition of tangentiality, we derive that $\langle Y_i(y), \nu \rangle = 0$ for every $i = 1, \ldots, N$. Since $\{Y_1(y), \ldots, Y_N(y)\}$ is a basis of $\mathbb{R}^N$, this yields $\nu = 0$, which is absurd (see Def. 10.21). This proves that, if $F(u) \ne \varnothing$, then $F(u) = \Omega$, i.e., L satisfies the Strong Maximum Principle on Ω. $\quad\sharp$

10.6.1 *Conclusions and a* résumé

We give a *résumé* of the contents of the chapter, showing their interrelations.

Let L be semielliptic and homogeneous on D, and let $A(x)$ be the matrix of its principal part. Suppose that $\Omega \subseteq D$ is open, X is a C^1 vector field on Ω and $F \subseteq \Omega$ is relatively closed in Ω.

Principality of X for L. A vector field X is said to be principal for L if for every $x \in D$ there exists $\lambda(x) > 0$ such that

$$\langle X(x), \xi \rangle^2 \leq \lambda(x) \langle A(x)\xi, \xi \rangle, \qquad \forall\, \xi \in \mathbb{R}^N.$$

X**-invariance of F.** F is called X-invariant if it "absorbs" the trajectory of every integral curve of X touching F; this is equivalent to requiring that, for every integral curve of X, say $\gamma : [a, b] \to \Omega$, one has $\gamma([a, b]) \subseteq F$ whenever $\gamma([a, b])$ intersects F in at least one point.

Maximum Propagation for L along principal v.f.s for L. By the above notions, we can state the Maximum Propagation Principle for L as follows: *for every L-subharmonic function u on Ω, the set $F(u) := \{x \in \Omega : u(x) = \max_\Omega u\}$ (provided it is non-empty) is invariant wrt every C^1 principal vector field for L;* in other words, an attained maximum of an L-subharmonic function "propagates" along the trajectories of the C^1 principal vector fields for L.

X **is tangent to F.** In order to prove the Maximum Propagation Principle, it is convenient to introduce the notion of tangentiality of X to F. Let $y \in \Omega \cap \partial F$; a non-null vector $\nu \in \mathbb{R}^N$ is (externally) orthogonal to F at y (written $\nu \perp F$ at y) if the closure of $B(y + \nu, \|\nu\|)$ intersects F only at the point y. The vector field X is said to be tangent to F if

$$\langle X(y), \nu \rangle = 0, \quad \text{whenever } \nu \perp F \text{ at } y.$$

Nagumo-Bony Corollary. This result ensures that *F is X-invariant if and only if X is tangent to F.*

Hopf-type Lemma. A keystone in the proof of the Maximum Propagation Principle is the following result: *let X be principal for L; then, for every L-subharmonic function u on Ω, X is tangent to the above set $F(u)$* (if this set is non-empty).

Proof of the Maximum Propagation Principle. Let u be L-subharmonic on the set Ω; the set $F(u)$ is relatively closed in Ω, and we assume it is non-empty. Let X be a principal v.f. for L; then X is tangent to $F(u)$ by the Hopf-type Lemma. Since the tangentiality of X to $F(u)$ is equivalent to the X-invariance of $F(u)$ (by the Nagumo-Bony Corollary), then $F(u)$ is X-invariant. This proves the Maximum Propagation Principle.

Application: a Strong Maximum Principle. Let $L = \sum_{j=1}^m X_j^2 + Y$, where $\{X_1, \ldots, X_m\}$ is a Hörmander system of v.f.s on the connected open set Ω and Y

is a continuous v.f. It is easily seen that $\pm X_1, \ldots, \pm X_m$ are principal for L. Hence, if u is an L-subharmonic function on Ω attaining its maximum on Ω, this maximum propagates along the trajectories of $\pm X_1, \ldots, \pm X_m$, in view of the Maximum Propagation Principle.

This shows that u is constant on Ω by any of the following arguments:

(1) by the Connectivity Theorem, since any two points can be joined by piece-wise trajectories of $\pm X_1, \ldots, \pm X_m$;

(2) if $F(u) \neq \varnothing$, one can prove that the set $\mathrm{Tg}(F(u))$ of the smooth v.f.s tangent to $F(u)$ is a Lie algebra of v.f.s (by exploiting the above Nagumo-Bony Corollary); since $\mathrm{Tg}(F(u))$ contains $X_1, \ldots, X_m$ (by the Hopf-type Lemma), it contains $\mathrm{Lie}\{X_1, \ldots, X_m\}$. By Hörmander's Rank Condition, this implies that there cannot exist any $\nu \neq 0$ which is orthogonal to $F(u)$ at any y in $\Omega \cap \partial F(u)$; therefore $F(u) = \Omega$, that is u is constant.

10.7 Exercises of Chap. 10

Exercise 10.1. Prove the assertion in Rem. 10.19.

Exercise 10.2. Let $f : [a, b] \to \mathbb{R}$ and assume that a is a local minimum point of f. Suppose that f is differentiable at a. Show that $f'(a) \geq 0$.

Exercise 10.3. For every $a, b, c \in \mathbb{R}^N$, prove the following inequality

$$\|a - b\| \leq \|c - b\| + \left\langle \frac{c - a}{\|c - a\|}, \frac{b - a}{\|b - a\|} \right\rangle \cdot \|c - a\|.$$

[*Hint:* Start from the following decomposition (for which it is required to give a geometric interpretation)

$$b - a = \left\langle c - a, \frac{b - a}{\|b - a\|} \right\rangle \cdot \frac{b - a}{\|b - a\|} - \left\langle c - b, \frac{b - a}{\|b - a\|} \right\rangle \cdot \frac{b - a}{\|b - a\|}.$$

Observe the analogy with:

- the inequality $\overline{AB} \leq \overline{CB} + \overline{AH}$ (an improved triangle inequality) in Fig. 10.9-(a), when $\langle c - a, b - a \rangle \geq 0$,

- the inequality $\overline{AB} + \overline{AH} \leq \overline{CB}$ in Fig. 10.9-(b), when $\langle c - a, b - a \rangle \leq 0$.]

Exercise 10.4. Prove Lem. 10.26 by completing the following argument:

(1) Let $\varepsilon > 0$ be fixed. We claim that (10.10) implies that the maximum of

$$[0, T] \ni t \mapsto G(t) := g(t) - g(0) - (M + \varepsilon)\, t$$

is attained at $t = 0$. Prove this claim by contradiction:

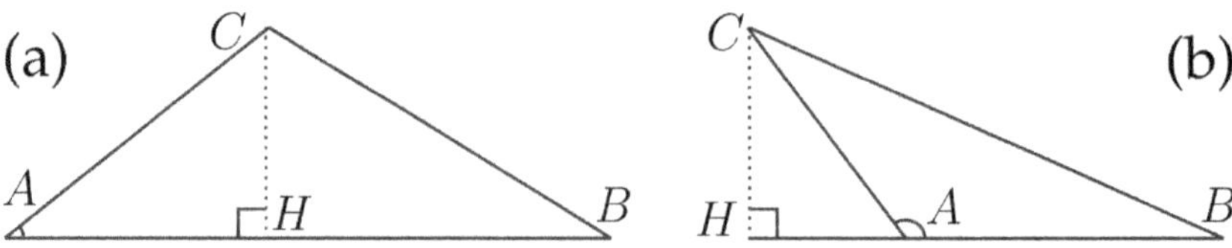

Fig. 10.9   The triangles in Exr. 10.3.

(2) Let $t_0 \in (0, T]$ be a maximum point of G; then

$$g(t) - g(0) - (M + \varepsilon)\, t \leq g(t_0) - g(0) - (M + \varepsilon)\, t_0, \quad \forall t \in [0, T].$$

We are allowed to take $t = t_0 + h$ with small $h < 0$ (why?); this gives

$$\frac{g(t_0 + h) - g(t_0)}{h} \geq M + \varepsilon.$$

By taking the $\lim \sup_{h \to 0^-}$, we contradict (10.10).
(3) Infer that $G(t) \leq G(0) = 0$ for every $t \in [0, T]$, that is,

$$g(t) - g(0) - (M + \varepsilon)\, t \leq 0 \quad \text{for every } t \in [0, T].$$

When $\varepsilon \to 0$, you obtain the proof of the lemma.

Exercise 10.5. Prove Gronwall's Lemma: *Let $g : [a, b] \to \mathbb{R}$ be a continuous function. Suppose that $C_0 \in \mathbb{R}$ and $L \geq 0$ are such that*

$$g(t) \leq C_0 + L \int_a^t g(x)\, \mathrm{d}x, \quad \forall\, t \in [a, b].$$

Then $g(t) \leq C_0 \exp(L(t - a))$ for every $t \in [a, b]$.

[*Hint:* Denoting $V(t) := C_0 + L \int_a^t g(x)\, \mathrm{d}x$, prove that $V'(t) \leq L\, V(t)$ and that $V(a) = C_0$; then derive that $V(t) \leq C_0 \exp(L(t - a))$...]

Deduce from Gronwall's Lemma the following fact (which is a special case of Lem. 10.27): *Let $g : [a, b] \to \mathbb{R}$ be differentiable; suppose that $g(a) = 0$, g is nonnegative and $g'(t) \leq L\, g(t)$ for any $t \in [a, b]$. Then $g \equiv 0$ on $[a, b]$.*

Exercise 10.6. Prove the assertion in Rem. 10.22 by the following argument.

- Since Ω is connected, $\Omega \cap \partial F$ is not empty.[3]

---

[3] Otherwise, if $\Omega \cap \partial F = \varnothing$ one would have

$$\Omega = [\Omega \cap \mathrm{Ext}(F)] \cup [\Omega \cap \mathrm{Int}(F)] \cup [\Omega \cap \partial F] = [\Omega \cap \mathrm{Ext}(F)] \cup [\Omega \cap \mathrm{Int}(F)] =: \Omega_1 \cup \Omega_2.$$

Ω_1, Ω_2 are obviously open and disjoint sets; Ω_1 is not empty, as F is a relatively closed *proper* subset of Ω; Ω_2 is not empty, otherwise any point of F would belong to ∂F, so that $F = \Omega \cap \partial F$ but this latter set is empty by assumption, whereas $F \neq \varnothing$ because F is a proper subset of Ω.

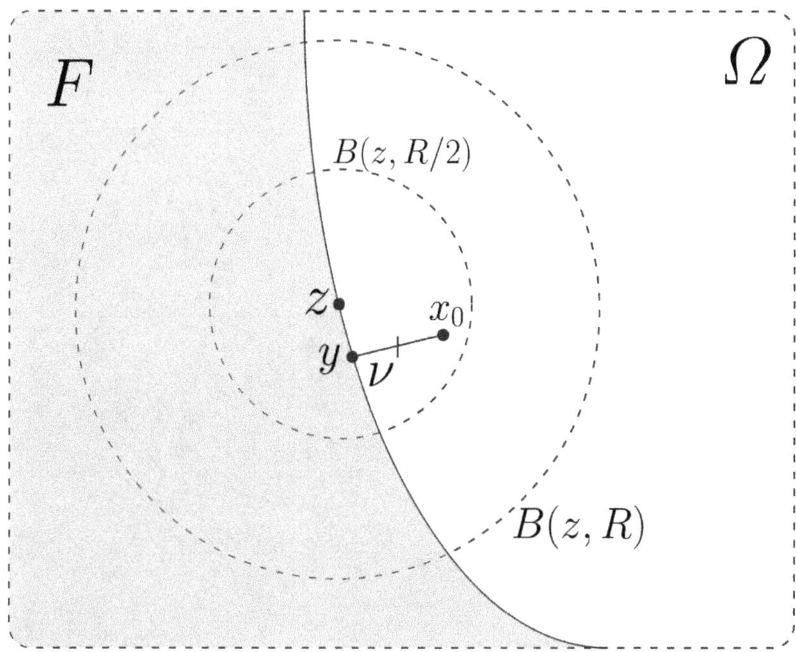

Fig. 10.10   The set F^* is non-empty, whenever F is a proper relatively closed subset of the connected open set Ω; see Rem. 10.22.

- With reference to Fig. 10.10, take a point $z \in \Omega \cap \partial F$, a ball $B(z, R) \subseteq \Omega$ and some $x_0 \in B(z, R/2)$ not belonging to F. Let $y \in \Omega \cap \partial F$ be such that

$$\|x_0 - y\| = \inf \left\{ \|x_0 - z\| \ : \ z \in \partial F \right\}. \tag{10.27}$$

  Then prove that $y \in F^*$, showing that $\nu := \frac{1}{2}\left(x_0 - y\right)$ is orthogonal to F at y. [*Hint:* use (10.27)...]

Prove also the following stronger fact: *Let F be a relatively closed proper subset of the open set Ω. Show that F^* is dense in $\Omega \cap \partial F$.*

[*Hint:* Use the above notation, and observe that, for every $z \in \Omega \cap \partial F$, we have $y \in \partial F \cap \overline{B(z, R)}$, and we can take R arbitrarily small...]

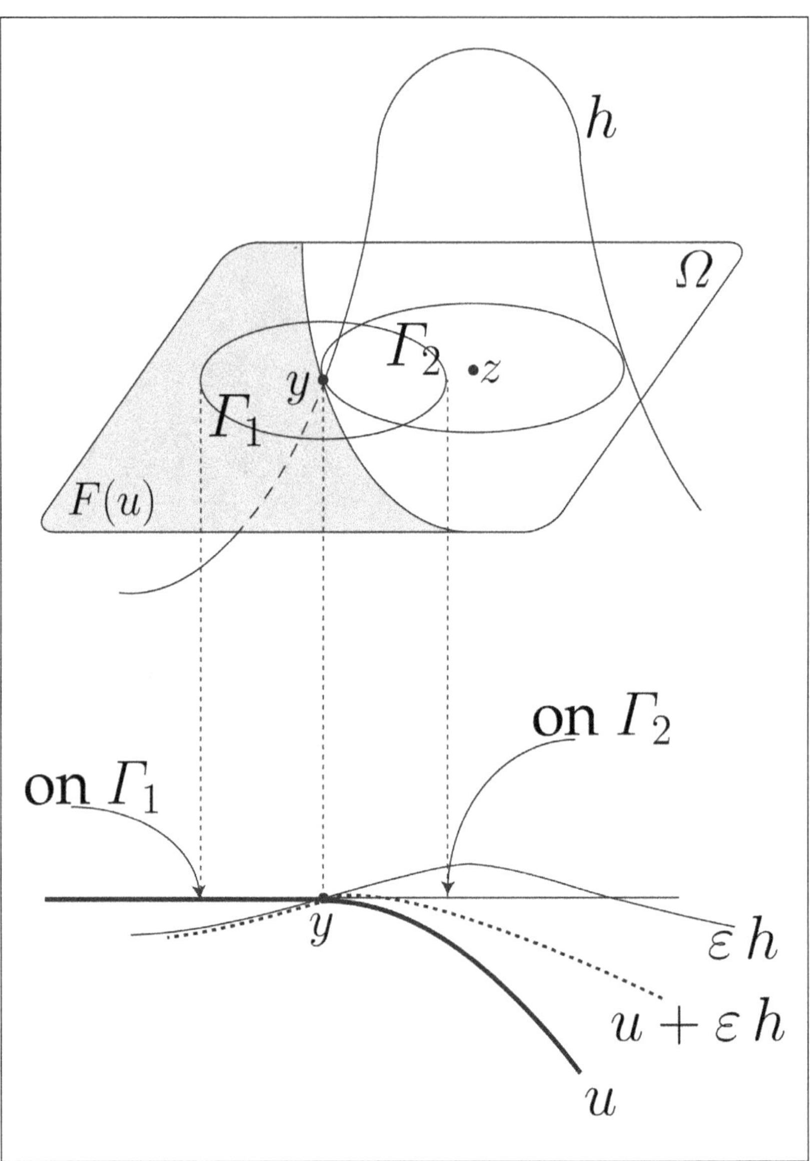

Fig. 10.11   The perturbation argument in the proof of the Hopf-type Lem. 10.31. On Γ_1: u is ≤ 0 and $\varepsilon\,h$ is negative; on Γ_2: u is smaller than a negative constant and $\varepsilon\,h$ is small. Thus $u + \varepsilon\,h$ is negative on $\Gamma_1 \cup \Gamma_2$, and null at y. This does not get along with the fact that u is L-subharmonic and h is strictly L-subharmonic on V, whose boundary is $\Gamma_1 \cup \Gamma_2$.

Chapter 11

The Maximum Propagation along the Drift

$\mathbf{T}$HE aim of this chapter is to complete the investigation of the Maximum Propagation Principle initiated in the previous chapter. The reader cannot have
missed that in Chap. 10, in dealing with operators of the form $L = \sum_{j=1}^{m} X_j^2 + X_0$,
nothing has been said about the maximum propagation along the integral curves
of the drift term X_0. This was not an oversight: the issue is that X_0 is not, in
general, a principal vector field for L.

The explicit example of the Heat operator $H = \sum_{j=1}^{N} \partial_j^2 - \partial_t$ proves that, not
only does the drift $-\partial_t$ miss to be principal for H, but neither can we expect (two-
sided) propagation of the maximum of an H-subharmonic function along the drift,
as the example of the fundamental solution Γ for H shows. Γ is indeed null in the
half-space $\{t < 0\}$ but elsewhere positive.

Nonetheless, a redeeming fact will be proved in this chapter: despite the lack of
X_0-invariance, we still have the *positive X_0-invariance* of the maximum-points set
$F(u)$ of an L-subharmonic function u. The proof of this fact is extremely delicate;
we follow the approach in [Amano (1979)]. Following [Amano (1979)], we consider
–more generally– the case of a second-order semielliptic PDO L (with no zero-
order term), rewritten as

$$Lu = \sum_{i=1}^{N} \frac{\partial}{\partial x_i}(X_i u) + X_0 u,$$

where $X_1, \ldots, X_N$ are the vector fields associated with the rows of the principal
matrix of L; in this case we say that X_0 is the *drift* of L, and the study of the positive
X_0-invariance is carried out.

At the end of the chapter we consider the *L-propagation set* of the maximum
points reachable (within positive times) from a given maximum point x_0. This is
clearly a Control Theory object, and we seize the opportunity to compare the reachability along a general set of v.f.s X to the reachability along the C^∞-module of X
(and, when possible, along the Lie algebra generated by X). This is a deep problem
to which Control Theory gives much more information than our overview.

Prerequisites for this chapter are the contents of Chap. 10.

237

11.1 Propagation along the drift

We go further in our study of the Maximum Propagation Principle by turning our attention to the role of the so-called drift vector field.

To be more precise let us consider a PDO of the form (10.1); we assume all the hypotheses in Sec. 10.1, plus the following one: the coefficient functions $a_{i,j}$ and b_i of L are required to be of class C^2 on D. From now on, we tacitly understand that Ω is a non-void open subset of D.

We introduce the following C^2 v.f.s associated with L:

$$X_i := \sum_{j=1}^{N} a_{i,j}(x)\, \frac{\partial}{\partial x_j}, \quad \text{for every } i = 1, \ldots, N;$$

$$X_0 := \sum_{j=1}^{N} \left(b_j(x) - \sum_{i=1}^{N} \frac{\partial a_{i,j}}{\partial x_i}(x) \right) \frac{\partial}{\partial x_j}. \tag{11.1}$$

By means of these vector fields, we can rewrite L in the following way:

$$L = \sum_{i=1}^{N} \frac{\partial}{\partial x_i} \left(\sum_{j=1}^{N} a_{i,j}(x)\, \frac{\partial}{\partial x_j} \right) + \sum_{j=1}^{N} \left(b_j(x) - \sum_{i=1}^{N} \frac{\partial a_{i,j}}{\partial x_i}(x) \right) \frac{\partial}{\partial x_j}$$

$$= \sum_{i=1}^{N} \frac{\partial}{\partial x_i} \circ X_i + X_0, \quad \text{on } D. \tag{11.2}$$

With an abuse of language borrowed from the case of the sums of squares of vector fields (see Exm. 8.3 on page 171), we say that X_0 is the **drift** of L.

Our aim is to show that, if $u \in C^2(\Omega)$ is L-subharmonic in Ω, then the maximum of u (if it exists) propagates *positively* along the drift X_0; in terms of Def. 10.18, we shall show that the set $F(u) := \{x \in \Omega : u(x) = \max_\Omega u\}$ is positively X_0-invariant. The notation $F(u)$ is henceforth tacitly understood.

Remark 11.1 (Principality of $X_1, \ldots, X_N$). We observe, for future reference, that the vector fields $X_1, \ldots, X_N$ defined in (11.1) are principal vector fields for L (see Def. 10.2): this is a direct consequence of Prop. 10.3 since $X_i(x)$ is precisely the i-th column of $A(x)$. Thus, if $u \in C^2(\Omega)$ is L-subharmonic in Ω, then $F(u)$ is X_i-invariant for every $i = 1, \ldots, N$ (provided it is $\neq \varnothing$).   ♯

The following is a maximum propagation result of independent interest; it will be useful in the sequel when considering propagation along the drift. The reader will acknowledge some analogy between the function f in the following statement and the use of L-barriers in Chap. 8 (see Def. 8.13).

Proposition 11.2. *Let $u \in C^2(\Omega)$ be an L-subharmonic function on Ω. Let us assume that u attains its maximum at a point $x_0 \in \Omega$, and that there exists a function $f \in C^2(\Omega)$ satisfying the following properties:*

(i) *$f(x_0) = 0$ and $\nabla f(x_0) \neq 0$;*

(ii) $Lf(x_0) > 0$.

Then, it is possible to find a neighborhood U_0 of x_0 such that, for every open neighborhood U of x_0 with closure contained in U_0, u attains its maximum not only at x_0 but also at some point of the set (see Fig. 11.1)

$$\partial U \cap \{f > 0\} := \{x \in \partial U : f(x) > 0\}.$$

Proof. If f is as in the assertion, we consider the C^2-function

$$F : \Omega \longrightarrow \mathbb{R}, \quad F(x) := f(x) - c\,\|x - x_0\|^2,$$

where $c > 0$ will be chosen in a moment. We have

- $F(x_0) = 0$;
- $\nabla F(x_0) = \nabla f(x_0)$;
- $LF(x_0) = Lf(x_0) - 2\,c\,\text{trace}(A(x_0))$.

By taking into account property (i) of f, we can choose a small $r > 0$ such that, setting $U_0 := B(x_0, r)$, we have $\overline{U_0} \subset \Omega$ and $\nabla F \neq 0$ on $\overline{U_0}$. By property (ii), we can choose $c > 0$ (depending on r) so small that $LF > 0$ on $\overline{U_0}$.

We now observe that, by definition of F and since $c > 0$, we have

$$\{x \in \Omega \setminus \{x_0\} : F(x) \geq 0\} \subseteq \{x \in \Omega : f(x) > 0\};$$

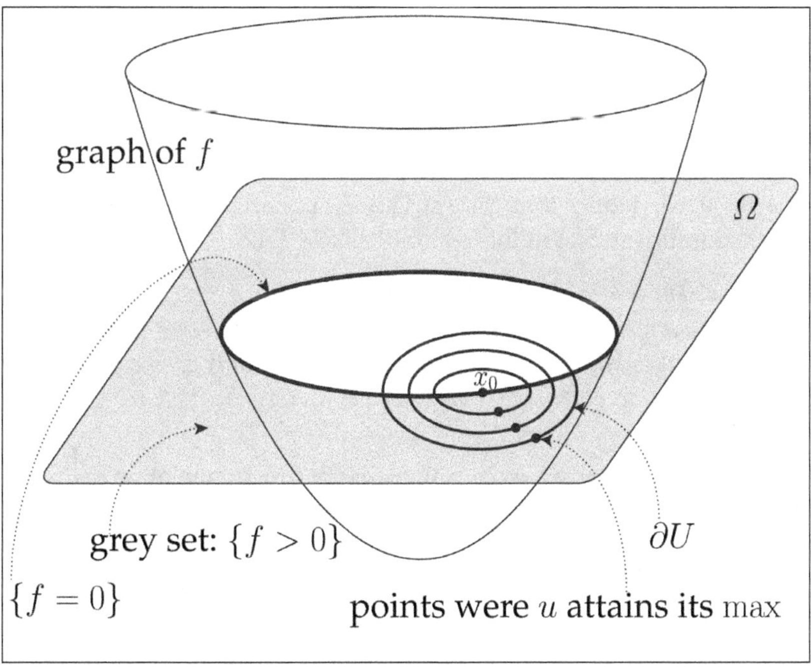

Fig. 11.1 The propagation of the maxima of u in Prop. 11.2.

hence, to prove the proposition it suffices to show that, if $U \Subset U_0$ is any open neighborhood of x_0, then u attains its maximum also at some point of

$$\Sigma = \partial U \cap \{F \geq 0\} := \{x \in \partial U : F(x) \geq 0\}.$$

We argue by contradiction and we assume that, for a certain open neighborhood $U \Subset U_0$ of x_0, the function u does not attain its maximum at any point of Σ. Since Σ is compact, this is equivalent to saying that

$$\sup\{u(x) : x \in \Sigma\} < u(x_0),$$

and thus it is possible to choose $\varepsilon > 0$ such that

$$0 < \varepsilon < \frac{u(x_0) - \sup_\Sigma u}{\sup_\Sigma F}. \tag{11.3}$$

We observe that $\sup_\Sigma F > 0$, since $F \geq 0$ on Σ and $\nabla F(x) \neq 0$ for every $x \in \overline{U_0}$. Setting $v := u + \varepsilon F$, we then have

- $Lv = Lu + \varepsilon\, LF \geq \varepsilon\, LF > 0$ on U;
- $v = u + \varepsilon\, F < u + u(x_0) - \sup_\Sigma u \leq u(x_0)$ on Σ (due to (11.3)).

On the other hand, since it obviously holds that

$$v < u \leq u(x_0) \text{ on } \partial U \cap \{F < 0\},$$

and since $v(x_0) = u(x_0)$, we deduce that $v \in C^2(U) \cap C(\overline{U})$ is a strictly L-subharmonic function on U which attains its maximum at some interior point of U, but this is in contradiction to Cor. 8.12 on page 176. □

Remark 11.3. Let the assumptions of Prop. 11.2 hold. If $U_0 \Subset \Omega$ is an open neighborhood of x_0 such that $Lf > 0$ on U_0, then $E := \{x \in U_0 : f(x) > 0\}$ cannot be empty. In fact, if this is not the case, then $f \leq 0$ on U_0; on the other hand, since $f(x_0) = 0$, we deduce that f is a strictly L-subharmonic function on U which attains its maximum at x_0, but this contradicts Cor. 8.12. ♯

Corollary 11.4. *Let $u \in C^2(\Omega)$ be L-subharmonic. Let us assume that u attains its maximum at a point $x_0 \in \Omega$, and that there exists a unit vector $\nu \in \mathbb{R}^N$ such that $\langle A(x_0)\,\nu, \nu \rangle > 0$. Then, for every $\rho > 0$, the function u attains its maximum not only at x_0 but also at some point of the set $\Omega \cap B(x_0 + \rho\nu, \rho)$. (See Fig. 11.2.)*

Proof. We consider, for any fixed $\rho > 0$, the Hopf function relative to the operator L, the point x_0 and the vector $\rho\nu$; we denote this function (see (10.22)) by h_ρ. Since $Lh_\rho(x_0) > 0$, we can apply Prop. 11.2, which ensures the existence of a neighborhood $U_0 \Subset \Omega$ such that, for any open neighborhood $U \Subset U_0$ of x_0, the function u attains its maximum somewhere on

$$\partial U \cap \{h_\rho > 0\} = \partial U \cap B(x_0 + \rho\nu, \rho) \subseteq \Omega \cap B(x_0 + \rho\nu, \rho),$$

and this is precisely what we wanted to prove. □

As the reader may have recognized, Prop. 11.2 and Cor. 11.4 can provide refined versions for the Hopf-type Lem. 10.31 in establishing propagation along principal v.f.s, as is explained in Exr. 11.1.

We can now prove the following proposition (whose proof is long and complicated!), a crucial step for the Maximum Propagation Principle along the drift.

Proposition 11.5. *Let $u \in C^2(\Omega)$ be an L-subharmonic function on Ω. Let X_0 be the drift of L, see (11.1). Let us assume that u attains its maximum at a point $x_0 \in \Omega$, and that there exists a unit vector $\nu \in \mathbb{R}^N$ such that*

$$\langle X_0(x_0), \nu \rangle > 0. \tag{11.4}$$

Then, for every $\rho > 0$, the function u attains its maximum not only at x_0 but also at some point of the set $\Omega \cap B(x_0 + \rho\nu, \rho)$. (See Fig. 11.2).

Proof. For the sake of clarity, we split the proof into three steps.

STEP I: First of all we observe that, if $\langle A(x_0)\,\nu, \nu \rangle > 0$, then the thesis follows from Cor. 11.4; hence, we can assume that $\langle A(x_0)\,\nu, \nu \rangle = 0$. This is equivalent to $A(x_0)\,\nu = 0$; thus, setting $r := \mathrm{rank}(A(x_0))$, we have $r < N$. By Sylvester's Theorem it is then possible to find a non-singular $N \times N$ matrix P such that[1]

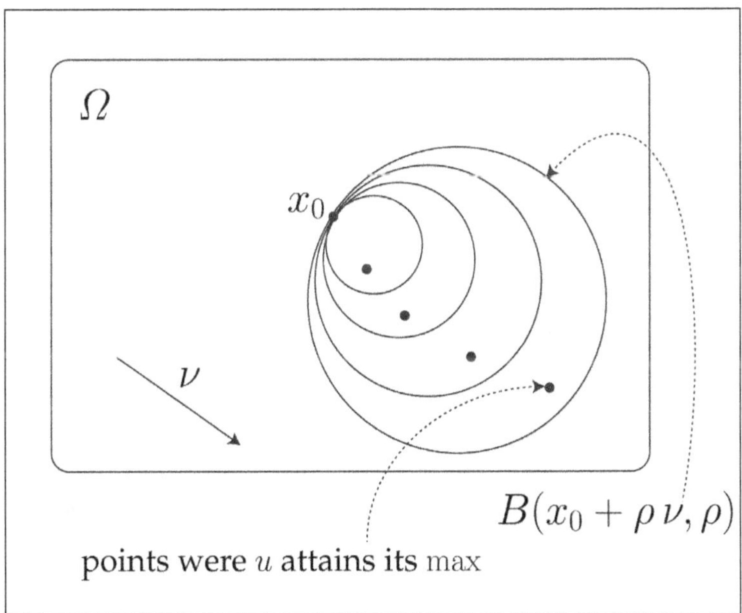

Fig. 11.2 The propagation of the maxima of u in Cor. 11.4 and in Prop. 11.5. Here ν is respectively: an ellipticity direction for the matrix of the quadratic form of L at x_0 (Cor. 11.4); a direction forming an acute angle with the drift of L at x_0 (Prop. 11.5).

---

[1] As usual, e_i is the i-th vector of the standard basis of $\mathbb{R}^N$, and $\mathbb{I}_r$ denotes the identity $r \times r$ matrix.

$$\text{(a):} \quad P^T e_N = \nu, \quad \text{and} \quad \text{(b):} \quad P\,A(x_0)\,P^T = \begin{pmatrix} \mathbb{I}_r & 0 \\ 0 & 0 \end{pmatrix}.$$

We then consider the linear change of variables Ψ defined as follows

$$\Psi : \mathbb{R}^N \to \mathbb{R}^N, \quad y = \Psi(x) := P(x - x_0).$$

According to Exr. 9.4 (page 209), L is turned by Ψ into a new linear PDO L', defined on the open set $D' := \Psi(D)$, which takes on the following form

$$L' = \sum_{i,j=1}^{N} \alpha_{i,j}(y) \frac{\partial^2}{\partial y_i \, \partial y_j} + \sum_{i=1}^{N} \beta_i(y) \frac{\partial}{\partial y_i},$$

where $A'(y) := (\alpha_{i,j}(y))_{i,j}$ and $\beta = (\beta_1, \ldots, \beta_N)^T$ are given, respectively, by

$$A'(y) = P\,A(\Psi^{-1}(y))\,P^T \quad \text{and} \quad \beta(y) = P\,b(\Psi^{-1}(y)).$$

Moreover, as we did for L, we introduce the C^2 v.f.s defined on D' as follows:

$$Y_i := \sum_{j=1}^{N} \alpha_{i,j}(y) \frac{\partial}{\partial y_j}, \quad \text{for every } i = 1, \ldots, N;$$

$$Y_0 := \sum_{j=1}^{N} \left(\beta_j(y) - \sum_{i=1}^{N} \frac{\partial \alpha_{i,j}}{\partial y_i}(y) \right) \frac{\partial}{\partial y_j}. \tag{11.5}$$

Since Ψ is an affine map, it is easy to check that X_0 and Y_0 are Ψ-related, i.e.,

$$Y_0(y) = J_\Psi\big(\Psi^{-1}(y)\big)\, X_0\big(\Psi^{-1}(y)\big) = P\,X_0\big(\Psi^{-1}(y)\big),$$

for every $y \in D'$; thus, by taking into account the above (a) (and by noticing that $\Psi(x_0) = 0$), from assumption (11.4) we infer that

$$\langle Y_0(0), e_N \rangle = \langle X_0(x_0), P^T e_N \rangle = \langle X_0(x_0), \nu \rangle > 0. \tag{11.6}$$

On the other hand, from property (b) we get

$$\alpha_{i,i}(0) = \langle A'(0)\,e_i, e_i \rangle = 0 \quad \text{for every } i = r+1, \ldots, N;$$

thus from Exr. 11.2 we obtain that (for $i, j \in \{r+1, \ldots, N\}$)

$$\frac{\partial \alpha_{i,j}}{\partial y_k}(0) = \left\langle \frac{\partial A'}{\partial y_k}(0)\,e_i, e_j \right\rangle = 0, \quad \text{for every } k = 1, \ldots, N. \tag{11.7}$$

As a consequence we get

$$\langle Y_0(0), e_N \rangle \overset{(11.5)}{=} \beta_N(0) - \sum_{i=1}^{N} \frac{\partial \alpha_{i,N}}{\partial y_i}(0) \overset{(11.7)}{=} \beta_N(0) - \sum_{i=1}^{r} \frac{\partial \alpha_{i,N}}{\partial y_i}(0).$$

From (11.6) we deduce that

$$\beta_N(0) - \sum_{i=1}^{r} \frac{\partial \alpha_{i,N}}{\partial y_i}(0) > 0. \tag{11.8}$$

We now set $\Omega' := \Psi(\Omega)$ and we consider the function $u' := u \circ \Psi^{-1}$. Obviously, $u' \in C^2(\Omega')$ and it attains its maximum at $\Psi(x_0) = 0$; furthermore, since L and L' are Ψ-related, we get $L'u'(y) \geq 0$ on Ω', that is, u' is L'-subharmonic in Ω'.

The aim of Step II is to show that the same statement of Prop. 11.5 holds true for L' and u', with the choice $x_0 = 0$ and $\nu = e_N$; that is, we prove that, for every $\rho > 0$, the function u' attains its maximum at some point of $\Omega' \cap B(\rho\, e_N, \rho)$.

Note that we are not replacing $B(x_0 + \rho\nu, \rho)$ with $\Psi(B(x_0 + \rho\nu, \rho))$ (which is an ellipsoid centred at 0); this would undoubtedly prove Prop. 11.5 for the original L and u by a simple change-of-coordinate argument, but it is more difficult to prove. We shall need Step III to show how Step II can be used all the same to prove Prop. 11.5 for L, for u and for $B(x_0 + \rho\nu, \rho)$.

STEP II: We consider the polynomial function defined by

$$f(y) := y_N - \frac{1}{2} \sum_{k=1}^{r} y_k \left(\frac{\partial \alpha_{k,N}(0)}{\partial y_k} y_k + 2 \sum_{j=k+1}^{N} \frac{\partial \alpha_{k,N}(0)}{\partial y_j} y_j \right)$$
$$- c \sum_{k=1}^{r} y_k^2 - C \sum_{k=r+1}^{N} y_k^2,$$

where c and C are suitable positive constants which we shall fix later on. Obviously, $f(0) = 0$ and $\nabla f(0) = e_N \neq 0$; moreover, a direct computation (based on (b) in Step I) shows that

$$L'f(0) = -2\,r\,c + \beta_N(0) - \sum_{k=1}^{r} \frac{\partial \alpha_{k,N}(0)}{\partial y_k},$$

and thus, taking into account (11.8), we can fix $c > 0$ in such a way that $L'f(0) > 0$. Owing to Prop. 11.2, it is then possible to find a closed ball $\overline{B(0, r_0)} \subseteq \Omega'$ such that, for every $0 < \varepsilon < r_0$, the function u' attains its maximum at some point $z \in B(0, \varepsilon) \cap \{f > 0\}$, that is, $\|z\| < \varepsilon$ and

$$z_N > \frac{1}{2} \sum_{k=1}^{r} z_k \left(\frac{\partial \alpha_{k,N}(0)}{\partial y_k} z_k + 2 \sum_{j=k+1}^{N} \frac{\partial \alpha_{k,N}(0)}{\partial y_j} z_j \right) + c \sum_{k=1}^{r} z_k^2 + C \sum_{k=r+1}^{N} z_k^2.$$
$$(11.9)$$

We claim that, for every $\rho > 0$, it is possible to choose ε and C in such a way that any maximum point $z \in B(0, \varepsilon) \cap \{f > 0\}$ of u' propagates as much as it can penetrate the set $B(\rho\, e_N, \rho)$.

To see this we consider, for every $y \in B(0, r_0)$, the integral curve $\gamma(t, Y_1, y)$ of Y_1 starting at y, which is defined and of class C^2 on a suitable open interval containing 0. Due to the continuity of $(t, y) \mapsto \gamma(t, Y_1, y)$, by shrinking r_0 if necessary, we can assume that, for every $y \in B(0, r_0)$,

$$y^{(1)} := \gamma(-y_1, Y_1, y) \quad \text{is well-defined and it belongs to } \Omega'.$$

Afterwards, we can consider the integral curve $\gamma(t, Y_2, y^{(1)})$ of Y_2 starting at $y^{(1)}$. Again by possibly shrinking r_0, we can assume that

$$y^{(2)} := \gamma(-y_2^{(1)}, Y_2, y^{(1)}) \quad \text{is well-defined and it belongs to } \Omega'.$$

By repeating this argument r times, we can suppose that $r_0 > 0$ is so small that, if $y \in B(0, r_0)$, then all the points

$$y^{(1)} = \gamma(-y_1, Y_1, y) \quad \cdots \quad y^{(r-1)} = \gamma\big(-y_{r-1}^{(r-2)}, Y_{r-1}, y^{(r-2)}\big)$$

are well-defined, and $\gamma(t, Y_r, y^{(r-1)})$ exists up to time $t = -y_r^{(r-1)}$.

We then consider the map

$$\theta : B(0, r_0) \longrightarrow \Omega' \qquad \theta(y) := \gamma\big(-y_r^{(r-1)}, Y_r, y^{(r-1)}\big).$$

Since the vector fields $Y_1, \ldots, Y_N$ are of class C^2, the map θ is of class C^2 on $B(0, r_0)$; moreover, since $Y_1, \ldots, Y_N$ are principal vector fields for L' (see Rem. 11.1), from Thm. 10.9 we deduce that, if $z \in B(0, \varepsilon) \cap \{f > 0\}$ is any maximum point of u', then the same is true of $\overline{z} = \theta(z)$.

We now aim to show that, given $\rho > 0$, if ε is sufficiently small and if C is sufficiently large, then $\theta(z) \in B(\rho e_N, \rho)$. To this end we first observe that, by Exr. 11.3, it is easy to recognize that the first r components $\theta_1, \ldots, \theta_r$ of θ admit the following second-order Taylor expansion near 0:

$$\theta_i(y) = -\frac{1}{2} \sum_{k=1}^{r} y_k \left(\frac{\partial \alpha_{k,i}(0)}{\partial y_k} y_k + 2 \sum_{j=k+1}^{N} \frac{\partial \alpha_{k,i}(0)}{\partial y_j} y_j \right) + o(\|y\|^2),$$

while for the last $N - r$ components $\theta_{r+1}, \ldots, \theta_N$ it holds that

$$\theta_i(y) = y_i - \frac{1}{2} \sum_{k=1}^{r} y_k \left(\frac{\partial \alpha_{k,i}(0)}{\partial y_k} y_k + 2 \sum_{j=k+1}^{N} \frac{\partial \alpha_{k,i}(0)}{\partial y_j} y_j \right) + o(\|y\|^2). \tag{11.10}$$

From this, by shrinking r_0 if necessary, we infer the existence of a positive constant $M > 0$ such that, for every $y \in B(0, r_0)$,

$$\sum_{k=1}^{N-1} \left(\theta_k(y) \right)^2 \le M \left(\sum_{i=1}^{r} y_i^4 + \sum_{i=r+1}^{N} y_i^2 \right); \tag{11.11}$$

moreover, by (11.10) with $i = N$ we can assume that, for every $y \in B(0, r_0)$,

$$\theta_N(y) > y_N - \frac{1}{2} \sum_{k=1}^{r} y_k \left(\frac{\partial \alpha_{k,N}(0)}{\partial y_k} y_k + 2 \sum_{j=k+1}^{N} \frac{\partial \alpha_{k,N}(0)}{\partial y_j} y_j \right) - \frac{c}{2} \|y\|^2. \tag{11.12}$$

In particular, if $z \in B(0, \varepsilon) \cap \{f > 0\}$ is a maximum point of u', by combining inequality (11.12) with (11.9) we obtain:

$$\overline{z}_N = \theta_N(z) > \frac{c}{2} \sum_{i=1}^{r} z_i^2 + \left(C - \frac{c}{2} \right) \sum_{i=r+1}^{N} z_i^2. \tag{11.13}$$

Therefore, if we choose $\varepsilon > 0$ in such a way that

 (i) $M \varepsilon^2 < \rho c / 2$,
 (ii) $\|\theta(y)\| < \rho$ for every $y \in B(0, \varepsilon)$,

and if let $C > 0$ be such that $C - c/2 > M/\rho$, by (11.11) and (11.13) we obtain

$$\bar{z}_N(2\rho - \bar{z}_N) = \theta_N(z)(2\rho - \theta_N(z)) \quad \text{(since } \|z\| < \varepsilon \text{ and owing to (ii))}$$

$$> \rho\,\theta_N(z) \overset{(11.13)}{>} \tfrac{\rho c}{2} \sum_{i=1}^r z_i^2 + \rho\,(C - c/2) \sum_{i=r+1}^N z_i^2 \quad \text{(owing to (i))}$$

$$> M \sum_{i=1}^r z_i^4 + \rho(C - c/2) \sum_{i=r+1}^N z_i^2 \quad \text{(by the choice of } C)$$

$$> M\Big(\sum_{i=1}^r z_i^4 + \sum_{i=r+1}^N z_i^2\Big) \overset{(11.11)}{\geq} \sum_{k=1}^{N-1} \big(\theta_k(z)\big)^2.$$

This proves that, if $z \in B(0,\varepsilon) \cap \{f > 0\}$ is a maximum point of u', then $\theta(z)$ (which is also a maximum point of u'), belongs to $\Omega' \cap B(\rho\,e_N,\rho)$.

STEP III: To conclude the demonstration, we are left to show that the results proved for L' and u' in Step II can be used to prove analogous facts for L and u. To this end, we choose $\rho > 0$ and we consider $E := \Psi\big(B(x_0 + \rho\nu,\rho)\big) \subseteq \Omega'$. Since Ψ is a diffeomorphism and since $B(x_0 + \rho\nu,\rho)$ has a smooth boundary, the same is true of E; moreover, since ν is the interior normal vector at x_0 to the ball $B(x_0 + \rho\nu,\rho)$, it is very easy to recognize that

$$(\mathcal{J}_{\Psi^{-1}}(0))^T\nu = \big(\mathcal{J}_\Psi(x_0)^T\big)^{-1}\nu = \big(P^T\big)^{-1}\nu = e_N$$

is the interior normal vector to E at $\Psi(x_0) = 0$. As a consequence, it is possible to find a positive $\delta > 0$ such that $B(\delta\,e_N,\delta) \subseteq E$. Owing to the results proved in Steps I and II, we know that u' attains its maximum at some $\bar{y} \in \Omega' \cap B(\delta\,e_N,\delta)$; hence $\bar{x} = \Psi^{-1}(\bar{y})$ is a maximum point of u lying in $\Omega \cap B(x_0 + \rho\nu,\rho)$, as desired. $\quad\square$

By means of Prop. 11.5 we can finally prove the following result.

Theorem 11.6 (Maximum Propagation Principle along the drift). *Let L be a second order PDO on D as in this section, and let X_0 be its drift (11.1). Let $u \in C^2(\Omega)$ be an L-subharmonic function on $\Omega \subseteq D$.*

Let us assume that the set of the maximum points $F(u)$ of u is non-empty. Then $F(u)$ is positively X_0-invariant, in the sense of Def. 10.18.

Proof. We can suppose that $F(u) \neq \Omega$, otherwise there is nothing to prove. Thus, owing to the Nagumo-Bony Thm. 10.28, we know that $F(u)$ turns out to be positively X_0-invariant if and only if

$$\langle X_0(y),\mu\rangle \leq 0, \quad \text{for every } y \in F(u)^* \text{ and every } \mu \perp F(u) \text{ at } y.$$

We argue by contradiction, assuming that there exist $y \in F(u)^*$ and $\mu \perp F(u)^*$ at y such that $\langle X_0(y),\mu\rangle > 0$ (see Fig. 11.3). From this, by applying Prop. 11.5 with $x_0 = y$, $\nu = \mu/\|\mu\|$ and $\rho = \|\mu\|$, we deduce that the function u attains its maximum at some point $\bar{y}$ belonging to $\Omega \cap B(y + \mu, \|\mu\|)$, that is,

$$F(u) \cap B(y + \mu, \|\mu\|) \neq \varnothing.$$

On the other hand, as μ is (externally) orthogonal to $F(u)$ at y, by definition we have $F(u) \cap B(y + \mu, \|\mu\|) = \varnothing$, but this is clearly a contradiction. $\quad\square$

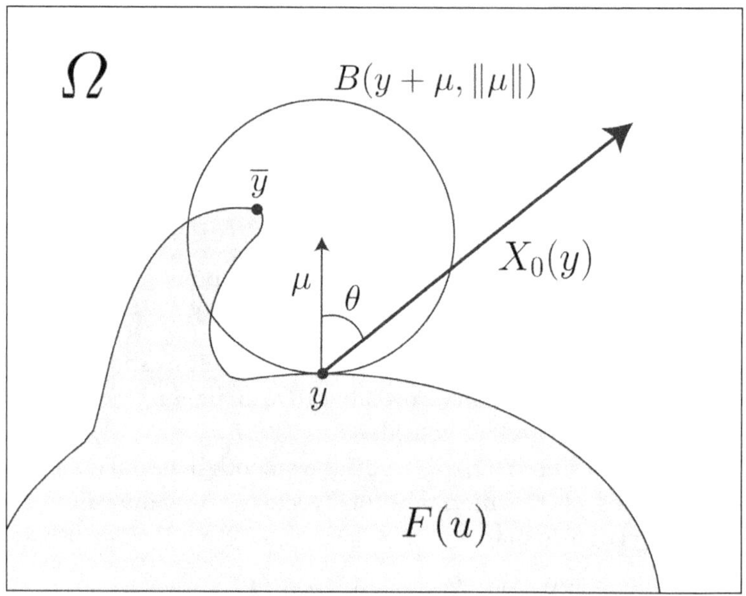

Fig. 11.3　The propagation of $F(u) = \{x : u(x) = \max_\Omega u\}$ along the drift X_0 in the proof of Thm. 11.6; the angle θ is acute if $\langle X_0(y), \mu \rangle > 0$.

11.2　A résumé of drift propagation

By combining Thm. 11.6 with the Propagation Principle in Thm. 10.9, we obtain the following *résumé* result.

Theorem 11.7 (Maximum propagation along principal/drift vector fields). *Let L be a homogeneous semielliptic PDO with $C^2(D)$ coefficients; let X_0 be the drift of L and let $X_1, \ldots, X_N$ be the v.f.s associated with the rows of the principal matrix of L, see (11.1).*

Suppose that $\Omega \subseteq D$ is an open set and that $u \in C^2(\Omega)$ is an L-subharmonic function on Ω. If u attains its maximum at some point of Ω, then the set

$$F(u) := \big\{ x \in \Omega : u(x) = \max_\Omega u \big\} \tag{11.14}$$

is positively invariant wrt any vector field X on Ω of the following form

$$X = \sum_{j=0}^{N} \xi_j(x)\, X_j, \tag{11.15}$$

for any choice of C^1 functions $\xi_0, \ldots, \xi_N$ with $\xi_0 \geq 0$ on Ω. If $\xi_0 \equiv 0$, there is also X-invariance, not only positive invariance.

Proof. We can suppose that $F(u) \neq \Omega$, otherwise there is nothing to prove. Thus, owing to the Nagumo-Bony Thm. 10.28 (see also Cor. 10.29), we know that $F(u)$ is positively X-invariant if and only if

$$\langle X(y), \nu \rangle \leq 0, \quad \text{for every } y \in F(u)^* \text{ and every } \nu \perp F(u) \text{ at } y.$$

On the other hand, since $F(u)$ is positively X_0-invariant (as ensured by Thm. 11.6) and since $F(u)$ is invariant wrt $X_1, \ldots, X_N$ (see Rem. 11.1), for every $y \in F(u)^*$ and every $\nu \perp F(u)$ at y we have

$$\langle X_0(y), \nu \rangle \leq 0 \quad \text{and} \quad \langle X_j(0), \nu \rangle = 0, \quad \text{for every } j = 1, \ldots, N.$$

Therefore, from our assumption (11.15) we immediately obtain

$$\langle X(y), \nu \rangle = \underbrace{\xi_0(y)}_{\geq 0} \cdot \underbrace{\langle X_0(y), \nu \rangle}_{\leq 0} \leq 0,$$

and this proves that $F(u)$ is positively X-invariant, as desired. $\qquad\square$

We close the section with the case of $L = \sum_{j=1}^{m} Z_j^2 + Z_0$. In principle, these could be seen as special cases of the operators considered so far, but it is more natural to consider the propagation along $Z_1, \ldots, Z_m$ and Z_0.

Theorem 11.8 (Maximum Propagation for sum of squares plus a drift). *Let D be an open subset of $\mathbb{R}^N$, let $Z_0, Z_1 \ldots, Z_m$ be C^2 vector fields on D, and let L be the sum of squares (plus a drift) PDO defined by*

$$L = \sum_{j=1}^{m} Z_j^2 + Z_0.$$

Suppose that $\Omega \subseteq D$ is open and that $u \in C^2(\Omega)$ is an L-subharmonic function. If u attains its maximum at some point of Ω, then the set $F(u)$ in (11.14) is positively invariant wrt any C^1 vector field on Ω of the form

$$X = \sum_{j=0}^{m} \xi_j(x) \, Z_j, \tag{11.16}$$

for any choice of C^1 functions $\xi_0, \ldots, \xi_m$ with $\xi_0 \geq 0$ on Ω. If $\xi_0 \equiv 0$, there is also X-invariance, not only positive invariance.

Proof. As usual, we can assume that $F(u) \neq \Omega$. Thus, owing to the Nagumo-Bony Thm. 10.28, we know that $F(u)$ is positively X-invariant iff

$$\langle X(y), \nu \rangle \leq 0, \quad \text{for every } y \in F(u)^* \text{ and every } \nu \perp F(u) \text{ at } y.$$

Now, $Z_1, \ldots, Z_m$ are principal vector fields for L owing to Exm. 10.6; thus the set $F(u)$ is invariant wrt $Z_1, \ldots, Z_m$, whence, for any $i = 1, \ldots, m$,

$$\langle Z_i(y), \nu \rangle = 0, \quad \text{for every } y \in F(u)^* \text{ and every } \nu \perp F(u) \text{ at } y. \tag{11.17}$$

On the other hand, by means of the results contained in Exr. 8.4 (on page 192), it is easy to recognize that L can be rewritten in the form (11.2); in particular, the matrix $A(x) = (a_{i,j}(x))$ and the drift vector field X_0 are respectively

$$A(x) = S(x) \cdot S(x)^T, \quad \text{with} \quad S(x) = (Z_1(x) \cdots Z_m(x)), \tag{11.18}$$
$$X_0 = Z_0 - \sum_{i=1}^{m} \operatorname{div}(Z_i I) \, Z_i. \tag{11.19}$$

Thus, since $F(u)$ is positively X_0-invariant (Thm. 11.6), for every $y \in F(u)^*$ and every $\nu \perp F(u)$ at y we obtain

$$0 \;\geq\; \langle X_0(y), \nu \rangle \stackrel{(11.19)}{=} \langle Z_0(y), \nu \rangle + \sum_{i=1}^{m} \operatorname{div}(Z_i I) \underbrace{\langle Z_i(y), \nu \rangle}_{=0} \stackrel{(11.17)}{=} \langle Z_0(y), \nu \rangle.$$

This gives

$$\langle Z_0(y), \nu \rangle \leq 0, \quad \text{for every } y \in F(u)^* \text{ and every } \nu \perp F(u) \text{ at } y. \tag{11.20}$$

Thus, by (11.17) and (11.20), we get (bearing in mind that $\xi_0 \geq 0$)

$$\langle X(y), \nu \rangle \stackrel{(11.16)}{=} \xi_0(y) \langle Z_0(y), \nu \rangle \leq 0,$$

and this proves that $F(u)$ is positively X-invariant. $\qquad\qquad\qquad\square$

A comparison of the results in Thms. 11.7 and 11.8 is in order:

Remark 11.9. Let $Z_0, \ldots, Z_m$ be as in Thm. 11.8 and let $L = \sum_{j=1}^{m} Z_j^2 + Z_0$. Let $X_1, \ldots, X_N$ be the v.f.s associated with the rows of $A(x)$ in (11.18) and let X_0 be the drift term, in the sense of (11.1). If $u \in \underline{S}(\Omega)$ and $F(u) \neq \varnothing$, by Thms. 11.7 and 11.8 we know that $F(u)$ is positively invariant wrt any v.f. X of the form

$$X = \sum_{i=0}^{N} \xi_i(x) \, X_i \quad \text{or} \quad X = \sum_{j=0}^{m} \eta_j(x) \, Z_j,$$

for C^1 functions ξ_i $(i = 0, \ldots, N)$ and η_j $(j = 0, \ldots, m)$, with $\xi_0(x), \eta_0(x) \geq 0$.

A natural question arises: what is the connection between the families of vector fields $\{X_i\}$ and $\{Z_j\}$? To give an answer we first observe that, due to (11.18), for every fixed $x \in D$ the columns of $A(x)$ (the X_i's) are linear combinations of the columns of $S(x)$ (the Z_j's), whence

$$\operatorname{span}\big(\{X_1(x), \ldots, X_N(x)\}\big) \subseteq \operatorname{span}\big(\{Z_1(x), \ldots, Z_m(x)\}\big);$$

on the other hand, since $S(x)$ and $A(x)$ have the same rank at every point of D (due to (11.18) again), we conclude that

$$\operatorname{span}\big(\{X_1(x), \ldots, X_N(x)\}\big) = \operatorname{span}\big(\{Z_1(x), \ldots, Z_m(x)\}\big).$$

Moreover, by (11.19) we have $X_0 = Z_0 - \sum_{i=1}^{m} f_i(x) Z_i$ for some functions f_i. This shows that a v.f. of the form $\sum_{i=0}^{N} \xi_i(x) \, X_i$ can always be written in the form $\sum_{j=0}^{m} \eta_j(x) \, Z_j$ with $\eta_0 = \xi_0$ and vice versa. $\qquad\qquad\qquad\sharp$

11.3   The point of view of reachable sets

Let $X = \{X_1, \ldots, X_m\}$ be C^1 v.f.s on the open set $D \subseteq \mathbb{R}^N$. Let us consider

$$\mathcal{P}_1(X) := \{X_1, \ldots, X_m\},$$

$$\mathcal{P}_2(X) := \{a_1 X_1 + \cdots + a_m X_m : a_1, \ldots, a_m \geq 0\}, \tag{11.21}$$

$$\mathcal{P}_3(X) := \Big\{ \sum_{i=1}^{m} a_i(x) X_i : a_i \in C^1(D), \ a_i \geq 0 \text{ on } D \text{ for } i = 1, \ldots, m \Big\}.$$

Clearly, $\mathcal{P}_1(X) \subseteq \mathcal{P}_2(X) \subseteq \mathcal{P}_3(X)$. Henceforth, we fix an open subset Ω of D.

Definition 11.10 (Propagation curve and set). Let $i \in \{1, 2, 3\}$ and let $\mathcal{P}_i(X)$ be as in (11.21). A continuous curve $\gamma : [0, T] \to \Omega$ (with $T > 0$) is said to be a $\mathcal{P}_i$-**propagation curve** (in Ω) if it satisfies the following property:

there exists a partition $\sigma = \{0 = t_0 < \ldots < t_p = T\}$ of $[0, T]$ such that the restriction of γ to every subinterval $[t_{k-1}, t_k]$ $(k = 1, \ldots, p)$, is an integral curve of a vector field in $\mathcal{P}_i(X)$. We denote by $\Lambda_i(\Omega)$ the set of all the $\mathcal{P}_i$-propagation curves.

Let $x_0 \in \Omega$. The $\mathcal{P}_i$-**propagation set of** x_0 in Ω, denoted by $\mathcal{P}_i(x_0, \Omega)$, is the set of all the end-points $\gamma(T)$ of the curves $\gamma : [0, T] \to \Omega$ in $\Lambda_i(\Omega)$ starting at x_0.

Clearly, $\mathcal{P}_1(x_0, \Omega) \subseteq \mathcal{P}_2(x_0, \Omega) \subseteq \mathcal{P}_3(x_0, \Omega)$, since $\Lambda_1(\Omega) \subseteq \Lambda_2(\Omega) \subseteq \Lambda_3(\Omega)$.

In the Control Theory literature, the set $\mathcal{P}_1(x_0, \Omega)$ is usually referred to as the **reachable set** of x_0 (in Ω, wrt X). Indeed, by unraveling the definition of $\mathcal{P}_1(x_0, \Omega)$, we see at once that

$$\mathcal{P}_1(x_0, \Omega) = \left\{ \Psi_{t_1}^{Z_1} \circ \Psi_{t_2}^{Z_2} \circ \cdots \circ \Psi_{t_n}^{Z_n}(x_0) \; : \; \begin{array}{l} n \in \mathbb{N}, \; t_1, \ldots, t_n \geqslant 0, \\ Z_1, \ldots, Z_n \in \{X_1, \ldots, X_m\} \end{array} \right\}.$$

Remark 11.11. Let $i \in \{1, 2, 3\}$; then Def. 11.10 gives at once:

(1) if $Z \in \mathcal{P}_i(X)$, any integral curve $\gamma : [0, T] \to \Omega$ of Z belongs to $\Lambda_i(\Omega)$;

(2) if $\gamma \in \Lambda_i(\Omega)$, any restriction of γ is itself in $\Lambda_i(\Omega)$;

(3) if $\gamma_1 : [0, T_1] \to \Omega$ and $\gamma_2 : [0, T_2] \to \Omega$ are two $\mathcal{P}_i$-propagation curves with $\gamma_1(T_1) = \gamma_2(0)$, the gluing of γ_1 and γ_2 is a $\mathcal{P}_i$-propagation curve.

As a consequence of the above facts, we have:

(4) if $x_0 \in \Omega' \subseteq \Omega$, then $\mathcal{P}_i(x_0, \Omega') \subseteq \mathcal{P}_i(x_0, \Omega)$;

(5) if $\gamma : [0, T] \to \Omega$ is in $\Lambda_i(\Omega)$ with $\gamma(0) = x_0$, then $\gamma([0, T]) \subseteq \mathcal{P}_i(x_0, \Omega)$;

(6) if $x_1 \in \mathcal{P}_i(x_0, \Omega)$, then $\mathcal{P}_i(x_1, \Omega) \subseteq \mathcal{P}_i(x_0, \Omega)$;

(7) as a consequence of (1) and (3), following Def. 10.18 (page 218), we infer that $\mathcal{P}_i(x_0, \Omega)$ is *positively* $Z|_\Omega$-*invariant*, for every $Z \in \mathcal{P}_i(X)$. ♯

Remark 11.12. We observe that, unlike the case of X-subunit curves (see Def. 6.13 on page 137), the $\mathcal{P}_i$-propagation curves are *oriented*.

More precisely, if $\gamma : [0, T] \to \Omega$ is in $\Lambda_i(\Omega)$, the curve

$$\psi : [0, T] \longrightarrow \Omega \qquad \psi(t) := \gamma(T - t)$$

may fail to be a $\mathcal{P}_i$-propagation curve. This is the case, e.g., of the Heat operator $L = \Delta_N - \partial_t$ on $\mathbb{R}^{N+1}$ (see Rem. 11.17). ♯

Rem. 11.11 implies the following simple topological property of $\mathcal{P}_i(x_0, \Omega)$.

Proposition 11.13. *Let* $i \in \{1, 2, 3, \}$. *If* $x_0 \in \Omega$, *the set* $\mathcal{P}_i(x_0, \Omega)$ *is path-connected. Thus, if* $\Omega_0 \subseteq \Omega$ *is the connected component of* x_0, *one has* $\mathcal{P}_i(x_0, \Omega) \subseteq \Omega_0$.

Proof. Let $y_1, y_2 \in \mathcal{P}_i(x_0, \Omega)$ and, according to Def. 11.10, let $\gamma_1 : [0, T_1] \to \Omega$ and $\gamma_2 : [0, T_2] \to \Omega$ be $\mathcal{P}_i$-propagation curves in Ω such that

$$\gamma_1(0) = \gamma_2(0) = x_0; \qquad \gamma_1(T_1) = y_1 \text{ and } \gamma_2(T_2) = y_2.$$

By Rem. 11.11-(5), both $\gamma_1([0, T_1])$ and $\gamma_2([0, T_2])$ are contained in Ω; thus, if we denote by ψ the curve defined by $\psi(t) := \gamma_1(T_1 - t)$ (with $t \in [0, T_1]$), then the gluing of ψ and γ_2 is a continuous curve in Ω joining y_1 and y_2.

We note that neither ψ nor its gluing with γ_2 are necessarily $\mathcal{P}_i$-propagation curves, but that does not affect path-connectedness. $\qquad\qquad\qquad\qquad\qquad\qquad\qquad\square$

We have the following redeeming feature of the $\mathcal{P}_i$-propagation sets:

Theorem 11.14. *Let $x_0 \in \Omega$ be fixed. Following Def. 11.10, we have that the sets*

$$F_i(x_0, \Omega) := \Omega \cap \overline{\mathcal{P}_i(x_0, \Omega)} \tag{11.22}$$

are all equal, as i ranges over $\{1, 2, 3\}$.

Proof. Let us drop the notation '(x_0, Ω)' in this proof. Clearly, $F_1 \subseteq F_2 \subseteq F_3$; so we are left to show that $F_3 \subseteq F_1$. We split the argument in three steps.

Step I. We prove a result of its own interest: we claim that

$$F_i \text{ is positively } Z|_\Omega\text{-invariant, for every } Z \in \mathcal{P}_i(X). \tag{11.23}$$

Indeed, let $y \in F_i$, $Z \in \mathcal{P}_i(X)$ and $\gamma : [0, T] \to \Omega$ be an integral curve of Z such that $\gamma(0) = y$; we need to prove that $\gamma(t) \in F_i$ for every $t \in [0, T]$. By definition of F_i, we know that there exists a sequence $y_k \in \mathcal{P}_i(x_0, \Omega)$ such that $\lim_k y_k = y$. Given any $t \in [0, T]$, since γ is an integral curve of Z defined on $[0, T]$, we have $t \in \mathcal{D}(Z, y)$. From $\lim_k y_k = y$, there exists $k_0 \in \mathbb{N}$ such that $t \in \mathcal{D}(Z, y_k)$ for every $k \geq k_0$; this follows from the semi-continuity results in Thm. B.33-(i). We consider $z_k := \Psi_t^Z(y_k)$, for $k \geq k_0$; these points are well-posed on account of the preceding argument. Any z_k belongs to $\mathcal{P}_i(x_0, \Omega)$, owing to Rem. 11.11-(7); thus, if we prove that $\lim_k z_k = \gamma(t)$, we shall deduce that $\gamma(t) \in F_i$, as desired.

Now, by construction, we have $\gamma(t) = \Psi_t^Z(y)$; therefore, from the continuity results for ODEs in Thm. B.33-(iii), we get

$$\lim_k z_k = \lim_k \Psi_t^Z(y_k) = \Psi_t^Z(y) = \gamma(t).$$

This proves the claimed (11.23).

Step II. Since F_i is relatively closed in Ω, we can use the theory developed in Sec. 10.4; in particular, by the characterization of positive invariance in the Nagumo-Bony Thm. 10.28, we infer that, for every $Z \in \mathcal{P}_i(X)$, we have

$$\langle Z(y), \nu \rangle \leq 0, \quad \text{for every } y \in (F_i)^* \text{ and every } \nu \perp F_i \text{ at } y. \tag{11.24}$$

Next we claim that F_1 is positively Z-invariant, for every $Z \in \mathcal{P}_3(X)$ (not only when $Z \in \mathcal{P}_1(X)$). Indeed, if $Z \in \mathcal{P}_3(X)$ we have $Z = \sum_{i=1}^m a_i(x) X_i$ for nonnegative C^1 functions a_i. Thus, for every $y \in (F_1)^*$ and every $\nu \perp F_1$ at y,

$$\langle Z(y), \nu \rangle = \sum_{i=1}^m \underbrace{a_i(y)}_{\geq 0} \cdot \underbrace{\langle X_i(y), \nu \rangle}_{\leq 0} \leq 0. \tag{11.25}$$

Here, we used the fact that $\langle X_i(y), \nu \rangle \leq 0$, as a consequence of (11.24) with $i = 1$ (and the fact that any X_i belongs to $\mathcal{P}_1(X)$). Again owing to the Nagumo-Bony Thm. 10.28, we see that (11.25) implies that F_1 is positively Z-invariant, as we claimed. By the very definition of invariance, this means that any integral curve (starting at any point of F_1) of any v.f. in $\mathcal{P}_3(X)$ remains in F_1 for positive times.

Step III. We are ready to prove that $F_3 \subseteq F_1$. To this aim, let $y \in F_3$; then there exist $y_k \in \mathcal{P}_3(x_0, \Omega)$ such that $\lim_k y_k = y$. By definition of $\mathcal{P}_3(x_0, \Omega)$, this means that any y_k can be connected to x_0 by a continuous curve, which is piecewise an integral curve of some v.f. in $\mathcal{P}_3(X)$. As x_0 is in F_1, we deduce from the last sentence of Step II that any y_k belongs to F_1. Thus $y = \lim_k y_k$ belongs to $\overline{F_1}$; as y is a point of Ω (bear in mind that $F_3 \subseteq \Omega$, see (11.22)), we derive that $y \in \Omega \cap \overline{F_1} = F_1$. This yields $y \in F_1$ and, from the arbitrariness of $y \in F_3$, the proof is complete. $\quad\square$

Remark 11.15. Suppose that the family X has the following special form

$$X = \{Y_1, \ldots, Y_m, \pm Z_1, \ldots, \pm Z_n\},$$

for some C^1 v.f.s Y_i and for some smooth v.f.s Z_j. Along with the sets in (11.21) we can consider the larger set

$$\mathcal{P}_4(X) := \left\{ a_0(x)Z + \sum_{i=1}^m a_i(x)Y_m : Z \in \mathrm{Lie}\{Z_1, \ldots, Z_n\} \right\},$$

where $a_1, \ldots, a_m$ are C^1 and nonnegative, and a_0 is C^1 (with no restriction on its sign). Arguing as in the proof of Thm. 11.14, one can show that

$$F_4(x_0, \Omega) := \Omega \cap \overline{\mathcal{P}_4(x_0, \Omega)}$$

is equal to the sets $F_i(x_0, \Omega)$ in (11.22), for $i = 1, 2, 3$ (and for any $x_0 \in \Omega$). We leave the proof to the reader, the main ingredients being:

- since both Z_k and $-Z_k$ belong to X, any $\mathcal{P}_i(x_0, \Omega)$ is Z_k-invariant;
- arguing as in the proof of Thm. 11.14, any $F_i(x_0, \Omega)$ is Z_k-invariant;
- using Prop. 10.38 (page 230), one infers that $F_i(x_0, \Omega)$ is invariant wrt any v.f. of the form $a(x)Z$, with $a \in C^1$ and $Z \in \mathrm{Lie}\{Z_1, \ldots, Z_n\}$;
- finally, one can retrace Steps II and III of the proof of Thm. 11.14, this time with the aid of the characterization of invariance in Cor. 10.29. $\quad\sharp$

11.3.1 *Examples of propagation sets for a PDO*

Let L be the linear PDO on D as in Sec. 11.1, and let $X_0, X_1, \ldots, X_N$ be the v.f.s in (11.1). Following [Amano (1979)], we can consider the set of $2N + 1$ vector fields $X := \{X_0, \pm X_1, \ldots, \pm X_N\}$. The construction of Sec. 11.3 can be performed in this case. We denote the associated sets $\mathcal{P}_3(X)$, $\Lambda_3(\Omega)$, $\mathcal{P}_3(x_0, \Omega)$ by

$$\mathcal{P}_L, \Lambda_L(\Omega), \mathcal{P}_L(x_0, \Omega), \text{ respectively.}$$

We use the naming 'L-propagation' instead of '$\mathcal{P}_3$-propagation'.

This naming is motivated by the following fact: Thm. 11.7 ensures that, if u is a L-subharmonic function on Ω and it attains its maximum at some point of Ω, then the set $F(u)$ in (11.14) is positively invariant with respect to any vector field $Z \in \mathcal{P}_3(X)$. This proves at once the following result.

Theorem 11.16. *Let $u \in C^2(\Omega)$ be an L-subharmonic function attaining its maximum at some point $x_0 \in \Omega$. Then*

$$u(x) = u(x_0) = \max_\Omega u, \qquad \text{for every } x \in \Omega \cap \overline{\mathcal{P}_L(x_0, \Omega)}. \tag{11.26}$$

On account of Thm. 11.14, we can replace $\mathcal{P}_L(x_0, \Omega)$ in (11.26) with $\mathcal{P}_1(x_0, \Omega)$ or $\mathcal{P}_2(x_0, \Omega)$, at our will.

We now furnish some explicit examples of L-propagation sets.

1. The Heat operator. Let us consider $\mathbb{R}^{1+N}$ (with $N \geq 1$) and let us denote its points by (t, x), with $t \in \mathbb{R}$ and $x \in \mathbb{R}^N$. We consider

$$H = \Delta_N - \frac{\partial}{\partial t} = \sum_{k=1}^{N} \frac{\partial^2}{\partial x_k^2} - \frac{\partial}{\partial t},$$

the classical heat operator on $\mathbb{R}^{1+N}$. Using the notation of Sec. 11.1,

$$X_0 = -\frac{\partial}{\partial t}, \qquad X_1 = 0, \ X_2 = \frac{\partial}{\partial x_1}, \ldots, X_{N+1} = \frac{\partial}{\partial x_N}.$$

As a consequence, we get (for suitable $\xi_0, \ldots, \xi_N \in C^1$ with $\xi_0 \geq 0$ on $\mathbb{R}^{1+N}$)

$$Z \in \mathcal{P}_H \iff Z = -\xi_0(t, x) \frac{\partial}{\partial t} + \sum_{j=1}^{N} \xi_j(t, x) \frac{\partial}{\partial x_j}. \tag{11.27}$$

We observe that the integral curves of $X_0, X_2, \ldots, X_{N+1}$ starting at $(t_0, x_0) \in \mathbb{R}^{1+N}$ are given, respectively, by

$$\gamma_{X_0, z}(s) = (t_0 - s, x_0),$$
$$\gamma_{X_i, z}(s) = (t_0, x_0 + s\, e_{i-1}) \qquad \text{(for } i = 2, \ldots, N+1),$$

where e_i denotes the i-th vector of the canonical basis of $\mathbb{R}^N$.

Furthermore, if $Z \in \mathcal{P}_H$ and if $\gamma : [a, b] \to \mathbb{R}^{1+N}$ is an integral curve of Z, for every $s \in [a, b]$ we have

$$\gamma = (\gamma_1, \ldots, \gamma_{N+1}) \quad \text{with} \quad \dot{\gamma}_1(s) = -\xi_0(\gamma(s)) \leq 0,$$

so that γ_1 is non-increasing on $[a, b]$. As a consequence, if $\gamma : [0, T] \to \Omega$ is an H-propagation curve, with the notation of Def. 11.10 we have

$$\gamma_1(0) = \gamma_1(t_0) \geq \gamma_1(t_1) \geq \ldots \geq \gamma_1(t_p) = \gamma_1(T). \tag{11.28}$$

Remark 11.17. Let $\gamma : [0, T] \to \Omega$ be in $\Lambda_H(\Omega)$ with $\gamma_1(0) > \gamma_1(T)$; this is the case, e.g., of any integral curve of $-\partial_t$. Then, the curve $\psi(t) = \gamma(T - t)$ is not an H-propagation curve, because $\psi_1(0) < \psi_1(T)$ (see (11.28)). ♯

By means of (11.28), we obtain at once the following fact.

Proposition 11.18. *Let $z_0 = (t_0, x_0) \in \Omega$ and let V_0 be the path-connected component of z_0 in $\Omega \cap \{t \leq t_0\}$. Then $\mathcal{P}_H(z_0, \Omega) \subseteq V_0 \subseteq \Omega \cap \{t \leq t_0\}$.*

When Ω is convex, Prop. 11.18 can be strengthened as follows.

Theorem 11.19. *Let Ω be convex and let $z_0 = (t_0, x_0) \in \Omega$. Then*

$$\mathcal{P}_H(z_0, \Omega) = \Omega \cap \{t \leq t_0\}.$$

Proof. On account of Prop. 11.18, it suffices to prove that $\Omega \cap \{t \leq t_0\}$ is in $\mathcal{P}_H(z_0, \Omega)$. To this end, let $z_1 = (t_1, x_1) \in \Omega \cap \{t \leq t_0\}$ and consider

$$\gamma : [0, 1] \longrightarrow \mathbb{R}^{1+N}, \qquad \gamma(s) := z_0 + s(z_1 - z_0).$$

We claim that $\gamma \in \Lambda_H(\Omega)$. Indeed, since Ω is convex, we have $\gamma([0, 1]) \subseteq \Omega$; moreover, since γ is an integral curve of

$$Z = -(t_0 - t_1)\frac{\partial}{\partial t} + \sum_{j=1}^{N} (x_1 - x_0)_j \frac{\partial}{\partial x_j},$$

and since, by (11.27), Z belongs to $\mathcal{P}_H$ (as $t_0 - t_1 \geq 0$), we conclude that $\gamma \in \Lambda_H(\Omega)$. As a consequence, $z_1 = \gamma(T) \in \mathcal{P}_H(z_0, \Omega)$, as desired. $\qquad\square$

The next theorem contains a non-obvious property of the H-propagation sets. The proof of this result is only sketched, as it rests on some technical properties of the H-propagation curves which we do not present.

Theorem 11.20. *Let $z_0 = (t_0, x_0) \in \Omega$. Then $\mathcal{P}_H(z_0, \Omega)$ is open in $\Omega \cap \{t \leq t_0\}$.*

Sketch. Let $z_1 = (t_1, x_1) \in \mathcal{P}_H(z_0, \Omega)$ and let $\rho > 0$ be such that $B(z_1, \rho) \subseteq \Omega$. If $t_1 = t_0$, it readily follows from Thm. 11.19 that

$$B(z_1, \rho) \cap \{t \leq t_0\} \subseteq \mathcal{P}_H(z_0, \Omega),$$

so that z_1 is in the interior of $\mathcal{P}_H(z_0, \Omega)$ (relative to $\Omega \cap \{t \leq t_0\}$).

If $t_1 < t_0$, it is possible to construct an H-propagation curve $\gamma : [0, T] \to \Omega$ such that γ_1 is strictly decreasing, at least near T. As a consequence, by possibly shrinking ρ, there exists $s \in [0, T]$ such that

$$\gamma(s) \in B(z_1, \rho) \subseteq \Omega \cap \{t < t_0\} \quad \text{and} \quad \tau := \gamma_1(s) > t_1.$$

Again from Thm. 11.19, we infer that $B(z_1, \rho) \cap \{t < \tau\} \subseteq \mathcal{P}_H(z_0, \Omega)$, whence z_1 is in the interior of $\mathcal{P}_H(z_0, \Omega)$. $\qquad\square$

See Fig. 11.4 for an explicit example of an H-propagation set.

2. A stationary Kolmogorov-type operator in $\mathbb{R}^2$. Let us consider

$$K = \frac{\partial^2}{\partial x_1^2} - x_1 \frac{\partial}{\partial x_2} \quad \text{in } \mathbb{R}^2, \tag{11.29}$$

a stationary *Kolmogorov-type operator* in $\mathbb{R}^2$. With the notation of Sec. 11.3, we have $X_0 = -x_1\,\partial/\partial x_2$, $X_1 = \partial/\partial x_1$ and $X_2 = 0$. As a consequence,

$$Z \in \mathcal{P}_K \quad \Longleftrightarrow \quad Z = -\xi_0(x)\,x_1\,\frac{\partial}{\partial x_2} + \xi_1(x)\,\frac{\partial}{\partial x_1},$$

for suitable $\xi_0, \xi_1 \in C^1(\mathbb{R}^2)$ with $\xi_0 \geq 0$ on $\mathbb{R}^2$. The integral curves of X_0 and X_1 starting at (x_0, y_0) are, respectively,

$$t \mapsto (x_0, y_0 - x_0\,t) \quad \text{and} \quad t \mapsto (x_0 + t, y_0).$$

From this, by arguing as in the case of the Heat operator, we infer that:

- if $\Omega \subseteq \{x_1 \geq 0\}$ and if $\gamma \in \Lambda_K(\Omega)$, then $\gamma_2(T) \leq \gamma_2(0)$, so that

$$\mathcal{P}_K((x_0, y_0), \Omega) \subseteq \Omega \cap \{x_2 \leq y_0\};$$

- if $\Omega \subseteq \{x_1 \leq 0\}$ and if $\gamma \in \Lambda_K(\Omega)$, then $\gamma_2(T) \geq \gamma_2(0)$, so that

$$\mathcal{P}_K((x_0, y_0), \Omega) \subseteq \Omega \cap \{x_2 \geq y_0\}.$$

This does not prevent the possibility of connecting any pair of points of $\mathbb{R}^2$ by means of integral curves, *via positive times*, of X_0 and $\pm X_1$; this means

$$\mathcal{P}_K((x_0, y_0), \mathbb{R}^2) = \mathbb{R}^2, \quad \text{for every } (x_0, y_0) \in \mathbb{R}^2.$$

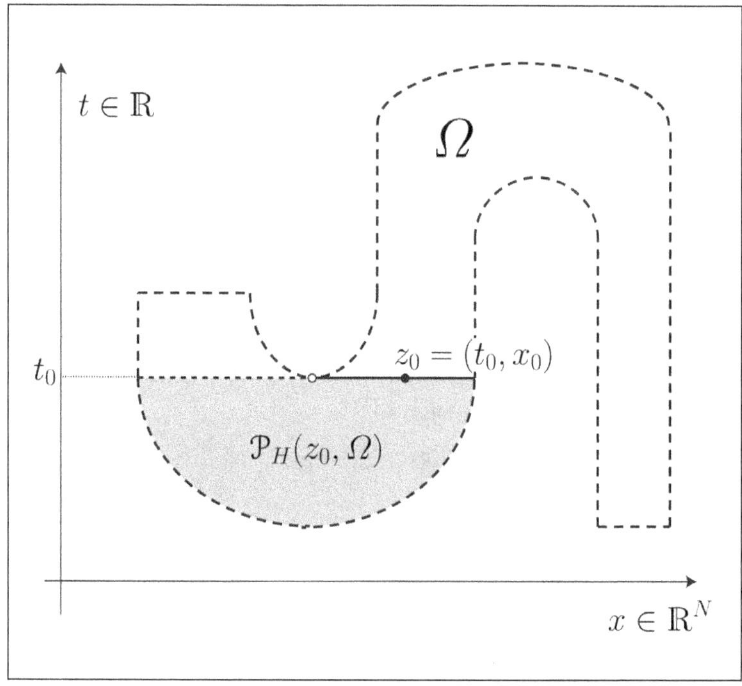

Fig. 11.4   An example of an H-propagation set, when H is the Heat operator.

See Fig. 11.5. The situation can drastically be different if we consider an open set $\Omega \neq \mathbb{R}^2$, as the next example shows.

Example 11.21. Even if Ω intersects the x_2-axis, the K-propagation set of a point in Ω can be strictly contained in Ω. For example, considering the convex open set

$$\Omega := \{(x_1, x_2) \in \mathbb{R}^2 : -2 < x_2 < 2, \ x_1 > x_2^2 - 1\},$$

then $\mathcal{P}_K((1,1), \Omega) = \Omega \cap \{x_2 \leq 1\} \neq \Omega.$ ♯

A more complicated geometry is involved with the PDO

$$K = \frac{\partial^2}{\partial x_1^2} + x_1 \frac{\partial}{\partial x_2} - \frac{\partial}{\partial x_3}, \tag{11.30}$$

the Kolmogorov operator in $\mathbb{R}^3$. See Exr. 11.4.

3. A Fediĭ operator. On $\mathbb{R}^2$ let us consider the PDO, studied in [Fediĭ (1971)],

$$F := \frac{\partial^2}{\partial x_1^2} + \left(a(x_1) \frac{\partial}{\partial x_2} \right)^2, \quad \text{where } a(x_1) = \begin{cases} 0 & \text{if } x_1 = 0, \\ \exp(-1/x_1^2) & \text{if } x_1 \neq 0. \end{cases}$$

Note that F is a sum of squares of C^∞ v.f.s, but *it is not a Hörmander PDO*, since the rank condition in violated on the line $x_1 = 0$. We have

$$X_0 = 0, \quad X_1 = \partial/\partial x_1, \quad X_2 = a(x_1) \partial/\partial x_2.$$

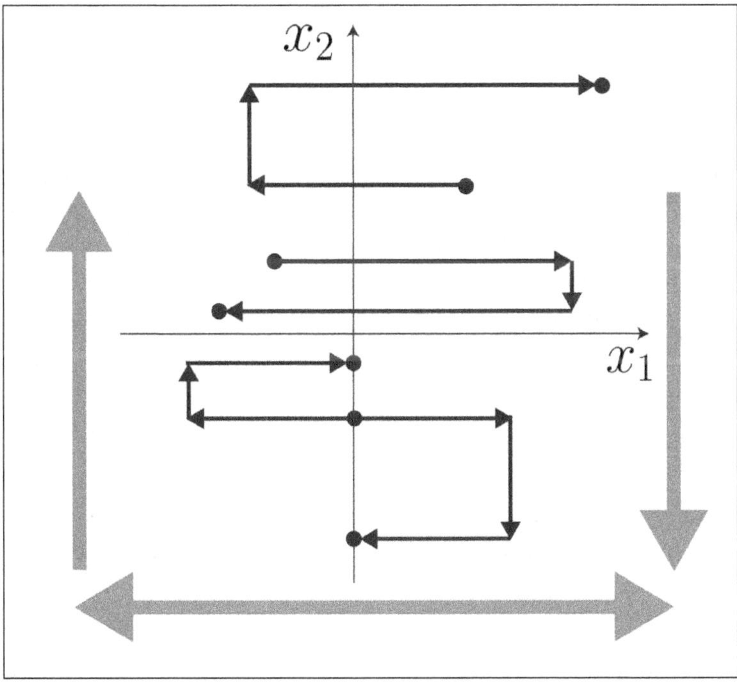

Fig. 11.5 Permissible directions (bold grey arrows) for the propagation related to the operator K in (11.29). Couple of points are shown (large black dots), and the arrowed polygonal chains picture some admissible K-propagation curves.

The integral curves of X_1 and X_2 starting at $x = (x_1, x_2)$ are, respectively,

$$t \mapsto (x_1 + t, x_2), \quad t \mapsto (x_1, x_2 + t\, a(x_1)).$$

This allows to prove the following fact.

Theorem 11.22. *Let Ω be a connected open set and let $x_0 \in \Omega$. Then*

$$\mathcal{P}_F(x_0, \Omega) = \Omega.$$

Proof. Since $x_0 \in \mathcal{P}_F(x_0, \Omega)$, it suffices to show that $\mathcal{P}_F(x_0, \Omega)$ is both open and closed in Ω.

$\mathcal{P}_F(x_0, \Omega)$ *is open.* Let $y \in \mathcal{P}_F(x_0, \Omega)$ and let $\rho > 0$ be such that

$$C(y, \rho) := (y_1 - \rho, y_1 + \rho) \times (y_2 - \rho, y_2 + \rho) \subseteq \Omega.$$

Via the integral curve of X_1 starting at y, we get that $\mathcal{P}_F(x_0, \Omega)$ contains

$$(y_1 - \rho, y_1 + \rho) \times \{y_2\};$$

thus, if z belongs to the latter segment and is such that $z_1 \neq 0$, running through the integral curve of X_2 starting at z we also deduce that $\{z_1\} \times (y_2 - \rho, y_2 + \rho)$ is in $\mathcal{P}_F(x_0, \Omega)$. Finally, choosing any point z' in the latter segment, and following again the integral curve of X_1 starting at z', we conclude that $C(y, \rho) \subseteq \mathcal{P}_F(x_0, \Omega)$.

$\mathcal{P}_F(x_0, \Omega)$ *is closed in Ω.* Let $y \in \overline{\mathcal{P}_F(x_0, \Omega)} \cap \Omega$ and let $(y_n)_n$ be a sequence in $\mathcal{P}_F(x_0, \Omega)$ converging to y as $n \to \infty$. Moreover, let $\rho > 0$ and $n \in \mathbb{N}$ be such that $y_n \in C(y, \rho) \subseteq \Omega$. By arguing as above, from $y_n \in \mathcal{P}_F(x_0, \Omega)$ we infer that $C(y, \rho)$ is contained in $\mathcal{P}_F(x_0, \Omega)$, whence $y \in \mathcal{P}_F(x_0, \Omega)$. $\qquad\square$

Remark 11.23. If $X = \{X_1, X_2\}$, it is contained in the proof of Thm. 11.22 the following (non-trivial) fact: *any open and connected subset Ω of $\mathbb{R}^2$ is X-connected, and any pair of points in Ω can be joined by an X-subunit curve, which is piecewise an integral curve of $\pm X_1, \pm X_2$.* We point out that this result cannot be deduced from Thm. 6.22, since X is not a Hörmander system in Ω (if Ω intersects the x_2 axis). $\qquad\sharp$

11.4   Exercises of Chap. 11

Exercise 11.1. Let $u \in \underline{S}(\Omega)$. Assume that $F(u) \neq \varnothing$. Prove that Cor. 11.4 gives another proof of the Maximum Propagation Principle, i.e., if X is a principal vector field for L, then $F(u)$ is X-invariant. Complete the following argument.

(1) By Cor. 10.29, $F(u)$ is X-invariant if and only if

$$\langle X(y), \mu \rangle = 0, \quad \text{for every } y \in F(u)^* \text{ and every } \mu \perp F(u) \text{ at } y.$$

(2) By Prop. 10.3 and the definition of $X_1, \ldots, X_N$ in (11.1), show that

$$X(x) \in \mathrm{span}\big(\{X_1(x), \ldots, X_N(x)\}\big), \quad \text{for every } x \in \Omega.$$

(3) Thus, to prove the X-invariance of $F(u)$ it suffices to show that, for every $j \leq N$, the displayed identity in (1) above holds with X replaced by X_j.

(4) Argue by contradiction: for some $y \in F(u)^*$ and some $\mu \perp F(u)$ at y, one has $\langle X_j(y), \mu \rangle \neq 0$. Since X_j is principal for L, infer that

$$\lambda(y) \langle A(y) \mu, \mu \rangle \geq \langle X_j(y), \mu \rangle^2 > 0.$$

Then use Cor. 11.4 with $x_0 = y$, $\nu = \mu / \|\mu\|$ and $\rho = \|\mu\|$, and deduce that u attains its maximum at some point in $B(y + \mu, \|\mu\|)$, i.e.,

$$F(u) \cap B(y + \mu, \|\mu\|) \neq \varnothing.$$

This is in plain contradiction to μ being orthogonal to $F(u)$ at y.

Exercise 11.2. For every $i, j \in \{1, \ldots, N\}$, let $m_{i,j}$ be differentiable functions on an open set $U \subseteq \mathbb{R}^N$. Let $M(x) = (m_{i,j}(x))_{i,j}$ be symmetric and positive semidefinite for every $x \in U$. Let us assume that there exists a point $x_0 \in U$ and two vectors $u, v \in \mathbb{R}^N$ such that $\langle M(x_0) u, u \rangle = \langle M(x_0) v, v \rangle = 0$.

Then, for every $k \in \{1, \ldots, N\}$, prove that

$$\left\langle \frac{\partial M(x_0)}{\partial x_k} u, v \right\rangle = 0.$$

Here we have denoted by $\dfrac{\partial M(x_0)}{\partial x_k}$ the matrix $\left(\dfrac{\partial m_{i,j}}{\partial x_k}(x_0) \right)_{i,j}$.

Complete the following argument.

(1) Use the Cauchy-Schwarz inequality to prove that

$$\langle M(x_0) u, v \rangle^2 \leq \langle M(x_0) u, u \rangle \cdot \langle M(x_0) v, v \rangle = 0;$$

deduce that $\langle M(x_0)(u + v), u + v \rangle = 0$.

(2) For $x \in U$, define $q_u(x) := \langle M(x) u, u \rangle$. Observe that x_0 is a minimum point of q_u; derive that (for $k = 1, \ldots, N$)

$$0 = \frac{\partial q_u(x_0)}{\partial x_k} = \left\langle \frac{\partial M(x_0)}{\partial x_k} u, u \right\rangle, \quad \text{so that}$$

$$\left\langle \frac{\partial M(x_0)}{\partial x_k} v, v \right\rangle = 0 = \left\langle \frac{\partial M(x_0)}{\partial x_k} (u + v), u + v \right\rangle.$$

(3) Deduce the thesis from the above identities ($\partial M / \partial x_k$ is symmetric).

Exercise 11.3. The next fact establishes the Maclaurin expansion of degree 2 of the flow map $\Psi_t^X(x)$ wrt (t, x) jointly.

Let X be a vector field of class C^2 on an open neighborhood U of 0; we denote by $\gamma(t, x)$ the map $(t, x) \mapsto \Psi_t^X(x)$ defined on the open set $\mathcal{D}(X)$.

Then, setting $X = \sum_{i=1}^N \alpha_i(x) \partial_{x_i}$, for every $i = 1, \ldots, N$ we have the following second-order expansion at $(t, x) = (0, 0)$:

$$\gamma_i(t, x) = \alpha_i(0) t + x_i + \frac{1}{2}\left(X\alpha_i(0) t^2 + 2t \langle \nabla \alpha_i(0), x \rangle \right) + r_i(t, x), \qquad (11.31)$$

where $r_i(t, x) = o(\|(t, x)\|^2)$ as $(t, x) \to (0, 0)$.

[*Hint:* Complete the following argument. Since X is of class C^2, infer that γ is of class C^2 on $\mathcal{D}(X)$. Let e_i be the i-th vector of the canonical basis of $\mathbb{R}^N$; take the i-th row vector in the matrix identity (1.29):

$$\nabla_x(\gamma_i(t, x)) = e_i + \int_0^t (\nabla\alpha_i)(\gamma(\tau, x)) \, \mathcal{J}_{\gamma(\tau, \cdot)}(x) \, d\tau. \tag{11.32}$$

Recognize that, by definition of flow, one has

$$\gamma(0, x) = x \quad \text{and} \quad \frac{\partial}{\partial t}(\gamma_i(t, x)) = \alpha_i(\gamma(t, x)). \tag{11.33}$$

Deduce that $\partial_t \gamma_i(0, 0) = \alpha_i(0)$ and $\nabla_x \gamma_i(0, 0) = e_i$, while the second-order derivatives of γ_i at the origin are given by

$$\frac{\partial^2 \gamma_i}{\partial t^2}(0, 0) \overset{(11.33)}{=} \frac{\partial}{\partial t}\Big|_{t=0} \alpha_i(\gamma(t, 0)) = (\nabla\alpha_i)(0) \, X(0) = X\alpha_i(0),$$

$$\mathrm{Hess}_x \gamma_i(0, 0) = \left(\frac{\partial^2}{\partial x_h \partial x_k}\Big|_{x=0} \gamma_i(0, x) \right)_{h,k} \overset{(11.33)}{=} 0,$$

$$\frac{\partial}{\partial t}\Big|_{t=0} \nabla_x\Big|_{x=0}(\gamma_i(t, x)) \overset{(11.32)}{=} \nabla\alpha_i(0) \, \mathcal{J}_{\gamma(0, \cdot)}(0) \overset{(11.33)}{=} \nabla\alpha_i(0).$$

Finally derive (11.31) from the Taylor Formula with a Peano remainder.]

Exercise 11.4. Consider the operator K in (11.30). Find the associated v.f.s $X_0, \ldots, X_3$ and their integral curves. Let $\Omega = (-1, 1)^{\times 3}$, and prove that

$$\mathcal{P}_K(0, \Omega) = \big\{ (x_1, x_2, x_3) \in \Omega : x_3 \leq 0, \ x_3 \leq x_2 \leq -x_3 \big\}.$$

Chapter 12

The Differential of the Flow wrt its Parameters

$\mathbf{T}$HE aim of this chapter is to study the flow map $\Psi_t^X(x)$ *as a function of the vector field* X, after we studied it as a function of t and of x in Chap. 1. Since X appears in the ODE defining $\Psi_t^X(x)$, we need some prerequisites of ODE Theory, namely the *non-autonomous equation of variation* ruling the following Cauchy problem (depending on the parameter ξ)

$$(\text{CP}): \qquad \begin{cases} \dot{y} = f(t,y,\xi) \\ y(\tau) = x. \end{cases}$$

If $t \mapsto \Phi_{t,\tau}^\xi(x)$ denotes the solution of the non-autonomous problem (CP), we obtain a remarkable integral identity involving the Jacobian matrices of $\Phi_{t,\tau}^\xi(x)$ wrt ξ and wrt x, and the Jacobian of $f(t,y,\xi)$ wrt ξ. This formula is

$$\frac{\partial}{\partial \xi} \Phi_{t,\tau}^\xi(x) = \int_\tau^t \left(\frac{\partial}{\partial x} \Phi_{t,s}^\xi \right)(\Phi_{s,\tau}^\xi(x)) \cdot \left(\frac{\partial f}{\partial \xi} \right)(s, \Phi_{s,\tau}^\xi(x), \xi) \, \mathrm{d}s.$$

Since this result is generally not presented in ODE textbooks, we provide the details in Sec. 12.1. Its autonomous counterpart is also obtained.

When the parametric second member f of the ODE comes, in an autonomous way, from a vector field of the form

$$f(t,y,\xi) = \xi_1 X_1(y) + \cdots + \xi_m X_m(y),$$

where $X_1, \ldots, X_m$ are C^1 vector fields on some open set in space, we can produce a formula for the partial derivatives wrt $\xi_1, \ldots, \xi_m$ of the function

$$\xi \mapsto \Psi_t^{\xi_1 X_1 + \cdots + \xi_m X_m}(x).$$

As a result, if $\{X_1, \ldots, X_m\}$ is the basis of a *finite-dimensional* Lie algebra $\mathfrak{g}$ of vector fields, we shall obtain in Chap. 13 a formula for the differential of the map

$$\mathfrak{g} \ni X \mapsto \Psi_t^X(x);$$

this is the case, e.g., when $\mathfrak{g}$ is the Lie algebra $\mathrm{Lie}(\mathbb{G})$ of some Lie group $\mathbb{G}$. We shall obtain applications (within or without the Lie group context) in Chaps. 13 and 14.

A prerequisite for this chapter is some basic ODE theory, given in App. B.

259

12.1   The non-autonomous equation of variation

In this section we use the following result.

Lemma 12.1 (Semi-group properties for non-autonomous ODEs). *Assume that* $f \in C(\Omega, \mathbb{R}^N)$ *is such that, for any* $(t_0, x_0) \in \Omega$, *the Cauchy problem*

$$\text{(CP)}: \qquad \dot{x} = f(t, x), \quad x(t_0) = x_0$$

has a unique maximal solution, denoted by $t \mapsto \gamma(t, t_0, x_0)$, *defined on its maximal domain* $\mathcal{D}(t_0, x_0)$. *Then the following facts hold:*

  (1) $t_0 \in \mathcal{D}(t_0, x_0)$ *and* $\gamma(t_0, t_0, x_0) = x_0$;
  (2) *if* $t \in \mathcal{D}(t_0, x_0)$ *then* $t_0 \in \mathcal{D}(t, \gamma(t, t_0, x_0))$ *and*

$$\gamma(t_0, t, \gamma(t, t_0, x_0)) = x_0; \tag{12.1}$$

  (3) *if* $\tau \in \mathcal{D}(t_0, x_0)$ *and if* $t \in \mathcal{D}(\tau, \gamma(\tau, t_0, x_0))$, *then* $t \in \mathcal{D}(t_0, x_0)$ *and*

$$\gamma(t, \tau, \gamma(\tau, t_0, x_0)) = \gamma(t, t_0, x_0). \tag{12.2}$$

Proof. (1): this is trivial.

(2): Let $t \in \mathcal{D}(t_0, x_0)$ so that $x_1 := \gamma(t, t_0, x_0)$ is well posed. At least on $\mathcal{D}(t_0, x_0)$, the map $\gamma(r) := \gamma(r, t_0, x_0)$ solves the following Cauchy problem:

$$\dot{x} = f(t, x), \quad x(t) = x_1.$$

In its turn, the maximal solution of the latter is $\gamma(r, t, x_1)$; by uniqueness of the maximal solution we infer that $\mathcal{D}(t_0, x_0) \subseteq \mathcal{D}(t, x_1)$, and $\gamma(r, t_0, x_0)$ must coincide with $\gamma(r, t, x_1)$ on $\mathcal{D}(t_0, x_0)$. This also shows that $t_0 \in \mathcal{D}(t, x_1)$ and if we take $r = t_0$ we get what we aimed to prove, i.e.,

$$x_0 \overset{(1)}{=} \gamma(t_0, t_0, x_0) = \gamma(t_0, t, x_1).$$

(3) Let $\tau \in \mathcal{D}(t_0, x_0)$; then $x_1 := \gamma(\tau, t_0, x_0)$ is well posed. By (2) we get that $t_0 \in \mathcal{D}(\tau, \gamma(\tau, t_0, x_0)) = \mathcal{D}(\tau, x_1)$. We consider the map $\gamma(r) := \gamma(r, \tau, x_1)$ defined on $\mathcal{D}(\tau, x_1)$, which obviously satisfies

$$\begin{cases} \dfrac{\mathrm{d}\gamma}{\mathrm{d}r}(r) = f(r, \gamma(r)) & \text{for any } r \in \mathcal{D}(\tau, x_1), \\ \gamma(t_0) = \gamma(t_0, \tau, x_1). \end{cases} \tag{12.3}$$

Now, owing to (2), we have $\gamma(t_0, \tau, x_1) = \gamma(t_0, \tau, \gamma(\tau, t_0, x_0)) = x_0$, so that (12.3) says that γ solves (CP) on $\mathcal{D}(\tau, x_1)$. By uniqueness, we infer that

$$\mathcal{D}(\tau, x_1) \subseteq \mathcal{D}(t_0, x_0), \text{ and} \tag{12.4}$$

$$\gamma(r) = \gamma(r, t_0, x_0) \text{ for any } r \in \mathcal{D}(\tau, x_1). \tag{12.5}$$

Let t belong to $\mathcal{D}(\tau, \gamma(\tau, t_0, x_0)) = \mathcal{D}(\tau, x_1)$; due to (12.4), t is an element of $\mathcal{D}(t_0, x_0)$ as well; finally, by taking $r = t$ in (12.5) we get

$$\gamma(t, \tau, \gamma(\tau, t_0, x_0)) = \gamma(t, \tau, x_1) = \gamma(t) \overset{(12.5)}{=} \gamma(t, t_0, x_0),$$

and (12.2) follows. This ends the proof. $\square$

Let $\Omega \subseteq \mathbb{R}^N$ be open; we denote the points of $\mathbb{R}^{1+N}$ by (t,y), with t in $\mathbb{R}$ and y in $\mathbb{R}^N$. Let V (the set of the 'parameters') be an open subset of $\mathbb{R}^m$, whose points are denoted by ξ. We assume that we are given an $\mathbb{R}^N$-valued function $f = f((t,y),\xi)$ which is continuous on $\Omega \times V$ and endowed with the partial derivatives

$$\frac{\partial f}{\partial y_i} \quad \text{and} \quad \frac{\partial f}{\partial \xi_j} \quad (1 \le i \le N, \ 1 \le j \le m),$$

which are required to be continuous on $\Omega \times V$. To simplify the notation, we write $f(t,y;\xi)$ in place of $f((t,y),\xi)$. Given $(\tau,x) \in \Omega$ we consider the Cauchy problem

$$\dot{y} = f(t,y;\xi), \quad y(\tau) = x,$$

whose maximal solution will be denoted by any of the notations

$$t \mapsto \gamma(t,\tau,x,\xi), \qquad t \mapsto \Phi^{\xi}_{t,\tau}(x).$$

Its maximal domain will be denoted by $\mathcal{D}(\tau,x,\xi)$. The full domain of γ wrt all its variables (t,τ,x,ξ) is $\mathcal{O} = \left\{(t,(\tau,x),\xi) \in \mathbb{R} \times \Omega \times V : t \in \mathcal{D}(\tau,x,\xi)\right\}$.

From Lem. 12.1 we get the **non-autonomous semigroup property**

$$\Phi^{\xi}_{t,t_0} \circ \Phi^{\xi}_{t_0,\tau}(x) = \Phi^{\xi}_{t,\tau}(x). \tag{12.6}$$

More precisely, if $t_0 \in \mathcal{D}(\tau,x,\xi)$ and if $t \in \mathcal{D}(t_0, \Phi^{\xi}_{t_0,\tau}(x),\xi)$, then it is true that $t \in \mathcal{D}(\tau,x,\xi)$ and (12.6) is fulfilled. By definition one has

$$\tfrac{d}{dt}\Phi^{\xi}_{t,\tau}(x) = f\left(t, \Phi^{\xi}_{t,\tau}(x);\xi\right), \qquad \Phi^{\xi}_{\tau,\tau}(x) = x. \tag{12.7}$$

In the sequel we use the notations $\frac{\partial}{\partial \xi}\Phi^{\zeta}_{t,\tau}(x)$ and $\frac{\partial}{\partial x}\Phi^{\xi}_{t,\tau}(x)$ to denote the $N \times m$ and $N \times N$ Jacobian matrices obtained by differentiating $\Phi^{\xi}_{t,\tau}(x)$ wrt ξ and wrt x.

From Thm. B.39 we know that $\gamma(t,\tau,x,\xi)$ admits second order mixed partial derivatives with respect to the couples of variables (t,τ), (t,x) and (t,ξ) and that the partial derivatives can be interchanged (even if we do not have, *a priori*, full C^2 regularity of γ wrt all the variables (t,τ,x,ξ)); see Rem. B.40.

If we differentiate both identities of (12.7) with respect to ξ (and we interchange derivatives), we get the matrix system

$$\begin{cases} \dfrac{d}{dt}\dfrac{\partial}{\partial \xi}\Phi^{\xi}_{t,\tau}(x) = \dfrac{\partial f}{\partial x}(t, \Phi^{\xi}_{t,\tau}(x);\xi) \cdot \dfrac{\partial}{\partial \xi}\Phi^{\xi}_{t,\tau}(x) + \dfrac{\partial f}{\partial \xi}(t, \Phi^{\xi}_{t,\tau}(x);\xi) \\[2ex] \dfrac{\partial}{\partial \xi}\Phi^{\xi}_{\tau,\tau}(x) = 0. \end{cases} \tag{12.8a}$$

If we differentiate both identities of (12.7) with respect to x (and we interchange derivatives), we get another matrix system

$$\begin{cases} \dfrac{d}{dt}\dfrac{\partial}{\partial x}\Phi^{\xi}_{t,\tau}(x) = \dfrac{\partial f}{\partial x}(t, \Phi^{\xi}_{t,\tau}(x);\xi) \cdot \dfrac{\partial}{\partial x}\Phi^{\xi}_{t,\tau}(x) \\[2ex] \dfrac{\partial}{\partial x}\Phi^{\xi}_{\tau,\tau}(x) = \mathbb{I}_N, \end{cases} \tag{12.8b}$$

where, as usual, $\mathbb{I}_N$ is the $N \times N$ identity matrix. Identities (12.8a) and (12.8b) are valid for every $(t, (\tau, x), \xi) \in \mathcal{O}$ (we shall tacitly understand this).

Let us introduce the following notations:

$$W_{t,\tau}^{\xi}(x) := \frac{\partial}{\partial \xi} \Phi_{t,\tau}^{\xi}(x), \qquad \Omega_{t,\tau}^{\xi}(x) := \frac{\partial}{\partial x} \Phi_{t,\tau}^{\xi}(x),$$

$$F_{t,\tau}^{\xi}(x) := \left(\frac{\partial f}{\partial \xi} \right)(t, \Phi_{t,\tau}^{\xi}(x); \xi), \qquad A_{t,\tau}^{\xi}(x) := \left(\frac{\partial f}{\partial x} \right)(t, \Phi_{t,\tau}^{\xi}(x); \xi). \tag{12.9}$$

With these notations, the systems (12.8a) and (12.8b) become

$$\frac{\mathrm{d}}{\mathrm{d}t} W_{t,\tau}^{\xi}(x) = A_{t,\tau}^{\xi}(x) \cdot W_{t,\tau}^{\xi}(x) + F_{t,\tau}^{\xi}(x), \quad W_{\tau,\tau}^{\xi}(x) = 0, \tag{12.10a}$$

$$\frac{\mathrm{d}}{\mathrm{d}t} \Omega_{t,\tau}^{\xi}(x) = A_{t,\tau}^{\xi}(x) \cdot \Omega_{t,\tau}^{\xi}(x), \quad \Omega_{\tau,\tau}^{\xi}(x) = \mathbb{I}_N. \tag{12.10b}$$

We are therefore in the presence of a (non-homogeneous) linear system of ODEs (12.10a) and its associated homogeneous system (12.10b), all depending on the parameters τ, x, ξ.

Remark 12.2. The reader must pay careful attention not to confuse (12.10b) with the system (see (12.25) in the dedicated appendix, Sec. 12.3)

$$\frac{\mathrm{d}}{\mathrm{d}t} \mathbf{X}(t, t_0) = A_{t,\tau}^{\xi}(x) \cdot \mathbf{X}(t, t_0), \quad \mathbf{X}(t_0, t_0) = \mathbb{I}_N, \tag{12.11}$$

characterizing the transition matrix $\mathbf{X}(t, t_0)$ associated with the system of ODEs $\dot{X}(t) = A_{t,\tau}^{\xi}(x) \cdot X(t)$, depending on the parameters τ, x, ξ. In other words, *this transition matrix $\mathbf{X}(t, t_0)$ is not $\Omega_{t,t_0}^{\xi}(x)$, as one can check.*[1]   ♯

We claim that the transition matrix $\mathbf{X}(t, t_0)$ characterized by (12.11) is

$$\mathbf{X}(t, t_0) := \Omega_{t,t_0}^{\xi}(\Phi_{t_0,\tau}^{\xi}(x)). \tag{12.12}$$

According to (12.11), we need to check that

$$\mathbf{X}(t_0, t_0) = \mathbb{I}_N \quad \text{and} \quad \frac{\mathrm{d}}{\mathrm{d}t} \mathbf{X}(t, t_0) = A_{t,\tau}^{\xi}(x) \cdot \mathbf{X}(t, t_0).$$

As for the first identity, we have

$$\mathbf{X}(t_0, t_0) \overset{(12.12)}{=} \Omega_{t_0,t_0}^{\xi}(\Phi_{t_0,\tau}^{\xi}(x)) \overset{(12.10b)}{=} \mathbb{I}_N.$$

---

[1]Indeed, since the system $\dot{X}(t) = A_{t,\tau}^{\xi}(x) \cdot X(t)$ depends on the parameters τ, x, ξ, its transition matrix should be denoted by $\mathbf{X}_{\tau,x,\xi}(t, t_0)$, so that (12.11) would read

$$(\bigstar): \qquad \frac{\mathrm{d}}{\mathrm{d}t} \mathbf{X}_{\tau,x,\xi}(t, t_0) = A_{t,\tau}^{\xi}(x) \cdot \mathbf{X}_{\tau,x,\xi}(t, t_0), \; \mathbf{X}_{\tau,x,\xi}(t_0, t_0) = \mathbb{I}_N.$$

Well, $\Omega_{t,t_0}^{\xi}(x)$ satisfies at least $\Omega_{t_0,t_0}^{\xi}(x) = \mathbb{I}_N$ as in the second identity of $(\bigstar)$, but its t-derivative is (see (12.10b) with $\tau = t_0$)

$$\frac{\mathrm{d}}{\mathrm{d}t} \Omega_{t,t_0}^{\xi}(x) = A_{t,t_0}^{\xi}(x) \cdot \Omega_{t,t_0}^{\xi}(x),$$

which is not the ODE in $(\bigstar)$, in general, but only when $\tau = t_0$.

The second needed identity comes from the following argument:

$$\frac{\mathrm{d}}{\mathrm{d}t}\mathbf{X}(t,t_0) \overset{(12.12)}{=} \frac{\mathrm{d}}{\mathrm{d}t}\left(\Omega^\xi_{t,t_0}(\Phi^\xi_{t_0,\tau}(x))\right) \overset{(12.10\mathrm{b})}{=} A^\xi_{t,t_0}(\Phi^\xi_{t_0,\tau}(x)) \cdot \Omega^\xi_{t,t_0}(\Phi^\xi_{t_0,\tau}(x))$$

$$\overset{(12.12)}{=} A^\xi_{t,t_0}(\Phi^\xi_{t_0,\tau}(x)) \cdot \mathbf{X}(t,t_0) \overset{(12.9)}{=} \frac{\partial f}{\partial x}(t,\Phi^\xi_{t,t_0}(\Phi^\xi_{t_0,\tau}(x));\xi) \cdot \mathbf{X}(t,t_0)$$

$$\overset{(12.6)}{=} \frac{\partial f}{\partial x}(t,\Phi^\xi_{t,\tau}(x);\xi) \cdot \mathbf{X}(t,t_0) \overset{(12.9)}{=} A^\xi_{t,\tau}(x) \cdot \mathbf{X}(t,t_0).$$

Upon verification of (12.12), we can use the linear ODE theory reviewed in Sec. 12.3; therefore, the solution of

$$\dot{X}(t) = A^\xi_{t,\tau}(x) \cdot X(t) + F(t), \qquad X(t_0) = x_0 \tag{12.13}$$

is given by (see (12.31))

$$\mathbf{X}(t,t_0)\,x_0 + \int_{t_0}^t \mathbf{X}(t,s) \cdot F(s)\,\mathrm{d}s$$

$$\overset{(12.12)}{=} \Omega^\xi_{t,t_0}(\Phi^\xi_{t_0,\tau}(x))\,x_0 + \int_{t_0}^t \Omega^\xi_{t,s}(\Phi^\xi_{s,\tau}(x)) \cdot F(s)\,\mathrm{d}s.$$

Taking into account that $W^\xi_{t,\tau}(x)$ is the solution of a system like (12.13) (see (12.10a) with $F(t) = F^\xi_{t,\tau}(x)$), from the above representation we get

$$W^\xi_{t,\tau}(x) = \int_\tau^t \Omega^\xi_{t,s}(\Phi^\xi_{s,\tau}(x)) \cdot F^\xi_{s,\tau}(x)\,\mathrm{d}s.$$

By the definitions in (12.9), we have therefore proved the relevant:

Theorem 12.3 (Equation of Variation for non-autonomous ODE systems). *Let $\Omega \subseteq \mathbb{R}^{1+N}$ and $V \subseteq \mathbb{R}^m$ be open sets. Let $f = f(t,y;\xi)$ be in $C(\Omega \times V, \mathbb{R}^N)$ and suppose that the partial derivatives $\frac{\partial f}{\partial y_i}$ and $\frac{\partial f}{\partial \xi_j}$ (for $1 \le i \le N, 1 \le j \le m$) exist and are continuous on $\Omega \times V$.*

If $t \mapsto \Phi^\xi_{t,\tau}(x)$ denotes the maximal solution (defined on its maximal domain $\mathcal{D}(\tau,x,\xi)$) of the non-autonomous ODE system (depending on the parameter ξ)

$$\dot{\gamma} = f(t,\gamma;\xi), \quad \gamma(\tau) = x, \tag{12.14}$$

we have the identity (involving three suitable Jacobian matrices)

$$\frac{\partial}{\partial \xi}\Phi^\xi_{t,\tau}(x) = \int_\tau^t \left(\frac{\partial}{\partial x}\Phi^\xi_{t,s}\right)(\Phi^\xi_{s,\tau}(x)) \cdot \left(\frac{\partial f}{\partial \xi}\right)(s,\Phi^\xi_{s,\tau}(x),\xi)\,\mathrm{d}s. \tag{12.15}$$

This identity is valid for any $(t,(\tau,x),\xi) \in \mathbb{R} \times \Omega \times V$ such that $t \in \mathcal{D}(\tau,x,\xi)$.

Even if this is non-standard, we shall refer to (12.15) as to the (integral form of the) **equation of variation** for the non-autonomous ODE system (12.14).

12.1.1 *The autonomous equation of variation*

In the particular case when the ODE system is autonomous (i.e., f does not depend on t), it is very simple to check that

$$\Phi_{t,\tau}^{\xi}(x) = \Phi_{t-\tau,0}^{\xi}(x). \tag{12.16}$$

In the case of an autonomous system, the vector-field notation is more effective: hence we suppose we are given a vector field X_{ξ} (on the open set $\Omega \subseteq \mathbb{R}^N$) depending (in a C^1 way) on a parameter ξ; this means that $X_{\xi} = \sum_{j=1}^{N} a_j(x,\xi) \frac{\partial}{\partial x_j}$, where $a_j(x,\xi)$ is C^1 in $(x,\xi) \in \Omega \times V$ for any $j = 1,\dots,N$. With our usual notation for the identity map I, we write $X_{\xi}I$ for the $\mathbb{R}^N$-valued map

$$x \mapsto (a_1(x,\xi),\dots,a_N(x,\xi))^T.$$

In the next result, we resume our usual notations $\Psi_t^X(x)$ and $\mathcal{D}(x,X)$.

Theorem 12.4 (Equation of Variation for the autonomous ODE systems). *Let $\Omega \subseteq \mathbb{R}^N$ and $V \subseteq \mathbb{R}^m$ be open sets. Suppose X_{ξ} is a C^1 vector field on Ω, depending in a C^1 way on the parameter $\xi \in V$. Let $\Psi_t^{X_{\xi}}(x)$ denote the flow of X_{ξ}, that is, the function $t \mapsto \Psi_t^{X_{\xi}}(x)$ is the maximal solution of*

$$\dot{\gamma}(t) = X_{\xi}(\gamma(t)), \quad \gamma(0) = x,$$

defined on the maximal domain $\mathcal{D}(x,X_{\xi})$. Then we have the identity

$$\frac{\partial}{\partial \xi}\Psi_t^{X_{\xi}}(x) = \int_0^t \left(\frac{\partial}{\partial x}\Psi_{t-s}^{X_{\xi}}\right)(\Psi_s^{X_{\xi}}(x)) \cdot \left(\frac{\partial}{\partial \xi}X_{\xi}I\right)(\Psi_s^{X_{\xi}}(x))\, \mathrm{d}s. \tag{12.17}$$

This identity is valid for any $(t,x,\xi) \in \mathbb{R} \times \Omega \times V$ such that $t \in \mathcal{D}(x,X_{\xi})$.

Proof. We apply Thm. 12.3: we take $\tau = 0$ and $f(t,y;\xi) := X_{\xi}(y)$ in (12.15), and we replace $\Phi_{t,s}^{\xi}$ with $\Psi_{t-s}^{X_{\xi}}$ (see (12.16)). $\qquad\square$

12.2 More on flow differentiation

The following result, concerning –roughly put– the differential of the flow wrt the v.f., will be applied many times in the sequel.

Theorem 12.5. *Suppose we are given a set of C^1 vector fields $\{X_1,\dots,X_m\}$ on the open set $\Omega \subseteq \mathbb{R}^N$. For any $\xi = (\xi_1,\dots,\xi_m) \in \mathbb{R}^m$, we let $\xi \cdot X := \sum_{i=1}^{m} \xi_i X_i$. Let also $x \in \Omega$ and $t \in \mathbb{R}$ be fixed. Finally, consider the map*

$$\mathbb{R}^m \ni \xi \mapsto E_{t,x}(\xi) := \Psi_t^{\xi \cdot X}(x) \in \mathbb{R}^N,$$

which is well-posed in a neighborhood of $0 \in \mathbb{R}^m$, at least. Then we have the identity

$$\frac{\partial E_{t,x}}{\partial \xi}(\xi) = \int_0^t \left[(X_1\Psi_{t-s}^{\xi \cdot X})(\Psi_s^{\xi \cdot X}(x)) \quad \cdots \quad (X_m\Psi_{t-s}^{\xi \cdot X})(\Psi_s^{\xi \cdot X}(x))\right] \mathrm{d}s. \tag{12.18}$$

In the above rhs, the brackets denote an $N \times m$ matrix whose i-th column-vector is

$$(X_i\Psi_{t-s}^{\xi \cdot X})(\Psi_s^{\xi \cdot X}(x));$$

as usual, a flow is thought of as a function valued in $\mathbb{R}^N$, written as an $N \times 1$ column vector, and v.f.s are allowed to operate componentwise on $\mathbb{R}^N$-valued functions.

With a pushforward notation, (12.18) is equivalent to stating that

$$\frac{\partial E_{t,x}(\xi)}{\partial \xi_i} = \int_0^t \left(\mathrm{d}\Psi_{t-s}^{\xi \cdot X} X_i \right)\left(\Psi_t^{\xi \cdot X}(x) \right) \mathrm{d}s \qquad (\text{for } i \in \{1, \ldots, m\}). \tag{12.19}$$

Proof. Formula (12.19) is a consequence of (12.18), taking into account that

$$\left(\mathrm{d}\Psi_{t-s}^{\xi \cdot X} X_i \right)\left(\Psi_t^{\xi \cdot X}(x) \right) = \left(X_i \Psi_{t-s}^{\xi \cdot X} \right)\left(\left(\Psi_{t-s}^{\xi \cdot X} \right)^{-1}\left(\Psi_t^{\xi \cdot X}(x) \right) \right) = \left(X_i \Psi_{t-s}^{\xi \cdot X} \right)\left(\Psi_s^{\xi \cdot X}(x) \right).$$

In the first identity we used the definition of the pushforward of a v.f. via a diffeomorphism (see (4.12) on page 101 with $f = I$); in the second identity we used

$$\left(\Psi_{t-s}^{\xi \cdot X} \right)^{-1} = \Psi_{s-t}^{\xi \cdot X} = \Psi_s^{\xi \cdot X} \circ \left(\Psi_t^{\xi \cdot X} \right)^{-1},$$

resulting from the semigroup properties of flows (Prop. 1.14 on page 11). Hence we are left to prove (12.18). We use (12.17) with $X_\xi = \xi \cdot X$. Thus,

$$\left(\frac{\partial}{\partial \xi} X_\xi I \right)(z) = \left[X_1(z) \cdots X_m(z) \right].$$

Thus i-th column of the matrix in the integrand in the rhs of (12.17) is

$$\left(\frac{\partial}{\partial x} \Psi_{t-s}^{\xi \cdot X} \right)\left(\Psi_s^{\xi \cdot X}(x) \right) \cdot X_i\left(\Psi_s^{\xi \cdot X}(x) \right). \tag{12.20}$$

Since $\left(\dfrac{\partial}{\partial x} \Psi_{t-s}^{\xi \cdot X} \right)(z)$ is the Jacobian matrix of $\Psi_{t-s}^{\xi \cdot X}$ at z, we infer that (on account of our usual compact notation for v.f.s acting on vector valued functions) (12.20) is simply $\left(X_i \Psi_{t-s}^{\xi \cdot X} \right)\left(\Psi_s^{\xi \cdot X}(x) \right)$. This gives the desired (12.18). $\qquad \square$

12.3   Appendix: A review of linear ODEs

Let $A(t) = (a_{i,j}(t))$ be an $N \times N$ matrix-valued map defined on an interval I, with $a_{i,j} \in C(I, \mathbb{R})$ for any i, j. Let also $b \in C(I, \mathbb{R}^N)$. We consider the linear homogeneous and non-homogeneous ODE systems

$$\dot{x} = A(t)\, x, \tag{12.21}$$
$$\dot{x} = A(t)\, x + b(t). \tag{12.22}$$

Any solution of (12.21) and of (12.22) is defined on I (Gronwall's Lem. B.19).

Definition 12.6. Let $V(t) = (v_{i,j}(t))_{i,j}$ be an $N \times N$ matrix-valued map, with $v_{i,j}$ in $C^1(I, \mathbb{R})$. Let us denote by v_i the $\mathbb{R}^N$-valued map associated with the i-th column of V. Finally, let $t_0 \in I$ be fixed.

We say that V is a **principal fundamental matrix** for (12.21) at t_0 if:

- $\{v_1, \ldots, v_N\}$ is a basis of the linear space of the solutions of (12.21),
- $V(t_0)$ is the $N \times N$ identity matrix $\mathbb{I}_N$.

In this case we also use the notation V_{t_0} instead of V.

Let $(t_0, x_0) \in I \times \mathbb{R}^N$ be fixed. The maximal solution $t \mapsto \gamma(t, t_0, x_0)$ of
$$\dot{x} = A(t)\, x, \qquad x(t_0) = x_0 \tag{12.23}$$
is given by $I \ni t \mapsto V_{t_0}(t)\, x_0$; hence, fixing $t \in I$ and letting x_0 vary, the map $x_0 \mapsto \gamma(t, t_0, x_0)$ is an endomorphism of $\mathbb{R}^N$. If we set
$$\mathbf{X}(t, t_0) := V_{t_0}(t) \qquad (t, t_0 \in I), \tag{12.24}$$
then $\mathbf{X}(t, t_0)$ is the unique $N \times N$ matrix representing (wrt the canonical basis of $\mathbb{R}^N$) this endomorphism: it is usually called the **transition matrix** (with initial time t_0 and final time t) of (12.21). It is easy to recognize that $\mathbf{X}(t, t_0)$ is characterized by the following (matrix-valued) Cauchy problem
$$\tfrac{\mathrm{d}}{\mathrm{d}t} \mathbf{X}(t, t_0) = A(t)\, \mathbf{X}(t, t_0), \qquad \mathbf{X}(t_0, t_0) = \mathbb{I}_N. \tag{12.25}$$
Any transition matrix is invertible, and it satisfies the remarkable identity
$$(\mathbf{X}(t, t_0))^{-1} = \mathbf{X}(t_0, t), \qquad t, t_0 \in I. \tag{12.26}$$
Indeed, $t \mapsto \mathbf{X}(t, t_0)\, x_0$ is the maximal solution of (12.23), i.e.,
$$\mathbf{X}(t, t_0)\, x_0 = \gamma(t, t_0, x_0), \qquad \forall\, t, t_0 \in I, \ \forall\, x_0 \in \mathbb{R}^N. \tag{12.27}$$
Thus, the identity $\gamma(t_0, t, \gamma(t, t_0, x_0)) = x_0$ in (12.1) is $\mathbf{X}(t_0, t)\, \mathbf{X}(t, t_0)\, x_0 = x_0$. The arbitrariness of $x_0 \in \mathbb{R}^N$ proves (12.26). More generally, we have
$$\mathbf{X}(t, \tau)\, \mathbf{X}(\tau, t_0) = \mathbf{X}(t, t_0), \qquad \forall\, t, \tau, t_0 \in I. \tag{12.28}$$
This is nothing but the semi-group property (12.2) rewritten in the language of transition matrices (via identity (12.27)).

Let us finally consider the non-homogeneous Cauchy problem
$$\dot{x} = A(t)\, x + b(t), \qquad x(t_0) = x_0. \tag{12.29}$$
Lagrange's constant-variation method ensures that, if $c \in C^1(I, \mathbb{R}^N)$ satisfies
$$V_{t_0}(t)\, \dot{c}(t) = b(t) \quad \text{for all } t \in I,$$
then $V_{t_0}(t)\, c(t)$ solves of (12.22). Thus, the maximal solution of (12.29) is
$$V_{t_0}(t) \left(x_0 + \int_{t_0}^{t} (V_{t_0}(s))^{-1} b(s)\, \mathrm{d}s \right), \quad \text{for all } t \in I. \tag{12.30}$$
This can be restated via transition matrices. Indeed, using the notation (12.24) in (12.30), the solution of (12.29) can be recast as
$$\mathbf{X}(t, t_0) \left(x_0 + \int_{t_0}^{t} (\mathbf{X}(s, t_0))^{-1} b(s)\, \mathrm{d}s \right) \overset{(12.26)}{=}$$
$$\mathbf{X}(t, t_0)\, x_0 + \int_{t_0}^{t} \mathbf{X}(t, t_0)\, \mathbf{X}(t_0, s)\, b(s)\, \mathrm{d}s \overset{(12.28)}{=} \mathbf{X}(t, t_0)\, x_0 + \int_{t_0}^{t} \mathbf{X}(t, s)\, b(s)\, \mathrm{d}s.$$
Summing up, the maximal solution of (12.29) is
$$\mathbf{X}(t, t_0)\, x_0 + \int_{t_0}^{t} \mathbf{X}(t, s)\, b(s)\, \mathrm{d}s, \quad \text{for all } t \in I. \tag{12.31}$$

12.4 Exercises of Chap. 12

Exercise 12.1. Prove all the assertions in Sec. 12.3 which were left unproved.

Chapter 13

The Exponential Theorem for ODEs

$\mathbf{T}$HE aim of this chapter is to prove a non-trivial analog, *in the realm of ODEs,* of the algebraic Exponential Theorem of Campbell-Baker-Hausdorff-Dynkin, previously investigated in Chap. 2. If compared to the algebraic setting of Chap. 2, the purpose of this chapter is much more ambitious, in that we want to replace the exponential function by the flow of a vector field.

Roughly put, since the flow of a vector field X is of an "exponential-type"

$$\Psi_t^X(x) = e^{tX}(x),$$

it is legitimate to presume that the composition of two flows $\Psi_t^Y \circ \Psi_t^X$ (an object that we studied in detail in Chap. 3) may be ruled by a CBHD-type formula. At the *finite* level of Taylor expansions, we have already proved that this is indeed the case (Thm. 3.6), modulo suitable remainders; the aim of this chapter is to show that much more is true, if we deal with *finite-dimensional Lie-algebras V of smooth vector fields.* We shall derive the formula

$$\binom{\textit{the CBHD Theorem}}{\textit{for ODEs}} \qquad \exp(Y)\big(\exp(X)(x)\big) = \exp(X \diamond Y)(x),$$

when $X, Y \in V$ are sufficiently close to the zero vector field. Here as usual $X \diamond Y$ is the CBHD series

$$X + Y + \frac{1}{2}[X, Y] + \cdots .$$

In accordance with the spirit of this book, we prove the above identity via an ODE argument: we show that

$$F(t) = \exp(tY)\big(\exp(X)(x)\big) \quad \text{and} \quad G(t) = \exp\big(X \diamond (tY)\big)(x)$$

solve the same ODE. In order to obtain $G'(t)$, we shall need many previous ingredients: Hadamard's Formula (Chap. 4), Poincaré's ODE (Chap. 5) and, obviously, the integral equation of variation derived in Chap. 12.

One can apply the CBHD Theorem for ODEs when V is the Lie algebra of a real Lie group (this will be done in Chap. 14); but it is also possible to use this result in the absence of any background Lie group structure: this will be our approach in Chap. 17 for *the very construction of Lie groups* starting from Lie algebras of vector fields V satisfying minimal assumptions.

Prerequisites for this chapter are some contents from Chaps. 4, 5 and 12.

13.1   Finite-dimensional algebras of vector fields

Convention. Let us fix a notation, once and for all in this chapter:

$$V \text{ is a } \textit{finite-dimensional} \text{ Lie subalgebra}$$
$$\text{of } \mathfrak{X}(\Omega), \text{ the smooth vector fields on the open set } \Omega \subseteq \mathbb{R}^N. \tag{13.1}$$

Theorem 13.1. *Suppose V satisfies (13.1). Let $t \in \mathbb{R}$ be such that $\Omega_t^X \neq \varnothing$. Then,*

$$\mathrm{d}\Psi_{-t}^X Y = e^{\mathrm{ad}\, tX} Y \qquad \begin{array}{l} \textit{as an equality of} \\ \textit{vector fields on } \Omega_t^X, \end{array} \tag{13.2}$$

for every choice of the vector fields X, Y in V.

The notation of (13.2) deserves an explanation: in its lhs, Y is thought of as a v.f. on Ω_{-t}^X (so that $\mathrm{d}\Psi_{-t}^X Y$ is a v.f. on Ω_t^X); in its rhs X and Y are thought of as v.f.s on Ω_t^X (and it is part of the proof to show that $e^{\mathrm{ad}\, tX} Y$ is a well posed v.f. on Ω_t^X). An equivalent formulation of (13.2) is: for any $x \in \Omega$ one has

$$(Y \Psi_{-t}^X)(\Psi_t^X(x)) = \sum_{j=0}^{\infty} ((\mathrm{ad}\, X)^j Y)(x)\, \frac{t^j}{j!}, \qquad t \in \mathcal{D}(X, x).$$

In particular, $t \mapsto \mathrm{d}\Psi_{-t}^X(Y)(x) = (Y \Psi_{-t}^X)(\Psi_t^X(x))$ is real-analytic on $\mathcal{D}(X, x)$.

Proof. The whole proof rests in the verification that the expansibility sufficient condition (4.27) in Thm. 4.27 (page 107) is fulfilled.

To this end, setting $m := \dim(V)$ (as a vector subspace of $\mathfrak{X}(\Omega)$), we fix a basis $\mathcal{B} = \{X_1, \ldots, X_m\}$ of V, and we denote by $c_{i,j,k}$ the structure constants of V (as a Lie algebra) with respect to $\mathcal{B}$ (see Sec. A.1). This means that

$$[X_i, X_j] = \sum_{k=1}^{m} c_{i,j,k} X_k, \quad \forall\, i, j = 1, \ldots, m.$$

Let $X, Y \in V$ be fixed. There exist unique constants α_j, β_j $(j \leq m)$ such that $X = \sum_{j=1}^{m} \alpha_j X_j$ and $Y = \sum_{j=1}^{m} \beta_j X_j$. Besides, it is easy to prove by induction on j that $(\mathrm{ad}\, X)^j Y$ equals

$$\sum_{k=1}^{m} \left(\sum_{i_1,\ldots,i_j=1}^{m} \alpha_{i_1} \cdots \alpha_{i_j} \sum_{k_1,\ldots,k_j=1}^{m} \beta_{k_1} c_{i_1,k_1,k_2} c_{i_2,k_2,k_3} \cdots c_{i_j,k_j,k} \right) X_k.$$

We fix any compact set $K \subset \Omega$; then, for any $j \in \mathbb{N}$,

$$\sup_{x \in K} \left\| ((\mathrm{ad}\, X)^j Y)(x) \right\| \leq$$

$$\sum_{k=1}^{m} \sup_{x \in K} \|X_k(x)\| \sum_{i_1,\ldots,i_j=1}^{m} |\alpha_{i_1}| \cdots |\alpha_{i_j}| \sum_{k_1,\ldots,k_j=1}^{m} |\beta_{k_1}|\, |c_{i_1,k_1,k_2}| \cdots |c_{i_j,k_j,k}|$$

$$\leq \alpha^j \beta\, c^j \sum_{k=1}^{m} \sup_{x \in K} \|X_k(x)\|,$$

where $\alpha = \max_{i \leq m} |\alpha_i|$, $\beta = \max_{k \leq m} |\beta_k|$, $c = \max_{i,j,k \leq m} |c_{i,j,k}|$.

Since the component functions of $X_1, \ldots, X_m$ are bounded on the compact subsets of Ω, we plainly recognize that condition (4.27) is fulfilled. $\qquad\square$

A direct consequence of Thm. 13.1 is the following fact.

Corollary 13.2. *Suppose V satisfies (13.1) and assume that any v.f. in V is global. Then, for every $X, Y \in V$ and every $f \in C^\infty(\Omega, \mathbb{R})$, one has*

$$(\mathrm{d}\Psi^X_{-t}Y)f(x) = \sum_{j=0}^{\infty} \frac{(\operatorname{ad} tX)^j Y}{j!} f(x) \qquad \forall\, x \in \Omega,\ t \in \mathbb{R}, \tag{13.3}$$

the power-series being convergent for any real t.

Since f can equivalently be replaced by the identity map I, this is equivalent:

- either, to an equality of vector fields on Ω:

$$\mathrm{d}\Psi^X_{-t}Y = e^{\operatorname{ad} tX} Y \quad \text{on } \Omega,$$

- or (fixing $x \in \Omega$), to an equality of $\mathbb{R}^N$-valued curves defined on $\mathbb{R}$:

$$(Y\Psi^X_{-t})(\Psi^X_t(x)) = \sum_{j=0}^{\infty}((\operatorname{ad} X)^j Y)(x)\frac{t^j}{j!} \quad \text{for } t \in \mathbb{R}.$$

Remark 13.3. We give an alternative ODE proof of Cor. 13.2. Suppose V satisfies (13.1) and assume that any v.f. in V is global. Then the noteworthy identity contained in the Hadamard's Formula (4.23) (page 104) is a genuine ODE in the finite-dimensional vector space V:

$$\dot{\gamma}(t) = (\operatorname{ad} X)(\gamma(t)), \quad \gamma(0) = Y. \tag{13.5}$$

Since $\operatorname{ad} X$ is a linear map on V, this is a *linear homogeneous Cauchy problem*, whose solution is therefore of the exponential type

$$\gamma(t) = \sum_{j=0}^{\infty} \frac{t^j (\operatorname{ad} X)^j}{j!} Y = e^{\operatorname{ad} tX} Y.$$

This gives back (13.3) since $t \mapsto \mathrm{d}\Psi^X_{-t}Y$ solves the same Cauchy problem (13.5), due to the mentioned Hadamard's Formula (4.23). ♯

13.2 The differential of the flow wrt the vector field

So far in this book we have studied $\Psi^X_t(x)$ sometimes as a function of t and sometimes as a function of x. We now take into account the non-trivial problem of considering the flow *as a function of the vector field X*. A very general (but very difficult) framework for this problem would involve the *infinite dimensional* vector space of the smooth v.f.s. This is definitely beyond our scope: since we shall be interested in finite-dimensional Lie algebras of smooth v.f.s (as in the case of Lie groups), we go on assuming that V satisfies assumption (13.1).

We want to give a representation formula for the "differential" of the map

$$V \ni X \mapsto \Psi(X) := \Psi^X_t(x) \in \mathbb{R}^N,$$

when $x \in \Omega$ and $t \in \mathcal{D}(X, x)$ are fixed. Since we investigated the notion of the differential of a function on a smooth (finite-dimensional) manifold in Chap. 4, it suffices to obtain a formula for the directional derivative

$$\frac{\partial \Psi}{\partial Y}(X) := \frac{\mathrm{d}}{\mathrm{d}\tau}\Big|_{\tau=0} \Psi(X + \tau Y) = \frac{\mathrm{d}}{\mathrm{d}\tau}\Big|_{\tau=0} \Psi_t^{X+\tau Y}(x),$$

for any X, Y in V (and with x and t as above). Since V is a finite-dimensional vector space (hence a smooth manifold in the usual way), the directional derivative $\frac{\partial \Psi}{\partial Y}(X)$ is equal to $\mathrm{d}_X \Psi(Y)$, where Y is thought of as a tangent vector to V at X (as is customary for finite-dimensional vector spaces).

First we show that the function $\Psi_t^{X+\tau Y}(x)$ is well posed. Let $X, Y \in V$, and let us fix $x \in \Omega$ and $t \in \mathcal{D}(X, x)$. We claim that there exists $\varepsilon > 0$ such that t also belongs to $\mathcal{D}(X + \tau Y, x)$, for every $\tau \in [-\varepsilon, \varepsilon]$. This will ensure that $\Psi_t^{X+\tau Y}(x)$ is well defined for any such τ. Now, the claim is a consequence of the continuous-dependence results for ODEs proved in App. B (see Thm. B.34 on page 373), taking into account that, as $\tau \to 0$, the coefficient vector of $X + \tau Y$ converges (locally uniformly) to the coefficient vector of X. The statement of the following result is therefore well posed.

Theorem 13.4 (Differential of the flow for finite-dimensional Lie-algebras). *Let V be a finite-dimensional Lie sub-algebra of $X(\Omega)$, and let $X, Y \in V$. Let us fix any $x \in \Omega$ and $t \in \mathcal{D}(X, x)$. Then, the following formula holds*

$$\frac{\mathrm{d}}{\mathrm{d}\tau}\Big|_{\tau=0} \Psi_t^{X+\tau Y}(x) = \sum_{j=0}^{\infty} \frac{t^{j+1}}{(j+1)!} \left((\mathrm{ad}\, X)^j Y\right) \Psi_t^X(x). \tag{13.6}$$

Formula (13.6) is equivalent to

$$\frac{\mathrm{d}}{\mathrm{d}\tau}\Big|_{\tau=0} \Psi_t^{X+\tau Y}(x) = \mathfrak{d}_{\Psi_t^X}(x) \cdot \sum_{j=0}^{\infty} \frac{t^{j+1} \left((\mathrm{ad}\, X)^j Y\right)(x)}{(j+1)!}. \tag{13.7}$$

If any vector field in V is global, the above formulae are valid for every $t \in \mathbb{R}$, and the series in the their rhs's are convergent for every $t \in \mathbb{R}$.

Formulae (13.6) and (13.7) can compactly be written respectively as

$$\frac{\mathrm{d}}{\mathrm{d}\tau}\Big|_{\tau=0} \Psi_t^{X+\tau Y}(x) = \left(\frac{e^{\mathrm{ad}\, tX} - 1}{\mathrm{ad}\, tX}(tY)\right) \Psi_t^X(x)$$
$$= \mathfrak{d}_{\Psi_t^X}(x) \cdot \left(\frac{e^{\mathrm{ad}\, tX} - 1}{\mathrm{ad}\, tX}(tY)\right)(x). \tag{13.8}$$

All of the above formulae are intended as equalities of $N \times 1$ column vectors, in that (as per usual) we let the v.f. $(\mathrm{ad}\, X)^j Y$ act componentwise on the vector-valued function Ψ_t^X (written as an $N \times 1$ function). Moreover, it is implicitly understood that the power series contained in these formulae are convergent.

Remark 13.5. When $t = 1$ is allowed (for example if all the v.f.s in V are global), (13.8) can compactly be written as

$$\frac{\mathrm{d}}{\mathrm{d}\tau}\Big|_{\tau=0}\Psi_1^{X+\tau Y}(x) = \left(\frac{e^{\mathrm{ad}\,X}-1}{\mathrm{ad}\,X}Y\right)\Psi_1^X(x)$$

$$= \mathcal{J}_{\Psi_1^X}(x)\cdot\left(\frac{e^{\mathrm{ad}\,X}-1}{\mathrm{ad}\,X}Y\right)(x). \tag{13.9}$$

Proof. We already discussed the fact that, under the assumption of the theorem, $\Psi_t^{X+\tau Y}(x)$ is well posed for small τ. By definition, $\gamma(s) := \Psi_s^{X+\tau Y}(x)$ solves on $\mathcal{D}(X+\tau Y, x)$ the Cauchy Problem

$$\dot\gamma(s) = (X+\tau Y)(\gamma(s)), \quad \gamma(0) = x.$$

We are entitled to apply the autonomous equation of variation (12.17) (page 264), which gives, for any $s \in \mathcal{D}(X+\tau Y, x)$,

$$\frac{\mathrm{d}}{\mathrm{d}\tau}\Psi_s^{X+\tau Y}(x) = \int_0^s \mathcal{J}_{\Psi_{s-r}^{X+\tau Y}}(\Psi_r^{X+\tau Y}(x))\cdot Y(\Psi_r^{X+\tau Y}(x))\,\mathrm{d}r$$

$$= \int_0^s \left(Y\Psi_{s-r}^{X+\tau Y}\right)(\Psi_r^{X+\tau Y}(x))\,\mathrm{d}r.$$

We now take $\tau = 0$; since t (which is a fixed time in $\mathcal{D}(X, x)$) belongs to any of the sets $\mathcal{D}(X+\tau Y, x)$ (as long as τ is conveniently small), we are contemporarily allowed to take $s = t$. Furthermore, we use the Chain Rule and the semigroup property $\Psi_{t-r}^X = \Psi_t^X \circ \Psi_{-r}^X$: we obtain

$$\frac{\mathrm{d}}{\mathrm{d}\tau}\Big|_{\tau=0}\Psi_t^{X+\tau Y}(x) = \int_0^t \left(Y\Psi_{t-r}^X\right)(\Psi_r^X(x))\,\mathrm{d}r$$

$$= \int_0^t \mathcal{J}_{\Psi_t^X}(\Psi_{-r}^X(\Psi_r^X(x)))\cdot(Y\Psi_{-r}^X)(\Psi_r^X(x))\,\mathrm{d}r$$

$$= \mathcal{J}_{\Psi_t^X}(x)\cdot\int_0^t (Y\Psi_{-r}^X)(\Psi_r^X(x))\,\mathrm{d}r$$

$$= \mathcal{J}_{\Psi_t^X}(x)\cdot\int_0^t (\mathrm{d}\Psi_{-r}^X Y)(x)\,\mathrm{d}r =: (\star).$$

Since we are dealing with X, Y in a finite-dimensional Lie sub-algebra of $\mathfrak{X}(\Omega)$, we can apply Thm 13.1, which ensures that (4.28a) holds true (see page 108); we are also entitled to integrate under the series sign, since the power series is convergent (see Thm. 4.27). We thus get

$$(\star) = \mathfrak{I}_{\Psi_t^X}(x) \cdot \int_0^t \sum_{j=0}^{\infty} ((\operatorname{ad} X)^j Y)(x) \frac{r^j}{j!} \, dr$$

$$= \mathfrak{I}_{\Psi_t^X}(x) \cdot \sum_{j=0}^{\infty} ((\operatorname{ad} X)^j Y)(x) \frac{t^{j+1}}{(j+1)!}$$

$$= \sum_{j=0}^{\infty} \frac{t^{j+1}}{(j+1)!} \, \mathfrak{I}_{\Psi_t^X}(x) \cdot ((\operatorname{ad} X)^j Y)(x)$$

$$= \sum_{j=0}^{\infty} \frac{t^{j+1}}{(j+1)!} ((\operatorname{ad} X)^j Y) \Psi_t^X(x).$$

This proves (13.6). $\qquad\qquad\qquad\qquad\qquad\qquad\qquad\qquad\qquad\qquad\square$

Definition 13.6 (Exponentiation of a finite-dim. Lie algebra of v.f.s). Let V be as in (13.1), and suppose that $\mathcal{B} = \{X_1, \ldots, X_m\}$ is a (linear) basis of V. For any $\xi \in \mathbb{R}^m$, let us briefly set

$$\xi \cdot X := \xi_1 X_1 + \cdots + \xi_m X_m, \quad \xi \in \mathbb{R}^m. \tag{13.10}$$

Let also $x \in \Omega$ and $t \in \mathbb{R}$ be fixed. Then, there exists $\varepsilon > 0$ (depending on V, $\mathcal{B}$, x and t) so small that the map

$$\xi \mapsto E_{t,x}(\xi) := \Psi_t^{\xi \cdot X}(x)$$

is well posed on the set $\{\xi \in \mathbb{R}^m : \|\xi\| \leq \varepsilon\}$. We say that $E_{t,x}$ is the **exponentiation of V** (at time t and starting point x) relative to the basis $\mathcal{B}$.

Remark 13.7. Suppose that $V \subset \mathfrak{X}(\Omega)$ is finite-dimensional, and that any v.f. in V is global. Let $X_1, \ldots, X_m$ be a (linear) basis of V, and let $x \in \Omega$ be fixed. Then the exponentiation function

$$F_x : \mathbb{R}^m \to \Omega, \quad F_x(\xi) := E_{1,x}(\xi) = \Psi_1^{\xi_1 X_1 + \cdots + \xi_m X_m}(x)$$

is well defined and smooth. By means of Thm. 13.4 we can compute the Jacobian matrix of F_x at any $\xi \in \mathbb{R}^m$. Indeed, using the notation of (13.10), given $i \in \{1, \ldots, m\}$, the i-th column-vector of $\mathfrak{I}_{F_x}(\xi)$ is

$$\frac{d}{d\tau}\bigg|_{\tau=0} \Psi_1^{\xi \cdot X + \tau X_i}(x) \stackrel{(13.9)}{=} \mathfrak{I}_{\Psi_1^{\xi \cdot X}}(x) \cdot \left(\frac{e^{\operatorname{ad}(\xi \cdot X)} - 1}{\operatorname{ad}(\xi \cdot X)} X_i \right)(x).$$

With a very abridged formalism, with the aid of the function $\varphi(\omega) := \frac{e^\omega - 1}{\omega}$, we get the matrix formula

$$\mathfrak{I}_{F_x}(\xi) = \mathfrak{I}_{\Psi_1^{\xi \cdot X}}(x) \cdot \varphi(\operatorname{ad}(\xi \cdot X)) \Big(X_1 \cdots X_m \Big)(x).$$

By means of Thm. 13.4, we can provide yet another formula for the differential of the exponentiation map for finite-dimensional Lie-algebras.

Corollary 13.8 (Differential of the exponentiation map). *Let V satisfy assumption (13.1), and let $E_{t,x}$ be the exponentiation map defined in Def. 13.6, which is well posed in a neighborhood U of $0 \in \mathbb{R}^m$, at least. When every vector field in V is global, $E_{t,x}$ is a C^∞ map defined on $U = \mathbb{R}^m$.*

Then for any $i \in \{1, \ldots, m\}$ and any $\xi \in U$, we have the identity

$$\frac{\partial E_{t,x}}{\partial \xi_i}(\xi) = \left(\sum_{j=0}^{\infty} \frac{t^{j+1}}{(j+1)!} \left(\mathrm{ad}\,(\xi \cdot X) \right)^j X_i \right) \Psi_t^{\xi \cdot X}(x). \tag{13.11}$$

If $\xi \in U$, the series is convergent for any $t \in \mathcal{D}(x, \xi \cdot X)$ (the latter coinciding with $\mathbb{R}$ if any v.f. in V is global). More compactly, we write this formula as

$$\frac{\partial E_{t,x}}{\partial \xi_i}(\xi) = \left(\frac{e^{t\,\mathrm{ad}\,(\xi \cdot X)} - 1}{\mathrm{ad}\,(\xi \cdot X)} X_i \right) \Psi_t^{\xi \cdot X}(x), \quad \xi \in U. \tag{13.12}$$

Proof. In the notation of the assertion, we have

$$\frac{\partial E_{t,x}}{\partial \xi_i}(\xi) = \frac{\mathrm{d}}{\mathrm{d}\tau}\bigg|_{\tau=0} \Psi_t^{\xi_1 X_1 + \cdots + (\xi_i + \tau)X_i + \cdots + \xi_m X_m}(x) = \frac{\mathrm{d}}{\mathrm{d}\tau}\bigg|_{\tau=0} \Psi_t^{\xi \cdot X + \tau X_i}(x).$$

As a consequence, (13.11) is a particular case of (13.6). $\qquad\qquad\square$

The rhs of (13.12) has to be intended as the componentwise action of

$$\frac{e^{t\,\mathrm{ad}\,(\xi \cdot X)} - 1}{\mathrm{ad}\,(\xi \cdot X)} X_i,$$

which is a vector field on Ω, on the $\mathbb{R}^N$-valued function (written as a column vector) $\Psi_t^{\xi \cdot X}$, the resulting function being computed at x.

13.3 The Exponential Theorem for ODEs

In the following statement –as usual– we denote by $\exp(X)(x)$ the integral curve, at time $t = 1$, of the vector field X starting at x when $t = 0$.

Theorem 13.9 (CBHD Theorem for ODEs). *Let V be a finite-dimensional Lie subalgebra of $\mathfrak{X}(\Omega)$, the smooth vector fields on the open set $\Omega \subseteq \mathbb{R}^N$, ant let $\| \cdot \|$ be a fixed norm on V. The following facts hold true.*

(a). There exists a positive real number ε, depending on $\|\cdot\|$, such that the homogeneous CBHD series (see Def. 2.24 on page 53)

$$Z(X,Y) := \sum_{n=1}^{\infty} Z_n(X,Y)$$

is convergent for every $X, Y \in V$ with $\|X\|, \|Y\| \leq \varepsilon$. Furthermore, for every $x \in \Omega$ there exists $\varepsilon(x) > 0$ (also depending on X, Y) such that the ODE identity

$$\exp(Y)\big(\exp(X)(x)\big) = \exp\big(Z(X,Y)\big)(x) \tag{13.13}$$

is fulfilled by any $X, Y \in V$ with $\|X\|, \|Y\| \le \varepsilon(x)$. If K is a compact subset of Ω, the same $\varepsilon(K) > 0$ can be taken in place of $\varepsilon(x)$, uniformly for any $x \in K$.

(b). If any vector field in V is a global vector field, then the above $\varepsilon(x)$ does not depend on $x \in \Omega$, but only on X and Y.

(c). Finally, if the CBHD series is convergent (in V) for any $X, Y \in V$ (this is the case, for example, if V is nilpotent) and if any vector field in V is a global vector field, then (13.13) holds true for every $X, Y \in V$ and every $x \in \Omega$, with no restriction on the smallness of $\|X\|, \|Y\|$.

Proof. Let $\| \cdot \|$ be a norm on V satisfying the Lie-sub-multiplicative property (5.1) (we can always assume that this is the case, by rescaling the existing norm on V; see Rem. 5.2 on page 114).

Owing to Thm. 5.3 (page 114) on the local convergence of the CBHD series, it is possible to find $\varepsilon > 0$ (depending on V and $\| \cdot \|$) such that the CBHD series $Z(X, Y) = \sum_{n=1}^{\infty} Z_n(X, Y)$ is convergent in V for $\|X\|, \|Y\| \le \varepsilon$. Obviously, if $Z(X, Y)$ converges for every $X, Y \in V$, then one can take $\varepsilon = \infty$. Again by the results in Thm. 5.3, it follows that the double series

$$Z(X, tY) = X + tY + \sum_{j=1}^{\infty} \left(\sum_{i=1}^{\infty} C_{i,j}(X, Y) \right) t^j$$

is also convergent in V whenever $\|X\|, \|Y\| \le \varepsilon$ and $t \in [0, 1]$.

Let now $x \in \Omega$ be fixed. By the results in App. B (see Thm. B.34 on page 373), there exists $\varepsilon(x) > 0$ (which we may assume to be $\le \varepsilon$) such that the maps

$$\begin{aligned} t &\mapsto F(t) := \exp(tY)\big(\exp(X)(x)\big) \\ t &\mapsto G(t) := \exp\big(Z(X, tY)\big)(x) \end{aligned} \tag{13.14}$$

are well posed for t in an open neighborhood of $[0, 1]$, whenever $\|X\|$ and $\|Y\|$ are $\le \varepsilon(x)$. Moreover, this $\varepsilon(x)$ does not depend on x but only on K, as long as K is a compact subset of Ω and $x \in K$. We skip any detail of this continuity argument for ODEs, but we observe that the estimate (see (5.6))

$$\|Z(X, Y)\| \le \log \left(\frac{1}{2 - e^{\|X\| + \|Y\|}} \right)$$

is also required to ensure the well posedness of the integral curve

$$s \mapsto \exp\big(s\, Z(X, tY)\big)(x),$$

up to $s = 1$, uniformly for any $t \in [0, 1]$ (and any X, Y with small norms).

In the case when any v.f. in V is global (hence also $Z(X, Y)$ is global for it belongs to V when it exists), then we can take $\varepsilon(x) = \varepsilon$ for any $x \in \Omega$ (where ε is as in the previous part of the proof), since $F(t)$ is defined for any $t \in \mathbb{R}$ and any $X, Y \in V$, whereas $G(t)$ is defined for any $t \in [0, 1]$ and whenever $\|X\|, \|Y\| \le \varepsilon$. In the further event that the series defining $Z(X, Y)$ converges for any $X, Y \in V$, then one can take $\varepsilon(x) = \varepsilon = \infty$, and $F(t)$ and $G(t)$ are defined for every $t \in \mathbb{R}$ and every $X, Y \in V$.

The proof will be complete if we show that $F(t) = G(t)$ for any $t \in [0,1]$: indeed the choice $t = 1$ will produce (13.13), while statements (b) and (c) of the theorem have already been justified. In order to prove that $F \equiv G$ on $[0,1]$, we show that they both solve the same Cauchy problem.

To this end, we trivially have $F(0) = \exp(X)(x) = G(0)$ (remember that $Z(X,0) = X$). The hardest part of the proof is obviously the differentiation of $F(t)$ and $G(t)$. We split the proof in two further steps.

Step I: obtaining $F'(t)$. This is a simple consequence of the definition of an integral curve (see also Thm. 1.25 on page 18), which gives

$$F'(t) = Y(F(t)), \quad t \in [0,1]. \tag{13.15}$$

Step II: obtaining $G'(t)$. For the differentiation of $G(t)$ we need the Poincaré-type ODEs in Thm. 5.8 and the differential of the flow in Thm. 13.8. Indeed, if we fix a basis $\{X_1, \ldots, X_m\}$ of V, there exist C^1 functions $\xi_1(t), \ldots, \xi_m(t)$ such that

$$Z(X,tY) = \sum_{j=1}^{m} \xi_j(t)\, X_j. \tag{13.16}$$

Using the notation of Thm. 13.8, we write $Z(X,tY) = \xi(t) \cdot X$, so that

$$G(t) = \Psi_1^{\xi(t) \cdot X}(x).$$

We have the following computation:

$$
\begin{aligned}
G'(t) &= \mathcal{J}_{\xi \,\mapsto\, \Psi_1^{\xi \cdot X}(x)}(\xi(t)) \cdot \xi'(t) \\[2mm]
&\overset{(13.12)}{=} \mathcal{J}_{\Psi_1^{\xi(t) \cdot X}}(x) \cdot \sum_{j=1}^{m}\left(\frac{e^{\operatorname{ad}(\xi(t) \cdot X)} - 1}{\operatorname{ad}(\xi(t) \cdot X)} X_j \right)(x) \cdot \xi'_j(t) \\[2mm]
&= \mathcal{J}_{\Psi_1^{\xi(t) \cdot X}}(x) \cdot \frac{e^{\operatorname{ad}(\xi(t) \cdot X)} - 1}{\operatorname{ad}(\xi(t) \cdot X)}\left(\sum_{j=1}^{m} \xi'_j(t)\, X_j \right)(x) =: (\star).
\end{aligned}
$$

We differentiate (13.16) wrt t; in doing this, we are entitled to use the notable Poincaré-type Thm. 5.8 (page 120), since V is (by assumption) finite-dimensional. We thus get (see precisely (5.20) in the mentioned theorem)

$$\sum_{j=1}^{m} \xi'_j(t)\, X_j = \frac{\mathrm{d}}{\mathrm{d}t} Z(X,tY) = \mathbf{b}(-\operatorname{ad} Z(X,tY))(Y).$$

With the short notation Z for $Z(X,tY)$, this gives

$$(\star) = \mathcal{J}_{\Psi_1^{Z}}(x) \cdot \frac{e^{\operatorname{ad} Z} - 1}{\operatorname{ad} Z}\Big(\mathbf{b}(-\operatorname{ad} Z) \Big) Y(x) =: (2\star).$$

Now we can invoke some basic tricks of formal-series manipulation, as explained in Sec. 2.8 (page 64): since (for $\omega \in \mathbb{C}$ near 0)

$$\frac{e^{\omega} - 1}{\omega} \cdot \mathbf{b}(-\omega) = \frac{e^{\omega} - 1}{\omega} \cdot \frac{-\omega}{e^{-\omega} - 1} = e^{\omega},$$

we can easily infer that

$$\frac{e^{\operatorname{ad} Z} - 1}{\operatorname{ad} Z}\Big(\mathbf{b}(-\operatorname{ad} Z)\Big) = e^{\operatorname{ad} Z}.$$

Therefore we obtain

$$(2\star) = \partial_{\Psi_1^Z}(x) \cdot \underbrace{(e^{\operatorname{ad} Z}Y)(x)}_{\mathrm{d}\Psi_{-1}^Z(Y)} = \partial_{\Psi_1^Z}(x) \cdot (\mathrm{d}\Psi_{-1}^Z Y)(x)$$

$$= (\mathrm{d}\Psi_{-1}^Z Y)\Psi_1^Z(x).$$

In the under-braced formula, we exploited (13.2). Summing up, we get

$$G'(t) = (\mathrm{d}\Psi_{-1}^Z Y)\Psi_1^Z(x) \overset{(4.12)}{=} Y\underbrace{(\Psi_1^Z \circ \Psi_{-1}^Z)}_{=I}(\Psi_1^Z(x)) = Y(\Psi_1^Z(x)).$$

Since $\Psi_1^Z(x) = \Psi_1^{Z(X,tY)}(x) = \exp(Z(X,tY))(x) = G(t)$ (see (13.14)), we deduce that $G'(t) = Y(G(t))$. By comparison with (13.15), we see that F and G satisfy the same ODE, which is what we intended to prove. $\qquad\square$

13.4   Exercises of Chap. 13

Exercise 13.1. Test the validity of formula (13.6) for the vector fields in $\mathbb{R}^3$

$$X = \partial_{x_1} + 2\,x_2\,\partial_{x_3}, \quad Y = \partial_{x_2} - 2\,x_1\,\partial_{x_3}.$$

First check that they belong to a finite dimensional Lie-subalgebra of $\mathcal{X}(\mathbb{R}^3)$.

[*Hint:* Prove that $\Psi_t^{X+\tau Y}(x) = (x_1 + t, x_2 + t\,\tau, x_3 + 2x_2 t - 2x_1 t\tau)$ and that both sides of (13.6) are equal to $(0, t, -2x_1 t)$.]

Chapter 14

The Exponential Theorem for Lie Groups

THE aim of this chapter is to derive an analog of the Exponential Theorem of Chap. 2 in the setting of Lie groups $(\mathbb{G}, *)$. We obtain a Campbell-Baker-Hausdorff-Dynkin Theorem for $\mathbb{G}$, concerning the group-product of two group-Exponentials:
$$\mathrm{Exp}(X) * \mathrm{Exp}(Y) = \mathrm{Exp}(X \diamond Y),$$
holding true for every $X, Y \in \mathrm{Lie}(\mathbb{G})$ sufficiently close to the null v.f. As usual, $X \diamond Y$ denotes the CBHD series $X + Y + \frac{1}{2}[X, Y] + \cdots$, which is well posed near 0 in the finite-dimensional Lie algebra $\mathrm{Lie}(\mathbb{G})$ (see Chap. 5).

It is time to fix a summary of the CBHD Theorems in this book:

(1) $e^x e^y = e^{x \diamond y}$ in the algebra of the formal power series in the indeterminates x, y (Chap. 2): here the product is that of formal power series;

(2) $e^Y(e^X(x)) = e^{X \diamond Y}(x)$, holding true for v.f.s X, Y close to the null v.f. in a finite-dimensional Lie algebra V of smooth v.f.s on some open set (see Chap. 13): here the product is the composition of flows;

(3) $\mathrm{Exp}(X) * \mathrm{Exp}(Y) = \mathrm{Exp}(X \diamond Y)$, which we prove in this chapter, for X, Y in $\mathrm{Lie}(\mathbb{G})$ and near 0; here the product is that of the Lie group $\mathbb{G}$.

The cornerstone in proving the latter CBHD identity is furnished by the following identity (a basic fact of Lie group theory; see App. C) valid for every $X, Y \in \mathrm{Lie}(\mathbb{G})$ and for time t close to 0,
$$\mathrm{Exp}(tX) * \mathrm{Exp}(tY) = \Psi_t^Y\big(\Psi_t^X(e)\big),$$
where e is the identity element of $\mathbb{G}$. Thus we recognize that the group product in (3) is given by a composition of flows as in (2): this allows us to reduce the CBHD Theorem on a Lie group to the CBHD Theorem for ODEs, fully investigated in Chap. 13. As a side result of the contents of Chap. 13, we also obtain a formula for the differential of the Exponential Map of $\mathbb{G}$.

The main prerequisite for this chapter is the knowledge of some basic Lie group theory (see App. C), plus some results on analytic functions. As remarked, the CBHD Theorem for ODEs of Chap. 13 is also needed.

277

14.1 The differential of the Exponential Map

Throughout the chapter, $\mathbb{G}$ will denote a Lie group, and N its dimension (as a manifold). We denote by $\mathfrak{g}$ the Lie algebra $\mathrm{Lie}(\mathbb{G})$ of $\mathbb{G}$.

Since we shall frequently perform computations in coordinates, we fix once and for all the notation. By $\mathcal{C} = (U, (x_i))$ we shall denote any (fixed) coordinate chart around the identity element e of $\mathbb{G}$, and $\partial/\partial x_1|_p, \ldots, \partial/\partial x_N|_p$ denote the associated coordinate (tangent) vectors at $p \in U$. It is known that there exist a uniquely-determined basis $\mathcal{J} = (J_1, \ldots, J_N)$ of left invariant v.f.s on $\mathbb{G}$ (depending on $\mathcal{C}$) such that

$$J_i|_e = \frac{\partial}{\partial x_i}\Big|_e, \quad i = 1, \ldots, N.$$

We say that $\mathcal{J}$ is the **Jacobian basis** for $\mathfrak{g}$ associated with the chart $\mathcal{C}$.

Proposition 14.1 (Local invertibility of Exp**).** *The map* $\mathrm{Exp} : \mathfrak{g} \to \mathbb{G}$ *is a local diffeomorphism of an open neighborhood of* $0 \in \mathfrak{g}$ *onto an open neighborhood of the neutral element* $e \in \mathbb{G}$.

Proof. Since the result is local, we can work in a local chart $(U, (x_i))$ centred at e, and identify U with an open neighborhood of 0 in $\mathbb{R}^N$. We further identify $\mathfrak{g}$ with $\mathbb{R}^N$ by fixing linear coordinates wrt the Jacobian basis $J_1, \ldots, J_N$ associated with this chart. In view of the Inverse Function Theorem, the assertion will follow if we show that the map

$$\Phi : \mathbb{R}^N \longrightarrow \mathbb{R}^N, \quad \Phi(\alpha_1, \ldots, \alpha_N) := \mathrm{Exp}(\alpha_1 J_1 + \cdots + \alpha_N J_N)$$

has an invertible Jacobian matrix at $\alpha = (\alpha_1, \ldots, \alpha_N) = 0$. To this end, consider the following computation:

$$\Phi(\alpha) = \exp\left(\sum_i \alpha_i J_i\right)(0) \qquad (\text{see Thm. } 1.25)$$

$$= \sum_i \alpha_i J_i(0) + \int_0^1 (1-s)\left(\sum_i \alpha_i J_i\right)^2 I\left(\gamma(s, \sum_i \alpha_i J_i, 0)\right) \mathrm{d}s.$$

It is not difficult[1] to prove that the integral remainder is a $\mathcal{O}(\|\alpha\|^2)$, when $\alpha \to 0$. Thus $\Phi(\alpha) = \alpha + \mathcal{O}(\|\alpha\|^2)$, as $\alpha \to 0$, which implies at once that $\mathcal{J}_\Phi(0)$ is the identity matrix of order N. $\qquad\square$

We now give an improvement of Prop. 14.1. We temporarily identify the Lie algebra $\mathfrak{g}$ with $T_e\mathbb{G}$, the tangent space at the identity element e of $\mathbb{G}$, via the isomorphism (of vector spaces):

$$\alpha : \mathfrak{g} \longrightarrow T_e\mathbb{G}, \qquad X \mapsto X_e. \tag{14.1}$$

---

[1] Beware that α also appears in $\gamma(s, \sum_i \alpha_i J_i, 0)$. In order to prove that

$$\left(\sum_i \alpha_i J_i\right)^2 I\left(\gamma(s, \sum_i \alpha_i J_i, 0)\right) = \mathcal{O}(\|\alpha\|^2), \quad \text{as } \alpha \to 0,$$

uniformly for $s \in [0, 1]$, one needs to prove first that $\gamma(s, \sum_i \alpha_i J_i, 0)$ lies in some compact subset of $\mathbb{R}^N$. This follows from general ODE results; see App. B.

Basic results of Lie group theory ensure that α^{-1} sends $\mathbf{v} \in T_e\mathbb{G}$ to the left invariant vector field acting on a smooth f as follows:

$$f \mapsto \alpha^{-1}(\mathbf{v})f(x) = \mathbf{v}(f \circ \tau_x), \quad \forall\, x \in \mathbb{G}. \tag{14.2}$$

The usual Lie bracket $[\cdot, \cdot]$ on $\mathfrak{g}$ (i.e., the Lie bracket of vector fields) is turned by α into an operation on $T_e\mathbb{G}$, obviously endowing $T_e\mathbb{G}$ with the structure of a Lie algebra isomorphic to $\mathfrak{g}$, and such that α is a Lie algebra isomorphism. We denote this operation by $[\cdot, \cdot]_e$: explicitly

$$[\mathbf{u}, \mathbf{v}]_e := \alpha\Big([\alpha^{-1}(\mathbf{u}), \alpha^{-1}(\mathbf{v})]\Big), \qquad \text{for every } \mathbf{u}, \mathbf{v} \in T_e\mathbb{G}. \tag{14.3}$$

If $(U, (x_i))$ is a fixed local chart around e, any element $\mathbf{v}$ of $T_e\mathbb{G}$ has the form

$$\mathbf{v} = v_1 \left.\frac{\partial}{\partial x_1}\right|_e + \cdots + v_N \left.\frac{\partial}{\partial x_N}\right|_e, \tag{14.4}$$

for a uniquely determined N-tuple of real numbers $(v_1, \ldots, v_N)$ (depending on the chart). Since we are about to provide a computation in the local chart $(U, (x_i))$, U is identified with an open set in $\mathbb{R}^N$, and (with a mild abuse of notation) e denotes the point of $\mathbb{R}^N$ corresponding to the identity element of $\mathbb{G}$ in this chart. Accordingly, m denotes the coordinate expression of the multiplication of $\mathbb{G}$ on U:

$$m(\alpha, \beta) := \alpha * \beta, \quad \alpha, \beta \in U.$$

We are ready to prove a useful lemma, showing that the mixed partial derivatives in the Hessian matrix of m at (e, e) suffice to determine (in a very precise way) the Lie algebra structure of $\mathfrak{g}$. The notion of structure constants can be found in Def. A.14 on page 345.

Lemma 14.2. *With all the above notation, we have*

$$\left[\sum_{i=1}^{N} u_i \left.\frac{\partial}{\partial x_i}\right|_e, \sum_{j=1}^{N} v_j \left.\frac{\partial}{\partial x_j}\right|_e \right]_e = \sum_{h=1}^{N} \left(\sum_{i,j=1}^{N} (u_i\, v_j - u_j\, v_i) \frac{\partial^2 m_h(e, e)}{\partial \alpha_i\, \partial \beta_j} \right) \left.\frac{\partial}{\partial x_h}\right|_e, \tag{14.5}$$

for every $u_i, v_i \in \mathbb{R}$ $(i = 1, \ldots, N)$.

Equivalently, the structure constants of the Lie algebra $(T_e\mathbb{G}, [\cdot, \cdot]_e)$ with respect to the basis $\partial/\partial x_1|_e, \ldots, \partial/\partial x_N|_e$ are given by

$$\left[\left.\frac{\partial}{\partial x_i}\right|_e, \left.\frac{\partial}{\partial x_j}\right|_e \right]_e = \sum_{h=1}^{N} \left(\frac{\partial^2 m_h(e, e)}{\partial \alpha_i\, \partial \beta_j} - \frac{\partial^2 m_h(e, e)}{\partial \alpha_j\, \partial \beta_i} \right) \left.\frac{\partial}{\partial x_h}\right|_e, \tag{14.6}$$

for any $i, j \in \{1, \ldots, N\}$.

Proof. For any smooth f near e we have (with the notation of (14.4) for $\mathbf{u}, \mathbf{v}$):

$$[\mathbf{u}, \mathbf{v}]_e f \overset{(14.3)}{=} \big[\alpha^{-1}(\mathbf{u}), \alpha^{-1}(\mathbf{v})\big] f(e)$$

$$= (\alpha^{-1}(\mathbf{u}))\big|_e (\alpha^{-1}(\mathbf{v})(f)) - \{\text{analogous, interchange } \mathbf{u}, \mathbf{v}\}$$

$$\overset{(14.2)}{=} \mathbf{u}\big(x \mapsto \mathbf{v}(f \circ \tau_x)\big) - \{\text{analogous, interchange } \mathbf{u}, \mathbf{v}\}$$

$$= \left(\sum_{i=1}^{N} u_i \left.\frac{\partial}{\partial x_i}\right|_e \right)\left\{ \left(\sum_{j=1}^{N} v_j \left.\frac{\partial}{\partial y_j}\right|_e \right) f(x * y) \right\} - \{\text{analogous} \ldots\}$$

$$= \sum_{i,j=1}^{N} u_i\, v_j \left.\frac{\partial}{\partial x_i}\right|_e \left.\frac{\partial}{\partial y_j}\right|_e \{f(x * y)\} - \left\{ \begin{array}{c} \text{analogous,} \\ \text{interchange } u_i \text{ with } v_i \end{array} \right\}.$$

We have, by definition of $\partial/\partial x_i$ and of m,

$$\frac{\partial}{\partial x_i}\Big|_e \frac{\partial}{\partial y_j}\Big|_e \{f(x*y)\} = \frac{\partial}{\partial \alpha_i}\Big|_e \frac{\partial}{\partial \beta_j}\Big|_e \{f(m(\alpha,\beta))\}$$

$$= \frac{\partial}{\partial \alpha_i}\Big|_e \sum_{h=1}^{N} \frac{\partial f}{\partial x_h}(m(\alpha,e)) \frac{\partial m_h}{\partial \beta_j}(\alpha,e)$$

(again by the chain rule, together with $m(\alpha,e) = \alpha$)

$$= \sum_{h=1}^{N} \frac{\partial^2 f}{\partial x_i\,\partial x_h}(e)\frac{\partial m_h}{\partial \beta_j}(e,e) + \sum_{h=1}^{N}\frac{\partial f}{\partial x_h}(e)\frac{\partial^2 m_h}{\partial \alpha_i\,\partial \beta_j}(e,e).$$

Now note that one has (as $m_h(e,\beta) = \beta_h$)

$$\frac{\partial m_h}{\partial \beta_j}(e,e) = \frac{\partial}{\partial \beta_j}\{m_h(e,\beta)\} = \frac{\partial}{\partial \beta_j}\beta_h = \delta_{j,h} \quad \text{(Kronecker symbol)}.$$

As a consequence we have the following formula

$$\frac{\partial}{\partial x_i}\Big|_e \frac{\partial}{\partial y_j}\Big|_e \{f(x*y)\} = \frac{\partial^2 f}{\partial x_i\,\partial x_j}(e) + \sum_{h=1}^{N}\frac{\partial f}{\partial x_h}(e)\frac{\partial^2 m_h}{\partial \alpha_i\,\partial \beta_j}(e,e).$$

Going back to the computation for $[\mathbf{u},\mathbf{v}]_e f$, we obtain

$$[\mathbf{u},\mathbf{v}]_e f = \sum_{i,j=1}^{N} u_i\, v_j \frac{\partial^2 f(e)}{\partial x_i\,\partial x_j} + \sum_{i,j,h=1}^{N} u_i\, v_j \frac{\partial f(e)}{\partial x_h}\frac{\partial^2 m_h(e,e)}{\partial \alpha_i\,\partial \beta_j} +$$

$$- \{\text{analogous, interchange } u_i \text{ with } v_i\}$$

$$= \sum_{i,j=1}^{N} u_i\, v_j \frac{\partial^2 f(e)}{\partial x_i\,\partial x_j} - \sum_{i,j=1}^{N} v_i\, u_j \frac{\partial^2 f(e)}{\partial x_i\,\partial x_j} +$$

$$+ \sum_{i,j,h=1}^{N} u_i\, v_j \frac{\partial f(e)}{\partial x_h}\frac{\partial^2 m_h(e,e)}{\partial \alpha_i\,\partial \beta_j} - \sum_{i,j,h=1}^{N} v_i\, u_j \frac{\partial f(e)}{\partial x_h}\frac{\partial^2 m_h(e,e)}{\partial \alpha_i\,\partial \beta_j}$$

(the first two sums cancel each other, by Schwarz's Theorem!)

$$= \sum_{h=1}^{N}\left(\sum_{i,j=1}^{N}(u_i\, v_j - u_j\, v_i)\frac{\partial^2 m_h(e,e)}{\partial \alpha_i\,\partial \beta_j}\right)\frac{\partial f(e)}{\partial x_h}.$$

This gives the desired (14.5), by the very definition of $(\partial/\partial x_h)|_e$. Obviously, (14.6) is a particular case of (14.5). $\qquad\square$

In the following statement, m denotes the coordinate expression of the multiplication of $\mathbb{G}$, using the notation preceding Lem. 14.2.

Corollary 14.3. *Let* $(U,(x_i))$ *be a chart around* e, *and let* $\mathfrak{J} = (J_1,\ldots,J_N)$ *be the associated Jacobian basis of* $\mathfrak{g}$. *The structure constants of* $\mathfrak{g}$ *wrt* $\mathfrak{J}$ *are given by*

$$[J_i, J_j] = \sum_{h=1}^{N}\left(\frac{\partial^2 m_h(e,e)}{\partial \alpha_i\,\partial \beta_j} - \frac{\partial^2 m_h(e,e)}{\partial \alpha_j\,\partial \beta_i}\right)J_h,$$

for any $i,j \in \{1,\ldots,N\}$.

Proof. This follows from (14.6), taking into account that

$$\alpha^{-1}\left(\frac{\partial}{\partial x_i}\Big|_e\right) = J_i, \quad i = 1, \ldots, N,$$

as one can derive from (14.1). $\qquad\square$

Theorem 14.4 (Differential of Exp**).** *Let* $\mathbb{G}$ *be a Lie group with Exponential Map* Exp *and identity element* e*. Let* $X, Y \in \mathfrak{g}$*. Then*

$$\frac{\mathrm{d}}{\mathrm{d}\tau}\Big|_{\tau=0} \mathrm{Exp}(X + \tau Y) = \mathrm{d}_e \rho_{\mathrm{Exp}(X)}\left(\frac{e^{\mathrm{ad}\,X} - 1}{\mathrm{ad}\,X}Y\right)_e.$$

The above identity is intended as an equality of two elements of $T_{\mathrm{Exp}(X)}\mathbb{G}$*.*

Proof. This is a direct consequence of Thm. 13.4, once one takes into account the following ingredients:

- Thm. 12.4 can be adapted to the smooth manifold case (see Appendix B in the book by [Duistermaat and Kolk (2000)]);
- $\rho_{\mathrm{Exp}(X)} = \Psi_1^X$ for any $X \in \mathfrak{g}$ (see (C.10) on page 391).

This justifies the following computation:

$$\frac{\mathrm{d}}{\mathrm{d}\tau}\Big|_{\tau=0} \Psi_1^{X+\tau Y}(e) = \mathrm{d}_e \Psi_1^X\left(\frac{e^{\mathrm{ad}\,X} - 1}{\mathrm{ad}\,X}Y\right)_e,$$

which ends the proof. $\qquad\square$

Remark 14.5. The (well-posed) vector field $\varphi(\mathrm{ad}\,X)(Y) = \frac{e^{\mathrm{ad}\,X}-1}{\mathrm{ad}\,X}Y$ is left invariant on $\mathbb{G}$; hence its value at the identity element e vanishes if and only if $\varphi(\mathrm{ad}\,X)(Y) = 0$ in $\mathfrak{g}$. The possible degeneracies of the Exponential Map are therefore characterized by the v.f.s $X \in \mathfrak{g}$ for which the linear map $\varphi(\mathrm{ad}\,X)$ has nontrivial kernel. The non-triviality of this kernel is equivalent to the fact that the (complex) non-vanishing eigenvalues of $\mathrm{ad}\,X$ belong to $\{2k\pi i : k \in \mathbb{Z} \setminus \{0\}\}$.

14.2 The Exponential Theorem for Lie groups

Throughout this section, if we make reference to a norm or to an ODE on $\mathfrak{g}$, or to real-analytic/smooth functions defined and/or valued in $\mathfrak{g}$, we shall always tacitly mean that we have identified $\mathfrak{g}$ with $\mathbb{R}^N$, via the choice of some fixed linear basis on $\mathfrak{g}$, the choice of the particular basis being immaterial.

We are ready for the main result of this chapter.

Theorem 14.6 (The CBHD Theorem for Lie Groups). *Let* $\mathbb{G}$ *be a Lie group. Then there exists a neighborhood* $\mathfrak{D} \subseteq \mathfrak{g}$ *of* 0 *(the zero vector field) such that the homogeneous Campbell-Baker-Hausdorff-Dynkin series* $X \diamond Y = \sum_{n=1}^{\infty} Z_n(X, Y)$ *is uniformly convergent on* $\mathfrak{D} \times \mathfrak{D}$*. Furthermore, by possibly shrinking* $\mathfrak{D}$*, we have the identity*

$$\mathrm{Exp}(X) * \mathrm{Exp}(Y) = \mathrm{Exp}(X \diamond Y), \qquad \text{for every } X, Y \in \mathfrak{D}. \tag{14.7}$$

If $\mathbb{G}$ is nilpotent, then the series $X \diamond Y$ converges for every $X, Y \in \mathfrak{g}$ (for it is a finite sum) and formula (14.7) holds for every $X, Y \in \mathfrak{g}$.

*We shall call (14.7) the **Campbell-Baker-Hausdorff-Dynkin Formula on the Lie group $\mathbb{G}$** (CBHD Formula, for short).*

Proof. Since $\mathfrak{g}$ is N-dimensional, the first assertion on the uniform (actually, normal) convergence of the CBHD series follows from Thm. 5.3 on page 114. Indeed, it suffices to fix any norm $\|\cdot\|$ on $\mathfrak{g}$, to fix $M > 0$ such that

$$\big\|[a, b]\big\| \leq M \, \|a\| \cdot \|b\|, \quad \text{for every } a, b \in \mathfrak{g},$$

and then we can define $\mathfrak{D}$ as the following ball centred at 0:

$$\mathfrak{D} := \left\{ X \in \mathfrak{g} : \|X\| \leq \frac{\log 2}{3\, M} \right\}. \tag{14.8}$$

This has been demonstrated in the proof of Thm. 5.3. The proof of (14.7) is crucially based on the following fact: by basic Lie group theory we know that left invariant v.f.s are global, and we have

$$\mathrm{Exp}(tX) * \mathrm{Exp}(tY) = \mathrm{Exp}(tX) * \Psi_t^Y(e) = \Psi_t^Y(\mathrm{Exp}(tX)) = \Psi_t^Y\big(\Psi_t^X(e)\big). \tag{14.9}$$

Hence, we recognize that the group product $\mathrm{Exp}(tX) * \mathrm{Exp}(tY)$ is given by a composition of flows: *this allows us to reduce the CBHD Theorem on a Lie group to the CBHD Theorem for ODEs!* Indeed, owing to Thm. 13.9 on page 273, we can choose $\varepsilon > 0$ so small that (14.7) holds true for every X, Y in $\mathfrak{D}_1 := \left\{ X \in \mathfrak{g} : \|X\| \leq \varepsilon \right\}$. The assertion for nilpotent groups is also a consequence of Thm. 13.9, since in this case the CBHD series $X \diamond Y$ is convergent for any X, Y. $\qquad\square$

14.3   An alternative approach with analytic functions

In this section we give a simple and self-contained proof of Thm. 14.6 for *analytic Lie groups*, and we add some comments in the C^∞ case.

For the sake of simplicity, we assume that the manifold underlying $\mathbb{G}$ is $\mathbb{R}^N$, so that we can ignore any problem in defining real analytic functions valued in $\mathbb{G}$. Let us suppose that $\mathbb{G} = (\mathbb{R}^N, *)$ and that the map

$$\mathbb{G} \times \mathbb{G} \ni (x, y) \mapsto x * y^{-1} \in \mathbb{G}$$

is of class C^ω (in the sense that its real-valued component functions are real analytic on $\mathbb{R}^{2N}$). Since $\mathfrak{g} = \mathrm{Lie}(\mathbb{G})$ is completely determined by the C^ω map $x \mapsto \mathrm{d}_e \tau_x$, we recognize that any vector field $X \in \mathfrak{g}$ has C^ω coefficients. By general results of ODE Theory, any integral curve of X is therefore real-analytic too, as an $\mathbb{R}^N$-valued curve. Furthermore, by the ODE-dependence facts, we deduce that $\mathrm{Exp} : \mathfrak{g} \to \mathbb{G}$ is of class C^ω as well.

This shows that, for every $X, Y \in \mathfrak{g}$, the map

$$\mathbb{R} \ni t \mapsto F(t) := \mathrm{Exp}(tX) * \mathrm{Exp}(tY)$$

is of class C^ω. Due to (14.9) we have $F(t) = \exp(tY)(\exp(tX)(e))$. We use the notation of the *incipit* of the proof of Thm. 14.6, and we take $X, Y \in \mathfrak{D}$, where $\mathfrak{D}$ is as in (14.8). Note that (by the homogeneity of $\|\cdot\|$) $tX, tY \in \mathfrak{D}$ for every $t \in [-1, 1]$, whence $(tX) \diamond (tY)$ is well defined. Due to the homogeneity of the polynomial function Z_h, one has

$$g(t) := (tX) \diamond (tY) = \sum_{h=1}^\infty Z_h(tX, tY) = \sum_{h=1}^\infty Z_h(X, Y)\, t^h.$$

Since the series $\sum_{h=1}^\infty \|Z_h(X, Y)\|$ is convergent (as we showed in the proof of Thm. 5.3), we deduce that $g(t)$ is a real-analytic function of $t \in [-1, 1]$ (or of t in some open neighborhood of $[-1, 1]$) valued in $\mathfrak{g}$.

Now we have another crucial fact: arguing as in the proof of Thm. 3.6 (page 78), one can obtain, for any arbitrary $n \in \mathbb{N}$,

$$\exp(tY)(\exp(tX)(e)) = \exp(g(t))(e) + \mathcal{O}_n(t^{n+1}), \quad \text{as } t \to 0.$$

Beware that $\mathcal{O}_n$ depends on n, so (even for $|t| < 1$) we cannot simply let n go to ∞ in the above identity!

By setting $G(t) := \mathrm{Exp}(g(t)) = \exp(g(t))(e)$, we therefore have

$$F(t) = G(t) + \mathcal{O}_n(t^{n+1}), \quad \text{as } t \to 0.$$

This yields the equality of $\dfrac{\mathrm{d}^k}{\mathrm{d}t^k}F(0)$ with $\dfrac{\mathrm{d}^k}{\mathrm{d}t^k}G(0)$, for every $k = 0, 1, \ldots, n$.

By the arbitrariness of $n \in \mathbb{N}$ we derive that

$$\frac{\mathrm{d}^k}{\mathrm{d}t^k}F(0) = \frac{\mathrm{d}^k}{\mathrm{d}t^k}G(0), \qquad \text{for every } k \geq 0.$$

Next, by the C^ω regularity of F and of G (note that G is the composition of two analytic functions, Exp and g) the equality of all the derivatives of F and of G at 0 implies that F and G coincide throughout their largest common domain of definition, which contains $[-1, 1]$.

In particular, $F(1) = G(1)$, and this identity is exactly (14.7).

Remark 14.7. Another proof of Thm. 14.6 can be given, via the following argument; roughly put, instead of the ODE identity

$$\exp(Y)\big(\exp(X)(x)\big) = \exp\big(X \diamond Y\big)(x),$$

we use the group-identity (of independent interest)

$$\mathrm{Log}\big(\mathrm{Exp}(tX) * \mathrm{Exp}(tY)\big) = (tX) \diamond (tY),$$

for X, Y and t specified below.

Indeed, let $X, Y \in \mathfrak{D}$ in (14.8). We also suppose that X, Y belong to a smaller neighborhood $\mathfrak{D}_1$ of 0 contained in $\mathfrak{D}$ in such a way that

$$[-1, 1] \ni t \mapsto Z(t) := \mathrm{Log}(\mathrm{Exp}(tX) * \mathrm{Exp}(tY)) \tag{14.10}$$

is well defined, where Log is the inverse function of the restriction of Exp to a neighborhood of $0 \in \mathfrak{g}$ (see Prop. 14.1). The existence of the above $\mathfrak{D}_1$ follows from a simple continuity argument. We think of the map $t \mapsto Z(t)$ as a smooth curve in $\mathfrak{g}$, passing through 0 when $t = 0$.

We can suppose that $\mathfrak{D}_1$ is contained in a small neighborhood $\mathfrak{D}_2$ of $0 \in \mathfrak{g}$ satisfying the following property: for every $X, Y \in \mathfrak{D}_2$ one has

$$|||\mathrm{ad}\, Z(t)||| < 2\pi, \quad \text{for every } t \in [-1, 1]. \tag{14.11}$$

The existence of such a $\mathfrak{D}_2$ follows from a continuity argument.[2] Then

$$\mathbf{b}(\mathrm{ad}\,(Z(t))) := \sum_{n=0}^{\infty} \frac{B_n}{n!} \left(\mathrm{ad}\, Z(t)\right)^n$$

is convergent in the Banach space $(\mathrm{End}(\mathfrak{g}), ||| \cdot |||)$ and it therefore defines an endomorphism of $\mathfrak{g}$ (see Exr. 14.1). As usual, the B_n's are the Bernoulli numbers (see Exrs. 5.7 and 5.8, page 129).

By the very definition of $Z(t)$ it follows that

$$\mathrm{Exp}(Z(t)) = \mathrm{Exp}(tX) * \mathrm{Exp}(tY), \quad t \in [-1, 1]. \tag{14.12}$$

By means of Thm. 13.4, we are able to take derivatives of both sides of (14.12); due to (14.11), we can further discover that $t \mapsto Z(t)$ is a solution, at least on $[-1, 1]$, of the following Cauchy problem in $\mathfrak{g}$:

$$(\mathrm{CP}): \quad Z' = \mathbf{b}(\mathrm{ad}\, Z)(X) + \mathbf{b}(-\mathrm{ad}\, Z)(Y), \quad Z(0) = 0, \tag{14.13}$$

where $\mathbf{b}$ is, as usual, the complex function $\mathbf{b}(\omega) = \frac{\omega}{e^\omega - 1}$. We recognize the notable Poincaré ODE, investigated in Thm. 5.7 on page 119. From that very theorem we infer that $Z(t)$ is of class C^ω wrt t on $[-1, 1]$, and (see (5.15))

$$Z(t) = \sum_{n=1}^{\infty} Z_n(X, Y)\, t^n = \sum_{n=1}^{\infty} Z_n(tX, tY) = (tX) \diamond (tY).$$

We highlight that the C^ω regularity of $Z(t)$ cannot be derived by its very definition (14.10) in the case of arbitrary (not necessarily C^ω) Lie groups. By taking $t = 1$ in the above formula we get

$$\mathrm{Log}(\mathrm{Exp}(X) * \mathrm{Exp}(Y)) = Z(1) = X \diamond Y,$$

for every $X, Y \in \mathfrak{D}_1$. By taking the Exponential on both sides we obtain (14.7). $\quad \natural$

---

[2]It is useful to observe that, fixing a norm $|| \cdot ||$ on $\mathfrak{g}$, by the continuity of $(a, b) \mapsto [a, b]$, we can choose $\delta_1 > 0$ so small that $||Z|| < \delta_1$ implies $|||\mathrm{ad}\, Z||| < 2\pi$ (see Exr. 14.2). Then one chooses $\delta_2 > 0$ so small that $||\mathrm{Log}(\mathrm{Exp}(X) * \mathrm{Exp}(Y))|| < \delta_1$ for every $X, Y \in \mathfrak{g}$ satisfying $||X||, ||Y|| < \delta_2$.

14.4 Exercises of Chap. 14

Exercise 14.1. Let $(V, \|\cdot\|)$ be a normed vector space over $\mathbb{K}$ ($= \mathbb{R}$ or $\mathbb{C}$). Let $\mathrm{End}(V)$ be the vector space of the endomorphisms of V.

(1). Prove that $|||A||| := \sup_{\|\xi\|\leq 1} \|A\xi\|$ defines a norm on $\mathrm{End}(V)$, which is called the *operator norm* associated with $\|\cdot\|$. Show that if $(V, \|\cdot\|)$ is complete then the same is true of the space $(\mathrm{End}(V), ||| \cdot |||)$.

(2). Prove that $||| \cdot |||$ is sub-multiplicative wrt the composition of maps $\circ$, i.e.,

$$|||A \circ B||| \leq |||A||| \cdot |||B|||, \quad \text{for all } A, B \in \mathrm{End}(V).$$

Deduce that the map $\mathrm{End}(V) \times \mathrm{End}(V) \ni (A, B) \mapsto A \circ B \in \mathrm{End}(V)$ is bilinear and continuous (on its domain we consider the product topology).

[*Hint:* It may be useful to use Exr. 5.2 on page 128.]

(3). Let $(V, \|\cdot\|)$ be complete. Let $\sum_{n=0}^{\infty} a_n z^n$ be a power series in $\mathbb{K}$. Suppose it has radius of convergence $\varrho \in (0, +\infty]$. Prove that, for every $A \in \mathrm{End}(V)$ satisfying $|||A||| < \varrho$, the series $\sum_{n=0}^{\infty} a_n A^n$ is convergent in $(\mathrm{End}(V), |||\cdot|||)$. Here A^n denotes the n-fold composition of A with itself.

(4). With the same hypothesis as in (3), if $f(z)$ denotes the function defined by $f(z) := \sum_{n=0}^{\infty} a_n z^n$ (for z belonging to the disc in $\mathbb{K}$ centred at 0 with radius ϱ), and if we denote by $f(A)$ the sum of the series $\sum_{n=0}^{\infty} a_n A^n$ in $\mathrm{End}(V)$ (for $A \in \mathrm{End}(V)$ satisfying $|||A||| < \varrho$), show that

$$f(A)\xi = \sum_{n=0}^{\infty} a_n A^n \xi \qquad (\xi \in V),$$

the latter series being convergent in V.

[*Hint:* Start showing that if $S_n \xrightarrow{n\to\infty} S$ in the space $(\mathrm{End}(V), ||| \cdot |||)$ then $S_n \xi \xrightarrow{n\to\infty} S$ in $(V, \| \cdot \|)$, for every $\xi \in V$.]

(5). With the above notation, prove that $|||f(A)||| \leq \sum_{n=0}^{\infty} |a_n| \cdot |||A|||^n$, for every $A \in \mathrm{End}(V)$ satisfying $|||A||| < \varrho$.

Exercise 14.2. Let $\mathfrak{g}$ be a real or complex Lie algebra and suppose that $\mathfrak{g}$ is equipped with a norm $\| \cdot \|$ such that

$$\big\|[a, b]\big\| \leq M \, \|a\| \cdot \|b\| \qquad \forall \, a, b \in \mathfrak{g},$$

for some constant $M > 0$. (By Exr. 5.2 this holds true iff $(a, b) \mapsto [a, b]$ is continuous; for example, if $\mathfrak{g}$ is finite-dimensional this is always the case.)

Prove that the linear map $\mathrm{ad} : \mathfrak{g} \to \mathrm{End}(\mathfrak{g})$ is continuous, where $\mathfrak{g}$ is equipped with the topology induced by the norm $\| \cdot \|$, while $\mathrm{End}(\mathfrak{g})$ is equipped with the topology induced by the operator norm induced by $\| \cdot \|$ (see the previous exercise).

[*Hint:* Show that $|||\mathrm{ad}\, a||| \leq M \, \|a\|$ and that this suffices to get the assertion.]

Deduce that for every $r > 0$ there exists $\delta > 0$ such that $|||\mathrm{ad}\, a||| < r$ whenever $a \in \mathfrak{g}$ satisfies $\|a\| < \delta$.

[*Hint:* Show that $\delta = r/M$ does the job!]

Exercise 14.3. Let $\mathscr{T}_1, \mathscr{T}_2$ be topologies on the same set X. Suppose that both of the topological spaces $(X, \mathscr{T}_1)$ and $(X, \mathscr{T}_2)$ are first countable.[3]

Prove that $\mathscr{T}_1 = \mathscr{T}_2$ if and only if $(X, \mathscr{T}_1)$ and $(X, \mathscr{T}_2)$ have the same convergent sequences; the latter condition means that, given a sequence $\{x_n\}_n$ in X and a point $x \in X$, one has $x_n \xrightarrow{n \to \infty} x$ in $(X, \mathscr{T}_1)$ if and only if $x_n \xrightarrow{n \to \infty} x$ in $(X, \mathscr{T}_2)$.

Show that the hypothesis of first countability is essential. Indeed, let $\mathscr{T}_1$ be the topology on $\mathbb{R}$ comprising $\varnothing$, $\mathbb{R}$ plus all the subsets of $\mathbb{R}$ having a countable complement (the so called *cocountable topology* on $\mathbb{R}$). Prove that $\mathscr{T}_1$ is different from the discrete topology $\mathscr{T}_2$ on $\mathbb{R}$ ($\mathscr{T}_2$ is the topology for which all sets are open), but these two topologies have the same convergent sequences.

Is either of $\mathscr{T}_1, \mathscr{T}_2$ first countable?

Exercise 14.4. Let V be a vector space over $\mathbb{K}$ ($= \mathbb{R}$ or $\mathbb{C}$), and let $\| \cdot \|_1$ and $\| \cdot \|_2$ be two equivalent norms on V, in the sense that there exist positive constants C_i ($i = 1, 2$) such that

$$(\bigstar) \qquad C_1 \|v\|_1 \leq \|v\|_2 \leq C_2 \|v\|_1, \qquad \forall\, v \in V.$$

Prove that $(V, \| \cdot \|_1)$ and $(V, \| \cdot \|_2)$ have the same convergent sequences. Deduce from Exr. 14.3 that these two metric spaces induce the same topology on V. Alternatively, prove directly from $(\bigstar)$ that the open sets of the metric topologies on $(V, \| \cdot \|_1)$ and on $(V, \| \cdot \|_2)$ are coincident.

[As a by-product of Exr. 5.3 and of the present one, we infer that *all norms on a finite-dimensional vector space induce the same topology.*]

Exercise 14.5. (1). Let $A : V \to W$ be a linear map, V, W being normed vector spaces over $\mathbb{K}$ ($= \mathbb{R}$ or $\mathbb{C}$). Prove that A is continuous if and only if

$$\|A(v)\| \leq M \|v\| \quad \text{for every } v \in V,$$

for some constant $M > 0$.

(2). Let V, W be as above. Suppose that V is finite-dimensional. Prove that any linear map $A : V \to W$ is continuous.

[*Hint:* For example, you may prove that, if $\{v_1, \ldots, v_n\}$ is a basis of V, then $\|A(x_1\, v_1 + \cdots + x_n\, v_n)\| \leq M \sum_{i=1}^{n} |x_i|$, for every $x_1, \ldots, x_n \in \mathbb{K}$ (for some constant $M > 0$). Then use the equivalence of all norms on V...]

(3). Let V be a finite dimensional vector space over $\mathbb{K}$. Let $\mathrm{End}(V)$ be the $\mathbb{K}$-vector space of all the endomorphisms of V. Let $\mathcal{B} = \{v_1, \ldots, v_n\}$ be a fixed basis

---

[3]$(X, \mathscr{T})$ is first countable if every point of X has a countable family of neighborhoods. For example, any metric space (hence any normed vector space) is first countable.

of V. Let $M_n(\mathbb{K})$ denote the $\mathbb{K}$-vector space of the $n \times n$ matrices with entries in $\mathbb{K}$. Consider the linear map $M : \mathrm{End}(V) \longrightarrow M_n(\mathbb{K})$ defined by $M(f) := \mathrm{Mat}_{\mathcal{B}}(f)$, where $\mathrm{Mat}_{\mathcal{B}}(f)$ is the matrix representing f wrt $\mathcal{B}$ (in domain and codomain).[4]

(3.a). Prove that $M(A \circ B) = M(A) \cdot M(B)$ for every $A, B \in \mathrm{End}(V)$ (here the symbol $\cdot$ denotes the usual matrix product).

(3.b). Let $\Psi : V \to \mathbb{K}^n$ be the linear isomorphism defined by

$$\Psi(x_1 v_1 + \cdots + x_n v_n) := (x_1, \ldots, x_n)^T, \qquad x_1, \ldots, x_n \in \mathbb{K}.$$

Prove that for every $v \in V$ and every $f \in \mathrm{End}(V)$ one has $\Psi(f(v)) = M(f)\,\Psi(v)$.

Note. If $\mathbb{K}$ is $\mathbb{R}$ or $\mathbb{C}$, from point (2) of the exercise we know that M *is continuous*, if $\mathrm{End}(V)$ and $M_n(\mathbb{K})$ are equipped with the topologies induced by any norm (all these topologies being coincident; see Exr. 14.4.)

(4). Let M be as above. Consider the notations in Exr. 14.1. Prove that, if we have $\varphi(z) = \sum_{n=0}^{\infty} a_n\, z^n$ (the series having radius of convergence $\varrho > 0$), then

$$M(\varphi(A)) = \sum_{n=0}^{\infty} a_n\, (M(A))^n = \varphi(M(A)),$$

for every $A \in \mathrm{End}(V)$ such that $|||A||| < \varrho$ (here $||| \cdot |||$ is the operator norm induced on $\mathrm{End}(V)$ by some arbitrarily fixed norm on V).

(5). Finally, we get to the part of the exercise which is crucial to understand that the ODE in the Cauchy Problem (14.13) is of class C^ω.

Let $\mathfrak{g}$ be a finite-dimensional real Lie algebra. Remember that any $Z \in \mathfrak{g}$ defines a linear map $\mathrm{ad}\, Z \in \mathrm{End}(\mathfrak{g})$. Let

$$\mathbf{b}(z) = \frac{z}{e^z - 1} = \sum_{n=0}^{\infty} \frac{B_n}{n!}\, z^n \quad \text{(where } B_n \text{ are the Bernoulli numbers)}.$$

Fix $X, Y \in \mathfrak{g}$. Finally consider the ODE in $\mathfrak{g}$

$$(\star) \qquad Z' = \mathbf{b}(\mathrm{ad}\, Z)(X) + \mathbf{b}(-\mathrm{ad}\, Z)(Y).$$

By fixing a basis $\mathcal{B} = \{v_1, \ldots, v_N\}$ of $\mathfrak{g}$, we can rewrite this ODE as an ordinary differential equation in $\mathbb{R}^N$: in other words we identify $\mathfrak{g}$ with $\mathbb{R}^N$ by applying the identification map Ψ introduced in point (3.b) of the exercise. Accordingly, the endomorphisms $\mathbf{b}(\pm \mathrm{ad}\, Z)$ must be replaced by the matrices representing them (as is described in (3.b)).

Deduce from point (4) of the exercise that these matrices are

$$M\big(\mathbf{b}(\pm \mathrm{ad}\, Z)\big) = \sum_{n=0}^{\infty} \frac{B_n}{n!} \left(\pm M(\mathrm{ad}\, Z) \right)^n = \mathbf{b}\big(\pm M(\mathrm{ad}\, Z) \big).$$

---

[4] We know that the j-th column of $\mathrm{Mat}_{\mathcal{B}}(f)$ is given by the coordinates wrt $\mathcal{B}$ of $f(v_j)$. That is, $\mathrm{Mat}_{\mathcal{B}}(f) = (a_{i,j})_{i,j \leq n}$ where $f(v_j) = \sum_{i=1}^{n} a_{i,j} v_i$, for every $j \leq n$.

Since $\mathrm{ad} : \mathfrak{g} \to \mathrm{End}(\mathfrak{g})$ is linear, we have $\mathrm{ad}\,(Z) = z_1\,\mathrm{ad}\,v_1 + \cdots + z_N\,\mathrm{ad}\,v_N$, where $\Psi(Z) = (z_1, \ldots, z_N)$. From the linearity of M we deduce that

$$M(\mathrm{ad}\,Z) = z_1 M_1 + \cdots + z_N M_N,$$

where $M_1, \ldots, M_N$ are fixed $N \times N$ matrices (obviously $M_i = M(\mathrm{ad}\,v_i)$).

Collecting together the above facts, we derive that $M\big(\mathbf{b}(\pm\,\mathrm{ad}\,Z)\big)$ is an analytic function of $z = (z_1, \ldots, z_N) \in \mathbb{R}^N$ valued in $M_N(\mathbb{R})$, convergent for

$$|||z_1 M_1 + \cdots + z_N M_N||| < 2\pi,$$

say, when $\|z\|$ is small. As a consequence, the application of Ψ to the ODE $(\star)$ turns it into $z' = M_+(z)(x) + M_-(z)(y)$, where

$$z = \Psi(Z),\, x = \Psi(X),\, y = \Psi(Y),\, M_\pm(z) = \mathbf{b}\big(\pm M(\mathrm{ad}\,Z)\big).$$

The above ODE is clearly defined on an open subset Ω of $\mathbb{R}^N$ and it is associated with a C^ω function on Ω, in the usual sense.

Exercise 14.6. Here we denote by $u^{(n)}$ the n-th derivative of a C^∞ function u (real-valued or valued in $\mathbb{R}^N$; in the latter case the derivative is taken component-wise).

Let $\Omega \subseteq \mathbb{R}^N$ be open and let $F \in C^\infty(\Omega, \mathbb{R})$. Let $u, v : [-\varepsilon, \varepsilon] \to \Omega$ be smooth curves. Suppose that $u^{(n)}(0) = v^{(n)}(0)$ for every $n \in \mathbb{N} \cup \{0\}$. Derive that

$$(F \circ u)^{(n)}(0) = (F \circ v)^{(n)}(0) \quad \text{for every } n \in \mathbb{N} \cup \{0\}.$$

[*Hint:* Prove by induction on n the following statement: for every $n \geq 1$,

$$(F \circ u)^{(n)}(t) = \sum_{k=1}^{n} \sum_{\substack{I=(i_1,\ldots,i_k) \\ J=(j_1,\ldots,j_k) \\ i_1+\cdots+i_k=n}} F_{k,I,J}(u(t))\, u_{j_1}^{(i_1)}(t) \cdots u_{j_k}^{(i_k)}(t),$$

where $j_1, \ldots, j_k \in \{1, \ldots, N\}$, $i_1, \ldots, i_k \geq 1$ and the functions $F_{k,I,J}$ are linear combinations of the partial derivatives of F (of orders $\leq n$).

Beware: there is no need to determine explicitly the functions $F_{k,I,J}$...]

Chapter 15

The Local Third Theorem of Lie

THE aim of this chapter is to prove the remarkable Third Fundamental Theorem of Lie in its *local* (real) form: for any finite-dimensional real Lie algebra $\mathfrak{g}$, there exists a local Lie group (on a neighborhood U of the origin of $\mathfrak{g}$) such that the smooth vector fields on U which are invariant under the (local) left translations form a Lie algebra isomorphic to $\mathfrak{g}$. This local Lie group is given by the Campbell-Baker-Hausdorff-Dynkin series $a \diamond b = \sum_{n=1}^{\infty} Z_n(a, b)$, which is well posed near 0, as we know from Chap. 5. The germ of associativity enjoyed by this local operation is the main ingredient: we proved it in Sec. 5.4, as a result of the algebraic machinery on the Exponential Theorem in Chap. 2.

When $\mathfrak{g}$ is nilpotent, this result can easily be globalized: the operation $\diamond$ is well posed throughout $\mathfrak{g}$, and $(\mathfrak{g}, \diamond)$ is a Lie group with Lie algebra isomorphic to $\mathfrak{g}$: this is the *global* version of Lie's Third Theorem for real and nilpotent Lie algebras. In Chap. 16, as a by-product, we shall derive a version of Lie's Third Theorem for (finite-dimensional) stratified Lie algebras $\mathfrak{g}$: for any such $\mathfrak{g}$ there always exists a

homogeneous Carnot group $\mathbb{G} = (\mathbb{R}^N, \diamond, \delta_\lambda)$

(in the sense of [Bonfiglioli *et al.* (2007), Definition 1.4.1]; here $N = \dim \mathfrak{g}$), such that $\mathrm{Lie}(\mathbb{G})$ is isomorphic to $\mathfrak{g}$.

The global form of the Third Theorem of Lie holds for general finite-dimensional Lie algebras, not necessarily nilpotent. Here, we restrict to considering the latter case for two reasons. Firstly, the proof of this theorem in the general case requires a deep knowledge of Differential Geometry, which is beyond the scope of this book; secondly –as remarked above– the local case and the global nilpotent case are sufficiently interesting for our scope, since their proofs can be carried out in a constructive way, and they profoundly exploit the Campbell-Baker-Hausdorff-Dynkin series introduced in Chap. 2 and studied in Chap. 5. Thus, this chapter furnishes a meaningful application of various topics of this book.

Prerequisites for this chapter are the contents of Chap. 5, and some basic Lie group theory (see App. C).

15.1 Local Lie's Third Theorem

We review some facts and notations from Sec. 5.4. Let $\mathfrak{g}$ be a real finite-dimensional Lie algebra, equipped with a Lie-sub-multiplicative norm $\|\cdot\|$ (which always exists; Rem. 5.2). By Thms. 5.3 and 5.9, there exist two balls centred at $0 \in \mathfrak{g}$, say

$$U := \Big\{ a \in \mathfrak{g} : \|a\| \leq h \Big\} \quad \text{and} \quad \Omega := \{ a \in \mathfrak{g} : \|a\| < \varepsilon \},$$

with the following properties (as usual $a \diamond b = \sum_{n=1}^{\infty} Z_n(a,b)$ is the homogeneous CBHD series):

(1) $\Omega \subset U$ (actually, $h \approx 0.231$ and $\varepsilon \approx 0.0937$);
(2) $a \diamond b$ is well posed for any $a, b \in U$;
(3) for any $a, b, c, \in \Omega$, one has $a \diamond b, b \diamond c \in U$, whence $(a \diamond b) \diamond c$ and $a \diamond (b \diamond c)$ are well posed; most importantly, they coincide. See Fig. 15.1.

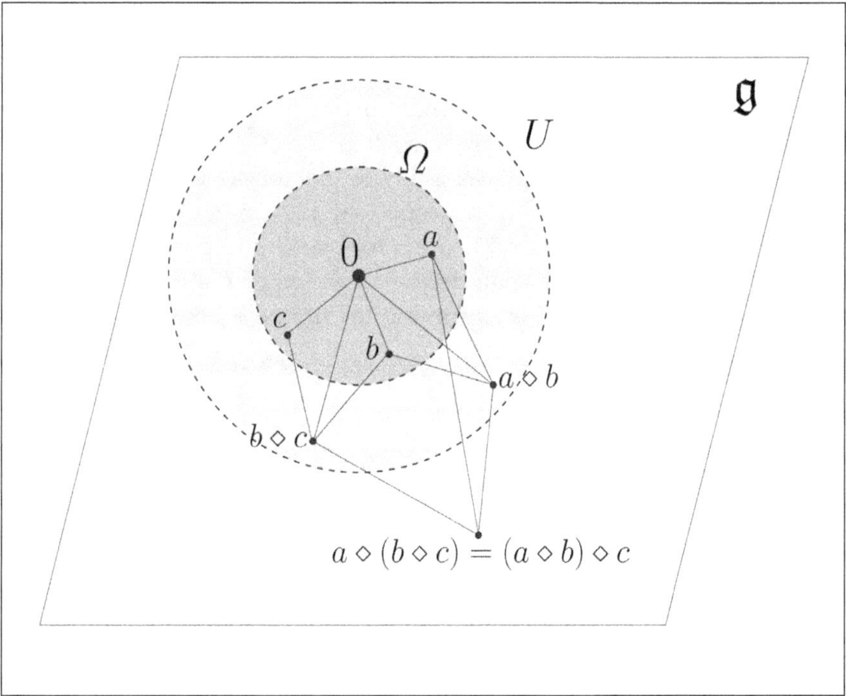

Fig. 15.1   The sets U and Ω constructed in Thms. 5.3 and 5.9, and the associativity of the CBHD operation $\diamond$ near 0.

We use the following notation, for the local "left-translations" of $\diamond$; for any fixed $a \in U$, we set

$$\tau_a : U \to \mathfrak{g}, \quad \tau_a(b) := a \diamond b. \tag{15.1}$$

Let us consider the family of functions (indexed by $a \in \Omega$) defined as follows

$$\mathfrak{F} := \left\{ \tau_a|_\Omega : \Omega \to U \right\}_{a \in \Omega}.$$

From property (3) we infer that

$$\tau_a \circ \tau_b \equiv \tau_{a \diamond b} \quad \text{on } \Omega, \text{ for every } a, b \in \Omega. \tag{15.2}$$

Beware that, in the previous formula, τ_a and τ_b are both maps from the family $\mathfrak{F}$, but their composition does not necessarily belong to $\mathfrak{F}$; from property (3) we only know that $\tau_{a \diamond b}$ is well posed as a function on U, in the sense of (15.1).

In what follows, we equip $\mathfrak{g}$ with the smooth manifold structure resulting from its being a real finite-dimensional vector space.

Remark 15.1. It is not difficult to check that, since $\diamond$ is defined by a totally convergent power series, the $\mathfrak{g}$-valued map $(a, b) \mapsto a \diamond b$ is real-analytic on $U \times U$ (hence smooth); in referring to the real-analyticity of a function defined on an open subset of $\mathfrak{g} \times \mathfrak{g}$, we are taking for granted the obvious identification of $\mathfrak{g} \times \mathfrak{g}$ with $\mathbb{R}^{2N}$, where $N = \dim \mathfrak{g}$, and the associated C^ω structure, which is compatible with the C^∞ one. We skip any further detail. ♯

We introduce the following family of smooth vector fields on U:

$$L(\mathfrak{g}) := \left\{ X \in \mathfrak{X}(U) \,\middle|\, \mathrm{d}_b \varphi(X_b) = X_{\varphi(b)}, \begin{array}{l} \text{for any } \varphi \in \mathfrak{F} \\ \text{and any } b \in \Omega \end{array} \right\}. \tag{15.3}$$

As for the well-posedness of this definition, we observe that:

- $\varphi(b) \in U$ for every $b \in \Omega$ and every $\varphi \in \mathfrak{F}$;
- for any $b \in \Omega$, the (point-wise) differential $\mathrm{d}_b \varphi$ is well posed on $T_b \mathfrak{g}$ since any $\varphi \in \mathfrak{F}$ is smooth on Ω (see Rem. 15.1).

Remark 15.2. Bearing in mind Def. 4.5, a smooth v.f. X on U belongs to $L(\mathfrak{g})$ if and only if the restriction of X to Ω is φ-related to X (as a v.f. on U), for every $\varphi \in \mathfrak{F}$. We shall avoid to try and rephrase this in terms of $\mathfrak{F}$-invariance (see Rem. 4.16), since the domain and codomain of the maps in $\mathfrak{F}$ are different. ♯

The following theorem holds true.

Theorem 15.3 (Local Third Theorem of Lie). *Let the above notations apply. The Lie algebra $L_\Omega(\mathfrak{g})$ obtained by restricting to Ω the vector fields belonging to $L(\mathfrak{g})$ in (15.3) form a Lie subalgebra of $\mathfrak{X}(\Omega)$ with the following properties:*

(i). $L_\Omega(\mathfrak{g})$ is finite dimensional, and its dimension is equal to $\dim(\mathfrak{g})$;

(ii). $L_\Omega(\mathfrak{g})$ and $\mathfrak{g}$ are isomorphic Lie algebras.

The proof of (ii) exhibits one of the nicest computations involving the CBHD operation, and we strongly recommend it to the reader's attention.

Proof. We split the proof in two steps.

Proof of (i): This argument is very similar to the proof that the Lie algebra of a Lie group is isomorphic to the tangent space at the neutral element; however, much care has to be taken due to the locality of the objects we are dealing with. Let us consider the following map:

$$\Lambda : L_\Omega(\mathfrak{g}) \longrightarrow T_0\mathfrak{g}, \quad X \mapsto X_0.$$

Since $T_0\mathfrak{g}$ has the same dimension of the manifold $\mathfrak{g}$ (and this is $\dim \mathfrak{g}$, in the sense of vector spaces), it is enough to show that Λ is a vector space isomorphism.

Clearly, Λ is linear. We prove it is injective: suppose $X \in L_\Omega(\mathfrak{g})$ is such that $X_0 = 0$. By definition, X is the restriction to Ω of a v.f. in $\mathfrak{X}(U)$ which we still denote by X; moreover, if we apply (15.3) with $b = 0$, $\varphi = \tau_a$ and $a \in \Omega$, we get

$$X_a = X_{\tau_a(0)} = \mathrm{d}_0\tau_a(X_0) = \mathrm{d}_0\tau_a(0) = 0.$$

This shows that X is null on Ω, so that Λ is injective.

We turn to the surjectivity of Λ. Let $\mathbf{v} \in T_0\mathfrak{g}$ be arbitrary. Let us define

$$X_a := \mathrm{d}_0\tau_a(\mathbf{v}) \in T_a\mathfrak{g}, \quad \text{for } a \in U. \tag{15.4}$$

This is meaningful, since τ_a is smooth on U (see (15.1)) for any $a \in U$. It is simple to show that X defines a smooth vector field on U (see also Rem. 15.1). We claim that $X \in L(\mathfrak{g})$. On account of (15.3), we need to prove that

$$\mathrm{d}_b\tau_a(X_b) = X_{a \diamond b}, \quad \text{for every } a, b \in \Omega.$$

In its turn, by the very definition (15.4) of X, this is equivalent to

$$\mathrm{d}_b\tau_a(\mathrm{d}_0\tau_b(\mathbf{v})) = \mathrm{d}_0\tau_{a \diamond b}(\mathbf{v}), \quad \text{for every } a, b \in \Omega.$$

In order to prove the latter, it is enough to differentiate at $0 \in \Omega$ both sides of (15.2) and to apply the resulting differentials at $\mathbf{v} \in T_0\mathfrak{g}$. This shows that $X \in L(\mathfrak{g})$. Finally we have $\Lambda(X|_\Omega) = \mathbf{v}$, i.e., $X_0 = \mathbf{v}$. Indeed

$$X_0 \overset{(15.4)}{=} \mathrm{d}_0\tau_0(\mathbf{v}) = \mathbf{v},$$

since τ_0 is the identity map of U. This ends the proof of part (i) of the theorem.

Proof of (ii): First we observe that $L_\Omega(\mathfrak{g})$ is a Lie sub-algebra of $\mathfrak{X}(U)$: this follows from Rem. 15.2 and the consistency of relatedness in Prop. 4.6. The proof is complete if we show that $L_\Omega(\mathfrak{g})$ and $\mathfrak{g}$ are isomorphic Lie algebras. On account of Lem. A.16 on page 345, it is sufficient to find linear bases for $L_\Omega(\mathfrak{g})$ and $\mathfrak{g}$ such that the associated structure constants (see Def. A.14) are the very same. To this end, we fix any basis $\mathcal{V} = \{v_1, \ldots, v_N\}$ for $\mathfrak{g}$ (with $N = \dim \mathfrak{g}$); let $\{c_{i,j}^k\}$ be the structure constants of $\mathfrak{g}$ wrt $\mathcal{V}$, i.e.,

$$[v_i, v_j] = \sum_{k=1}^{N} c_{i,j}^k \, v_k, \quad i,j = 1, \ldots, N. \tag{15.5}$$

Every v_i defines a derivation $\mathbf{v}_i$ at 0 in the obvious way:

$$\mathbf{v}_i f = \frac{\mathrm{d}}{\mathrm{d}t}\Big|_{t=0}\{f(tv_i)\}, \quad \forall\, f \in C^\infty(\mathfrak{g}). \tag{15.6}$$

Accordingly, we can define a unique smooth vector field X^i on Ω via the isomorphism $\Lambda : L_\Omega(\mathfrak{g}) \to T_0\mathfrak{g}$ in the first part of the proof. Owing to (15.4) and (15.6), we see that X_a^i acts as follows:

$$X_a^i f = \frac{\mathrm{d}}{\mathrm{d}t}\Big|_{t=0}\{f(a \diamond (tv_i))\}, \quad \text{for } a \in U \text{ and } f \in C^\infty(\mathfrak{g}). \tag{15.7}$$

We already know that the restrictions to Ω of the v.f.s $X^1,\ldots,X^N$ give a basis, say $\mathcal{B}$, of $L_\Omega(\mathfrak{g})$. We claim that the structure constants of $L_\Omega(\mathfrak{g})$ (equipped with the usual commutator of v.f.s) wrt $\mathcal{B}$ coincide with $\{c_{i,j}^k\}$. Since Λ is an isomorphism of vector spaces, this claim is equivalent to

$$\Lambda([X^i, X^j]) = \sum_{k=1}^{N} c_{i,j}^k \Lambda(X^k), \quad \text{for any } i,j = 1,\ldots,N. \tag{15.8}$$

Since Λ is the evaluation at 0, we can check (15.8) upon the verification that both its sides equally act at 0 on smooth functions f:

$$[X^i, X^j]_0 f = \sum_{k=1}^{N} c_{i,j}^k X_0^k f, \quad \text{for any } f \in C^\infty(\mathfrak{g}) \text{ and } i,j \leq N. \tag{15.9}$$

We fix $f \in C^\infty(\mathfrak{g})$ and $i,j \in \{1,\ldots,N\}$. Due to (15.7), the rhs of (15.9) is

$$\sum_{k=1}^{N} c_{i,j}^k \frac{\mathrm{d}}{\mathrm{d}t}\Big|_0 \{f(tv_k)\}. \tag{15.10}$$

The lhs of (15.9) is $X_0^i(X^j f) - X_0^j(X^i f)$. We compute $X_0^i(X^j f)$ (and then we interchange i and j); we have the following computation, based on (15.7):

$$X_0^i(X^j f) = \frac{\mathrm{d}}{\mathrm{d}t}\Big|_0\left\{X_{tv_i}^j f\right\} = \frac{\mathrm{d}}{\mathrm{d}t}\Big|_0 \frac{\mathrm{d}}{\mathrm{d}s}\Big|_0\left\{f\big((tv_i) \diamond (sv_j)\big)\right\}. \tag{15.11}$$

Bearing in mind the first two summands of the CBHD series, one has

$$(tv_i) \diamond (sv_j) = tv_i + sv_j + \frac{st}{2}[v_i, v_j] + \mathcal{O}(|s| + |t|)^3$$

$$\stackrel{(15.5)}{=} tv_i + sv_j + \frac{st}{2}\sum_{k=1}^{N} c_{i,j}^k v_k + \mathcal{O}(|s| + |t|)^3, \quad \text{as } (s,t) \to (0,0). \tag{15.12}$$

We now perform a computation in a (global) chart for $\mathfrak{g}$: let $\varphi : \mathfrak{g} \to \mathbb{R}^N$ be the linear map sending (for any $i = 1,\ldots,N$) the vector v_i to the i-th vector of the standard basis of $\mathbb{R}^N$, e_i say. If we use the notation $\widetilde{f} := f \circ \varphi^{-1}$, then (15.10) becomes (via standard multivariate calculus)

$$\sum_{k=1}^{N} c_{i,j}^k \frac{\mathrm{d}}{\mathrm{d}t}\Big|_0 \{\widetilde{f}(te_k)\} = \sum_{k=1}^{N} c_{i,j}^k \partial_k \widetilde{f}(0). \tag{15.13}$$

Analogously, taking into account (15.11) and (15.12),

$$
\begin{aligned}
X_0^i(X^j f) &= \frac{\mathrm{d}}{\mathrm{d}t}\Big|_0 \frac{\mathrm{d}}{\mathrm{d}s}\Big|_0 \left\{ \widetilde{f}\left(te_i + se_j + \frac{st}{2}\sum_{k=1}^N c_{i,j}^k\, e_k + \widetilde{\mathcal{O}}(|s|+|t|)^3 \right) \right\} \\
&= \frac{\mathrm{d}}{\mathrm{d}t}\Big|_0 \left\{ \left\langle (\nabla \widetilde{f})\big(te_i + \widetilde{\mathcal{O}}(t^3)\big),\, e_j + \frac{t}{2}\sum_{k=1}^N c_{i,j}^k\, e_k + \widetilde{\mathcal{O}}(t^2) \right\rangle \right\} \\
&= \frac{\mathrm{d}}{\mathrm{d}t}\Big|_0 \left\{ \partial_j \widetilde{f}\big(te_i + \widetilde{\mathcal{O}}(t^3)\big) + \frac{t}{2}\sum_{k=1}^N c_{i,j}^k\, \partial_k \widetilde{f}\big(te_i + \widetilde{\mathcal{O}}(t^3)\big) + \widetilde{\mathcal{O}}(t^2) \right\} \\
&= \partial_{i,j}\widetilde{f}(0) + \frac{1}{2}\sum_{k=1}^N c_{i,j}^k\, \partial_k \widetilde{f}(0).
\end{aligned}
$$

As a result, by interchanging i and j (and by Schwarz Theorem) we get

$$
X_0^i(X^j f) - X_0^j(X^i f) = \frac{1}{2}\sum_{k=1}^N c_{i,j}^k\, \partial_k \widetilde{f}(0) - \frac{1}{2}\sum_{k=1}^N c_{j,i}^k\, \partial_k \widetilde{f}(0) = \sum_{k=1}^N c_{i,j}^k\, \partial_k \widetilde{f}(0),
$$

since $c_{j,i}^k = -c_{i,j}^k$ by skew-symmetry of the bracket. Comparing to (15.13), we see that the rhs and the lhs of (15.9) are equal, and the proof is complete. $\qquad\square$

15.2   Global Lie's Third Theorem in the nilpotent case

In this section we shall prove the *global* Third Theorem of Lie in the particular case of nilpotent Lie algebras. To begin with, the reader may want to review the notion of a nilpotent Lie algebra and of the descending central series, both given in Def. A.7 on page 342.

We want to prove the following result.

Theorem 15.4 (Global Lie's Third Theorem for nilpotent Lie algebras). *Let $\mathfrak{g}$ be a finite-dimensional real Lie algebra, and suppose that $\mathfrak{g}$ is nilpotent. If $r \in \mathbb{N}$ is the step of nilpotence of $\mathfrak{g}$, consider the **CBHD operation** on $\mathfrak{g}$, which is the map*

$$
\diamond : \mathfrak{g} \times \mathfrak{g} \longrightarrow \mathfrak{g} \qquad (a,b) \mapsto a \diamond b = \sum_{n=1}^r Z_n(a,b). \tag{15.14}
$$

Then $(\mathfrak{g}, \diamond)$ is a Lie group, and its Lie algebra is isomorphic to $\mathfrak{g}$. Moreover, 0 is the neutral element of $\diamond$, while the group inverse of $a \in \mathfrak{g}$ is $-a$.

Proof. This is a consequence of Thm. 15.3. Indeed, no issue of convergence troubles the CBHD series, since this is a finite sum, due to the nilpotence of $\mathfrak{g}$; hence the series $\sum_{n=1}^\infty Z_n(a,b)$ converges for any $a,b \in \mathfrak{g}$, and its sum is the CBHD operation $a \diamond b$ in (15.14). In any system of linear coordinates on $\mathfrak{g}$, $\diamond$ is a polynomial.

    The local associativity of $\diamond$ established in Thm. 5.9 turns into a global associativity on $\mathfrak{g}$, since (in linear coordinates) the identity $a \diamond (b \diamond c) = (a \diamond b) \diamond c$ (holding true for a, b, c near 0) becomes an identity involving two polynomials, so that its validity near 0 ensures is global validity.

As a consequence, the proof of Thm. 15.3 holds true *verbatim* by replacing the small neighborhoods U and Ω of the origin with the whole of $\mathfrak{g}$. Accordingly, the vector fields in the Lie algebra $L_\Omega(\mathfrak{g})$ in Thm. 15.3 are precisely the left invariant vector fields on $(\mathfrak{g}, \diamond)$, since the maps τ_a in (15.1) are the left translations of the group. This ends the proof. $\qquad\square$

Motivated by what we shall do in Chap. 16, we realize the group $G = (\mathfrak{g}, \diamond)$ as a Lie group on $\mathbb{R}^N$, where $N = \dim \mathfrak{g}$, by fixing some linear coordinates on $\mathfrak{g}$; we are particularly interested in simplifying the Exponential Map of G as much as possible, and this is particularly detectable in coordinates. It is understood from now that $\mathfrak{g}$ satisfies the assumption of Thm. 15.4.

We arbitrarily fix a basis $\mathcal{E} := \{E_1, \ldots, E_N\}$ of $\mathfrak{g}$ and we consider the coordinates on $\mathfrak{g}$ wrt $\mathcal{E}$, i.e., we set

$$\pi_\mathcal{E} : \mathbb{R}^N \longrightarrow \mathfrak{g}, \qquad \pi_\mathcal{E}(x) = x \cdot E := \sum_{k=1}^{N} x_k\, E_k. \tag{15.15}$$

Obviously, $\pi_\mathcal{E}$ is a linear isomorphism of vector spaces; hence, in particular, $\pi_\mathcal{E}$ is a bijection and we can transform $\diamond$ via $(\pi_\mathcal{E})^{-1}$ as follows:

Definition 15.5. Let the above notations apply. We define

$$*_\mathcal{E} : \mathbb{R}^N \times \mathbb{R}^N \longrightarrow \mathbb{R}^N, \qquad x *_\mathcal{E} y := \pi_\mathcal{E}^{-1}\big(\pi_\mathcal{E}(x) \diamond \pi_\mathcal{E}(y)\big).$$

By Thm. 15.4, $\mathbb{G} := (\mathbb{R}^N, *_\mathcal{E})$ is a Lie group isomorphic to $(\mathfrak{g}, \diamond)$ via $\pi_\mathcal{E}$:

$$\pi_\mathcal{E}(x *_\mathcal{E} y) = \pi_\mathcal{E}(x) \diamond \pi_\mathcal{E}(y), \quad \text{for every } x, y \in \mathbb{R}^N. \tag{15.16}$$

Remark 15.6. Let $x, y \in \mathbb{R}^N$; $x *_\mathcal{E} y$ is characterized by the identity

$$\Big(\sum_{k=1}^{N} x_k\, E_k\Big) \diamond \Big(\sum_{k=1}^{N} y_k\, E_k\Big) = \sum_{k=1}^{N} (x *_\mathcal{E} y)_k\, E_k,$$

where we have set $x *_\mathcal{E} y = \big((x *_\mathcal{E} y)_1, \ldots, (x *_\mathcal{E} y)_N\big)$. $\qquad\sharp$

Since we are now dealing with usual Euclidean space $\mathbb{R}^N$, we resume our usual notations on vector fields X on $\mathbb{R}^N$ and their coefficient vectors $X(x)$. We denote by $\mathcal{J} := \{J_1, \ldots, J_N\}$ the set of the left invariant vector fields in $\mathrm{Lie}(\mathbb{G})$ such that (for any $i = 1, \ldots, N$) the column vector $J_i(0)$ coincides with the i-th element of the canonical basis of $\mathbb{R}^N$. It is well known from basic Lie group theory that such a J_i is uniquely determined, that $\mathcal{J}$ is a linear basis of $\mathrm{Lie}(\mathbb{G})$, and that J_i acts on functions as follows (see Exm. C.27 in App. C):

$$J_i f(x) = \frac{\partial}{\partial y_i}\Big|_{y=0} \big\{f(x *_\mathcal{E} y)\big\}, \qquad \forall f \in C^\infty(\mathbb{R}^N), \quad x \in \mathbb{R}^N. \tag{15.17}$$

Occasionally,[1] we say that $\mathcal{J}$ is the **Jacobian basis** of $\mathrm{Lie}(\mathbb{G})$.

---

[1] Beware that this is not a coordinate-free notion: this basis heavily depends on the standard coordinates of $\mathbb{R}^N$ that we have fixed.

We summarize what we have proved in Thms. 15.3 and 15.4:

Theorem 15.7. *Let $\mathfrak{g}$ be finite-dimensional real Lie algebra, with dimension N and nilpotent of step $r \in \mathbb{N}$. Let $*_{\mathcal{E}}$ be the multiplication on $\mathbb{R}^N$ (associated with the basis $\mathcal{E} = \{E_1, \ldots, E_N\}$) introduced in Def. 15.5. Then we have:*

(1) $\mathbb{G} = (\mathbb{R}^N, *_{\mathcal{E}})$ is a nilpotent Lie group;

(2) $*_{\mathcal{E}}$ has polynomial component functions;

(3) if $\{J_1, \ldots, J_N\}$ is the basis of $\mathrm{Lie}(\mathbb{G})$ introduced in (15.17), then the linear map

$$\varphi : \mathrm{Lie}(\mathbb{G}) \longrightarrow \mathfrak{g} \qquad \varphi(J_i) = E_i \quad (\text{for } i = 1, \ldots, N),$$

is an isomorphism of Lie algebras.

Remark 15.8. Let $\mathfrak{g}$ be as in Thm. 15.7. Let $\mathfrak{h}$ be a Lie algebra isomorphic to $\mathfrak{g}$. By Lem. A.16, it is possible to find a basis $\mathcal{F} = \{F_1, \ldots, F_N\}$ of $\mathfrak{h}$ such that the structure constants of $\mathfrak{h}$ wrt $\mathcal{F}$ coincide with those of $\mathfrak{g}$ wrt $\mathcal{E}$; hence, taking into account the universal expression of the Dynkin polynomials, it is not difficult to see that the operations $*_{\mathcal{E}}$ and $*_{\mathcal{F}}$ on $\mathbb{R}^N$ do coincide.   ♯

15.2.1   *The Exponential Map of* $\mathbb{G}$

We study the Exponential Map of the group $\mathbb{G}$ in Thm. 15.7. We warn the reader that, in the proof of the next result, some results of basic Lie group theory (concerning homomorphisms and their differentials) are needed.

Theorem 15.9. *Let $\mathbb{G}$ be the Lie group in Thm. 15.7. If $\mathcal{J} = \{J_1, \ldots, J_N\}$ is the basis of $\mathrm{Lie}(\mathbb{G})$ introduced in (15.17), the Exponential Map of $\mathbb{G}$ acts as follows*

$$\mathrm{Exp}\Big(\sum_{k=1}^{N} \xi_k\, J_k \Big) = \xi, \quad \text{for every } \xi \in \mathbb{R}^N. \tag{15.18}$$

In other words, if we identify $\mathrm{Lie}(\mathbb{G})$ with $\mathbb{R}^N$ via the basis $\mathcal{J}$, then Exp is the identity map, and it is therefore globally invertible.

A few words about the meaning of this result. In the case of nilpotent Lie groups $(G, \cdot)$, it is known that the Exponential Map $\mathrm{Exp}_G : \mathrm{Lie}(G) \to G$ is globally invertible, and that the CBHD-type identity

$$\mathrm{Exp}_G(X) \cdot \mathrm{Exp}_G(Y) = \mathrm{Exp}_G\big(\textstyle\sum_n Z_n(X, Y) \big)$$

holds globally for every $X, Y \in \mathrm{Lie}(G)$ (see e.g., [Corwin and Greenleaf (1990)]). Thus it is possible to identify G with $\mathrm{Lie}(G)$ via Exp_G. In principle, this process is possible also in our case when $(G, \cdot)$ is our $\mathbb{G} = (\mathbb{R}^N, *_{\mathcal{E}})$. Now, Thm. 15.9 states that no further simplification is brought by the identification of our $\mathbb{G}$ with $\mathrm{Lie}(\mathbb{G})$, since Exp is ...already as simple as one may expect.

Roughly put, the group $\mathbb{G}$ that we have constructed gives the best simplification (at least at the level of the Exp map) amongst the possible isomorphic connected and simply connected Lie groups that one may attach to $\mathfrak{g}$.

Proof of Thm. 15.9. Let $\mathfrak{g}$ be as in Thm. 15.7. We firstly observe that, since $\mathrm{Lie}(\mathbb{G})$ is isomorphic to $\mathfrak{g}$, then $\mathrm{Lie}(\mathbb{G})$ is N-dimensional and nilpotent; hence, we are entitled to re-apply Thm. 15.7 to the new Lie algebra $\mathrm{Lie}(\mathbb{G})$: if $\diamond$ is the CBHD multiplication on $\mathrm{Lie}(\mathbb{G})$ and if we define (see the above (15.15))

$$*_{\mathfrak{J}} : \mathbb{R}^N \times \mathbb{R}^N \longrightarrow \mathbb{R}^N, \quad \xi *_{\mathfrak{J}} \eta := \pi_{\mathfrak{J}}^{-1}\big(\pi_{\mathfrak{J}}(\xi) \diamond \pi_{\mathfrak{J}}(\eta)\big),$$

then $\mathbb{F} := (\mathbb{R}^N, *_{\mathfrak{J}})$ is yet another Lie group. On the other hand, since the structure constants of $\mathrm{Lie}(\mathbb{G})$ and of $\mathfrak{g}$, respectively associated with $\mathfrak{J}$ and $\mathcal{E}$, are the same (by point (3) in Thm. 15.7), it follows from Rem. 15.8 that

$$\xi *_{\mathfrak{J}} \eta = \xi *_{\mathcal{E}} \eta, \quad \text{for every } \xi, \eta \in \mathbb{R}^N. \tag{15.19}$$

We then consider the following map

$$E : \mathbb{G} \longrightarrow \mathbb{G}, \qquad E(\xi) := (\mathrm{Exp} \circ \pi_{\mathfrak{J}})(\xi) = \mathrm{Exp}\big(\textstyle\sum_{k=1}^{N} \xi_k J_k\big).$$

Since $\mathbb{G}$ is a Lie group and since $\mathrm{Lie}(\mathbb{G})$ is nilpotent, by the CBHD formula on $\mathbb{G}$ (see Thm. 14.7 on page 281) we have

$$\mathrm{Exp}(X \diamond Y) = \mathrm{Exp}(X) *_{\mathcal{E}} \mathrm{Exp}(Y), \quad \text{for every } X, Y \in \mathrm{Lie}(\mathbb{G});$$

as a consequence, by exploiting identity (15.19), we deduce that E is a homomorphism from $\mathbb{G}$ to itself: indeed, for every $\xi, \eta \in \mathbb{R}^N$ we have

$$E(\xi *_{\mathcal{E}} \eta) \overset{(15.19)}{=} E(\xi *_{\mathfrak{J}} \eta) = \mathrm{Exp}\big(\pi_{\mathfrak{J}}(\xi) \diamond \pi_{\mathfrak{J}}(\eta)\big)$$
$$= \mathrm{Exp}\big(\pi_{\mathfrak{J}}(\xi)\big) *_{\mathcal{E}} \mathrm{Exp}\big(\pi_{\mathfrak{J}}(\eta)\big) = E(\xi) *_{\mathcal{E}} E(\eta).$$

Now, taking into account that 0 is the identity of $\mathbb{G}$, it is contained in the proof of Prop. 14.1 on page 278 that $\mathfrak{J}_E(0)$ is the $N \times N$ identity matrix. This shows that the differential of the homomorphism E is the identity map. Since $\mathbb{G}$ is connected, basic results of Lie group theory (see Thm. C.22-(1), page 397) imply that E is necessarily the identity map, which is precisely (15.18). $\qquad\square$

Remark 15.10. Let the notations in the proof of Thm. 15.9 apply for the map E. Since, by (15.18), E coincides with the identity map on $\mathbb{G}$, we deduce that

$$\mathrm{Exp} = \pi_{\mathfrak{J}}^{-1}. \tag{15.20}$$

In other words, the map Exp is the *linear bijective map* sending the k-th element of the Jacobian basis of $\mathrm{Lie}(\mathbb{G})$ into the k-th element of the canonical basis of $\mathbb{R}^N$. Furthermore, from (15.20) we also deduce that $\mathrm{Log} = \pi_{\mathfrak{J}}$. $\qquad\sharp$

15.3 Exercises of Chap. 15

In the next two exercises, the notation $\diamond_{\mathcal{E}}$ from Def. 15.5 is used.

Exercise 15.1. Consider on $\mathbb{R}^3$ the v.f.s

$$X_1 = \partial_{x_1} + 2\,x_2\,\partial_{x_3}, \quad X_2 = \partial_{x_2} - 2\,x_1\,\partial_{x_3}.$$

Let $\mathfrak{g} = \mathrm{Lie}\{X_1, X_2\}$; if $\{\mathfrak{g}_k\}_k$ is the descending central series, prove that

- $\mathfrak{g}_1 = \mathfrak{g} = \mathrm{span}\{X_1, X_2, X_3\}$, where $X_3 := [X_1, X_2] = -4\,\partial_{x_3}$;
- $\mathfrak{g}_2 = \mathrm{span}\{X_3\}$; $\mathfrak{g}_k = \{0\}$ for every $k \geq 3$.

Infer that $\dim(\mathfrak{g}) = 3$ and that $\mathfrak{g}$ is nilpotent of step 2. If $\mathcal{E}$ is the basis of $\mathfrak{g}$ given by $\mathcal{E} = \{X_1, X_2, X_3\}$, prove that

$$\left(\textstyle\sum_{k=1}^{3} x_k\, X_k\right) \diamond \left(\textstyle\sum_{k=1}^{3} y_k\, X_k\right)$$
$$= (x_1 + y_1)\, X_1 + (x_2 + y_2)\, X_2 + \left(x_3 + y_3 + \tfrac{1}{2}\,(x_1 y_2 - x_2 y_1)\right) X_3,$$

so that $x *_{\mathcal{E}} y = \left(x_1 + y_1, x_2 + y_2, x_3 + y_3 + \tfrac{1}{2}\,(x_1 y_2 - x_2 y_1)\right)$.

The group $\mathbb{G} = (\mathbb{R}^3, *_{\mathcal{E}})$ is called **the first Heisenberg group** $\mathbb{H}^1$.

Exercise 15.2. Consider on $\mathbb{R}^2$ the vector fields $X_1 = \partial_{x_1}$ and $X_2 = x_1\,\partial_{x_2}$, and set $\mathfrak{g} = \mathrm{Lie}\{X_1, X_2\}$. If $\{\mathfrak{g}_k\}_k$ is the descending central series, prove that

- $\mathfrak{g}_1 = \mathfrak{g} = \mathrm{span}\{X_1, X_2, X_3\}$, where $X_3 := [X_1, X_2] = \partial_{x_2}$;
- $\mathfrak{g}_2 = \mathrm{span}\{X_3\}$; $\mathfrak{g}_k = \{0\}$ for every $k \geq 3$.

Infer that $\dim(\mathfrak{g}) = 3$ and that $\mathfrak{g}$ is nilpotent of step 2. Consider the basis $\mathcal{E}$ of $\mathfrak{g}$ given by $\mathcal{E} = \{X_1, X_2, X_3\}$; prove that

$$\left(\textstyle\sum_{k=1}^{3} x_k\, X_k\right) \diamond \left(\textstyle\sum_{k=1}^{3} y_k\, X_k\right)$$
$$= (x_1 + y_1)\, X_1 + (x_2 + y_2)\, X_2 + \left(x_3 + y_3 + \tfrac{1}{2}\,(x_1 y_2 - x_2 y_1)\right) X_3.$$

Therefore, the group $\mathbb{G} = (\mathbb{R}^3, *_{\mathcal{E}})$ coincides with the one constructed in Exr. 15.1. We observe that, in this case, there cannot exist a Lie group on $\mathbb{R}^2$ with Lie algebra precisely equal to $\mathfrak{g}$: in fact, the dimension of $\mathfrak{g}$ is strictly greater than the dimension of the ambient space $\mathbb{R}^2$.

Chapter 16

Construction of Carnot Groups

$\mathbf{T}$HE aim of this chapter is to construct the *homogeneous Carnot groups* (HCGs, for short) starting from the (finite-dimensional) *stratified Lie algebras*, with the technique that we employed in Chap. 15, based on the CBHD operation on finite-dimensional Lie algebras.

A Lie algebra $\mathfrak{s}$ is stratified when it admits a decomposition of the form

$$\mathfrak{s} = V \oplus [V, V] \oplus [V, [V, V]] \oplus [V, [V, [V, V]]] \oplus \cdots,$$

where V is a subspace of $\mathfrak{s}$. If $\mathfrak{s}$ is finite-dimensional, this decomposition eventually becomes trivial (i.e., successive summands are $\{0\}$), so that $\mathfrak{s}$ is nilpotent. We can therefore equip $\mathfrak{s}$ with a Lie group structure, namely $(\mathfrak{s}, \diamond)$, as we did in Chap. 15. It turns out that this group can be further endowed with a homogeneous structure turning it into an HCG, whose Lie algebra is isomorphic to $\mathfrak{s}$, hence stratified.

HCGs (and their sub-Laplacian operators) are fully studied in the monograph by [Bonfiglioli *et al.* (2007)]. The way HCGs are presented in this chapter is more intrinsic than what is done in [Bonfiglioli *et al.* (2007)], in that the focus is here shifted to the stratified Lie algebra as a *datum*, and the associated HCG is obtained by the constructive global Lie's Third Thm. 15.4.

If one chooses a global chart for $\mathfrak{s}$ given some linear basis adapted (in a suitable sense) to the stratification

$$V \oplus [V, V] \oplus [V, [V, V]] \oplus \cdots$$

of $\mathfrak{s}$, the Lie group multiplication takes on a "pyramidal" (and polynomial) form, making the study of the group $(\mathfrak{s}, \diamond)$ in these coordinates particularly simple. For example, one can define on $\mathfrak{s}$ a group of diagonalizable automorphisms δ_λ, called *anisotropic dilations*, which –in the cited coordinates– have the form

$$\delta_\lambda(x_1, \ldots, x_N) = (\lambda^{\sigma_1} x_1, \ldots, \lambda^{\sigma_N} x_N),$$

with successive integer exponents $1 = \sigma_1 \leq \cdots \leq \sigma_N$ (here $N = \dim(\mathfrak{s})$). As a by-product, the many well-behaved properties of HCGs (mostly, the existence of dilations) are consequences of the Lie-algebra properties.

Prerequisites for this chapter are the contents of Chap. 15, plus some basic Lie group theory (given in App. C).

299

16.1 Finite-dimensional stratified Lie algebras

The aim of this section is to establish a refinement of the global Lie's Third Thm. 15.4 in the particular case of (finite-dimensional) *stratified* Lie algebras (a subclass of the nilpotent Lie algebras). As a by-product, this will provide the construction of the so-called homogeneous Carnot groups (in the sense of [Bonfiglioli *et al.* (2007), Section 4.1]) in Sec. 16.2. We begin with the relevant definition.

Definition 16.1 (Stratified Lie algebra). A Lie algebra $\mathfrak{s}$ is called **stratified** if it admits a **stratification** $\{V_i\}_{i\in\mathbb{N}}$, that is if there exists a family of vector subspaces $\{V_i\}_{i\in\mathbb{N}}$ of $\mathfrak{s}$ such that

$$\mathfrak{s} = \bigoplus_{i=1}^{\infty} V_i \quad \text{and } V_{i+1} = [V_1, V_i] \ \text{ for every } i \geq 1. \tag{16.1}$$

This is the same as asking for the existence of a subspace V_1 of $\mathfrak{s}$ such that

$$\mathfrak{s} = V_1 \oplus [V_1, V_1] \oplus [V_1, [V_1, V_1]] \oplus [V_1, [V_1, [V_1, V_1]]] \oplus \cdots$$

For example, any nilpotent Lie algebra $\mathfrak{g}$ of step 2 is stratified (Exr. 16.2).

Let $(\mathfrak{g}, \{V_n\}_n)$ be a fixed real *finite-dimensional* stratified Lie algebra, with dimension N. According to Rem. A.18 in App. A, $\mathfrak{g}$ is nilpotent of step r, say:

$$\mathfrak{g} = \bigoplus_{n=1}^{r} V_n, \quad \text{and} \quad \begin{cases} V_{n+1} = [V_1, V_n], & \text{for } n = 1, \ldots, r-1, \\ [V_1, V_r] = \{0\}. \end{cases}$$

These notations will be fixed for the rest of the section. Moreover, p_n will denote the canonical projection of $\mathfrak{g}$ onto V_n (see Rem. A.19)

We now define a family of dilations $\{\Delta_\lambda\}_{\lambda>0}$ on $\mathfrak{g}$ in the following way.

Definition 16.2 (Dilations on $\mathfrak{g}$). For every $\lambda > 0$, we denote by Δ_λ the unique linear map from $\mathfrak{g}$ to itself such that

$$\Delta_\lambda\big|_{V_i} = \lambda^i \, \mathrm{id}_{V_i}, \quad \text{for every } i = 1, \ldots, r. \tag{16.2}$$

In other words, for every $a \in \mathfrak{g}$ and every $\lambda > 0$ we have

$$\Delta_\lambda(a) = \Delta_\lambda\big(\textstyle\sum_{n=1}^{r} \mathrm{p}_n(a)\big) = \textstyle\sum_{i=1}^{r} \lambda^i \, \mathrm{p}_i(a).$$

Note that this definition is well-posed, since $\mathrm{p}_1(a), \ldots, \mathrm{p}_r(a)$ are uniquely defined (as $\mathfrak{g}$ is the direct sum of $V_1, \ldots, V_r$).

Lemma 16.3. *For every $\lambda > 0$, the map Δ_λ is a Lie algebra automorphism of $\mathfrak{g}$.*

Proof. Obviously, Δ_λ is an automorphism of $\mathfrak{g}$ as a vector space; moreover, for every $a, b \in \mathfrak{g}$ we have (see Rem. A.19)

$$[\Delta_\lambda(a), \Delta_\lambda(b)] \overset{(16.2)}{=} \textstyle\sum_{i,j=1}^{r} \lambda^{i+j} \, [\mathrm{p}_i(a), \mathrm{p}_j(b)] \qquad (\text{since } [\mathrm{p}_i(a), \mathrm{p}_j(b)] \in V_{i+j})$$
$$= \Delta_\lambda\big(\textstyle\sum_{i,j=1}^{r} [\mathrm{p}_i(a), \mathrm{p}_j(b)]\big) = \Delta_\lambda([a, b]).$$

This ends the proof. $\qquad\square$

Clearly, it is possible to define dilations Δ_λ also for infinite-dimensional stratified Lie algebras, and Lem. 16.3 holds as well (Exr. 16.3)

As a simple consequence of Lem. 16.3, we obtain the following result.

Proposition 16.4. *For every $\lambda > 0$, the map Δ_λ is an automorphism of the group $(\mathfrak{g}, \diamond)$ (where $\diamond$ is the CBHD multiplication on $\mathfrak{g}$), that is,*

$$\Delta_\lambda(a \diamond b) = \Delta_\lambda(a) \diamond \Delta_\lambda(b), \quad \text{for every } a, b \in \mathfrak{g}. \tag{16.3}$$

Proof. For every $n = 1, \ldots, r$, let Z_n be the n-th Dynkin polynomial (relative to $\mathfrak{g}$). Since Z_n is a homogeneous Lie-polynomial of degree n, by repeatedly applying Lem. 16.3 we get $\Delta_\lambda\big(Z_n(a, b)\big) = Z_n\big(\Delta_\lambda(a), \Delta_\lambda(b)\big)$, for every $a, b \in \mathfrak{g}$; from this, by the very definition of the operation $\diamond$, we conclude that

$$\Delta_\lambda(a \diamond b) = \sum_{n=1}^{r} \Delta_\lambda\big(Z_n(a, b)\big) = \sum_{n=1}^{r} Z_n\big(\Delta_\lambda(a), \Delta_\lambda(b)\big) = \Delta_\lambda(a) \diamond \Delta_\lambda(b),$$

and this is precisely the desired identity (16.3). $\qquad\square$

16.2 Construction of Carnot groups

We now transfer the family $\{\Delta_\lambda\}_{\lambda>0}$ of Def. 16.2 to $\mathbb{R}^N$, by fixing a suitable basis $\mathcal{E}$ of $\mathfrak{g}$. To this end, we first give the following definition.

Definition 16.5 (Adapted basis to the stratification). Let $(\mathfrak{g}, \{V_n\}_n)$ be a stratified Lie algebra. We say that a basis $\mathcal{E} = \{E_1, \ldots, E_N\}$ of $\mathfrak{g}$ is **adapted to the stratification** $\{V_n\}_n$ if it can be decomposed as

$$\mathcal{E} = \{E_1^{(1)}, \ldots, E_{N_1}^{(1)}, \ldots, E_1^{(r)}, \ldots, E_{N_r}^{(r)}\},$$

where, for every $i = 1, \ldots, r$, we have $N_i = \dim(V_i)$ (hence, $N_1 + \cdots + N_r = N$), and $\{E_1^{(i)}, \ldots, E_{N_i}^{(i)}\}$ is a basis of V_i.

We fix $\mathcal{E}$ as above, and we let $\pi_\mathcal{E}$ be the associated coordinate function as in (15.15). We define a family of dilations $\{\delta_\lambda\}_\lambda$ on $\mathbb{R}^N$ in the following way.

Definition 16.6. For every $\lambda > 0$, we set

$$\delta_\lambda : \mathbb{R}^N \longrightarrow \mathbb{R}^N, \qquad \delta_\lambda(x) := \pi_\mathcal{E}^{-1}\big(\Delta_\lambda(\pi_\mathcal{E}(x))\big). \tag{16.4}$$

The following result is a refinement of Thm. 15.7, along with a definition of *homogeneous Carnot groups*, in the sense of [Bonfiglioli *et al.* (2007), Sec. 1.4].

Theorem 16.7. *Let $(\mathfrak{g}, V)$ be a (real) stratified Lie algebra, with finite dimension N. Let $\mathcal{E} = \{E_1, \ldots, E_N\}$ be a basis of $\mathfrak{g}$ adapted to V. Let $*_\mathcal{E}$ be the CBHD multiplication on $\mathbb{R}^N$ (associated with $\mathcal{E}$) introduced in Def. 15.5, and let $\{\delta_\lambda\}_\lambda$ be the family of dilations introduced in Def. 16.6.*

*Then $\mathbb{G} = (\mathbb{R}^N, *_\mathcal{E}, \delta_\lambda)$ is a **homogeneous Carnot group**, that is:*

(1) in the standard coordinates of $\mathbb{R}^N$, δ_λ has the diagonal form

$$\delta_\lambda(x_1,\ldots,x_N) = (\lambda^{\sigma_1} x_1,\ldots,\lambda^{\sigma_N} x_N), \qquad (16.5)$$

with $\sigma_1 \leq \cdots \leq \sigma_N$, and $\{\sigma_1,\ldots,\sigma_N\} = \{1,2,\ldots,r\}$, where r is the step of nilpotence of $\mathfrak{g}$;

(2) $\{\delta_\lambda\}_\lambda$ is a family of automorphisms of the group $(\mathbb{R}^N, *_{\mathcal{E}})$;

(3) if N_1 is the cardinality of $\{i : \sigma_i = 1\}$, then the first N_1 elements of the Jacobian basis $\{J_1,\ldots,J_N\}$ of $\mathrm{Lie}(\mathbb{G})$ (see (15.17)) Lie-generate $\mathrm{Lie}(\mathbb{G})$:

$$\mathrm{Lie}\{J_1,\ldots,J_{N_1}\} = \mathrm{Lie}(\mathbb{G});$$

(4) for any $i \in \{1,\ldots,N\}$, J_i is δ_λ-homogeneous of degree σ_i, that is

$$J_i(f \circ \delta_\lambda) = \lambda^{\sigma_i}(J_i f) \circ \delta_\lambda, \qquad \forall f \in C^\infty(\mathbb{R}^N), \quad \lambda > 0.$$

Further properties hold true:

- *The neutral element of $\mathbb{G}$ is 0 and the group inversion is $x \mapsto -x$. The (polynomial) group operation $*_{\mathcal{E}}$ is nothing but the CBHD operation $\diamond$ on $\mathfrak{g}$, read in a system of coordinates associated with the linear basis $\mathcal{E}$ of $\mathfrak{g}$. Thus, $*_{\mathcal{E}}$ is uniquely determined by the structure constants of $\mathfrak{g}$ wrt $\mathcal{E}$ and the 'universal' operation $\diamond$.*

- *Furthermore, the linear map*

$$\varphi : \mathrm{Lie}(\mathbb{G}) \longrightarrow \mathfrak{g} \qquad \varphi(J_i) = E_i \quad (\text{for } i = 1,\ldots,N), \qquad (16.6)$$

  is a Lie-algebra isomorphism.

- *Finally, $\mathrm{Exp} : \mathrm{Lie}(\mathbb{G}) \to \mathbb{G}$ is globally invertible and*

$$\mathrm{Exp}\big(\textstyle\sum_{k=1}^N \xi_k J_k\big) = \xi, \quad \text{for every } \xi \in \mathbb{R}^N. \qquad (16.7)$$

Proof. First of all we observe that, since $\mathfrak{g}$ is stratified and finite-dimensional, $\mathfrak{g}$ is nilpotent (see Cor. A.21). Hence, we can apply to $\mathfrak{g}$ the results in Thm. 15.7: $\mathbb{G} = (\mathbb{R}^N, *_{\mathcal{E}})$ is a Lie group on $\mathbb{R}^N$, and the map φ in (16.6) is an isomorphism of Lie algebras; moreover, Thm. 15.9 ensures that (16.7) holds.

To complete the demonstration, we are left to prove that $\{\delta_\lambda\}_\lambda$ endows $\mathbb{G}$ with a structure of homogeneous Carnot group, in the sense of the statement. To this end, we split $\mathbb{R}^N$ according to the stratification of $\mathfrak{g}$, i.e.,

$$\mathbb{R}^N = \mathbb{R}^{N_1} \times \cdots \times \mathbb{R}^{N_r},$$

where, for every $i = 1,\ldots,r$, we have $N_i := \dim(V_i)$. Accordingly, we denote by $x = (x^{(1)},\ldots,x^{(r)})$ the points of $\mathbb{R}^N$ and, for $i = 1,\ldots,r$, we set

$$x^{(i)} = (x_1^{(i)},\ldots,x_{N_i}^{(i)}) \in \mathbb{R}^{N_i}.$$

By using these notations, for every $\lambda > 0$ and every $x \in \mathbb{R}^N$ we have

$$\delta_\lambda(x) \overset{(16.4)}{=} \pi_{\mathcal{E}}^{-1}\Big(\Delta_\lambda\Big(\textstyle\sum_{i=1}^r \sum_{k=1}^{N_i} x_k^{(i)} E_k^{(i)}\Big)\Big) \quad (\text{since } E_k^{(i)} \in V_i \text{ for } i \leq r)$$

$$\overset{(16.2)}{=} \pi_{\mathcal{E}}^{-1}\Big(\textstyle\sum_{i=1}^r \sum_{k=1}^{N_i} \lambda^i x_k^{(i)} E_k^{(i)}\Big) = (\lambda x^{(1)}, \lambda^2 x^{(2)},\ldots,\lambda^r x^{(r)});$$

this proves (16.5) and point (1) of the assertion. Moreover, since Δ_λ is an automorphism of $(\mathfrak{g}, \diamond)$ (by Prop. 16.4) and since $\pi_{\mathcal{E}}$ is a Lie-group homomorphism between $(\mathfrak{g}, \diamond)$ and $\mathbb{G}$ (see (15.16)), we deduce that $\delta_\lambda = \pi_{\mathcal{E}}^{-1} \circ \Delta_\lambda \circ \pi_{\mathcal{E}}$ is an automorphism of $\mathbb{G}$ for every $\lambda > 0$. This gives point (2).

Finally, let $J_1, \ldots, J_{N_1}$ be the first N_1 elements of the Jacobian basis of $\mathrm{Lie}(\mathbb{G})$. Since $\mathfrak{g}$ is generated (as a Lie algebra) by V_1 (see Prop. A.17) and since φ is a Lie algebra isomorphism (Thm. 15.7), we deduce that

$$\mathrm{Lie}(\mathbb{G}) = \mathrm{Lie}\{\varphi^{-1}(V_1)\}. \tag{16.8}$$

Recalling that $\{E_1^{(1)}, \ldots, E_{N_1}^{(1)}\}$ is a basis of V_1 ($\mathcal{E}$ being adapted to $\mathcal{V}$), by the definition of φ we have $\{J_1, \ldots, J_{N_1}\} = \varphi^{-1}\{E_1^{(1)}, \ldots, E_{N_1}^{(1)}\}$, so that (16.8) proves point (3) of the theorem. Finally we prove point (4) of the assertion. We fix any $i \in \{1, \ldots, N\}$; we know that J_i acts as follows:

$$J_i f(x) = \frac{\partial}{\partial y_i}\Big|_{y=0} \{f(x * y)\}, \qquad \forall f \in C^\infty(\mathbb{R}^N), \quad x \in \mathbb{G}.$$

Hence the following computation holds true:

$$J_i(f \circ \delta_\lambda)(x) = \frac{\mathrm{d}}{\mathrm{d}t}\Big|_{t=0}\Big\{(f \circ \delta_\lambda)(x * (te_i))\Big\} = \frac{\mathrm{d}}{\mathrm{d}t}\Big|_{t=0}\Big\{f\big(\delta_\lambda(x) * \delta_\lambda(te_i)\big)\Big\}$$

$$= \frac{\mathrm{d}}{\mathrm{d}t}\Big|_{t=0}\Big\{f\big(\delta_\lambda(x) * (\lambda^{\sigma_i} te_i)\big)\Big\} = \lambda^{\sigma_i}\frac{\mathrm{d}}{\mathrm{d}r}\Big|_{r=0}\Big\{f\big(\delta_\lambda(x) * (r\,e_i)\big)\Big\}$$

$$= \lambda^{\sigma_i}\frac{\partial}{\partial y_i}\Big|_{y=0}\Big\{f\big(\delta_\lambda(x) * y\big)\Big\} = \lambda^{\sigma_i}\,(J_i f)(\delta_\lambda(x)).$$

This proves that J_i is δ_λ-homogeneous of degree σ_i. $\qquad\square$

Remark 16.8. The above argument provides the explicit coordinate expression of $\{\delta_\lambda\}_\lambda$: more precisely, for every $x \in \mathbb{R}^N$ and every $\lambda > 0$ we have

$$\delta_\lambda(x) = \big(\lambda x^{(1)}, \lambda^2 x^{(2)}, \ldots, \lambda^r x^{(r)}\big),$$

where we have used the notation $x = (x^{(1)}, \ldots, x^{(r)})$, with $x^{(i)} \in \mathbb{R}^{N_i}$. $\qquad\sharp$

Example 16.9. Let us consider the Lie algebra $\mathfrak{g}$ introduced in Exr. 15.2. It is easy to see that, setting $V_1 := \mathrm{span}\{X_1, X_2\}$ and $V_2 := \mathrm{span}\{X_3\}$, the family $\mathcal{V} = \{V_1, V_2\}$ is a stratification of $\mathfrak{g}$, and $N_1 = \dim(V_1) = 2$ and $N_2 = \dim(V_2) = 1$; moreover, $\mathcal{E} = \{X_1, X_2, X_3\}$ is a basis for $\mathfrak{g}$ adapted to the stratification $\mathcal{V}$. Then, if $\{\delta_\lambda\}_\lambda$ is the family of dilations defined in (16.4), for every $\lambda > 0$ and every $x = (x^{(1)}, x^{(2)})$ in $\mathbb{R}^3 = \mathbb{R}^2 \times \mathbb{R}$, we have (see Rem. 16.8)

$$\delta_\lambda(x) = \big(\lambda x^{(1)}, \lambda^2 x^{(2)}\big) = \big(\lambda x_1^{(1)}, \lambda x_2^{(1)}, \lambda^2 x^{(2)}\big);$$

furthermore, Thm. 16.7 ensures that $\mathbb{G} = (\mathbb{R}^3, *_{\mathcal{E}}, \delta_\lambda)$ is a Carnot group. $\qquad\sharp$

Remark 16.10. A Lie group $(G, \cdot)$ is called a **Carnot group** if it is connected, simply connected and if its Lie algebra $\mathrm{Lie}(G)$ is stratified (in the sense of Def. 16.1). Since $\mathrm{Lie}(G)$ is finite-dimensional, then it is nilpotent too (see Cor. A.21).

It is not difficult to recognize that G is isomorphic to a homogeneous Carnot group as in Thm. 16.7. Indeed, by general results on nilpotent Lie groups (see e.g., [Corwin and Greenleaf (1990)]), one knows that $\mathrm{Exp}_G : (\mathrm{Lie}(G), \diamond) \to (G, \cdot)$ is a group isomorphism since G is connected, simply connected and nilpotent. Here $\diamond$ is the CBHD multiplication on $\mathrm{Lie}(G)$.

Now we can apply our Lie group construction related to the (finite-dimensional and nilpotent) Lie algebra $\mathfrak{g} := \mathrm{Lie}(G)$. We chose any basis $\mathcal{E} = \{E_1, \ldots, E_N\}$ of $\mathrm{Lie}(G)$ and, via Thm. 15.7, we know that $\mathbb{G} = (\mathbb{R}^N, *_{\mathcal{E}})$ is a Lie group isomorphic to $(\mathrm{Lie}(G), \diamond)$. As a consequence $\mathbb{G} \equiv \mathbb{R}^N$ and G are isomorphic Lie groups, via the map $(x_1, \ldots, x_N) \mapsto \mathrm{Exp}_G(x_1 E_1 + \cdots + x_N E_N)$. Most importantly, $\mathbb{G}$ brings along its *homogeneous* Carnot group structure, in the sense of the properties in Thm. 16.7. This shows that *any Carnot group is isomorphic to a homogeneous Carnot group.*   ♯

16.3   Exercises of Chap. 16

Exercise 16.1. Assume the following general result on vector spaces (a consequence of Zorn's Lemma): let $V \neq \{0\}$ be a vector space. Let also I, G be subsets of V such that $I \subseteq G$, I is linearly independent and G generates V; then there exists a basis $\mathcal{B}$ of V with $I \subseteq \mathcal{B} \subseteq G$.

Prove that, if $V \neq \{0\}$ is a vector space and $W \subseteq V$ is any vector subspace, then there exists a direct complement U of W in V, that is U is a vector subspace of V such that[1] $V = U \oplus W$.

Exercise 16.2. Prove that any Lie algebra $\mathfrak{g}$ which is nilpotent of step 2 is stratified. Complete the following argument: Let $V_2 := [\mathfrak{g}, \mathfrak{g}]$ and let V_1 be any direct complement of V_2 in $\mathfrak{g}$, that is $\mathfrak{g} = V_1 \oplus V_2$ (see Exr. 16.1). Show that $[V_1, V_2] = \{0\}$. Finally show that $[V_1, V_1] = V_2$.

[*Hint:* the inclusion $\subseteq$ is trivial; vice versa, any bracket $[w_1, w_2]$ belongs to $[V_1, V_1]$ for every $w_1, w_2 \in \mathfrak{g}$: indeed, $w_i = a_i + b_i$ with $a_i \in V_1$ and $b_i \in V_2$; thus $[w_1, w_2] = [a_1, a_2] \in [V_1, V_1]$, for $[a_1, b_2] = [b_1, a_2] = [b_1, b_2] = 0\ldots$]

Exercise 16.3. Let $(\mathfrak{s}, \mathcal{V} = \{V_n\}_n)$ be a stratified Lie algebra (not necessarily of finite dimension). Define a family of dilations $\{\Delta_\lambda\}_{\lambda>0}$ on $\mathfrak{s}$ as follows: Δ_λ is the unique endomorphism of $\mathfrak{s}$ such that $\Delta_\lambda\big|_{V_n} = \lambda^n \, \mathrm{id}_{V_n}$, for every $n \in \mathbb{N}$.

Prove that Δ_λ is well-defined, is a Lie algebra automorphism of $\mathfrak{s}$, and, for every $a \in \mathfrak{s}$ and every $\lambda > 0$, one has (see the notation in Rem. A.19)

$$\Delta_\lambda(a) = \Delta_\lambda\Big(\sum_{n=1}^{\infty} \mathrm{p}_n(a)\Big) = \sum_{n=1}^{\infty} \lambda^n \, \mathrm{p}_n(a).$$

---

[1] $V = U \oplus W$ means that any $v \in V$ can be *uniquely* written as $v = u + w$ where $u \in U$, $w \in W$.

Chapter 17

Exponentiation of Vector Field Algebras into Lie Groups

$\mathbf{T}$HE main aim of this chapter is to provide an exhaustive answer to the following natural question:

(Q) *Given a Lie subalgebra* $\mathfrak{g}$ *of the smooth vector fields on* $\mathbb{R}^N$, *is it possible to find a Lie group* $\mathbb{G} = (\mathbb{R}^N, *)$ *(whose underlying manifold is* $\mathbb{R}^N$ *endowed with its usual differentiable structure) such that* $\mathrm{Lie}(\mathbb{G}) = \mathfrak{g}$?

It is clear that, if we do not assume any hypothesis on $\mathfrak{g}$, the answer is negative: for example, if a vector field in $\mathfrak{g}$ is not global, then $\mathfrak{g}$ cannot be the Lie algebra of any Lie group on $\mathbb{R}^N$. Taking into account what is known in Lie group theory, it is not difficult to find some *necessary*[1] conditions for (Q) to have a positive answer:

(1) every $X \in \mathfrak{g}$ must be a global vector field (see Def. 1.17);

(2) $\mathfrak{g}$ must satisfy Hörmander's rank condition (see Def. 6.5):
$$\dim\left(\{X(x) \in \mathbb{R}^N \,:\, X \in \mathfrak{g}\}\right) = N, \quad \text{for every } x \in \mathbb{R}^N;$$

(3) the dimension of $\mathfrak{g}$, as a linear subspace of $\mathfrak{X}(\mathbb{R}^N)$, must be equal to N.

The main result of this chapter shows that the above conditions are also a set of *independent and sufficient* conditions on $\mathfrak{g}$ for a positive answer to (Q).

The means we shall use in order to prove this fact are the following:

• the Campbell-Baker-Hausdorff-Dynkin Theorem (for the composition of flows of v.f.s) in order to equip $\mathbb{R}^N$ with a local Lie-group structure;

• the use of a prolongation argument for ODEs in order to globalize this local Lie group. To this end, the uniqueness of the solution of a Cauchy problem will play a fundamental role as a powerful "globalizing" tool.

Prerequisites for this chapter are a few elements of Lie group theory, given in App. C, and –most importantly– the CBHD Thm. 13.9 for ODEs.

---

[1]The necessity of (1) is shown in Rem. C.3; the necessity of (2) is contained in Exm. C.27-(4); for the necessity of (3), see the identification in (C.4).

17.1   The assumptions for the exponentiation

The main goal of this section is to introduce, for a selected class of Lie algebras $\mathfrak{g} \subset \mathfrak{X}(\mathbb{R}^N)$, the exponential map and the logarithmic map. Such maps will be fundamental to answer question (Q) posed above.

Convention. Since in this chapter we shall only deal with smooth v.f.s on $\mathbb{R}^N$, in the sequel it is tacitly understood that $\mathfrak{g}$ *is a Lie sub-algebra of* $\mathfrak{X}(\mathbb{R}^N)$.

Definition 17.1. We shall say that $\mathfrak{g}$ satisfies hypothesis

(G) if every $X \in \mathfrak{g}$ is a global vector field;

(H) if Hörmander's rank condition holds:

$$\dim\left(\{X(x) \in \mathbb{R}^N \, : \, X \in \mathfrak{g}\}\right) = N \quad \text{for every } x \in \mathbb{R}^N; \tag{17.1}$$

(ND) if $\mathfrak{g}$ is N-dimensional, as a linear subspace of $\mathfrak{X}(\mathbb{R}^N)$.

Here '(G)' stands for 'global'; '(H)' stands for 'Hörmander v.f.s'; '(ND)' stands for 'N-dimensional'.

Remark 17.2. To distinguish the two dimensions appearing in conditions (H) and (ND) we observe that, for every subspace V of $\mathfrak{X}(\mathbb{R}^N)$ and every $x \in \mathbb{R}^N$, one has

$$\dim\left(\{X(x) \in \mathbb{R}^N \, : \, X \in V\}\right) \leq \dim(V).$$

Indeed, $V \ni X \mapsto \Lambda(X) := X(x) \in \mathbb{R}^N$ is linear and $\Lambda(V) = \{X(x) : X \in V\}$.   ♯

     We have already remarked that conditions (G), (H) and (ND) are necessary for question (Q) to have a positive answer. We now highlight the independence[2] of these conditions with the aid of the following examples.

Example 17.3 ((H)+(ND)$\nRightarrow$(G)). Consider in $\mathfrak{X}(\mathbb{R})$ the v.f. $X := (1 + x_1^2)\frac{\partial}{\partial x_1}$, and let $\mathfrak{g} := \mathrm{Lie}\{X\}$. It is easy to recognize that $\mathfrak{g}$ satisfies (H) and (ND) with $N = 1$, but it violates (G): indeed, the integral curve of X starting at 0 is the function $t \mapsto \tan t$, which is not defined on the whole of $\mathbb{R}$.   ♯

Example 17.4 ((G)+(ND)$\nRightarrow$(H)). Consider, in $\mathfrak{X}(\mathbb{R})$, the v.f. $X := x_1 \frac{\partial}{\partial x_1}$, and let $\mathfrak{g} := \mathrm{Lie}\{X\}$. Then $\mathfrak{g}$ satisfies (G) and (ND) with $N = 1$. On the other hand, condition (H) does not hold, since (17.1) is not satisfied at $x = 0$.   ♯

Example 17.5 ((G)+(H)$\nRightarrow$(ND)). Let us consider, in $\mathfrak{X}(\mathbb{R})$, the v.f.s

$$X := x_1 \frac{\partial}{\partial x_1}, \quad Y := \frac{\partial}{\partial x_1},$$

and let $\mathfrak{g} := \mathrm{Lie}\{X, Y\}$. Since $[X, Y] = X$, condition (ND) does not hold: in fact X, Y being linearly independent in $\mathfrak{X}(\mathbb{R})$, we have

$$\mathfrak{g} = \mathrm{span}_{\mathbb{R}}\{X, Y\}, \text{ whence } \dim(\mathfrak{g}) = 2.$$

---

[2]This means that the joint validity of two of them does not imply the validity of the third.

On the other hand, $\mathfrak{g}$ satisfies conditions (G) and (H) with $N = 1$.

We remark that a Lie algebra $\mathfrak{g}$ can satisfy conditions (G) and (H) without being finite-dimensional (as a subspace of $\mathfrak{X}(\mathbb{R}^N)$). This is the case, e.g., of the Lie algebra generated by the v.f.s in $\mathbb{R}^2$

$$X := \frac{\partial}{\partial x_1}, \quad Y := \frac{1}{1 + x_1^2} \frac{\partial}{\partial x_2}.$$

Owing to Def. 17.1, we can state the main theorem of this chapter, which provides a complete answer to question (Q). The proof of this theorem is accomplished in Secs. 17.2 and 17.3.

Theorem 17.6. *Let $\mathfrak{g}$ be a Lie algebra of smooth vector fields on $\mathbb{R}^N$ satisfying conditions (G), (H) and (ND) in Def. 17.1.*

Then, there exists a Lie group $\mathbb{G} = (\mathbb{R}^N, *)$ (with underlying manifold given by $\mathbb{R}^N$ with its usual differentiable structure), with neutral element 0, and such that $\mathrm{Lie}(\mathbb{G}) = \mathfrak{g}$. This equality is not intended up to some isomorphism, but as an equality of two sets of linear first-order differential operators.

The first ingredient to prove Thm. 17.6 is the definition of exponentiation of a Lie algebra $\mathfrak{g} \subset \mathfrak{X}(\mathbb{R}^N)$ satisfying condition (G), having a strong analogy with the definition of Exponential Map on a Lie group (Sec. C.1.2).

Definition 17.7 (Exponentiation of $\mathfrak{g}$). Let $\mathfrak{g}$ satisfy condition (G). We set
$$\mathrm{Exp}_{\mathfrak{g}} : \mathfrak{g} \longrightarrow \mathbb{R}^N, \quad \mathrm{Exp}_{\mathfrak{g}}(X) := \gamma(1, X, 0),$$
where $t \mapsto \gamma(t, X, 0)$ denotes the integral curve of X starting at 0. With an abuse of language borrowed from Lie group theory, we shall sometimes call this map the **exponential map** of $\mathfrak{g}$. We also denote $\mathrm{Exp}_{\mathfrak{g}}(X)$ by $\exp(X)(0)$.

Assumption (G) on $\mathfrak{g}$ is essential for Def. 17.7 to make sense: indeed, if $X \in \mathfrak{g}$ is not complete, the integral curve of X starting at 0 may not be defined for $t = 1$.

Remark 17.8. Let $\mathfrak{g}$ satisfy condition (G), and let us suppose that there exists a Lie group $\mathbb{G} = (\mathbb{R}^N, *)$ with Lie algebra equal to $\mathfrak{g}$. If the neutral element of $\mathbb{G}$ is 0, then the reader acquainted with Lie group theory knows that the above map $\mathrm{Exp}_{\mathfrak{g}}$ is nothing but the Exponential Map of $\mathbb{G}$ (see Def. C.7). ♯

Our next purpose is to investigate the regularity of the map $\mathrm{Exp}_{\mathfrak{g}}$: to this end, we need an additional structure on $\mathfrak{g}$ allowing us to talk about open sets and smooth functions. Hence, we assume that $\mathfrak{g}$ also satisfies condition (ND): if this is the case, the vector space $\mathfrak{g}$ can be endowed with the topological-differentiable structure by identifying it with $\mathbb{R}^N$ via the choice of a basis.

Lemma 17.9. *Let $\mathfrak{g}$ satisfy conditions (ND) and (H). Then there exists a basis of $\mathfrak{g}$ as a subspace of $\mathfrak{X}(\mathbb{R}^N)$, say $\{J_1, \ldots, J_N\}$, such that*

$$\det(J_1(x) \cdots J_N(x)) \neq 0 \quad \text{for all } x \in \mathbb{R}^N, \tag{17.2}$$

$$(J_1(0) \cdots J_N(0)) = \mathbb{I}_N, \tag{17.3}$$

where $\mathbb{I}_N$ is the $N \times N$ identity matrix.

In the sequel, when we refer to coordinates on $\mathfrak{g}$, we tacitly mean wrt the above basis $\{J_1, \ldots, J_N\}$; the same is true if we mention a differentiable structure on $\mathfrak{g}$ (this time the choice of a basis being immaterial).

Proof. First of all, since $\mathfrak{g}$ satisfies condition (ND), there exist $Z_1, \ldots, Z_N$ in $\mathfrak{g}$ such that $\mathcal{Z} := \{Z_1, \ldots, Z_N\}$ is a basis of $\mathfrak{g}$. We claim that, for every $x \in \mathbb{R}^N$, the vectors $Z_1(x), \ldots, Z_N(x)$ are linearly independent in $\mathbb{R}^N$.

Indeed, since $\mathfrak{g}$ also satisfies condition (H), there exist $W_1, \ldots, W_N \in \mathfrak{g}$ such that the vectors $W_1(x), \ldots, W_N(x)$ are linearly independent (in $\mathbb{R}^N$); on the other hand, since $\mathcal{Z}$ is a basis of $\mathfrak{g}$, we have

$$\mathrm{span}_{\mathbb{R}}\{W_1(x), \ldots, W_N(x)\} \subseteq \mathrm{span}_{\mathbb{R}}\{Z_1(x), \ldots, Z_N(x)\},$$

and this shows that $Z_1(x), \ldots, Z_N(x)$ are linearly independent, as claimed.

A simple linear change of coordinates drives from $\mathcal{Z}$ to a basis $\{J_1, \ldots, J_N\}$ satisfying (17.2) and (17.3). $\qquad\square$

Remark 17.10. If $\mathfrak{g}$ satisfies condition (ND) and if it is possible to find a basis $\{J_1, \ldots, J_N\}$ of $\mathfrak{g}$ (as a subspace of $\mathcal{X}(\mathbb{R}^N)$) such that $J_1(x), \ldots, J_N(x)$ are linearly independent for all $x \in \mathbb{R}^N$, then it is easy to see that $\mathfrak{g}$ also satisfies condition (H).

Thus, due to Lem. 17.9, if $\mathfrak{g}$ satisfies (ND), then it fulfills condition (H) if and only if it fulfills (H$'$), where

(H$'$): there exists a basis $\{J_1, \ldots, J_N\}$ of $\mathfrak{g}$ such that $J_1(x), \ldots, J_N(x)$ are linearly independent for every $x \in \mathbb{R}^N$. $\qquad\sharp$

Remark 17.11. Let $\mathfrak{g}$ satisfy (H) and (ND), and let us assume that $\mathbb{G} = (\mathbb{R}^N, *)$ is a Lie group on $\mathbb{R}^N$, with neutral element 0, such that $\mathrm{Lie}(\mathbb{G})$ is equal to $\mathfrak{g}$. Then, a basis of $\mathfrak{g}$ as in Lem. 17.9 is unique, since a left invariant vector fields X is completely determined by its value $X(0)$. $\qquad\sharp$

Proposition 17.12. *Let $\mathfrak{g}$ satisfy conditions* (G), (H) *and* (ND). *Then* $\mathrm{Exp}_{\mathfrak{g}}$ *is a smooth map on $\mathfrak{g}$ with non-singular differential at $0 \in \mathfrak{g}$.*

Consequently, there exists an open and connected neighborhood $\mathfrak{U}$ of 0 in $\mathfrak{g}$ such that $(\mathrm{Exp}_{\mathfrak{g}})|_{\mathfrak{U}}$ *is a diffeomorphism.*

Proof. The regularity of $\mathrm{Exp}_{\mathfrak{g}}$ follows from Thm. B.41 in App. B. We now turn to show that the differential of $\mathrm{Exp}_{\mathfrak{g}}$ at $X = 0$ in non-singular. To this end, we consider the coordinate representation of $\mathrm{Exp}_{\mathfrak{g}}$ given by

$$E : \mathbb{R}^N \to \mathbb{R}^N, \quad E(\xi) := \mathrm{Exp}_{\mathfrak{g}}\left(\textstyle\sum_{k=1}^{N} \xi_k J_k\right),$$

and we compute its Jacobian matrix at $\xi = 0$. From the Maclaurin expansion in (1.20) (page 18), it is easy to show that

$$E(\xi) = \xi + \mathcal{O}(\|\xi\|^2), \qquad \text{as } \xi \to 0,$$

whence $\mathcal{J}_E(0) = \mathbb{I}_N$, which ends the proof. $\qquad\square$

Remark 17.13. It is in the proof of Prop. 17.12 the following fact: under conditions (G), (H), (ND), if $\{J_1, \ldots, J_N\}$ is a basis of $\mathfrak{g}$ as in Lem. 17.9, and if

$$\pi(\xi) := \sum_{k=1}^{N} \xi_k J_k \quad \text{and} \quad E := \mathrm{Exp}_{\mathfrak{g}} \circ \pi,$$

then E is a smooth map and $\mathcal{J}_E(0) = \mathbb{I}_N$, where $\mathbb{I}_N$ is the identity $N \times N$ matrix. ♯

Definition 17.14 (Logarithmic map on $\mathfrak{g}$). Let $\mathfrak{g}$ satisfy (G), (H) and (ND), and let $\mathfrak{U}$ be as in Prop. 17.12. We set $V := \mathrm{Exp}_{\mathfrak{g}}(\mathfrak{U})$ and we denote by $\mathrm{Log}_{\mathfrak{g}} : V \to \mathfrak{U}$ the inverse map of $\mathrm{Exp}_{\mathfrak{g}} : \mathfrak{U} \to V$.

We call this map the **logarithmic map** of $\mathfrak{g}$ (relative to $\mathfrak{U}$).

17.2   Construction of the local Lie group

In this section we shall show that, if $\mathfrak{g}$ satisfies (G), (H) and (ND), it is possible to endow $\mathbb{R}^N$ with a *local* Lie-group structure in such a way that the vector fields in $\mathfrak{g}$ are (locally) left invariant.

To begin with, to keep the exposition clear, we fix once and for all the main notations used in the sequel:

- we denote by $\mathfrak{g}$ a fixed Lie algebra of smooth vector fields on $\mathbb{R}^N$ satisfying conditions (G), (H) and (ND) in Def. 17.1;
- we denote by Exp the exponential map $\mathrm{Exp}_{\mathfrak{g}}$ of $\mathfrak{g}$ and we let $\mathrm{Log} : V \to \mathfrak{U}$ denote its local inverse (with $V := \mathrm{Exp}(\mathfrak{U})$) as in Def. 17.14;
- we fix a basis $\mathcal{J} = \{J_1, \ldots, J_N\}$ of $\mathfrak{g}$ as in Lem. 17.9 and we introduce the map $\pi : \mathbb{R}^N \to \mathfrak{g}$ defined by $\pi(\xi) := \sum_{k=1}^{N} \xi_k J_k$.

17.2.1   *The local Lie-group multiplication*

We let the above notations apply; then we give the following definition.

Definition 17.15 (The local Lie-group multiplication). We set

$$m : \mathbb{R}^N \times V \longrightarrow \mathbb{R}^N, \qquad m(x, y) := \exp(\mathrm{Log}(y))(x). \tag{17.4}$$

As usual, if $X \in \mathfrak{X}(\mathbb{R}^N)$ and if $x \in \mathbb{R}^N$, we denote by $t \mapsto \exp(t\,X)(x)$ the maximal integral curve of X starting at x.

By the results in App. B it follows that m is smooth on $\mathbb{R}^N \times V$.

Remark 17.16. Let us assume that there exists a Lie group $\mathbb{G} = (\mathbb{R}^N, *)$, with neutral element 0, and such that $\mathrm{Lie}(\mathbb{G}) = \mathfrak{g}$. As pointed out in Rem. 17.8, the exponential map Exp on $\mathfrak{g}$ coincides with the Exponential Map of $\mathbb{G}$; as a consequence, if $y \in V$ and if $Y = \mathrm{Log}(y) \in \mathfrak{g}$, the reader familiar with Lie group theory knows that (see Thm. C.10)

$$m(x, y) = \exp(Y)(x) = \gamma(1, Y, x) = x * y.$$

We want to show that m in (17.4) is locally associative near 0, and that 0 is a neutral element for m. To this end, we first remind the following notable result, deriving from Thm. 13.9 on page 273.

Theorem 17.17. *Let $\mathfrak{h}$ be a Lie algebra of smooth vector fields on $\mathbb{R}^N$ satisfying conditions (G) and (ND), ant let $\| \cdot \|$ be a fixed norm on $\mathfrak{h}$. There exists a positive real number ε, depending on $\| \cdot \|$, such that the CBHD series (Def. 2.24)*

$$Z(X,Y) := \sum_{h=1}^{\infty} Z_h(X,Y)$$

is totally convergent on $\mathfrak{B}(0,\varepsilon) \times \mathfrak{B}(0,\varepsilon)$, where $\mathfrak{B}(0,\varepsilon) := \{V \in \mathfrak{h} : \|V\| < \varepsilon\}$. Furthermore, for every $X,Y \in \mathfrak{B}(0,\varepsilon)$, we have the following ODE identity

$$\exp(Y)\big(\exp(X)(x)\big) = \exp\big(Z(X,Y)\big)(x), \quad \textit{for every } x \in \mathbb{R}^N. \tag{17.5}$$

As usual, we also use the notation $X \diamond Y := Z(X,Y)$.

In order to apply the remarkable identity (17.5) to our setting, we need to fix a norm on $\mathfrak{g}$; for simplicity, we consider the Euclidean norm obtained by identifying $\mathfrak{g}$ with $\mathbb{R}^N$ via the basis $\mathcal{J}$, that is,

$$\left\| \sum_{k=1}^{N} \xi_k J_k \right\|_{\mathfrak{g}} := \sqrt{\xi_1^2 + \cdots + \xi_N^2}.$$

For balls wrt $\| \cdot \|_{\mathfrak{g}}$ we use the letter $\mathfrak{B}$, whereas B is used for Euclidean balls in $\mathbb{R}^N$. By means of Thm. 17.17, we are able to derive a powerful representation for the map m as in the next theorem.

Theorem 17.18. *Let $\varepsilon > 0$ be as in Thm. 17.17 and let us suppose (by possibly shrinking ε) that $\mathfrak{B}(0,\varepsilon) \subseteq \mathfrak{U}$. It is then possible to find a real $\rho_0 > 0$ such that*

$$\mathfrak{Z} : B(0,\rho_0) \times B(0,\rho_0) \longrightarrow \mathfrak{B}(0,\varepsilon) \qquad \mathfrak{Z}(x,y) := \mathrm{Log}(x) \diamond \mathrm{Log}(y)$$

is well-defined and, for every $x,y \in B(0,\rho_0)$, the following identity holds true

$$m(x,y) = \mathrm{Exp}(\mathfrak{Z}(x,y)). \tag{17.6}$$

Proof. Let $Z : \mathfrak{B}(0,\varepsilon) \times \mathfrak{B}(0,\varepsilon) \to \mathfrak{g}$ be defined by $Z(X,Y) := X \diamond Y$. Since, by Thm. 17.17 (and by the choice of ε), the CBHD series $X \diamond Y$ is totally convergent on $\mathfrak{B}(0,\varepsilon) \times \mathfrak{B}(0,\varepsilon)$, Z is well-defined and continuous on its domain; as a consequence, it is possible to find $0 < \varepsilon_1 < \varepsilon$ such that

$$Z(X,Y) \in \mathfrak{B}(0,\varepsilon), \quad \text{for all } X,Y \in \mathfrak{B}(0,\varepsilon_1). \tag{17.7}$$

Analogously, since Log is continuous on V and $\mathrm{Log}(0) = 0$, there exists $\delta > 0$ such that $B(0,\delta) \subseteq V$ and

$$\mathrm{Log}(x) \in \mathfrak{B}(0,\varepsilon_1), \quad \text{for all } x \in B(0,\delta). \tag{17.8}$$

We then set $\rho_0 := \delta$ and we show that it satisfies all the properties in the statement of the theorem. To this end, let $x,y \in B(0,\rho_0)$ be fixed.

By (17.8), we have $\mathrm{Log}(x), \mathrm{Log}(y) \in \mathfrak{B}(0, \varepsilon_1)$; therefore, as $\varepsilon_1 < \varepsilon$, by (17.7) we see that the CBHD series $\mathrm{Log}(x) \diamond \mathrm{Log}(y)$ is convergent, whence $\mathfrak{z}$ is well-defined, and $\mathfrak{z}(x, y)$ belongs to $\mathfrak{B}(0, \varepsilon)$.

As for identity (17.6) we observe that, since $B(0, \rho_0) \subseteq V$, we have

$$m(x, y) \overset{(17.4)}{=} \exp(\mathrm{Log}(y))(x) = \exp\big(\mathrm{Log}(y)\big)\big(\mathrm{Exp}(\mathrm{Log}(x))\big)$$
$$= \exp\big(\mathrm{Log}(y)\big)\big(\exp(\mathrm{Log}(x))(0)\big).$$

Thus, by gathering together (17.8) and (17.5), we conclude that

$$m(x, y) = \exp(\mathrm{Log}(x) \diamond \mathrm{Log}(y))(0) = \mathrm{Exp}(\mathfrak{z}(x, y)),$$

which is exactly what we wanted to prove. $\qquad\square$

Remark 17.19. Let $\rho_0 > 0$ be as in Thm. 17.18. If $x \in B(0, \rho_0)$, we have

$$\|\mathrm{Log}(x)\|_{\mathfrak{g}} < \varepsilon. \tag{17.9}$$

Indeed, if $x \in B(0, \rho_0)$, we have $\mathrm{Log}(x) = \mathfrak{z}(x, 0)$; thus, from Thm. 17.18 we infer that $\mathrm{Log}(x) \in \mathfrak{B}(0, \varepsilon)$, which is exactly (17.9). In particular, since $\mathfrak{B}(0, \varepsilon)$ is symmetric wrt multiplication by -1, we have

$$-\mathrm{Log}(x) \in \mathfrak{B}(0, \varepsilon), \quad \text{for all } x \in B(0, \rho_0). \tag{17.10}$$

By Thm. 17.18, we can provide a simple proof of the local associativity of m.

Theorem 17.20. *Let $\rho_0 > 0$ be as in Thm. 17.18. Then $m(a, b) \in V$ for every fixed $a, b \in B(0, \rho_0)$ and m is associative near the origin; more precisely, we have*

$$m(x, m(y, z)) = m(m(x, y), z) \quad \forall\, x, y, z \in \mathbb{R}^N \text{ with } \|y\|, \|z\| < \rho_0. \tag{17.11}$$

Furthermore, the point $0 \in \mathbb{R}^N$ provides a local neutral element for m, that is,

$$m(x, 0) = x, \quad \text{for all } x \in \mathbb{R}^N, \tag{17.12}$$
$$m(0, y) = y, \quad \text{for all } y \in V. \tag{17.13}$$

Proof. Let $a, b \in B(0, \rho_0)$. Since $\mathfrak{z}$ takes values in $\mathfrak{B}(0, \varepsilon) \subseteq \mathfrak{U}$, we have

$$m(a, b) \overset{(17.6)}{=} \mathrm{Exp}(\mathfrak{z}(a, b)) \in \mathrm{Exp}(\mathfrak{U}) = V. \tag{17.14}$$

We now prove (17.11). To this end, let $x \in \mathbb{R}^N$ and let $y, z \in B(0, \rho_0)$. Firstly, by (17.14) (and since $B(0, \rho_0) \subseteq V$), both sides of (17.11) are well-defined; moreover, by means of Thm. 17.18 we can write

$$\mathrm{Log}(m(y, z)) \overset{(17.6)}{=} \mathrm{Log}(\mathrm{Exp}(\mathfrak{z}(y, z))) \qquad (\text{since } \mathfrak{z}(y, z) \in \mathfrak{U})$$
$$= \mathrm{Log}\Big(\mathrm{Exp}|_{\mathfrak{U}}(\mathfrak{z}(y, z))\Big) = \mathfrak{z}(y, z). \tag{17.15}$$

As a consequence, the lhs of (17.11) can be rewritten as follows:

$$m(x, m(y, z)) = \exp(\mathrm{Log}(m(y, z)))(x) \overset{(17.15)}{=} \exp(\mathfrak{z}(y, z))(x). \tag{17.16}$$

As for the rhs we observe that, by definition of m, we have

$$m(m(x,y),z) = \exp(\mathrm{Log}(z))(m(x,y)) = \exp\big(\mathrm{Log}(z)\big)(\exp(\mathrm{Log}(y)(x));$$

therefore, since $\mathrm{Log}(y), \mathrm{Log}(z) \in \mathcal{B}(0, \varepsilon)$ (by the choice of ρ_0, see (17.9)), we can apply identity (17.5), which gives

$$m(m(x,y),z) = \exp(\mathrm{Log}(y) \diamond \mathrm{Log}(z))(x) = \exp(\mathfrak{Z}(y,z))(x). \tag{17.17}$$

Finally, by comparing (17.16) and (17.17), we derive (17.11). As for identity (17.12), it is an easy consequence of the definition of m: indeed, if $x \in \mathbb{R}^N$,

$$m(x,0) \overset{(17.4)}{=} \exp(\mathrm{Log}(0))(x) = \exp(0)(x) = x.$$

On the other hand, if $y \in V$, by definition of Exp we have

$$m(0,y) = \exp(\mathrm{Log}(y))(0) = \mathrm{Exp}(\mathrm{Log}(y)) = y,$$

and this is precisely the desired (17.13). This ends the proof. $\qquad\square$

Remark 17.21. It is noteworthy that, in the proof of the local associativity of m, *we have not used the local associativity of the CBHD operation $\diamond$, but only its existence!* Indeed, a closer inspection to the previous proof shows that all that we used was the existence of *some vector-field valued function F* such that

$$\exp\big(Y\big)(\exp(X)(x)) = \exp(F(X,Y))(x), \tag{17.18}$$

for any x and for X, Y close to the zero vector field. *De facto*, this function F inherits an associative-like property from *the associativity of the composition of functions*: indeed (assuming a repeated application of (17.18) is legitimate)

$$\exp(Z)\Big(\exp(Y)\big(\exp(X)(0)\big)\Big) = \exp(Z)\Big(\exp(F(X,Y))(0)\Big)$$
$$= \exp\Big(F(F(X,Y),Z)\Big)(0),$$

and analogously (this time proceeding from the outer exp)

$$\exp(Z)\Big(\exp(Y)\big(\exp(X)(0)\big)\Big) = \exp(F(Y,Z))\big(\exp(X)(0)\big)$$
$$= \exp\Big(F(X,F(Y,Z))\Big)(0);$$

thus, provided we can apply Log to the far right-hand members, we infer

$$F(F(X,Y),Z) = F(X,F(Y,Z)).$$

Definition 17.22. We define

$$\iota : V \longrightarrow \mathbb{R}^N, \qquad \iota(x) := \mathrm{Exp}(-\mathrm{Log}(x)). \tag{17.19}$$

As in the case of m, the smoothness of the v.f.s in $\mathfrak{g}$ implies that $\iota \in C^\infty(V, \mathbb{R}^N)$; moreover, we have the following remark.

Remark 17.23. Let us assume that there exists a Lie group $\mathbb{G} = (\mathbb{R}^N, *)$, with neutral element 0 and such that $\mathrm{Lie}(\mathbb{G}) = \mathfrak{g}$. Then, for every $x \in V$, we have

$$\iota(x) = x^{-1}.$$

Indeed, since the map Exp is precisely the Exponential Map on $\mathbb{G}$, if $x \in V$ and if $X = \mathrm{Log}(x) \in \mathfrak{g}$, Prop. 1.14 on page 11 implies that

$$x * \iota(x) = x * \mathrm{Exp}(-\mathrm{Log}(x)) = x * \exp(-X)(0) = \exp(-X)(x)$$
$$= \exp(-X)(\mathrm{Exp}(X)) = \exp(-X)(\exp(X)(0)) = 0,$$

and this proves that $\iota(x) = x^{-1}$, as claimed. In the third equality we have used a known property in Lie group theory (see (C.9), page 391). $\qquad\sharp$

We now prove that the map ι provides a local inverse for m.

Theorem 17.24. *Let $\rho_0 > 0$ be as in Thm. 17.18. Then the map ι in (17.19) provides a local inverse for m on $B(0, \rho_0)$, that is,*

$$m(x, \iota(x)) = 0, \quad \text{for all } x \in B(0, \rho_0), \tag{17.20}$$
$$m(\iota(x), x) = 0, \quad \text{for all } x \in B(0, \rho_0). \tag{17.21}$$

Proof. Let $x \in B(0, \rho_0)$ be fixed and let $X = \mathrm{Log}(x) \in \mathfrak{g}$. By Rem. 17.19, we have $-X \in \mathcal{B}(0, \varepsilon) \subseteq \mathfrak{U}$, whence

$$\iota(x) = \mathrm{Exp}(-\mathrm{Log}(x)) = \mathrm{Exp}(-X) \in \mathrm{Exp}(\mathfrak{U}) = V;$$

as a consequence, since $\mathcal{B}(0, \varepsilon) \subseteq \mathfrak{U}$, we also have

$$\mathrm{Log}(\iota(x)) = \mathrm{Log}\big(\mathrm{Exp}(-X)\big) = -X = -\mathrm{Log}(x). \tag{17.22}$$

From these identities we deduce that $m(x, \iota(x))$ is well-defined (since $\iota(x)$ belongs to V) and that

$$m(x, \iota(x)) = \exp\big(\mathrm{Log}(\iota(x))\big)(x) \overset{(17.22)}{=} \exp(-X)(x) = \exp(-X)\big(\exp(X)(0)\big).$$

As a consequence, since $X, -X \in \mathcal{B}(0, \varepsilon)$, we have

$$m(x, \iota(x)) \overset{(17.5)}{=} \exp(X \diamond (-X))(0). \tag{17.23}$$

The desired (17.20) now follows from (17.23), by noticing that

$$X \diamond (-X) = X + (-X) + \sum_{h=2}^{\infty} Z_h(X, -X) = 0.$$

As for identity (17.21) we observe that, by definition of ι, we have

$$m(\iota(x), x) = \exp(\mathrm{Log}(x))(\iota(x)) = \exp\big(X\big)(\exp(-X)(0));$$

thus, by arguing as above, we conclude

$$m(\iota(x), x) \overset{(17.5)}{=} \exp((-X) \diamond X)(0) = 0.$$

This ends the proof. $\qquad\square$

17.2.2 *The local left invariance of* $\mathfrak{g}$

By gathering the results in Thms. 17.20 and 17.24, we see that m actually defines a local Lie-group structure on $\mathbb{R}^N$; we end this section by showing that the Lie algebra $\mathfrak{g}$ is deeply connected to this structure.

Theorem 17.25 (Local left-invariance of $\mathfrak{g}$). *For every $X \in \mathfrak{g}$ it holds that*

$$X(m(x,y)) = \frac{\partial m}{\partial y}(x,y)\, X(y), \quad \text{for every } (x,y) \in \mathbb{R}^N \times B(0, \rho_0), \tag{17.24}$$

where $\rho_0 > 0$ is as in Thm. 17.18.

Proof. We first prove that identity (17.24) holds for $y = 0$, that is,

$$X(x) = \frac{\partial m}{\partial y}(x,0)\, X(0), \quad \text{for all } x \in \mathbb{R}^N. \tag{17.25}$$

To this end, let $x \in \mathbb{R}^N$ and let $\eta > 0$ be such that $tX \in \mathfrak{U}$ for all $t \in \mathbb{R}$ with $|t| < \eta$. For these values of t, we have $\mathrm{Exp}(tX) \in \mathrm{Exp}(\mathfrak{U}) = V$, whence

$$\exp(tX)(x) = \exp\left(\mathrm{Log}(\mathrm{Exp}(tX))\right)(x) \overset{(17.4)}{=} m\left(x, \mathrm{Exp}(tX)\right). \tag{17.26}$$

By taking the derivative w.r.t. t of both sides of identity (17.26) and evaluating at $t = 0$, we get (since $t \mapsto \exp(tX)(x)$ is an integral curve of X)

$$\begin{aligned}
X(x) &= \frac{\mathrm{d}}{\mathrm{d}t}\Big|_{t=0}\big\{\exp(tX)(x))\big\} = \frac{\mathrm{d}}{\mathrm{d}t}\Big|_{t=0}\big\{m(x, \mathrm{Exp}(tX))\big\} \\
&= \frac{\partial m}{\partial y}(x,0)\,\frac{\mathrm{d}}{\mathrm{d}t}\Big|_{t=0}\mathrm{Exp}(tX) = \frac{\partial m}{\partial y}(x,0)\, X(0),
\end{aligned}$$

which is exactly the desired (17.25).

We now turn to prove identity (17.24). To this end we first notice that, since m is associative near the origin (as ensured by Thm. 17.20), we have

$$m(m(x,y),z) = m(x, m(y,z)), \quad \forall\, x \in \mathbb{R}^N \text{ with } \|y\|, \|z\| < \rho_0;$$

thus, by differentiating w.r.t. z the above identity and evaluating at $z = 0$, we get (setting, to avoid ambiguities, $m = m(\alpha, \beta)$)

$$\frac{\partial m}{\partial \beta}(m(x,y),0) = \frac{\partial m}{\partial \beta}(x, m(y,0))\,\frac{\partial m}{\partial \beta}(y,0). \tag{17.27}$$

From this, by multiplying both sides of (17.27) by the column vector $X(0)$, we obtain (remind that $m(y,0) = y$)

$$\frac{\partial m}{\partial \beta}(m(x,y),0)\, X(0) = \frac{\partial m}{\partial \beta}(x,y)\,\frac{\partial m}{\partial \beta}(y,0)\, X(0),$$

which gives (returning to the $m = m(x,y)$ notation),

$$\begin{aligned}
X(m(x,y)) &\overset{(17.25)}{=} \frac{\partial m}{\partial y}(m(x,y),0)\, X(0) = \frac{\partial m}{\partial y}(x,y)\,\frac{\partial m}{\partial y}(y,0)\, X(0) \\
&\overset{(17.25)}{=} \frac{\partial m}{\partial y}(x,y)\, X(y), \quad \text{for all } (x,y) \in \mathbb{R}^N \times B(0, \rho_0).
\end{aligned}$$

This is precisely the desired (17.24), and the proof is complete. $\qquad\square$

Throughout what follows, we tacitly understand that $\mathcal{J} = \{J_1, \ldots, J_N\}$ is a basis of $\mathfrak{g}$ as in Lem. 17.9, and we define the following $N \times N$ matrix

$$J(x) := \big(J_1(x), \ldots, J_N(x)\big), \quad x \in \mathbb{R}^N.$$

Due to (17.25) (with $X = J_i$) in the proof of Thm. 17.25, a notable differential identity links this matrix $J(x)$ to our local operation $m = m(x, y)$:

$$J(x) = \frac{\partial m}{\partial y}(x, 0), \quad \text{for every } x \in \mathbb{R}^N.$$

Consequently, identity (17.24) in Thm. 17.25 becomes

$$\frac{\partial m}{\partial y}(m(x, y), 0) = \frac{\partial m}{\partial y}(x, y) \frac{\partial m}{\partial y}(y, 0), \tag{17.28}$$

valid for every $x \in \mathbb{R}^N$ and every $y \in B(0, \rho_0)$. Equivalently, we can reconstruct the full Jacobian matrix of m wrt y in terms of J, as follows:

$$\frac{\partial m}{\partial y}(x, y) = J(m(x, y)) \, J(y)^{-1}, \tag{17.29}$$

for every $x \in \mathbb{R}^N$ and every $y \in B(0, \rho_0)$.

In Lie group theory, a similar identity is solved by the multiplication map m defined by $(x, y) \mapsto \tau_x(y)$ (where τ_x is the group left translation by x), and in this framework it is known as **Lie's First Theorem** (see e.g., [Cohn (1957), p. 95]). This is plainly not sheer chance since our map $m(x, y)$ is destined to be a left translation on a Lie group! (For some details, see Sec. C.1.4.)

Remark 17.26. *A surprising loop connects the associativity of m and the "left-invariance" identity (17.29), rewritten as*

$$J(m(x, y)) = \frac{\partial m}{\partial y}(x, y) \, J(y). \tag{17.30}$$

Indeed, we have proved that the local associativity of m implies (17.30); vice versa, as the hardest part of the next section, we shall prove that we can prolong m in such a way that (17.30) remains true for every x and y, and that (17.30) *implies the associativity of this prolongation of m.* $\sharp$

17.3   Local to global

The aim of this last section is to show that the local-group structure constructed in Sec. 17.2 can be (uniquely) continued to be global. In what follows, we fix all the notations introduced so far.

To begin with, we prove that the map m admits a C^∞-extension to the whole of $\mathbb{R}^N \times \mathbb{R}^N$. Our idea is the following: for every fixed $x, y \in \mathbb{R}^N$, we consider the curve $\gamma_{x,y}$ defined by

$$\gamma_{x,y}(t) := m(x, t\,y).$$

Since m is defined on $\mathbb{R}^N \times V$, there exists a (possibly small) open neighborhood of $0 \in \mathbb{R}$ on which $\gamma_{x,y}$ is well-defined. We show that $\gamma_{x,y}$ satisfies a suitable Cauchy problem which possesses a (unique) global maximal solution, say $\mathbb{R} \ni t \mapsto \varphi_{x,y}(t)$. Then it is natural to extend m as follows

$$x * y := \varphi_{x,y}(1).$$

This is accomplished in the next section.

17.3.1 *Schur's ODE on $\mathfrak{g}$ and prolongation of solutions*

We decided to name the (crucial!) ODE appearing in the next result as *Schur's ODE*, since it was [Schur (1889)] who first studied the local exponentiation of Lie algebras of vector fields by means of ODE identities.

Theorem 17.27 (Schur's ODE). *Let $x, y \in \mathbb{R}^N$ be fixed and let $I_y \subseteq \mathbb{R}$ be a real interval containing 0 such that $ty \in B(0, \rho_0)$ for all $t \in I_y$. We set*

$$\gamma_{x,y} : I_y \longrightarrow \mathbb{R}^N, \qquad \gamma_{x,y}(t) := m(x, ty).$$

Then the curve $t \mapsto \gamma_{x,y}(t)$ is a solution on I_y of the following Cauchy problem

$$\dot{z}(t) = \sum_{k=1}^{N} a_k(t, y) J_k(z(t)), \quad z(0) = x, \tag{17.31}$$

where $A(t, y) = (a_1(t, y), \ldots, a_N(t, y))^T$ is defined by[3]

$$A(t, y) := (J(t\,y))^{-1}\, y. \tag{17.32}$$

Equivalently, the ODE in (17.31) can be written in the following way

$$\dot{z}(t) = J(z(t))\, J(ty)^{-1}\, y. \tag{17.33}$$

*We say that (17.33) is **Schur's ODE** (for $\mathfrak{g}$ and y) and that (17.31) is **Schur's (Cauchy) problem** (for $\mathfrak{g}$ and x, y).*

Note that Schur's ODE is non-autonomous.

Proof. Firstly, since m is smooth on $\mathbb{R}^N \times V$, then $\gamma_{x,y} \in C^\infty(I, \mathbb{R}^N)$; moreover, by exploiting (17.29), we get

$$\dot{\gamma}_{x,y}(t) = \frac{\partial m}{\partial y}(x, ty)\, y \overset{(17.29)}{=} J(m(x, ty))\, J(ty)^{-1}\, y$$

$$= J(\gamma_{x,y}(t))\, A(t, y) = \sum_{k=1}^{N} a_k(t, y) J_k(\gamma_{x,y}(t)).$$

Thus $\gamma_{x,y}$ satisfies the ODE in (17.31); $\gamma_{x,y}(0) = x$ follows from (17.12). $\square$

Our next task is to prove that (17.31) admits a maximal solution which is defined on the whole of $\mathbb{R}$. To this end, we first establish the following result.

---

[3]In the rhs of (17.32), y is intended as an $N \times 1$ column vector; the same is true of the function $A(t, y)$ in the lhs.

Theorem 17.28 (Prolongation of the solution of Schur's ODE).
Suppose that $X_1, \ldots, X_n \in \mathfrak{g}$ and let $\alpha_1, \ldots, \alpha_n \in C(\mathbb{R})$. Then, for every $\xi \in \mathbb{R}^N$, the maximal solution of the Cauchy problem

$$\dot{z}(t) = \sum_{k=1}^{n} \alpha_k(t) X_k(z(t)), \quad z(0) = \xi, \tag{17.34}$$

is defined on the whole of $\mathbb{R}$.

As a consequence, for any $x, y \in \mathbb{R}^N$, the maximal solution, say $t \mapsto \varphi_{x,y}(t)$, of Schur's Cauchy problem (17.31) is defined on the whole of $\mathbb{R}$.

Proof. Let $\varphi : \mathcal{D} \to \mathbb{R}^N$ be the unique maximal solution of (17.34) and let us assume, by contradiction, that $\mathcal{D} \neq \mathbb{R}$. To fix ideas, we suppose that

$$0 < T := \sup(\mathcal{D}) < \infty,$$

and we set $K := [0, T]$. By Exr. 17.4, there exists $\varepsilon > 0$ such that the (unique) maximal solution u_s of the parametric problem

$$\dot{x}(t) = \sum_{k=1}^{n} \alpha_k(t + s) X_k(x(t)), \quad x(0) = 0, \tag{17.35}$$

is defined at least on $[-\varepsilon, \varepsilon]$, uniformly for $s \in K$, and it satisfies

$$|u_s(t)| \le \rho_0, \quad \text{for all } t \in [-\varepsilon, \varepsilon] \text{ and every } s \in K. \tag{17.36}$$

Let now $\tau \in (0, T)$ be such that $T - \tau < \varepsilon$ and let $x := \varphi(\tau)$ (note that x is well-defined, since $\tau \in (0, T) \subseteq \mathcal{D}$). We then define

$$\nu : [0, \varepsilon] \longrightarrow \mathbb{R}^N, \quad \nu(t) := m(x, u_\tau(t)), \tag{17.37}$$

where u_τ is the maximal solution of the Cauchy problem (17.35) with $s = \tau$.

We observe that ν is well-defined and smooth on $[0, \varepsilon]$, since m belongs to $C^\infty(\mathbb{R}^N \times V, \mathbb{R}^N)$ and, by (17.36), we have

$$u_\tau(t) \in B(0, \rho_0) \subseteq V, \quad \text{for all } t \in [0, \varepsilon].$$

We claim that ν solves on $[0, \varepsilon]$ the following Cauchy problem

$$\dot{z}(t) = \sum_{k=1}^{n} a_k(t + \tau) X_k(z(t)), \quad z(0) = x.$$

Indeed, since $m(x, 0) = x$, we have $\nu(0) = m(x, u_\tau(0)) = m(x, 0) = x$; moreover, by Thm. 17.25 (and since $u_\tau(t) \in B(0, \rho_0)$ for all $t \in [0, \varepsilon]$), one has

$$\begin{aligned}
\dot{\nu}(t) &= \frac{\partial m}{\partial y}(x, u_\tau(t))\, \dot{u}_\tau(t) \\[4pt]
&\overset{(17.35)}{=} \sum_{k=1}^{n} a_k(t + \tau) \left(\frac{\partial m}{\partial y}(x, u_\tau(t))\, X_k(u_\tau(t)) \right) \\[4pt]
&\overset{(17.24)}{=} \sum_{k=1}^{n} a_k(t + \tau) X_k\big(m(x, u_\tau(t))\big) \overset{(17.37)}{=} \sum_{k=1}^{n} a_k(t + \tau) X_k(\nu(t)).
\end{aligned}$$

We then consider the gluing of φ and ν, that is, the map

$$\Phi : [0, \tau + \varepsilon] \longrightarrow \mathbb{R}^N, \quad \Phi(t) := \begin{cases} \varphi(t), & t \in [0, \tau] \\ \nu(t - \tau) & t \in (\tau, \tau + \varepsilon]. \end{cases}$$

It is easy to prove that $\Phi \in C^1([0, \tau + \varepsilon], \mathbb{R}^N)$, and that Φ is a solution of (17.34). As a consequence, since $\tau + \varepsilon > T$ (by the choice of τ), Φ turns out to be a prolongation of φ beyond $[0, T)$; this is clearly in contradiction to the maximality of φ. $\qquad\square$

By means of Thm. 17.28, we are able the extend the map m.

Definition 17.29. Let $x, y \in \mathbb{R}^N$, and let $t \mapsto \varphi_{x,y}(t)$ be the (unique) maximal solution of the related Schur's Cauchy problem (17.31). We define

$$M : \mathbb{R}^N \times \mathbb{R}^N \longrightarrow \mathbb{R}^N, \qquad M(x, y) := \varphi_{x,y}(1). \tag{17.38}$$

We also use the notation

$$x * y := M(x, y), \quad x, y \in \mathbb{R}^N. \tag{17.39}$$

Moreover, for every fixed $x \in \mathbb{R}^N$ we set

$$\tau_x, \rho_x : \mathbb{R}^N \longrightarrow \mathbb{R}^N \qquad \tau_x(y) := x * y, \quad \rho_x(y) := y * x.$$

Warning. The reader should not be misled by the $*$ notation, resemblant to a group operation: we do not even know that $*$ is associative, and this will require much work! The same warning holds for the notations τ_x, ρ_x, resemblant to those used in the Lie group setting.

As is natural to expect, M turns out to be a smooth extension of m.

Theorem 17.30. *The function M defined in (17.38) is smooth on $\mathbb{R}^N \times \mathbb{R}^N$, and it extends the function m, that is,*

$$M(x, y) = m(x, y) = \exp(\mathrm{Log}(y))(x), \quad \forall\, x \in \mathbb{R}^N,\ y \in B(0, \rho_0). \tag{17.40}$$

Proof. The smoothness of $*$ is a consequence of ODE theory results on C^∞-dependence on parameters. We now prove (17.40). To this end, we fix $x \in \mathbb{R}^N$ and $y \in B(0, \rho_0)$, and we define

$$\gamma_{x,y} : [0, 1] \longrightarrow \mathbb{R}^N, \qquad \gamma_{x,y}(t) := m(x, ty).$$

By Thm. 17.27, $\gamma_{x,y}(t)$ solves (17.31) for all $t \in [0, 1]$; therefore, since $\varphi_{x,y}$ is the maximal solution of the same problem, we have

$$\gamma_{x,y}(t) = \varphi_{x,y}(t), \quad \text{for all } 0 \leq t \leq 1.$$

In particular, choosing $t = 1$, we get $m(x, y) = \gamma_{x,y}(1) = \varphi_{x,y}(1) = x * y$, and this proves that $*$ coincides with m on $\mathbb{R}^N \times B(0, \rho_0)$, as desired. $\qquad\square$

Now that we have extended the map m to $\mathbb{R}^N \times \mathbb{R}^N$, we devote the rest of this section to prove that $*$ inherits all the local-group properties of m and it turns them into global ones. To begin with, we prove the following simple lemma.

Lemma 17.31. *Let $x, y \in \mathbb{R}^N$ be fixed and let $t \mapsto \varphi_{x,y}(t)$ be the maximal solution of Schur's problem (17.31). Then*

$$\varphi_{x,y}(t) = M(x, ty), \qquad \text{for every } t \in \mathbb{R}. \tag{17.41}$$

With equivalent notations, $\varphi_{x,y}(t) = \varphi_{x,ty}(1) = x * (ty) = \tau_x(ty)$.

Proof. First of all we observe that, by definition of M, we have

$$M(x, ty) = \varphi_{x,ty}(1), \quad \text{for every } t \in \mathbb{R},$$

where $\varphi_{x,ty}$ is the maximal solution of the parametric problem

$$\dot{z}(s) = J(z(s))\, J(sty)^{-1}\, ty, \quad z(0) = x. \tag{17.42}$$

We then fix $t \in \mathbb{R}$ and we consider the curve $\gamma : \mathbb{R} \to \mathbb{R}^N$ defined as follows:

$$\gamma(s) := \varphi_{x,y}(ts) \qquad (s \in \mathbb{R}).$$

We claim that γ solves (17.42) on $\mathbb{R}$. In fact, since $\varphi_{x,y}$ solves (17.31), one has $\gamma(0) = \varphi_{x,y}(0) = x$; moreover, by the chain rule, for every $t \in \mathbb{R}$, we have

$$\frac{\mathrm{d}}{\mathrm{d}s}\gamma(s) = t\,\frac{\mathrm{d}}{\mathrm{d}u}\Big|_{u=ts}\{\varphi_{x,y}(u)\} \stackrel{(17.33)}{=} tJ(\varphi_{x,y}(ts))\, J(tsy)^{-1}y.$$

Thus γ solves the Cauchy problem (17.42) as $\varphi_{x,ty}$ does; by uniqueness, we conclude that $\varphi_{x,y}(t) = \gamma(1) = \varphi_{x,ty}(1)$, which is (17.41). $\qquad\square$

From Lem. 17.31 we straightforwardly obtain the following result.

Corollary 17.32. *If M is the map in (17.38), then*

$$\begin{cases} \dfrac{\mathrm{d}}{\mathrm{d}t}\{M(x, ty)\} = J(M(x, ty))\, J(ty)^{-1}\, y \\[2mm] \qquad\quad M(x, 0) = x, \end{cases} \tag{17.43}$$

for every $x, y \in \mathbb{R}^N$ and every $t \in \mathbb{R}$. As a consequence, for every $x, y \in \mathbb{R}^N$,

$$\frac{\partial M}{\partial y}(x, y)\, y = J(M(x, y))\, J(y)^{-1}\, y, \tag{17.44}$$

$$\frac{\partial M}{\partial y}(x, 0) = J(x). \tag{17.45}$$

Proof. (17.43) follows from Lem. 17.31 and Schur's problem for $\varphi_{x,y}(t)$. Obviously, the ODE in (17.43) is equivalent to

$$\frac{\partial M}{\partial y}(x, ty)\, y = J(M(x, ty))\, J(ty)^{-1}\, y. \tag{17.46}$$

When $t = 1$, (17.46) produces (17.44); when $t = 0$ it produces

$$\frac{\partial M}{\partial y}(x, 0)\, y = J(x)\, y \qquad \forall\, x, y \in \mathbb{R}^N,$$

remembering that $J(0)$ is the identity matrix, and $M(x, 0) = x$. The latter formula is (17.45), due to the arbitrariness of y. $\qquad\square$

Our next major effort lies in the proof that the right multiplication by y in (17.44) can be dropped; this is not obvious, since the matrices

$$A(y) = \begin{pmatrix} 0 & 0 \\ 0 & 0 \end{pmatrix} \quad \text{and} \quad B(y) = \begin{pmatrix} y_2 & -y_1 \\ 0 & 0 \end{pmatrix}$$

satisfy $A(y)\, y = B(y)\, y$, but they are equal only at $y = 0$.

We need a technical lemma.

Lemma 17.33. *Let $\mathfrak{J} = \{J_1, \ldots, J_N\}$ and let us consider the structure constants of $\mathfrak{g}$ wrt $\mathfrak{J}$, that is the constants $\{C_{i,j}^k\}_{i,j,k \leq N}$ such that*

$$[J_i, J_j] = \sum_{k=1}^{N} C_{i,j}^k\, J_k, \quad \forall\, i,j \in \{1, \ldots, N\}. \tag{17.47}$$

Let $j \in \{1, \ldots, N\}$ be fixed. Then, if $C(j) := (C_{i,j}^k)_{k,i \leq N}$, one has

$$C(j)\, J(z)^{-1} z \tag{17.48}$$

$$= J(z)^{-1} \frac{\partial J_j}{\partial z}(z)\, z - J(z)^{-1} \left(\frac{\partial J_1}{\partial z}(z)\, J_j(z) \cdots \frac{\partial J_N}{\partial z}(z)\, J_j(z) \right) J(z)^{-1} z,$$

for every $z \in \mathbb{R}^N$.

Proof. Taking the coefficient column-vectors of the v.f.s in (17.47) we get

$$J_i(J_j(z)) - J_j(J_i(z)) = \sum_{k=1}^{N} C_{i,j}^k\, J_k(z) \quad \forall\, z.$$

If $C_{i,j}$ is the column vector whose components are $C_{i,j}^1, \ldots, C_{i,j}^N$, this is just

$$\frac{\partial J_j}{\partial z}(z)\, J_i(z) - \frac{\partial J_i}{\partial z}(z)\, J_j(z) = J(z)\, C_{i,j}. \tag{17.49}$$

Let us fix j and let us create, from this identity between $N \times 1$ column matrices, the identity between the $N \times N$ matrices whose columns are, orderly, the two sides of (17.49) when $i = 1, \ldots, N$:

$$\frac{\partial J_j}{\partial z}(z)\, J(z) - \left(\frac{\partial J_1}{\partial z}(z)\, J_j(z) \cdots \frac{\partial J_N}{\partial z}(z)\, J_j(z) \right) = J(z)\, C(j).$$

Let us left multiply by the invertible matrix $J(z)^{-1}$; we get

$$J(z)^{-1} \frac{\partial J_j}{\partial z}(z)\, J(z) - J(z)^{-1} \left(\frac{\partial J_1}{\partial z}(z)\, J_j(z) \cdots \frac{\partial J_N}{\partial z}(z)\, J_j(z) \right) = C(j).$$

Finally, if we right multiply by $J(z)^{-1} z$, we get (17.48). $\qquad\square$

Theorem 17.34. *For every $x, y \in \mathbb{R}^N$ we have*

$$\frac{\partial M}{\partial y}(x, y) = J(M(x, y))\, J(y)^{-1}. \tag{17.50}$$

Equivalently, with the $*$ notation, this gives the "left-invariance" property

$$\mathfrak{J}_{\tau_x}(y) = J(x * y)\, J(y)^{-1}. \tag{17.51}$$

As a consequence, one also has

$$X(x * y) = \mathfrak{J}_{\tau_x}(y)\, X(y), \quad \text{for every } X \in \mathfrak{g}. \tag{17.52}$$

Proof. Clearly, (17.51) is a restatement of (17.50), which implies

$$J(x * y) = \vartheta_{\tau_x}(y)\, J(y).$$

Since the v.f.s related to the columns of J form a basis for $\mathfrak{g}$, (17.52) follows from the latter by a linearity argument.

Thus we are left with the non-trivial and lengthy proof of (17.50).

The rough idea is to put ty in place of y in each members of (17.50), and to show that the lhs and rhs solve (as functions of t) the same Cauchy problem. We follow this line, but only after a trick has been performed: we also multiply both sides of (17.50) by t. We now explain why.

For the rest of the proof, $x, y \in \mathbb{R}^N$ are fixed and $t \in \mathbb{R}$. If we take the Jacobian matrix wrt y in the ODE in (17.43), we get the matrix ODE

$$\frac{\mathrm{d}}{\mathrm{d}t}\left\{ t\,\frac{\partial M}{\partial y}(x, ty) \right\} = \frac{\partial}{\partial y}\left\{ J(M(x, ty))\, J(ty)^{-1}\, y \right\}. \tag{17.53}$$

Thus it is more natural to concoct a Cauchy problem for the curve

$$t \mapsto A(t) := t\,\frac{\partial M}{\partial y}(x, ty),$$

which is valued in the $N \times N$ real matrices. Our thesis (17.50) will then follow from the eventual equality of $A(t)$ with

$$B(t) := t\, J(M(x, ty))\, J(ty)^{-1}, \tag{17.54}$$

and by taking $t = 1$. Note that there are no problems with the initial datum: $A(0) = 0 = B(0)$. So, firstly, we need to obtain an ODE for $A(t)$, starting from (17.53); namely, we are interested in a linear ODE.

Before embarking on a tedious computation, we need some notation:

$$J(z) = (J_1(z) \cdots J_N(z)), \quad \text{where} \quad J_k(z) = \begin{pmatrix} J_k^1(z) \\ \vdots \\ J_k^N(z) \end{pmatrix};$$

furthermore, since x is never involved, with a small abuse we write $M(y)$ in place of $M(x, y)$. Helped by some basic differential calculus (see Exr. 17.6), we are ready to perform the y-Jacobian at the rhs of (17.53):

$$\frac{\partial}{\partial y}\left\{ J(M(ty))\, J(ty)^{-1}\, y \right\}$$

$$= J(M(ty))\, \frac{\partial}{\partial y}\left(J(ty)^{-1}\, y \right) + \sum_{k=1}^{N} (J(ty)^{-1}\, y)_k\, \frac{\partial}{\partial y}\left\{ J_k(M(ty)) \right\}$$

$$= J(M(ty))\, \frac{\partial}{\partial y}\left(J(ty)^{-1}\, y \right) + \sum_{k=1}^{N} (J(ty)^{-1}\, y)_k\, \frac{\partial J_k}{\partial y}(M(ty))\, \frac{\partial M}{\partial y}(x, ty)\, t.$$

This shows that $A(t)$ solves the linear ODE system

$$A'(t) = J(M(ty))\, \frac{\partial}{\partial y}\left(J(ty)^{-1}\, y \right) + \sum_{k=1}^{N} (J(ty)^{-1}\, y)_k\, \frac{\partial J_k}{\partial y}(M(ty))\, A(t).$$

We need to show that $B(t)$ does the same, i.e., we need to prove the following

  - CLAIM 1:

$$B'(t) = J(M(ty)) \frac{\partial}{\partial y}\left(J(ty)^{-1} y\right) + \sum_{k=1}^{N} (J(ty)^{-1} y)_k \frac{\partial J_k}{\partial y}(M(ty)) \, B(t).$$

Let us compare the two sides of CLAIM 1; taking into account (17.54) we see that the lhs of CLAIM 1 is

$$B'(t) = J(M(ty)) \frac{\mathrm{d}}{\mathrm{d}t}\{t J(ty)^{-1}\} + t \frac{\mathrm{d}}{\mathrm{d}t}\{J(M(ty))\} \, J(ty)^{-1}.$$

Thus CLAIM 1 holds true iff the following claim is fulfilled

  - CLAIM 2:

$$0 = J(M(ty))\left(\frac{\mathrm{d}}{\mathrm{d}t}\{t\, J(ty)^{-1}\} - \frac{\partial}{\partial y}\left(J(ty)^{-1} y\right)\right)$$
$$+ \left(t \frac{\mathrm{d}}{\mathrm{d}t}\{J(M(ty))\} - \sum_{k=1}^{N} (J(ty)^{-1} y)_k \frac{\partial J_k}{\partial y}(M(ty)) \, t\, J(M(ty))\right) J(ty)^{-1}.$$

We write this identity as $0 = J(M(ty))\,(1^{\text{st}}) + (2^{\text{nd}})J(ty)^{-1}$, with a clear meaning of 1st and 2nd. After simple computations, we discover that

$$2^{\text{nd}} = t \sum_{k=1}^{N} (J(ty)^{-1}y)_k \left(J_k(J_j^i)(M(ty)) - J_j(J_k^i)(M(ty))\right)_{i,j}.$$

By inserting the structure constants of $\mathfrak{g}$ wrt $\mathcal{J}$ (see (17.47)), we have

$$2^{\text{nd}} = t \sum_{k=1}^{N} (J(ty)^{-1}y)_k \, J(M(ty)) \left(C_{k,j}^i\right)_{i,j}$$
$$= t J(M(ty)) \left(\sum_{k=1}^{N} C_{k,j}^i (J(ty)^{-1}y)_k\right)_{i,j}.$$

Thus, our CLAIM 2 is equivalent to

  - CLAIM 3: $0 = J(M(ty))\left(\frac{\mathrm{d}}{\mathrm{d}t}\{t\, J(ty)^{-1}\} - \frac{\partial}{\partial y}\left(J(ty)^{-1} y\right)\right)$
$$+ t J(M(ty)) \left(\sum_{k=1}^{N} C_{k,j}^i (J(ty)^{-1}y)_k\right)_{i,j} J(ty)^{-1}.$$

Since $J(M(ty))$ is invertible, this is equivalent to

$$0 = \left(\frac{\mathrm{d}}{\mathrm{d}t}\{t\, J(ty)^{-1}\} - \frac{\partial}{\partial y}\left(J(ty)^{-1} y\right)\right) J(ty) + \left(\sum_{k=1}^{N} C_{k,j}^i (J(ty)^{-1}(ty))_k\right)_{i,j}.$$

After simple calculations, the latter is equivalent to (setting $z = ty$)

  - CLAIM 4:

$$J(z)^{-1}\left(\nabla(J_j^i)(z)z\right)_{i,j} - \left(e_i^T J(z)^{-1}(\partial_j J)(z)J(z)^{-1}z\right)_{i,j} J(z)$$
$$= \left(\sum_{k=1}^{N} C_{k,j}^i (J(z)^{-1}z)_k\right)_{i,j}.$$

As an identity between the column vectors of the matrices from both sides, this boils down to (see the notation for $C(j)$ in Lem. 17.33)

$$J(z)^{-1} \frac{\partial J_j}{\partial z}(z)\, z - \sum_{i=1}^{N} J_j^i(z)J(z)^{-1}(\partial_j J)(z)J(z)^{-1}z$$
$$= C(j)\, J(z)^{-1}z, \qquad \text{for } j = 1, \ldots, N.$$

Now the last identity comes from (17.48). $\square$

Remark 17.35. We observe that (17.52) is the global version of (17.24). In particular, once we have proved that $\mathbb{G} := (\mathbb{R}^N, *)$ is a Lie group on $\mathbb{R}^N$, (17.52) will imply that $\mathfrak{g}$ is made of left-invariant v.f.s, that is, $\mathfrak{g} \subseteq \mathrm{Lie}(\mathbb{G})$. ♮

Theorem 17.36. *For every fixed $x \in \mathbb{R}^N$, the map τ_x in Def. 17.29 is a smooth local diffeomorphism of $\mathbb{R}^N$, hence an open map.*

Proof. This is a direct consequence of (17.51) and of the Inverse Function Theorem, since $J(z)$ is non-singular for every $z \in \mathbb{R}^N$. $\square$

With Thms. 17.34 and 17.36 at hand, we can prove that $*$ globalizes all the local-group properties of m.

Theorem 17.37. *The origin is a global neutral element for $*$, that is,*

$$x * 0 = 0 * x = x, \quad \text{for all } x \in \mathbb{R}^N.$$

Proof. We first prove that 0 is a right neutral element. To this end, let $x \in \mathbb{R}^N$ be fixed. By definition of $*$ we know that $x * 0 = \varphi_{x,0}(1)$, where $\varphi_{x,0}$ is the maximal solution of Schur's problem with $y = 0$. Taking a look at (17.33), we infer that $\varphi_{x,0} \equiv x$; thus $x * 0 = \varphi_{x,0}(1) = x$, as desired.

As for 0 being the left neutral element, we observe that $0 * x = \varphi_{0,x}(1)$, where $\varphi_{0,x}(t)$ solves Schur's problem

$$\dot{z}(t) = J(z(t)) \, J(tx)^{-1}x, \quad z(0) = 0.$$

It is clear that $z(t) = tx$ is the solution of this Cauchy problem. As a consequence, $0 * x = \varphi_{0,x}(1) = z(1) = x$, as desired. $\square$

We now prove that $*$ is globally associative on $\mathbb{R}^N$.

Theorem 17.38. *The map $*$ is globally associative on $\mathbb{R}^N$, that is*

$$x * (y * z) = (x * y) * z, \quad \text{for all } x, y, z \in \mathbb{R}^N.$$

Proof. By definition of $*$, we have $(x * y) * z = \varphi_{x*y,z}(1)$, where $t \mapsto \varphi_{x*y,z}(t)$ is the maximal solution of

$$\dot{\varphi}(t) = J(\varphi(t)) \, J(tz)^{-1}z, \quad \varphi(0) = x * y, \tag{17.55}$$

We consider the curve on $\mathbb{R}$ defined by

$$\gamma(t) := x * \big(y * (tz)\big) = \tau_x\big(y * (tz)\big),$$

and we claim that it is a solution (on $\mathbb{R}$) of (17.55). By uniqueness, this will give

$$x * (y * z) = \gamma(1) = \varphi_{x*y,z}(1) = (x * y) * z.$$

So we are left with the proof of the claim. Since 0 is a global neutral element for $*$ (Thm. 17.37), we have $\gamma(0) = x * (y * 0) = x * y$; moreover, by crucially exploiting Thm. 17.34, for every $t \in \mathbb{R}$ we have the following computation

$$
\begin{aligned}
\dot{\gamma}(t) \;&=\; \frac{\mathrm{d}}{\mathrm{d}t}\big\{\tau_x\big(y * (tz)\big)\big\} = \partial\tau_x\big(y * (tz)\big)\,\frac{\mathrm{d}}{\mathrm{d}t}\{y * (tz)\} \\
&\overset{(17.43)}{=} \partial\tau_x\big(y * (tz)\big)\, J(y * (tz))\, J(tz)^{-1}\, z \\
&\overset{(17.51)}{=} J\big(x * (y * (tz))\big)\, J(tz)^{-1}\, z = J(\gamma(t))\, J(tz)^{-1}\, z.
\end{aligned}
$$

This ends the proof. $\qquad\square$

We now turn our attention to the existence of a global inversion map for $*$. A first result in this direction is contained in the following lemma.

Lemma 17.39. *There exists an open and connected neighborhood W of 0 such that*

$$
x * \iota(x) = \iota(x) * x = 0, \quad \text{for every } x \in W.
$$

Here, $\iota : V \to \mathbb{R}^N$ is the map introduced in Def. 17.22.

Proof. Let $\rho_0 > 0$ be as in Thm. 17.18. Since ι is continuous on V, there exists an open and connected neighborhood $W \subseteq B(0, \rho_0)$ of 0 such that

$$
\iota(x) \in B(0, \rho_0) \subseteq V, \quad \text{for every } x \in W.
$$

From this, since $*$ coincides with m on $\mathbb{R}^N \times B(0, \rho_0)$ and ι provides a local inverse for m on $B(0, \rho_0)$, we obtain

$$
x * \iota(x) = m(x, \iota(x)) = 0 = m(\iota(x), x) = \iota(x) * x, \quad \forall\, x \in W.
$$

This ends the proof. $\qquad\square$

The following simple topological result will allow us to extend the map ι to the whole of $\mathbb{R}^N$; it is left as an exercise (Exr. 17.5).

Proposition 17.40. *Let $W \subseteq V$ be as in Lem. 17.39. Then we have*

$$
\mathbb{R}^N = \bigcup_{n=1}^{\infty} \big\{ w_1 * \cdots * w_n : w_1, \ldots, w_n \in W \big\}. \tag{17.56}
$$

Remark 17.41. If $\mathbb{G} = (\mathbb{R}^N, *)$ is a Lie group with neutral element 0, the result contained in Prop. 17.40 is fully expected as a general fact of Lie group (actually, of topological group) theory; see Lem. C.1. $\qquad\sharp$

From Prop. 17.40, we easily deduce the following crucial result.

Proposition 17.42. *For every $x \in \mathbb{R}^N$, there exists a unique $y_x \in \mathbb{R}^N$ such that*

$$
x * y_x = y_x * x = 0. \tag{17.57}
$$

Proof. Let $W \subseteq V$ be as in Lem. 17.39. By Prop. 17.40, it is possible to find $w_1, \ldots, w_n \in W$ (not necessarily unique) such that $x = w_1 * \cdots * w_n$; hence, we define $y := \iota(w_n) * \cdots * \iota(w_1)$. Since $W \subseteq V$, y is well defined; moreover, by the associativity of $*$ and Lem. 17.39, we have $x * y = 0$. Analogously, $y * x = 0$. The uniqueness part follows from the associativity of $*$. $\qquad\square$

The result contained in Prop. 17.42 provides a very natural way to extend the map ι to the whole of $\mathbb{R}^N$.

Definition 17.43. For every $x \in \mathbb{R}^N$, let $y_x \in \mathbb{R}^N$ be the unique point satisfying (17.57). We define

$$\tilde{\iota} : \mathbb{R}^N \longrightarrow \mathbb{R}^N, \quad \tilde{\iota}(x) := y_x. \tag{17.58}$$

As we did for $*$, we prove that $\tilde{\iota}$ is a smooth extension of ι.

Theorem 17.44. *Let W be as in Lem. 17.39. The map $\tilde{\iota}$ defined in (17.58) is smooth on $\mathbb{R}^N$, and it extends ι beyond W, that is,*

$$\tilde{\iota}(x) = \iota(x), \quad \text{for all } x \in W. \tag{17.59}$$

Proof. We first prove that $\tilde{\iota} \in C^\infty(\mathbb{R}^N, \mathbb{R}^N)$. To this end, let $x_0 \in \mathbb{R}^N$ and let $y_0 := \tilde{\iota}(x_0)$. Then, by definition, we have $x_0 * y_0 = 0$. Let us set $M(x, y) := x * y$. We claim that the Jacobian matrix of M at (x_0, y_0) has full rank. This follows from the identity (in block form)

$$\mathfrak{J}_M(x_0, y_0) = \left(\mathfrak{J}_{\rho_{y_0}}(x_0) \quad \mathfrak{J}_{\tau_{y_0}}(x_0) \right),$$

where τ_{x_0} and ρ_{x_0} are as in Def. 17.29, remembering that $\mathfrak{J}_{\tau_{y_0}}(x_0)$ is non-singular (see Thm. 17.36). By the Inverse Function Theorem, we can then find two open neighborhoods U, U' of x_0 and y_0 resp., and a smooth function $f : U \to U'$ such that $f(x_0) = y_0 = \tilde{\iota}(x_0)$ and

$$\{(x, y) \in U \times U' : M(x, y) = 0\} = \{(x, f(x)) : x \in U\}.$$

Now, by the uniqueness statement in Prop. 17.42, we get

$$f(x) = \tilde{\iota}(x), \quad \text{for every } x \in U,$$

and this proves that $\tilde{\iota}$ is of class C^∞ on U. From the arbitrariness of x_0, we infer that $\tilde{\iota}$ is smooth on $\mathbb{R}^N$. As for identity (17.59) we know that, if $x \in W$, we have

$$x * \iota(x) = \iota(x) * x = 0;$$

hence, again by Prop. 17.42 we get $\iota(x) = \tilde{\iota}(x)$. $\qquad\square$

We are finally ready for the proof of Thm. 17.6.

Proof of Thm. 17.6. Let $*$ be the map defined in (17.39) and let $\mathbb{G} := (\mathbb{R}^N, *)$. By Thm. 17.30, $*$ is of class C^∞ on $\mathbb{R}^N \times \mathbb{R}^N$; moreover, Thms. 17.37 and 17.38 show that 0 is a neutral element for $*$ and that $*$ is associative. Finally, if $\tilde{\iota}$ is the map in (17.58), by Thm. 17.44 we know that $\tilde{\iota}$ is smooth and that it provides an inversion map for $*$. Summing up, $\mathbb{G}$ is a Lie group on $\mathbb{R}^N$ with neutral element 0.

To conclude the demonstration of the theorem, we turn to show that $\mathrm{Lie}(\mathbb{G}) = \mathfrak{g}$. To this end we observe that, since Thm. 17.34 ensures that

$$X(x * y) = \mathfrak{J}_{\tau_x}(y)\, X(y) \quad \text{for all } x, y \in \mathbb{R}^N \text{ and } X \in \mathfrak{g},$$

then $\mathfrak{g} \subseteq \mathrm{Lie}(\mathbb{G})$; from this, bearing in mind that $\mathfrak{g}$ has dimension N (see (ND)), we conclude that $\mathfrak{g} = \mathrm{Lie}(\mathbb{G})$, and the proof is complete. $\qquad\square$

17.4   Exercises of Chap. 17

Exercise 17.1. Let us consider the vector fields on $\mathbb{R}^3$ defined by

$$J_1 := \partial_{x_1}, \quad J_2 := \partial_{x_2}, \quad J_3 := \partial_{x_3} - x_1 \partial_{x_2},$$

and let $\mathfrak{g}$ be the vector subspace of $\mathcal{X}(\mathbb{R}^3)$ generated by J_1, J_2, J_3.

Prove that $\mathfrak{g}$ is a Lie subalgebra of $\mathcal{X}(\mathbb{R}^3)$ which satisfies conditions (G), (H) and (ND) in Def. 17.1. Write down explicitly the ODEs involved in the definition of the maps m and $*$, by completing the following argument:

(1) For every $\xi = (\xi_1, \xi_2, \xi_3) \in \mathbb{R}^3$ show that

$$\mathrm{Exp}\big(\xi_1 J_1 + \xi_2 J_2 + \xi_3 J_3\big) = \begin{pmatrix} \xi_1 \\ \xi_2 - \tfrac{1}{2}\xi_1\xi_3 \\ \xi_3 \end{pmatrix},$$

    and prove that Exp is a bijection from $\mathfrak{g}$ to $\mathbb{R}^3$.

(2) For every $y = (y_1, y_2, y_3) \in \mathbb{R}^3$ show that

$$\mathrm{Log}(y) := \big(\mathrm{Exp}\big)^{-1}(y) = y_1 J_1 + \left(y_2 + \frac{1}{2} y_1 y_3 \right) J_2 + y_3 J_3.$$

(3) For every $x, y \in \mathbb{R}^3$ prove that the Cauchy problem involved in the definition of $m(x,y)$ is the following autonomous system

$$\gamma = \gamma_{x,y} : \quad \begin{cases} \dot{\gamma}_1 = y_1 \\ \dot{\gamma}_2 = y_2 + \tfrac{1}{2} y_1 y_3 - \gamma_1 y_3 \\ \dot{\gamma}_3 = y_3 \\ \gamma(0) = (x_1, x_2, x_3). \end{cases}$$

Find the solution and define $m(x,y) = \gamma_{x,y}(1)$.

(4) For every $x, y \in \mathbb{R}^3$ prove that the Cauchy problem involved in the definition of $x * y$ is the following non-autonomous system

$$\varphi = \varphi_{x,y} : \begin{cases} \dot{\varphi}_1(t) = a_1(t, y) \\ \dot{\varphi}_2(t) = a_2(t, y) - \varphi_1 a_3(t, y) \\ \dot{\varphi}_3(t) = a_3(t, y) \\ \varphi(0) = (x_1, x_2, x_3), \end{cases}$$

where

$$\begin{pmatrix} a_1(t, y) \\ a_2(t, y) \\ a_3(t, y) \end{pmatrix} = \begin{pmatrix} y_1 \\ y_2 + t y_1 y_3 \\ y_3 \end{pmatrix}.$$

Find the solution and define $x * y = \varphi_{x,y}(1)$.

Verify that the Lie group on $(\mathbb{R}^3, *)$ with Lie algebra equal to $\mathfrak{g}$ (and neutral element equal to 0) is given by

$$m(x, y) = x * y = \left(x_1 + y_1, x_2 + y_2 - x_1 y_3, x_3 + y_3\right).$$

Exercise 17.2. Let us consider the vector fields on $\mathbb{R}^4$ defined by

$$J_1 := \partial_{x_1}, \quad J_2 := \partial_{x_2} + x_1 \partial_{x_3} + \frac{1}{2} x_1^2 \partial_{x_4}, \quad J_3 := \partial_{x_3} + x_1 \partial_{x_4}, \quad J_4 := \partial_{x_4},$$

and let $\mathfrak{g}$ be the vector subspace of $X(\mathbb{R}^4)$ generated by J_1, J_2, J_3, J_4.

Prove that $\mathfrak{g}$ is a Lie subalgebra of $X(\mathbb{R}^4)$ which satisfies conditions (G), (H) and (ND) in Def. 17.1. Write down explicitly the ODEs involved in the definitions of the maps m and $*$, by completing the following argument:

(1) For every $\xi = (\xi_1, \xi_2, \xi_3, \xi_4) \in \mathbb{R}^4$ show that

$$\mathrm{Exp}\left(\sum_{i=1}^{4} \xi_i J_i\right) = \begin{pmatrix} \xi_1 \\ \xi_2 \\ \xi_3 + \frac{1}{2}\xi_1 \xi_2 \\ \xi_4 + \frac{1}{2}\xi_1 \xi_3 + \frac{1}{6}\xi_1^2 \xi_2 \end{pmatrix},$$

and prove that the map Exp is a bijection from $\mathfrak{g}$ to $\mathbb{R}^4$.

(2) For every $y = (y_1, y_2, y_3, y_4) \in \mathbb{R}^4$ show that

$$\mathrm{Log}(y) = y_1 J_1 + y_2 J_2 + \left(y_3 + \frac{1}{2}y_1 y_2\right) J_3 + \left(y_4 - \frac{1}{2}y_1 y_3 + \frac{1}{12}y_1^2 y_2\right) J_4.$$

(3) For every $x, y \in \mathbb{R}^4$ prove that the Cauchy problem involved in the definition of $m(x, y)$ is the following

$$\gamma = \gamma_{x,y} : \begin{cases} \dot{\gamma}_1 = y_1 \\ \dot{\gamma}_2 = y_2 \\ \dot{\gamma}_3 = y_3 + \gamma_1 y_2 - \frac{1}{2}y_1 y_2 \\ \dot{\gamma}_4 = y_4 - \frac{1}{2}y_1 y_3 + \frac{1}{12}y_1^2 y_2 + \frac{1}{2}\gamma_1^2 y_2 + \gamma_1\left(y_3 - \frac{1}{2}y_1 y_2\right) \\ \gamma(0) = (x_1, x_2, x_3, x_4). \end{cases}$$

(4) For every $x, y \in \mathbb{R}^3$ prove that the Cauchy problem involved in the definition of $x * y$ is the following

$$\varphi = \varphi_{x,y} : \quad \begin{cases} \dot{\varphi}_1(t) = a_1(t, y) \\ \dot{\varphi}_2(t) = a_2(t, y) \\ \dot{\varphi}_3(t) = a_3(t, y) + \varphi_1 a_2(t, y) \\ \dot{\varphi}_4(t) = a_4(t, y) + \varphi_1 a_3(t, y) + \frac{1}{2}\varphi_1^2 a_2(t, y) \\ \varphi(0) = (x_1, x_2, x_3, x_4), \end{cases}$$

where

$$\begin{pmatrix} a_1(t, y) \\ a_2(t, y) \\ a_3(t, y) \\ a_4(t, y) \end{pmatrix} = \begin{pmatrix} y_1 \\ y_2 \\ y_3 - ty_1 y_2 \\ y_4 - ty_1 y_3 + \frac{1}{2}t^2 y_1^2 y_2 \end{pmatrix}.$$

Verify that $m(x, y) = \gamma_{x,y}(1) = \varphi_{x,y}(1) = x * y$, where

$$x * y = \left(x_1 + y_1, x_2 + y_2, x_3 + y_3 + x_1 y_2, x_4 + y_4 + \frac{1}{2}x_1^2 y_2 + x_1 y_3 \right).$$

$(\mathbb{R}^4, *)$ is called the Engel group on $\mathbb{R}^4$.

Exercise 17.3. Let us consider the vector fields on $\mathbb{R}^3$ defined by

$$J_1 := \partial_{x_1}, \quad J_2 := \cos(x_1)\partial_{x_2} + \sin(x_1)\partial_{x_3}, \quad J_3 := -\sin(x_1)\partial_{x_2} + \cos(x_1)\partial_{x_3},$$

and let $\mathfrak{g}$ be the vector subspace of $\mathfrak{X}(\mathbb{R}^3)$ generated by J_1, J_2, J_3.
　　Show that the associated Exp map is

$$\mathrm{Exp}\big(\xi_1 J_1 + \xi_2 J_2 + \xi_3 J_3\big) = \begin{pmatrix} \xi_1 \\ \xi_2 f(\xi_1) + \xi_3 g(\xi_1) \\ \xi_2 g(\xi_1) + \xi_3 f(\xi_1) \end{pmatrix},$$

where the functions f and g are defined by

$$f(t) := \frac{\sin(t)}{t}, \quad g(t) := \frac{\cos(t) - 1}{t}, \quad \text{for all } t \in \mathbb{R}.$$

Write down explicitly the Cauchy problems involved in the definition of the maps m and $*$, by completing the following argument:

(1) Prove that, setting $\Omega := \{X = \sum_{i=1}^{3} \xi_i J_i : |\xi_1| < 2\pi\} \subseteq \mathfrak{g}$, Exp is a bijection from Ω to the open subset V of $\mathbb{R}^3$ defined by

$$V = \{x = (x_1, x_2, x_3) \in \mathbb{R}^3 : |x_1| < 2\pi\}.$$

　　Note that, in this case, the map Exp *is not globally injective nor surjective.*

(2) For every $y = (y_1, y_2, y_3) \in V$ show that

$$\mathrm{Log}(y) = y_1 J_1 + \big(y_2 \alpha(y_1) - y_3 \beta(y_1)\big) J_2 + \big(y_2 \beta(y_1) + y_3 \alpha(y_1)\big) J_3,$$

　　where the functions α and β are defined by

$$\alpha(t) := \frac{t \sin(t)}{2(1 - \cos(t))}, \quad \beta(t) := -\frac{t}{2}, \quad \text{for all } |t| < 2\pi.$$

(3) For every $x, y \in \mathbb{R}^3$ prove that the Cauchy problem involved in the definition of $m(x, y) = \gamma_{x,y}(1)$ is the following

$$\begin{cases} \dot{\gamma}_1 = y_1 \\ \dot{\gamma}_2 = \big(y_2\alpha(y_1) - y_3\beta(y_1)\big)\cos(\gamma_1) - \big(y_2\beta(y_1) + y_3\alpha(y_1)\big)\sin(\gamma_1) \\ \dot{\gamma}_3 = \big(y_2\alpha(y_1) - y_3\beta(y_1)\big)\sin(\gamma_1) + \big(y_2\beta(y_1) + y_3\alpha(y_1)\big)\cos(\gamma_1) \\ \gamma(0) = (x_1, x_2, x_3). \end{cases}$$

(4) For every $x, y \in \mathbb{R}^3$ prove that the Cauchy problem involved in the definition of $x * y = \varphi_{x,y}(1)$ is the following

$$\begin{cases} \dot{\varphi}_1 = a_1(t, y) \\ \dot{\varphi}_2 = a_2(t, y)\cos(\varphi_1) - a_3(t, y)\sin(\varphi_1) \\ \dot{\varphi}_3 = a_2(t, y)\sin(\varphi_1) + a_3(t, y)\cos(\varphi_1) \\ \varphi(0) = (x_1, x_2, x_3), \end{cases}$$

where

$$\begin{pmatrix} a_1(t, y) \\ a_2(t, y) \\ a_3(t, y) \end{pmatrix} = \begin{pmatrix} y_1 \\ y_2\cos(ty_1) + y_3\sin(ty_1) \\ y_3\cos(ty_1) - y_2\sin(ty_1)) \end{pmatrix}.$$

Verify that the group $(\mathbb{R}^3, *)$ with Lie algebra $= \mathfrak{g}$ (and identity 0) is

$$x * y = \big(x_1 + y_1, x_2 + y_2\cos x_1 - y_3\sin x_1, x_3 + y_2\sin x_1 + y_3\cos x_1\big).$$

Exercise 17.4. Prove the following ODE result:

Let $X_1, \ldots, X_n$ be locally Lipschitz v.f.s on $\mathbb{R}^N$ and let $a_1, \ldots, a_n \in C(\mathbb{R}^N, \mathbb{R})$. For every $s \in \mathbb{R}$, let $\gamma(\cdot\,; s) \in C^1(\mathcal{D}(s), \mathbb{R}^N))$ be the unique maximal solution of the parametric Cauchy problem

$$\begin{cases} \dot{x}(t) = \displaystyle\sum_{j=1}^{n} a_j(t + s)\, X_j(x(t)) \\ x(0) = 0. \end{cases} \tag{17.60}$$

Then, for every compact set $K \subseteq \mathbb{R}$ and every real $h > 0$ there exists $\varepsilon > 0$, only depending on K and h, such that $[-\varepsilon, \varepsilon] \subseteq \mathcal{D}(s)$ for every $s \in K$ and

$$\|\gamma(t; s)\| \leq h, \qquad \text{for every } t \in [-\varepsilon, \varepsilon] \text{ and every } s \in K.$$

[*Hint:* For $s \in \mathbb{R}$, let $\mathcal{D}(s) = (\underline{\omega}(s), \overline{\omega}(s))$. Since, by Thm. B.34, $\underline{\omega}$ and $\overline{\omega}$ are, resp., upper and lower semi-continuous on $\mathbb{R}$, there exist $a, b \in \mathbb{R}$ such that $a < 0 < b$ and $\underline{\omega}(s) < a < b < \overline{\omega}(s)$ for every $s \in K$; thus $\gamma(\cdot\,; s)$ is defined at least on $[a, b]$, uniformly for $s \in K$.

On the other hand, since $(t, s) \mapsto \gamma(t; s)$ is (defined and) uniformly continuous on $[a, b] \times K$ (again by Thm. B.34), there exists $\varepsilon > 0$, only depending on K and h, such that, for every $s \in K$, one has $[-\varepsilon, \varepsilon] \subseteq [a, b] \subseteq \mathcal{D}(s)$ and

$$\|\gamma(t; s) - \gamma(0; s)\| \overset{(17.60)}{=} \|\gamma(t; s)\| \leq h, \quad \forall\, t \in [-\varepsilon, \varepsilon].]$$

Exercise 17.5. Prove Prop. 17.40 by completing the following argument.

Denote by A the set in rhs of (17.56) and prove it is open and closed:

- *A is open:* prove that $A_n = \{w_1 * \cdots * w_n : w_1, \ldots, w_n \in W\}$ is open. Indeed $A_1 = W$ and (since $*$ is associative) $A_n = \bigcup_{x \in A_{n-1}} \tau_x(W)$; then use the fact that any τ_x is open (Thm. 17.36).

- *A is closed:* Let $x_0 \in \overline{A}$. By continuity of ι, there exists an open neighborhood $U \subseteq W$ of 0 such that $\iota(U) \subseteq W$; recognize that $\tau_{x_0}(U)$ is an open neighborhood of $x_0 = \tau_{x_0}(0)$; deduce that $\tau_{x_0}(U) \cap A \neq \varnothing$. As a consequence, there exist $w_1, \ldots, w_n \in W$ and $u \in U$ such that
$$w_1 * \cdots * w_n = \tau_{x_0}(u) = x_0 * u.$$

Use the associativity of $*$ and derive that $(w_1 * \cdots * w_n) * \iota(u) = x_0$; since $w_1, \ldots, w_n, \iota(u) \in W$, this gives $x_0 \in A$.

Exercise 17.6. Supposing the involved functions are differentiable and well posed, prove the following formula ($\frac{\partial}{\partial x}$ denotes the Jacobian operator, and $x \in \mathbb{R}^N$):

if $A(x)$ is an $N \times N$ matrix-valued function, and if $v(x)$ is an $N \times 1$ vector-valued function, show that

$$\frac{\partial}{\partial x}(A(x)\,v(x)) = A(x)\,\frac{\partial v}{\partial x}(x) + \sum_{k=1}^{N} v_k(x)\frac{\partial A_k}{\partial x}(x),$$

where A_k is the k-th column vector of A, and v_k is the k-th entry of v.

Chapter 18

On the Convergence of the CBHD Series

$\mathbf{T}$HE aim of this chapter is to furnish an application of the formal PDEs of Sec. 2.6.2 (page 59): we provide a domain of convergence for the homogeneous CBHD series $Z(a,b) = \sum_{n=1}^{\infty} Z_n(a,b)$ on a finite-dimensional Lie algebra $\mathfrak{g}$. The argument is inspired to that in [Blanes and Casas (2004)] (see also [Mérigot (1974)]); in doing this, we show how the algebraic computations (involving formal power series) obtained in Sec. 2.8 (page 64) can profitably be used in order to find the cited convergence domain.

More precisely, in Chap. 2 we established an algebraic machinery in order to prove that $C = C(s,t) := Z(sa, tb)$ solves the Poincaré PDE

$$\frac{\partial C}{\partial s} = \frac{\operatorname{ad} C}{e^{\operatorname{ad} C} - 1}(a).$$

Since the Taylor expansion of

$$\frac{z}{e^z - 1} \quad \text{is} \quad \sum_{n=0}^{\infty} \frac{B_n}{n!} z^n$$

(B_n being the Bernoulli numbers), this PDE enabled us to infer that, turning to the non-homogenous notation $C = \sum_{i,j} C_{i,j} \, s^i t^j$, one has recursive identities

$$(i+1)\, C_{i+1,j} = \sum_{\substack{1 \le h \le i+j \\ (i_1,j_1),\dots,(i_h,j_h) \ne (0,0) \\ i_1 + \cdots + i_h = i, \ j_1 + \cdots + j_h = j}} \frac{B_h}{h!} \, [C_{i_1,j_1} \dots [C_{i_h,j_h}, x] \dots].$$

Now, if we consider the majorant series $\sum_{n=0}^{\infty} \frac{|B_n|}{n!} z^n$ (which can explicitly be summed), and if we solve a suitable Cauchy problem associated to this function, then we can easily obtain an upper bound for $\sum_{i,j \ge 0} \|C_{i,j}(a,b)\|$, hence an upper bound for $\sum_{n \ge 1} \|Z_n(a,b)\|$.

Repeating the same argument wrt t (for which we also have a suitable Poincaré PDE), we get a symmetric convergence domain for the CBHD series. Since they are based on algebraic techniques only, the arguments of this chapter also apply to the more general case of the *infinite-dimensional* Banach-Lie algebras (see [Biagi and Bonfiglioli (2014)]).

A prerequisite for this chapter is a careful reading of Sec.s 2.1, 2.2 and 2.8, plus some basic facts of holomorphic functions.

18.1   A domain of convergence for the CBHD series

The aim of this section is to show how the formal PDEs for C obtained in Sec. 2.6.2 can be used in order to find a domain of convergence for the CBHD series. We tacitly inherit all the notations in Sec. 2.8.

To this aim, we begin with fixing once and for all a Lie algebra $(\mathfrak{g}, [\cdot, \cdot])$ over $\mathbb{R}$ with *finite dimension* (as a real vector space), say N. Due to this assumption, $\mathfrak{g}$ can be equipped with topological and differentiable structures by identifying it with $\mathbb{R}^N$ via the choice of a basis. Indeed, let $\mathcal{B} := \{v_1, \ldots, v_N\}$ be a fixed basis of $\mathfrak{g}$ and let $\pi_{\mathcal{B}} : \mathbb{R}^N \longrightarrow \mathfrak{g}$ be the linear map $\pi_{\mathcal{B}}(x_1, \ldots, x_N) := \sum_{k=1}^{N} x_k v_k$.

Since $\pi_{\mathcal{B}}$ is an isomorphism of vector spaces, we can set $\|a\|_{\mathcal{B}} := \|\pi_{\mathcal{B}}^{-1}(a)\|$ (for $a \in \mathfrak{g}$), where $\| \cdot \|$ on the rhs denotes the Euclidean norm on $\mathbb{R}^N$. It is very easy to prove the following facts:

  (i) the map $\| \cdot \|_{\mathcal{B}}$ defines a norm on $\mathfrak{g}$ and $(\mathfrak{g}, \| \cdot \|_{\mathcal{B}})$ is a Banach space;

 (ii) the map $\pi_{\mathcal{B}}$ is a smooth isomorphism between $\mathbb{R}^N$ and $\mathfrak{g}$ (the latter being equipped with the structure of a differentiable manifold resulting from the global chart $\pi_{\mathcal{B}}^{-1} : \mathfrak{g} \to \mathbb{R}^N$);

(iii) the choice of a different basis $\mathcal{B}'$ in $\mathfrak{g}$ gives rise to an equivalent norm $\| \cdot \|_{\mathcal{B}'}$, so that the Banach spaces $(\mathfrak{g}, \| \cdot \|_{\mathcal{B}})$ and $(\mathfrak{g}, \| \cdot \|_{\mathcal{B}'})$ are equivalent (and they have the same convergent sequences).

We remark that, since $\mathfrak{g}$ is a finite-dimensional vector space, all the norms on $\mathfrak{g}$ (not necessarily associated with a basis $\mathcal{B}$) are in fact equivalent.

We now turn our attention to the Lie bracket $[\cdot, \cdot]$ in $\mathfrak{g}$. Since $\mathfrak{g}$ is a finite-dimensional Banach space (wrt any fixed norm $\| \cdot \|$) and $[\cdot, \cdot]$ is a bilinear map, then it is a continuous map from $\mathfrak{g} \times \mathfrak{g}$ to $\mathfrak{g}$, so that there exists a positive real constant M such that

$$\|[a, b]\| \leq M \|a\| \, \|b\|, \quad \text{for all } a, b \in \mathfrak{g}. \tag{18.1}$$

By rescaling the norm $\| \cdot \|$ (replacing it with $M \| \cdot \|$), it is not restrictive to assume that identity (18.1) holds with $M = 1$, that is,

$$\|[a, b]\| \leq \|a\| \, \|b\|, \quad \text{for all } a, b \in \mathfrak{g}. \tag{18.2}$$

From now on, we assume that a norm on $\mathfrak{g}$ has been fixed is such a way that identity (18.2) holds. For instance, this norm can always be taken to be a scalar multiple of the previous $\| \cdot \|_{\mathcal{B}}$.

We begin with some preliminary results of independent interest.

Lemma 18.1. *Let $\{b_n\}_{n \geq 0}$ be the sequence of the coefficients of the Maclaurin series of the function $\mathbf{b}$ introduced in (2.43) (page 61). Then the complex power series*

$$F(z) := \sum_{n=0}^{\infty} |b_n| \, z^n$$

has radius of convergence equal to 2π, and for all $z \in B(0, 2\pi)$ one has

$$F(z) = \begin{cases} 1 & \text{if } z = 0 \\ 2 + \dfrac{z}{2}\left(1 - \cot\left(\dfrac{z}{2}\right)\right) & \text{if } 0 < |z| < 2\pi. \end{cases} \tag{18.3}$$

Proof. First, since b is defined and holomorphic on $B(0, 2\pi)$ (see (2.43)), we have[1]

$$\mathbf{b}(z) = \sum_{n=0}^{\infty} b_n z^n, \quad \text{for all } z \in B(0, 2\pi),$$

and thus the complex power series $T(z) := \sum_{n=0}^{\infty} b_n z^n$ has a radius of convergence ρ which is greater than or equal to 2π. On the other hand, since b is singular for $z = 2\pi i$, then ρ cannot be greater than 2π, and this proves that the radius of convergence of $F(z)$ (which is the same as that of $T(z)$) is exactly equal to 2π.

We now prove identity (18.3). It can be proved that the sequence $\{b_n\}_{n\geq 0}$ satisfies the following properties (see Ex. 5.8)

$$\begin{cases} b_0 = 1, \quad b_1 = -\tfrac{1}{2}, \quad b_{2n+1} = 0 \quad \text{for all } n \geq 1, \\ b_{2n} = (-1)^{n-1}|b_{2n}| \quad \text{for all } n \geq 1. \end{cases} \tag{18.4}$$

For all $z \in B(0, 2\pi)$, by taking into account (18.4) and by definition of $F(z)$, we get

$$\mathbf{b}(iz) = \frac{iz}{e^{iz} - 1} = \sum_{n=0}^{\infty} b_n (iz)^n = 1 - \frac{iz}{2} - \sum_{n=1}^{\infty} |b_{2n}| z^{2n} = 2 + \frac{z - iz}{2} - F(z),$$

which yields, after some tedious computations,

$$F(z) = 2 + \frac{z - iz}{2} - \frac{iz}{e^{iz} - 1} = [\cdots] = 2 + \frac{z}{2}\left(1 - \cot\left(\frac{z}{2}\right)\right).$$

This is exactly the desired (18.3), and the proof is complete. $\qquad\square$

Proposition 18.2. *Let α be any fixed nonnegative real number, and let F be the function defined in (18.3). We set*

$$F_\alpha : (-2\pi, 2\pi) \longrightarrow \mathbb{R}, \qquad F_\alpha(t) := \alpha F(t).$$

Moreover, let β be a real number in $[0, 2\pi)$ and let γ be the maximal solution of

$$y' = F_\alpha(y), \quad y(0) = \beta. \tag{18.5}$$

Then the following identities hold true, for all $n \geq 1$:

$$\gamma'(0) = \alpha\left(1 + \sum_{m=1}^{\infty} |b_m|\beta^m\right), \tag{18.6a}$$

$$(n+1)\frac{\gamma^{(n+1)}(0)}{(n+1)!} = \alpha \sum_{m=1}^{\infty} |b_m| \left(\sum_{\substack{i_1,\ldots,i_m \geq 0 \\ i_1+\cdots+i_m = n}} \frac{\gamma^{(i_1)}(0)\cdots\gamma^{(i_m)}(0)}{i_1!\cdots i_m!} \right). \tag{18.6b}$$

---

[1]We use the short notation $b_n := \frac{B_n}{n!}$, where the B_n are the Bernoulli numbers in (2.18), page 47.

Proof. From Lem. 18.1 we know that F_α is real-analytic on its domain and that, for every $|t| < 2\pi$, we have $F_\alpha(t) = \alpha \sum_{n=0}^{\infty} |b_n| t^n$. On the other hand, since γ solves the Cauchy problem (18.5), it follows from general results of ODE Theory that γ is real-analytic on its maximal domain I, so that it is possible to find a positive real number $\delta > 0$ such that

$$\gamma(t) = \sum_{n=0}^{\infty} \frac{\gamma^{(n)}(0)}{n!}\, t^n, \quad \text{for all } t \in (-\delta, \delta) \subseteq I.$$

From (18.5) we have $\gamma'(t) = F_\alpha(\gamma(t))$, for all $|t| < \delta$, so that the Maclaurin expansions of these two functions must coincide. This gives at once (18.6a) and (18.6b) after some trivial computations on power series. $\qquad\square$

From the recursive identities (18.6a) and (18.6b) we obtain:

Corollary 18.3. *Let α, F_α, β and γ be as in Prop. 18.2; then we have*

$$\gamma^{(n)}(0) \geq 0, \quad \text{for all } n \in \mathbb{N} \cup \{0\}.$$

Owing to the preliminaries presented so far, we can now turn our attention to the problem of finding a domain of convergence for the homogeneous CBHD series in the finite dimensional real Lie algebra $\mathfrak{g}$; this is the series

$$a \diamond b = \sum_{n=1}^{\infty} Z_n(a, b) = \sum_{n=1}^{\infty} \left(\sum_{i,j \geq 0:\ i+j=n} C_{i,j}(a, b) \right).$$

The following theorem is the core of this section, and it provides an estimate for the non-homogeneous summands $C_{i,j}$ of the CBHD series in $\mathfrak{g}$, by combining the abstract PDEs for C in Sec. 2.6 with the recursive identities (18.6a) and (18.6b).

Theorem 18.4. *Let $a, b \in \mathfrak{g}$ be fixed elements of $\mathfrak{g}$ such that $\|a\|, \|b\| < 2\pi$, and let F be the function defined in (18.3). Moreover, let γ and ψ be the two maximal solutions of the following Cauchy problems in $\mathbb{R}$, respectively:*

$$\begin{cases} y' = \|b\|\, F(y) \\ y(0) = \|a\| \end{cases} \quad \text{and} \quad \begin{cases} y' = \|a\|\, F(y) \\ y(0) = \|b\|. \end{cases}$$

Then we have the following estimates for the summands $C_{i,j}$:

$$\sum_{i=0}^{\infty} \|C_{i,j}(a, b)\| \leq \frac{\gamma^{(j)}(0)}{j!} \qquad \text{for all } j \geq 0, \tag{18.7}$$

$$\sum_{j=0}^{\infty} \|C_{i,j}(a, b)\| \leq \frac{\psi^{(i)}(0)}{i!} \qquad \text{for all } i \geq 0. \tag{18.8}$$

Proof. We prove the estimate (18.7), since (18.8) can be proved analogously. We proceed by induction on $j \in \mathbb{N} \cup \{0\}$.

If $j = 0$, since $C_{1,0}(a, b) = a$ and $C_{i,0}(a, b) = 0$ for all $i \geq 2$, we have

$$\sum_{i=0}^{\infty} \|C_{i,0}(a, b)\| = \|C_{1,0}\| = \|a\| = \gamma(0),$$

which is exactly (18.7) for $j = 0$. Let as now assume that the inequality (18.7) holds for all $p \in \mathbb{N} \cup \{0\}$ which are less than or equal to a certain $j \geq 0$, and let us prove that it also holds for $p = j + 1$.

If $j = 0$, by using the recursive identity (2.51) (page 64) together with (18.6a), we get (since $C_{i,0}(a,b) = 0$ if $i \geq 2$)

$$\sum_{i=0}^{\infty} \|C_{i,1}(a,b)\| = \|C_{0,1}(a,b)\| + \sum_{i=1}^{\infty} \|C_{i,1}(a,b)\| = \|b\| + \sum_{i=1}^{\infty} \|C_{i,1}(a,b)\|$$

$$\overset{(2.51)}{=} \|b\| + \sum_{i=1}^{\infty} \Big\| \sum_{\substack{1 \leq h \leq i \\ (i_1,j_1),\ldots,(i_h,j_h) \neq (0,0) \\ i_1 + \cdots + i_h = i \\ j_1 + \cdots + j_h = 0}} (-1)^h b_h \, [C_{i_1,j_1}(a,b), \ldots, [C_{i_h,j_h}(a,b), b] \ldots] \Big\|$$

$$\overset{(18.2)}{\leq} \|b\| + \sum_{i=1}^{\infty} \sum_{\substack{1 \leq h \leq i \\ i_1,\ldots,i_h \neq 0 \\ i_1 + \cdots + i_h = i}} |b_h| \, \|C_{i_1,0}(a,b)\| \cdots \|C_{i_h,0}(a,b)\| \, \|b\|$$

$$= \|b\| \Big(1 + \sum_{i=1}^{\infty} |b_i| \, \|a\|^i \Big) \overset{(18.6a)}{=} \gamma'(0),$$

which is the desired (18.7) for $j = 1$.

On the other hand, if $j \geq 1$, by using once again (2.51), identity (18.6b) and the inductive hypothesis, we obtain

$$\sum_{i=0}^{\infty} \|C_{i,j+1}(a,b)\|$$

$$\overset{(2.51)}{=} \frac{1}{j+1} \sum_{i=0}^{\infty} \Big\| \sum_{\substack{1 \leq h \leq i+j \\ (i_1,j_1),\ldots,(i_h,j_h) \neq (0,0) \\ i_1 + \cdots + i_h = i \\ j_1 + \cdots + j_h = j}} (-1)^h b_h \, [C_{i_1,j_1}, \ldots, [C_{i_h,j_h}, b] \ldots] \Big\|$$

$$\overset{(18.2)}{\leq} \frac{1}{j+1} \sum_{i=0}^{\infty} \sum_{\substack{1 \leq h \leq i+j \\ (i_1,j_1),\ldots,(i_h,j_h) \neq (0,0) \\ i_1 + \cdots + i_h = i, \quad j_1 + \cdots + j_h = j}} |b_h| \, \|C_{i_1,j_1}\| \cdots \|C_{i_h,j_h}\| \, \|b\|$$

$$\leq \frac{\|b\|}{j+1} \sum_{h=1}^{\infty} |b_h| \sum_{\substack{j_1,\ldots,j_h \geq 0 \\ j_1 + \cdots + j_h = j}} \Big(\sum_{i_1 \geq 0} \|C_{i_1,j_1}\|\Big) \cdots \Big(\sum_{i_h \geq 0} \|C_{i_h,j_h}\|\Big)$$

$$\overset{(18.7)}{\leq} \frac{\|b\|}{j+1} \sum_{h=1}^{\infty} |b_h| \sum_{\substack{j_1,\ldots,j_h \geq 0 \\ j_1 + \cdots + j_h = j}} \frac{\gamma^{(j_1)}(0) \cdots \gamma^{(j_h)}(0)}{i_1! \cdots i_h!} \overset{(18.6b)}{=} \frac{\gamma^{(j+1)}(0)}{(j+1)!},$$

which is inequality (18.7) with j replaced by $j + 1$. $\qquad\square$

We can finally provide a domain of convergence for the CBHD series.

Theorem 18.5 (A domain of convergence for the CBHD series). *Let $\mathfrak{g}$ be a finite dimensional real Lie algebra. Let F be the function defined in (18.3) and let u be the integral function defined by*

$$u : [0, 2\pi) \longrightarrow \mathbb{R}, \qquad u(r) := \int_r^{2\pi} \frac{1}{F(y)} \, \mathrm{d}y.$$

We denote by $\mathcal{G}(u)$ the hypograph of u, that is,

$$\mathcal{G}(u) := \Big\{ (r, \rho) \in [0, 2\pi) \times \mathbb{R} : \rho < u(r) \Big\}.$$

Then the homogeneous CBHD series in $\mathfrak{g}$ is absolutely convergent on the set

$$\Delta := \Big\{ (a, b) \in \mathfrak{g} \times \mathfrak{g} \ \Big| \ (\|a\|, \|b\|) \in \mathcal{G}(u) \ \text{ or } \ (\|b\|, \|a\|) \in \mathcal{G}(u) \Big\}.$$

(See also Fig. 18.3 on page 339.)

Proof. We begin by remarking some simple facts concerning the function F and its integral function u. First of all, it is simple to show that the function F attains its minimum on $(-2\pi, 2\pi)$ which is *strictly positive*, so that the reciprocal function $1/F$ is well-defined, bounded and continuous on $(-2\pi, 2\pi)$. (See Ex. 2.13 on page 68; see also Fig. 18.2 on page 338.) Due to this fact, the function u is well-defined (and finite) for all $r \in [0, 2\pi)$.

Let us now fix $a, b \in \mathfrak{g}$ such that $(\|a\|, \|b\|) \in \mathcal{G}(u)$ and let us prove that the CBHD series $a \diamond b$ is absolutely convergent. First of all, from the estimate (18.7) in Thm. 18.4 we get

$$\sum\nolimits_{i=0}^{\infty} \|C_{i,j}(a, b)\| \leq \frac{\gamma^{(j)}(0)}{j!}, \quad \text{ for all } j \geq 0, \tag{18.9}$$

where γ is the maximal solution of the Cauchy problem

$$y' = \|b\| \, F(y), \quad y(0) = \|a\|.$$

We know from Prop. 18.2 that γ is real analytic on its maximal domain $I = (\alpha, \beta)$, and since the ODE solved by γ is *separable*, we have the explicit expression

$$\beta = \frac{1}{\|b\|} \int_{\|a\|}^{2\pi} \frac{1}{F(y)} \, \mathrm{d}y = \frac{1}{\|b\|} \, u(\|a\|). \tag{18.10}$$

Moreover, it follows from Cor. 18.3 that the Maclaurin series of γ has *nonnegative coefficients*, so that we can apply the so-called Vivanti-Pringsheim Theorem (see e.g., [Titchmarsh (1939), Theorem 7.21, p. 214]) to derive that the Maclaurin series of γ has radius of convergence exactly equal to β, that is,

$$\gamma(t) = \sum\nolimits_{j=0}^{\infty} \frac{\gamma^{(j)}(0)}{j!} \, t^j, \quad \text{ for all } t \in [0, \beta).$$

We are now ready to conclude: since $(\|a\|, \|b\|) \in \mathcal{G}(u)$ it follows from (18.10) that $\beta > 1$, so that the Maclaurin series of γ is convergent for $t = 1$. By taking into account the estimates (18.9), we get

$$\sum\nolimits_{n=0}^{\infty} \|Z_n(a, b)\| \leq \sum\nolimits_{i,j=0}^{\infty} \|C_{i,j}(a, b)\| \overset{(18.9)}{\leq} \sum\nolimits_{j=0}^{\infty} \frac{\gamma^{(j)}(0)}{j!} = \gamma(1),$$

and this proves that the homogeneous CBHD series $a \diamond b$ is absolutely convergent. Let us now assume that $a, b \in \mathfrak{g}$ are such that $(\|b\|, \|a\|) \in \mathcal{G}(u)$ and let us prove that $a \diamond b$ is again absolutely convergent. By Ex. 2.9 on page 67, we have

$$\sum_{n=1}^{\infty} \|Z_n(a, b)\| = \sum_{n=1}^{\infty} \|(-1)^{n+1} Z_n(b, a)\| = \sum_{n=1}^{\infty} \|Z_n(b, a)\|, \qquad (18.11)$$

and since $(\|b\|, \|a\|) \in \mathcal{G}(u)$, then the CBHD series $b \diamond a = \sum_{n=1}^{\infty} Z_n(b, a)$ is absolutely convergent, as we proved above. From (18.11) it follows immediately that the CBHD series $a \diamond b = \sum_{n=1}^{\infty} Z_n(a, b)$ is absolutely convergent as well. $\qquad \square$

We end this section with the following remarks.

Remark 18.6.

(a) The set Δ introduced in Thm. 18.5 may be strictly contained in the domain of convergence of the CBHD, that is, the CBHD series $a \diamond b$ can be (absolutely) convergent even if (a, b) does not belong to Δ. This happens, for instance, in any finite dimensional nilpotent Lie algebra, where the series expressing $a \diamond b$ is in fact a finite sum. Moreover, if $a \in \mathfrak{g}$ is *any* element of $\mathfrak{g}$ (with norm not necessarily smaller that 2π), then the CBHD series $a \diamond 0$ is absolutely convergent, since $\sum_{n=1}^{\infty} \|Z_n(a, 0)\| = \|a\|$.

(b) Let us consider the function u defined in the statement of Thm. 18.5, and let r_0 be the (unique) fixed point of u on $[0, 2\pi)$, that is, the unique solution of the equation $u(r) = r$. Then the CBHD series is absolutely convergent on *any* square $[0, s] \times [0, s]$ with $s < r_0$.

(c) The arguments presented in Sec. 18.1 can be extended to the case of *infinite-dimensional Banach-Lie algebras*: these are infinite-dimensional (real or complex) Banach spaces A endowed with a Lie-algebra structure such that

$$\|[a, b]\| \leq \|a\| \, \|b\|,$$

for any $a, b \in A$. See [Biagi and Bonfiglioli (2014)] for all the details. $\qquad \sharp$

18.2 Exercises of Chap. 18

Exercise 18.1. Determine the graphs of the (real version of the) functions met in this chapter (see Fig. 18.1, 18.2, 18.3 below):

(1) $F : (-2\pi, 2\pi) \longrightarrow \mathbb{R}$, $F(x) = 2 + \dfrac{x}{2} \left(1 - \cot\left(\dfrac{x}{2}\right)\right)$ (setting $F(0) = 1$); prove that F has a minimum point on the interval $(-2\pi, 2\pi)$ which is the solution of $\cos x + \sin x = x + 1$ different from $x = 0$;

(2) $1/F$, if F is as above;

(3) $u : [0, 2\pi] \longrightarrow \mathbb{R}$, $u(x) = \displaystyle\int_{x}^{2\pi} \dfrac{1}{F(y)} \, \mathrm{d}y$.

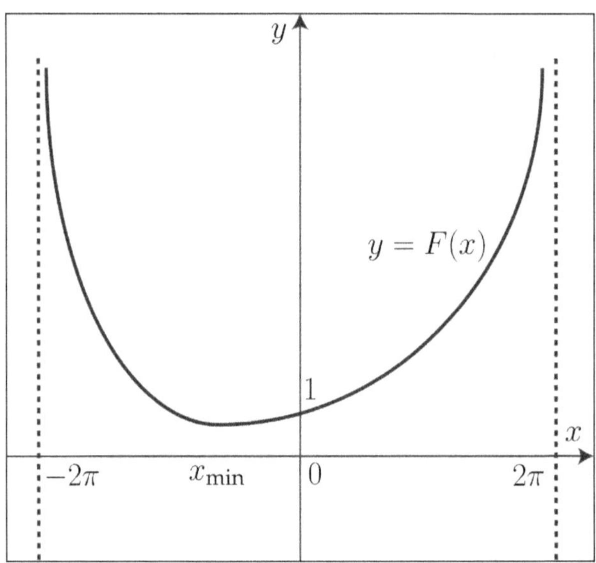

Fig. 18.1   Graph of the function $F(x) = 2 + \dfrac{x}{2}\left(1 - \cot\left(\dfrac{x}{2}\right)\right)$ on the interval $(-2\pi, 2\pi)$. This function has a minimum point at $x_{\min} \approx -2.41201114$ with value $F(x_{\min}) \approx 0.333442$. The point $x_{\min}$ is the solution of $\cos x + \sin x = x + 1$ different from $x = 0$.

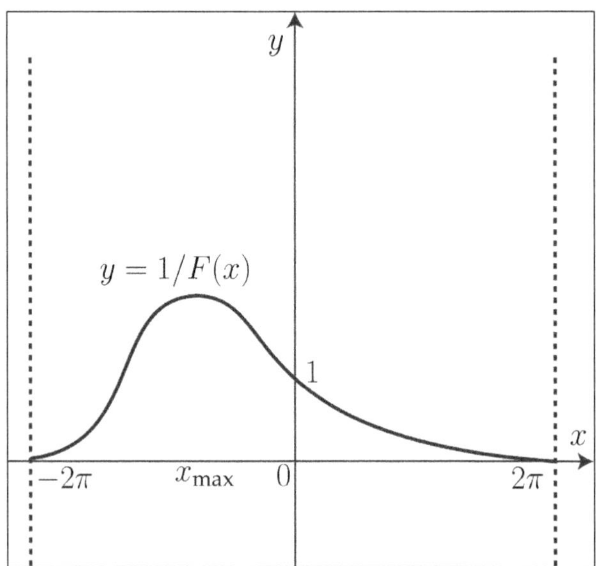

Fig. 18.2   Graph of the function $1/F(x)$ on the interval $(-2\pi, 2\pi)$. This function has a maximum point at $x_{\max} \approx -2.41201114$ with value $F(x_{\max}) \approx 2.99902$.

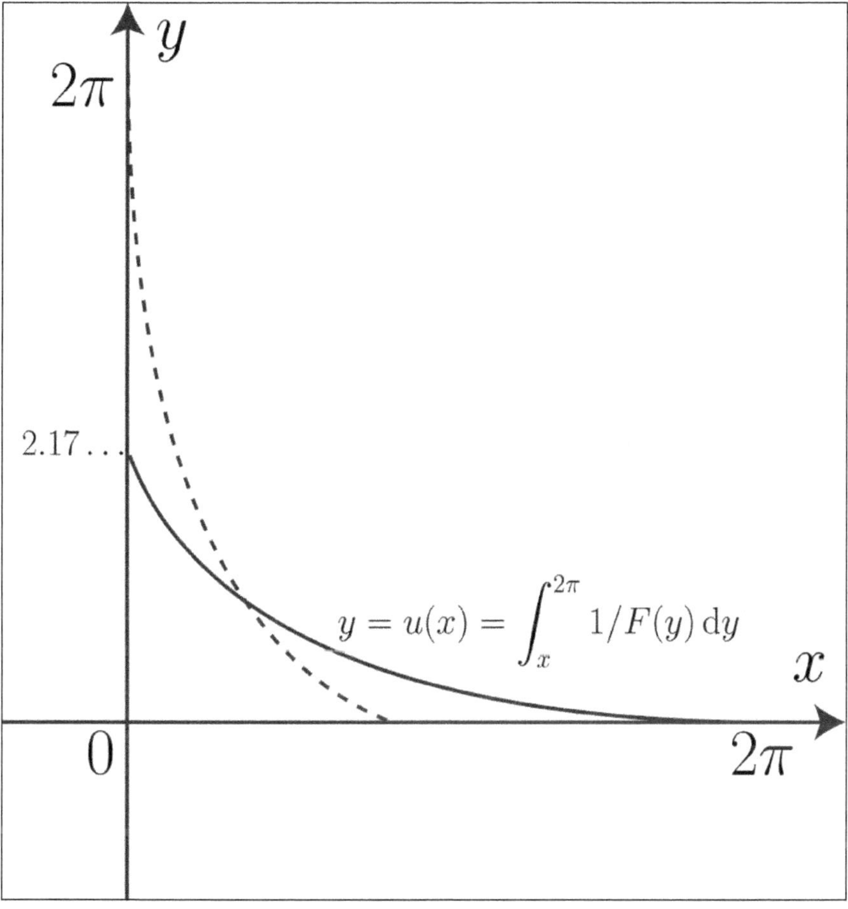

Fig. 18.3 Graph of the function $u(x) = \int_x^{2\pi} \dfrac{1}{F(y)}\,\mathrm{d}y$ on the interval $[0, 2\pi]$. The maximum value is approximately 2.17374. The dashed line represents the graph of the inverse function u^{-1} of u. The intersection of the two graphs is the point (x_0, x_0) with $x_0 \approx 1.2357$. The union of the hypographs of u and of u^{-1} provides the convergence set Δ in Thm. 18.5.

Appendix A

Some prerequisites of Linear Algebra

$\mathbf{T}$HE aim of this appendix is to collect, in a clear way, the basic definitions and results of Algebra and of Linear Algebra that we used in the book.

A.1 Algebras and Lie algebras

We begin with the main algebraic definitions that we used.

Definition A.1 (Algebra). Let A be a vector space over the field $\mathbb{K}$. Suppose $*$ is a binary operation on A, that is, $*$ is a function of the form

$$A \times A \longrightarrow A, \quad (a, a') \mapsto a * a'.$$

We say that $*$ endows A with the structure of:

    (1) an **algebra**, if $*$ is $\mathbb{K}$-bilinear;

    (2) an **associative algebra**, if $(A, *)$ is an algebra and $*$ is associative:

$$a * (b * c) = (a * b) * c, \qquad \text{for every } a, b, c \in A;$$

    (3) a **unital associative algebra**, if $(A, *)$ is an associative algebra and $*$ is endowed with a unit element $e \in A$:

$$a * e = a = e * a, \quad \text{for every } a \in A;$$

    (4) a **Lie algebra**, if $(A, *)$ is an algebra, $*$ is skew-symmetric and the following so-called **Jacobi identity** holds

$$a * (b * c) + b * (c * a) + c * (a * b) = 0, \quad \text{for every } a, b, c \in A.$$

Given an algebra $(A, *)$ we say that $B \subseteq A$ is a **subalgebra** of A if B is a vector subspace of B closed under the operation $*$, hence $(B, *|_{B \times B})$ is itself an algebra. In the case of a Lie algebra we speak of a **Lie subalgebra**.

As is customary, in the setting of Lie algebras the operation $*$ will be denoted by $(a, a') \mapsto [a, a']$ and it will be called the **Lie bracket** of A. Thus the Jacobi identity can be rewritten as follows:

$$[a, [b, c]] + [b, [c, a]] + [c, [a, b]] = 0, \quad \text{for every } a, b, c \in A.$$

Remark A.2. If $(A, *)$ is an associative algebra, A can naturally be endowed with the structure of a Lie algebra by setting

$$[a, b] := a * b - b * a, \quad \text{for every } a, b \in A.$$

We say that the above $[\cdot, \cdot]$ (occasionally denoted by $[\cdot, \cdot]_*$) is the **commutator related to** $*$. We leave to the reader the verification that the commutator related to $*$ is skew-symmetric and it satisfies the Jacobi identity (Exr. A.1).   ♮

Example A.3. Observe that, if V is a vector space over the field $\mathbb{K}$ and if $\mathrm{End}(V)$ is the vector space of the endomorphisms of V, then $(\mathrm{End}(V), \circ)$ is a unital associative algebra over $\mathbb{K}$ (whose identity is the identity map of V).   ♮

Definition A.4 (Derivation of an algebra). Let $(A, *)$ be an algebra (associative or not). A map $D : A \to A$ is called a **derivation** of A if D is linear and it satisfies

$$D(a * b) = (Da) * b + a * (Db), \quad \text{for every } a, b \in A.$$

When $(A, [\cdot, \cdot])$ is a Lie algebra, the above condition becomes

$$D[a, b] = [Da, b] + [a, Db], \quad \text{for every } a, b \in A.$$

Left as an exercise is to show that if D is a derivation of the associative algebra $(A, *)$ then it is also a derivation of the Lie algebra $(A, [\cdot, \cdot]_*)$ (see Exr. A.2).

Definition A.5 (Adjoint map). Let $(\mathfrak{g}, [\cdot, \cdot])$ be a Lie algebra. Given $x \in \mathfrak{g}$, we set

$$\mathrm{ad}\, x : \mathfrak{g} \longrightarrow \mathfrak{g}, \qquad \mathrm{ad}\, x(y) := [x, y] \quad (y \in \mathfrak{g}).$$

We say that $\mathrm{ad}\, x$ is the **adjoint** (map) of x.

It is easy to prove that $\mathrm{ad}\, x$ is a derivation of the Lie algebra $\mathfrak{g}$ (Exr. A.3). The above definition makes sense, e.g., in any associative algebra $(A, *)$ since we know that A can be equipped with a Lie algebra structure by the commutator defined by $[a, b] := a * b - b * a$ (Rem. A.2).

Notation A.6. Let $(\mathfrak{g}, [\cdot, \cdot])$ be a Lie algebra. If U, V are subsets of $\mathfrak{g}$ we set

$$[U, V] := \mathrm{span}\{[u, v] : u \in U, \ v \in V\}.$$

Note that $[U, V]$ is not merely the set of the brackets $[u, v]$ with $u \in U$ and $v \in V$, but the *span* of these brackets.

Definition A.7 (Nilpotent Lie algebra). Let $(\mathfrak{g}, [\cdot, \cdot])$ be a Lie algebra. The following sequence of subspaces of $\mathfrak{g}$ is called the **descending central series** of $\mathfrak{g}$:

$$\mathfrak{g}_1 := \mathfrak{g}, \quad \mathfrak{g}_2 := [\mathfrak{g}, \mathfrak{g}], \quad \mathfrak{g}_3 = [\mathfrak{g}, [\mathfrak{g}, \mathfrak{g}]]$$
$$\mathfrak{g}_{n+1} := [\mathfrak{g}, \mathfrak{g}_n] = \mathrm{span}\{[g, g_n] : g \in \mathfrak{g}, \ g_n \in \mathfrak{g}_n\} \quad (n \in \mathbb{N}, \ n \geq 1). \tag{A.1}$$

We say that $\mathfrak{g}$ is **nilpotent of step** r if and only if $\mathfrak{g}_r \neq \{0\}$ and $\mathfrak{g}_{r+1} = \{0\}$.

Let $(\mathfrak{g}, [\cdot, \cdot])$ be a Lie algebra and let $U \subseteq \mathfrak{g}$. We say that the elements of U are **brackets of length** 1 of U. Let now $k \geq 2$. Inductively, once the brackets of lengths $1, \ldots, k-1$ of U have been defined, we say that $[u, v]$ is a bracket of **length** k of U if and only if u and v are brackets of lengths i and j of U, with $i + j = k$. We shall also allow **height** or **order** as synonyms for length.

To fix the notation, we set:

$$B_1(U) := \mathrm{span}\{U\}, \qquad B_2(U) := [B_1(U), B_1(U)] \quad \text{and, for } n \geq 2,$$
$$B_n(U) := \mathrm{span}\Big\{ [u, v] \; : \; u \in B_i(U), \; v \in B_j(U) \text{ and } i + j = n \Big\}. \tag{A.2}$$

In other words $B_n(U)$ is the vector space spanned by the brackets of length n of U. For example, if $u_1, \ldots, u_7 \in U$, then

$$[[u_1, u_2], [[[u_3, [u_4, u_5]], u_6], u_7]], \qquad [[[u_1, [[u_2, u_3], u_4]], u_5], [u_6, u_7]]$$

are elements of $B_7(U)$. Note that, given $U \subseteq \mathfrak{g}$, *an element of $\mathfrak{g}$ may have more than one length* (wrt U) or even infinitely many, or even no length at all (if it does not belong to any of the $B_n(U)$ above); see the following example.

Example A.8. In the Lie algebra of the C^∞ v.f.s on $\mathbb{R}$, we take $U = \{\partial_x, x\, \partial_x\}$. As

$$\partial_x = [\cdots [\partial_x, \underbrace{x\, \partial_x] \cdots x\, \partial_x}_{k \text{ times}}] \quad \text{for every } k \in \mathbb{N},$$

we see that ∂_x belongs to $B_n(U)$ for every $n \in \mathbb{N}$, and it is therefore a bracket of length n of U, for every $n \in \mathbb{N}$ (Exr. A.5).   ♯

When $u_1, \ldots, u_k \in U$, brackets of the form

$$[u_1, [u_2 \cdots [u_{k-1}, u_k] \cdots]], \qquad [[\cdots [u_1, u_2] \cdots u_{k-1}], u_k]$$

are called **nested** (respectively, **right-nested** and **left-nested**). Observe that right-nested brackets can be written by means of compositions of ad maps:

$$[u_1, [u_2 \cdots [u_{k-1}, u_k] \cdots]] = (\mathrm{ad}\, u_1) \circ (\mathrm{ad}\, u_2) \circ \cdots \circ (\mathrm{ad}\, u_{k-1})(u_k).$$

Analogously, by skew-symmetry, the same is true of left-nested brackets since one has $[a, b] = [-b, a] = (-\mathrm{ad}\, b)(a) = \mathrm{ad}\, (-b)(a)$, whence:

$$[[\cdots [u_1, u_2] \cdots u_{k-1}], u_k] = (-1)^{k-1} [u_k, [u_{k-1} \cdots [u_2, u_1] \cdots]]. \tag{A.3}$$

The following Thm. A.11 shows that the set of the right-nested brackets span the set of the brackets of any length. An analogous statement holds for left-nested ones, owing to (A.3). First we give a definition.

Definition A.9 (Lie subalgebra generated by a set). Let $\mathfrak{g}$ be a Lie algebra and let $U \subseteq \mathfrak{g}$. We denote by $\mathrm{Lie}\{U\}$ the smallest Lie subalgebra of $\mathfrak{g}$ containing U, and we call it **the Lie algebra generated by** U (in $\mathfrak{g}$). More precisely, $\mathrm{Lie}\{U\} = \bigcap \mathfrak{h}$, where $\mathfrak{h}$ runs over the set of all the Lie subalgebras of $\mathfrak{g}$ which contain U.

If $\mathrm{Lie}\{U\} = \mathfrak{g}$ we say that U is a set of **Lie-generators** of $\mathfrak{g}$ (or, equivalently, we say that U Lie-generates $\mathfrak{g}$).

Remark A.10. With the above notation, it is easy to see that Lie$\{U\}$ *coincides with the span of the brackets of all lengths of the elements of U.* Indeed (see (A.2)),

$$\mathrm{Lie}\{U\} = \biguplus_{n \in \mathbb{N}} B_n(U),$$

where $\biguplus$ denotes a sum of vector subspaces[2] of $\mathfrak{g}$. Note that, by means of this notation, we can compactly write

$$B_n(U) = \biguplus_{i+j=n} [B_i(U), B_j(U)], \quad n \in \mathbb{N}. \tag{A.4}$$

Theorem A.11 (Nested brackets). *Let $\mathfrak{g}$ be a Lie algebra and let $U \subseteq \mathfrak{g}$. We set*

$$R_1(U) := \mathrm{span}\{U\}, \qquad R_n(U) := [U, R_{n-1}(U)], \quad n \geq 2. \tag{A.5}$$

In other words $R_n(U)$ is the vector space spanned by the right-nested brackets of length n of U. Then we have $R_n(U) = B_n(U)$ for every $n \in \mathbb{N}$, where $B_n(U)$ has been introduced in (A.2). As a consequence, Lie$\{U\} = \biguplus_{n \in \mathbb{N}} R_n(U)$.

The same assertion holds using left-nested brackets instead of right-nested ones.

The above theorem states that every element of Lie$\{U\}$ is a linear combination of right-nested brackets (an analogous statement is valid for left-nested ones). Observe that the central series $\mathfrak{g}_n$ in (A.1) satisfies $\mathfrak{g}_n = R_n(\mathfrak{g})$, for every n.

To show the idea behind the proof of Thm. A.11 (a consequence of the Jacobi identity and of the skew-symmetry), we take $u_1, u_2, v_1, v_2 \in U$ and we prove that $[[u_1, u_2], [v_1, v_2]]$ is a linear combination of right-nested brackets of length 4.

By the Jacobi identity $[X, [Y, Z]] = -[Y, [Z, X]] - [Z, [X, Y]]$ one has

$$[\underbrace{[u_1, u_2]}_{X}, [\underbrace{v_1}_{Y}, \underbrace{v_2}_{Z}]] = -[v_1, [v_2, [u_1, u_2]]] - [v_2, [[u_1, u_2], v_1]]$$

$$= -[v_1, [v_2, [u_1, u_2]]] + [v_2, [v_1, [u_1, u_2]]] \in R_4(U).$$

Proof of Thm. A.11. We argue by induction on n. Clearly

$$R_1(U) = B_1(U) = \mathrm{span}\{U\} \quad \text{and} \quad R_2(U) = B_2(U) = [U, U].$$

We now suppose that $R_n(U) = B_n(U)$ for $n = 1, \ldots, k$ and we prove it for $n = k+1$. It is obvious that $R_{k+1}(U) \subseteq B_{k+1}(U)$. To prove the reverse inclusion, taking into account (A.4) and the inductive hypothesis, we can restrict to proving that, if $i + j = k + 1$ and if $u_1, \ldots, u_i, v_1, \ldots, v_j \in U$, then

$$\Big[[u_1[u_2[\cdots[u_{i-1}, u_i]\cdots]]]; [v_1[v_2[\cdots[v_{j-1}, v_j]\cdots]]]\Big] \in R_{i+j}(U).$$

---

[2] We review the relevant definition. Let V be a vector space and let $\{W_i\}_{i \in I}$ be any family of vector subspaces of V. We denote by $\biguplus_{i \in I} W_i$ the set of the finite sums of the elements of the spaces W_i, that is, the set of the vectors of the form

$$w_{i_1} + \cdots + w_{i_n},$$

where n varies in $\mathbb{N}$, $\{i_1, \ldots, i_n\} \subseteq I$ and $w_{i_1} \in W_{i_1}, \ldots, w_{i_n} \in W_{i_n}$.

Exploiting repeatedly the induction hypothesis, the Jacobi identity and the skew-symmetry, we have (setting $u = [u_1[u_2[\cdots[u_{i-1}, u_i]\cdots]]]$)

$$\Big[u, [v_1[v_2[\cdots[v_{j-1}, v_j]\cdots]]]\Big]$$

$$= -[v_1, \underbrace{[[v_2, [v_3, \cdots]], u]]}_{\text{length } k} - [[v_2, [v_3, \cdots]], [u, v_1]]$$

$$= \{\text{element of } R_{k+1}(U)\} - [[v_1, u], [v_2, [v_3, \cdots]]]$$

$$= \{\text{element of } R_{k+1}(U)\} + [v_2, \underbrace{[[v_3, \cdots], [v_1, u]]}_{\text{length } k}] + [[v_3, \cdots], [[v_1, u], v_2]]$$

$$= \{\text{element of } R_{k+1}(U)\} + [[v_2, [v_1, u]], [v_3, \cdots]]$$

(after finitely many steps...)

$$= \{\text{element of } R_{k+1}(U)\} + (-1)^j \, [v_j, [v_{j-i}, [v_{j-2}, \cdots [v_1, u]]]],$$

and the latter belongs to $R_{k+1}(U) = R_{i+j}(U)$. This ends the proof. $\qquad\square$

Remark A.12. Let $\mathfrak{g}$ be a Lie algebra and let $\{\mathfrak{g}_n\}_n$ be the descending central series of $\mathfrak{g}$. By definition, each set $\mathfrak{g}_n$ is a vector subspace of $\mathfrak{g}$; moreover, if $n \geq 2$, the elements in $\mathfrak{g}_n$ are linear combinations of nested brackets

$$[a_1, [a_2, [a_3, [\cdots [a_{k-1}, a_n]\cdots]]]], \quad \text{with } a_1, \ldots, a_n \in \mathfrak{g}.$$

We also know that $\mathfrak{g}_{n+1} \subseteq \mathfrak{g}_n$, for every $n \in \mathbb{N}$. $\qquad\sharp$

Remark A.13. Let $\mathfrak{g}$ be a nilpotent Lie algebra of step r, and let $\{\mathfrak{g}_n\}_n$ be the descending central series of $\mathfrak{g}$. Since $\mathfrak{g}_n = \{0\}$ for every $n \geq r + 1$, from Rem. A.12 we deduce that any nested bracket of elements in $\mathfrak{g}$ containing more than r terms is equal to 0. Thus, since *any* bracket of length n is a linear combination of nested brackets of the same length (Thm. A.11), we deduce that any bracket (not necessarily nested) of elements in $\mathfrak{g}$ containing more than r terms is 0. $\qquad\sharp$

We now give a very general definition concerning Lie algebras.

Definition A.14 (Structure constants of a Lie algebra wrt a basis). Let $\mathfrak{g}$ be an N-dimensional Lie algebra, and let $\mathcal{F} = \{F_1, \ldots, F_N\}$ be a basis of $\mathfrak{g}$.

For every $i, j \in \{1, \ldots, N\}$ there exist uniquely defined scalars $c_{i,j}^k$ such that

$$[F_i, F_j] = \sum_{k=1}^{N} c_{i,j}^k \, F_k. \tag{A.6}$$

We call such numbers the **structure constants** of $\mathfrak{g}$ wrt the basis $\mathcal{F}$.

Remark A.15. Let the above notations apply. By skew-symmetry, we have $c_{i,i}^k = 0$ and $c_{i,j}^k = -c_{j,i}^k$, for every choice of i, j and k in $\{1, \ldots, N\}$. $\qquad\sharp$

The structure constants of a Lie algebra completely determine its Lie-algebraic structure, as the following lemma shows.

Lemma A.16. *Let $\mathfrak{g}_1, \mathfrak{g}_2$ be Lie algebras of finite dimensions. The following are equivalent:*

(i) $\mathfrak{g}_1$ *and* $\mathfrak{g}_2$ *are isomorphic (as Lie algebras);*

(ii) *for every basis of* $\mathfrak{g}_1$ *there exists a basis of* $\mathfrak{g}_2$ *(of the same cardinality) such that the associated structure constants coincide.*

The simple proof is left as an exercise.

A.1.1 *Stratified Lie algebras*

In this section we collect some results (used in Chap. 16) about stratified Lie algebras, defined in Def. 16.1 on page 300.

From a set-theoretic point of view, a same Lie algebra $\mathfrak{g}$ may admit many different stratifications. In the sequel, when we deal with a stratified Lie algebra we tacitly refer to a *couple* $(\mathfrak{g}, \mathcal{V})$, where $\mathcal{V} = \{V_i\}_{i \in \mathbb{N}}$ is the datum of a stratification of $\mathfrak{g}$. In the sequel, we tacitly understand that all linear structures are over $\mathbb{R}$.

Proposition A.17. *Let* $(\mathfrak{g}, \{V_i\}_i)$ *be a stratified Lie algebra. Then* $\mathfrak{g}$ *is Lie-generated by the set* V_1, *that is,* $\mathfrak{g} = \mathrm{Lie}\{V_1\}$.

Proof. Let $\mathfrak{h} := \mathrm{Lie}\{V_1\}$. By Thm. A.11 we infer that $\mathfrak{h} = \biguplus_{n=1}^{\infty} R_n(V_1)$, where $R_1(V_1) = \mathrm{span}\{V_1\}$ and, for every $n \geq 2$, $R_n(V_1) = [V_1, R_{n-1}(V_1)]$. On the other hand, since $R_1(V_1) = V_1$, an induction argument shows that $R_n(V_1) = V_n$ for every $n \in \mathbb{N}$; therefore, since $\mathfrak{g}$ is stratified, we get $\mathfrak{h} = \biguplus_{n=1}^{\infty} V_n = \bigoplus_{n=1}^{\infty} V_n = \mathfrak{g}$. $\qquad \square$

Remark A.18. Let $\mathfrak{g}$ be a stratified Lie algebra and let $\mathcal{V} = \{V_i\}_i$ be a stratification of $\mathfrak{g}$. If $\mathfrak{g}$ is *finite-dimensional*, then only finitely many V_i in (16.1) (page 300) can be different from $\{0\}$, say $V_1, \ldots, V_n$; we can obviously discard any V_i which coincides with $\{0\}$. As a consequence $\dim(\mathfrak{g}) = \sum_{i=1}^{n} \dim(V_i) \geq n$; hence, it is possible to find a natural r with the following properties:

(i) $V_i \neq \{0\}$ for every $i = 1, \ldots, r$;
(ii) $V_i = \{0\}$ for every $i \geq r + 1$.

Therefore, since $V_{i+1} = [V_1, V_i]$ for every $i \in \mathbb{N}$, we can write

$$\mathfrak{g} = V_1 \oplus \underbrace{[V_1, V_1]}_{V_2} \oplus \underbrace{[V_1, [V_1, V_1]]}_{V_3} \oplus \cdots \oplus \underbrace{[V_1, \ldots, [V_1, V_1] \ldots]}_{V_r},$$

and $V_1, \ldots, V_r$ are all different from $\{0\}$. $\qquad\qquad\qquad\qquad\qquad \sharp$

Remark A.19. Let $(\mathfrak{g}, \{V_n\}_n)$ be a stratified Lie algebra. Any element $a \in \mathfrak{g}$ can be uniquely written as $a = \sum_{n \in \mathbb{N}} a_n$, where $a_n \in V_n$ for every $n \in \mathbb{N}$, and only finitely-many (possibly none) a_n are different from 0. As a consequence, for every natural n we can define a map $\mathrm{p}_n : \mathfrak{g} \longrightarrow V_n$ by requiring that $\mathrm{p}_n(a) := a_n$.

Obviously, p_n is a linear surjective map, referred to as the **canonical projection** of $\mathfrak{g}$ onto V_n. If $a \in \mathfrak{g}$, there exists a finite set $I_a \subseteq \mathbb{N}$ such that $\mathrm{p}_n(a) = 0$, for every $n \notin I_a$. Hence

$$a = \sum_{n=1}^{\infty} \mathrm{p}_n(a) = \sum_{n \in I_a} \mathrm{p}_n(a).$$

In particular, if $\mathfrak{g}$ has finite dimension and if $r \in \mathbb{N}$ is such that $V_n = \{0\}$ for every $n \geq r+1$ (see Rem. A.18), we can write $a = \sum_{n=1}^{r} p_n(a)$. ♯

The following result characterizes the nilpotent, stratified Lie algebras.

Lemma A.20. *Let $(\mathfrak{g}, \{V_n\}_n)$ be a stratified Lie algebra. The following are equivalent:*

(i) $\mathfrak{g}$ *is nilpotent of step $r \in \mathbb{N}$;*

(ii) $V_r \neq \{0\}$ *and* $V_n = \{0\}$ *for every $n \geq r+1$.*

Proof. (i) $\Rightarrow$ (ii): Let $\{\mathfrak{g}_n\}_n$ be the descending central series of $\mathfrak{g}$. Since, by assumption, $\mathfrak{g}$ is nilpotent of step r, we have (see Def. A.7)

$$\mathfrak{g}_r \neq \{0\} \quad \text{and} \quad \mathfrak{g}_n = \{0\} \quad \text{for every } n \geq r+1;$$

as a consequence, for every $n \in \mathbb{N}$ such that $n \geq r+1$, one has

$$V_n \overset{(16.1)}{=} \underbrace{[V_1, \ldots [V_1, V_1]]}_{n \text{ times}} \subseteq \underbrace{[\mathfrak{g}, \ldots [\mathfrak{g}, \mathfrak{g}]]}_{n \text{ times}} \overset{(A.1)}{=} \mathfrak{g}_n = \{0\},$$

and $\mathfrak{g}$ can be decomposed as $\mathfrak{g} = V_1 \oplus \cdots \oplus V_r$ (see also (16.1)).

On the other hand, since $\mathfrak{g}_r \neq \{0\}$ and since $\mathfrak{g}_r$ is generated by nested brackets of length r (see Rem. A.12), there exist $a_1, \ldots, a_r \in \mathfrak{g}$ such that

$$a = [a_1, \ldots, [a_{r-1}, a_r] \ldots] \neq 0;$$

therefore, observing that $p_n \equiv 0$ for every $n \geq r+1$, we obtain

$$0 \neq a = \sum_{i_1, \ldots, i_r = 1}^{r} \underbrace{[p_{i_1}(a_1), \ldots, [p_{i_{r-1}}(a_{r-1}), p_{i_r}(a_r)] \ldots]}_{\in V_{i_1 + \cdots + i_r}}$$

$$\big(\text{since } V_n = \{0\} \text{ for every } n \geq r+1\big)$$

$$= [p_1(a_1), \ldots, [p_1(a_{r-1}), p_1(a_r)] \ldots] \in V_r,$$

and this proves that $V_r \neq \{0\}$, as desired.

(ii) $\Rightarrow$ (i): Since, by assumption, $V_r \neq \{0\}$ and since $V_r \subseteq \mathfrak{g}_r$, we have $\mathfrak{g}_r \neq \{0\}$; moreover, as $V_n = \{0\}$ for every $n \geq r+1$, we obtain

$$[a_1, \ldots, [a_r, a_{r+1}] \ldots] = \sum_{i_1, \ldots, i_{r+1} = 1}^{r} \underbrace{[p_{i_1}(a_1), \ldots, [p_{i_r}(a_r), p_{i_{r+1}}(a_{r+1})] \ldots]}_{\in V_{i_1 + \cdots + i_{r+1}}}$$

$$= 0 \quad \big(\text{since } i_1 + \cdots + i_{r+1} \geq r+1\big).$$

Since $\mathfrak{g}_{r+1}$ is generated by nested brackets of length $r+1$, we get $\mathfrak{g}_n = \{0\}$ for every $n \geq r+1$; thus $\mathfrak{g}$ is nilpotent of step r, as desired. □

By combining Rem. A.18 and Lem. A.20, we obtain the following fact.

Corollary A.21. *Any finite-dimensional stratified Lie algebra is nilpotent.*

A.2   Positive semidefinite matrices

Convention. *In the sequel, it is understood that all matrices are real-valued. When we mention a positive (or negative) semidefinite matrix A, we understand that A is symmetric.*

We fix another convention: we identify any $m \times N$ matrix B with the linear map from $\mathbb{R}^N$ to $\mathbb{R}^m$ mapping $x \in \mathbb{R}^N$ to $Bx \in \mathbb{R}^m$. Given a set $U \subseteq \mathbb{R}^N$, $U^\perp$ denotes the sets of the vectors ξ in $\mathbb{R}^N$ such that $\langle \xi, u \rangle = 0$ for every $u \in U$. Moreover, following the identification between matrices and linear maps, we denote by $\ker(B)$ the kernel of B, and by range(B) we mean the range of the matrix B, that is, the span of the column-vectors of B.

It is well known that A is positive semidefinite if and only if $A = R^T R$, for some square matrix R of the same order of A (Exr. A.8). As in the decomposition of the principal matrix of a sum-of-square operator (see Sec. 8.3 on page 171), it can be relevant to consider positive semidefinite matrices of the form $B^T B$ where B is not necessarily a square matrix. This is the main reason why, in the sequel, we also consider decompositions of this form.

Let us fix a notation: given a symmetric matrix A of order N, we write

$$\mathrm{Isotr}(A) := \big\{ \xi \in \mathbb{R}^N : \langle A\xi, \xi \rangle = 0 \big\}, \tag{A.7}$$

and we say that any element of $\mathrm{Isotr}(A)$ is an **isotropic vector for** A (or an isotropic vector for the quadratic form associated with A). If A is positive semidefinite, it is easy to check (Exr. A.9) that $\ker(A) = \mathrm{Isotr}(A)$.

Moreover, if $A = B^T B$ for some $m \times N$ matrix B (with m not necessarily equal to N), one has $\ker(A) = \mathrm{Isotr}(A) = \ker(B)$. Moreover, another identity concerns with the ranges of these matrices (Exr. A.11), namely range$(A) =$ range(B^T).

In Chap. 10, in the characterization of the principality of v.f.s, we used:

Proposition A.22 (Principality). *Given an $N \times N$ positive semidefinite matrix A and a vector $v \in \mathbb{R}^N$, the following conditions are equivalent:*

   (a) *there exists $\lambda > 0$ such that $\langle v, \xi \rangle^2 \leq \lambda \langle A\xi, \xi \rangle$, for every $\xi \in \mathbb{R}^N$;*
   (b) *$v \in (\ker A)^\perp$;*
   (c) *v is a linear combination of the rows (or of the columns) of A.*

If, moreover, B is any real $m \times N$ matrix such that $A = B^T B$, we have yet another equivalent condition:

   (d) *v is a linear combination of the columns of B^T (i.e., the rows of B).*

Proof. The equivalence of (b) and (c) is obvious,[3] whereas (d) follows from the identity range$(B^T B) =$ range(B^T) (see Exr. A.11). The only non-trivial implication is (b) $\Rightarrow$ (a): for the proof of this, see the hint to Exr. A.12. $\square$

---

[3]For any matrix B one has that $(\ker B)^\perp$ is the range of B^T, i.e., the span of the row vectors of B: this follows from the very definition of $\ker B$, and it can also be seen as the trivial version of the so-called Fredholm's Alternative; see Exr. A.10.

In studying Picone's WMP, in Chap. 8 we used the following result.

Lemma A.23 (Féjer). *Let A, B be real symmetric matrices of the same dimension. Let A be positive semidefinite and B negative semidefinite. Then $\mathrm{trace}(AB) \leq 0$.*

Actually (Exr. A.17), given a symmetric matrix B, the family of inequalities

$$\mathrm{trace}(AB) \leq 0 \quad \text{(as } A \text{ ranges in the positive definite matrices)}$$

characterizes the negative semidefinite matrices B.

Proof. Write $A = R R^T$, with some matrix R of the same order as A (Exr. A.8). Thus (as $\mathrm{trace}(HK) = \mathrm{trace}(KH)$ for any square matrices H, K), one has

$$\mathrm{trace}(AB) = \mathrm{trace}(R R^T B) = \mathrm{trace}(R^T B R) = \sum_i (R^T)^i B R_i,$$

where $(R^T)^i$ denotes the i-th row of R^T and R_i denotes the i-th column of R. (For the last equality, see also Exr. A.16.) As $(R^T)^i = (R_i)^T$, we have

$$(R^T)^i B R_i = (R_i)^T B R_i = \langle R_i, B R_i \rangle \leq 0.$$

The above inequality follows from the assumption that B is negative semidefinite. Thus $\mathrm{trace}(AB) \leq 0$, which ends the proof. $\qquad\square$

A.3 The Moore-Penrose pseudo-inverse

Let V, W be *finite-dimensional* vector spaces on the same field $\mathbb{K}$, and suppose that $T : V \to W$ is a $\mathbb{K}$-linear map. Let $\ker T$ be the kernel of T, and let U be any direct-sum complement of $\ker T$ in V, i.e.,

$$V = U \oplus \ker T. \tag{A.8}$$

This means that, for any $v \in V$, there exist unique $\overline{u} \in U$ and $\overline{k} \in \ker T$ such that $v = \overline{u} + \overline{k}$. Consequently, there exist uniquely defined maps $u, k : V \to V$ such that, for every $v \in V$,

$$v = u(v) + k(v), \quad \text{with } u(v) \in U \text{ and } k(v) \in \ker T. \tag{A.9}$$

By means of (A.8), it is very easy to show that

$$T|_U : U \to T(V) \quad \text{is an isomorphism of vector spaces.} \tag{A.10}$$

Incidentally, from (A.8) we get $\dim(U) + \dim(\ker T) = \dim(V)$; therefore, since (A.10) gives $\dim(U) = \dim(T(V))$, we get the well-known fact

$$\dim(T(V)) + \dim(\ker T) = \dim(V).$$

The inverse map of $T|_U$ is obviously characterized by the following property: for any $w \in T(V)$ one has that

$$v := (T|_U)^{-1}(w) \text{ is the unique vector such that } \begin{cases} T(v) = w \\ v \in U. \end{cases} \tag{A.11}$$

Equivalently, if $w = T(v)$ for some $v \in V$ (which is not uniquely defined by w whenever T is not injective), then

$$(T|_U)^{-1}(T(v)) = u(v), \qquad (A.12)$$

where $u(\cdot)$ is as in (A.9). This identity is well posed (that is, $T(v) = T(v')$ implies that $u(v) = u(v')$) owing to (A.8).

Let now $\mathbb{K} = \mathbb{R}$ (the complex case can be treated analogously), and suppose that V is finite-dimensional and is endowed with an inner product $\langle \cdot, \cdot \rangle$. Due to the finite-dimensionality of V, we have (Exr. A.13) $V = (\ker T)^{\perp} \oplus \ker T$, where orthogonality is meant wrt the fixed inner product $\langle \cdot, \cdot \rangle$. Hence we can choose $U := (\ker T)^{\perp}$ in the previous paragraphs.

By (A.10) we infer that the restriction of T to $(\ker T)^{\perp}$ is an isomorphism of vector spaces from $(\ker T)^{\perp}$ to $T(V)$. Its inverse map

$$T^{\dagger} := (T|_{(\ker T)^{\perp}})^{-1} : T(V) \longrightarrow (\ker T)^{\perp}$$

will be referred to as the **(Moore-Penrose) pseudo-inverse of** T. By (A.11) we infer that $T^{\dagger}$ is characterized by the following fact: for any $w \in T(V)$,

$$v := T^{\dagger}(w) \text{ is the unique element of } V \text{ such that } \quad \begin{cases} T(v) = w \\ v \perp \ker T. \end{cases} \qquad (A.13)$$

If we use (A.12) instead, we infer that $T^{\dagger}$ is also characterized by the following fact: for every $v \in V$, one has

$$T^{\dagger}(T(v)) = u(v), \quad \text{where } v = u(v) + k(v) \text{ and } \quad \begin{cases} u(v) \perp \ker T \\ k(v) \in \ker T. \end{cases} \qquad (A.14)$$

We have yet another characterization of $T^{\dagger}$ (see Fig. A.1):

Remark A.24. Under the previous notations, the pseudo-inverse $T^{\dagger}$ of T enjoys the following characterizing property:

For any $w \in T(V)$ one has

$$T(T^{\dagger}(w)) = w \text{ and } \quad \|T^{\dagger}(w)\| = \min\{\|v\| : T(v) = w\}; \qquad (A.15)$$

the minimum is attained at $T^{\dagger}(w)$ only.

Indeed, if $T(v) = w$ then $v = u(v) + k(v)$, with $u(v) \perp \ker T$, $k(v) \in \ker T$; by Pythagorean Theorem we have

$$\|v\|^2 = \|u(v)\|^2 + \|k(v)\|^2 \geq \|u(v)\|^2 \overset{(A.14)}{=} \|T^{\dagger}(w)\|^2.$$

As for uniqueness, due to (A.13) it suffices to prove that, if $\bar{v} \in V$ is such that $T(\bar{v}) = w$ and if

$$\|\bar{v}\| = \min\{\|v\| : T(v) = w\}, \qquad (A.16)$$

then $\overline{v} \perp \ker T$. To this end, let $\{v_1, \ldots, v_n\}$ be an orthonormal basis of $\ker T$; then, again by Pythagorean Theorem, we have

$$\|\overline{v}\|^2 = \left\|\overline{v} - \sum_{j=1}^{n} \langle \overline{v}, v_j \rangle \, v_j \right\|^2 + \sum_{j=1}^{n} \langle \overline{v}, v_j \rangle^2. \tag{A.17}$$

Since $u(\overline{v}) = \overline{v} - \sum_{j=1}^{n} \langle \overline{v}, v_j \rangle \, v_j$ and $T(\overline{v}) = w$, then $T(u(\overline{v})) = w$ so that $u(\overline{v})$ is one of the vectors v in (A.16). Due to the minimality assumption on $\overline{v}$,

$$\left\|\overline{v} - \sum_{j=1}^{n} \langle \overline{v}, v_j \rangle \, v_j \right\|^2 = \|u(\overline{v})\|^2 \overset{(A.16)}{\geq} \|\overline{v}\|^2.$$

By (A.17) we infer that $\langle \overline{v}, v_j \rangle^2 = 0$ for any $j = 1, \ldots, n$; thus $\overline{v} \perp \ker T$. ♯

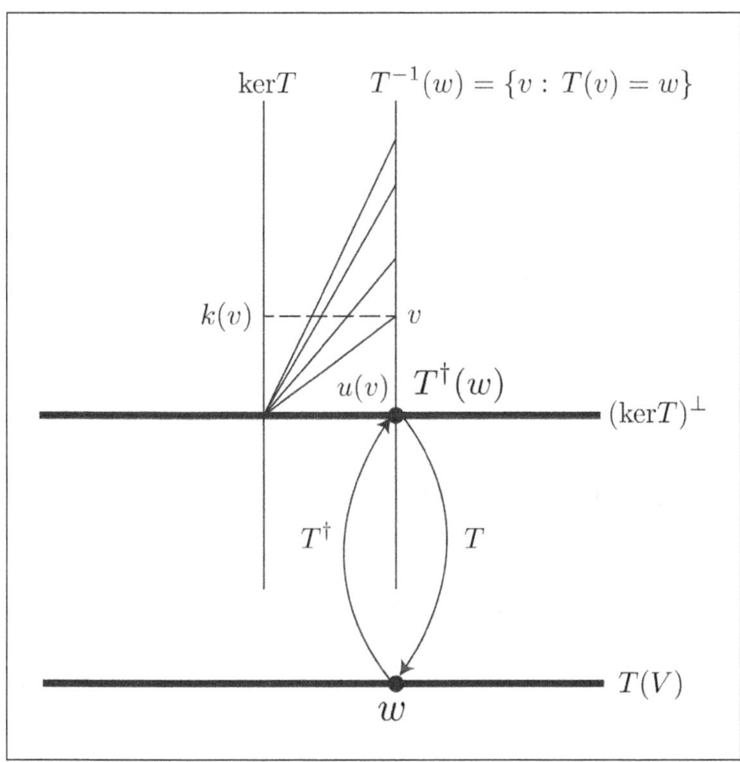

Fig. A.1   The pseudo-inverse $T^{\dagger}$ of T.

In the sequel, as usual, when a vector of $\mathbb{R}^m$ is involved in a matrix computation, we agree that it is written as an $m \times 1$ column matrix. We shall tacitly assume this throughout. In the next result we make no assumptions on m and N.

Corollary A.25. *Let S be an $N \times m$ real matrix. For every $x \in \mathbb{R}^m$ there exists at least one $\eta \in \mathbb{R}^N$ such that*

$$Sx = SS^T \eta. \tag{A.18}$$

Given $x \in \mathbb{R}^m$, all the solutions η of (A.18) are precisely those $\eta \in \mathbb{R}^N$ such that

$$S^T \eta = T^\dagger(Sx), \tag{A.19}$$

where $T : \mathbb{R}^m \to \mathbb{R}^N$ is the linear map canonically[4] associated with S, and $T^\dagger$ is its Moore-Penrose pseudo-inverse.

In particular, given $x \in \mathbb{R}^m$, a solution $\eta \in \mathbb{R}^N$ of (A.18) may not be uniquely defined by x, but $S^T \eta$ is uniquely determined by x via (A.19).

Proof. Given $x \in \mathbb{R}^m$, one has $Sx \in \mathrm{range}(S)$; hence, the existence of a solution $\eta \in \mathbb{R}^N$ of (A.18) is a consequence of $\mathrm{range}(S) = \mathrm{range}(SS^T)$ (see Exr. A.11). Let T be as in the statement of Cor. A.25, and let η solve (A.18). We apply $T^\dagger$ to both sides of (A.18), and we get

$$T^\dagger(S\,x) = T^\dagger(SS^T \eta). \tag{A.20}$$

The lhs of (A.20) is the rhs of (A.19); thus, (A.19) will follow if we show that $T^\dagger(SS^T \eta) = S^T \eta$. In its turn, the latter is a consequence of the following identities: if u denotes the orthogonal projection onto $(\ker T)^\perp$, we have

$$T^\dagger(SS^T \eta) = T^\dagger(T(S^T \eta)) \overset{(A.14)}{=} u(S^T \eta) = S^T \eta.$$

The last identity follows from $S^T \eta \in \mathrm{range}(S^T) = (\ker T)^\perp$ (Exr. A.10).

Vice versa, given $x \in \mathbb{R}^m$, let $\eta \in \mathbb{R}^N$ satisfy (A.19); if we apply T to both sides of the latter, then we get

$$SS^T \eta = T(S^T \eta) \overset{(A.19)}{=} T(T^\dagger(Sx)) \overset{(A.15)}{=} Sx.$$

In the last identity we used the fact that $Sx \in \mathrm{range}(S) = T(\mathbb{R}^m)$. The above identities yield (A.18) and this ends the proof. $\qquad\square$

We are finally ready to give the proof of Prop. 6.12 on page 137:

Proof of Prop. 6.12. Suppose that $v = \sum_{j=1}^m c_j v_j$ with $\sum_{j=1}^m c_j^2 \le 1$. Then

$$|\langle v, \xi \rangle| = \left| \left\langle \sum_{j=1}^m c_j v_j, \xi \right\rangle \right| \le \sum_{j=1}^m |c_j| \cdot |\langle v_j, \xi \rangle| \le \qquad \text{(Cauchy-Schwarz)}$$

$$\le \sqrt{\sum_{j=1}^m c_j^2} \cdot \sqrt{\sum_{j=1}^m |\langle v_j, \xi \rangle|^2} \le \sqrt{\sum_{j=1}^m |\langle v_j, \xi \rangle|^2}.$$

By squaring we get (6.2), so that $v \in \mathrm{Sub}(\mathcal{V})$.

Vice versa, let $v \in \mathrm{Sub}(\mathcal{V})$. By Rem. 6.10 there exists $x \in \mathbb{R}^m$ such that

$$v = \sum_{j=1}^m x_j v_j. \tag{A.21}$$

Since the scalars x_j satisfying (A.21) are not necessarily unique (if the vector v_j are not linearly independent), *we cannot infer that* $\sum_j x_j^2 \le 1$ (and this may hardly be the case). It is then natural to look for some $c = (c_1, \dots, c_m)$ with minimal norm,

---

[4]By this we mean that $T(x) = Sx$ for every $x \in \mathbb{R}^m$.

satisfying $v = \sum_{j=1}^{m} c_j v_j$, hence we make use of the pseudo-inverse $T^\dagger$ of the map $T : \mathbb{R}^m \to \mathbb{R}^N$ in (6.4).

To this end, let S be the $N \times m$ matrix representing T (with respect to the standard bases in $\mathbb{R}^m$ and $\mathbb{R}^N$). Note that (A.21) means that $v = Sx$. By Cor. A.25, if $x \in \mathbb{R}^m$ is as in (A.21), there exists at least one $\xi_x \in \mathbb{R}^N$ satisfying

$$Sx = SS^T \xi_x. \tag{A.22}$$

We test the inequality (6.2) with ξ replaced by ξ_x: we get

$$\langle Sx, \xi_x \rangle^2 \leq \sum_{j=1}^{m} \langle v_j, \xi_x \rangle^2. \tag{A.23}$$

In view of (A.22), the lhs is $\langle SS^T \xi_x, \xi_x \rangle^2 = \langle S^T \xi_x, S^T \xi_x \rangle^2 = \|S^T \xi_x\|^4$; in its turn, the rhs is (if e_j is the j-th vector of the canonical basis of $\mathbb{R}^m$)

$$\sum_{j=1}^{m} \langle v_j, \xi_x \rangle^2 = \sum_{j=1}^{m} \langle Se_j, \xi_x \rangle^2 = \sum_{j=1}^{m} \langle e_j, S^T \xi_x \rangle^2 = \|S^T \xi_x\|^2.$$

Thus, (A.23) produces the information $\|S^T \xi_x\|^2 \leq 1$. If we set $c := S^T \xi_x$, we have just obtained $\|c\|^2 \leq 1$, and

$$v = \sum_{j=1}^{m} x_j v_j = Sx \overset{\text{(A.22)}}{=} Sc = \sum_{j=1}^{m} c_j v_j,$$

and this is what we desired. $\qquad\qquad\qquad\qquad\qquad\qquad\qquad\qquad\qquad\square$

A.4   Exercises of App. A

Exercise A.1. Prove that the commutator associated with an associative algebra $(A, *)$ endows A with the structure of a Lie algebra.

Exercise A.2. Let $(A, *)$ be an associative algebra and suppose D is a derivation of A. Prove that D is also a derivation of the *Lie algebra* $(A, [\cdot, \cdot]_*)$, where $[\cdot, \cdot]_*$ is the commutator associated with $*$.

Exercise A.3. Let $(\mathfrak{g}, [\cdot, \cdot])$ be a Lie algebra. Prove that, for every $x \in \mathfrak{g}$, $\operatorname{ad} x$ is a derivation of the Lie algebra $\mathfrak{g}$.

Exercise A.4. If $\{\mathfrak{g}_n\}_n$ is the descending central series of a Lie algebra $\mathfrak{g}$ (see Def. A.7), prove the following facts:

   (1) any $\mathfrak{g}_n$ is a Lie subalgebra of $\mathfrak{g}$ (use also Thm. A.11);
   (2) $\mathfrak{g}_n \supseteq \mathfrak{g}_{n+1}$ for every $n \in \mathbb{N}$.

Exercise A.5. Let $\mathfrak{g}$ be the Lie algebra of the smooth vector fields on $\mathbb{R}^1$. Let $X = \partial_x$ and $Y = x \, \partial_x$. Prove that

$$X = [\cdots [X, \underbrace{Y] \cdots Y}_{k \text{ times}}], \quad \forall \, k \in \mathbb{N}.$$

Deduce that X is a bracket of length k of $U = \{X, Y\}$, *for every* $k \in \mathbb{N}$. Note that $\mathfrak{a} := \operatorname{span}\{X, Y\}$ is a Lie subalgebra of $\mathfrak{g}$. What is the descending central series $\{\mathfrak{a}_n\}_n$ of $\mathfrak{a}$? Observe that $\{\mathfrak{a}_n\}_n$ is constant for $n \geq 2$, but $\mathfrak{a}$ is not nilpotent.

Exercise A.6. Prove (A.3) (showing that, by reversing orders and modulo a factor ± 1) left-nested brackets can be converted into right-nested ones and vice versa.

Exercise A.7. Prove the assertion in Rem. A.10.

[*Hint:* Prove that $\biguplus_{k\in\mathbb{N}} B_k(U)$ is a Lie subalgebra of $\mathfrak{g}$ and that it is necessarily contained in every Lie subalgebra $\mathfrak{h}$ of $\mathfrak{g}$ containing U.]

Prove also that (with reference to the notation in (A.2))

$$[B_i(U), B_j(U)] \subseteq B_{i+j}(U), \quad \text{for every } i, j \in \mathbb{N}.$$

Due to Thm. A.11, deduce that $[R_i(U), R_j(U)] \subseteq R_{i+j}(U)$, for every $i, j \in \mathbb{N}$ (see the notation in (A.5)).

Exercise A.8. Let A be an $N \times N$ real symmetric matrix; assume that A is positive semidefinite. Prove that $A = R^T R$, for some $N \times N$ real matrix R.

[*Hint:* By Sylvester's Theorem, A is congruent to the diagonal matrix P having as many 1's on the main diagonal as the rank of A, and 0's otherwise. Hence, there exists a square matrix M such that $A = M^T P M$. Note that $P = P^2 = P^T P$, whence $A = (PM)^T (PM)$. Then take $R = PM$.]

Exercise A.9. (a) Let A be an $N \times N$ real symmetric matrix; assume that A is positive semidefinite. Prove that $\ker (A) = \mathrm{Isotr}(A)$ (see (A.7)).

[*Hint:* If R is as in Exr. A.8, $\ker (R) \subseteq \ker (A) \subseteq \mathrm{Isotr}(A) \subseteq \ker (R)$.]

Considering the case of $A = \left(\begin{smallmatrix} 0 & 1 \\ 0 & 0 \end{smallmatrix}\right)$, observe that the hypothesis of symmetry cannot be removed (in this case $\mathrm{Isotr}(A)$ is not even a vector space).

(b) Prove that, if $A = B^T B$ for some $m \times N$ real matrix B (with m not necessarily equal to N), one has $\ker (A) = \mathrm{Isotr}(A) = \ker (B)$.

(c) Moreover, if A is an $N \times N$ real positive semidefinite symmetric matrix, prove that the following facts are equivalent:

(1) $A \neq 0$;

(2) $\ker (A) \neq \mathbb{R}^N$;

(3) $\mathrm{rank}(A) \geq 1$;

(4) $\mathrm{Isotr}(A) \neq \mathbb{R}^N$;

(5) A has a positive eigenvalue;

(6) $\mathrm{trace}(A) > 0$;

(7) there exists $i \in \{1, \ldots, N\}$ such that $a_{i,i} \neq 0$.

Exercise A.10. Let B be an $N \times m$ matrix. Recognize that the very definition of the kernel of B gives the representation

$$\ker (B) = (\mathrm{span}\{B^1, \ldots, B^N\})^\perp,$$

where $B^1, \ldots, B^N$ are the row-vectors of B. Deduce the following identities:

(1) $(\ker (B))^\perp = \mathrm{span}\{B^1, \ldots, B^N\}$;
(2) $(\ker (B))^\perp = \mathrm{range}(B^T)$;

(3) $\ker(B) = (\mathrm{range}(B^T))^\perp$;
(4) $\mathrm{range}(B) = (\ker(B^T))^\perp$;
(5) $(\mathrm{range}(B))^\perp = \ker(B^T)$.

Derive from (4) the following simple form of the so-called *Fredholm's Alternative*: given a vector $b \in \mathbb{R}^N$, one and only one of the following facts holds:

(a) there exists a solution $x \in \mathbb{R}^m$ of $Bx = b$,
(b) there exists $y \in \mathbb{R}^N$ such that $B^T y = 0$ with $\langle y, b \rangle \neq 0$.

Exercise A.11. Prove that, for any $N \times m$ matrix B, one has

(i) $\ker(B^T) = \ker(B\,B^T)$;
(ii) $\mathrm{range}(B) = \mathrm{range}(B\,B^T)$.

[*Hint:* (i) follows from (b) of Exr. A.9; deduce (ii) from (i) and Exr. A.10:

$$\mathrm{range}(B) = (\ker(B^T))^\perp = \left(\ker(B\,B^T)\right)^\perp = \mathrm{range}((B\,B^T)^T) = \mathrm{range}(B\,B^T).]$$

Exercise A.12. Let A be an $N \times N$ positive semidefinite matrix, and let $\mathrm{Isotr}(A)$ be as in (A.7). Finally let $v \in \mathbb{R}^N$.
   Prove that the following conditions are equivalent:

(1) there exists $\lambda > 0$ such that $\langle v, \xi \rangle^2 \leq \lambda \langle A\xi, \xi \rangle$, for every $\xi \in \mathbb{R}^N$;
(2) $\mathrm{Isotr}(A) \subseteq (\mathrm{span}\{v\})^\perp$;
(3) $v \in (\ker A)^\perp$;
(4) v is a linear combination of the rows of A (i.e., $v \in \mathrm{range}(A^T)$);
(5) v is a linear combination of the columns of A (i.e., $v \in \mathrm{range}(A)$).

If, moreover, B is any real $N \times m$ matrix such that $A = B\,B^T$, show that we have yet another equivalent condition:

(6) v is a linear combination of the columns of B (i.e., $v \in \mathrm{range}(B)$).

[*Hint:* It is easy to see (by also using Exrs. A.9 and A.10) that conditions (2), (3), (4), (5), (6) are equivalent and that (1) implies (2). Prove that (3) implies (1) by giving the details of the following argument: when $A \neq 0$, consider

$$\Lambda : S \to \mathbb{R}, \quad \Lambda(\eta) := \frac{\langle v, \eta \rangle^2}{\langle A\eta, \eta \rangle},$$

where $S := (\ker A)^\perp \cap \{\eta : \|\eta\| = 1\}$. Complete the following argument:

- show that Λ is well defined (it may help that $\mathrm{Isotr}(A) = \ker A$);
- show that $\lambda := \max_S \Lambda$ is finite and it satisfies (1) when $\xi \in S$;
- show that if (1) holds true for $\xi \in S$ then it holds true for any $\xi \in (\ker A)^\perp$;
- in order to prove (1) for any $\xi \in \mathbb{R}^N$, write $\xi = \xi_1 + \xi_2$ where $\xi_1 \in (\ker A)^\perp$ and $\xi_2 \in \ker A$; since $v \in (\ker A)^\perp$, we have $\langle v, \xi \rangle^2 = \langle v, \xi_1 \rangle^2$, while

$$\langle A\xi, \xi \rangle = \langle A\xi_1, \xi \rangle = \langle A\xi_1, \xi_1 \rangle + \langle A\xi_1, \xi_2 \rangle = \langle A\xi_1, \xi_1 \rangle,$$

the latter inequality deriving from $\langle A\xi_1, \xi_2\rangle = \langle \xi_1, A\xi_2\rangle = 0$, as $\xi_2 \in \ker A$. This shows that it suffices to prove (1) for $\xi \in (\ker A)^\perp$.]

Exercise A.13. Let V be a finite-dimensional vector space over $\mathbb{R}$, endowed with an inner product $\langle \cdot, \cdot \rangle$. Let $U \subseteq V$ be a vector subspace.

Prove that $V = U \oplus U^\perp$, where $U^\perp = \{v \in V : \langle v, u\rangle = 0 \quad \forall\, u \in U\}$.

[*Hint*: Let $u_1, \ldots, u_n$ be an orthonormal basis of U, and if $v \in V$ is arbitrary, show that $u(v) := \sum_{i=1}^n \langle v, u_i\rangle\, u_i$ is such that $v - u(v) \in U^\perp$...]

Exercise A.14. Let $v \in \mathbb{R}^N$. Prove that

$$\|v\| = \sup_{\|\xi\|<1} \langle v, \xi\rangle = \max_{\|\xi\|\leq 1} \langle v, \xi\rangle = \max_{\|\xi\|=1} \langle v, \xi\rangle,$$

and that

$$\|v\| = \sup_{\|\xi\|<1} |\langle v, \xi\rangle| = \max_{\|\xi\|\leq 1} |\langle v, \xi\rangle| = \max_{\|\xi\|=1} |\langle v, \xi\rangle|.$$

[*Hint:* Use: Cauchy-Schwarz's inequality; $\langle v, \pm\xi\rangle = \pm\langle v, \xi\rangle$; if $v \neq 0$, then take $\xi = (1 - \varepsilon)\, v/\|v\|$ with $\varepsilon = 0$ or $\varepsilon > 0$ small...]

Exercise A.15. Let A be a real $p \times q$ matrix, with $p \leq q$; suppose that the p rows of A are linearly independent vectors of $\mathbb{R}^q$. Prove that AA^T is invertible.

Let $T : \mathbb{R}^q \to \mathbb{R}^p$ be the linear map canonically associated with A. Prove that the matrix, say $A^\dagger$, representing the linear map $T^\dagger$ (the Moore-Penrose pseudo-inverse of T) is explicitly given by $A^\dagger = A^T(AA^T)^{-1}$.

[*Hint:* Remember that $(\ker T)^\perp$ is spanned by the rows of A...]

Exercise A.16. In Féjer's Lem. A.23 we used the following simple fact: for any triple of square matrices (of the same dimension) A, B, C we have

$$\mathrm{trace}(ABC) = \sum_i A^i BC_i,$$

where A^i is the i-th row of A and C_i is the i-th column of C. Prove this identity.

Exercise A.17. Let B be a real $m \times m$ symmetric matrix. Prove that B is negative semidefinite if and only if the inequality $\mathrm{trace}(AB) \leq 0$ is satisfied for every real symmetric positive-definite matrix A of dimension $m \times m$.

Show that, in the last sentence, 'positive-definite' can be replaced by 'positive semidefinite' as well.

[*Hint:* The 'only if' part is contained in Féjer's Lem. A.23. For the 'if' part complete the following argument:

(1) Given any $\xi \in \mathbb{R}^m$ and any $\varepsilon > 0$ prove that $A := \xi \cdot \xi^T + \varepsilon\,\mathbb{I}_m$ is positive definite (here $\mathbb{I}_m$ is the identity matrix of order m).
(2) Test the inequality in the 'if'-hypothesis with this matrix...]

Appendix B

Dependence Theory for ODEs

$\mathbf{T}$HE aim of this appendix is to collect and to clearly expose the definitions and the results concerning ODE Theory that we used in this book. Only a few selected results will be proved, since it is understood that the reader already knows many basic facts on ODEs. So our main intention is fixing the notation and stating the results for the sake of future reference.

We start by reviewing the basic facts about ODEs (existence and uniqueness theorems and maximal solutions); then we shall face the problem of C^0, C^k and C^ω dependence of the solutions of an ODE wrt the parameters and the initial conditions. We often follow the approach in the comprehensive monograph by [Hartman (1982)]; however we also introduce some non-standard definitions (such as those of a *control cylinder* and of a *covering control triple*).

B.1   Review of basic ODE Theory

This fist section is totally devoted to presenting basic facts concerning ODE Theory. We begin by reviewing well-known existence and uniqueness theorems; then we shall review the problem of the existence of maximal solutions. Only a couple of proofs will be given.

B.1.1   *Preliminaries*

Let us consider Euclidean space $\mathbb{R}^{1+N}$ and let us denote the points of $\mathbb{R}^{1+N}$ by (t, x), where $t \in \mathbb{R}$ and $x \in \mathbb{R}^N$. We consider the Cauchy problem:

$$(\mathrm{CP}): \qquad \dot{x} = f(t, x), \quad x(t_0) = x_0, \tag{B.1}$$

where f is an $\mathbb{R}^N$-valued function defined and (at least) continuous on a subset E of $\mathbb{R}^{1+N}$ and (t_0, x_0) is a fixed point in E. A function γ defined on an interval $I \subseteq \mathbb{R}$ (not reduced to a singleton) is a solution of the ordinary differential equation (ODE, for brevity) $\dot{x} = f(t, x)$ iff:

(i) $\gamma \in C^1(I, \mathbb{R}^N)$;

(ii) $(t, \gamma(t)) \in E$ for all $t \in I$;

357

(iii) $\dot{\gamma}(t) = f(t, \gamma(t))$ for all $t \in I$.

Moreover, a function γ defined on an interval $I \subseteq \mathbb{R}$ containing t_0 is a solution of (CP) if it is a solution of the ODE $\dot{x} = f(t, x)$ and if $\gamma(t_0) = x_0$.

Equivalently, a function γ (defined on an interval $I \subseteq \mathbb{R}$ containing t_0) is a solution of (CP) if and only if γ is continuous on I, the graph $\{(t, \gamma(t)) : t \in I\}$ of γ is contained in E, and γ solves the associated Volterra's integral equation

$$\gamma(t) = x_0 + \int_{t_0}^{t} f(s, \gamma(s)) \, ds, \quad \text{for all } t \in I.$$

There are essentially two fundamental results about the existence of a (local) solution of (CP): the *Peano's Theorem* and the *Cauchy-Peano-Picard's Theorem*. Let us fix some notation and definitions, used throughout.

- We call **closed cylinder** of $\mathbb{R}^{1+N}$ any set C of the form[5]

$$C = C(h, r) := [t_0 - h, t_0 + h] \times \overline{B(x_0, r)}, \tag{B.2}$$

  where $(t_0, x_0) \in \mathbb{R}^{1+N}$, $r > 0$, $h > 0$ and $\overline{B(x_0, r)}$ denotes the closed Euclidean ball with centre x_0 and radius r.

- If $E \subseteq \mathbb{R}^{1+N}$ is any set and if $f \in C(E, \mathbb{R}^N)$, we say that a closed cylinder $C = C(h, r) \subseteq E$ of the form (B.2) is a **control cylinder** for f iff

$$\max_{(t,x) \in C} \|f(t, x)\| \cdot h \le r. \tag{B.3}$$

- Finally, if $E \subseteq \mathbb{R}^{1+N}$ is any set and $f : E \longrightarrow \mathbb{R}^N$, we say that f is *locally Lipschitz continuous on E wrt x* if, for every compact set $K \subseteq E$, there exists a constant $L_K > 0$ such that

$$\|f(t, x_1) - f(t, x_2)\| \le L_K \|x_1 - x_2\| \quad \text{for every } (t, x_1), (t, x_2) \in K. \tag{B.4}$$

  If there exists a constant L such that (B.4) holds on E with L_K replaced by L, then we say that f is Lipschitz continuous on E wrt x.

Remark B.1. (1). If $E \subseteq \mathbb{R}^{1+N}$, $f \in C(E, \mathbb{R}^N)$ and $C = C(h, r)$ is a closed cylinder contained in E, then there always exists a closed cylinder C_1 contained in C which is a control cylinder for f and with the same x-projection of C: in fact, if $f \equiv 0$ on $C(h, r)$ then (B.3) is trivially fulfilled by any h and any $r > 0$; otherwise, setting

$$M = \max_{C(h,r)} \|f\| \quad \text{and} \quad a := \min\left\{h, \frac{r}{M}\right\},$$

then it is immediate to check that $C_1 = C(a, r)$ is a control cylinder contained in $C(h, r)$. See Fig. B.1.

(2). Moreover, if $C(h, r)$ is a control cylinder for f, then $C(h, r)$ is a control cylinder for any function g such that $\|g\| \le \|f\|$. ♯

---

[5]Clearly, $C(h, r)$ also depends on its centre (t_0, x_0) but we shall avoid expressing this dependence when the centre is understood (typically, in dealing with a Cauchy problem with a given initial datum).

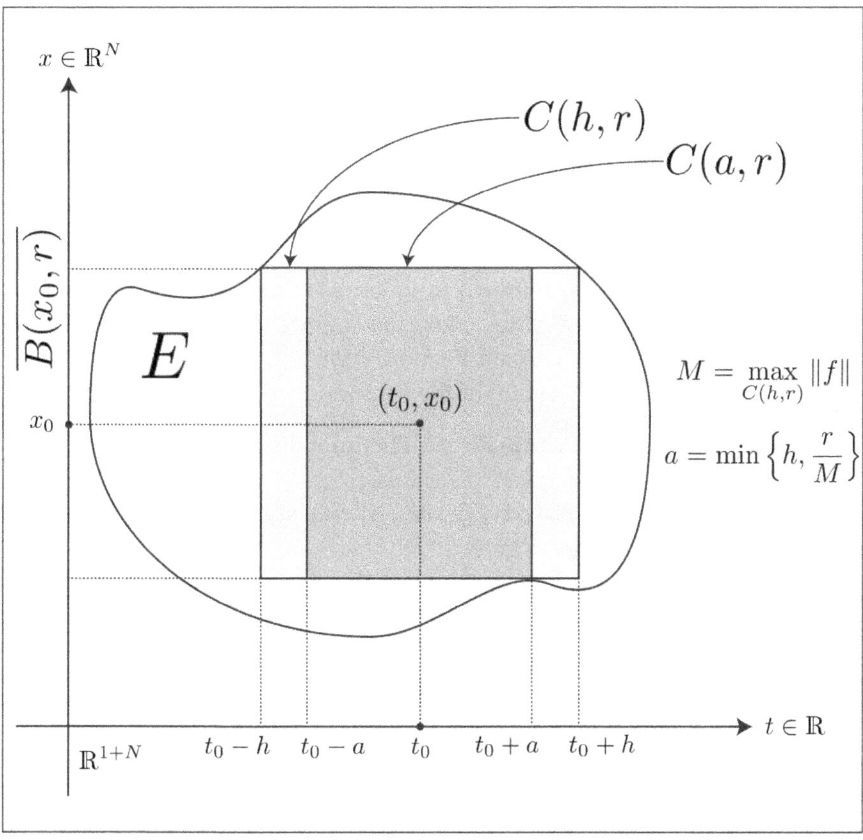

Fig. B.1   Construction of a control cylinder for a Cauchy Problem inside a set E.

Theorem B.2 (Peano's Theorem). *Let $(t_0, x_0) \in \mathbb{R}^{1+N}$, and let us consider the closed cylinder $C_0 := [t_0 - h, t_0 + h] \times \overline{B(x_0, r)}$, with $h, r > 0$. Let $f \in C(C_0, \mathbb{R}^N)$ and let us assume that C_0 is control cylinder for f, see (B.3).*

Then there exists a solution $\gamma \in C^1([t_0 - h, t_0 + h], \mathbb{R}^N)$ of the Cauchy problem (CP), whose graph $\{(t, \gamma(t)) : t \in [t_0 - h, t_0 + h]\}$ is contained in C_0.

Note that this theorem gives an *a priori* knowledge of the domain of the solution, which is the whole projection on the t-axis of the control cylinder C_0. *This is the most important feature of control cylinders.*

It is well known that, under the sole continuity of f, we cannot expect any uniqueness result, as the following example shows.

Example B.3. Let us consider the closed cylinder in $\mathbb{R}^2$ centred at $(0,0)$ given by $C_0 := [-1, 1] \times [-1, 1]$. It is immediate to recognize that C_0 is a control cylinder for the continuous function $f(t, x) = \sqrt{|x|}$. One can directly check that, for every fixed $c \in [0, 1]$, the function γ_c defined by

$$\gamma_c : [-1, 1] \longrightarrow \mathbb{R} \quad \gamma_c(t) := \begin{cases} 0, & \text{for } -1 \le t \le c \\ (t - c)^2/4, & \text{for } c \le t \le 1, \end{cases} \tag{B.5}$$

is a solution of the Cauchy problem $\dot{x} = \sqrt{|x|}$, $x(0) = 0$. Thus, the latter possesses a "continuous family" of solutions, all defined on I. (Clearly they can be thought of as solutions of (CP) on the whole of $\mathbb{R}$.) Note that f does not satisfy any Lipschitz continuity assumption at $(0, 0)$.   ♯

We have the following well-known uniqueness result. At this stage, we mean 'uniqueness' *on a fixed control cylinder*; another uniqueness result (related to open domains) is given in Thm. B.8. As in Peano's Thm. B.2, the following result gives an *a priori* knowledge of the domain of the solution.

Theorem B.4 (The Cauchy–Peano–Picard Theorem on control cylinders). *Let* $(t_0, x_0) \in \mathbb{R}^{1+N}$, *and let us consider the cylinder* $C_0 = [t_0 - h, t_0 + h] \times \overline{B(x_0, r)}$, *with* $h, r > 0$. *Let* $f \in C(C_0, \mathbb{R}^N)$ *be Lipschitz continuous on* C_0 *wrt* x. *Finally, let us assume that* C_0 *is a control cylinder for* f.

Then, setting $I = [t_0 + h, t_0 + h]$, *there exists a unique solution* $\gamma \in C^1(I, \mathbb{R}^N)$ *of* (CP), *whose graph is contained in* C_0. *More precisely, if* $\psi \in C^1(J, \mathbb{R}^N)$ *is another solution of* (CP), *then* $\psi \equiv \gamma$ *on* $J \cap I$.

There exist several proofs of this theorem, usually based on the Banach-Caccioppoli Fixed Point Theorem. Indeed, one can easily show that the map

$$u \mapsto F(u)(t) := x_0 + \int_{t_0}^{t} f(s, u(s))\,\mathrm{d}s \qquad (t \in [t_0 - h, t_0 + h])$$

is such that (for a suitable n) F^n is a contraction of the complete metric space

$$X = \left\{ u \in C([t_0 - h, t_0 + h], \mathbb{R}^N) : \|u(t) - x_0\| \le r \text{ if } |t - t_0| \le h \right\},$$

equipped with the supremum norm. The fixed point u of F^n is a fixed point of F, so that u solves (CP) under its equivalent Volterra form. The fact that $C(h, r)$ is a control cylinder ensures that F maps X into X:

$$\|F(u)(t) - x_0\| \le \left| \int_{t_0}^{t} \|f(s, u(s))\|\,\mathrm{d}s \right| \le M|t - t_0| \le Mh \overset{(\text{B.3})}{\le} r.$$

Local-Lipschitzianity proves that F^n is a contraction, since, for any $u, v \in X$,

$$\max_{t \in I} \left\| F^n(u)(t) - F^n(v)(t) \right\| \le \frac{(L_{C_0} h)^n}{n!} \cdot \max_{t \in I} \|u(t) - v(t)\|.$$

Uniqueness is achieved since, if $\psi \in C^1(J, \mathbb{R}^N)$ solves (CP) then

$$\|F^n(x_0)(t) - \psi(t)\| \le M\frac{(L_{C_0})^n |t - t_0|^{n+1}}{(n + 1)!}, \qquad \text{for any } n \ge 0 \text{ and } t \in J.$$

Thms. B.2 and B.4 ensure the existence-uniqueness of the solution of any Cauchy problem defined on a control cylinder of $\mathbb{R}^{1+N}$. However, in many situations one has to deal with Cauchy problems defined on open subsets of $\mathbb{R}^{1+N}$, so that it is important to have existence-uniqueness results for such case as well.

From now on, Ω will tacitly denote a non-empty open subset of $\mathbb{R}^{1+N}$.

Lemma B.5 (A uniform covering lemma with control cylinders). *Let f belong to $C(\Omega, \mathbb{R}^N)$, and let $K \subset \Omega$ be a fixed compact set. There exist real numbers $h, r > 0$ and a bounded open set Ω_0 (all depending on $\Omega, \|f\|, K$) such that:*

(i) $K \subset \Omega_0 \subset \overline{\Omega_0} \subset \Omega$;
(ii) *for any $(t, x) \in K$ the closed cylinder $C = [t - h, t + h] \times \overline{B(x, r)}$ is contained in Ω_0 and it is a control cylinder for f.*

Due to the importance of the above lemma (whose proof follows from a simple topological exercise based on compactness), we give –for future reference– the following (non-standard but useful) definition.

Definition B.6 (Covering control triple). Let $f \in C(\Omega, \mathbb{R}^N)$ and let $K \subseteq \Omega$ be a compact set. We say that a triple $\mathcal{C} = (\Omega_0, h, r)$ is a **covering control triple** for K and f (in Ω) if Ω_0, h, r satisfy all the requirements in Lem. B.5.

By this definition, we can rephrase Lem. B.5 as follows: *if $f \in C(\Omega, \mathbb{R}^N)$ and $K \subseteq \Omega$ is compact, there exists a covering triple for K and f (in Ω).*

Remark B.7. Due to Rem. B.1-(2), if $\mathcal{C}$ is a covering control triple for K and f, then it is a covering control triple for K and any g such that $\|g\| \leq \|f\|$. ♯

The following result ensures the *solvability of a family of Cauchy problems whose Cauchy data (t_0, x_0) vary in a compact set K*, and it shows that the solution domains have lengths depending on K only, and their graphs lie entirely in a fixed compact set. It is therefore the first result (of a list destined to grow) giving a good *a priori* stability of the domains and of the graphs of the solutions of a parametric Cauchy problem. See Fig. B.2.

Theorem B.8 (Existence-Uniqueness for Cauchy Problems on open sets). *Let $f \in C(\Omega, \mathbb{R}^N)$ and let $K \subset \Omega$ be compact. The following facts hold.*

(I). *There exist real numbers $h, r > 0$, only depending on $\Omega, \|f\|$ and K, such that for every $(t_0, x_0) \in K$ the Cauchy problem (CP) has a solution γ defined on $I = [t_0 - h, t_0 + h]$, further satisfying $\|\gamma(t) - x_0\| \leq r$, for all $t \in I$.*

(II). *The same assertion holds if h, r come from any covering control triple $\mathcal{C}(\Omega_0, h, r)$ for K and f (in Ω), according to Def. B.6.*

(III). *The same h and r work also for any $g \in C(\Omega, \mathbb{R}^N)$ satisfying $\|g\| \leq \|f\|$.*

(IV). *If f is assumed to be locally Lipschitz continuous on Ω, then we have the following uniqueness property of the solution:*
if $\gamma_1 \in C^1(I_1, \mathbb{R}^N)$ and $\gamma_2 \in C^1(I_2, \mathbb{R}^N)$ solve (CP), then $\gamma_1 \equiv \gamma_2$ on $I_1 \cap I_2$.

The proof of this result is based on Lem. B.5 together with Thms. B.2 and B.4; statement (III) follows from Rem. B.7.

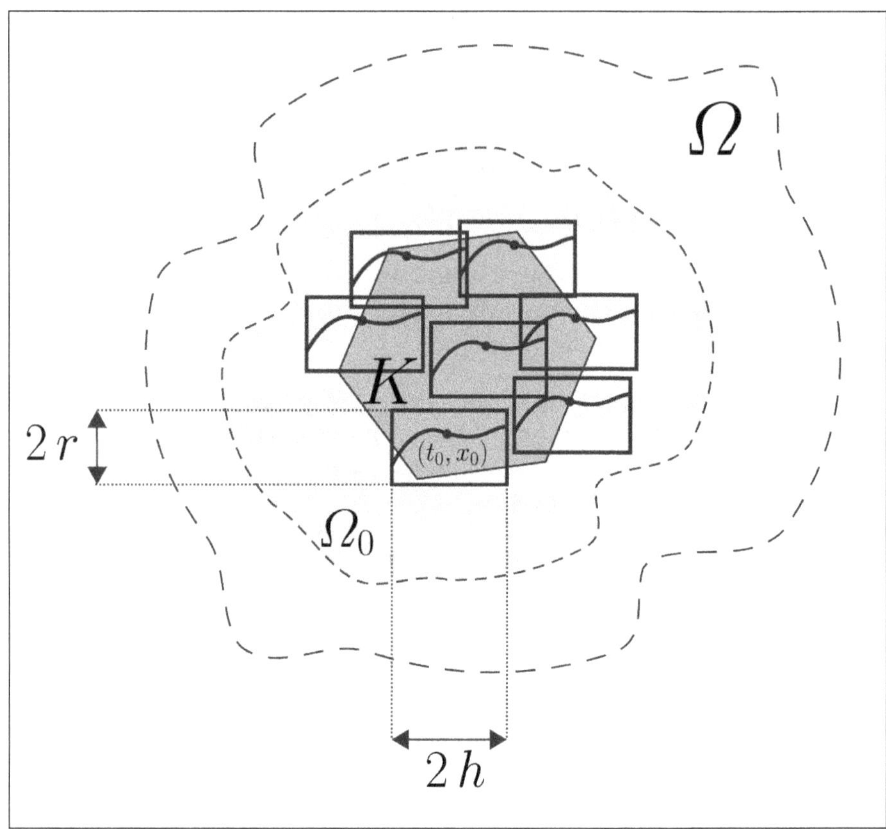

Fig. B.2   Thm. B.8: An *a priori* control on the size of the domains and on the graphs of the solutions, when the initial datum lies in a compact K. Here $\mathcal{C}(\Omega_0, h, r)$ is a covering control triple for K, f (in Ω).

B.1.2 *Maximal solutions*

Definition B.9 (Prolongation of a solution of an ODE). Let $E \subseteq \mathbb{R}^{1+N}$ be any set, and let $f \in C(E, \mathbb{R}^N)$. Moreover, let $\gamma \in C^1(I, \mathbb{R}^N)$ be a solution of the ODE $\dot{x} = f(t, x)$. We say that another solution $\psi \in C^1(J, \mathbb{R}^N)$ of the same ODE is a **proper prolongation** of γ if the following conditions hold:

$$\text{(i): } I \subseteq J \text{ and } I \neq J; \qquad \text{(ii): } \psi \text{ coincides with } \gamma \text{ on } I.$$

Example B.10. As is shown in Exm. B.3, the null function on $[-1, 1]$ solves the Cauchy problem $\dot{x} = \sqrt{|x|}$, $x(0) = 0$, and it admits infinitely many *different* prolongations defined on $\mathbb{R}$ (see (B.5)).   ♯

Definition B.11 (Maximal solution of an ODE). Let $E \subseteq \mathbb{R}^{1+N}$ and suppose that $f \in C(E, \mathbb{R}^N)$. Let $\gamma \in C^1(I, \mathbb{R}^N)$ be a solution of the ODE $\dot{x} = f(t, x)$. We say that γ is a **maximal solution** of this ODE if there does not exist any proper prolongation of γ. In this case, I is said to be the **maximal domain** of γ.

We can define the notion of a maximal solution of the Cauchy Problem (CP) in the obvious way.

Remark B.12. Beware that the notion of a maximal solution implicitly depends on E as well. For instance if $E = \mathbb{R} \times (0, \infty)$ and $f : E \to \mathbb{R}$ is defined by $f(t, x) = 1$, then $\gamma : (0, \infty) \to \mathbb{R}$ defined by $\gamma(t) := t$ is a maximal solution of $\dot{x} = f(t, x)$, despite γ can be prolonged to a map (namely, $\mathbb{R} \ni t \mapsto t$) defined on $\mathbb{R}$ and solution of the same ODE, when f is understood as a map on $\mathbb{R}^2$ rather than on E. ♯

The key result for the existence of a maximal solution is the following:

Remark B.13 (Merging of solutions). Let $\gamma_i \in C^1(I_i, \mathbb{R}^N)$ $(i = 1, 2)$ be solutions of the ODE $\dot{x} = f(t, x)$. If $I_1 \cap I_2 \neq \varnothing$ and if $\gamma_1 \equiv \gamma_2$ on $I_1 \cap I_2$, then

$$(\gamma_1 \vee \gamma_2) : I_1 \cup I_2 \longrightarrow \mathbb{R}^N \qquad (\gamma_1 \vee \gamma_2)(t) := \begin{cases} \gamma_1(t) & \text{if } t \in I_1 \\ \gamma_2(t) & \text{if } t \in I_2 \end{cases} \qquad \text{(B.6)}$$

is a solution of $\dot{x} = f(t, x)$ on $I_1 \cup I_2$. ♯

Throughout, we understand that $\gamma_1 \vee \gamma_2$ denotes the **merging** of γ_1 and γ_2 as defined in (B.6). With this notion one can prove the following result.

Theorem B.14 (Existence of a maximal solution of Cauchy Problem). *Let us suppose that $f \in C(\Omega, \mathbb{R}^N)$ and $(t_0, x_0) \in \Omega$. Then the Cauchy problem (CP) has at least one maximal solution γ, which is defined on an open interval containing t_0.*

Proof. Let $\{K_n\}_n$ be a family of compact subsets of Ω satisfying the following properties: $K_n \subset \text{int}(K_{n+1})$ for all $n \in \mathbb{N}$ and $\Omega = \bigcup_{n \in \mathbb{N}} K_n$. An iterative argument based on Thm. B.8 and the merging technique in Rem. B.13 leads either to the construction of a solution on $[t_0, \infty)$, or to the construction of an increasing sequence $\{n_k\}_k$ of natural numbers, an increasing sequence $\{t_k\}_k$ in (t_0, ∞) and a sequence of functions $\{\gamma_k\}_k$ satisfying:

(i) γ_k is defined and C^1 on $[t_0, t_k]$, and it solves (CP) on this same interval;
(ii) γ_{k+1} is a prolongation of γ_k, that is $\gamma_{k+1} \equiv \gamma_k$ on $[t_0, t_k]$;
(iii) $(t_k, \gamma_k(t_k)) \notin K_{n_k}$.

Setting $\overline{\omega} := \lim_{k \to \infty} t_k$, we obtain a solution $\psi_+ \in C^1([t_0, \overline{\omega}), \mathbb{R}^N)$ of (CP) prolonging any γ_k; ψ_+ cannot be further prolonged to the right beyond $\overline{\omega}$, since the sequence $\{(t_k, \psi_+(t_k))\}_k$ is either unbounded or possesses a cluster point on $\partial\Omega$. The same procedure can be performed to the left of t_0. □

A simple argument on control cylinders proves the following fact.

Lemma B.15. *Let $f \in C(\Omega, \mathbb{R}^N)$ and let $I = [a, b)$ be a bounded interval of $\mathbb{R}$. Moreover, let $\gamma \in C^1(I, \mathbb{R}^N)$ be a solution of the ODE $\dot{x} = f(t, x)$. We make the assumption* *that there exists a sequence $\{t_n\}_n \subseteq I$ converging to b and a point $x_0 \in \mathbb{R}^N$ such that $\lim_{n\to\infty} \gamma(t_n) = x_0$. If $(b, x_0) \in \Omega$, then one has*

$$\lim_{t \to b^-} \gamma(t) = x_0,$$

and the function γ can be prolonged (as a solution of $\dot{x} = f(t, x)$) on $[a, b]$.

By means of Lem. B.15 one can easily prove the following result.

Proposition B.16 (Boundary behavior of a maximal solution). *Suppose that f is in $C(\Omega, \mathbb{R}^N)$ and let ψ be any maximal solution of $\dot{x} = f(t, x)$ with maximal domain $\mathcal{D}$. Then $\mathcal{D}$ is an open interval, say $\mathcal{D} = (\underline{\omega}, \overline{\omega})$, and*

$$\psi(t) \text{ tends to } \partial\Omega \quad \text{as } t \to \underline{\omega} \text{ and as } t \to \overline{\omega}.$$

More precisely, for every compact set $K \subseteq \Omega$ there exist real numbers $\alpha, \beta \in \mathcal{D}$ (depending on K) such that

$$(t, \psi(t)) \notin K, \quad \text{for all } t \in (\underline{\omega}, \alpha) \text{ and for all } t \in (\beta, \overline{\omega}). \tag{B.7}$$

If $\Omega = (a_0, b_0) \times \mathbb{R}^N$, then one necessarily has

- *$\overline{\omega} = b_0$ or $\|\psi(t)\| \longrightarrow \infty$ as $t \to \overline{\omega}-$;*
- *$\underline{\omega} = a_0$ or $\|\psi(t)\| \longrightarrow \infty$ as $t \to \underline{\omega}+$.*

From Thms. B.8 and B.14, and Prop. B.16 we obtain at once the following:

Theorem B.17 (Existence-Uniqueness of the maximal solution of (CP)). *Suppose that $f \in C(\Omega, \mathbb{R}^N)$ is locally Lipschitz continuous wrt x, and let $(t_0, x_0) \in \Omega$.*

Then (CP) admits a unique maximal solution $\psi \in C^1(\mathcal{D}, \mathbb{R}^N)$, whose maximal domain $\mathcal{D}$ is an open interval $(\underline{\omega}, \overline{\omega})$ containing t_0.

Moreover, ψ tends to $\partial\Omega$ as $t \to \underline{\omega}$ and $t \to \overline{\omega}$ in the sense of (B.7).

The *a priori* control on domains/graphs of solutions described in Thm. B.8 gives the following improvement of Thm. B.17.

Corollary B.18. *Let $f \in C(\Omega, \mathbb{R}^N)$ be locally Lipschitz continuous on Ω wrt x, and let $K \subseteq \Omega$ be a compact set. Let $\mathcal{C} = (\Omega_0, h, r)$ be any covering control triple for K and f in Ω (at least one such $\mathcal{C}$ exists due to Lem. B.5).*

Then, for every point $(t_0, x_0) \in K$, the unique maximal solution ψ_0 of (CP) is (at least) defined on $I = [t_0 - h, t_0 + h]$ and it satisfies the estimate

$$\|\psi_0(t) - x_0\| \leq r, \quad \text{for all } t \in I.$$

Furthermore, the set of the trajectories

$$\mathcal{T}(K) := \left\{ (t, \psi_0(t)) : t \in I \text{ and } (t_0, x_0) \in K \right\}$$

lies entirely in Ω_0 (thus its closure is a compact subset of Ω).

Finally, if $g \in C(\Omega, \mathbb{R}^N)$ is locally Lipschitz continuous on Ω wrt x, and it satisfies $\|g\| \leq \|f\|$, then the same uniform Ω_0, h and r work for g as well.

The following result, concerning *sub-solutions of a linear integral equation*, has a prominent role in the *a priori* knowledge of the maximal domain in many relevant cases (namely for linear ODEs).

Lemma B.19 (Gronwall's lemma). *Let $[a, b] \subseteq \mathbb{R}$ and let $u, v \in C([a, b], \mathbb{R})$, with u nonnegative, and let $C \geq 0$ be a constant. Let us assume that*

$$v(t) \leq C + \int_a^t u(s)\, v(s)\, \mathrm{d}s, \quad \text{for all } t \in [a, b]. \tag{B.8}$$

Then the following inequality holds true:

$$v(t) \leq C \exp\left(\int_a^t u(s)\, \mathrm{d}s \right), \quad \text{for all } t \in [a, b]. \tag{B.9}$$

Proof. Let us consider the function $t \mapsto F(t)$ in the rhs of (B.8). Since u and v are continuous, then $F \in C^1([a, b], \mathbb{R})$; moreover, assumption (B.8) says that $v \leq F$ so that (by multiplying by $u \geq 0$)

$$F'(t) = u(t)\, v(t) \leq u(t)\, F(t), \quad \text{for all } t \in [a, b].$$

From the differential inequality $F' \leq u\, F$ we immediately get (for $t \in [a, b]$)

$$w'(t) \leq 0, \quad \text{where} \quad w(t) := F(t) \exp\left(- \int_a^t u(s)\, \mathrm{d}s \right),$$

and this implies that w is non-increasing on $[a, b]$. Therefore,

$$\exp\left(- \int_a^t u(s)\, \mathrm{d}s \right) v(t) \stackrel{\text{(B.8)}}{\leq} \exp\left(- \int_a^t u(s)\, \mathrm{d}s \right) F(t) = w(t) \leq w(a) = C,$$

for every $t \in [a, b]$, and this is equivalent to (B.9). $\qquad\square$

B.1.3 *ODEs depending on parameters*

From now on we are interested in ODEs depending on some parameters ξ, namely $\dot{x} = f_\xi(t, x)$. It is convenient to consider the second member of the ODE as a function of the parameter ξ; hence we make the following assumption.

We consider continuous functions f of the following form:

$$f : \Omega \times V \longrightarrow \mathbb{R}^N, \qquad f = f((t, x), \xi) \quad \text{(with } (t, x) \in \Omega \text{ and } \xi \in V\text{)},$$

where $\Omega \subseteq \mathbb{R}^N$ and $V \subseteq \mathbb{R}^m$ are open sets. To ease the notation, we write

$$f(t, x; \xi) \quad \text{in place of } f((t, x), \xi),$$

where the semicolon serves as a separator between time-space variables (t, x) and the parameters ξ. Furthermore, it is a beneficial idea to "incorporate" the parameters in the spatial variable, as is formalized in the following result:

Proposition B.20. *Let f be as above, and take any $(t_0, x_0) \in \Omega$ and $\xi_0 \in V$. Then the following Cauchy problems are equivalent*

$$(\mathrm{CP})_{\xi_0} : \begin{cases} \dot{x} = f(t, x; \xi_0) \\ x(t_0) = x_0 \end{cases} \quad \text{and} \quad (\mathrm{CP})_* : \begin{cases} \dot{y} = f_*(t, y) \\ y(t_0) = y_0, \end{cases}$$

where $y_0 := (x_0, \xi_0)$, and we introduced the $\mathbb{R}^{N+m}$-valued function

$$f_*(t, y) := \Big(f\big(t, y_1, \ldots, y_N; y_{N+1}, \ldots, y_{N+m}\big), 0 \Big),$$

which is defined for $(t, y_1, \ldots, y_N) \in \Omega$ and $(y_{N+1}, \ldots, y_{N+m}) \in V$.

More precisely, a function $\psi \in C^1(I, \mathbb{R}^{N+m})$ is a solution of $(\mathrm{CP})_*$ if and only if $\psi(t) = (\gamma(t), \xi_0)$ and $\gamma \in C^1(I, \mathbb{R}^N)$ is a solution of $(\mathrm{CP})_{\xi_0}$.

With no regularity assumption wrt ξ (except for continuity), gathering the results proved so far, we have the following first result on parametric (CP)'s.

Theorem B.21. *Let $f \in C(\Omega \times V, \mathbb{R}^N)$ and suppose that $K \subseteq \Omega$ and $K' \subseteq V$ are compact sets. We assume that, for every fixed $\xi \in V$, the function*

$$\Omega \ni (x, t) \mapsto f(t, x; \xi)$$

is locally Lipschitz continuous on Ω wrt x (the associated local Lipschitz constant as in (B.4) is allowed to depend on ξ).

Then there exist positive numbers h and r (only depending on $\Omega, V, K, K', \|f\|$) such that, uniformly for every $(t_0, x_0) \in K$ and every $\xi \in V$, the (unique) maximal solution γ_ξ of the parametric Cauchy problem

$$(\mathrm{CP})_\xi : \qquad \dot{x} = f(t, x; \xi), \quad x(t_0) = x_0$$

is defined at least on $I := [t_0 - h, t_0 + h]$, and it satisfies the a priori estimate

$$\|\gamma_\xi(t) - x_0\| \le r, \quad \text{for all } t \in I.$$

Furthermore, the set of the trajectories

$$\mathcal{T}(K, K') := \big\{ \big(t, \gamma_\xi(t)\big) : t \in I, (t_0, x_0) \in K \text{ and } \xi \in K' \big\}$$

is contained in a compact subset K_0 of Ω.

Finally, if $g \in C(\Omega \times V, \mathbb{R}^N)$ enjoys the same assumptions as f, and it also satisfies $\|g\| \le \|f\|$, then the same uniform K_0, h and r work for g as well.

Proof. Using the notation of Prop. B.20, since f_* is continuous on the open set $\Omega_* := \Omega \times V$, and as $K \times K' \subseteq \Omega_*$ is compact, Lem. B.5 proves the existence of a covering control triple $\mathcal{C} := (W, h, r)$ for $K \times K'$ and f_* in Ω_*. The proof of the theorem now follows from Cor. B.18 and Prop. B.20. $\qquad\square$

Here is an example of how we applied Thm. B.21 in this book, concerning exponentiation of vector fields.

Example B.22. Let $O \subseteq \mathbb{R}^N$ be an open set and let $X = \{X_1, \ldots, X_m\}$ be a set of locally Lipschitz continuous vector fields on O. Fixing $x_0 \in O$, we consider, for $\xi \in \mathbb{R}^m$, the following parametric Cauchy problem

$$(\mathrm{CP})_\xi : \qquad \dot{\gamma} = \sum_{i=1}^m \xi_i X_i(\gamma), \quad \gamma(0) = x_0.$$

Obviously,[6] the function

$$f : \mathbb{R} \times O \times \mathbb{R}^m \longrightarrow \mathbb{R}^N \qquad f(t, x; \xi) := \sum_{i=1}^{m} \xi_i X_i(x)$$

is locally Lipschitz continuous wrt $(x, \xi) \in O \times \mathbb{R}^m$. Given a compact set $K \subset O$, we claim that there exists $\varepsilon = \varepsilon(K, X) > 0$ such that the maximal domain of the maximal solution of $(CP)_\xi$ contains $[-1, 1]$, for every $x_0 \in K$, and every $\xi \in \mathbb{R}^m$ such that $\|\xi\| \leq \varepsilon$.

Indeed, since $K \subset O$ is compact, there exists a compact set $K' \subset O$ and a small $r > 0$ such that $\bigcup_{x_0 \in K} \overline{B(x_0, r)} \subset K'$. Fixing $\xi \in \mathbb{R}^m$, the cylinder

$$C_{x_0}(1, r) := [-1, 1] \times \overline{B(x_0, r)}$$

is a control cylinder for $(t, x) \mapsto \sum_{i=1}^{m} \xi_i X_i(x)$, provided that

$$\max_{(t, x) \, \in \, C_{x_0}(1, r)} \left\| \sum_{i=1}^{m} \xi_i X_i(x) \right\| \leq r.$$

If $M := \max\{\sum_{i=1}^{m} \|X_i(x)\| : x \in K'\}$, the above lhs is bounded from above by $\|\xi\| M$, uniformly for $x_0 \in K$. We choose $\varepsilon > 0$ such that $\varepsilon < r/M$ and we take any $\xi \in \mathbb{R}^m$ such that $\|\xi\| \leq \varepsilon$. Due to Thm. B.4, we then know that $(CP)_\xi$ is solvable at least on $[-1, 1]$, and the solution is prolonged by the unique maximal solution referred to in Thm. B.21. This proves our claim. ♯

B.2 Continuous dependence

The aim of this section is to address the issue of the continuous dependence of the solutions of a Cauchy problem wrt the data and the equation. A crucial role in continuous dependence is played by the Arzelà-Ascoli Theorem.

B.2.1 *The Arzelà-Ascoli Theorem*

We state a few results for the sake of later reference, mainly without proofs.

To begin with, we consider a compact metric space (T, d) and we denote by $C(T, \mathbb{R}^p)$ the Banach space of the $\mathbb{R}^p$-valued continuous functions on T, equipped with the usual norm $\|u\| := \max_{t \in T} |u(t)|$, for every $u \in C(T, \mathbb{R}^p)$. Convergence in this norm is called **uniform convergence**. A subset $\mathcal{F}$ of $C(T, \mathbb{R}^p)$ is said to be **uniformly-bounded** if there exists $M > 0$ such that[7]

$$|u(t)| \leq M, \quad \text{for every } u \in \mathcal{F} \text{ and every } t \in T,$$

---

[6]We note that, if $K \subset O$ and $H \subset \mathbb{R}^m$ are compact, then

$$\|f(t, x; \xi) - f(t, x'; \xi')\| \leq \sum_{j=1}^{m} |\xi_j - \xi'_j| \max_{x \in K} \|X_j(x)\| +$$

$$+ \sum_{j=1}^{m} \max_{\xi' \in H} |\xi'_j| \cdot \|X_j(x) - X_j(x')\| \leq \alpha(K) \|\xi - \xi'\| + \beta(H) L_K \|x - x'\|.$$

[7]Up to this moment, we denoted Euclidean norm on $\mathbb{R}^p$ by $\|\cdot\|$; to avoid confusion, we pass to the notation $|\cdot|$, while using $\|\cdot\|$ for the norm on $C(T, \mathbb{R}^p)$.

that is, $\mathcal{F}$ is bounded wrt the norm $\|\cdot\|$; moreover, the set $\mathcal{F}$ is said to be **equi-continuous** if the following condition holds: for every $\varepsilon > 0$ there exists $\rho_\varepsilon > 0$, only depending on ε, such that

$$|u(t) - u(s)| < \varepsilon, \quad \text{for any } u \in \mathcal{F} \text{ and any } t, s \in T \text{ with } d(t, s) < \rho_\varepsilon.$$

Throughout this section, (T, d) will tacitly denote a compact metric space.

One of the most remarkable application of the notion of equi-continuity is represented by the following result (Exr. B.2).

Proposition B.23. *Let $\{u_n\}_n$ be a sequence in $C(T, \mathbb{R}^p)$ and let E be a dense subset of T. Let us assume that $\{u_n\}_n$ is point-wise convergent on E.*

If the set $\mathcal{F} = \{u_n \mid n \in \mathbb{N}\}$ is equi-continuous, then the sequence $\{u_n\}_n$ is uniformly convergent on T. The converse is also true.

The Bolzano-Weierstrass Theorem and a Cantor diagonal argument give the following result (Exr. B.3).

Proposition B.24. *Let $\{u_n\}_n$ be a uniformly-bounded sequence in $C(T, \mathbb{R}^p)$ and let E be a countable subset of T. Then, there exists a subsequence $\{u_{n_k}\}_k$ of $\{u_n\}_n$ which is point-wise convergent on the set E.*

Props. B.23 and B.24 easily give the following theorem (Exr. B.4).

Theorem B.25 (Ascoli Theorem). *Let $\{u_n\}_n \subset C(T, \mathbb{R}^p)$ be uniformly-bounded and equi-continuous. There exists $\{u_{n_k}\}_k$ uniformly convergent on T.*

From Thm. B.25 and the well-known characterization of the compact sets in metric spaces, we immediately obtain the following corollary.

Corollary B.26. *Let $\mathcal{F}$ be a subset of $C(T, \mathbb{R}^p)$ uniformly-bounded and equi-continuous. Then $\mathcal{F}$ is totally bounded. In particular, if $\mathcal{F}$ is also closed, then it is compact.*

The next theorem, usually attributed to Arzelà, shows that the statement of Cor. B.26 can be reversed (see Exr. B.5).

Theorem B.27 (Arzelà Theorem). *Let $\mathcal{F}$ be a totally bounded subset of $C(T, \mathbb{R}^p)$. Then $\mathcal{F}$ is uniformly-bounded and equi-continuous.*

By Thms. B.25 and B.27, we immediately derive the following characterization of the compact sets of $C(T, \mathbb{R}^p)$.

Theorem B.28 (Arzelà-Ascoli Theorem). *Let T be a compact metric space and let $\mathcal{F}$ be a subset of $C(T, \mathbb{R}^p)$. Then $\mathcal{F}$ is compact (in $C(T, \mathbb{R}^p)$) if and only if it is closed, uniformly-bounded and equi-continuous.*

The following consequences of the theorems of Arzelà-Ascoli will later be useful.

Corollary B.29. *Let $\{u_n\}_n$ be a sequence in $C(T, \mathbb{R}^p)$, uniformly-bounded and equi-continuous. Moreover, let $t_0 \in T$ and let $u_0 \in \mathbb{R}$ be a cluster point of the sequence of real numbers $\{u_n(t_0)\}_n$.*

Then, there exists a subsequence $\{u_{n_k}\}_k$ of $\{u_n\}_n$ which is uniformly convergent on T, further satisfying

$$\lim_{k \to \infty} u_{n_k}(t_0) = u_0. \tag{B.10}$$

Proof. Since u_0 is a cluster point of $\{u_n(t_0)\}_n$, we can assume (by possibly extracting a subsequence) that $u_n(t_0) \longrightarrow u_0$ as $n \to \infty$. On the other hand, since $\{u_n\}_n$ is uniformly-bounded and equi-continuous, we know from the Ascoli Thm. B.25 that there exists a subsequence $\{u_{n_k}\}_k$ of $\{u_n\}_n$ which is uniformly convergent on T; as a consequence, we have

$$\lim_{k \to \infty} u_{n_k}(t_0) = \lim_{n \to \infty} u_n(t_0) = u_0,$$

which is precisely (B.10). This ends the proof. $\qquad\square$

Corollary B.30. *Let $\{u_n\}_n$ be a sequence in $C(T, \mathbb{R}^p)$, uniformly-bounded and equi-continuous. Let us assume, in addition, that every (uniformly) convergent subsequence of $\{u_n\}_n$ has the same limit, say u_0.*

Then $\{u_n\}_n$ is uniformly convergent to u_0 on T.

Proof. Let us assume, by contradiction, that $\{u_n\}_n$ does not converge uniformly to u_0 on T. Hence, by possibly extracting a subsequence, it is possible to find a positive $\varepsilon > 0$ such that

$$\|u_n - u_0\| \geq \varepsilon, \quad \text{for every } n \in \mathbb{N}. \tag{B.11}$$

On the other hand, since $\{u_n\}_n$ is uniformly-bounded and equi-continuous, the Ascoli Thm. B.25 ensures the existence of a subsequence $\{u_{n_k}\}_k$ of $\{u_n\}_n$ which is uniformly convergent on T. Then, by assumption, $u_{n_k} \longrightarrow u_0$ uniformly on T as $k \to \infty$, but this contradicts (B.11). $\qquad\square$

B.2.2 *Dependence on the equation*

In the sequel, we tacitly understand that $\Omega \subseteq \mathbb{R}^{1+N}$ is a non-empty open set.

To begin with, we prove the following general and very useful result.

Proposition B.31. *Let Ω be bounded, let $\{f_n\}_n$ be a sequence in $C(\overline{\Omega}, \mathbb{R}^N)$ and let $(t_n, x_n) \in \Omega$ for any n. We assume that:*

(1) *there exists a function $f_0 \in C(\overline{\Omega}, \mathbb{R}^N)$ such that*

$$\lim_{n \to \infty} f_n = f_0 \quad \text{uniformly on } \overline{\Omega};$$

(2) *there exists $(t_0, x_0) \in \Omega$ such that $(t_n, x_n) \longrightarrow (t_0, x_0)$ as $n \to \infty$;*

(3) *for every $n \in \mathbb{N} \cup \{0\}$, the Cauchy problem*

$$(\mathrm{CP})_n : \qquad \dot{x} = f_n(t, x), \quad x(t_n) = x_n$$

admits a unique maximal solution γ_n with maximal domain $\mathcal{D}_n$;
(4) *there exists $h > 0$ such that $I = [t_0 - h, t_0 + h] \subseteq \mathcal{D}_n$ for every $n \in \mathbb{N}$.*

Then one has $\lim_{n \to \infty} \gamma_n = \gamma_0$ uniformly on I.

In the following Thm. B.32 we shall show that assumption (4) essentially follows from (1), (2) and a slightly stronger form of (3).

Proof. We first show that $\{\gamma_n\}_n$ is uniformly-bounded and equi-continuous in $C(I, \mathbb{R}^N)$. To this end we observe that, since γ_n is a solution of $(\mathrm{CP})_n$ on $I \subseteq \mathcal{D}_n$, by definition we have $(t, \gamma_n(t)) \subseteq \Omega$, for every $t \in I$ and every n; hence, since Ω is bounded, we infer that $\{\gamma_n\}_n$ is uniformly-bounded.

As for the equi-continuity we notice that, since $\{f_n\}_n$ uniformly converges to f_0 on the compact set $\overline{\Omega}$, there exists a constant $M > 0$ such that

$$\sup_\Omega \|f_n\| \leq M, \quad \text{for every } n \in \mathbb{N} \cup \{0\};$$

from this we infer that, for every $t, s \in I$,

$$\|\gamma_n(t) - \gamma_n(s)\| = \left\| \int_s^t \dot{\gamma}_n(\tau) \, \mathrm{d}\tau \right\| = \left\| \int_s^t f_n(\tau, \gamma_n(\tau)) \, \mathrm{d}\tau \right\| \leq M \, |t - s|,$$

and this obviously implies that $\{\gamma_n\}_n$ is equi-continuous on I.

We now prove that, if $\{\gamma_{n_k}\}_k$ is any subsequence of $\{\gamma_n\}_n$ which is (uniformly) convergent on I to a function u, then $u = \gamma_0$. To this end we observe that, as $\{f_n\}_n$ is uniformly convergent on $\overline{\Omega}$, the Arzelà Thm. B.27 ensures that $\{f_n\}_n$ is equi-continuous on $\overline{\Omega}$; hence, we have

$$\lim_{k \to \infty} f_{n_k}(t, \gamma_{n_k}(t)) = f(t, u(t)), \quad \text{uniformly for } t \in I.$$

From this, by passing to the limit as $k \to \infty$ in the Volterra integral equation

$$\gamma_{n_k}(t) = \gamma_{n_k}(t_0) + \int_{t_0}^t f_{n_k}(s, \gamma_{n_k}(s)) \, \mathrm{d}s \qquad (t \in I, \ k \in \mathbb{N}),$$

we plainly see that u is actually a solution of $(\mathrm{CP})_0$. Our uniqueness assumption (3) then implies that $u = \gamma_0$ on I, as claimed. We now deduce from Cor. B.30 that $\{\gamma_n\}_n$ uniformly converges to γ_0 (on I), as desired. $\qquad \square$

Theorem B.32. *Let $\{f_n\}_n$ be a sequence in $C(\Omega, \mathbb{R}^N)$ and let $\{(t_n, x_n)\}_n$ be a sequence in Ω. We make the assumptions:*

(1) *there exists a function $f_0 \in C(\Omega, \mathbb{R}^N)$ such that*

$$\lim_{n \to \infty} f_n = f \quad \text{uniformly on every compact subset of } \Omega;$$

(2) *there exists $(t_0, x_0) \in \Omega$ such that $(t_n, x_n) \longrightarrow (t_0, x_0)$ as $n \to \infty$;*

(3) *for every $n \in \mathbb{N} \cup \{0\}$ and every $(t^*, x^*) \in \Omega$, the Cauchy problem*

$$(\mathrm{CP})_{n,*}: \qquad \dot{x} = f_n(t, x), \quad x(t^*) = x^*$$

admits a unique maximal solution.

If, for every natural $n \geq 0$, we denote by $\gamma_n \in C^1(\mathcal{D}_n, \mathbb{R}^N)$ the unique maximal solution of $(\mathrm{CP})_{n,*}$ relative to the choice $(t^*, x^*) = (t_n, x_n)$, then:

(i) *for every compact interval $I \subseteq \mathcal{D}_0$, there exists a natural $\overline{n}$ such that*

$$I \subseteq \mathcal{D}_n, \quad \text{for every } n \geq \overline{n};$$

(ii) *the sequence $\{\gamma_n\}_n$ is uniformly convergent to γ_0 on I.*

In particular, if we set $\mathcal{D}_n = (\underline{\omega}_n, \overline{\omega}_n)$ for every $n \geq 0$, then one has

$$\limsup_{n \to \infty} \underline{\omega}_n \leq \underline{\omega}_0 < \overline{\omega}_0 \leq \liminf_{n \to \infty} \overline{\omega}_n. \tag{B.12}$$

Proof. Let $\{\Omega_n\}_n$ be a family of open sets in $\mathbb{R}^{1+N}$ such that

$$\overline{\Omega_n} \text{ is compact}; \quad \overline{\Omega_n} \subseteq \overline{\Omega_{n+1}} \text{ for every } n; \quad \Omega = \bigcup_n \Omega_n.$$

By (2) there exists $n_1 \gg 1$ such that $(t_n, x_n) \in \overline{\Omega_{n_1}}$ for every $n \geq 0$. Since $\overline{\Omega_{n_1}}$ is compact and since $\{f_n\}_n$ uniformly converges to f_0 on every compact subset of Ω, it is possible to construct a triple $\mathcal{C} = (O, 2h_0, 2r_0)$ which is a covering control triple for $\overline{\Omega_{n_1}}$ and f_n (in Ω_{n_1+1}), and this can be done *uniformly*[8] for every $n \geq 0$.

Thus, each solution γ_n is defined at least on $[t_n - 2h_0, t_n + 2h_0]$, and

$$\big(t, \gamma_n(t)\big) \in \Omega_{n_1+1}, \quad \text{for every } |t - t_n| \leq 2h_0.$$

Now, since $t_n \longrightarrow t_0$ as $n \to \infty$, we can suppose that $|t_n - t_0| \leq h_0$ for every natural n; hence, $I_0 := [t_0 - h_0, t_0 + h_0] \subseteq \mathcal{D}_n$ and

$$\big(t, \gamma_n(t)\big) \in \Omega_{n_1+1}, \quad \text{for every } t \in I_0.$$

We are then entitled to apply Prop. B.31, which ensures that $\{\gamma_n\}_n$ uniformly converges to γ_0 on I_0. Setting $t_1 = t_0 + h_0$, if $(t_1, \gamma_0(t_1))$ still belongs to Ω_{n_1} we can repeat the above argument, with $\{(t_n, x_n)\}_n$ replaced by $\{(t_1, \gamma_n(t_1)\}_n$, obtaining the uniform convergence of $\{\gamma_n\}_n$ to γ_0 on $[t_0, t_0 + 2h_0]$. By proceeding likewise finitely many times, we find $t_1 > t_0$ such that:

$$(t_1, \gamma_0(t_1)) \notin \Omega_{n_1}; \quad \{\gamma_n\}_n \text{ uniformly converges to } \gamma_0 \text{ on } [t_0, t_1].$$

An iteration of this procedure allows us to construct an increasing sequence $\{t_k\}_k$ in $[t_0, \infty)$ and an increasing sequence $\{n_k\}_k$ in $\mathbb{N}$ with the following properties:

(a) $(t_k, \gamma_0(t_k)) \notin \Omega_{n_k}$ for every k;

---

[8] *It is important to observe that we can choose $\mathcal{C}$ independently of n: indeed, one can always construct a covering control triple for $\overline{\Omega_{n_1}}$ and f_n, depending only on $\sup \|f_n\|$ over a suitable bounded open set containing $\overline{\Omega_{n_1}}$, and $\sup \|f_n\|$ is bounded from above by a finite constant independent of n, since $\{f_n\}_n$ is uniformly-bounded on the compact subsets of Ω.*

(b) $\{t_k\}_k \subseteq \mathcal{D}_n$ for every n;

(c) $\{\gamma_n\}_n$ uniformly converges to γ_0 on $[t_0, t_k]$ for every k.

As a consequence, setting $\alpha := \lim_{k\to\infty} t_k$, we see from property (a) that α must coincide with $\sup(\mathcal{D}_0) = \overline{\omega}_0$; moreover, properties (b) and (c) imply that, for every $t^* \in [t_0, \overline{\omega}_0)$, there exists a natural $\overline{k}$ such that

$$t^* < t_k < \overline{\omega}_k, \quad \text{for every } k \geq \overline{k}, \tag{B.13}$$

and $\{\gamma_n\}_n$ uniformly converges to γ_0 on $[t_0, t^*]$.

By proceeding in the same way, we can construct an analogous t_* on the left. Then, if $I = [t_*, t^*] \subseteq \mathcal{D}_0$, we deduce from (B.13) the existence of $\overline{n}$ such that $I \subseteq \mathcal{D}_n$ for every $n \geq \overline{n}$, and $\{\gamma_n\}_n$ uniformly converges to γ_0 on I. Hence, in particular, we have $\underline{\omega}_n < t_* < t^* < \overline{\omega}_n$, for every $n \geq \overline{n}$, and this gives (B.12) in the statement of the theorem. This ends the proof. $\qquad\square$

B.2.3 *Dependence on the datum*

We now briefly review some basic facts concerning lower and upper semi-continuous functions.

Let (E, d) be any metric space, and let $f : E \to \overline{\mathbb{R}} = [-\infty, \infty]$. We say that f is **lower semi-continuous** (l.s.c., for short) on E if

$$\liminf_{x \to x_0} f(x) \geq f(x_0), \quad \text{for every } x_0 \in E.$$

Similarly, f is said to be **upper semi-continuous** (u.s.c., for short) on E if

$$\limsup_{x \to x_0} f(x) \leq f(x_0), \quad \text{for every } x_0 \in E.$$

It is not difficult to see that l.s.c. and u.s.c. functions can be characterized in the following way: a function f is l.s.c. on E if and only if

$$\{f > t\} := \{x \in E : f(x) > t\} \quad \text{is open for every } t \in \overline{\mathbb{R}},$$

whereas f is u.s.c. on E if and only if

$$\{f < t\} := \{x \in E : f(x) < t\} \quad \text{is open for every } t \in \overline{\mathbb{R}}.$$

It is known that, if (E, d) is compact, then a l.s.c. function $f : E \to (-\infty, \infty]$ attains its minimum on E, whereas an u.s.c. function $f : E \to [-\infty, \infty)$ attains its maximum on E.

Theorem B.33 (Continuous dependence on the Cauchy datum). *Suppose that f is in $C(\Omega, \mathbb{R}^N)$. We assume that, for every $(t_0, x_0) \in \Omega$, the problem*

$$(\mathrm{CP}): \qquad \dot{x} = f(t, x), \quad x(t_0) = x_0 \tag{B.14}$$

admits a unique maximal solution $\gamma(\cdot\, ; t_0, x_0) \in C^1(\mathcal{D}(t_0, x_0), \mathbb{R}^N)$.

Then, setting $\mathcal{D}(t_0, x_0) = (\underline{\omega}(t_0, x_0), \overline{\omega}(t_0, x_0))$, the following facts hold true:

(i) *the function $\underline{\omega}$ is u.s.c. on Ω, while $\overline{\omega}$ is l.s.c. on Ω;*

(ii) *the set $\mathcal{O} := \{(t, (t_0, x_0)) \in \mathbb{R} \times \Omega : t \in \mathcal{D}(t_0, x_0)\}$ is open;*

(iii) *the function $(t, t_0, x_0) \mapsto \gamma(t, t_0, x_0)$ is continuous on $\mathcal{O}$.*

Proof. (i) follows from (B.12) in Thm. B.32 (with $f_n = f$ for every $n \geq 0$).

(ii) Let $(t_0, x_0) \in \Omega$ and let $t \in \mathcal{D}(t_0, x_0) = (\underline{\omega}(t_0, x_0), \overline{\omega}(t_0, x_0))$. Moreover, let $[a, b] \subseteq \mathcal{D}(t_0, x_0)$ be such that $t \in [a, b]$. Since we know from (i) that $\underline{\omega}$ is u.s.c. and $\overline{\omega}$ is l.s.c. on Ω, it is possible to find an open neighborhood $\mathcal{U} \subseteq \Omega$ of (t_0, x_0) such that

$$\underline{\omega}(t^*, x^*) < a < b < \overline{\omega}(t^*, x^*), \quad \text{for every } (t^*, x^*) \in \mathcal{U};$$

hence, by definition, the product $[a, b] \times \mathcal{U}$ (which is a neighborhood of $(t, (t_0, x_0)))$ is wholly contained in $\mathcal{O}$, and this proves that $\mathcal{O}$ is open.

(iii) Let $z_0 = (t_0, x_0) \in \Omega$ and let $I := [a, b] \subseteq \mathcal{D}(t_0, x_0)$. If $\{(t_n, x_n)\}_n$ is a sequence in Ω converging to (t_0, x_0) as $n \to \infty$, it follows from Thm. B.32 that $I \subseteq \mathcal{D}(t_n, x_n)$ if n is sufficiently large, and $\gamma(\cdot\,; t_n, x_n)$ uniformly converges to $\gamma(\cdot\,; t_0, x_0)$ on I as $n \to \infty$; hence, for every $\varepsilon > 0$ there exists a positive $\rho_1 > 0$, depending on ε, z_0 and I, such that $I \times B(z_0, \rho_1) \subseteq \mathcal{O}$ and

$$\|\gamma(t; t^*, x^*) - \gamma(t; t_0, x_0)\| < \varepsilon, \text{ for } t \in I \text{ and } (t^*, x^*) \in B(z_0, \rho_1). \tag{B.15}$$

On the other hand, since $\gamma(\cdot\,; t_0, x_0)$ is of class C^1 on $I \subseteq \mathcal{D}(t_0, x_0)$, there exists a positive ρ_2, again depending on ε, z_0 and I, such that

$$\|\gamma(t; t_0, x_0) - \gamma(s; t_0, x_0)\| < \varepsilon, \text{ for } t, s \in I \text{ with } |t - s| < \rho_2. \tag{B.16}$$

Therefore, setting $\rho = \min\{\rho_1, \rho_2\}$, by (B.15) and (B.16) we get

$$\|\gamma(t; t_0, x_0) - \gamma(s; t^*, x^*)\| < 2\varepsilon,$$

for every $t, s \in I$ with $|t - s| < \rho$ and every $(t^*, x^*) \in B(z_0, \rho)$. $\qquad\square$

B.2.4 *Dependence on the parameters*

With the trick in Prop. B.20, the following result follows from Thm. B.33

Theorem B.34 (Continuous dependence on the parameters). *Let $\Omega \subseteq \mathbb{R}^{1+N}$ and $V \subseteq \mathbb{R}^m$ be open sets, and let $f \in C(\Omega \times V, \mathbb{R}^N)$. Let us assume that, for every (t_0, x_0) in Ω and every ξ_0 in V, the parametric Cauchy problem*

$$(CP)_{\xi_0} : \qquad \dot{x} = f(t, x; \xi_0), \quad x(t_0) = x_0 \tag{B.17}$$

admits a unique maximal solution $\gamma(\cdot\,; t_0, x_0, \xi_0) \in C^1(\mathcal{D}(t_0, x_0, \xi_0), \mathbb{R}^N)$.

Then, setting $\mathcal{D}(t_0, x_0, \xi_0) = (\underline{\omega}(t_0, x_0, \xi_0), \overline{\omega}(t_0, x_0, \xi_0))$, we have:

(1) *$\underline{\omega}$ is u.s.c. and $\overline{\omega}$ is l.s.c. on $\Omega \times V$;*

(2) *the set $\mathcal{O} := \{(t, (t_0, x_0), \xi_0) \in \mathbb{R} \times \Omega \times V : t \in \mathcal{D}(t_0, x_0, \xi_0)\}$ is open;*

(3) *the function* $(t, t_0, x_0, \xi_0) \mapsto \gamma(t; t_0, x_0, \xi_0)$ *is continuous on* $\mathcal{O}$.

Corollary B.35. *In the same assumption of Thm. B.34, let $K \subseteq \Omega$ and $K' \subseteq V$ be compact sets. Then, it is possible to find real numbers $h_1, h_2 > 0$ such that, for any $(t_0, x_0) \in K$ and any $\xi_0 \in K'$, every maximal solution of the Cauchy problem $(\mathrm{CP})_{\xi_0}$ in (B.17) is defined at least on $I := (t_0 - h_1, t_0 + h_2)$.*

Moreover, if $I \subseteq \mathbb{R}$ is compact and $I \times K \times K' \subseteq \mathcal{O}$, then the set of the trajectories

$$\mathcal{T}(I, K, K') := \left\{ (t, \gamma(t; t_0, x_0, \xi_0)) : t \in I, (t_0, x_0) \in K \text{ and } \xi_0 \in K' \right\}$$

is a compact subset of $\mathbb{R} \times \Omega$.

B.3 C^k dependence

In Sec. B.2 we studied continuous dependence. We now prove a stronger regularity result: roughly put, γ is of class C^k (with $k = 1, \ldots, \infty$) on $\mathcal{O}$ if $f \in C^k(\Omega, \mathbb{R}^N)$. We begin with a technical lemma.

Lemma B.36 (Integral Mean Value Theorem). *Let $I = (a, b)$, let $V \subseteq \mathbb{R}^N$ be a convex open set, and let $f = f(t, y)$ be a real-valued continuous function on $I \times V$. Let us assume that, for every fixed $t \in I$, the map $y \mapsto f(t, y)$ is differentiable on V and that $\nabla_y f$ is continuous on $I \times V$.*

Then, the function $g \in C(I \times V \times V, \mathbb{R}^N)$ defined by

$$g(t, y_1, y_2) := \int_0^1 (\nabla_y f)(t, u\, y_2 + (1 - u)\, y_1)\, \mathrm{d}u$$

has the following properties: for any $(t, y), (t, y_1), (t, y_2) \in I \times V$

$$g(t, y, y) = (\nabla_y f)(t, y) \quad \text{and} \quad f(t, y_1) - f(t, y_2) = \langle g(t, y_1, y_2), y_1 - y_2 \rangle.$$

Proof. This identity is simply $F(1) - F(0) = \int_0^1 F'(u)\, \mathrm{d}u$, applied to the function $F(u) := f(t, uy_2 + (1 - u)y_1)$ defined on $[0, 1]$. $\qquad\qquad\square$

Remark B.37. The integral mean value Lem. B.36 can easily be extended to *vector-valued functions*. More precisely, if I, V are as above, assume that f belongs to $C(I \times V, \mathbb{R}^p)$. Suppose also that, for any fixed $t \in I$, the $\mathbb{R}^p$-valued map $y \mapsto f(t, y)$ is differentiable on V and that $\mathcal{J}_y f$ is continuous on $I \times V$.

Then, by applying Lem. B.36 to each component of f, we are led to the $(p \times N)$-matrix-valued function g, continuous in $I \times V \times V$, given by

$$g(t, y_1, y_2) = \int_0^1 \frac{\partial f}{\partial y}(t, uy_2 + (1 - u)y_1)\, \mathrm{d}u,$$

satisfying, for any $(t, y), (t, y_1), (t, y_2) \in I \times V$,

$$g(t, y, y) = \frac{\partial f}{\partial y}(t, y), \quad \text{and} \quad f(t, y_1) - f(t, y_2) = g(t, y_1, y_2)\,(y_2 - y_1). \qquad \sharp$$

In the next fundamental result we need to consider the components of the point $x_0 \in \mathbb{R}^N$; this is why we take the liberty of denoting x_0 also by

$$x_0 = (x_1^0, \dots, x_N^0).$$

Theorem B.38 (C^1-dependence on the data). *Let $\Omega \subseteq \mathbb{R}^{1+N}$ be an open set and let $f = f(t, x) \in C(\Omega, \mathbb{R}^N)$ be such that the partial derivatives $\partial_{x_1} f, \dots, \partial_{x_N} f$ exist and are continuous on Ω. For every $(t_0, x_0) \in \Omega$, let $\gamma(\cdot\,; t_0, x_0) \in C^1(\mathcal{D}(t_0, x_0), \mathbb{R}^N)$ be the unique maximal solution of the Cauchy problem (CP) in (B.14).*

Then the function $(t, t_0, x_0) \mapsto \gamma(t; t_0, x_0)$ is of class C^1 on the open set

$$\mathcal{O} = \big\{ (t, (t_0, x_0)) \in \mathbb{R} \times \Omega : t \in \mathcal{D}(t_0, x_0) \big\}.$$

Moreover, the following facts hold true.

(i) *For every $(t_0, x_0) \in \Omega$ and every $j = 1, \dots, N$, the function*

$$\mathcal{D}(t_0, x_0) \ni t \mapsto \frac{\partial \gamma}{\partial x_j^0}(t; t_0, x_0)$$

   is the maximal solution of the linear Cauchy problem

$$\begin{cases} \dot{x} = A_{t_0, x_0}(t)\, x \\ x(t_0) = e_j, \end{cases} \qquad \text{where} \quad A_{t_0, x_0}(t) := \frac{\partial f}{\partial x}(t, \gamma(t; t_0, x_0)). \qquad \text{(B.18)}$$

(ii) *For every $(t_0, x_0) \in \Omega$ and every $t \in \mathcal{D}(t_0, x_0)$, we have*

$$\frac{\partial \gamma}{\partial t_0}(t; t_0, x_0) = -\frac{\partial \gamma}{\partial x_0}(t; t_0, x_0) \cdot f(t_0, x_0). \qquad \text{(B.19)}$$

Proof. First of all we observe that, since $\gamma(\cdot\,; t_0, x_0)$ solves (CP), one has

$$\frac{\partial \gamma}{\partial t}(t; t_0, x_0) = f(t, \gamma(t; t_0, x_0)), \quad \text{for every } t \in \mathcal{D}(t_0, x_0);$$

thus, since $\gamma \in C(\mathcal{O})$ (by Thm. B.34) and since $f \in C(\Omega, \mathbb{R}^N)$, we infer that the map $(t, t_0, x_0) \mapsto \frac{\partial \gamma}{\partial t}(t; t_0, x_0)$ is continuous on $\mathcal{O}$. We then turn to prove the existence and the continuity of the partial derivatives of γ wrt t_0 and x_0.

To this end, let $(s, z^*) = (s, (t^*, x^*)) \in \mathcal{O}$ and let $I = [a, b] \subseteq \mathcal{D}(t^*, x^*)$ be such that $s, t^* \in I$. Moreover, let us choose $r > 0$ in such a way that

(a) $K := [a, b] \times \overline{B(z^*, r)} \subseteq \mathcal{O}$ (owing to Thm. B.33);

(b) for every $(t_0, x_0) \in B(z^*, r)$, one has $t_0 \in I$.

Due to the local nature of the results we aim to show, we can limit ourselves to work in the interior $\mathcal{U}$ of K, which is a neighborhood of $(s, (t^*, x^*))$.

(i) We first prove the existence and the regularity of the partial derivatives of γ wrt x_0. To this aim, we choose a point $(t_0, x_0) \in B(z^*, r)$, an index j in $\{1, \dots, N\}$ and, for every $h \in \mathbb{R}$ such that $(t_0, x_0 + h e_j) \in B(z^*, r)$, we set

$$\gamma_h(t) := \gamma(t; t_0, x_0 + h e_j), \quad \text{with } t \in \mathcal{D}(t_0, x_0 + h e_j).$$

Since $K \subseteq \mathcal{O}$, γ_h is defined at least on I (which is a neighborhood of t_0); moreover, since γ is continuous on $\mathcal{O}$ (and K is compact), one has

$$\lim_{h \to 0} \gamma_h(t) = \gamma_0(t) = \gamma(t; t_0, x_0), \quad \text{uniformly for } t \in I. \tag{B.20}$$

We set $T := \{(t, \gamma(t; t_0, x_0)) : a \le t \le b\}$ and we choose $\eta, \delta > 0$ and $\Omega_0 \subseteq \Omega$ such that (see also Lem. B.5):

(1) Ω_0 is open, bounded and compactly contained in Ω;

(2) $[t - \eta, t + \eta] \times \overline{B(\gamma_0(t), \delta)} \subseteq \Omega_0$ for every $t \in [a, b]$.

Due to (B.20), it is possible to find $\varepsilon > 0$ such that

$$\sup_{t \in I} \|\gamma_h(t) - \gamma_0(t)\| < \delta, \quad \text{for every } h \in \mathbb{R} \text{ with } |h| \le \varepsilon. \tag{B.21}$$

We now consider, for every $0 < |h| < \varepsilon$, the function u_h defined as follows

$$u_h : I \longrightarrow \mathbb{R}^N, \quad u_h(t) := \big(\gamma_h(t) - \gamma_0(t)\big)/h.$$

Since γ_h is the solution of (CP) starting from $(t_0, x_0 + he_j)$, we see that $u_h(t_0) = e_j$ and that, for every $t \in I$,

$$\dot{u}_h(t) = \big(f(t, \gamma_h(t)) - f(t, \gamma_0(t))\big)/h.$$

On the other hand, since (B.21) implies that $\gamma_h(t) \in B(\gamma_0(t), \delta)$, we are entitled to apply Lem. B.36, which gives

$$\dot{u}_h(t) = A_h(t)\, u_h(t), \quad \text{for every } t \in I \text{ and every } 0 < |h| < \varepsilon,$$

where we have set

$$A_h(t) := \int_0^1 \frac{\partial f}{\partial x}\Big(t, u\gamma_h(t) + (1 - u)\gamma_0(t)\Big)\, du, \qquad |h| < \varepsilon.$$

Summing up, we can conclude that, for every $0 < |h| < \varepsilon$, the map u_h is the unique maximal solution of the linear Cauchy problem (on $I \times \mathbb{R}^N$)

$$(\text{CP})_h : \qquad \dot{x} = A_h(t)\, x, \quad x(t_0) = e_j.$$

Now, by exploiting the uniform continuity of f on $\overline{\Omega}_0 \subseteq \Omega$ and the identities (B.20) and (B.21), it is not difficult to see that:

- $(t, h) \mapsto A_h(t)$ is continuous on $I \times (-\varepsilon, \varepsilon)$;
- $A_h(t) \longrightarrow A(t) = \dfrac{\partial f}{\partial x}(t, \gamma(t; t_0, x_0))$ as $h \to 0$, uniformly for $t \in I$.

By applying to $(\text{CP})_h$ the continuous-dependence results in Thm. B.34, we deduce the following non-trivial fact: the limit

$$u_0(t) := \lim_{h \to 0} u_h(t) = \lim_{h \to 0} \frac{1}{h}\big(\gamma_h(t) - \gamma_0(t)\big)$$

exists for every $t \in I$, and u_0 is the unique maximal solution of

$$(\text{CP})_0 : \qquad \dot{x} = A(t)\, x, \quad x(t_0) = e_j.$$

From the arbitrariness of $(t_0, x_0) \in B(z^*, r)$, we know that $\frac{\partial \gamma}{\partial x_j^0}(t; t_0, x_0) = u_0(t)$ exists for every $(t, t_0, x_0) \in \mathcal{U}$, and it is the maximal solution of (B.18).

To prove that such a derivative is continuous on $\mathcal{U}$, we rewrite the Cauchy problem (B.18) in the following more explicit way

$$\begin{cases} \dot{x} = A(t, t_0, x_0)\, x, \\ x(t_0) = e_j, \end{cases} \quad \text{where} \quad A(t, t_0, x_0) = \frac{\partial f}{\partial x}(t, \gamma(t; t_0, x_0)). \tag{B.22}$$

Since $f \in C(\Omega, \mathbb{R}^N)$ and $\gamma \in C(\mathcal{O}, \mathbb{R}^N)$, the map $(t, t_0, x_0) \mapsto A(t, t_0, x_0)$ is continuous on $\mathcal{U}$; hence (Thm. B.34) we get that $(t, t_0, x_0) \mapsto \frac{\partial \gamma}{\partial x_j^0}(t; t_0, x_0)$ is continuous on $\mathcal{U}$, and, as a function of t, this is the maximal solution of (B.22).

(ii) We now turn to show the existence (and the continuity) of the partial derivative of γ wrt t_0. To this end, we choose again a point $(t_0, x_0) \in B(z^*, r)$ and, for every $h \in \mathbb{R}$ such that $(t_0 + h, x_0) \in B(z^*, r)$, we set

$$\psi_h(t) := \gamma(t; t_0 + h, x_0).$$

By arguing as in the previous part, we see that ψ_h is defined at least on I (which is a neighborhood of t_0 and of $t_0 + h$, owing to property (b) on page 376) and that

$$\lim_{h \to 0} \psi_h(t) = \psi_0(t) = \gamma(t; t_0, x_0), \quad \text{uniformly for } t \in I. \tag{B.23}$$

Thus, if T, η, δ and Ω_0 are as above, it is possible to find $\varepsilon > 0$ such that

$$\sup_{t \in I} \|\psi_h(t) - \psi_0(t)\| < \delta, \quad \text{for every } h \in \mathbb{R} \text{ with } |h| \leq \varepsilon. \tag{B.24}$$

We also notice that, since (CP) has a unique maximal solution (namely, $\gamma(\cdot\,; t_0, x_0)$), we have the following identities (see Lem. 12.1):

(I) $\mathcal{D}(t_0 + h, x_0) = \mathcal{D}(t_0, \gamma(t_0; t_0 + h, x_0))$;

(II) $\gamma(t; t_0 + h, x_0) = \gamma(t; t_0, \gamma(t_0; t_0 + h, x_0))$ for every $t \in I \subseteq \mathcal{D}(t_0 + h, x_0)$.

We now consider, for every $0 < |h| < \varepsilon$, the function v_h defined as follows

$$v_h : I \longrightarrow \mathbb{R}^N \quad v_h(t) := \big(\psi_h(t) - \psi_0(t)\big)/h,$$

and we observe that, owing to (II), we can write (for every $t \in I$)

$$\begin{aligned} v_h(t) &= \frac{1}{h}\big(\psi_h(t) - \psi_0(t)\big) = \frac{1}{h}\big(\gamma(t; t_0 + h, x_0) - \gamma(t; t_0, x_0)\big) \\ &\overset{(II)}{=} \frac{1}{h}\big(\gamma(t; t_0, \gamma(t_0; t_0 + h, x_0)) - \gamma(t; t_0, x_0)\big). \end{aligned} \tag{B.25}$$

Since $\gamma(t_0; t_0, x_0) = x_0$ and since ψ_h uniformly converges to ψ_0 on I as $h \to 0$, by possibly shrinking ε we can assume that (see also (B.24))

$$(t_0, \gamma(t_0; t_0 + h, x_0)) \in B(z^*, r), \quad \text{for every } |h| \leq \varepsilon.$$

Therefore, since the function γ has continuous partial derivatives wrt x_0, we are entitled to apply Lem. B.36 to (B.25), obtaining

$$v_h(t) = J_h(t) \cdot \frac{\gamma(t_0; t_0 + h, x_0) - x_0}{h}, \quad \forall\, t \in I,\ \forall\, 0 < |h| < \varepsilon,$$

where we have set

$$J_h(t) := \int_0^1 \frac{\partial \gamma}{\partial x_0}\Big(t; t_0, u\,\gamma(t_0; t_0 + h, x_0) + (1 - u)x_0\Big)\,du, \qquad |h| < \varepsilon.$$

Now, from the continuity of $\partial\gamma/\partial x_0$ on $\mathcal{O}$ (and since $K = [a, b] \times \overline{B(z^*, r)}$ is compact and contained in $\mathcal{O}$), we easily infer that

$$\lim_{h\to 0} J_h(t) = J_0(t) = \frac{\partial\gamma}{\partial x_0}(t; t_0, x_0), \quad \text{uniformly for } t \in I. \tag{B.26}$$

On the other hand, since $x_0 = \gamma(t_0 + h; t_0 + h, x_0)$, we have

$$\gamma(t_0; t_0 + h, x_0) - x_0 = \gamma(t_0; t_0 + h, x_0) - \gamma(t_0 + h; t_0 + h, x_0)$$

$$= -h \int_0^1 \dot\gamma(t_0 + (1 - u)h; t_0 + h, x_0)\,du$$

$$= -h \int_0^1 f\big(t_0 + (1 - u)h, \gamma(t_0 + (1 - u)h; t_0 + h, x_0)\big)\,du$$

$$= -h \int_0^1 f\big(t_0 + (1 - u)h, \psi_h(t_0 + (1 - u)h)\big)\,du.$$

By the uniform continuity of f on $\overline{\Omega}_0 \subseteq \Omega$, by (B.23) and (B.24), we get

$$\lim_{h\to 0} \frac{\gamma(t_0; t_0 + h, x_0) - x_0}{h} = -f(t_0, x_0). \tag{B.27}$$

By gathering together identities (B.26) and (B.27), we then obtain

$$\lim_{h\to 0} v_h(t) = -\frac{\partial\gamma}{\partial x_0}(t; t_0, x_0)\,f(t_0, x_0), \quad \text{for every } t \in I;$$

hence, from the arbitrariness of $(t_0, x_0) \in B(z^*, r)$, we can conclude that

$$\frac{\partial\gamma}{\partial t_0}(t; t_0, x_0) = -\frac{\partial\gamma}{\partial x_0}(t; t_0, x_0)\,f(t_0, x_0),$$

for every $(t, t_0, x_0) \in \mathcal{U}$, and identity (B.19) holds true.

Finally, from the continuity of f on Ω and from the continuity of $\partial\gamma/\partial x_0$ on $\mathcal{O}$, we infer from (B.19) that the function $(t, t_0, x_0) \mapsto \frac{\partial\gamma}{\partial t_0}(t; t_0, x_0)$ is continuous on $\mathcal{O}$, and this completes the proof of the theorem. $\qquad\square$

B.3.1 *The equation of variation*

The results in Thm. B.38 can be extended to the parametric case by using the strategy described in Prop. B.20:

Theorem B.39 (C^1-dependence wrt the parameters). *Suppose that $\Omega \subseteq \mathbb{R}^{1+N}$ and $V \subseteq \mathbb{R}^m$ are open sets, and let $f \in C(\Omega \times V, \mathbb{R}^N)$ be such that*

$$\frac{\partial f}{\partial x_i} \quad \text{and} \quad \frac{\partial f}{\partial \xi_j} \quad (1 \le i \le N,\ 1 \le j \le m)$$

exist and are continuous on $\Omega \times V$.

For every $((t_0, x_0), \xi_0) \in \Omega \times V$, let $\gamma(\,\cdot\,; t_0, x_0, \xi_0) \in C^1(\mathcal{D}(t_0, x_0, \xi_0), \mathbb{R}^N)$ be the unique maximal solution of the Cauchy problem

$$\dot{x} = f(t, x; \xi_0), \quad x(t_0) = x_0. \tag{B.28}$$

Then the function $(t, t_0, x_0, \xi_0) \mapsto \gamma(t; t_0, x_0, \xi_0)$ is of class C^1 on the open set

$$\mathcal{O} = \left\{ (t, (t_0, x_0), \xi_0) \in \mathbb{R} \times \Omega \times V : t \in \mathcal{D}(t_0, x_0, \xi_0) \right\}.$$

More precisely, the following facts hold true:

(1) *For every $((t_0, x_0), \xi_0) \in \Omega \times V$ and every $i = 1, \ldots, N$, the function*

$$\mathcal{D}(t_0, x_0, \xi_0) \ni t \mapsto \frac{\partial \gamma}{\partial x_i^0}(t; t_0, x_0, \xi_0)$$

is the maximal solution of the linear homogeneous Cauchy problem

$$\begin{cases} \dot{x} = A_{t_0, x_0, \xi_0}(t)\, x \\ x(t_0) = e_j, \end{cases} \quad \text{where} \quad A_{t_0, x_0, \xi_0}(t) := \frac{\partial f}{\partial x}\Big(t, \gamma(t; t_0, x_0, \xi_0); \xi_0\Big).$$

$$\tag{B.29}$$

(2) *For every $((t_0, x_0), \xi_0) \in \Omega \times V$ and every $t \in \mathcal{D}(t_0, x_0, \xi_0)$, we have*

$$\frac{\partial \gamma}{\partial t_0}(t; t_0, x_0, \xi_0) = -\frac{\partial \gamma}{\partial x_0}(t; t_0, x_0, \xi_0)\, f(t_0, x_0; \xi_0). \tag{B.30}$$

(3) *For every $((t_0, x_0), \xi_0) \in \Omega \times V$ and every $j = 1, \ldots, m$, the function*

$$\mathcal{D}(t_0, x_0, \xi_0) \ni t \mapsto \frac{\partial \gamma}{\partial \xi_j^0}(t; t_0, x_0, \xi_0)$$

is the maximal solution of the linear non-homogeneous Cauchy problem

$$\begin{cases} \dot{x} = A_{t_0, x_0, \xi_0}(t)\, x + B_{t_0, x_0, \xi_0}^j(t) \\ x(t_0) = 0, \end{cases} \tag{B.31}$$

where $B_{t_0, x_0, \xi_0}^j(t) := \dfrac{\partial f}{\partial \xi_j}\Big(t, \gamma(t; t_0, x_0, \xi_0); \xi_0\Big).$

*The ODE in (B.31) is the so-called **equation of variation** associated with (B.28).*

Remark B.40. Let the assumptions of Thm. B.39 apply. We observe that, since γ is of class C^1 on $\mathcal{O}$, it follows from (B.29), (B.30) and (B.31) that

$$\frac{\partial}{\partial t}\left(\frac{\partial \gamma}{\partial t_0}(t; t_0, x_0, \xi_0)\right), \quad \frac{\partial}{\partial t}\left(\frac{\partial \gamma}{\partial x_0}(t; t_0, x_0, \xi_0)\right), \quad \frac{\partial}{\partial t}\left(\frac{\partial \gamma}{\partial \xi_0}(t; t_0, x_0, \xi_0)\right)$$

exist and are continuous wrt all the variables; on the other hand, since γ is a solution of the Cauchy problem (B.28), for every $(t, (t_0, x_0), \xi_0) \in \mathcal{O}$ we have

$$\frac{\partial \gamma}{\partial t}(t; t_0, x_0, \xi_0) = f\big(t, \gamma(t; t_0, x_0, \xi_0); \xi_0\big).$$

Hence, as f admits continuous partial derivatives (on $\Omega \times V$) wrt x and ξ, we deduce that also the partial derivatives

$$\frac{\partial}{\partial t_0}\left(\frac{\partial \gamma}{\partial t}(t; t_0, x_0, \xi_0)\right), \quad \frac{\partial}{\partial x_0}\left(\frac{\partial \gamma}{\partial t}(t; t_0, x_0, \xi_0)\right), \quad \frac{\partial}{\partial \xi_0}\left(\frac{\partial \gamma}{\partial t}(t; t_0, x_0, \xi_0)\right)$$

exist and are continuous on $\mathcal{O}$. By the Schwarz Theorem, we can interchange the order of differentiation for all the above second-order derivatives. ♯

An inductive argument on k proves the following crucial result, starting from the case $k = 1$ in Thm. B.39. We skip all the details, except for underlining that, for the inductive argument to work, one must exploit the linear equations in (B.29), (B.30) and (B.31).

Theorem B.41 (C^k-dependence of solutions). *Let $\Omega \subseteq \mathbb{R}^{1+N}$ and $V \subseteq \mathbb{R}^m$ be open sets, and let $f \in C^k(\Omega \times V, \mathbb{R}^N)$ (with $k \geq 1$ or $k = \infty$).*

For every $((t_0, x_0), \xi_0)$ in $\Omega \times V$, let $\gamma(\,\cdot\,; t_0, x_0, \xi_0) : \mathcal{D}(t_0, x_0, \xi_0) \to \mathbb{R}^N$ be the unique maximal solution of (B.28). *Then the function*

$$(t, t_0, x_0, \xi_0) \mapsto \gamma(t; t_0, x_0, \xi_0)$$

is of class C^k on the open set

$$\mathcal{O} = \left\{ (t, (t_0, x_0), \xi_0) \in \mathbb{R} \times \Omega \times V : t \in \mathcal{D}(t_0, x_0, \xi_0) \right\}.$$

B.4 C^ω dependence

We conclude our investigation on the regularity of the maximal solution of a Cauchy problem by turning our attention to the real-analytic case. We review only a few basic facts about the real-analytic functions of several variables (see e.g., the book by [Krantz and Parks (2002)] for an exhaustive treatment).

Let $U \subseteq \mathbb{R}^N$ be open, let $f \in C^\infty(U, \mathbb{R})$ and let $x_0 \in U$. We say that f is **real-analytic** at x_0 if there exists $\rho > 0$ such that $\overline{B(x_0, \rho)} \subseteq U$ and:

(i) the series $\displaystyle\sum_{n=0}^{\infty} \left(\sum_{|\alpha|=n} \frac{|D^\alpha f(x_0)|}{\alpha!} \right) \rho^n$ is convergent;

(ii) f coincides on $B(x_0, \rho)$ with the sum of its Taylor series centred at x_0:

$$f(x) = \sum_{n=0}^{\infty} \left(\sum_{|\alpha|=n} \frac{D^\alpha f(x_0)}{\alpha!} (x - x_0)^\alpha \right) \quad \text{for every } x \in B(x_0, \rho).$$

We say that f is real-analytic in U (and we write $f \in C^\omega(U, \mathbb{R})$) if f is real-analytic at every point of U. For vector-valued functions $f = (f_1, \ldots, f_m)$, real-analyticity is defined component-wise.

Assumption (i) implies that the Taylor series of f centred at x_0 is normally convergent on $\overline{B(x_0, \rho)}$, i.e.,

$$\sum_{n=0}^{\infty} \sup_{x \in \overline{B(x_0, \rho)}} \left| \sum_{|\alpha|=n} \frac{D^\alpha f(x_0)}{\alpha!} (x - x_0)^\alpha \right| < \infty.$$

Assumption (i) also guarantees the existence of a constant $M > 0$ such that

$$\left| \frac{D^\alpha f(x_0)}{\alpha!} \right| \leq \frac{M}{\rho^{|\alpha|}}, \quad \text{for every } \alpha \in (\mathbb{N} \cup \{0\})^N.$$

Finally, if $f \in C^\infty(U, \mathbb{R}^k)$ is real-analytic at $x_0 \in U$, then there exists $\rho > 0$ such that f belongs to $C^\omega(B(x_0, \rho), \mathbb{R}^k)$.

After these preliminaries, we now state the main theorem of this section.

Theorem B.42 (C^ω-regularity of γ). *Let $\Omega \subseteq \mathbb{R}^{1+N}, V \subseteq \mathbb{R}^m$ be open sets and let $f \in C^\omega(\Omega \times V, \mathbb{R}^N)$. Moreover, for every $((t_0, x_0), \xi_0) \in \Omega \times V$, let*

$$\gamma(\cdot\,; t_0, x_0, \xi_0) \in C^\infty(\mathcal{D}(t_0, x_0, \xi_0), \mathbb{R}^N)$$

be the unique maximal solution of (B.28).

Then the map $(t; t_0, x_0, \xi_0) \mapsto \gamma(t; t_0, x_0, \xi_0)$ is real-analytic on the open set

$$\mathcal{O} = \Big\{ (t, (t_0, x_0), \xi_0) \in \mathbb{R} \times \Omega \times V : t \in \mathcal{D}(t_0, x_0, \xi_0) \Big\}.$$

The proof of Thm. B.42 requires some preliminary results. In the sequel it is tacitly understood that Ω, V and f are as above.

Lemma B.43. *Suppose (to fix ideas) that $(0,0) \in \Omega$ and $0 \in V$. Moreover, let $f^{(1)}, f^{(2)} \in C^\infty(\Omega \times V, \mathbb{R}^N)$ and, for every $\xi \in V$, let $t \mapsto \gamma^{(i)}(t; 0, 0, \xi)$ (which is smooth on $\mathcal{D}^{(i)}(0, 0, \xi)$) be the unique maximal solution of*

$$\dot{x} = f^{(i)}(t, x; \xi), \quad x(0) = 0.$$

Let us assume that, for all multi-indices

$$p \in \mathbb{N} \cup \{0\}, \quad \alpha \in (\mathbb{N} \cup \{0\})^N, \quad \beta \in (\mathbb{N} \cup \{0\})^m,$$

and, for every $j = 1, \ldots N$, one has the estimates

$$\left| D_t^p D_x^\alpha D_\xi^\beta f_j^{(1)}(0, 0; 0) \right| \le D_t^p D_x^\alpha D_\xi^\beta f_j^{(2)}(0, 0; 0). \tag{B.32}$$

Then, for any $p \in \mathbb{N} \cup \{0\}$, any $\beta \in (\mathbb{N} \cup \{0\})^m$ and any $j \in \{1, \ldots, N\}$, one has

$$\left| D_t^p\big|_{t=0} D_\xi^\beta\big|_{\xi=0} \gamma_j^{(1)}(t; 0, 0, \xi) \right| \le D_t^p\big|_{t=0} D_\xi^\beta\big|_{\xi=0} \gamma_j^{(2)}(t; 0, 0, \xi). \tag{B.33}$$

Proof. Apart from the case $p = 0$ (trivial, since $\gamma^{(i)}(0; 0, 0, \xi) = 0$), the general case can be treated via the following ingredients: bearing in mind the ODE

$$D_t^p \gamma_j^{(i)}(t; 0, 0, \xi) = D_t^{p-1}\Big\{ f_j^{(i)}(t, \gamma^{(i)}(t; 0, 0, \xi); \xi) \Big\}, \tag{B.34}$$

the Chain Rule (and a recursive use of (B.34) itself) ensures the existence of a "universal" polynomial Q_p, with nonnegative integer coefficients, such that

$$D_t^p \gamma_j^{(i)}(t; 0, 0, \xi) = Q_p\left\{ \frac{\partial^{k+|\alpha|} f^{(i)}}{\partial t^k\, \partial x^\alpha}(t, \gamma^{(i)}(t; 0, 0, \xi); \xi) : \ k + |\alpha| \le p - 1 \right\}.$$

As Q_p does not depend on i, one can prove (B.33) starting from (B.32). $\qquad\square$

Lemma B.44. *Let $(t_0, x_0) \in \Omega$ be fixed and let $f \in C^\omega(\Omega, \mathbb{R}^N)$. Moreover, for every $\xi \in V$, let $t \mapsto \gamma(t; t_0, x_0, \xi)$ be the unique maximal solution of*

$$(\mathrm{CP})_\xi : \qquad \dot{x} = f(t, x; \xi), \quad x(t_0) = x_0.$$

Then, for any $\xi_0 \in V$, the map $(t; \xi) \mapsto \gamma(t; t_0, x_0, \xi)$ is real-analytic at $(t_0; \xi_0)$.

Proof. We can suppose, without loss of generality, that (t_0, x_0) and ξ_0 vanish; moreover, to simplify the notations, we write $\gamma(t; \xi)$ in place of $\gamma(t; 0, 0, \xi)$ and we denote the multi-indices $\sigma \in (\mathbb{N} \cup \{0\})^{1+N+m}$ by

$$\sigma = (p, \alpha, \beta), \quad \text{with } p \geq 0, \, \alpha \in (\mathbb{N} \cup \{0\})^N \text{ and } \beta \in (\mathbb{N} \cup \{0\})^m.$$

Since f is real-analytic at $((0,0), 0) \in \Omega \times V$, we can find $M, \rho > 0$ such that

$$U_\rho := B((0,0), \rho) \times B(0, \rho) \subseteq \Omega \times V,$$

and, for every $j \in \{1, \ldots, N\}$,

$$\left| \frac{1}{\sigma!} D^\sigma f_j(0, 0; 0) \right| \leq \frac{M}{\rho^{|\sigma|}}, \quad \text{for every } \sigma \in (\mathbb{N} \cup \{0\})^{1+N+m}. \tag{B.35}$$

Thus, if we define $g : U_\rho \to \mathbb{R}$ as

$$g(t, x; \xi) := \frac{M}{1 - t/\rho} \prod_{k=1}^{N} \frac{1}{1 - x_k/\rho} \prod_{h=1}^{m} \frac{1}{1 - \xi_h/\rho},$$

we have (for any σ as above and any $j \in \{1, \ldots, N\}$)

$$D_t^p D_x^\alpha D_\xi^\beta g(0, 0; 0) = \sigma! \frac{M}{\rho^{|\sigma|}} \overset{\text{(B.35)}}{\geq} \left| D_t^p D_x^\alpha D_\xi^\beta f_j(0, 0; 0) \right|. \tag{B.36}$$

As a crucial trick, we now denote by $G(t, x; \xi)$ the $\mathbb{R}^N$-valued map with N components all equal to $g(t, x; \xi)$; if $\psi(\cdot\,; \xi)$ is the maximal solution of

$$\dot{x} = G(t, x; \xi), \quad x(0) = 0,$$

we obtain, from (B.36) and Lem. B.43, the following estimates:

$$\left| D_t^p D_\xi^\beta \gamma_j(0; 0) \right| \leq D_t^p D_\xi^\beta \psi_j(0; 0). \tag{B.37}$$

Now, the key fact is that $\psi_1(\cdot\,; \xi), \ldots, \psi_N(\cdot\,; \xi)$ are all equal to a certain function $\psi_0(\cdot\,; \xi)$ which can explicitly be determined: in fact, we have

$$\psi_0(t; \xi) = \rho \left\{ 1 - \left(1 + \frac{2M(1+N)}{c(\xi)} \log \left(1 - t/\rho \right) \right)^{1/(N+1)} \right\},$$

where $c(\xi) = \prod_{h=1}^{m}(1 - \xi_h/\rho)$. In particular, we see that $(t; \xi) \mapsto \psi_0(t; \xi)$ is real-analytic on an open neighborhood O of $(0; 0)$ in $\mathbb{R} \times V$.

From this, by exploiting (B.37), we infer the existence of a real $r > 0$ such that $C_r := [-r, r] \times \overline{B(0, r)} \subseteq O$ and, for any $j \in \{1, \ldots, N\}$,

$$\sum_{n=0}^{\infty} \left(\sum_{p+|\beta|=n} \frac{|D_t^p D_\xi^\beta \gamma_j(0; 0)|}{p! \, \beta!} \right) r^n \leq \sum_{n=0}^{\infty} \left(\sum_{p+|\beta|=n} \frac{D_t^p D_\xi^\beta \psi_0(0; 0)}{p! \, \beta!} \right) r^n < \infty.$$

These estimates ensure the well-posedness and the real-analyticity of

$$v : C_r \to \mathbb{R}^N, \quad v(t; \xi) := \sum_{n=0}^{\infty} \left(\sum_{p+|\beta|=n} \frac{D_t^p D_\xi^\beta \gamma(0; 0)}{p! \, \beta!} t^p \xi^\beta \right).$$

We are left to prove that $\gamma \equiv v$ in a neighborhood of $(0;0)$. First of all, we can assume that $(t, v(t;\xi)) \in \Omega$ for all $(t,\xi) \in C_r$; moreover, by a classical induction argument, and the fact that $\gamma(\cdot\,;\xi)$ solves $(\text{CP})_\xi$, we see that

$$
\begin{aligned}
D_t^p D_\xi^\beta v(0;0) &= D_t^p D_\xi^\beta \gamma(0;0) \\
&= D_t^{p-1}\big|_{t=0} D_\xi^\beta\big|_{\xi=0} \left\{ f(t, \gamma(t);\xi) \right\} \\
&= D_t^{p-1}\big|_{t=0} D_\xi^\beta\big|_{\xi=0} \left\{ f(t, v(t);\xi) \right\},
\end{aligned}
$$

for every $p \geq 1$ and every $\beta \in (\mathbb{N} \cup \{0\})^m$. From this, by the Unique Continuation Principle, we infer that also $v(\cdot\,;\xi)$ solves $(\text{CP})_\xi$ on $[-r, r]$, whence $\gamma(t;\xi) = v(t;\xi)$ for all $(t;\xi) \in C_r$. This ends the proof. $\qquad\square$

From Lem. B.44 we obtain the following result, by means of an argument similar to that in Sec. B.1.3 in order to treat the data as parameters.

Lemma B.45. *Let the assumptions and the notation of Thm. B.42 apply.*
Then, for every $((\bar{t}, \bar{x}), \xi_0) \in \Omega \times V$, the map $(t; t_0, x_0, \xi) \mapsto \gamma(t; t_0, x_0, \xi)$ is real-analytic at $(\bar{t}; \bar{t}, \bar{x}, \xi_0)$.

Proof. We choose $\rho > 0$ in such a way that $B((\bar{t}, \bar{x}), 2\rho) \subseteq \Omega$, and we denote the points of $V' := B((\bar{t}, \bar{x}), 2\rho) \times V$ by $\xi' = ((t_0, x_0), \xi)$; we define

$$
f' : B((0,0), \rho) \times V' \longrightarrow \mathbb{R}^N, \qquad f'(t, x; \xi') := f(t + t_0, x + x_0; \xi).
$$

Since f' is real-analytic on its domain, if we let $t \mapsto \psi(t;\xi') \in C^\infty(\mathcal{D}'(\xi'), \mathbb{R}^N)$ be the maximal solution of the Cauchy problem

$$
\dot{x} = f'(t, x; \xi'), \quad x(0) = 0,
$$

we infer from Lem. B.44 that the map $(t; \xi') \mapsto \psi(t;\xi')$ is real-analytic at the point $(0; ((\bar{t}, \bar{x}), \xi_0)) \in \mathbb{R} \times V'$. As a consequence, since

$$
\gamma(t; t_0, x_0, \xi) = x_0 + \psi(t - t_0; \xi') = x_0 + \psi(t - t_0; ((t_0, x_0), \xi)),
$$

for all $(t; \xi') \in \mathbb{R} \times V'$ such that $t - t_0 \in \mathcal{D}'(\xi')$, we conclude that γ is real-analytic at $(\bar{t}; \bar{t}, \bar{x}, \xi_0)$, as required. $\qquad\square$

Remark B.46. Let the assumptions of Lem. B.45 apply. Let $(\bar{t}, \bar{x}) \in \Omega$ and let $\xi_0 \in V$. Since γ is real-analytic at $(\bar{t}; \bar{t}, \bar{x}, \xi_0)$, it is possible to find $\rho > 0$ such that γ is real-analytic at every point in $\mathcal{O}$ of the form $(t; s, x, \xi)$ with

$$
t \in (\bar{t} - \rho, \bar{t} + \rho), \quad (s, x) \in B((\bar{t}, \bar{x}), \rho), \quad \xi \in B(\xi_0, \rho).
$$

With Lem. B.45 at hand, we can finally prove Thm. B.42. We shall use, in a crucial way, the semi-group property for non-autonomous ODEs, established in Lem. 12.1 (page 260).

Proof of Thm. B.42. For simplicity, we denote (t_0, x_0) in Ω by z_0; moreover, given any $(z_0, \xi) \in \Omega \times V$ and any $\delta > 0$, we denote by $P_\delta(z_0, \xi)$ the open poly-disk

$$P_\delta(z_0, \xi) := B(z_0, \delta) \times B(\xi, \delta) \subset \mathbb{R}^{1+N} \times \mathbb{R}^m.$$

Let us fix $(s; \bar{t}, \bar{x}, \xi_0) \in \mathcal{O}$ (and let us set $\bar{z} := (\bar{t}, \bar{x})$). If $s = \bar{t}$, we know from Lem. B.45 that γ is real-analytic at $(\bar{t}; \bar{z}, \xi_0)$; we can thus suppose that $s \neq \bar{t}$ and, to fix ideas, we assume that $s > \bar{t}$.

Chosen $a, b \in \mathcal{D}(\bar{z}, \xi_0)$ in such a way that $[\bar{t}, s] \subseteq (a, b)$, we let $r > 0$ be such that $[a, b] \times \overline{P_r(\bar{z}, \xi_0)} \subseteq \mathcal{O}$ (see Thm. B.34), and we define

$$K := \left\{ \left((t, \gamma(t; z_0, \xi)), \xi \right) : t \in [a, b] \text{ and } (z_0, \xi) \in \overline{P_r(\bar{z}, \xi_0)} \right\} \subseteq \Omega \times V.$$

Since K is compact, from Rem. B.46 we infer the existence of a real $\rho \in (0, r)$, only depending on K, such that $[\bar{t} - \rho, \bar{t} + \rho] \subseteq (a, b)$ and we get that, for every point $(\zeta, \xi) = ((t, \gamma(t; z_0, \xi)), \xi) \in K$ (with $t \in [a, b]$ and $(z_0, \xi) \in \overline{P_r(\bar{z}, \xi_0)}$), one has

(a) $[t - \rho, t + \rho] \times \overline{P_\rho(\zeta, \xi)} \subseteq \mathcal{O}$;

(b) γ is real-analytic at every point of $(t - \rho, t + \rho) \times P_\rho(\zeta, \xi)$.

If $s \in [\bar{t}, \bar{t} + \rho)$, we immediately conclude that γ is real-analytic at $(s; \bar{z}, \xi_0)$; if, instead, $s \geq \bar{t} + \rho$, we have $[\bar{t} - \rho, \bar{t} + \rho] \subseteq (a, b)$ and

$$\gamma\big(u; t, \gamma(t; z_0, \xi), \xi\big) \overset{(12.2)}{=} \gamma(u; z_0, \xi)$$

is real-analytic at every point $(u; z_0, \xi)$ such that

$$u \in (t - \rho, t + \rho), \quad t \in (\bar{t} - \rho, \bar{t} + \rho), \quad (z_0, \xi) \in P_\rho(\bar{z}, \xi_0).$$

In particular, γ is real-analytic on $(\bar{t} - \rho, \bar{t} + 2\rho) \times P_\rho(\bar{z}, \xi_0)$.

By repeating this argument finitely many times, we conclude that there exists an open neighborhood $\mathcal{U}$ of $[\bar{t}, s]$ such that γ is real-analytic at every point of the set $\mathcal{U} \times P_\rho(\bar{z}, \xi_0) \subseteq \mathcal{O}$, whence γ is real-analytic at $(s, \bar{z}, \xi_0)$. $\square$

B.5 Exercises of App. B

Exercise B.1. Let $(t_0, x_0) \in \mathbb{R}^{1+N}$ and let $C_0 = [t_0 - h, t_0 + h] \times \overline{B(x_0, r)}$ (for $h, r > 0$). Moreover, let $f \in C(C_0, \mathbb{R}^N)$ and let us assume that C_0 is a control cylinder for f. Setting $I := [t_0, t_0 + h]$, let $\sigma = \{t_0, \ldots, t_p\}$ be a partition of I and let $u_\sigma : I \longrightarrow \mathbb{R}^N$ be the function defined as follows:

$$u_\sigma(t) := \begin{cases} x_0, & \text{if } t = t_0 \\ u_\sigma(t_{i-1}) + (t - t_{i-1}) f(t_i, u_\sigma(t_{i-1})), & \text{if } t_{i-1} < t \leq t_i \ (1 \leq i \leq p). \end{cases}$$

Prove the following facts:

(1) $u_\sigma \in C(I, \mathbb{R}^N)$ for every partition σ;

(2) the family $\{u_\sigma\}_\sigma$ is uniformly-bounded and equi-continuous in $C(I, \mathbb{R}^N)$;

(3) there exists a cluster point $\gamma \in C(I, \mathbb{R}^N)$ for the family $\{u_\sigma\}_\sigma$, which is a solution of (CP) on I.

Exploit this argument to provide another proof of Peano's Theorem.

Exercise B.2. Prove Prop. B.23 by completing the following argument.

(1) For any $\varepsilon > 0$ there exists $\rho_\varepsilon > 0$ such that
$$|u_n(t) - u_n(s)| < \varepsilon, \quad \text{for any } n \in \mathbb{N} \text{ and } t, s \in T \text{ with } d(t, s) < \rho_\varepsilon.$$

(2) The family[9] $\mathcal{U} := \{B_{\rho_\varepsilon}(s) : s \in E\}$ is an open cover of T. Infer the existence of $s_1, \ldots, s_p \in E$ such that
$$T = \bigcup_{j=1}^{p} B_{\rho_\varepsilon}(s_j).$$

(3) Prove that there exists $\bar{n}$ (only depending on ε) such that
$$|u_n(s_j) - u_m(s_j)| < \varepsilon, \quad \text{for any } n, m \geq \bar{n} \text{ and any } \forall j = 1, \ldots, p.$$

(4) Infer that $|u_n(t) - u_m(t)| < 3\,\varepsilon$, for every $n, m \geq \bar{n}$ and every $t \in T$ (uniform convergence is gained).

(5) Vice versa, let u_n converge uniformly to u on T. Use the Heine-Cantor Theorem to show that for any $\varepsilon > 0$ there exists $\rho_\varepsilon > 0$ such that
$$|u(t) - u(s)| < \varepsilon, \quad \text{for any } t, s \in T \text{ with } d(t, s) < \rho_\varepsilon.$$

(6) Deduce that you can find $\bar{n} \in \mathbb{N}$ such that
$$|u_n(t) - u(t)| < \varepsilon, \quad \text{for any } n \geq \bar{n} \text{ and any } t \in T.$$

(7) Prove that for any $n \geq \bar{n}$ and any $t, s \in T$ satisfying $d(t, s) < \rho_\varepsilon$ we have
$$|u_n(t) - u_n(s)| < |u_n(t) - u(t)| + |u(t) - u(s)| + |u(s) - u_n(s)| < 3\,\varepsilon.$$

Since $u_1, \ldots, u_{\bar{n}-1}$ are uniformly continuous on T, the same inequality holds for $n < \bar{n}$, by shrinking ρ_ε if necessary...

Exercise B.3. Prove Prop. B.24 by completing the following argument.

(1) Let $E = \{s_j : j \in \mathbb{N}\}$. Use the Bolzano-Weierstrass Theorem to construct a family of sequences $\mathcal{F} = \{\{u_n^{(j)}\}_n : j \in \mathbb{N}\}$ with the following properties:

- $\{u_n^{(1)}\}_n$ is a subsequence of $\{u_n\}_n$;
- $\{u_n^{(j+1)}\}_n$ is a subsequence of $\{u_n^{(j)}\}_n$ for every $j \geq 1$;
- $\{u_n^{(j)}(s_j)\}_n$ is convergent for every $j \in \mathbb{N}$.

(2) Applying a suitable Cantor diagonal argument to obtain a subsequence $\{u_{n_k}\}_k$ such that $\{u_{n_k}(s_k)\}_k$ is convergent for every k...

Exercise B.4. Prove Thm. B.25 by the following argument.

(1) There exists a countable set $E \subseteq T$ which is dense in T.

---

[9] $B_r(t)$ denotes the metric open ball with centre t and radius r.

(2) There exists $\{u_{n_k}\}_k$ point-wise convergent on E (Prop. B.24...).

(3) $\{u_{n_k}\}_k$ is uniformly convergent on T (Prop. B.23...).

Exercise B.5. Prove Thm. B.27 by completing the argument below.

(1) Since $\mathcal{F}$ is totally bounded, for every $\varepsilon > 0$ it is possible to find sets $\mathcal{F}_1, \ldots, \mathcal{F}_p$ contained in $C(T, \mathbb{R}^p)$ such that

(i) $\mathcal{F} = \bigcup_{k=1}^{p} \mathcal{F}_k$;

(ii) $\mathrm{diam}(\mathcal{F}_k) < \varepsilon$ for every $k = 1, \ldots, p$. We choose $u_k \in \mathcal{F}_k$ for $k \leq p$.

(2) Show that $\mathcal{F}$ is uniformly-bounded (if $u \in \mathcal{F}$ belongs to $\mathcal{F}_k$, then show that $\|u\| \leq \varepsilon + \max\{\|u_1\|, \ldots, \|u_p\|\}$).

(3) Show that $\mathcal{F}$ is equi-continuous: the Heine-Cantor theorem ensures the existence of $\rho > 0$, only depending on ε, such that $|u_k(t) - u_k(s)| < \varepsilon$, for any $t, s \in T$ with $d(t, s) < \rho$ and any $k = 1, \ldots, p$. Thus, for such t, s deduce that $|u(t) - u(s)| < 3\,\varepsilon$ for every $u \in \mathcal{F}$.

Appendix C

A brief review of Lie Group Theory

$\mathbf{T}$HE aim of this appendix is to collect notations and basic facts about Lie groups that have been used in the book. This appendix is not meant as a comprehensive introduction to Lie groups, but only as a companion of statements for the rest of the book. Proofs will be kept to a minimum amount; the reader is referred to e.g., [Lee (2013)] or [Varadarajan (1984)] for complete demonstrations.

C.1 A short review of Lie groups

A Lie group is a smooth manifold $\mathbb{G}$ equipped with a group law $*$ (in the algebraic sense) such that the multiplication and the inversion maps
$$\mathbb{G} \times \mathbb{G} \ni (x, y) \mapsto x * y \in \mathbb{G} \quad \text{and} \quad \mathbb{G} \ni x \mapsto \iota(x) := x^{-1} \in \mathbb{G}$$
are both smooth. Equivalently $(\mathbb{G}, *)$ is a Lie group iff $(x, y) \mapsto x * y^{-1}$ is smooth. Usually, the group law is denoted by juxtaposition; we take the liberty of using $*$ to make it resemblant to the operations constructed in other parts of the book. Throughout, e (or $e_{\mathbb{G}}$ if more than one group is involved) will denote the identity element of $\mathbb{G}$. For $x \in \mathbb{G}$ fixed, we use the notations
$$\tau_x, \rho_x : \mathbb{G} \longrightarrow \mathbb{G}, \qquad \tau_x(y) := x * y, \quad \rho_x(y) := y * x$$
to denote, respectively, the left-translation and the right-translation by x (occasionally $\tau_x^{\mathbb{G}}$ and $\rho_x^{\mathbb{G}}$). Restatements of the associativity of $*$ are:
$$\tau_x \circ \rho_y = \rho_y \circ \tau_x, \quad \tau_{x*y} = \tau_x \circ \tau_y, \quad \rho_{x*y} = \rho_y \circ \rho_x \qquad \forall \; x, y \in \mathbb{G}.$$
Moreover, $\tau_e = \rho_e$ is the identity map of $\mathbb{G}$. Clearly, τ_x and $\tau_{x^{-1}}$ are inverse maps, hence C^∞-diffeomorphisms of $\mathbb{G}$; the same holds true for $\rho_x, \rho_{x^{-1}}$. The map ι is also a C^∞-diffeomorphism of $\mathbb{G}$, and an involution as well. Thus, any differential $\mathrm{d}_x \tau_\alpha : T_x \mathbb{G} \to T_{\alpha * x} \mathbb{G}$ is an isomorphism of vector spaces.

Later, we shall use the following fact, whose proof is a topology exercise.

Lemma C.1. *Let $(\mathbb{G}, *)$ be a connected Lie group. Let W be any open neighborhood of the identity element of $\mathbb{G}$. Then we have*
$$\mathbb{G} = \bigcup_{n=1}^{\infty} \left\{ w_1 * \cdots * w_n : w_1, \ldots, w_n \in W \right\}. \tag{C.1}$$

For a proof of this fact, one can mimic Exr. 17.5, page 330.

387

C.1.1   *The Lie algebra of* $\mathbb{G}$

Let $(\mathbb{G}, *)$ be a Lie group; a smooth[10] vector field X on $\mathbb{G}$ is called left invariant if X is τ_x-related to itself, for every left translation τ_x; this means that

$$X(f \circ \tau_x) = (Xf) \circ \tau_x, \quad \forall\, x \in \mathbb{G}, \quad \forall\, f \in C^\infty(\mathbb{G}).$$

Equivalently, in terms of point-wise differentials, X is left invariant iff

$$d_y \tau_x(X_y) = X_{x*y}, \quad \text{for every } x, y \in \mathbb{G}. \tag{C.2}$$

According to Def. 4.14, this is equivalent to saying that X is τ_x-invariant for every $x \in \mathbb{G}$. whence the name 'left invariance'. A left invariant v.f. X is determined by its value at the identity e, since (C.2) with $y = e$ gives

$$d_e \tau_x(X_e) = X_x, \quad \text{for every } x \in \mathbb{G}. \tag{C.3}$$

The vector subspace of $\mathfrak{X}(\mathbb{G})$ of all the left invariant v.f.s on $\mathbb{G}$ is called the Lie algebra of $\mathbb{G}$; we denote it by $\mathrm{Lie}(\mathbb{G})$ or by the corresponding ('Fraktur') symbol $\mathfrak{g}$.

By Rem. 4.16 (page 98), $\mathrm{Lie}(\mathbb{G})$ is a Lie sub-algebra of the smooth v.f.s on $\mathbb{G}$. It is finite-dimensional, with $\dim(\mathrm{Lie}(\mathbb{G})) = \dim(\mathbb{G})$ (the latter is meant in the sense of smooth manifolds), since the evaluation map

$$\Lambda : \mathrm{Lie}(\mathbb{G}) \to T_e\mathbb{G} \qquad X \mapsto X_e \tag{C.4}$$

is an isomorphism (this follows by arguing as in the 'local' proof of Thm. 15.3-(i), page 291). Hence, we shall sometimes adopt the identification

$$\mathrm{Lie}(\mathbb{G}) \equiv T_e\mathbb{G}, \quad X \equiv X_e. \tag{C.5}$$

Remark C.2. The inverse of Λ in (C.4) is the map sending $\mathbf{v}_e \in T_e\mathbb{G}$ to the vector field X such that $X_x = d_e\tau_x(\mathbf{v}_e)$, for every $x \in \mathbb{G}$. The reader can easily check that this X belongs to $\mathrm{Lie}(\mathbb{G})$ and that X acts as follows:

$$C^\infty(\mathbb{G}) \ni f \mapsto X_x f = \mathbf{v}_e(f \circ \tau_x), \quad x \in \mathbb{G}. \tag{C.6}$$

If $X \in \mathrm{Lie}(\mathbb{G})$, since any τ_α is a smooth diffeomorphism, we have $d\tau_\alpha X = X$ (in the sense of push-forwards) for every $\alpha \in \mathbb{G}$; thus, by Cor. 4.11 (page 96),

$$\tau_\alpha(\Psi_t^X(x)) = \Psi_t^X(\alpha * x), \quad \forall\, x, \alpha \in \mathbb{G}, \tag{C.7}$$

for every $t \in \mathcal{D}(X, x) = \mathcal{D}(X, \alpha * x)$. Hence, all maximal domains $\mathcal{D}(X, x)$ of X are equal, as x varies in $\mathbb{G}$; for instance, $\mathcal{D}(X, x) = \mathcal{D}(X, e)$ for every x.

Remark C.3. Starting from the latter information, it is simple to show that *any left-invariant vector field is global*: indeed, if $[-\varepsilon, \varepsilon] \subseteq \mathcal{D}(X, e)$, then

$$[-\varepsilon, \varepsilon] \subseteq \mathcal{D}(X, \Psi_{\pm\varepsilon}^X(e)).$$

Thus we can glue (see page 139 for the notion of gluing) the three curves

$$[-\varepsilon, 0] \ni t \mapsto \gamma\Big(t, X, \Psi_{-\varepsilon}^X(e)\Big), \quad [-\varepsilon, \varepsilon] \ni t \mapsto \gamma\Big(t, X, e\Big),$$

$$\text{and} \quad [0, \varepsilon] \ni t \mapsto \gamma\Big(t, X, \Psi_\varepsilon^X(e)\Big),$$

---

[10]Smoothness could be omitted, since it is regained by the definition of left invariance; see (C.3).

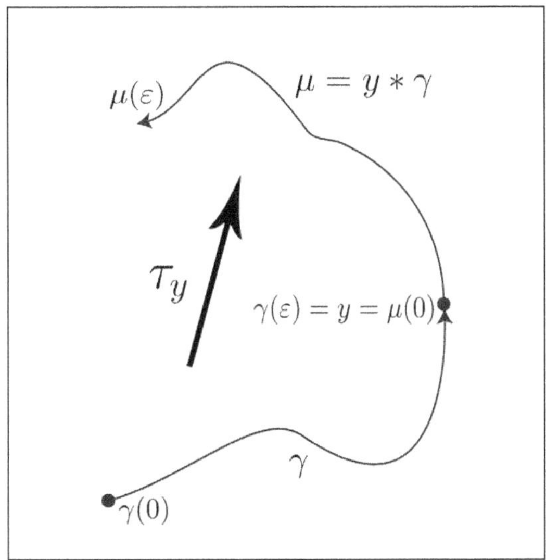

Fig. C.1 "Globalization" of the integral curve of a left invariant vector field X. Here $\gamma(t) = \Psi_t^X(e)$, with $t \in [0, \varepsilon] \subseteq \mathcal{D}(X, e)$ and $y := \Psi_\varepsilon^X(e)$; we then glue to γ the curve $\mu(t) := \Psi_t^X(y) = \tau_y(\gamma(t))$, defined for $t \in [0, \varepsilon]$. This prolongs γ to $[0, 2\varepsilon]$.

obtaining an integral curve of X starting at e, defined on $[-2\varepsilon, 2\varepsilon]$. By an inductive argument, this shows that $\mathcal{D}(X, e) = \mathbb{R}$. See Fig. C.1. $\qquad\qquad\qquad\sharp$

Definition C.4 (Nilpotent Lie group). A Lie group $\mathbb{G}$ is said to be nilpotent of step r if its Lie algebra $\mathrm{Lie}(\mathbb{G})$ is nilpotent of step r, according to Def. A.7.

We know from Exm. 6.2 on page 134 that there is, *a priori*, no relation between the linear independence of the v.f.s $X^1, \ldots, X^m$ as *linear differential operators*, and the linear independence of the derivations at x given by $X_x^1, \ldots, X_x^m$, as *elements of $T_x M$*. This is not the case, however, for left invariant v.f.s. Indeed, given a family of vector fields $X^1, \ldots, X^m \in \mathrm{Lie}(\mathbb{G})$, we show that the dimension of the subset of $T_x \mathbb{G}$ spanned by $\{X_x^1, \ldots, X_x^m\}$ is independent of $x \in \mathbb{G}$ and it equals the dimension of $\mathrm{span}\{X^1, \ldots, X^m\}$ as a subspace of $\mathrm{Lie}(\mathbb{G})$:

Proposition C.5 (Constant rank). *Let $\mathbb{G}$ be a Lie group with Lie algebra $\mathfrak{g}$ and identity e. Let $X^1, \ldots, X^m \in \mathfrak{g}$. Then the following statements are equivalent:*

  (i) $X^1, \ldots, X^m$ *are linearly independent in $\mathfrak{g}$ as a subspace of $\mathfrak{X}(\mathbb{G})$;*
  (ii) $X_e^1, \ldots, X_e^m$ *are linearly independent in $T_e \mathbb{G}$;*
  (iii) $\exists\, x_0 \in \mathbb{G}$ *such that $X_{x_0}^1, \ldots, X_{x_0}^m$ are linearly independent in $T_{x_0} \mathbb{G}$;*
  (iv) $X_x^1, \ldots, X_x^m$ *are linearly independent in $T_x \mathbb{G}$, for every $x \in \mathbb{G}$.*

Proof. By identity (C.3), we have, for $i = 1, \ldots, m$,
$$\mathrm{d}_e \tau_x(X_e^i) = X_x^i, \quad \text{for every } x \in \mathbb{G}.$$

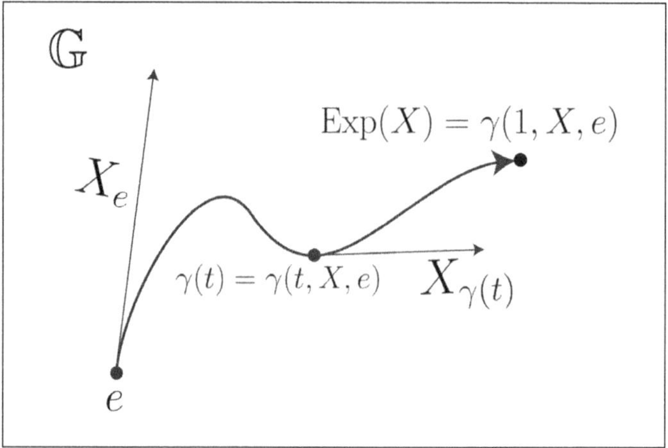

Fig. C.2   The Exponential Map as the solution of a Cauchy problem.

On the other hand, $\mathrm{d}_e \tau_x : T_e \mathbb{G} \to T_x \mathbb{G}$ is obviously a linear isomorphism. This clearly proves that (ii), (iii) and (iv) are equivalent. The equivalence of (i) and (ii) follows from the fact that Λ in (C.4) is an isomorphism. $\qquad\square$

Example C.6. Consider on $\mathbb{R}^2$ the v.f.s $X^1 = \partial_{x_1}$, $X^2 = x_1 \partial_{x_2}$. Note that, since X^1 and X^2 are independent in $\mathcal{X}(\mathbb{R}^2)$ but X_0^1, X_0^2 are dependent in $T_0 \mathbb{R}^2 \equiv \mathbb{R}^2$, we can deduce that X^1 and X^2 *cannot* belong to the Lie algebra of any Lie group whose underlying manifold is $\mathbb{R}^2$ (see Prop. C.5). $\qquad\sharp$

C.1.2   *The exponential map of* $\mathbb{G}$

Rem. C.3 shows that the following definition is well posed. See Fig. C.2.

Definition C.7 (Exponential Map). Let $\mathbb{G}$ be a Lie group on $\mathbb{R}^N$ with identity e and Lie algebra $\mathrm{Lie}(\mathbb{G})$. We set

$$\mathrm{Exp} : \mathrm{Lie}(\mathbb{G}) \longrightarrow \mathbb{G}, \qquad \mathrm{Exp}(X) := \Psi_t^X(e)\big|_{t=1}.$$

We say that Exp is the **Exponential Map** of $\mathbb{G}$.

In other words, given $X \in \mathrm{Lie}(\mathbb{G})$, we have $\mathrm{Exp}(X) = \gamma(1, X, e)$ where the map $\gamma(t) := \gamma(t, X, e)$ denotes –as usual– the maximal solution of

$$\dot{\gamma}(t) = X_{\gamma(t)}, \quad \gamma(0) = e. \tag{C.8}$$

General results on smooth dependence of the solution for ODEs (See Sec. B.3) ensure that Exp is a smooth map. Here we equip $\mathrm{Lie}(\mathbb{G})$ with the differentiable structure deriving from its being a finite dimensional vector space. Explicitly, if $\{X_1, \ldots, X_N\}$ is any basis of $\mathrm{Lie}(\mathbb{G})$, the following map is smooth:

$$\mathbb{R}^N \ni \xi \mapsto \mathrm{Exp}(\xi_1 X_1 + \cdots + \xi_N X_N) \in \mathbb{G}.$$

Moreover, we know that Exp is a local diffeomorphism of a neighborhood of $0 \in \mathfrak{g}$ onto a neighborhood of $e \in \mathbb{G}$ (Prop. 14.1 on page 278); indeed $\mathrm{d}_0\mathrm{Exp} : T_0\mathfrak{g} \to T_e\mathbb{G}$ is the identity map, if we canonically identify $T_0\mathfrak{g}$ and $T_e\mathbb{G}$ with $\mathfrak{g}$ itself.

Remark C.8. Given $x \in \mathbb{G}$ and $X \in \mathrm{Lie}(\mathbb{G})$ one has

$$\gamma(1, X, x) = x * \mathrm{Exp}(X). \tag{C.9}$$

Indeed, unraveling the definitions, we see that

$$\gamma(1, X, x) = \Psi_1^X(x * e) \overset{(C.7)}{=} \tau_x(\Psi_1^X(e)) = x * \mathrm{Exp}(X).$$

In particular, this gives an alternative expression for the flow of a left-invariant vector field on a Lie group:

$$\Psi_t^X = \rho_{\mathrm{Exp}(tX)}, \qquad \text{for every } X \in \mathrm{Lie}(\mathbb{G}) \text{ and } t \in \mathbb{R}. \tag{C.10}$$

This follows from (C.9), and $\gamma(1, tX, x) = \gamma(t, X, x) = \Psi_t^X(x)$; in the first identity we used Prop. 1.14-(3) and the fact that a left invariant v.f. is global. Summing up, we have the identities

$$\gamma(t, X, x) = \Psi_t^X(x) = x * \mathrm{Exp}(tX) = \rho_{\mathrm{Exp}(tX)}(x),$$

holding true for every $t \in \mathbb{R}$, $X \in \mathrm{Lie}(\mathbb{G})$ and $x \in \mathbb{G}$.

Example C.9. Let $M(n, \mathbb{R})$ denote the vector space of the $n \times n$ real matrices; we equip it with its natural smooth structure (as a finite dimensional vector space). As usual, $\mathrm{GL}(n, \mathbb{R})$ denotes the subset of $M(n, \mathbb{R})$ of the non-singular matrices: it is an open subset of $M(n, \mathbb{R})$, namely $\{A \in M(n, \mathbb{R}) : \det A \neq 0\}$, hence a manifold in its own right. The Lie algebra of $\mathrm{GL}(n, \mathbb{R})$, denoted by $\mathfrak{gl}(n, \mathbb{R})$, can be identified via (C.5) with the tangent space to $\mathrm{GL}(n, \mathbb{R})$ at the identity matrix $\mathbb{I}_n$. By (4.1) (page 91) and (C.6), any $A \in M(n, \mathbb{R})$ gives rise to one and only one $X^A \in \mathfrak{gl}(n, \mathbb{R})$ such that $(X^A)_{\mathbb{I}_n} = A$ acting as follows:

$$C^\infty(\mathrm{GL}(n, \mathbb{R})) \ni f \mapsto (X^A)_G f = \frac{\mathrm{d}}{\mathrm{d}s}\Big|_0 f(G + s\,G\,A), \quad G \in \mathrm{GL}(n, \mathbb{R}). \tag{C.11}$$

The mapping $A \mapsto X^A$ is not only a vector-space isomorphism, but it can be proved that it is also a Lie-algebra isomorphism, if $M(n, \mathbb{R})$ is equipped with the usual bracket of matrices $[A, B] = AB - BA$.

Due to (C.11), if both members of the EDO (C.8) defining $\mathrm{Exp}(X^A)$ act on a smooth f, then we get the identity

$$\frac{\mathrm{d}}{\mathrm{d}t}f(\gamma(t)) = \frac{\mathrm{d}}{\mathrm{d}s}\Big|_0\Big\{f(\gamma(t) + s\,\gamma(t)\,A)\Big\}. \tag{C.12}$$

Then, $\gamma(1) = \mathrm{Exp}(X^A)$. If we replace f with the function mapping a matrix into its (i, j)-entry (and we let i, j vary in $\{1, \ldots, n\}$), then (C.12) becomes a genuine matrix ODE, namely $\dot{\gamma}(t) = \gamma(t)\,A$. Bearing in mind the initial condition $\gamma(0) = \mathbb{I}_n$, the solution is $\gamma(t) = e^{tA} = \sum_{k=0}^\infty (tA)^k/k!$. Thus, one can identify the Exponential map $\mathrm{Exp} : \mathfrak{gl}(n, \mathbb{R}) \to \mathrm{GL}(n, \mathbb{R})$ with the usual matrix exponential $A \mapsto e^A$. $\qquad \sharp$

Theorem C.10. *Let* $(\mathbb{G}, *)$ *be a Lie group. Let* $x, y \in \mathbb{G}$ *and suppose there exists a vector field* $Y \in \mathrm{Lie}(\mathbb{G})$ *such that* $\mathrm{Exp}(Y) = y$*. Then*

$$x * y = \gamma(1, Y, x). \tag{C.13}$$

Proof. This is a particular case of (C.9): indeed, we have $x * y = x * \mathrm{Exp}(Y)$ (by hypothesis), and the latter is $\gamma(1, Y, x)$ owing to (C.9). $\qquad\square$

As we proved in Prop. 14.1 (page 278), since Exp is invertible on a suitable neighborhood U of $0 \in \mathrm{Lie}(\mathbb{G})$, then for every $y \in \mathrm{Exp}(U)$ (which is a neighborhood of $e \in \mathbb{G}$) there exists $Y \in U$ such that $y = \mathrm{Exp}(Y)$ and formula (C.13) makes sense for any such y and for every $x \in \mathbb{G}$.

Remark C.11 (Again on Lie group construction). Suppose that Exp is globally invertible, and let us denote its inverse map by $\mathrm{Log} : \mathbb{G} \to \mathrm{Lie}(\mathbb{G})$. Then, by applying Thm. C.10 to the left invariant v.f. $Y = \mathrm{Log}(y)$, we derive

$$x * y = \gamma(1, \mathrm{Log}(y), x), \quad \text{for every } x, y \in \mathbb{G}. \tag{C.14}$$

Now, observe that the rhs of this identity is an "ODE object" in the following sense: the Exp map is obtained by solving a system of ODEs (in fact, by definition, $\mathrm{Exp}(X) = \gamma(1, X, e)$), hence its inverse map Log (when it exists) is completely determined by this ODE system solely; furthermore, $\gamma(1, \mathrm{Log}(y), x)$ is the point of arrival at time 1 of the integral curve of $\mathrm{Log}(y)$ starting at x, another ODE object.

Thus formula (C.14) shows that the group composition $(x, y) \mapsto x * y$ is in some sense an ODE object too. This evocative fact suggests that it may be possible to attach a Lie group structure to suitable classes of Lie algebras of smooth vector fields simply by setting $x * y := \gamma(1, \mathrm{Log}(y), x)$. This has been our view-point in Chap. 17. Note that this construction, based on solutions of ODEs, may be much simpler than the one that we shall describe in Sec. C.1.4, based on solutions of suitable first order PDEs.

C.1.3   *Right invariant vector fields*

Let $(\mathbb{G}, *)$ be a Lie group; a smooth v.f. X on $\mathbb{G}$ is called **right invariant** if X is ρ_x-related to itself, for every right translation ρ_x. One can obtain analogous results for right invariance as in Sec. C.1.1. We prefer to follow an alternative way, showing that right invariant v.f.s are left invariant on another Lie group with the same manifold of $\mathbb{G}$, and isomorphic to $\mathbb{G}$ itself.

Definition C.12 (Inverse group). Let $(\mathbb{G}, *)$ be a Lie group with inversion map $\iota(x) = x^{-1}$. We define a new Lie group, denoted by $\widehat{\mathbb{G}}$ and referred to as the **inverse group** of $\mathbb{G}$, by pushing forward the group structure of $\mathbb{G}$ via the involution ι. Explicitly, if we denote by $\widehat{*}$ the composition of $\widehat{\mathbb{G}}$, this is defined by

$$x \,\widehat{*}\, y := y * x, \qquad x, y \in \mathbb{G}. \tag{C.15}$$

It is a simple exercise to verify that $\widehat{\mathbb{G}}$ is isomorphic to $\mathbb{G}$ and that $\iota : \mathbb{G} \to \widehat{\mathbb{G}}$ is a Lie group isomorphism (see Sec. C.2). Moreover, $\widehat{\mathbb{G}}$ and $\mathbb{G}$ have the same inversion map and identity element. Finally, it is almost tautological that the inverse group of $\widehat{\mathbb{G}}$ is $\mathbb{G}$ itself. Note that identity (C.15) is equivalent to

$$\widehat{\tau_x^*} \equiv \rho_x^*, \qquad \widehat{\rho_y^*} \equiv \tau_y^*. \tag{C.16}$$

These identities prove at once that $X \in \mathfrak{X}(\mathbb{G})$ is left invariant on $\widehat{\mathbb{G}}$ if and only if it is right invariant on $\mathbb{G}$, and vice versa, X is right invariant on $\widehat{\mathbb{G}}$ if and only if it is left invariant on $\mathbb{G}$. This proves at once the following fact:

Proposition C.13. *Let $\mathbb{G}$ be a Lie group. Any left invariant vector field on $\mathbb{G}$ commutes with any right invariant vector field on $\mathbb{G}$.*

Proof. Indeed, suppose that R, L are smooth v.f.s on $\mathbb{G}$, the first is right invariant and the second is left invariant. Hence, by (C.10) and (C.16), one has

$$\Psi_t^R = \widehat{\rho_{\mathrm{Exp}_{\widehat{\mathbb{G}}}(tR)}^*} = \tau_{\mathrm{Exp}_{\widehat{\mathbb{G}}}(tR)}^* \quad \text{and} \quad \Psi_s^L = \rho_{\mathrm{Exp}_{\mathbb{G}}(sL)}^*.$$

Since right and left translations of $\mathbb{G}$ commute, we infer that the flows of R commute with those of L; thus, on account of Thm. 4.25 (page 106) on commuting v.f.s, we deduce that R and L commute. $\qquad\qquad\square$

C.1.4 *Lie's First Theorem*

Let $\mathbb{G} = (\Omega, *)$ be a Lie group on the open set $\Omega \subseteq \mathbb{R}^N$ (with its usual differentiable structure), with identity e. Let $J_1, \dots, J_N$ be the basis of $\mathrm{Lie}(\mathbb{G})$ such that (for any $i = 1, \dots, N$) J_i is the unique left invariant v.f. on $\mathbb{G}$ coinciding at e with $\partial/\partial x_i|_e$. We introduce the notation

$$J(x) := \mathfrak{J}_{\tau_x}(e), \qquad x \in \Omega.$$

Note that $J(x)$ is invertible for every $x \in \Omega$ since τ_x is a smooth diffeomorphism for any x. It is not difficult to recognize that one has

$$\Big(J_1(x) \cdots J_N(x)\Big) = J(x). \tag{C.17}$$

By applying identity (C.2) to $J_1, \dots, J_N$ we obtain at once the matrix identity

$$\mathfrak{J}_{\tau_x}(y)\, J(y) = J(x * y), \qquad x, y \in \Omega.$$

This can also be obtained by differentiating the identity expressing associativity, that is $x * (y * z) = (x * y) * z$, with respect to z at $z = e$.

In its turn, this produces the following identity, often referred to as **Lie's First Theorem** (see e.g., [Cohn (1957), p. 95]):

$$\mathfrak{J}_{\tau_x}(y) = J(\tau_x(y)) \left(J(y)\right)^{-1}, \qquad x, y \in \Omega. \tag{C.18}$$

Note that, roughly put, given the matrix-valued function $y \mapsto J(y)$, if we fix an arbitrary $x \in \Omega$ as a parameter, and if we set $f(y) := \tau_x(y)$, (C.18) is a *system of first order PDEs* in the indeterminate functions $f_1(y), \ldots, f_N(y)$:

$$\mathcal{J}_f(y) = J(f(y)) \left(J(y) \right)^{-1}.$$

In some cases, the knowledge of the sole v.f.s $J_1, \ldots, J_N$ and of the neutral element e (giving the initial condition $f(e) = \tau_x(e) = x$) allows to solve these PDEs: in this fashion, *we can construct a group operation starting from a set of v.f.s* (which turn out to form the basis of the associated Lie algebra) and from the choice of a point (the candidate neutral element). For an example of this construction, see Exm. C.14.

We have exhaustively studied the problem of the construction of a Lie group on $\mathbb{R}^N$, starting from vector fields satisfying suitable conditions, in Chap. 17. See also Rem. C.11. Thus, *Lie's First Theorem gives yet another way to obtaining a Lie group starting from its candidate Lie algebra.*

Example C.14. Suppose that we know that the vector fields

$$J_1 = \partial_1, \qquad\qquad J_2 = \partial_2 + x_1 \partial_3 + \frac{x_1^2}{2} \partial_4$$

$$J_3 = \partial_3 + x_1 \partial_4, \qquad J_4 = \partial_4$$

are the v.f.s of the Jacobian basis (see Exm. C.27-(2) for this notion) of some group $\mathbb{G} = (\mathbb{R}^4, *)$ with identity $e = 0$, *but we ignore this Lie group.* Fixing $x \in \mathbb{R}^4$, the group composition $f(y) := x * y$ is a solution of the system of PDEs (C.18), with initial condition $f(0) = x$. With the notation in (C.17), since

$$J(x) = \begin{pmatrix} 1 & 0 & 0 & 0 \\ 0 & 1 & 0 & 0 \\ 0 & x_1 & 1 & 0 \\ 0 & \frac{x_1^2}{2} & x_1 & 1 \end{pmatrix},$$

after some simple computations, system (C.18) turns out to be

$$\mathcal{J}_f(y) = \begin{pmatrix} 1 & 0 & 0 & 0 \\ 0 & 1 & 0 & 0 \\ 0 & f_1(y) - y_1 & 1 & 0 \\ 0 & \dfrac{(f_1(y) - y_1)^2}{2} & f_1(y) - y_1 & 1 \end{pmatrix}.$$

The system for the gradient of f_1 is therefore

$$\partial_1 f_1(y) = 1, \quad \partial_2 f_1(y) = 0, \quad \partial_3 f_1(y) = 0, \quad \partial_4 f_1(y) = 0.$$

It is simple to solve this system: taking into account that $f_1(0) = x_1$, one obtains $f_1(y) = x_1 + y_1$. The system for ∇f_2 is analogous and it leads to $f_2(y) = x_2 + y_2$. Taking into account the value of $f_1(y)$, the system for ∇f_3 is

$$\partial_1 f_3(y) = 0, \; \partial_2 f_3(y) = x_1, \; \partial_3 f_3(y) = 1, \; \partial_4 f_3(y) = 0, \quad f_3(0) = x_3.$$

Simple computations give $f_3(y) = x_3 + y_3 + x_1\, y_2$. Finally, for ∇f_4 we have
$$\partial_1 f_4(y) = 0, \ \partial_2 f_4(y) = x_1^2/2, \ \partial_3 f_4(y) = x_1, \ \partial_4 f_4(y) = 1, \quad f_4(0) = x_4.$$
Simple computations give $f_4(y) = x_4 + y_4 + \frac{1}{2}\, x_1^2 y_2 + x_1 y_3$. The group composition is therefore *necessarily* given by
$$x * y = f(y) = \left(x_1 + y_1, x_2 + y_2, x_3 + y_3 + x_1 y_2, x_4 + y_4 + \tfrac{1}{2}\, x_1^2 y_2 + x_1 y_3 \right).$$
The group $\mathbb{G} = (\mathbb{R}^4, *)$ is the so-called **Engel's group**. For another example of the reconstruction of the group starting from a set of v.f.s, see Exr. C.15. ♯

C.2 Homomorphisms

Let $(\mathbb{G}, *)$, $(\mathbb{F}, \star)$ be Lie groups; a map $\varphi : \mathbb{G} \to \mathbb{F}$ is called a (Lie-group) **homomorphism** (or briefly, a **morphism**) if φ is of class C^∞ and it is also a group homomorphism (in the algebraic sense):
$$\varphi(x * y) = \varphi(x) \star \varphi(y), \quad \text{for every } x, y \in \mathbb{G}.$$
A restatement of this is
$$\varphi \circ \tau_x^{\mathbb{G}} = \tau_{\varphi(x)}^{\mathbb{F}} \circ \varphi, \quad \text{for any } x \in \mathbb{G}. \tag{C.19}$$
If φ is not only a Lie group morphism, but also a diffeomorphism of $\mathbb{G}$ onto $\mathbb{F}$, φ is called a Lie-group isomorphism; in this case we say that $\mathbb{G}$ and $\mathbb{F}$ are isomorphic; we speak of an **automorphism** if $\mathbb{G} = \mathbb{F}$. For example, if $x \in \mathbb{G}$, the **conjugation** C_x by x (i.e., the map $y \mapsto C_x(y) = x * y * x^{-1}$) is an automorphism of $\mathbb{G}$.

Example C.15 (One-parameter subgroups). A one-parameter subgroup of a Lie group $\mathbb{G}$ is any Lie group homomorphism $\gamma : \mathbb{R} \to \mathbb{G}$, where $\mathbb{R}$ is considered as a Lie group under addition. It is easy to show by differentiating (wrt s at $s = 0$) the identity $\gamma(t) * \gamma(s) = \gamma(t + s)$ that a one-parameter subgroup of $\mathbb{G}$ is the maximal integral curve of a left-invariant v.f. on $\mathbb{G}$; the converse is also very simple to show, owing to the semigroup property of the flows.

On account of Exm. C.9, the one-parameter subgroups of $\mathrm{GL}(n, \mathbb{R})$ are the maps $t \mapsto e^{tA}$, with $A \in \mathfrak{gl}(n, \mathbb{R}) \equiv M(n, \mathbb{R})$. ♯

Remark C.16. *Every Lie group homomorphism has constant rank.*[11] Indeed, taking the differentials at $e_{\mathbb{G}}$ of both sides of (C.19) gives
$$\mathrm{d}_x \varphi \circ \mathrm{d}_{e_{\mathbb{G}}} \tau_x^{\mathbb{G}} = \mathrm{d}_{e_{\mathbb{F}}} \tau_{\varphi(x)}^{\mathbb{F}} \circ \mathrm{d}_{e_{\mathbb{G}}} \varphi,$$
so that $\mathrm{d}_x \varphi$ has the same rank as $\mathrm{d}_{e_{\mathbb{G}}} \varphi$, since the composition by linear isomorphisms does not alter the rank. As a consequence of the Global Rank Theorem,[12] *a bijective Lie group homomorphism is a diffeomorphism.* ♯

---

[11] A smooth map $F : M \to N$ (of smooth manifolds M, N) is said to have constant rank if the ranks of the linear maps $\mathrm{d}_x F : T_x M \to T_{F(x)} N$ are constant as x varies in M; it is understood that the rank of a linear map $L : V \to W$ between vector spaces V, W is the dimension of $L(V)$ as a subspace of W.

[12] See [Lee (2013), Theorem 4.14]; this theorem states that, if $F : M \to N$ is a smooth map having constant rank (in the sense of the previous footnote), then:

    (1) if F is surjective, it is a smooth submersion (i.e., $\mathrm{d}_x F$ is surjective for any x);

    (2) if F is injective, it is a smooth immersion (i.e., $\mathrm{d}_x F$ is injective for any x);

    (3) if F is bijective, it is a smooth diffeomorphism.

Let $\mathbb{G}$ and $\mathbb{H}$ be Lie groups with neutral elements $e_{\mathbb{G}}$ and $e_{\mathbb{H}}$, respectively; let $\varphi : \mathbb{G} \to \mathbb{H}$ be a Lie group morphism. Since $\varphi(e_{\mathbb{G}}) = e_{\mathbb{H}}$, we have

$$d_{e_{\mathbb{G}}}\varphi(X_{e_{\mathbb{G}}}) \in T_{e_{\mathbb{H}}}\mathbb{H},$$

for every v.f. X on $\mathbb{G}$, and in particular for any $X \in \mathrm{Lie}(\mathbb{G})$. According to (C.5), we can associate with φ a linear map from $\mathrm{Lie}(\mathbb{G})$ to $\mathrm{Lie}(\mathbb{H})$, which we denote by $d\varphi$.

Definition C.17 (Differential of a homomorphism). In the above notations, we define $d\varphi : \mathrm{Lie}(\mathbb{G}) \to \mathrm{Lie}(\mathbb{H})$ as the unique linear function sending $X \in \mathrm{Lie}(\mathbb{G})$ to the vector field $d\varphi X$ which is the unique element of $\mathrm{Lie}(\mathbb{H})$ satisfying

$$(d\varphi X)_{e_{\mathbb{H}}} = d_{e_{\mathbb{G}}}\varphi(X_{e_{\mathbb{G}}}).$$

Observe that $d\varphi$ only depends on $d_{e_{\mathbb{G}}}\varphi$; thereby, if φ and ψ are Lie-group morphisms with the same differential at $e_{\mathbb{G}}$, then $d\varphi$ and $d\psi$ coincide.

Remark C.18. As a consequence, by the very definition of $d\varphi$, and bearing in mind Rem. C.2, if $X \in \mathrm{Lie}(\mathbb{G})$, then $d\varphi X$ is uniquely defined by

$$(d\varphi X)_y = d_{e_{\mathbb{G}}}(\tau_y^{\mathbb{H}} \circ \varphi)(X_{e_{\mathbb{G}}}), \quad \text{for every } y \in \mathbb{H}.$$

If we take $y = \varphi(x)$ (with $x \in \mathbb{G}$) and we take into account (C.19), we infer

$$(d\varphi X)_{\varphi(x)} = d_{e_{\mathbb{G}}}(\varphi \circ \tau_x^{\mathbb{G}})(X_{e_{\mathbb{G}}}) = d_x\varphi\Big(d_{e_{\mathbb{G}}}\tau_x^{\mathbb{G}}(X_{e_{\mathbb{G}}})\Big) \overset{(C.3)}{=} d_x\varphi(X_x),$$

for every $x \in \mathbb{G}$. This means that X and $d\varphi X$ are φ-related; this is the reason of the pushforward-type notation '$d\varphi$' (applicable even if φ may not be a diffeomorphism). This abused notation will cause no confusion. ♯

We have proved the following fact.

Theorem C.19. *Let $\mathbb{G}$ and $\mathbb{H}$ be Lie groups and let $\mathfrak{g}$ and $\mathfrak{h}$ be, respectively, their Lie algebras. Moreover, let $\varphi : \mathbb{G} \to \mathbb{H}$ be a morphism of Lie groups; finally let $d\varphi : \mathfrak{g} \to \mathfrak{h}$ be as in Def. C.17. Then we have the following facts:*

(i) *for every $X \in \mathfrak{g}$, the vector fields X and $d\varphi X$ are φ-related, i.e.,*

$$(d\varphi X)_{\varphi(x)} = d_x\varphi(X_x), \quad \text{for every } x \in \mathbb{G}. \tag{C.20}$$

Condition (C.20) characterizes $d\varphi X$, that is, $d\varphi X$ is the unique left invariant vector field on $\mathbb{H}$ which is φ-related to X;

(ii) *$d\varphi : \mathfrak{g} \to \mathfrak{h}$ is a morphism of Lie algebras.*

Statement (ii) stems from (i) and the consistency of φ-relatedness in Prop. 4.6 (page 94). If $\varphi : \mathbb{G} \to \mathbb{H}$ is an isomorphism of Lie groups, then $d\varphi$ as defined in Def. C.17 coincides with the pushforward φ_* (also denoted by $d\varphi$) introduced in Def. 4.3 on page 92 (see also Rem. 4.4).

The next lemma shows that the correspondence $\varphi \mapsto d\varphi$ is natural.

Lemma C.20 (Functorial properties of $d\varphi$). *The following facts hold true:*

(1) If $\mathbb{G}$ is a Lie group with Lie algebra $\mathfrak{g}$, then $\mathrm{d}(\mathrm{id}_{\mathbb{G}}) = \mathrm{id}_{\mathfrak{g}}$.

(2) If $\varphi : \mathbb{G} \to \mathbb{H}$ and $\psi : \mathbb{H} \to \mathbb{F}$ are Lie-group morphisms, $\mathrm{d}(\psi \circ \varphi) = \mathrm{d}\psi \circ \mathrm{d}\varphi$.

(3) If $\varphi : \mathbb{G} \to \mathbb{H}$ is an isomorphism of Lie groups, then $\mathrm{d}\varphi$ is an isomorphism of Lie algebras and $(\mathrm{d}\varphi)^{-1} = \mathrm{d}(\varphi^{-1})$.

The interaction of the Lie-group morphisms and the Exp map is also as natural as one may expect:

Proposition C.21. *Let $\varphi : \mathbb{G} \to \mathbb{H}$ be a morphism of Lie groups. Then we have*

$$\varphi\big(\mathrm{Exp}_{\mathbb{G}}(X)\big) = \mathrm{Exp}_{\mathbb{H}}(\mathrm{d}\varphi(X)), \quad \text{for every } X \in \mathrm{Lie}(\mathbb{G}).$$

In other words, the following diagram is commutative:

$$
\begin{array}{ccc}
\mathbb{G} & \xrightarrow{\;\;\varphi\;\;} & \mathbb{H} \\[4pt]
{\scriptstyle \mathrm{Exp}_{\mathbb{G}}}\big\uparrow & & \big\uparrow{\scriptstyle \mathrm{Exp}_{\mathbb{H}}} \\[4pt]
\mathrm{Lie}(\mathbb{G}) & \xrightarrow{\;\;\mathrm{d}\varphi\;\;} & \mathrm{Lie}(\mathbb{H})
\end{array}
\tag{C.21}
$$

The proof of this result follows from Lem. 4.8 (page 94), the relatedness of X and $\mathrm{d}\varphi X$, and the fact that $\mathrm{Exp}(X) = \Psi_1^X(e)$ when X is left invariant.

We close this section with the following profound result, showing that any Lie group morphism is completely characterized by its differential.

Theorem C.22. *Let $\mathbb{G}$ and $\mathbb{H}$ be Lie groups, with associated Lie algebras $\mathfrak{g}$ and $\mathfrak{h}$. Then the following facts hold true:*

(1) Suppose that $\mathbb{G}$ is connected; assume that $\varphi, \psi : \mathbb{G} \to \mathbb{H}$ are Lie group morphisms such that $\mathrm{d}\varphi, \mathrm{d}\psi : \mathfrak{g} \to \mathfrak{h}$ are coincident; then $\varphi = \psi$.

(2) If $\mathbb{G}$ is connected and simply connected, and if $T : \mathfrak{g} \to \mathfrak{h}$ is a Lie algebra morphism, there exists a unique Lie group morphism $\varphi : \mathbb{G} \to \mathbb{H}$ such that $\mathrm{d}\varphi = T$.

(3) Suppose $\mathbb{G}$ and $\mathbb{H}$ are both connected and simply connected; if $\mathfrak{g}$ and $\mathfrak{h}$ are isomorphic Lie algebras, then $\mathbb{G}$ and $\mathbb{H}$ are isomorphic Lie groups.

Remark C.23. If $\mathbb{G}$ is simply connected, the unique homomorphism $\varphi : \mathbb{G} \to \mathbb{G}$ whose differential is the identity map of $\mathfrak{g}$ is the identity map of $\mathbb{G}$. ♯

For the proof of (2), see [Lee (2013), Theorem 20.19]; (3) is a simple consequence of (2) (and Rem. C.23). As for (1), we argue as follows: with the notation in (1), Prop. C.21 ensures that $\varphi \circ \mathrm{Exp}_{\mathbb{G}} \equiv \psi \circ \mathrm{Exp}_{\mathbb{G}}$; in particular φ and ψ coincide on an open neighborhood of e, hence they coincide throughout (by Lem. C.1).

Remark C.24. In the spirit of this book (indeed we are about to use the CBHD Theorem), we can give a proof of the *local* construction of φ in (2) of Thm. C.22. Indeed, if $T : \mathfrak{g} \to \mathfrak{h}$ is a Lie algebra morphism, we can perform the following (local) construction; we use the notations $(\mathbb{G}, *)$, $(\mathbb{H}, \star)$ and Exp and Log with subindices $\mathbb{G}$ or $\mathbb{H}$ with the obvious meanings (se Fig. C.3):

- by the CBHD Theorem for Lie groups (Thm. 14.6 on page 281), there exists a neighborhood $\mathfrak{D}_1$ of $0 \in \mathfrak{h}$ such that the CBHD series $h \diamond h' = \sum_n Z_n(h, h')$ converges on $\mathfrak{D}_1$ and such that

$$\mathrm{Exp}_{\mathbb{H}}(h \diamond h') = \mathrm{Exp}_{\mathbb{H}}(h) \star \mathrm{Exp}_{\mathbb{H}}(h'), \quad h, h' \in \mathfrak{D}_1;$$

- we can find a neighborhood $\mathcal{E}_1$ of $0 \in \mathfrak{h}$ such that $\mathcal{E}_1 \subseteq \mathfrak{D}_1$, and such that $\mathrm{Exp}_{\mathbb{H}}|_{\mathcal{E}_1}$ is injective with inverse $\mathrm{Log}_{\mathbb{H}} : E_1 \to \mathcal{E}_1$, where $E_1 := \mathrm{Exp}_{\mathbb{H}}(\mathcal{E}_1)$;

- arguing as above, there exists a neighborhood $\mathfrak{D}_2$ of $0 \in \mathfrak{g}$ such that the CBHD series $g \diamond g' = \sum_n Z_n(g, g')$ converges on $\mathfrak{D}_2$ and such that

$$\mathrm{Exp}_{\mathbb{G}}(g \diamond g') = \mathrm{Exp}_{\mathbb{G}}(g) * \mathrm{Exp}_{\mathbb{G}}(g'), \quad g, g' \in \mathfrak{D}_2;$$

- since T is a linear map of finite-dimensional vector spaces, it is continuous; hence there exists an open Ω with $0 \in \Omega \subseteq \mathfrak{D}_2$ and $T(\Omega) \subseteq \mathcal{E}_1$;

- we can find an open $\mathcal{E}_2$ such that $0 \in \mathcal{E}_2 \subseteq \Omega$, and such that $\mathrm{Exp}_{\mathbb{G}}|_{\mathcal{E}_2}$ is injective with inverse $\mathrm{Log}_{\mathbb{G}} : E_2 \to \mathcal{E}_2$, where $E_2 := \mathrm{Exp}_{\mathbb{G}}(\mathcal{E}_2)$.

Now we define the map

$$\varphi : E_2 \subseteq \mathbb{G} \longrightarrow E_1 \subseteq \mathbb{H}, \quad \varphi := \mathrm{Exp}_{\mathbb{H}} \circ T \circ \mathrm{Log}_{\mathbb{G}}.$$

This is clearly inspired by the diagram (C.21). *We claim that φ behaves like a Lie-group morphism in E_1, and the differential of φ at $e_{\mathbb{G}}$ coincides with T.* Thus φ is a local[13] solution of (2) in Thm. C.22.

The second statement of the claim is quite simple, since, up to the identifications $T_0 \mathfrak{g} \equiv \mathfrak{g}$, $T_0 \mathfrak{h} \equiv \mathfrak{h}$ and $\mathrm{d}_0 T \equiv T$ (as T is linear), one has

$$\mathrm{d}_{e_{\mathbb{G}}} \varphi = \mathrm{d}_0 \mathrm{Exp}_{\mathbb{H}} \circ \mathrm{d}_0 T \circ \mathrm{d}_{e_{\mathbb{G}}} \mathrm{Log}_{\mathbb{G}} = T.$$

The first statement of the claim is a very nice computation involving the CBHD series: indeed, if $g, g' \in E_2$ we have

$$\varphi(g * g') = \mathrm{Exp}_{\mathbb{H}}\left(T\left(\mathrm{Log}_{\mathbb{G}}(g * g')\right)\right) = \mathrm{Exp}_{\mathbb{H}}\left(T\left(\mathrm{Log}_{\mathbb{G}}(g) \diamond \mathrm{Log}_{\mathbb{G}}(g')\right)\right)$$

$$= \mathrm{Exp}_{\mathbb{H}}\left(T\left(\mathrm{Log}_{\mathbb{G}}(g) + \mathrm{Log}_{\mathbb{G}}(g')\right) + \tfrac{1}{2}[\mathrm{Log}_{\mathbb{G}}(g), \mathrm{Log}_{\mathbb{G}}(g')] + \cdots\right)$$

(T is linear, continuous and a Lie-algebra morphism)

$$= \mathrm{Exp}_{\mathbb{H}}\left(T\mathrm{Log}_{\mathbb{G}}(g) + T\mathrm{Log}_{\mathbb{G}}(g') + \tfrac{1}{2}[T\mathrm{Log}_{\mathbb{G}}(g), T\mathrm{Log}_{\mathbb{G}}(g')] + \cdots\right)$$

$$= \mathrm{Exp}_{\mathbb{H}}\left((T\mathrm{Log}_{\mathbb{G}}(g)) \diamond (T\mathrm{Log}_{\mathbb{G}}(g'))\right)$$

$$= \mathrm{Exp}_{\mathbb{H}}(T\mathrm{Log}_{\mathbb{G}}(g)) \star \mathrm{Exp}_{\mathbb{H}}(T\mathrm{Log}_{\mathbb{G}}(g')) = \varphi(g) \star \varphi(g').$$

---

[13]It is simple imagining how a globalization of φ to $\mathbb{G}$ should look like: see Lem. C.1; however, the well posedness of this prolongation is non-trivial. Simply connectedness of $\mathbb{G}$ must play a role here.

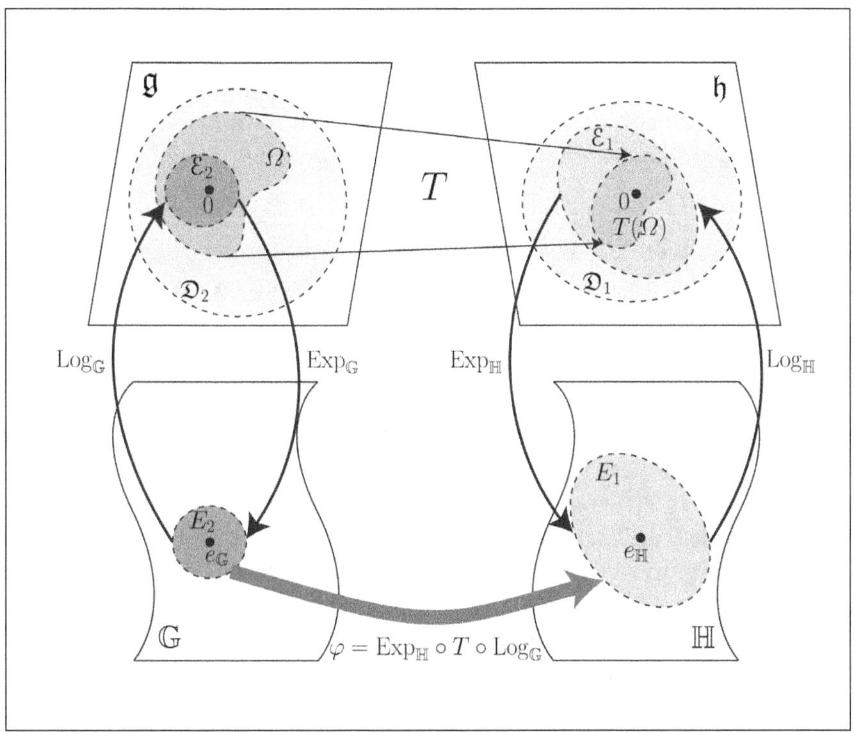

Fig. C.3   The construction in Rem. C.24: given a morphism $T : \mathfrak{g} \to \mathfrak{h}$ (with $\mathfrak{g} = \mathrm{Lie}(\mathbb{G}), \mathfrak{h} = \mathrm{Lie}(\mathbb{H})$), we construct, via the CBHD operation, a local Lie-group morphism φ satisfying $\mathrm{d}\varphi = T$.

Note that in the fifth equality we have used the universality of the summands Z_n of the CBHD series as defined on $\mathfrak{g}$ or on $\mathfrak{h}$, indifferently. ♯

We end the section with a short review of Lie group actions. Let $(\mathbb{G}, *)$ be a Lie group and let M be a smooth manifold. Let $\theta : \mathbb{G} \times M \to M$ be a map and, given $g \in \mathbb{G}$, we set

$$\theta_g : M \to M, \quad \theta_g(x) := \theta(g, x) \quad \text{for any } x \in M.$$

If θ is smooth, we say that it is a (smooth) **left action of $\mathbb{G}$ on M** iff

$$\theta_e = \mathrm{id}_M, \qquad \theta_g \circ \theta_{g'} = \theta_{g*g'} \quad \forall\ g, g' \in \mathbb{G}.$$

For each $x \in M$, the set $\{\theta_g(x) : g \in \mathbb{G}\}$ is called the orbit of x.

Example C.25. Let X be a smooth v.f. on the manifold M; we suppose X is global. Then the flow map

$$\theta : \mathbb{R} \times M \to M \quad \theta(t, x) := \Psi_t^X(x)$$

defines a smooth left action of the additive group $(\mathbb{R}, +)$ on M. The orbit of $x \in M$ is the trajectory of the maximal integral curve of X starting at x. ♯

Suppose that, along with $(\mathbb{G}, *)$, we have another Lie group $(\mathbb{H}, \star)$ and that $\theta : \mathbb{G} \times \mathbb{H} \to \mathbb{H}$ is a smooth left action of $\mathbb{G}$ on (the manifold of) $\mathbb{H}$. If, for every $g \in \mathbb{G}$, the map $\theta_g : \mathbb{H} \to \mathbb{H}$ is an automorpshim of the group $(\mathbb{H}, \star)$, we say that θ is an **action by automorphisms**. If θ is an action by automorphisms, it is not difficult to prove that the following binary operation on $\mathbb{G} \times \mathbb{H}$

$$(g, h) \bullet (g', h') := (g * g', h \star \theta_g(h')) \qquad (g, g' \in \mathbb{G}, \ h, h' \in \mathbb{H})$$

defines a Lie group on $\mathbb{G} \times \mathbb{H}$, called the **semidirect product** of $\mathbb{G}$ and $\mathbb{H}$, denoted by $\mathbb{G} \rtimes_\theta \mathbb{H}$. The identity of $\mathbb{G} \rtimes_\theta \mathbb{H}$ is $(e_\mathbb{G}, e_\mathbb{H})$, and the inversion is

$$(g, h) \mapsto (g^{-1}, \theta_{g^{-1}}(h^{-1})).$$

Example C.26. Let B be a fixed real $n \times n$ matrix. We consider the vector field $\mathbf{B}$ on $\mathbb{R}^n$ whose coefficient vector at $x \in \mathbb{R}^n$ is Bx. From Exm. 1.19 (page 13) we know that the flow of $\mathbf{B}$ is $\Psi_t^\mathbf{B}(x) = e^{tB}x$.

On account of Exm. C.25, we know that $\theta(t, x) = e^{tB}x$ is a smooth left action of the additive group $\mathbb{G} = (\mathbb{R}, +)$ on $\mathbb{R}^n$. Moreover, If we consider the additive group $\mathbb{H} = (\mathbb{R}^n, +)$, we recognize that θ is an action by automorphisms of $\mathbb{G}$ on $\mathbb{H}$, since the map $x \mapsto e^{tB}x$ is linear. Consequently, the Lie group operation (which we denote by $*_B$) of the semidirect product $\mathbb{G} \rtimes_\theta \mathbb{H}$ is

$$(t, x) *_B (t', x') = (t + t', \ x + \exp(t\,B)\,x'),$$

where $t, t' \in \mathbb{R}$ and $x, x' \in \mathbb{R}^n$. The class of the Lie groups of the form $(\mathbb{R}^{1+n}, *_B)$ has interesting connections with Ornstein-Uhlenbeck and Kolmogorov-Fokker-Planck PDOs, and it has been studied in [Bonfiglioli and Lanconelli (2012b)].   ♯

C.3   A few examples

Since they often intervene in the Analysis of linear PDOs, in this section we collect some explicit examples of Lie groups whose underlying manifold is an open subset Ω of Euclidean space $\mathbb{R}^N$. Thus, a single chart (the identity map!) covers the whole manifold, and chart-computations become particularly effective. We leave it as an exercise the verification of what is stated in the next example.

Example C.27. In the following statements, we suppose that the open set $\Omega \subseteq \mathbb{R}^N$ (with its usual differentiable structure) is equipped with a Lie group structure by the operation $*$, and we set $\mathbb{G} = (\Omega, *)$ and $\mathfrak{g} := \mathrm{Lie}(\mathbb{G})$. We use the standard notation e for the identity of $\mathbb{G}$.

(1) $X \in \mathfrak{X}(\Omega)$ belongs to $\mathfrak{g}$ iff one of the following matrix identities hold:

$$J_{\tau_x}(y)\,X(y) = X(x * y), \quad \text{or equivalently} \quad J_{\tau_x}(e)\,X(e) = X(x),$$

for any $x, y \in \Omega$. Given any $\eta \in \mathbb{R}^N$ there exists (a unique) left invariant v.f. X such that $X(e) = \eta$; this is precisely the v.f. associated with $J_{\tau_x}(e)\,\eta$.

(2) A (linear) basis for $\mathfrak{g}$ is given by the N v.f.s associated with the N column vectors of the matrix $\mathcal{J}_{\tau_x}(e)$, the Jacobian matrix of the left translation τ_x at the identity e of $\mathbb{G}$. This basis, say $\mathcal{J} = \{J_1, \ldots, J_N\}$, is called **the Jacobian basis**. This is a non-canonical, chart-dependent notion, meaningful in this setting where we can work in the global chart given by the standard coordinates of $\mathbb{R}^N$. J_i acts in the following way

$$J_i f(x) = \frac{\partial}{\partial y_i}\Big|_{y=e} \{f(x*y)\}, \qquad \text{for } f \in C^\infty(\Omega) \text{ and } x \in \Omega.$$

(3) By Cor. 14.3 on page 280, the structure constants of $\mathfrak{g}$ wrt $\mathcal{J}$ are

$$[J_i, J_j] = \sum_{h=1}^{N} \left(\frac{\partial^2 m_h(e,e)}{\partial \alpha_i \, \partial \beta_j} - \frac{\partial^2 m_h(e,e)}{\partial \alpha_j \, \partial \beta_i} \right) J_h,$$

for any $i, j \in \{1, \ldots, N\}$, where $m(\alpha, \beta) := \alpha * \beta$.

(4) Important examples of Hörmander vector fields are given in the setting of Lie groups of the type $\mathbb{G} = (\Omega, *)$. Indeed, let $X := \{X_1, \ldots, X_m\} \subseteq \mathfrak{g}$ be a set of Lie-generators of $\mathfrak{g}$, i.e., $\mathrm{Lie}\{X\} = \mathfrak{g}$. Then X is a Hörmander system of smooth vector fields on Ω. In particular, this is true of any linear basis of $\mathfrak{g}$. Vice versa, if $X = \{X_1, \ldots, X_m\}$ is a Hörmander system of left-invariant v.f.s on $\mathbb{G}$, then X is a set of Lie-generators of $\mathfrak{g}$.

(5) Let $\mathbb{G} = (\Omega, *)$ and $\mathbb{H} = (\Omega', \star)$ be Lie groups (with Ω, Ω' open sets in $\mathbb{R}^n$ and $\mathbb{R}^m$, resp.), with neutral elements $e_{\mathbb{G}}$ and $e_{\mathbb{H}}$, and let $\varphi : \mathbb{G} \to \mathbb{H}$ be a morphism of Lie groups. Moreover, let $\mathcal{J}$ and $\mathcal{Z}$ be the associated Jacobian bases of $\mathrm{Lie}(\mathbb{G})$ and $\mathrm{Lie}(\mathbb{H})$, respectively. Then, the matrix of the linear map $\mathrm{d}\varphi$ (w.r.t. the bases $\mathcal{J}$ and $\mathcal{Z}$) is the Jacobian matrix of φ at $e_{\mathbb{G}}$.

(6) As for the left invariant v.f.s, one can define the Jacobian basis of the right invariant v.f.s on a Lie group of the form $(\Omega, *)$. Namely, the i-th element of this basis is the vector field, R_i say, associated with the i-th column of $\mathcal{J}_{\rho_x}(e)$; with this notation one has

$$R_i f(x) = \frac{\partial}{\partial z_i}\Big|_{z=e} \{f(z*x)\}, \qquad \text{for } f \in C^\infty(\Omega) \text{ and } x \in \Omega.$$

Example C.28. $\mathbb{R}^3$ is endowed with the structure of a Lie group, say $\mathbb{H}^1$, by

$$x * y = \left(x_1 + y_1, \quad x_2 + y_2, \quad x_3 + y_3 + 2x_2 y_1 - 2x_1 y_2\right).$$

$\mathbb{H}^1$ is called the **Heisenberg group** (or the *first* Heisenberg group). The identity element is 0, whilst the inversion map is given by $\iota(x) = -x$. Since

$$\mathcal{J}_{\tau_x}(0) = \begin{pmatrix} 1 & 0 & 0 \\ 0 & 1 & 0 \\ 2x_2 & -2x_1 & 1 \end{pmatrix},$$

the v.f.s $\partial_1 + 2x_2\partial_3$, $\partial_2 - 2x_1\partial_3$, ∂_3 form a basis for $\mathrm{Lie}(\mathbb{H}^1)$ (see Exm. C.27-(2)). Hence, attached to this group we can consider the left-invariant operator

$$\mathcal{L}_1 = \left(\frac{\partial}{\partial x_1} + 2x_2 \frac{\partial}{\partial x_3}\right)^2 + \left(\frac{\partial}{\partial x_2} - 2x_1 \frac{\partial}{\partial x_3}\right)^2,$$

the so-called *Kohn-Laplacian* on first Heisenberg group. ♯

Example C.29. $\mathbb{R}^3$ is endowed with the structure of a Lie group by

$$x * y = \big(x_1 + y_1, \quad x_2 + y_2 - x_1 y_3, \quad x_3 + y_3\big).$$

The famous Kolmogorov's operator in $\mathbb{R}^3$, namely

$$\mathcal{K} = \Big(\frac{\partial}{\partial x_1}\Big)^2 - \Big(\frac{\partial}{\partial x_3} - x_1 \frac{\partial}{\partial x_2}\Big),$$

is left-invariant on $(\mathbb{R}^3, *)$. The identity element is 0, whilst the inversion map is given by $\iota(x) = (-x_1, -x_2 - x_1 x_3, -x_3)$. This group is isomorphic to the one in Exm. C.28, as the reader can check (Exr. C.3). ♮

Example C.30. $\mathbb{R}^3$ is endowed with the structure of a Lie group by the map

$$x * y = \big(x_1 + y_1, \quad x_2 + y_2 \cos x_1 - y_3 \sin x_1, \quad x_3 + y_2 \sin x_1 + y_3 \cos x_1\big).$$

The identity element is 0, whilst

$$\iota(x) = \big(-x_1, -x_2 \cos x_1 - x_3 \sin x_1, x_2 \sin x_1 - x_3 \cos x_1\big).$$

If, in place of $\mathbb{R}^3$, we consider $\mathbb{S}^1 \times \mathbb{R}^2$, and we denote the points of the latter by (θ, x) (with $\theta \in \mathbb{S}^1$ and $x = (x_1, x_2) \in \mathbb{R}^2$), we can rewrite $*$ as

$$(\theta, x) * (\theta', x')$$
$$= \big(\theta + \theta', \quad x_1 + x_1' \cos \theta - x_2' \sin \theta, \quad x_2 + x_1' \sin \theta + x_2' \cos \theta\big)$$
$$= \big(\theta + \theta', x + R(\theta)x'\big), \qquad \text{where } R(\theta) = \begin{pmatrix} \cos \theta & -\sin \theta \\ \sin \theta & \cos \theta \end{pmatrix}.$$

The latter is called the **roto-translation group**. The following PDO, often referred to as Mumford's operator, is left-invariant on $(\mathbb{S}^1 \times \mathbb{R}^2, *)$:

$$\mathcal{M} = \Big(\frac{\partial}{\partial \theta}\Big)^2 - \Big(\cos \theta \frac{\partial}{\partial x_1} + \sin \theta \frac{\partial}{\partial x_2}\Big).$$

Example C.31. Given $n \in \mathbb{N}$, we denote the points of $\mathbb{R}^{2n+1}$ by (x, y, t), where $x, y \in \mathbb{R}^n$ and $t \in \mathbb{R}$. Consider the operation

$$(x, y, t) * (x', y', t') = \big(x + x', y + y', t + t' + 2\langle y, x'\rangle - 2\langle x, y'\rangle\big),$$

where $\langle \cdot, \cdot \rangle$ is the usual inner product of $\mathbb{R}^n$. $\mathbb{H}^n := (\mathbb{R}^{2n+1}, *)$ is a Lie group, called the **Heisenberg group on** $\mathbb{R}^{2n+1}$ (or the n-th Heisenberg group). The identity of $\mathbb{H}^n$ is 0 and the inversion is $-x$. The Jacobian basis of $\mathrm{Lie}(\mathbb{H}^n)$ is

$$X_1 = \partial_{x_1} + 2\,y_1\,\partial_t, \qquad \ldots \qquad X_n = \partial_{x_n} + 2\,y_n\,\partial_t,$$
$$Y_1 = \partial_{y_1} - 2\,x_1\,\partial_t, \qquad \ldots \qquad Y_n = \partial_{y_n} - 2\,x_n\,\partial_t, \qquad T = \partial_t.$$

The only non-trivial commutators are given by $[X_i, Y_i] = -4T$, for $i \leq n$. A remarkably important left-invariant operator on $\mathbb{H}^n$ is

$$\mathcal{L}_n = \sum_{j=1}^{n} (X_j^2 + Y_j^2),$$

the so-called *Kohn-Laplacian* on the n-th Heisenberg group. ♮

The operators $\mathcal{L}_1, \mathcal{K}, \mathcal{M}, \mathcal{L}_n$ in the previous examples
are semielliptic operators (in the sense of Chap. 8);
they are also *Hörmander operators* (in the sense of Def. 6.7).

Example C.32. Let us consider the group $(\mathbb{R}^3, *)$ in Exm. C.30. It is easy to see that the associated Jacobian basis is

$$J_1 = \partial_1, \quad J_2 = \cos x_1 \partial_2 + \sin x_1 \partial_3, \quad J_3 = -\sin x_1 \partial_2 + \cos x_1 \partial_3.$$

We compute the associated Exponential Map. Setting $X = \xi_1 J_1 + \xi_2 J_2 + \xi_3 J_3$ (for $\xi_1, \xi_2, \xi_3 \in \mathbb{R}$), we have to solve $\dot{\gamma} = X(\gamma)$, $\gamma(0) = 0$, i.e.,

$$\begin{cases} \dot{\gamma}_1(t) = \xi_1 & \gamma_1(0) = 0 \\ \dot{\gamma}_2(t) = \xi_2 \cos(\gamma_1(t)) - \xi_3 \sin(\gamma_1(t)) & \gamma_2(0) = 0 \\ \dot{\gamma}_3(t) = \xi_2 \sin(\gamma_1(t)) + \xi_3 \cos(\gamma_1(t)) & \gamma_3(0) = 0. \end{cases}$$

The first equation gives $\gamma_1(t) = \xi_1 t$. If we insert this in the other equations, a direct quadrature gives out the solutions:

$$\gamma_2(t) = \xi_2 \frac{\sin(\xi_1 t)}{\xi_1} + \xi_3 \frac{\cos(\xi_1 t) - 1}{\xi_1},$$

$$\gamma_3(t) = -\xi_2 \frac{\cos(\xi_1 t) - 1}{\xi_1} + \xi_3 \frac{\sin(\xi_1 t)}{\xi_1}.$$

Here we assign the value t to $\frac{\sin(\xi_1 t)}{\xi_1}$ when $\xi_1 = 0$ and the value 0 to $\frac{\cos(\xi_1 t)-1}{\xi_1}$ when $\xi_1 = 0$. Taking $t = 1$ we get

$$\mathrm{Exp}(\xi_1 J_1 + \xi_2 J_2 + \xi_3 J_3) = \begin{pmatrix} \xi_1 \\ \xi_2 \dfrac{\sin \xi_1}{\xi_1} + \xi_3 \dfrac{\cos \xi_1 - 1}{\xi_1} \\ -\xi_2 \dfrac{\cos \xi_1 - 1}{\xi_1} + \xi_3 \dfrac{\sin \xi_1}{\xi_1} \end{pmatrix}.$$

Note that Exp is a C^ω function of (ξ_1, ξ_2, ξ_3). We remark that Exp *is not injective nor surjective*. Indeed, on the one hand, for every fixed k in $\mathbb{Z} \setminus \{0\}$,

$$\mathrm{Exp}(2k\pi\, J_1 + \xi_2\, J_2 + \xi_3\, J_3) = (2k\pi, 0, 0), \quad \forall\, \xi_2, \xi_3 \in \mathbb{R},$$

which proves that Exp is not injective. On the other hand, for every fixed k in $\mathbb{Z} \setminus \{0\}$, the point $(2k\pi, x_2, x_3)$ belongs to the image set of Exp only if $x_2 = x_3 = 0$, which proves that Exp is not surjective. $\sharp$

Example C.33. Let B be a real $n \times n$ matrix. We denote the points of $\mathbb{R}^{1+n}$ by (x, y), with $x \in \mathbb{R}$ and $y \in \mathbb{R}^n$. We consider the group operation $*_B$ on $\mathbb{R}^{1+n}$ associated with the semidirect product in Exm. C.26. With the present notations, this becomes

$$(x, y) *_B (x', y') = (x + x', \; y + \exp(x\, B)\, y'),$$

where $x, x' \in \mathbb{R}$ and $y, y' \in \mathbb{R}^n$. We set $\mathbb{G}(B) := (\mathbb{R}^{1+n}, *_B)$; we know that this is a Lie group with identity $(0,0)$ and inversion $(x,y)^{-1} = (-x, -e^{-xB}y)$. We next find the Jacobian basis. Since (in matrix block form)

$$J_{\tau_{(x,y)}}(0,0) = \begin{pmatrix} 1 & 0 \\ 0 & e^{xB} \end{pmatrix},$$

the Jacobian basis is given by $X := \partial_x$ and $Y_1, \ldots, Y_n$, where

$$Y_j := \sum_{k=1}^{n} (e^{xB})_{k,j} \frac{\partial}{\partial y_k}, \quad j = 1, \ldots, n.$$

We now turn to compute the associated Exp map. If $\xi \in \mathbb{R}$ and $\eta \in \mathbb{R}^n$, the integral curve $(\gamma(t), \Gamma(t))$ starting at $(0,0)$ (with γ real-valued and Γ valued in $\mathbb{R}^n$) of $\xi X + \sum_{j=1}^{n} \eta_j Y_j$ is the solution of

$$\begin{cases} \dot{\gamma}(t) = \xi, & \gamma(0) = 0 \\ \dot{\Gamma}(t) = \exp(\gamma(t)B)\, \eta, & \Gamma(0) = 0. \end{cases}$$

We solve the first equation and we integrate by series the second one; we get

$$\Gamma(1) = \sum_{k=0}^{\infty} \frac{\xi^k B^k}{(k+1)!}\, \eta =: \varphi(\xi B)\, \eta,$$

where $\varphi(\xi B)$ is the matrix defined by substitution of ξB into the power series decomposition of the entire function $\varphi(z) = (e^z - 1)/z$. As a consequence,

$$\mathrm{Exp}\big(\xi\, X + \eta_1\, Y_1 + \cdots + \eta_n\, Y_n\big) = (\xi, \varphi(\xi B)\, \eta), \qquad \xi, \eta_1, \ldots, \eta_n \in \mathbb{R}.$$

It is easy (Exr. C.9) to see that the following conditions are equivalent for $\mathbb{G}(B)$:

(i) Exp is injective,
(ii) Exp is surjective,
(iii) Exp is a diffeomorphism,
(iv) the complex eigenvalues of B are contained in $\{0\} \cup \{a + ib \in \mathbb{C} : a \neq 0\}$.

Example C.34. Let us consider the matrix $B = \left(\begin{smallmatrix} 0 & 0 \\ 1 & 0 \end{smallmatrix}\right)$, which is nilpotent of degree 2. Hence the group $\mathbb{G}(B)$ is $(\mathbb{R}^3, *_B)$, where

$$(x, y_1, y_2) *_B (x', y_1', y_2') = (x + x', y_1 + y_1', y_2 + y_2' + x\, y_1').$$

Since the only eigenvalue of B is 0, the Exp map is a diffeomorphism. It is easy to see that $\mathbb{G}(B)$ is isomorphic to the Heisenberg group on $\mathbb{R}^3$. The group $\mathbb{G}(B)$ is sometimes called *polarized Heisenberg group*. ♮

Example C.35. It is not difficult to recognize that the group $\mathbb{G}(B)$ associated with $B = \left(\begin{smallmatrix} 0 & -1 \\ 1 & 0 \end{smallmatrix}\right)$ is the group $(\mathbb{R}^3, *)$ in Exm. C.30 (Exr. C.11). ♮

C.4 Exercises of App. C

Exercise C.1. Suppose that $(\mathbb{R}^N, *)$ is a group (in the algebraic sense) and that

$$f : \mathbb{R}^N \times \mathbb{R}^N \longrightarrow \mathbb{R}^N \qquad f(x, y) = x * y$$

is smooth. (Nothing is asked about the smoothness of the group inversion, apart from its sole existence.) For $x \in \mathbb{R}^N$ set as usual $\tau_x : \mathbb{R}^N \to \mathbb{R}^N$, $\tau_x(y) = x * y$. Show that any τ_x is a C^∞-diffeomorphism. Deduce that $\mathcal{J}_{\tau_x}(y)$ is invertible for every x, y. If e is the identity of the group, note that x^{-1} is implicitly defined by the equation $f(x, y) - e = 0$. By exploiting the Implicit Function Theorem, derive that the function $x \mapsto x^{-1}$ is C^∞ on $\mathbb{R}^N$.

Exercise C.2. Prove that $\mathbb{R}^3$ is endowed with a group structure by the operations introduced in Exms. C.28, C.29, C.30. What kind of trigonometric formulae are involved in the proof of the associativity of the composition law in Exm. C.30?

Exercise C.3. Prove that the groups in Exms. C.28 and C.29 are isomorphic Lie groups, by showing that

$$\phi(x_1, x_2, x_3) = \left(x_1, \frac{x_3}{4} - \frac{x_1 x_2}{2}, x_2 \right)$$

is a Lie group isomorphism from $\mathbb{H}^1$ onto the group of Exm. C.29.

Exercise C.4. Find the Jacobian basis $\{J_1, J_2\}$ of the group $(\mathbb{R}^2, *)$, where

$$(x_1, x_2) * (y_1, y_2) = \left(y_1 + x_1 \, e^{y_2}, x_2 + y_2 \right).$$

Prove the commutator relations (where J_2 appears n times)

$$[\cdots [J_1, J_2] \cdots J_2] = J_1, \qquad \forall n \in \mathbb{N}.$$

This proves that J_1 is a higher-order commutator of J_1, J_2 of *any order* $n \in \mathbb{N}$. Finally, show that the Exp map is

$$\mathrm{Exp}(\alpha_1 J_1 + \alpha_2 J_2) = \left(\alpha_1 \, \frac{e^{\alpha_2} - 1}{\alpha_2}, \alpha_2 \right), \qquad \alpha_1, \alpha_2 \in \mathbb{R},$$

where $\frac{e^{\alpha_2} - 1}{\alpha_2}$ is understood $= 1$ when $\alpha_2 = 0$. Prove that Exp is bijective.

Exercise C.5. Let us consider the so-called Engel's group $\mathbb{G} = (\mathbb{R}^4, *)$:

$$x * y = \left(x_1 + y_1, x_2 + y_2, x_3 + y_3 + x_1 y_2, x_4 + y_4 + \tfrac{1}{2} x_1^2 y_2 + x_1 y_3 \right).$$

Find the associated Jacobian basis $J_1, \ldots, J_4$ and the commutator relations among its elements. Prove that the related Exp map is given by

$$\mathrm{Exp}\big(\xi_1 \, J_1 + \xi_2 \, J_2 + \xi_3 \, J_3 + \xi_4 \, J_4\big) =$$
$$\big(\xi_1, \xi_2, \xi_3 + \tfrac{1}{2} \xi_1 \xi_2, \xi_4 + \tfrac{1}{2} \xi_1 \xi_3 + \tfrac{1}{6} \xi_1^2 \xi_2\big), \quad \text{for } \xi_1, \xi_2, \xi_3, \xi_4 \in \mathbb{R}.$$

Exercise C.6. With reference to the computations in Exm. C.32, determine explicitly the image set of the map Exp of the group $(\mathbb{R}^3, *)$ in Exm. C.30, and observe that this set is dense in $\mathbb{R}^3$. Find also a connected open neighborhood of $0 \in \mathrm{Lie}(\mathbb{G})$ on which Exp is injective.

If $\{J_1, J_2, J_3\}$ is the Jacobian basis, prove that (at least in a neighborhood of $0 \in \mathbb{R}^3$) the inverse map of Exp is

$$\mathrm{Log}(y) = y_1 J_1 + \frac{1}{2}\left(\frac{y_2\, y_1 \sin y_1}{1 - \cos y_1} + y_1\, y_3\right) J_2 + \frac{1}{2}\left(\frac{y_3\, y_1 \sin y_1}{1 - \cos y_1} - y_1\, y_2\right) J_3.$$

Observe that the complex map $z \mapsto \frac{z \sin z}{1 - \cos z}$ is holomorphic in a neighborhood of $0 \in \mathbb{C}$. Determine all of its poles; compare to the information obtained in the investigation of the image set of Exp.

Exercise C.7. Consider the Heisenberg group $\mathbb{H}^n$ in Exm. C.31 (we follow the notation therein). Prove, by explicitly solving the ODE defining it, that its Exponential Map is $\mathrm{Exp}\big(\sum_{j=1}^{n}(a_j X_j + b_j Y_j) + cT\big) = (a, b, c)$.

Exercise C.8. Let B be an $n \times n$ matrix with real (or complex) entries. Let $\lambda_1, \ldots, \lambda_n$ be its *complex* eigenvalues (counted with multiplicities). We know that B is similar (via a complex matrix M) to its Jordan canonical form J, say $B = M\,J\,M^{-1}$, where J has diagonal blocks of the form

$$\begin{pmatrix} \lambda_i & 1 & \cdots & 0 \\ 0 & \lambda_i & \ddots & \vdots \\ \vdots & \vdots & \ddots & 1 \\ 0 & 0 & \cdots & \lambda_i \end{pmatrix}.$$

Let $\varphi(z) = \sum_{k=0}^{\infty} a_k z^k$ be an entire complex function. Show that the matrix $\varphi(B) := \sum_{k=0}^{\infty} a_k B^k$ is well defined as a convergent series of matrices (with the usual Banach space structure on the set of the square real or complex matrices) and that $\varphi(B) = M\,\varphi(J)\,M^{-1}$. Prove that $\varphi(J)$ is upper triangular (i.e., the entries below the main diagonal are 0) and that the entries on the main diagonal of $\varphi(J)$ are $\varphi(\lambda_1), \ldots, \varphi(\lambda_n)$. Deduce that the latter are the complex eigenvalues of $\varphi(B)$.

Exercise C.9. Resume the notations in Exm. C.33. Denote $(\xi, \varphi(\xi B)\,\eta)$ by $\Psi(\xi, \eta)$. Let $\lambda_1, \ldots, \lambda_n \in \mathbb{C}$ be the eigenvalues of B (counted with multiplicities); prove that $\varphi(\xi B)$ is invertible iff the non-vanishing elements of $\{\xi\lambda_1, \ldots, \xi\lambda_n\}$ are not of the form $2k\pi\, i$ (with $k \in \mathbb{Z}$), that is, $\{\xi\lambda_1, \ldots, \xi\lambda_n\} \setminus \{0\} \cap \{2k\pi\, i : k \in \mathbb{Z}\} = \varnothing$.

[*Hint:* by Exr. C.8, prove that the eigenvalues of $\varphi(\xi B)$ are

$$\varphi(\xi\,\lambda_k) = \begin{cases} \dfrac{e^{\xi\lambda_k} - 1}{\xi\lambda_k} & \text{if } \xi\lambda_k \neq 0 \\[2mm] 1 & \text{if } \xi\lambda_k = 0. \end{cases}$$

Then provide the details of the following two cases:

(1) *The eigenvalues of B are contained in $\{0\} \cup \{a + ib \in \mathbb{C} : a \neq 0\}$. Desume that Ψ is invertible, with inverse $\Psi^{-1}(x, y) = (x, (\varphi(x\,B))^{-1}\,y)$.*

(2) *B possesses at least one non-zero eigenvalue on the imaginary axis.* Let ib be any such eigenvalue (with $b \in \mathbb{R} \setminus \{0\}$). Prove that Ψ is not injective nor surjective. Indeed, if we fix any $k \in \mathbb{Z} \setminus \{0\}$ and we set $A_k := \varphi\left(\frac{2\pi k}{b} B\right)$, then every point of $\mathbb{R}^{1+n}$ of the form[14]

$$\left(\frac{2\pi k}{b}, y\right), \quad \text{with } y \in \mathbb{R}^n \setminus \mathrm{range}(A_k)$$

lies outside the image set of Ψ. This proves that Ψ is not surjective. In order to prove that it is not even injective, it suffices to observe that (with the same notation) all the points $(2\pi k/b, y)$ (with $y \in \ker(A_k)$) are mapped by Ψ into $(2\pi k/b, 0)$.]

Exercise C.10. Consider the group $\mathbb{G}(B)$ in Exm. C.33.

(1) With reference to Def. C.12, prove that the composition of the associated inverse group $\widehat{\mathbb{G}}$ (which we denote by $\widehat{*}_B$) is the following one:

$$(x, y) \,\widehat{*}_B\, (x', y') = (x + x', y' + \exp(x'B)\, y), \qquad x, x' \in \mathbb{R}, \ y, y' \in \mathbb{R}^n.$$

(2) Prove that the Jacobian matrix at $(0,0)$ of the map

$$(x', y') \mapsto (x, y) \,\widehat{*}_B\, (x', y') = (x', y') *_B (x, y)$$

is $\left(\begin{smallmatrix} 1 & 0 \\ By & \mathbb{I}_n \end{smallmatrix}\right)$, where $\mathbb{I}_n$ is the identity matrix of order n. Deduce that a basis for the Lie algebra of $\widehat{\mathbb{G}}$ is $\{X, Y_1, \ldots, Y_n\}$, where

$$X := \frac{\partial}{\partial x} + \sum_{i,j=1}^n b_{i,j}\, y_j\, \frac{\partial}{\partial y_i}, \qquad Y_1 := \frac{\partial}{\partial y_1}, \ldots, Y_n := \frac{\partial}{\partial y_n}.$$

(3) Show that the Exp Map of $\widehat{\mathbb{G}}$ can be computed by solving this system of ODEs: given $\xi \in \mathbb{R}, \eta \in \mathbb{R}^n$ we have $\mathrm{Exp}(\xi X + \sum_{j=1}^n \eta_j Y_j) = (\gamma(1), \Gamma(1))$, where

$$\text{(CP)} \qquad \begin{cases} \dot\gamma(t) = \xi & \gamma(0) - 0 \\ \dot\Gamma(t) = \eta + \xi B\, \Gamma(t) & \Gamma(0) = 0. \end{cases}$$

Prove that the following curve is the solution of (CP):

$$\gamma(t) = \xi\, t, \qquad \Gamma(t) = \int_0^t \exp(\xi\, (t - s)\, B)\, \eta\, ds.$$

(4) Use the same arguments of expansion by series and integration used in Exm. C.33 to show that the Exp map is

$$\mathrm{Exp}(\xi X + \eta_1 Y_1 + \cdots + \eta_n Y_n) = \left(\xi, \varphi(\xi B)\, \eta\right).$$

Exercise C.11. Prove that the group $\mathbb{G}(B)$ associated with $B = \left(\begin{smallmatrix} 0 & -1 \\ 1 & 0 \end{smallmatrix}\right)$ is the group in Exm. C.30. Show that the Jacobian basis of its inverse group is

$$\partial_{x_1} - x_3\, \partial_{x_2} + x_2\, \partial_{x_3}, \quad \partial_{x_2}, \quad \partial_{x_3}.$$

Observe how different the vector fields of $\mathrm{Lie}(\mathbb{G}(B))$ and of $\mathrm{Lie}(\widehat{\mathbb{G}(B)})$ are (nonetheless these Lie algebras are isomorphic since the associated Lie groups are isomorphic). Verify the validity of what is stated in Prop. C.13 proving that the vector fields in $\mathrm{Lie}(\mathbb{G}(B))$ commute with those in $\mathrm{Lie}(\widehat{\mathbb{G}(B)})$.

---

[14]$\mathrm{range}(A_k)$ is the image set of the linear function associated with the matrix A_k.

Exercise C.12. Consider the group in Exm. C.29. Compute the Exp map and prove that this is invertible. Denote by Log its inverse map. Show that (as ensured by (C.14)) $\gamma(1, \mathrm{Log}(y), x)$ coincides with $x * y$ for every $x, y \in \mathbb{R}^3$.

Exercise C.13. Consider the Lie group on $\mathbb{R}^2 \setminus \{0\}$ defined by the composition

$$x * y = (x_1 y_1 - x_2 y_2, x_1 y_2 + x_2 y_1).$$

Find the inversion map and the identity element. Prove that, if $\{J_1, J_2\}$ is the associated Jacobian basis, then the Exponential Map is

$$\mathrm{Exp}(\xi_1 J_1 + \xi_2 J_2) = (e^{\xi_1} \cos \xi_2, e^{\xi_1} \sin \xi_2).$$

Observe that Exp is not injective, but it is surjective onto $\mathbb{R}^2 \setminus \{0\}$. Find the inverse function of the restriction of Exp to the set

$$\left\{ \xi_1 J_1 + \xi_2 J_2 : \ \xi_1 \in \mathbb{R}, \ \xi_2 \in (-\pi, \pi] \right\}.$$

[*Hint:* It may help observing that the composition $*$ is nothing but the usual multiplication in $\mathbb{C}$, rewritten via the identification $\mathbb{C} \equiv \mathbb{R}^2$...]

Exercise C.14. Consider the Lie group on $\Omega := \{x \in \mathbb{R}^2 : \ x_1 \neq 0, \ x_2 \neq 0\}$ defined by the composition $x * y = (x_1 y_1, x_2 y_2)$. Find the inversion map and the identity element. Prove that, if $\{J_1, J_2\}$ is the associated Jacobian basis, then the Exponential Map is $\mathrm{Exp}(\xi_1 J_1 + \xi_2 J_2) = (e^{\xi_1}, e^{\xi_2})$. Observe that Exp is injective, but it is not surjective on the whole of Ω, but only onto $\{x \in \mathbb{R}^2 : \ x_1 > 0, \ x_2 > 0\}$.

Exercise C.15. Suppose we know that $J_1 = x_1 \partial_{x_1} + x_2 \partial_{x_2}$ and $J_2 = \partial_{x_2}$ form the Jacobian basis of some unknown group $\mathbb{G} = (\Omega, *)$ with identity $(1, 0)$, where $\Omega \subseteq \mathbb{R}^2$ is an open and connected set containing $(1, 0)$. Arguing as in Exm. C.14, show that, fixing $x \in \mathbb{R}^2$ and setting $f(y) := x * y$, then

$$J_f(y) = \begin{pmatrix} \dfrac{f_1(y)}{y_1} & 0 \\ \dfrac{f_2(y) - y_2}{y_1} & 1 \end{pmatrix}.$$

Prove that Ω can be taken as $\{x \in \mathbb{R}^2 : \ x_1 \neq 0\}$ and the group law is

$$x * y = (x_1 \, y_1, \ x_2 \, y_1 + y_2).$$

This is referred to as the *group of the affine transformations on a line.*

[*Hint:* First show that $f_1(y) = u(y_1)$ (for some smooth u), where u satisfies the Cauchy problem $u'(y_1) = u(y_1)/y_1$, $u(1) = x_1$. Then show that $f_2(y) = y_2 + v(y_1)$, for a smooth v satisfying $v'(y_1) = v(y_1)/y_1$, $v(1) = x_2$...]

Exercise C.16. Consider the Lie group $\mathbb{G} = (\mathbb{R}^3, *)$, where

$$x * y = \left(\mathrm{arcsinh}(\sinh(x_1) + \sinh(y_1)), x_2 + y_2 + \sinh(x_1) y_3, x_3 + y_3 \right).$$

Determine the Jacobian basis of $\mathrm{Lie}(\mathbb{G})$ and prove that $\mathbb{G}$ is nilpotent. Observe that the group law is not polynomial.

Further Readings

- FURTHER REFERENCES FOR CHAP. 1

For the sake of simplicity in an introductory chapter, the contents of Chap. 1 have been introduced in $\mathbb{R}^N$ space. The generalization to the smooth manifold setting is quite standard and simple, and is left to the interested reader; see e.g., [Lee (2013)].

- FURTHER REFERENCES FOR CHAP. 2

A comprehensive exposition of the history, of the proofs, of the applications and of the bibliographical references concerning the Campbell-Baker-Hausdorff-Dynkin Theorem can be found in the monograph by [Bonfiglioli and Fulci (2012)]. In particular, we refer the interested reader to [Bonfiglioli and Fulci (2012), Section 5.7, p. 359] for a detailed and commented list of bibliographical references. See also [Achilles and Bonfiglioli (2012)] for an historical investigation.

The proof of the Exponential Theorem given in Chap. 2 follows the line of the remarkably transparent argument given in [Djoković (1975)] (see also [Bonfiglioli and Fulci (2012), Section 4.3] for the detailed algebraic background).

The argument in Sec. 2.4 reducing the determination of $Z(t)$ to a "quadrature" problem can be found in [Duistermaat and Kolk (2000), Section 1.7, p. 29]. The algebraic technique in Sec. 2.6 has been used in the paper by [Biagi and Bonfiglioli (2014)] for the study of a convergence domain of the CBHD series in (possibly infinite-dimensional) Banach-Lie algebras.

- FURTHER REFERENCES FOR CHAP. 3

Chap. 3 contains the version of the CBHD Theorem mostly referred to by analysts. For instance, Thm. A in Exr. 3.11 is used in the recent papers by [Christ *et al.* (1999)] and by [Morbidelli (2000)]; it is also a restatement of Prop. 4.3 in the pioneering paper by [Nagel *et al.* (1985)] or, moreover, an analogue of a result exploited for the celebrated Lifting Procedure by [Rothschild and Stein (1976), Sec. 10]. Both in [Nagel *et al.* (1985); Rothschild and Stein (1976)] this version of the CBHD Theorem appears without a proof. Approximations for higher order commutators are frequently used in PDEs, see e.g., [Folland (1975), §5, p. 193]; [Varopoulos *et al.* (1992), §III.3, p. 34]; see also the recent papers by [Bonfiglioli and Lanconelli (2012a); Citti and Manfredini (2006); Magnani (2006); Morbidelli (2000)].

- FURTHER REFERENCES FOR CHAP. 4

Some references for Chap. 4, concerning smooth manifold theory, are: [Boothby (1986)]; [Gallot *et al.* (1987)]; [Lee (2013)]; [Tu (2011)]; [Warner (1983)].

The reader is referred to the paper by [Rampazzo and Sussmann (2007)] for generalizations of Thms. 1.32 and 4.25 for Lipschitz vector fields.

- FURTHER REFERENCES FOR CHAP. 5

The argument for the local convergence of the CBHD series in the proof of Thm. 5.3 is due to [Dynkin (1947)]: it is indeed based on the explicit presentation for the commutator series of $\log(e^x\, e^y)$, first given in the paper by [Dynkin (1947)].

The generalization to the infinite-dimensional case of the so-called Banach-Lie algebras was investigated again by [Dynkin (1950)]. The papers [Dynkin (1947, 1950)] (see also [Dynkin (2000)]) paved the way for the intensively studied problems concerning other possible presentations of $\log(e^x\, e^y)$ and concerning the improved domains of convergence of the CBHD series. See [Bonfiglioli and Fulci (2012)] for a comprehensive list of references.

- FURTHER REFERENCES FOR CHAP. 6

The celebrated paper [Hörmander (1967)] is the seminal paper on what were later to be called 'Hörmander operators'. For a thorough introduction to Hörmander vector fields and operators, see the recent monograph by [Bramanti (2014)]; the references in [Bramanti (2014)] are strongly recommended to the interest reader.

We gave the definition of X-subunit path in connection with a family X of vector fields (see Def. 6.13); one can obtain a more general definition of L-subunit path by replacing the rhs of (6.5) with the characteristic form of a second-order differential operator L (with nonnegative characteristic form). This more general definition is originally due to [Fefferman and Phong (1983)]. Contemporarily, [Franchi and Lanconelli (1983)] independently introduced subunit paths for a family X of diagonal (non-smooth) vector fields. Since then, these ideas have proved to be tremendously fruitful in the theory of sub-elliptic PDEs.

- FURTHER REFERENCES FOR CHAP. 7

Some pertinent references for Chap. 7 are: the comprehensive book by [Agrachev *et al.* (2016)]; the book by [Busemann (1955)]; the paper by [Garofalo and Nhieu (1996)]; the Memoir by [Hajłasz and Koskela (2000)].

See also [Agrachev and Sachkov (2004); Ambrosio and Tilli (2004); Gromov (1996, 2007); Monti (2001)]. References for Geometric Measure Theory are the books by [Federer (1969); Morgan (2009); Mattila (1995)].

- FURTHER REFERENCES FOR CHAPS. 8 AND 9

The second-order PODs which we named 'semielliptic' are usually called '*elliptic-parabolic*', using the terminology introduced in the pioneering paper by [Picone (1913)]. We allowed ourselves permission to avoid this (old but traditional) terminology, with the aim to reserve the adjective 'elliptic' only to the stronger meaning given to it in Def. 8.1. In this sense, our use of the word 'semielliptic' makes explicit mention to the possibility that the quadratic form of the PDO be degenerate.

A systematic use of the Green operator (see Chap. 9) for Potential Theoretic purposes was first introduced in the seminal paper by [Bony (1969)].

The reader interested in further investigations (and applications) of maximum principles for second-order operators is referred to the comprehensive monographs by [Fraenkel (2000); Kresin and Maz'ya (2012); Protter and Weinberger (1967); Pucci and Serrin (2007); Sperb (1981)]. See also [Nirenberg (1992)]; [Gilbarg and Trudinger (2001), Chapter 3]; [Taira (1988), Chapter 7].

Quoting [Brezis and Browder (1998), p. 84]:

> *"the maximum principle for second order elliptic and parabolic equations has played a central role throughout the 20th century."*

Indeed, maximum principles for linear and non-linear PDOs have been incessantly investigated in the literature. An exhaustive list of references and a description of the countless settings where maximum principles have been investigated so far is beyond our scope.

- FURTHER REFERENCES FOR CHAP. 10

For our exposition of the Maximum Propagation Principle, we basically followed the scheme in [Bonfiglioli *et al.* (2007), Section 5.13] (see also [Lanconelli (2010)]), where sub-Laplace operators on Carnot groups are considered; this is however not the case with the proof of the Hopf-type Lem. 10.31, for the proof of which we followed [Bony (1969)], which does *not* make use of the Weak Maximum Principle.

Thm. 10.28 is usually attributed to [Nagumo (1942)] and [Bony (1969)]; other contributions are due to [Brezis (1970)] and [Redheffer (1972)].

The study of flow-invariant sets is occasionally to be found in textbooks on ODEs: see, e.g., that by [Wolfgang (1998)]. For a survey on the notion of invariant set in Control Theory, see the paper by [Blanchini (1999)]; in this paper the interested reader can also find many other references related to Thm. 10.28.

For recent generalizations of the Strong Maximum Principle and for Hopf's original contribution to it, see the book by [Pucci and Serrin (2004)].

- FURTHER REFERENCES FOR CHAP. 11

In Sec. 11.1 we followed the approach by [Amano (1979)]; see also the papers by [Bony (1969); Hill (1970); Redheffer (1971)].

See [Brockett (2014)] for a thorough historical survey on differential geometric control theory in the 1970-80's; the list of references in [Brockett (2014)] is particularly recommended to the reader's attention. See in particular: [Haynes and Hermes (1970); Hermann (1962); Lobry (1970); Krener (1974)].

Other references for the Control Theory point of view in reachability are: [Hirschorn (1975); Jurdjevic (1997); Jurdjevic and Sussmann (1972b,a)]. See also: [Burago *et al.* (2001); Sussmann (1972)].

- FURTHER REFERENCES FOR CHAP. 12

The equation of variation in its differential form (as an ODE solved by the differential of the solution wrt its parameters) is a classical topic in ODEs (see [Hartman (1982)]). The integral form of this identity (as presented in Thm. 12.3) is less common in textbooks (we were unable to locate a precise reference). In the autonomous case, however, this appears e.g., in [Duistermaat and Kolk (2000)].

- FURTHER REFERENCES FOR CHAP. 13

Applications of Thm. 13.9 to the analysis of some PDEs appear in the papers by [Biagi and Bonfiglioli (2015); Bonfiglioli and Lanconelli (2012a)]. See also Chap. 17 for an application to the construction of Lie groups.

- FURTHER REFERENCES FOR CHAPS. 14 AND 15

The CBHD Theorem for Lie groups is a classical topic in Lie group theory; for the analytic case (Sec. 14.3) we followed the approach by [Varadarajan (1984)]. For the smooth case, our approach is less common, as it is based on the ODE version of the CBHD Theorem, treated in Chap. 13.

In Lie group textbooks, the identity $\mathrm{Exp}(X) * \mathrm{Exp}(Y) = \mathrm{Exp}(X \diamond Y)$ is usually proved as a consequence, when $t = 1$, of

$$\mathrm{Exp}(tX) * \mathrm{Exp}(tY) = \mathrm{Exp}((tX) \diamond (tY)),$$

and the latter is proved by showing that both sides (as functions of t) solve the same Cauchy problem. In doing this, it is first required the knowledge of the differential of Exp. As our focus is on the ODE version of the CBHD Theorem as an autonomous ODE fact, this is not the path that we followed.

In classic textbooks of Lie group theory, the Third Theorem of Lie is presented in its global form; a proof of this result is onerous, and it requires a lot of prerequisites of Lie group theory and topology (see e.g., [Varadarajan (1984)]). For the local version, see e.g., [Cohn (1957)]. More recently, growing interest has been given in infinite-dimensional Lie groups, see [Neeb (2006)] and the references therein.

- FURTHER REFERENCES FOR CHAP. 16

The same procedure performed in the construction of Carnot groups starting from stratified Lie algebras can be carried out, more generally, if one starts from a homogeneous Lie algebra (in the sense of [Folland and Stein (1982)]). This leads to a coordinate model for any homogeneous Lie group; for a thorough analysis of the homogeneous Lie groups (and, more generally, of the nilpotent Lie groups) the reader is referred to the book by [Fischer and Ruzhansky (2016)].

Owing to what is proved in Rem. 16.10 (p. 303), we established a bridge between abstract Carnot groups and homogeneous Carnot groups (HCG, for short). A thorough analysis of HCGs (and of their sub-Laplacian operators) is carried out in the monograph by [Bonfiglioli *et al.* (2007)]. As a consequence, the results of Chap. 16 may serve as a link to the contents of [Bonfiglioli *et al.* (2007)].

We observe that the perspective of Chap. 16 is somewhat reversed if compared to the approach in [Bonfiglioli *et al.* (2007)]: we introduce HCGs not as Lie groups on $\mathbb{R}^N$ satisfying properties 1-to-4 in Thm. 16.7 (p. 301), as is done in [Bonfiglioli *et al.* (2007)]; instead, we first introduced finite-dimensional stratified Lie algebras, then we proved that any such Lie algebra can be exponentiated to an HCG.

This provides a more intrinsic view of HCGs, in that the focus is shifted to their Lie algebras, and the many properties of HCGs (mostly, their homogeneous structure) are consequences of the Lie-algebra properties, instead of a datum.

- FURTHER REFERENCES FOR CHAP. 17

For the results in Chap. 17, we followed the paper by [Biagi and Bonfiglioli (2015)].

We highlight that question (Q) introduced in the *incipit* of Chap. 17 can also be answered by means of advanced tools of Lie group theory and of differential geometry, such as the (global version of the) Third Theorem of Lie and the properties of the action of a Lie group on an analytic manifold. See the Memoir by [Palais (1957)]. We did not follow this approach, since it requires a deep topological machinery, which is beyond the scope of this book.

Sec. 17.3 contains some new results, see [Biagi *et al.* (2017)].

- FURTHER REFERENCES FOR CHAP. 18

Chap. 18 furnishes a detailed exposition of some results in the recent paper [Biagi and Bonfiglioli (2014)]. The literature on the convergence domain for CBHD-type series is extremely vast; see the paper by [Blanes and Casas (2004)] and the references therein (see also [Biagi and Bonfiglioli (2014)]).

- FURTHER REFERENCES FOR APPS. A, B, C

The results of App. B are classical results of ODE Theory (see e.g., [Hartman (1982)]), but it is difficult to find them collected together in the same textbook. For example, it is quite rare to find a thorough exposition for the C^ω dependence; an exception is the Appendix B in the book by [Duistermaat and Kolk (2000)], where the C^ω case is sketched. For Sec. B.4 we also made profitable use of some results in the book by [Cartan (1995)]. For finer regularity and dependence results, see also the book by [Parenti and Parmeggiani (2010)].

As regards App. C, it is not an aim of this book furnishing any exhaustive introduction to the (immense) theory of Lie groups. The reader will be more profitably introduced to this theory by the following classical references:

[Abbaspour and Moskowitz (2007); Bröcker and tom Dieck (1985); Chevalley (1946); Cohn (1957); Duistermaat and Kolk (2000); Godement (1982); Gorbatsevich *et al.* (1997); Hall (2003); Hausner and Schwartz (1968); Helgason (1978); Hilgert and Neeb (1991); Hochschild (1965); Hofmann and Morris (2006); Jacobson (1962); Rossmann (2002); Sagle and Walde (1973); Sepanski (2007); Serre (2006); Varadarajan (1984)].

List of abbreviations:

App.	Appendix
Chap.	Chapter
Exm.	Example
Exr.	Exercise
iff	if and only if
Lem.	Lemma
lhs	left-hand side
Not.	Notation
Prop.	Proposition
Rem.	Remark
rhs	right-hand side
Sec.	Section
Thm.	Theorem
wrt	with respect to

Bibliography

Abbaspour, H. and Moskowitz, M. (2007). *Basic Lie theory* (World Scientific Publishing Co. Pte. Ltd., Hackensack, NJ).

Achilles, R. and Bonfiglioli, A. (2012). The early proofs of the theorem of Campbell, Baker, Hausdorff, and Dynkin, *Arch. Hist. Exact Sci.* **66**, pp. 295–358.

Agrachev, A., Barilari, D., and Boscain, U. (2016). Introduction to geodesics in sub-Riemannian geometry, in *Geometry, analysis and dynamics on sub-Riemannian manifolds. Vol. II*, EMS Ser. Lect. Math. (Eur. Math. Soc., Zürich), pp. 295–358.

Agrachev, A. A. and Sachkov, Y. L. (2004). *Control theory from the geometric viewpoint, Encyclopaedia of Mathematical Sciences*, Vol. 87 (Springer-Verlag, Berlin).

Amano, K. (1979). Maximum principles for degenerate elliptic-parabolic operators, *Indiana Univ. Math. J.* **28**, pp. 545–557.

Ambrosio, L. and Tilli, P. (2004). *Topics on analysis in metric spaces, Oxford Lecture Series in Mathematics and its Applications*, Vol. 25 (Oxford University Press, Oxford).

Biagi, S. and Bonfiglioli, A. (2014). On the convergence of the Campbell-Baker-Hausdorff-Dynkin series in infinite-dimensional Banach-Lie algebras, *Linear Multilinear Algebra* **62**, pp. 1591–1615.

Biagi, S. and Bonfiglioli, A. (2015). A completeness result for time-dependent vector fields and applications, *Commun. Contemp. Math.* **17**, pp. 1–26.

Biagi, S., Bonfiglioli, A., and Matone, M. (2017). On the baker-campbell-hausdorff theorem: non-convergence and prolongation issues, .

Blanchini, F. (1999). Set invariance in control, *Automatica J. IFAC* **35**, pp. 1747–1767.

Blanes, S. and Casas, F. (2004). On the convergence and optimization of the Baker-Campbell-Hausdorff formula, *Linear Algebra Appl.* **378**, pp. 135–158.

Bonfiglioli, A. and Fulci, R. (2012). *Topics in noncommutative algebra. The theorem of Campbell, Baker, Hausdorff and Dynkin, Lecture Notes in Math.*, Vol. 2034 (Springer, Heidelberg).

Bonfiglioli, A. and Lanconelli, E. (2012a). Lie groups related to Hörmander operators and Kolmogorov-Fokker-Planck equations, *Commun. Pure Appl. Anal.* **11**, pp. 1587–1614.

Bonfiglioli, A. and Lanconelli, E. (2012b). Matrix-exponential groups and Kolmogorov-Fokker-Planck equations, *J. Evol. Equ.* **12**, pp. 59–82.

Bonfiglioli, A., Lanconelli, E., and Uguzzoni, F. (2007). *Stratified Lie groups and potential theory for their sub-Laplacians*, Springer Monographs in Mathematics (Springer, Berlin).

Bony, J.-M. (1969). Principe du maximum, inégalite de Harnack et unicité du problème de Cauchy pour les opérateurs elliptiques dégénérés, *Ann. Inst. Fourier (Grenoble)* **19**, pp. 277–304.

Boothby, W. M. (1986). *An introduction to differentiable manifolds and Riemannian geometry, Pure and Applied Mathematics*, Vol. 120, 2nd edn. (Academic Press, Inc., Orlando, FL).

Bramanti, M. (2014). *An invitation to hypoelliptic operators and Hörmander's vector fields*, SpringerBriefs in Mathematics (Springer, Cham).

Brezis, H. (1970). On a characterization of flow-invariant sets, *Comm. Pure Appl. Math.* **23**, pp. 261–263.

Brezis, H. and Browder, F. (1998). Partial differential equations in the 20th century, *Adv. Math.* **135**, pp. 76–144.

Bröcker, T. and tom Dieck, T. (1985). *Representations of compact Lie groups, Graduate Texts in Mathematics*, Vol. 98 (Springer-Verlag, New York).

Brockett, R. (2014). The early days of geometric nonlinear control, *Automatica J. IFAC* **50**, pp. 2203–2224.

Burago, D., Burago, Y., and Ivanov, S. (2001). *A course in metric geometry, Graduate Studies in Mathematics*, Vol. 33 (American Mathematical Society, Providence, RI).

Busemann, H. (1955). *The geometry of geodesics* (Academic Press Inc., New York, N. Y.).

Carathéodory, C. (1909). Untersuchungen über die Grundlagen der Thermodynamik, *Math. Ann.* **67**, pp. 355–386.

Cartan, H. (1995). *Elementary theory of analytic functions of one or several complex variables* (Dover Publications, Inc., New York).

Castaing, C. and Valadier, M. (1977). *Convex analysis and measurable multifunctions*, Lecture Notes in Mathematics, Vol. 580 (Springer-Verlag, Berlin-New York).

Chevalley, C. (1946). *Theory of Lie Groups. I*, Princeton Mathematical Series, vol. 8 (Princeton University Press, Princeton, N. J.).

Chow, W.-L. (1939). über Systeme von linearen partiellen Differentialgleichungen erster Ordnung, *Math. Ann.* **117**, pp. 98–105.

Christ, M., Nagel, A., Stein, E. M., and Wainger, S. (1999). Singular and maximal Radon transforms: analysis and geometry, *Ann. of Math. (2)* **150**, pp. 489–577.

Citti, G. and Manfredini, M. (2006). Implicit function theorem in Carnot-Carathéodory spaces, *Commun. Contemp. Math.* **8**, pp. 657–680.

Cohn, P. M. (1957). *Lie groups, Cambridge Tracts in Mathematics and Mathematical Physics*, Vol. 46 (Cambridge University Press, New York, N.Y.).

Cohn-Vossen, S. (1936). Existenz kürzester Wege, *Compositio Math.* **3**, pp. 441–452.

Corwin, L. J. and Greenleaf, F. P. (1990). *Representations of nilpotent Lie groups and their applications. Part I: Basic theory and examples, Cambridge Studies in Advanced Mathematics*, Vol. 18 (Cambridge University Press, Cambridge).

Djoković, D. v. Z. (1975). An elementary proof of the Baker-Campbell-Hausdorff-Dynkin formula, *Math. Z.* **143**, pp. 209–211.

Duistermaat, J. J. and Kolk, J. A. C. (2000). *Lie groups*, Universitext (Springer-Verlag, Berlin).

Dynkin, E. B. (1947). Calculation of the coefficients in the Campbell-Hausdorff formula, *Doklady Akad. Nauk SSSR (N.S.)* **57**, pp. 323–326.

Dynkin, E. B. (1950). Normed Lie algebras and analytic groups, *Uspehi Matem. Nauk (N.S.)* **5**, pp. 135–186.

Dynkin, E. B. (2000). *Selected papers of E. B. Dynkin with commentary* (American Mathematical Society, Providence, RI; International Press, Cambridge, MA), edited by A. A. Yushkevich, G. M. Seitz and A. L. Onishchik.

Federer, H. (1969). *Geometric measure theory, Die Grundlehren der mathematischen Wissenschaften*, Vol. 153 (Springer-Verlag New York Inc., New York).

Fediĭ, V. S. (1971). On a criterion for hypoellipticity, *Mat. USSR Sb.* **14**, pp. 15–45.

Fefferman, C. and Phong, D. H. (1983). Subelliptic eigenvalue problems, in *Conference on harmonic analysis in honor of Antoni Zygmund, Vol. I, II (Chicago, Ill., 1981)*, Wadsworth Math. Ser. (Wadsworth, Belmont, CA), pp. 590–606.

Fischer, V. and Ruzhansky, M. (2016). *Quantization on nilpotent Lie groups, Progress in Mathe-*

matics, Vol. 314 (Birkhäuser/Springer, Cham).

Folland, G. B. (1975). Subelliptic estimates and function spaces on nilpotent Lie groups, *Ark. Mat.* **13**, pp. 161–207.

Folland, G. B. and Stein, E. M. (1982). *Hardy spaces on homogeneous groups, Mathematical Notes*, Vol. 28 (Princeton University Press; University of Tokyo Press, Princeton, N.J.; Tokyo).

Fraenkel, L. E. (2000). *An introduction to maximum principles and symmetry in elliptic problems, Cambridge Tracts in Mathematics*, Vol. 128 (Cambridge University Press, Cambridge).

Franchi, B. and Lanconelli, E. (1983). Une métrique associée à une classe d'opérateurs elliptiques dégénérés, *Rend. Sem. Mat. Univ. Politec. Torino* , pp. 105–114.

Gallot, S., Hulin, D., and Lafontaine, J. (1987). *Riemannian geometry*, Universitext (Springer-Verlag, Berlin).

Garofalo, N. and Nhieu, D.-M. (1996). Isoperimetric and Sobolev inequalities for Carnot-Carathéodory spaces and the existence of minimal surfaces, *Comm. Pure Appl. Math.* **49**, pp. 1081–1144.

Gilbarg, D. and Trudinger, N. S. (2001). *Elliptic partial differential equations of second order*, Classics in Mathematics (Springer-Verlag, Berlin).

Godement, R. (1982). *Introduction à la théorie des groupes de Lie. Tome 1, Publications Mathématiques de l'Université Paris VII [Mathematical Publications of the University of Paris VII]*, Vol. 11 (Université de Paris VII, U.E.R. de Mathématiques, Paris).

Gorbatsevich, V. V., Onishchik, A. L., and Vinberg, E. B. (1997). *Foundations of Lie theory and Lie transformation groups* (Springer-Verlag, Berlin).

Gromov, M. (1996). Carnot-Carathéodory spaces seen from within, in *Sub-Riemannian geometry, Progr. Math.*, Vol. 144 (Birkhäuser, Basel), pp. 79–323.

Gromov, M. (2007). *Metric structures for Riemannian and non-Riemannian spaces*, Modern Birkhäuser Classics (Birkhäuser Boston, Inc., Boston, MA).

Hajłasz, P. and Koskela, P. (2000). Sobolev met Poincaré, *Mem. Amer. Math. Soc.* **145**.

Hall, B. C. (2003). *Lie groups, Lie algebras, and representations: an elementary introduction, Graduate Texts in Mathematics*, Vol. 222 (Springer-Verlag, New York).

Hartman, P. (1982). *Ordinary differential equations*, 2nd edn. (Birkhäuser, Boston, Mass.).

Hausner, M. and Schwartz, J. T. (1968). *Lie groups; Lie algebras* (Gordon and Breach Science Publishers, New York-London-Paris).

Haynes, G. W. and Hermes, H. (1970). Nonlinear controllability via Lie theory, *SIAM J. Control* **8**, pp. 450–460.

Helgason, S. (1978). *Differential geometry, Lie groups, and symmetric spaces, Pure and Applied Mathematics*, Vol. 80 (Academic Press, Inc., New York-London).

Hermann, R. (1962). The differential geometry of foliations. II, *J. Math. Mech.* **11**, pp. 303–315.

Hermann, R. (1968). *Differential geometry and the calculus of variations*, Mathematics in Science and Engineering, Vol. 49 (Academic Press, New York-London).

Hermes, H. and LaSalle, J. P. (1969). *Functional analysis and time optimal control, Mathematics in Science and Engineering*, Vol. 56 (Academic Press, New York-London).

Hilgert, J. and Neeb, K.-H. (1991). *Lie-Gruppen und Lie-Algebren* (Vieweg, Braunschweig).

Hill, C. D. (1970). A sharp maximum principle for degenerate elliptic-parabolic equations, *Indiana Univ. Math. J.* **20**, pp. 213–229.

Hirschorn, R. M. (1975). Controllability in nonlinear systems, *J. Diff. Equations* **19**, pp. 46–61.

Hochschild, G. (1965). *The structure of Lie groups* (Holden-Day, Inc., London).

Hofmann, K. H. and Morris, S. A. (2006). *The structure of compact groups. A primer for the student - a handbook for the expert, De Gruyter Studies in Mathematics*, Vol. 25, second revised and augmented edn. (Walter de Gruyter & Co., Berlin).

Hörmander, L. (1967). Hypoelliptic second order differential equations, *Acta Math.* **119**, pp. 147–171.

Jacobson, N. (1962). *Lie algebras*, Interscience Tr1acts in Pure and Applied Mathematics, No. 10 (Interscience Publishers (a division of John Wiley & Sons), New York-London).

Jerison, D. and Sánchez-Calle, A. (1987). Subelliptic, second order differential operators, in *Complex analysis, III (College Park, Md., 1985–86), Lecture Notes in Math.*, Vol. 1277 (Springer, Berlin), pp. 46–77.

Jurdjevic, V. (1997). *Geometric control theory, Cambridge Studies in Advanced Mathematics*, Vol. 52 (Cambridge University Press, Cambridge).

Jurdjevic, V. and Sussmann, H. J. (1972a). Control systems on Lie groups, *J. Differential Equations* **12**, pp. 313–329.

Jurdjevic, V. and Sussmann, H. J. (1972b). Controllability of nonlinear systems, *J. Differential Equations* **12**, pp. 95–116.

Krantz, S. G. and Parks, H. R. (2002). *A primer of real analytic functions*, 2nd edn., Birkhäuser Advanced Texts: Basler Lehrbücher. (Birkhäuser Boston, Inc., Boston, MA).

Krener, A. J. (1974). A generalization of Chow's theorem and the bang-bang theorem to non-linear control problems, *SIAM J. Control* **12**, pp. 43–52.

Kresin, G. and Maz'ya, V. (2012). *Maximum principles and sharp constants for solutions of elliptic and parabolic systems, Mathematical Surveys and Monographs*, Vol. 183 (American Mathematical Society, Providence, RI).

Lanconelli, E. (2010). Maximum principles and symmetry results in sub-Riemannian settings, in *Symmetry for elliptic PDEs, Contemp. Math.*, Vol. 528 (Amer. Math. Soc., Providence, RI), pp. 17–33.

Lee, J. M. (2013). *Introduction to smooth manifolds, Graduate Texts in Mathematics*, Vol. 218, 2nd edn. (Springer, New York).

Lobry, C. (1970). Contrôlabilité des systèmes non linéaires, *SIAM J. Control* **8**, pp. 573–605.

Magnani, V. (2006). Lipschitz continuity, Aleksandrov theorem and characterizations for H-convex functions, *Math. Ann.* **334**, pp. 199–233.

Mattila, P. (1995). *Geometry of sets and measures in Euclidean spaces, Cambridge Studies in Advanced Mathematics*, Vol. 44 (Cambridge University Press, Cambridge).

Mérigot, M. (1974). Domaine de convergence de la série de campbell-hausdorff, .

Monti, R. (2001). *Distances, boundaries and surface measures in Carnot-Carathéodory spaces*, Ph.D. thesis, University of Trento, Trento, Italy.

Morbidelli, D. (2000). Fractional Sobolev norms and structure of Carnot-Carathéodory balls for Hörmander vector fields, *Studia Math.* **139**, pp. 213–244.

Morgan, F. (2009). *Geometric measure theory: A beginner's guide*, 4th edn. (Elsevier/Academic Press, Amsterdam).

Nagel, A., Stein, E. M., and Wainger, S. (1985). Balls and metrics defined by vector fields. I. Basic properties, *Acta Math.* **155**, pp. 103–147.

Nagumo, M. (1942). über das Randwertproblem der nicht linearen gewöhnlichen Differentialgleichungen zweiter Ordnung, *Proc. Phys.-Math. Soc. Japan* **24**, pp. 845–851.

Neeb, K.-H. (2006). Towards a Lie theory of locally convex groups, *Jpn. J. Math.* **1**, pp. 291–468.

Nirenberg, L. (1992). *On the maximum principle*, AMS-MAA Joint Lecture Series (American Mathematical Society, Providence, RI).

Palais, R. S. (1957). A global formulation of the Lie theory of transformation groups, *Mem. Amer. Math. Soc. No.* **22**.

Parenti, C. and Parmeggiani, A. (2010). *Algebra Lineare ed Equazioni Differenziali Ordinarie* (Springer, Milan, Italy).

Picone, M. (1913). Teoremi di unicità nei problemi dei valori al contorno per le equazioni ellittiche e paraboliche, *Rend. Mat. Acc. Lincei* **22**, pp. 275–282.

Poincaré, H. (1900). Sur les groupes continus, *Cambr. Trans.* **18**, pp. 220–255.

Protter, M. H. and Weinberger, H. F. (1967). *Maximum principles in differential equations* (Prentice-Hall, Inc., Englewood Cliffs, N.J.).

Pucci, P. and Serrin, J. (2004). The strong maximum principle revisited, *J. Differential Equations* **196**, pp. 1–66.

Pucci, P. and Serrin, J. (2007). *The maximum principle, Progress in Nonlinear Differential Equations and their Applications*, Vol. 73 (Birkhäuser Verlag, Basel).

Rampazzo, F. and Sussmann, H. J. (2007). Commutators of flow maps of nonsmooth vector fields, *J. Differential Equations* **232**, pp. 134–175.

Rashevskiĭ, P. K. (1938). On the connectability of two arbitrary points of a totally nonholonomic space by an admissible curve, *Uchen. Zap. Mosk. Ped. Inst. Ser. Fiz.-Mat. Nauk* **3**, pp. 83–94.

Redheffer, R. M. (1971). The sharp maximum principle for nonlinear inequalities, *Indiana Univ. Math. J.* **21**, pp. 227–248.

Redheffer, R. M. (1972). The theorems of Bony and Brezis on flow-invariant sets, *Amer. Math. Monthly* **79**, pp. 740–747.

Rossmann, W. (2002). *Lie groups. An introduction through linear groups, Oxford Graduate Texts in Mathematics*, Vol. 5 (Oxford University Press, Oxford).

Rothschild, L. P. and Stein, E. M. (1976). Hypoelliptic differential operators and nilpotent groups, *Acta Math.* **137**, pp. 247–320.

Rudin, W. (1987). *Real and complex analysis*, 3rd edn. (McGraw-Hill Book Co., New York).

Sagle, A. A. and Walde, R. E. (1973). *Introduction to Lie groups and Lie algebras, Pure and Applied Mathematics*, Vol. 51 (Academic Press, New York-London).

Schur, F. (1889). Neue Begründung der Theorie der endlichen Transformationsgruppen, *Math. Ann.* **35**, pp. 161–197.

Sepanski, M. R. (2007). *Compact Lie groups, Graduate Texts in Mathematics*, Vol. 235 (Springer, New York).

Serre, J.-P. (2006). *Lie algebras and Lie groups, Lecture Notes in Mathematics*, Vol. 1500 (Springer-Verlag, Berlin).

Sperb, R. P. (1981). *Maximum principles and their applications, Mathematics in Science and Engineering*, Vol. 157 (Academic Press, Inc., New York-London).

Sussmann, H. J. (1972). The "bang-bang" problem for certain control systems in $GL(n, R)$, *SIAM J. Control* **10**, pp. 470–476.

Taira, K. (1988). *Diffusion processes and partial differential equations* (Academic Press, Inc., Boston, MA).

Titchmarsh, E. C. (1939). *The theory of functions*, 2nd edn. (Oxford University Press, Oxford).

Tu, L. W. (2011). *An introduction to manifolds*, 2nd edn., Universitext (Springer, New York).

Varadarajan, V. S. (1984). *Lie groups, Lie algebras, and their representations, Graduate Texts in Mathematics*, Vol. 102 (Springer-Verlag, New York).

Varopoulos, N. T., Saloff-Coste, L., and Coulhon, T. (1992). *Analysis and geometry on groups, Cambridge Tracts in Mathematics*, Vol. 100 (Cambridge University Press, Cambridge).

Wang, Z. X. and Guo, D. R. (1989). *Special functions* (World Scientific Publishing Co., Inc., Teaneck, NJ).

Warner, F. W. (1983). *Foundations of differentiable manifolds and Lie groups, Graduate Texts in Mathematics*, Vol. 94 (Springer-Verlag, New York-Berlin).

Wei, J. (1963). Note on the global validity of the Baker-Hausdorff and Magnus theorems, *J. Mathematical Phys.* **4**, pp. 1337–1341.

Wolfgang, W. (1998). *Ordinary differential equations, Graduate Texts in Mathematics*, Vol. 182 (Springer-Verlag, New York).

Index